2nd International Conference on Ship Manoeuvring in Shallow and Confined Water: Ship to Ship Interactions 2011

Trondheim, Norway
18-20 May 2011

Editors:

Bjornar Pettersen
Katrien Eloot
Tor Einar Berg
Marc Vantorre

ISBN: 978-1-62276-119-7

Printed from e-media with permission by:

Curran Associates, Inc.
57 Morehouse Lane
Red Hook, NY 12571

Some format issues inherent in the e-media version may also appear in this print version.

Printed by Curran Associates, Inc. (2012)

For permission requests, please contact The Royal Institution of Naval Architects]
at the address below.

The Royal Institution of Naval Architects
10 Upper Belgrave Street
London SW1X 8BQ UK

Phone: 44 207 235 4622
Fax: 44 207 259 5912

publications@rina.org.uk

Additional copies of this publication are available from:

Curran Associates, Inc.
57 Morehouse Lane
Red Hook, NY 12571 USA
Phone: 845-758-0400
Fax: 845-758-2634
Email: curran@proceedings.com
Web: www.proceedings.com

2nd International Conference on Ship Manoeuvring in Shallow and Confined Water:

Ship to Ship Interaction

May 18 - 20, 2011,
Trondheim, Norway

Editors

Bjørnar Pettersen (NTNU)
Tor Einar Berg (MARINTEK)
Katrien Eloot (Flanders Hydraulics Research)
Marc Vantorre (Ghent University)

These proceedings are sponsored by: MANAGING RISK DNV

2nd International Conference on

Ship Manoeuvring in Shallow and Confined Water:

Ship to Ship Interaction

INTERNATIONAL ORGANIZING AND PAPER COMMITTEE (OPC)

Reviewer	Affiliation
Prof. Knud Benedict	Univ. of Technology, Business and Design, Germany
Dr. Tor Einar Berg	MARINTEK, Norway
Dr. Larry Daggett	Waterway Simulation Technology, USA
Dr. Jonathan Duffy	Australian Maritime College, Australia
Prof. Katrien Eloot	Flanders Hydraulics Research, Belgium
Dr. Andreas Gronarz	Development Centre for Ship Technology and Transport Systems, Germany
Prof. Katsuro Kijima	Nagasaki Institute of Applied Science, Japan
Prof. Egil Pedersen	Norwegian University of Science and Technology, Norway
Prof. Bjørnar Pettersen	Norwegian University of Science and Technology, Norway
Prof. Jo Pinkster	Pinkster Marine Hydrodynamics, The Netherlands
Prof. K-P Rhee	Seoul National University, Korea
Dr. Renato Skejic	MARINTEK, Norway
Prof. Marc Vantorre	Ghent University, Belgium
Prof. Kamlesh Varyani	University of Strathclyde, UK

Review co-ordination: Dr. Maxim Candries, Ghent University, Belgium

PREFACE

Since the first STS conference in Antwerp two years ago, the world fleet has grown, the question of qualification of ship's crews has been discussed at the Manila meeting and the number of ship groundings and collisions is still showing a growing trend. Modern ships are equipped with more and more instruments. Operations are becoming more complex, while the competence of ship's officers may be reduced as a consequence of reduced time at sea, even while the quality of basic maritime education is rising. Insurance statistics show that the number of incidents and accidents is increasing all over the world. In Norway we are registering a growing number of groundings with serious oil spill as a consequence.

Norway is a leading nation in the development of advanced vessels for offshore petroleum operations. These vessels are used in complex marine operations such as anchor handling, providing supplies and stand-by services to floating platforms, towing offshore structures and pipe-laying. During the production phase, tankers are used in offshore loading and to some extent in side-by-side cargo transfer operations. As exploitation of our petroleum resources in deep, hostile and arctic waters increase in scope, new solutions for bringing the product to market will be needed.

Naval operations will become ever more dependent on replenishment at sea operations between supply ship and naval vessels. The duration of such operations should be kept as short as possible in order to minimize the vulnerability of the vessels involved.

Simulators are important tools for training bridge staff on interaction effects between ships in close proximity or in confined waters. Simulators are used for various purposes, including research, training and vessel design. For training purposes, some centres provide both general ship handling and specialist training. In connection with these training courses it is important to understand and possess a correct description and model of the fluid flow around the trainee's own vessel and to understand how interaction forces are influenced by a number of important parameters. Results based on model tests, analytical analysis and computational fluid dynamic simulations can be used to document forces on the vessels involved.

CFD has still some way to go. Even simulating the bare hull in a manoeuvring situation is challenging. Propeller and rudder interactions complicate the situation even more. There is therefore a need for close cooperation with experimental fluid dynamics (EFD) and new measurement techniques.

For several years, MARINTEK and the Department of Marine Technology at NTNU have been involved in studies of ship-to-ship operations. We are now approaching the end of a four-year international R&D programme supported by the Research Council of Norway. The aim of the project has been to generate new knowledge and develop new tools for studies of complex ship-to-ship operations. Project work packages were designed to improve computational fluid dynamic tools for multiple bodies in close proximity and models of interaction forces and moments. The final activity focused on operational experience as an input for studies of future operational guidance tools for ship-to-ship operations. Model tests were run at Flanders Hydraulics Research, and ship data were made available for validation studies by several research organisations and universities.

The first conference in Antwerp in 2009 demonstrated its value in bringing together people with different backgrounds and competence to improve communication and discussion on the subject of Bank Effects. This second conference has chosen to focus on Ship-to-Ship Interactions. The 45 papers that will be presented in the course of the next three days will cover a wide range of topics and research approaches. The authors and participants come from all over the world. We hope that your time in Trondheim will be useful for sharing operational experience and scientific results related to all types of ship-to-ship and ship-to-shore interaction, and for further developing personal and scientific relationships with your colleagues.

On behalf of Flanders Hydraulics Research, Ghent University, RINA and the local organisers, NTNU and MARINTEK, we bid you all welcome to Trondheim and hope you will enjoy a pleasant and fruitful stay in our city.

Prof. Bjørnar Pettersen and Dr. Tor Einar Berg
Local organising committee

IN COOPERATION WITH:

The Royal Institution of Naval Architects

THE ROYAL INSTITUTION OF NAVAL ARCHITECTS
10 Upper Belgrave Street
London SWIX 8BQ
United Kingdom

http://www.rina.org.uk

department
Mobility and Public Works

Flemish Government
Department Mobility and Public Works
Flanders Hydraulics Research
Berchemlei 115 - 2140 Antwerp, Belgium

http://www.watlab.be/en

MARITIME TECHNOLOGY
GHENT UNIVERSITY

Ghent University -
Maritime Technology Division
Technologiepark 904
9052 Ghent, Belgium

http://www.maritiem.ugent.be/english.htm

NTNU
Norwegian University of Science and Technology

NTNU
Department of Marine Technology
Marine Technology Centre
NO-7491 Trondheim, Norway
http://www.ntnu.edu/imt/english

MARINTEK
SINTEF

Norsk Marinteknisk Forskningsinstitutt AS
Norwegian Marine Technology Research Institute
P.O.Box 4125 Valentinlyst
NO-7450 Trondheim, Norway

http://www.marintek.sintef.no/

Gold sponsors:

department
Mobility and
Public Works

Flemish Government
Department Mobility and Public Works
Flanders Hydraulics Research
Berchemlei 115 - 2140 Antwerp, Belgium

http://www.watlab.be/en

MANAGING RISK DNV

DNV (Det Norske Veritas),
Veritasveien 1,
NO-1363 Høvik, Norway

http://www.dnv.com

NTNU
Norwegian University of
Science and Technology

NTNU
Department of Marine Technology
Marine Technology Centre
NO-7491 Trondheim, Norway

http://www.ntnu.edu/imt/english

MARINTEK
SINTEF

Norsk Marinteknisk Forskningsinstitutt AS
P.O.Box 4125 Valentinlyst
NO-7450 Trondheim, Norway

http://www.marintek.sintef.no/

Silver sponsors:

SPT, THE MARINE TRANSFER SOLUTION

SPT Inc.
Two Houston Center
900 Fannin St. Suite 3300
Houston, TX 77010

http://www.sptmts.com/

Bronze sponsors:

Promotie Binnenvaart Vlaanderen
Armand Hertzstraat 23
3500 Hasselt, Belgium

http://www.binnenvaart.be/index_eng.asp

Ghent University -
Maritime Technology Division
Technologiepark 904
9052 Ghent, Belgium

http://www.maritiem.ugent.be/english.htm

Knowledge Centre
Ship Manoeuvring
in Shallow and Confined Water

T +32 (0) 3 224 60 35
E info@shallowwater.be

www.shallowwater.be

Second International Conference on Ship Manoeuvring in Shallow and Confined Water: Ship-to-Ship Interaction

Marine Technology Centre,
Otto Nielsens vei 10, Trondheim, Norway

CONFERENCE PROGRAMME

FIRST DAY: 18 May 2011 (Wednesday)

8:00-9:00	**Registration**	
9:00-9:15	**Opening address**: Ingvald Strømmen, Professor at NTNU, Dean of Faculty of Engineering Science and Technology	
9:15-10.00	**Invited Keynote Speech** by Cal Hayden, Vice-President, Fleet Operations, SPT Inc. *Ship to Ship Transfers – a global service*	
	SESSION J1: Interaction Ship – Floating bodies (Chair: Prof. E. Pedersen)	
10:00-10.30	*Seabase model tests and surface effect ship code evaluation for T-craft* A.L. Silver, M.J. Hughes, R.C. Bishop, S.S. Lee, D.W. Wundrow, J.F. O'Dea	
10:30-10:50	**Refreshment break**	
	SESSION A1: Lightering (Chair: Prof. E. Pedersen)	
10:50-11:20	*Ship to ship interaction forces during lightering operations* E. Lataire, M. Vantorre, J. Vandenbroucke, K. Eloot	
11:20-11:50	*Hydrodynamic interaction loads between two ships during lightering operation in calm water* X. Xiang, R. Skejic, O. Faltinsen, T. E. Berg	
11:50-12:20	*A new approach in modelling the interaction forces* A. Gronarz	
12:20-13:30	**Lunch**	
	SESSION F1: Overtaking and meeting in channels and canals (Chair: Dr. J. Duffy)	**SESSION I1: Squat** (Chair: Prof. K.-P. Rhee)
13:30-14:00	*The influence of different vessel speeds of navigation on meeting in navigable channels of hydro system Danube-Tisa-Danube* V. Škiljaica, I. Škiljaica, M. Živković	*Optimisation of trim with respect to squat and under-keel-clearance* T. Berndt, J. Reinking, A. Laupichler, A. Härting
14:00-14:30	*Improved simulation model and methods for determination of the required passing distance between ships in restricted fairways* K. Benedict, W. Köpnick, M. Baldauf, A. Gronarz, B. Friedhoff, F. de Mello Petey	*Ship Squat Comparison and Validation Using PIANC, Ankudinov and BNT Predictions* M. J. Briggs, P. J. Kopp, V. Ankudinov, A. L. Silver
14:30-15:00	*Interaction-A practical viewpoint* G. Maxwell	*Squat during ship-to-ship interactions in shallow water* K. Eloot, M. Vantorre, J. Verwilligen
15:00-15:30	**Refreshment break**	

	SESSION L1: Experimental investigations of interaction (Chair: Prof. K. Benedict)	SESSION H1: Shallow and confined water effects (Chair: Prof. K. Eloot)
15:30-16:00	*Hydrodynamic interaction of two thin ships with rudder in close proximity* H. Yasukawa, S. Yoshida	*CFD Prediction and validation of ship-bank interaction in a canal* L. Zou, L. Larsson, G. Delefortrie, E. Lataire
16:00-16:30	*Integration and testing of STS decision-support system in a ship maneuvering simulator* D. Husjord, T. A. Øritsland, E. Pedersen	*An experimental research on the hydrodynamic forces acting on a twin-screw ferry during unberthing* S. Lee, J. Park, K.-P. Rhee, W. Yoo
16:30-17:00	*A comparative study of PIV experiments and numerical simulations of flow fields around two interacting ships* T. Arslan, B. Pettersen, J. Visscher, C. Muthanna, H.I. Andersson	*The importance of the form of time domain forces and moments on berthed ship interaction* J. T. Duffy, M. R. Renilson

SECOND DAY: 19 May 2011 (Thursday)

8:15-09:15	**Registration**	
9:15-10.00	**Invited Keynote Speech** by Odd Faltinsen, Professor at NTNU, *Modeling of ship-ship interaction*	
	SESSION F2: Overtaking and meeting in channels and canals (Chair: Prof. M. Vantorre)	
10:00-10.30	*Post widening ship/ship interaction and operational decisions in the Panama Canal* L.L. Daggett; J.C. Hewlett; R.S. Gray	
10:30-11:00	*Numerical investigation on ship-ship hydrodynamic interaction in restricted waters* C. Zhang, Z. Zou	
11:00:11.20	**Refreshment break**	
	SESSION E1: Moored and passing vessels (Chair: Prof. B. Pettersen)	
11:20-11:50	*Prediction of hydrodynamic interaction effects of vessels in ports* J. A. Pinkster	
11:50-12:20	*Calculation of forces on moored ships due to passing ships* W. van der Molen, P. Swiegers, J. Moes, M. Vantorre	
12:20-12:50	*The use of a Boussinesq-type wave model to determine ship-induced waves over non-uniform bathymetries* A. van der Hout, O. Weiler, M. Borsboom	
13:00-14:20	**Lunch**	
	SESSION A2: Lightering – Operational aspects (Chair: Dr. T.E. Berg)	**SESSION H2: Shallow and confined water effects** (Chair: Prof. J. Pinkster)
14:30-15:00	*Relative distances and speeds measurement system for STS lightering operations using stereo-cameras* E. Shimizu, E. Pedersen	*Prediction of linear hydrodynamic derivatives in shallow water* Y. Furukawa; Y. Nakiri; K. Kijima
15:00-15:30	*Decision support system in ship-to-ship operations using high accuracy DGPS* E. Okuyama, H. Oda, E. Shimizu	*A study on estimation of hydrodynamic forces acting on a ship hull in shallow water by CFD* Y. Kimura, E. Kobayashi, Y. Tahara, S. Koshimura
15:30-16:00	*Critical elements of a lightering operation* G. Gudmundseth	*New thinking on ship generated hydrodynamic fields: Introducing concepts for predicting bank suction and rejection* P. J. McArthur
16:00-16:15	**Refreshment break**	
	SESSION A3: Lightering – Operational aspects (Chair: Dr. T.E. Berg)	**SESSION H3: Shallow and confined water effects** (Chair: Dr. A. Gronarz)
16:15-16:45	*Decision-support system for STS training on ship handling simulator by means of visualization of danger zone due to hydrodynamic interaction* K. Kirimoto, E. Pedersen	*Numerical investigation of bank influence on a ship motion* A.M. Chetvertakov, M.P. Lebedeva, D.V. Nikushchenko
16:45-17:15	*Advanced ship manoeuvring and mooring support system at ship-to-ship transfer operations by fender monitoring system of pneumatic fenders* S. Sakakibara, S. Yamada, E. Shimizu, E. Okuyama, H. Oda	*URANS studies of ship-ship interactions in shallow water* H. Sadat-Hosseini, P.-C. Wu, Y. Toda, P. Carrica, F. Stern

THIRD DAY: 20 May 2011 (Friday)

08:00-08:30	**Registration**	
	SESSION M1: Interaction – CFD studies (Chair: Dr. R. Skejic)	**SESSION G1: Ship – tug interaction** (Chair: Prof. K. Kijima)
08:30-09:00	*Numerical study of ship-to-ship interaction forces on the basis of various flow models* J.M.A. Fonfach; S. Sutulo; C. Guedes Soares	*Interaction forces in tug operations* S. Geerts, M. Vantorre, K. Eloot, R. Huijsmans, N. Fierens
09:00-09:30	*URANS studies of ship-ship interactions in calm water and oblique waves for replenishment and overtaking conditions* S.M. Mousaviraad, S.H. Sadat-Hosseini, P. Carrica, F. Stern	*Slack towline on tow-towed ships interaction during maneuvering* A. Fitriadhy, H. Yasukawa, Y. Masaki
09.30-10:00	*Study on the effects of unsteady ship to ship interaction by CFD method* H. Yang, B.S. Wu, Q.M. Miao, X. Xiang , T. E. Berg, X.F. Kuang	*Dynamic interaction between tow and towed ships in head seas* Y. Nakayama, J. Watanabe, H. Yasukawa
10:00-10:30	**Refreshment break**	
	SESSION J2: Interaction Ship – Floating bodies / Session B1 : Replenishment (Chair: Dr. L. Daggett)	**SESSION G2: Ship – tug interaction / SESSION I2: Squat** (Chair: Prof. M. Vantorre)
10:30-11:00	*Investigating ship maneuvers around a floating structure under the influence of a uniform current in deep and calm water* R. Skejic, M. Breivik, T. E. Berg	*CFD based prediction of ship-ship interaction forces on a tug beside a tanker* C. D. Simonsen, C. K. Nielsen, J. F. Otzen, K. Agdrup
11:00-11:30	*Using dynamic positioning for side-by-side cargo transfer operations* F.H.H.A. Quadvlieg, R. Hallmann, G. Hughes, D.H. Harris	*Squat computation for container ships in restricted waterways* A. von Graefe, B. El Moctar, V. Shigunov, H. Söding, T. Zorn
11:30-12:00	*Simulation of the Launch/Recovery of a Submersible from a Host ship* D. Cassidy, P. Carrica, M. Hyman	*Squat of ships sailing in a non centered track in a canal* H. J. de Koning Gans
12:00-12:15	**Closing Remarks** by Marc Vantorre, Professor at Ghent University	
12:30-13:45	**Lunch**	
14:00-17:00	**Afternoon visits**	

TABLE OF CONTENTS

All papers, except the keynote papers, are in alphabetic order according to the family name of first author.

MODELLING OF MANOEUVRING WITH ATTENTION TO SHIP-SHIP INTERACTION AND WIND WAVES

O M Faltinsen, Centre for Ships and Ocean Structures (CeSOS) and Department of Marine Technology, Norwegian University of Science and Technology, Norway

SUMMARY

CFD and simplified modular manoeuvring models are discussed with emphasis on ship-ship interaction. The importance of ship end effects on ship-ship interaction necessitates a 3D analysis. A limited error analysis for a modular manoeuvring model is applied to a single ship in calm water. How to account for wave effects by means of mean wave loads is briefly described. A method to predict mean wave loads on a single ship as well as on interacting ships at forward speed is presented. Satisfactory agreement between numerical simulations and experiments of the trajectory of the SR-108 container vessel in a turning manoeuvre in regular waves is demonstrated. The presence of resonant water motion in the gap between two ships is discussed and a simplified method to account for flow separation at the gap entrance is presented. It is discussed how to generalize the method to ship manoeuvring.

1. INTRODUCTION

Analysis of ship manoeuvring requires integrated knowledge about resistance, propulsion, ship machinery, seakeeping, steering and automatic control. Manoeuvring is analyzed traditionally in calm-water conditions. However, replenishment and lightering operations are influenced by environmental conditions. Further, broaching in waves represents a potentially dangerous situation for smaller vessels and must be accompanied with proper steering strategies. It is most often sufficient for displacement vessels to consider only the effect of lateral and longitudinal velocities and "yaw rate". Higher speeds need to incorporate the effect of heel. Experimental results for trimarans have shown that linear manoeuvring analysis is more limited than for monohulls and catamarans. Hydrofoil vessels require an analysis of six degrees of freedom during manoeuvring. There is an increased importance of dynamic instabilities with increasing speed of high-speed vessels. For instance, calm-water broaching of semi-displacement round-bilge monohulls may suddenly happen beyond a certain Froude number. Spinout is a serious consequence of bad fast-boat driving and can occur when the boat is altering course and slows down. The bow will as a consequence drop. Directional stability is lost as a consequence of the reduced draft at transom. A scenario can be rounding a mark in a race. Other spinout scenarios involving waves are described by Pike [1].

CFD, simplified modular methods and experiments are used in manoeuvring analysis and represent all valuable tools. Simplified modular methods have the advantage in doing real-time analysis and can either be based on a combination of empirical results and potential-flow theory or by obtaining hydrodynamic coefficients from experiments such as PMM tests. If model tests are used, attention should be paid to scaling of Reynolds number effects associated with flow separation. Further, results from PMM tests assume a functional form of the nonlinear part of forces and moments that can be questioned. Recommended error analysis by ITTC ought to be a must for numerical and experimental tools.

Our final focus will be on ship-ship interactions in waves. We will limit ourselves to unrestricted water of infinite depth and consider a speed range where the Froude number (Fn) effect is secondary, i.e. $Fn \lessapprox 0.2$. How the Froude number and depth affect hydrodynamic manoeuvring terms are discussed by Faltinsen [2]. We start with giving a brief summary of CFD. Even though the development of CFD is promising, we are still not yet to the point that CFD can be the only tool. It is difficult to imagine that experiments will be replaced by CFD in the future. We will use our own developed simplified modular manoeuvring model for calm-water conditions of a single ship as a basis for a limited error analysis to illustrate factors influencing manoeuvring. Our model for incorporating wave effects by a two-time scale method will be shortly presented. The important wave effects are speed and heading dependent mean wave loads. The method is a generalization of state-of-the art slow-drift analysis of moored or dynamic positioned floating structures in waves. The ship-ship interactions are associated with both calm-water and wave effects. Since the hull ends play an important role in the interaction, the necessity of a 3D analysis is emphasized. Potential-flow theory can to a large extent be used in seakeeping analysis. However, flow separation plays an important role in piston-mode wave resonance in the gap between ships when the hulls have sharp corners at the lower gap entrance. A computational time-efficient numerical method based on using a Navier-Stokes solver in the bilge area together with a domain-decomposition method is presented. It is discussed how to generalize the method to calm-water manoeuvring of a single ship in unrestricted water.

2. COMPUTATIONAL FLUID DYNAMICS

CFD is commonly used in marine hydrodynamics and is often associated with methods solving the Navier-Stokes equations. There is a broad variety of numerical methods based on CFD. An overview for application with interface flow is given in Figure 1. An advantage of CFD methods is flow visualization providing details such as vorticity distribution similarly as it could be done through model tests. Navier-Stokes CFD methods are usually robust and their flow simulations may look convincing by themselves, but reliability is still an issue and must be assessed with verification and validation.

The required computational time by CFD is a limiting factor. The computational time prohibits presently solutions of the fully nonlinear seakeeping problem during the duration of a sea state. Another matter is that many realizations of the same sea state are needed to get reliable extreme value predictions. Further, all physical effects are not completely understood for violent fluid motions. Examples are modelling of air cavities and the effect of mixing between air and water. Description of impact between water and a blunt surface is demanding and requires small time steps and small local grid sizes. An example of successful seakeeping studies of limited time duration is by Oberhagemann and el Moctar [4]. They studied global hydroelastic behaviour of a ship in waves associated with springing and whipping which are associated with steady-state and transient oscillations, respectively. Whipping is caused by slamming and springing is excited by both linear and nonlinear wave effects.

CFD is well established in predictions of calm-water ship resistance. Durante et al. [5] solved the unsteady RANS equations for the turning manoeuvre of a fully appended ship in calm water. A finite volume pseudo-compressible formulation was applied and the free surface was captured by a single-phase level set approach. Local time stepping, multi-grid technique and parallelization were used to reduce the computational time. Dynamic overlapping grid was applied to handle a movable rudder. An approximation is that propellers loads are based on an actuator disc model. Their method was validated by comparing with model tests of turning circle manoeuvre carried out at the lake of Nemi close to Rome. Three test series were performed with large discrepancy between test series I and test series II and III. The two latter test series were therefore the basis for the comparisons with the CFD calculations. Fine (with 6.2 M cells) and medium (every other cell in each direction is removed) grids were selected for comparisons. Only the fine grid gave satisfactory agreement with the experimental turning circle trajectory and was selected for further detailed studies. Both the drift angle and the yaw rate agreed well with the experiments while the speed drop was overestimated. Examples of error sources are the modelling of turbulence and the propeller model.

3. CALM-WATER MODULAR MANOEUVRING MODEL

Real time simulations of ship manoeuvring require a modular manoeuvring simulation model. A calm-water modular manoeuvring model has been developed at CeSOS through the doctoral studies by Skejic [6] and

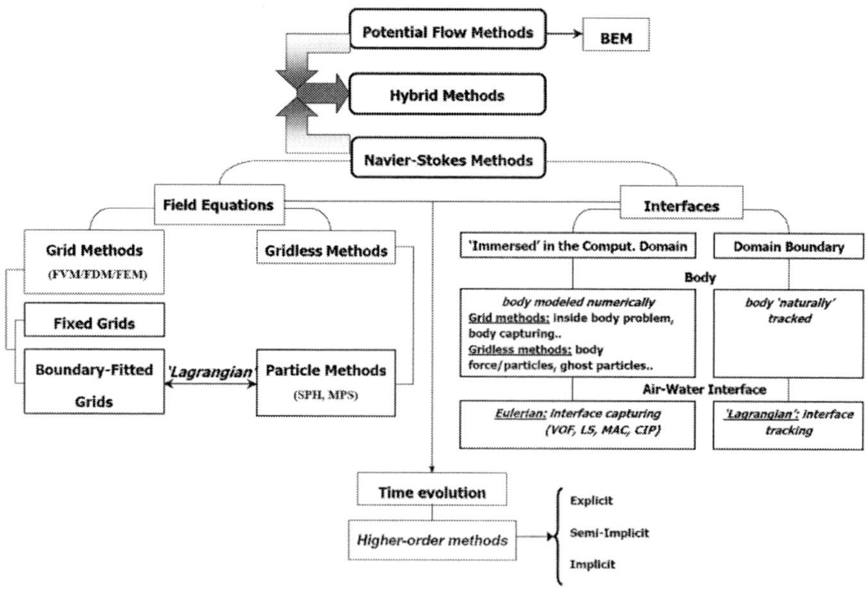

Figure 1: Overview of numerical methods in fluid dynamics with emphasis on possible solution strategies within Navier-Stokes solvers. It is, for instance, illustrated that the field equations in Navier-Stokes methods are solved by either grid methods or gridless methods. Further, it is indicated in the figure how interfaces between air (gas) and water (liquid) are numerically handled as well as different ways of accounting for the presence of a body (Faltinsen and Timokha [3]).

Xiang. The model solves for surge, sway, roll and yaw and can consider two interacting ships. The resistance is either calculated by Holtrop's [7] formula or by using the ITTC formulation for the viscous resistance combined with direct calculations of the wave resistance. Air resistance can be added by using empirical drag coefficients.

The nonlifting dynamic hull loads accounting for ship interactions can be calculated by either a 3D potential flow method using BEM (Xiang and Faltinsen [8]) or by Tuck and Newman's [9] far-field slender-body theory. Both methods are based on a rigid free-surface condition which implies that the Froude number roughly speaking should be less than 0.2. The 3D method allows for arbitrary relative configurations of the ships up to the point of collision. Further, more than two ships and other interacting structures can be considered. The expressions involve added mass coefficients accounting for body interactions. One term is expressed as the time derivative of a sum of terms that are products of added mass coefficients and velocities. There are also terms with added mass coefficients that are not time differentiated. Finally, there is a term involving integration of velocity-square terms over the mean wetted hull surface. Tuck and Newman's theory assumes the two ships to be on parallel course. The ships must not be too close to each other. Since the ship ends have important contribution to the interaction loads, a slender-body theory is only adequate for slender ship ends. So far, we have only applied the 3D method to unrestricted water of infinite depth. However, no problems are expected in using the method for finite depth with a solid bottom and in channels as long as ship-wave generation does not matter and cross-flow separation is secondary. The interaction model by Xiang and Faltinsen [8] has been validated by comparing with model tests of the transverse interaction force between two oil tankers at parallel course with the same speed and different staggers. How to account for wave generation is discussed in section 4. Even though one only considers a single ship, it is advisable to use a 3D potential-flow method instead of strip theory to calculate added mass coefficients, in particular, for angular accelerations.

The hull-lift is evaluated by Söding's [10] modification of slender-body theory and does not account for hull interactions. The slender-body hull-lift model is only rational as long as a trailing edge with flow separation and the associated longitudinal vortex sheet is clearly defined. The latter fact is normally not true for displacement vessels. The viscous cross-flow model is based on the cross-flow principle with constant drag coefficient along the ship. A cross-flow principle can be questioned at forward speed. The latter fact can, for instance, be shown by using a 2D+t theory as outlined by Faltinsen [2]. A 2D+t theory illustrates the fact that flow separation is more dominant in the aft part than in the fore part of a ship during manoeuvring at forward speed. The cross-flow principle does not account for this fact. The drift angle becomes an important parameter for the transverse loads due to flow separation at forward speed. The Keulegan-Carpenter (KC) number influences clearly the drag coefficients at zero forward speed at small and moderate KC numbers. The ship may at forward speed partly run away from the returning cross-flow-generated vortices and hence cause a reduced KC-number influence. The viscous effect on the "centrifugal" hydrodynamic force is given by empirical formula. The latter force acts in a similar way as the centrifugal force in causing ship resistance during turning. An alternative would be to use PMM tests to obtain hydrodynamic hull coefficients. Proper care must be shown in ensuring turbulent boundary layer condition. Flow separation at curved surfaces depends strongly on laminar or turbulent boundary layer condition in a local upstream distance from a separation line. Open water diagram for the propulsion are input. The wake is found by Holtrop's formula [6] while the thrust deduction coefficients are given by Holtrop and Mennen [11]. The rudder forces are predicted according to Söding [10]. The fact that the propeller race increases the longitudinal inflow flow velocity over part of a rudder is not presently considered. Ankudinov et al.'s [12] flow-rectification factors of the inflow to the rudder caused by the hull and the propeller is implemented. The inflow velocity to the rudder is expressed as $V_{RS} = (u^2_{RS} + v^2_{RS})^{0.5}$ where $v_{RS} = \gamma_v v + \gamma_r x_d r$, γ_v and γ_r are the flow rectification factors, v is the transverse velocity of the COG of the ship, x_d is the longitudinal x-coordinate of the rudder centre relative to COG and r is the yaw rate. The modelling of the propulsion machinery is avoided by assuming a constant number of propeller revolutions per second. In addition, automatic control procedures can be used such as in lightering and replenishment operations. An autopilot is then implemented by using rudder and propulsion control to achieve the desired (safe) transverse/longitudinal clearances and heading angles for operation by compensating for the deviations caused by the hydrodynamic interacting forces and moments between ships.

3.1 ERROR ANALYSIS OF MANOEUVRING OF A SINGLE SHIP

Skejic [6] validated the described modular manoeuvring model by comparing with full scale trials of "MARINER" and "Esso Osaka". The speed loss of "Esso Osaka" during 35 degrees port turn manoeuvre with approach forward speed 7.7 knots is well predicted. The trajectories differ more. Similar trends appear for the trajectories and speed loss of "Esso Osaka" for 20/20 degrees zigzag manoeuvre with approach forward speed 7.8 knots. The trajectory of the "MARINER" hull during 20.9 degrees starboard turn manoeuvre with approach forward speed 15.4 knots was well predicted. The latter case is the basis for a limited error analysis. The error sources with base value are:

- Drag coefficient C_d associated with hull cross-flow is 1.35

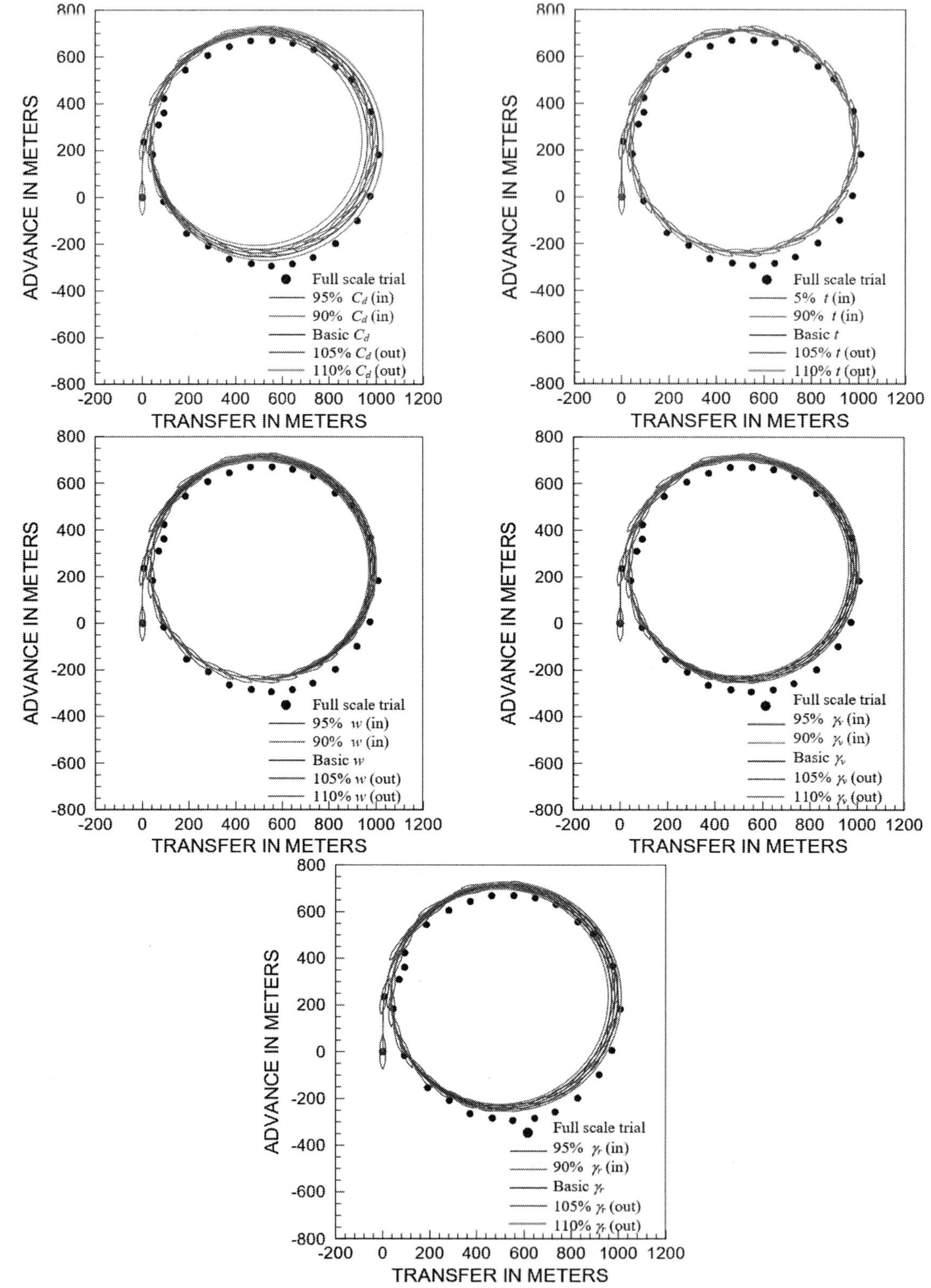

Figure 2: "MARINER" hull 20.9° Starboard turn manoeuvre with approach forward speed 15.4 knots in calm water. A modular calm-water manoeuvring model is compared with full scale trial. The influence of variation of drag coefficient, wake factor, thrust deduction factor and flow rectification coefficients is examined (Skejic, personal communication 2011).

- Thrust deduction factor $t = 0.1732$
- Wake factor $w = 0.267$
- Flow rectification factors γ_v and γ_r are 1.0.

The base values of the error sources were varied with ± 10%. The justification for this variation can be questioned. We ought to have considered other error sources such as choice of "trailing" cross-section in the hull-lift formulation by Söding [10] and the propeller race effect on the rudder. Further, varying only the constant drag coefficient in the cross-flow drag formulation does not reflect the fact that flow separation is more pronounced in the aft part than in the fore part of the vessel. The turning trajectories following from varying drag coefficient, thrust deduction factor, wake factor and flow rectification factors are presented in Figure 2. The variation in the drag coefficient has the largest influence while the influences of the wake factor and thrust deduction factor are not large. The flow rectification factors matter.

The advance and the tactical diameter are measures of ship manoeuvrability according to IMO manoeuvring criteria from 2002 for ships longer than 100 m at service speed. The advance and the tactical diameter in the ship trial and by the modular manoeuvring model are close and presented in Table 1. Table 2 and 3 presents the error analysis for the advance and the tactical diameter, respectively. The error associated with each error source is assumed independent of the error associated with other error source. The resulting error of the advance and the tactical diameter by this limited error analysis is small and equal to $0.13L_{pp}$ and $0.37L_{pp}$, respectively.

	Advance - A	**Tactical diameter - T**
Full scale trial (L_{pp} = 160.934 m)	$4.11L_{pp}$	$6.15L_{pp}$
Base value	$A^{Basic} = 4.37L_{pp}$	$T^{Basic} = 6.06L_{pp}$

Table 1: IMO manoeuvring criteria. Comparison between full scale trial and modular manoeuvring model (see Figure 2) (Skejic, personal communication 2011).

Advance error Δ_A (m)				
$A_{C_d}^{90\%} = 4.30L_{pp}$	$A_t^{90\%} = 4.37L_{pp}$	$A_w^{90\%} = 4.31L_{pp}$	$A_{\gamma_v}^{90\%} = 4.33L_{pp}$	$A_{\gamma_r}^{90\%} = 4.31L_{pp}$
$A_{C_d}^{110\%} = 4.44L_{pp}$	$A_t^{110\%} = 4.37L_{pp}$	$A_w^{110\%} = 4.44L_{pp}$	$A_{\gamma_v}^{110\%} = 4.42L_{pp}$	$A_{\gamma_r}^{110\%} = 4.44L_{pp}$
$\Delta A_{C_d}^{90\%} = \lvert A_{C_d}^{90\%} - A^{Basic} \rvert$	$\Delta A_t^{90\%} = \lvert A_t^{90\%} - A^{Basic} \rvert$	$\Delta A_w^{90\%} = \lvert A_w^{90\%} - A^{Basic} \rvert$	$\Delta A_{\gamma_v}^{90\%} = \lvert A_{\gamma_v}^{90\%} - A^{Basic} \rvert$	$\Delta A_{\gamma_r}^{90\%} = \lvert A_{\gamma_r}^{90\%} - A^{Basic} \rvert$
$\Delta A_{C_d}^{110\%} = \lvert A_{C_d}^{110\%} - A^{Basic} \rvert$	$\Delta A_t^{110\%} = \lvert A_t^{110\%} - A^{Basic} \rvert$	$\Delta A_w^{110\%} = \lvert A_w^{110\%} - A^{Basic} \rvert$	$\Delta A_{\gamma_v}^{110\%} = \lvert A_{\gamma_v}^{110\%} - A^{Basic} \rvert$	$\Delta A_{\gamma_r}^{110\%} = \lvert A_{\gamma_r}^{110\%} - A^{Basic} \rvert$
$\Delta A_{C_d} = (\Delta A_{C_d}^{90\%} + \Delta A_{C_d}^{110\%})/2$	$\Delta A_t = (\Delta A_t^{90\%} + \Delta A_t^{110\%})/2$	$\Delta A_w = (\Delta A_w^{90\%} + \Delta A_w^{110\%})/2$	$\Delta A_{\gamma_v} = (\Delta A_{\gamma_v}^{90\%} + \Delta A_{\gamma_v}^{110\%})/2$	$\Delta A_{\gamma_r} = (\Delta A_{\gamma_r}^{90\%} + \Delta A_{\gamma_r}^{110\%})/2$
$\Delta_A = \sqrt{(\Delta A_{C_d})^2 + (\Delta A_t)^2 + (\Delta A_w)^2 + (\Delta A_{\gamma_v})^2 + (\Delta A_{\gamma_r})^2} = 0.13L_{pp}$				

Table 2: Estimation of error in Advance IMO manoeuvring criterion (Skejic, personal communication 2011).

Tactical diameter error Δ_T (m)				
$T_{C_d}^{90\%} = 5.79L_{pp}$	$T_t^{90\%} = 6.06L_{pp}$	$T_w^{90\%} = 5.96L_{pp}$	$T_{\gamma_v}^{90\%} = 5.92L_{pp}$	$T_{\gamma_r}^{90\%} = 5.90L_{pp}$
$T_{C_d}^{110\%} = 6.34L_{pp}$	$T_t^{110\%} = 6.08L_{pp}$	$T_w^{110\%} = 6.18L_{pp}$	$T_{\gamma_v}^{110\%} = 6.21L_{pp}$	$T_{\gamma_r}^{110\%} = 6.24L_{pp}$
$\Delta T_{C_d}^{90\%} = \lvert T_{C_d}^{90\%} - T^{Basic} \rvert$	$\Delta T_t^{90\%} = \lvert T_t^{90\%} - T^{Basic} \rvert$	$\Delta T_w^{90\%} = \lvert T_w^{90\%} - T^{Basic} \rvert$	$\Delta T_{\gamma_v}^{90\%} = \lvert T_{\gamma_v}^{90\%} - T^{Basic} \rvert$	$\Delta T_{\gamma_r}^{90\%} = \lvert T_{\gamma_r}^{90\%} - T^{Basic} \rvert$
$\Delta T_{C_d}^{110\%} = \lvert T_{C_d}^{110\%} - T^{Basic} \rvert$	$\Delta T_t^{110\%} = \lvert T_t^{110\%} - T^{Basic} \rvert$	$\Delta T_w^{110\%} = \lvert T_w^{110\%} - T^{Basic} \rvert$	$\Delta T_{\gamma_v}^{110\%} = \lvert T_{\gamma_v}^{110\%} - T^{Basic} \rvert$	$\Delta T_{\gamma_r}^{110\%} = \lvert T_{\gamma_r}^{110\%} - T^{Basic} \rvert$
$\Delta T_{C_d} = (\Delta T_{C_d}^{90\%} + \Delta T_{C_d}^{110\%})/2$	$\Delta T_t = (\Delta T_t^{90\%} + \Delta T_t^{110\%})/2$	$\Delta T_w = (\Delta T_w^{90\%} + \Delta T_w^{110\%})/2$	$\Delta T_{\gamma_v} = (\Delta T_{\gamma_v}^{90\%} + \Delta T_{\gamma_v}^{110\%})/2$	$\Delta T_{\gamma_r} = (\Delta T_{\gamma_r}^{90\%} + \Delta T_{\gamma_r}^{110\%})/2$
$\Delta_T = \sqrt{(\Delta T_{C_d})^2 + (\Delta T_t)^2 + (\Delta T_w)^2 + (\Delta T_{\gamma_v})^2 + (\Delta T_{\gamma_r})^2} = 0.37L_{pp}$				

Table 3: Estimation of error in Tactical diameter IMO manoeuvring criterion (Skejic, personal communication 2011).

4. MANOEUVRING IN A SEAWAY

The CeSOS manoeuvring model of a ship in waves assumes different time scales for the manoeuvring and the linear wave-induced loads. The fact that manoeuvring models operate with a body-fixed accelerated coordinate system while linear and weakly nonlinear seakeeping problems are traditionally solved in an inertial system must be recognized in the analysis. The effect of regular waves on ship manoeuvring is expressed in terms of mean wave forces and moments acting on the ships that are functions of the slowly varying ship heading relative to the incident waves, ship speed and the frequency of encounter. The procedure may be generalized to a short-term irregular sea by creating a time history of the wave elevation and approximating the wave elevation between two successive zero-crossings as a regular wave. The calculations of mean wave loads are done interactively with the ship manoeuvring. The procedure is described by Skejic and Faltinsen [13] who based the mean wave load expressions for any wave directions and ship speeds on Faltinsen et al. [14], which presents an asymptotic small wavelength formulation and a direct pressure integration method in the wave-frequency range of non-negligible ship motions. Mean wave load results depend on how the linear wave-induced ship motions are evaluated. Skejic and Faltinsen [13] used the Salvesen-Tuck-Faltinsen strip theory. Some manoeuvring models in waves are incomplete in the way that mean wave loads are accounted for, e.g. the only considered effect is due to integration of the linear pressure over the instantaneous wetted surface. It is well known from direct pressure integration methods of mean wave loads that accounting only partly for terms can lead to erroneous answers. The latter fact is illustrated in Faltinsen [15] by applying the direct pressure integration method in the estimation of the mean wave loads on a 2D vertical wall.

When the frequency of encounter is small, one has to solve simultaneously the seakeeping and manoeuvring problems. Examples are in analyzing the effect of long-periodic swell and steering due to broaching in following waves. If the wave-generation effect of the ship can be neglected, we can for small frequency of encounter evaluate in a simple way the effect of the incident waves as the sum of Froude-Kriloff loads, added mass loads due to wave diffraction and the effects of incident wave velocity on hull lift as well as propeller and rudder loads. The Froude-Kriloff loads can easily be done by a nonlinear formulation in combination with nonlinear hydrostatic terms. The latter matters in analyzing capsize due to broaching in waves.

A further development of the mean wave load calculations is undergoing at CeSOS. The linear wave-induced motions are found by a 3D Rankine source method that solves the Neumann-Kelvin problem in the time domain. Xiang and Faltinsen [16] validated the method for the two interacting independent ships studied experimentally by Ronæss [17]. The two ships were towed side-by-side in regular head waves at the MARINTEK towing tank in Trondheim which is 270 m long and 10.5 m wide. The distance between the centerlines of the two models was maintained as 2.47 times the average beam of the two ships (1.25 m) during the tests. The comparisons were made when the two ships advanced with amid-ship sections approximately side-by-side (i.e. non-staggered), and the forward speed corresponds to Froude number 0.1 when using the length of the largest ship. The heave and pitch motions of the two ships and the roll motion of the largest ship are adequately predicted by the present method, while the roll motion prediction of the smallest ship is clearly larger than the model test results. This is consistent with Ronæss [17] which commented that the reason was from the experimental setup. Further, the predictions and model tests show that the hydrodynamic interaction effects have much greater influence on the motions of the smallest ship than the largest ship. The described method neglects the local interaction between the steady and unsteady flow which matters, in particular, for non-slender ships. When local interaction between steady and unsteady flow is accounted for in methods formulated in an inertial coordinate system, a difficulty occurs at body surface areas with high curvature. The reason is the so-called m_j - terms in the unsteady body boundary conditions containing terms involving second-derivatives of the steady velocity potential. Actually the m_j - terms are singular at sharp corners. If a body-fixed coordinate system is used instead of an inertial coordinate system in formulating the boundary value problem, the m_j - terms do not appear (Shao and Faltinsen [18]). The latter method has so far only been validated for single ships.

A special case of the described analysis is for calm water when the waves are generated due to slowly varying ship motions associated with manoeuvring and the forward speed. If the two ships are in the far-field of each other, one can consider the waves generated by one ship as incident waves to the other ship and calculate the resulting responses. The procedure can, for instance, be applied to a lightering operation and combined with a database based method to reduce the computational time.

The new method for calculating mean wave loads are based on time averaging expressions for the time rate of change of momentum and angular momentum of the flow. The effect of the second-order velocity potential is neglected which was shown by Grue and Palm [19] to matter at forward speed, in particular, for the mean yaw moment. It is more challenging to evaluate the second-order potential than the first-order potential. Part of the mean wave-load expressions involves integrations over control surfaces that can be chosen at an arbitrary distance from a ship. However, it is advantageous from a computational point of view to choose the control surface not too far away from the considered ship. The arbitrary choice of control-surface position is important when studying two interacting ships. The expressions have been validated by Xiang (personal communication, 2011) for single advancing ships by comparing with the added resistance experiments for the S-175 ship by Fujii [20] and Nakamura [21] at Fn = 0.15 and Fn = 0.25 in head sea (180°) and bow seas with 150° and 120° headings. Satisfactory agreement with the experimental results for the container ship SR-108 reported by Yasukawa [22] was also documented. Yasukawa [22] presented results for added resistance in head and beam sea and mean transverse wave force and yaw moment in beam sea at forward speed. There is a need for more experimental results of mean sway force and yaw moment on single ships at forward speed as a function of wave heading and frequency. Further, we are not aware of reliable experimental results for mean wave loads on interacting ships at forward speed.

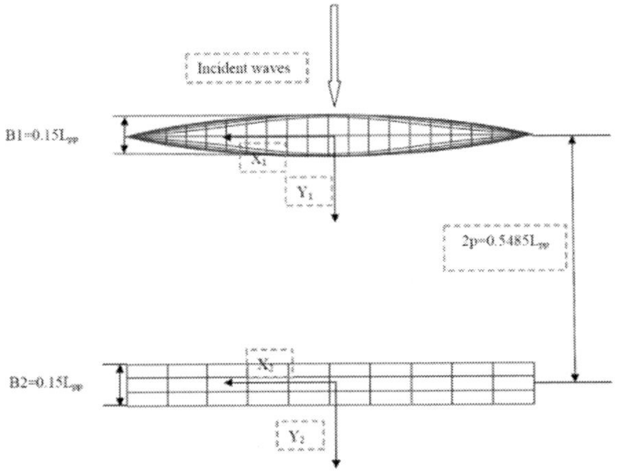

Figure 3: Arrangement of a fixed Wigley hull and a fixed barge in the beam sea regular wave tests by Kashiwagi et al. [23].

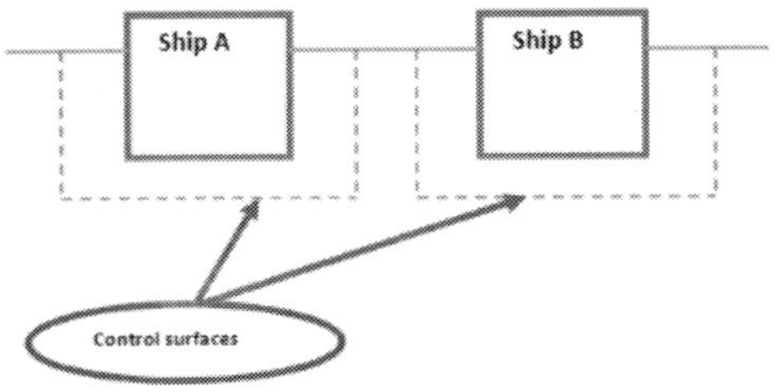

Figure 4: Illustration of arrangement of control surfaces used in calculating mean wave loads on two interacting ships.

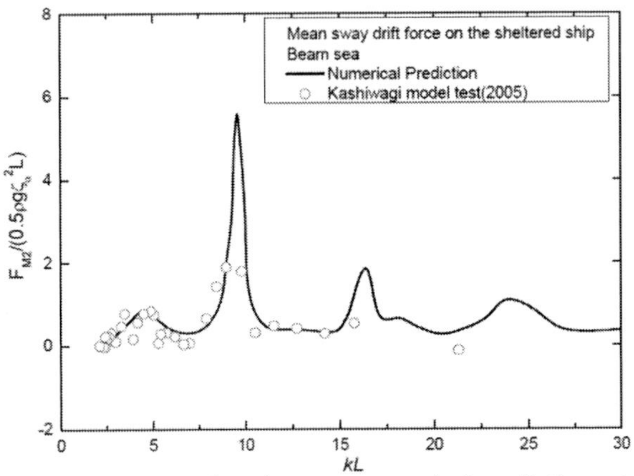

Figure 5: Comparison between numerical predictions at CeSOS and experimental results by Kashiwagi et al. [23] of mean sway force F_{M2} on the sheltered barge shown in Figure 3. k - wave number of incident beam sea waves. ς_a - incident wave amplitude. L - ship length (Xiang, personal communication 2011).

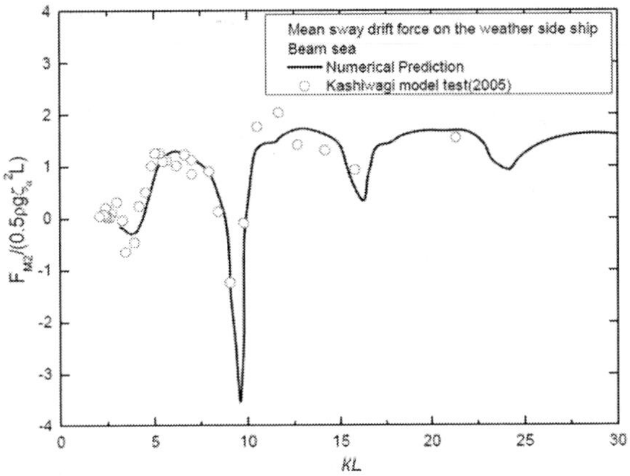

Figure 6: Comparison between numerical predictions at CeSOS and experimental results by Kashiwagi et al. [23] of mean sway force F_{M2} on the weatherside ship shown in Figure 3. k - wave number of incident beam sea waves. ς_a - incident wave amplitude. L - ship length (Xiang, personal communication 2011).

At zero speed we can compare with experimental results for a fixed Wigley hull ship arranged side-by-side with a fixed barge presented by Kashiwagi et al. [23]. Incident beam sea regular waves of different wavelengths were examined. The set up is shown in Figure 3.

The control surfaces used in calculating the mean wave loads are illustrated in Figure 4. The theoretical and experimental mean sway forces on the sheltered and weatherside ships are shown in Figure 5 and Figure 6, respectively. The local peaks and rapid change in values in certain frequency domains indicate piston mode and sloshing modes resonance occurring between the two hulls (Faltinsen and Timokha [2]). If the bodies have sharp corners at the lower entrance to the gap, flow separation can strongly affect the resonant piston mode wave elevation in the gap and thereby influence the mean wave loads.

The latter fact is well known for piston mode resonance in moonpools and was studied for 2D flow conditions by Kristiansen and Faltinsen [24]. A hybrid method that combines a potential flow and a CFD solver by domain decomposition was presented. Linear free-surface conditions according to potential flow were assumed. A viscous domain is placed in the bilge area of the ship as illustrated with two rectangular rigidly- connected barges (only half of the domain was simulated due to symmetry) in Figure 7. The viscous (CFD) domain is red, while the potential-flow domain is blue. There are two matching criteria along the boundary that separates the two domains; the normal velocity and the pressure are required to be continuous. This is achieved in this hybrid method by solving a combined Laplace and Poisson equation for the whole water domain at each time-step. The finite volume method is used for the whole water domain, and the Poisson equation appears in a projection step of the Navier-Stokes solver.

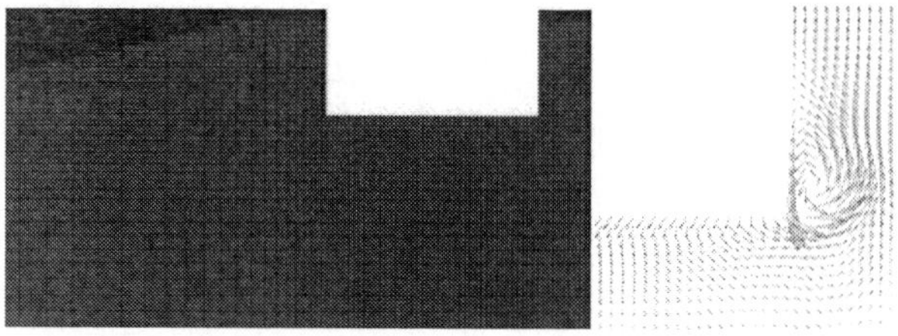

Figure 7: Left: Illustration of part of the computational grid in Kristiansen and Faltinsen's [24] hybrid method for forced heave oscillation of a 2D catamaran with rectangular side hulls. Only the grid next to the left side hull and in half of the gap is shown. Red: CFD domain. Blue: Potential-flow domain. Right: Snapshot of the velocity field (Kristiansen, personal communication 2011).

At every time step, a sparse matrix system must be solved. Thus, the method is quite easily parallelized on a computer cluster. This feature makes it feasible for 3D implementation. From numerical experiments with a 2D implementation, the present hybrid method is orders of magnitude faster than traditional CFD methods. The hybrid method with 7000 unknowns and 30 forced oscillation periods took 73 s on a 2.4 GHz laptop with 4 GB RAM. The CFD code OpenFOAM with 8 x 2.0 GHz processor in parallel was also applied with 380709 grid cells. Many of the cells were needed in the free-surface zone due to the used free-surface capturing method (VOF). The computational time was one hour to reach steady state for one forced oscillation period. The difference with respect to the experiments was up to 15%. Increasing number of grid cells lacked convergence and did not improve the results.

Figure 8 presents a validation of the hybrid method by comparing with experimental data for forced heave oscillations of a catamaran cross-section with rectangular side hulls. Around the resonant piston-mode frequency, the linear potential-flow theory clearly over predicts the experimental values. Using nonlinear free-surface conditions in a potential-flow method had a small influence. There are no existing physically based methods that can account for the viscous flow separation in potential-flow methods associated with gap resonance, i.e. we cannot add empirical viscous damping coefficients in a similar way as done in roll predictions.

Both current and forward speed will affect the gap resonance. A general tendency is that the gap resonance effect diminishes with increasing forward speed, as e.g. indicated by the numerical results presented in the right plot of Figure 8 for small Froude numbers. A special gap resonance effect was noted in the head-sea regular-wave experiments by Ronæss [17] with two independent ships on parallel course. At $\tau = \omega_e U/g$ slightly above 1/4 large amplitude waves were building up between the hulls.

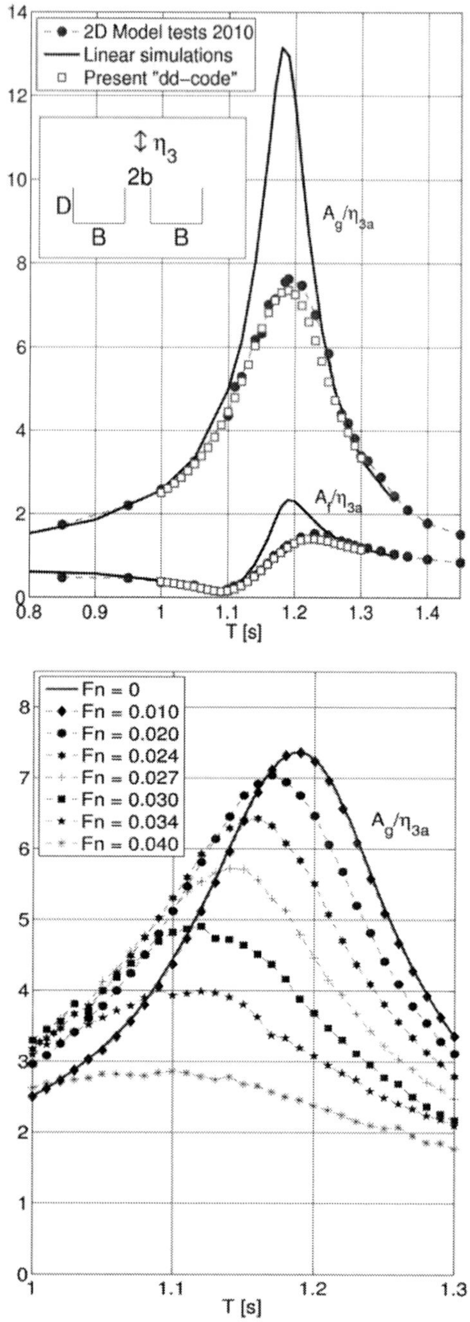

Figure 8: Validation of hybrid method for forced heave motion of a catamaran section with rectangular side hulls in unrestricted water with infinite depth. A_g - wave amplitude in the gap. A_f - far-field wave amplitude. Side-hull beam B = 0.36 m. Side-hull draft D = 0.18 m. Gap ($2b$) between side hulls = 0.18 m. Forced heave amplitude η_{3a} = 0.005 m. Left: No current, comparison between simulations using the hybrid method and experiments. Right: Effect of current, simulations only. $Fn = U/[2g(B+b)]^{0.5}$ (Kristiansen and Faltinsen [24]).

Here ω_e is the frequency of encounter, U the ship speed and g the gravitational acceleration. The phenomenon was observed for both studied staggers. The waves were propagating forwards and appeared 2D in the longitudinal vertical plane, as sketched in Figure 9. On the outside of largest hull, M1862, no such wave pattern was observed, while a weak tendency of this pattern was observed outside the smallest hull. The wave front appeared to move longer and longer forwards during the run but died out before it reached the bows. The rear waves were shorter and steeper than the first waves. The rear waves were almost breaking and reached the deck level of model M1862. It took some time to build up the large waves. Since the towing length was limited, it is not known how large the waves could have become. The wave trapping occurred in a very narrow τ range and may not be of concern from a practical point of view. The wave trapping is partly associated with the fact that there are no unsteady linear waves generated upstream by a ship when $\tau > 1/4$.

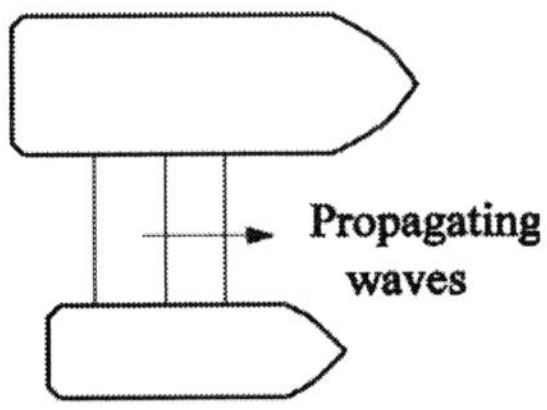

Figure 9: 2D trapped waves between the hulls (Ronæss [17]).

The two-time scale method for manoeuvring in regular waves has been validated by comparing with model test results of a turning circle manoeuvre reported by Yasukawa [22].

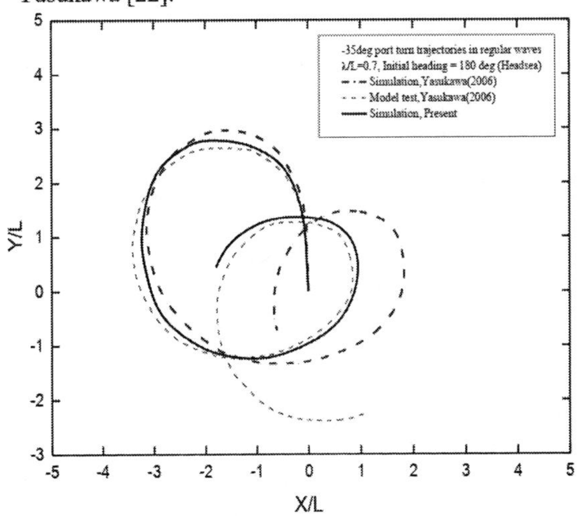

Figure 10: Comparison between numerical simulations and experiment of the trajectory of the SR-108 container vessels in a turning manoeuvre with 35 degrees port turn in regular waves with $\lambda/L = 0.7$. Initial wave heading is head sea. Initial speed corresponds to $Fn = 0.15$ (Xiang, personal communication 2011).

Yasukawa [25] did also zigzag and stopping manoeuvres. Free-running model tests were carried out in regular and irregular waves using the SR-108 container ship model. The model tests were conducted at Seakeeping and Maneuvering Basin, Nagasaki R & D Center, MHI. The regular wave tests were carried out in various wavelengths in initially head and beam sea waves.

The comparison with the model test in regular waves with wavelength $\lambda/L = 0.7$, initially head sea waves and Froude number 0.15 are presented in Figure 10 together with numerical simulation results by Yasukawa [22]. Our simulation results agree better with the experiments than Yasukawa's simulations. Since the two simulation methods agreed well for calm water, the reason must be how the mean wave loads are estimated. Yasukawa [22] uses a database based method. Two sets of added resistance are prepared, i.e., $Fn = 0$ and $Fn = 0.15$, for different wave headings. Then the added resistance is obtained via linear interpolation during the simulation. The calculation of added resistance at $Fn = 0.15$ is strip-theory based, by use of formulations by Maruo [26] and Fujii and Takahashi [27] while the calculations at zero speed are based on a 3D panel method. No speed effects are considered in evaluating the mean sway force and yaw moment. The latter fact is an error source to be further investigated.

5. SIMPLIFIED TIME-EFFICIENT NUMERICAL MANOEUVRING MODEL

Can we learn from the hybrid method described in the previous section for gap resonance and develop a rational and time-efficient 3D numerical manoeuvring model that is not based on a complete CFD method as illustrated with an example in section 2? The reasons why the described hybrid method was time-efficient was due to

- A fixed grid system
- Domain decomposition where viscous effects were accounted for where it mattered
- Linear free-surface conditions based on potential flow.

The consequence of the above-mentioned features is that we can only consider a single ship in unrestricted water in calm-water conditions. What we cannot account for with a fixed grid system is movable rudders and rotating propellers. Linear free-surface conditions with gravity wave effects imply that slender ships at limited Froude numbers must be analyzed. However, when $Fn <\approx 0.2$, we can apply the rigid free-surface condition for both slender and non-slender ships.

The described hybrid method was based on laminar flow and fixed separation points. When flow separation from continuously curved surfaces happens, the necessity of accounting for turbulent boundary layer locally upstream of a separation line is necessary. Further, turbulent boundary layers must obviously be considered in evaluating the viscous resistance. Free-vorticity flow will be turbulent even in model tests and matters for calculating wakes and thereby inflow to rudders and propellers. Hull resistance, the hull rectification of the inflow to the rudder, the hull-lift and viscous cross-flow effects as well as the viscous resistance reduction due to the "centrifugal" hydrodynamic force, are described in a rational way by the method.

The author's experience with describing separated flow is unexpected surprises. Since the described method is not a complete manoeuvring model and one has to separately account for rudders, propellers and thrust deduction factors, one should have in mind the fact that "the weakest link breaks the chain" and that one must not have imbalance in how accurately that sub problems are described. However, it is the accuracy of the relevant response that matters such as described by procedures for lightering and replenishment operations and by IMO manoeuvring criteria for ships longer than 100 m at service speed in terms of turning circle and zigzag manoeuvres and stopping ability. An important consideration is that some applications such as manoeuvring simulators require real time simulations. The latter constraint is hard to achieve with the described method.

6. CONCLUSIONS

CFD and simplified modular manoeuvring models are discussed with emphasis on ship-ship interaction. The importance of ship end effects on ship interaction necessitates a 3D analysis. A limited error analysis for a modular manoeuvring model is applied to a single ship in calm water. The variation in the drag coefficients has the largest influence while the influences of the wake factor and thrust deduction factor are not large. The flow rectification factors matter. The influence in the studied variations of the wake and thrust deduction factors is negligible. It is emphasized that the effect of a seaway on ship manoeuvring is sometimes wrongly modeled. When the frequency of encounter is not small, a two-time scale model for ship manoeuvring in waves is recommended. A 3D method to predict mean wave loads on a single ship as well as on interacting ships at forward speed is presented. Satisfactory agreement between numerical simulations and experiments of the trajectory of the SR-108 container vessel in a turning manoeuvre in regular waves is demonstrated. The presence of resonant water motion in the gap between two ships is discussed. If the ships have sharp corners at the lower gap entrance, viscous flow separation strongly limits the resonant piston mode amplitude with the consequence that potential-flow methods have limited applicability and cannot in a rational way be modified by using empirical viscous coefficients. A highly CPU saving approach using a local CFD solver at the flow separation area is presented. It is discussed how to generalize the method to ship manoeuvring. Future developments of the manoeuvring model in waves should include the effect of wind gusts.

7. ACKNOWLEDGEMENTS

The contributions from Dr. Renato Skejic (MARINTEK), Dr. Trygve Kristiansen (MARINTEK and CeSOS) and Xu Xiang (CeSOS) are appreciated.

8. REFERENCES

1. PIKE, D., '*Fast Powerboat Seamanship. The complete Guide to Boat Handling, Navigation, and Safety*', International Marine/ McGraw-Hill, 2004.

2. FALTINSEN, O. M. '*Hydrodynamics of High-Speed Marine Vehicles*', Cambridge University Press, 2005.

3. FALTINSEN, O. M. and TIMOKHA, A. N., '*Sloshing*', Cambridge University Press, 2009.

4. OBERHAGEMANN, J, and EL MOCTAR, O., 'Numerical and Experimental Investigations of whipping and springing of ship structures', *Proc. ISOPE*, Maui, 2011.

5. DURANTE, D., BROGLIA, R., MUSCARI, R. and DI MASCIO, A., 'Numerical simulations of a turning circle manoeuvre for a fully appended hull', *Proc. 28th ONR Symposium on Naval Hydrodynamics*, Pasadena, 12-17 Sept. 2010.

6. SKEJIC, R. , '*Maneuvering and Seakeeping of a Single Ship and of Two Ships in Interaction*', Doctoral thesis, Department of Marine Technology, Faculty of Engineering Science and Technology, Norwegian University of Science and Technology, Trondheim, Norway, 2008.

7. HOLTROP, J., 'A statistical re–analysis of resistance and propulsion data', *Int. Shipbuild. Progr.*, 31, 1984.

8. XIANG, X. and FALTINSEN, O. M., 'Maneuvering of two Interacting Ships in Calm Water', Proc. *PRADS2010, the 11th International Symposium on Practical Design of Ships and other Floating Structures*, September 19-24, Rio de Janeiro, Brazil, 2010.

9. TUCK, E. O. and NEWMAN, J. N., 'Hydrodynamic interaction between ships', *Proc. 10th Symp. on Naval Hyd.*, Cambridge, Mass., USA, pp. 35–70, 1974.

10. SÖDING, H., 'Prediction of ship steering capabilities', *Schiffstechnik*, 29:3–29, 1982.

11. HOLTROP, J. and MENNEN, G. G. J., 'An approximate power prediction method', *Int. Shipbuild. Progr.* 29, 1982.

12. ANKUDINOV, V., KAPLAN, P. and JACOBSEN, B. K., 'Assessment and principal structure of the modular mathematical model for ship maneuverabilty prediction and real–time maneuvering simulations',

MARSIM'93, St. Johns, Newfoundland, Canada, 1993.

13. SKEJIC, R. and FALTINSEN, O. M., 'A unified seakeeping and maneuvering analysis of ships in regular waves', *Journal of Marine Science and Technology,* 13:371–394, 2008. http://www.springerlink.com/content/88373xm3g12h4067/fulltext.pdf

14. FALTINSEN, O. M., MINSAAS, K., LIAPIS, N. and SKJØRDAL, S. O., 'Prediction of resistance and propulsion of a ship in a seaway', *Proc. 13th Symposium on Naval Hydrodynamics*, Tokyo, pp 505-529, 1980.

15. FALTINSEN, O. M., '*Sea Loads on Ships and Offshore Structures* ', Cambridge University Press, 1990.

16. XIANG, X. and FALTINSEN, O. M., 'Time domain simulation of two interacting ships advancing parallel in waves ', *Proc. 30th Int. Conf. on Ocean, Offshore and Arctic Engineering (OMAE2011)*, Rotterdam, The Netherlands, 2011

17. RONÆSS, M., '*Wave induced motions of two ships advancing on parallel course* ', Dr. Ing. thesis, Dept. of Marine Hydrodynamics, NTNU, Trondheim, 2002.

18. SHAO, Y-L. and FALTINSEN, O. M. , 'Numerical study of the second-order wave loads on a ship with forward speed', *26th Int. Workshop on Water Waves and Floating Bodies*, Athens, Greece, 2011.

19. GRUE, J. AND PALM, E., The mean drift force and yaw moment on marine structures in waves and current. *J. Fluid Mech.* 250:121-142, 1993.

20. FUJII, H. and TAKAHASHI, T. , 'Experimental study on the resistance increase of a ship in regular oblique waves', *Proc.14th ITTC*, Vol. 4, p.351, 1975.

21. NAKAMURA, S. and NAITO, S., 'Propulsive performance of a container ship in waves', *The Society of Naval Architects of Japan. Naval Architecture and Ocean Engineering*, Vol. 15, 1977.

22. YASUKAWA, H., 'Simulations of ship maneuvering in waves (first report: turning circle maneuvers)' (In Japanese), *Paper collections of Japanese Society of Naval Architects and Marine Engineers*, December, 2006.

23. KASHIWAGI, M., ENDO, K. and YAMAGUSHI, H., 'Wave drift forces and moments on two ships arranged side by side in waves', *Ocean Engineering*, 32 (5-6), pp. 529-555, 2005.

24. KISTIANSEN, T. and FALTINSEN, O. M., 'Gap resonances analyzed by a domain-decomposition method', *26th Int. Workshop on Water Waves and Floating Bodies*, Athens, Greece, 2011.

25. YASUKAWA, H. 'Simulations of ship maneuvering in waves (second report: zigzag and stopping maneuvers)' (In Japanese), *Paper collections of Japanese Society of Naval Architects and Marine Engineers*, June 2008.

26. MARUO, H., 'Wave Resistance of a Ship in Regular Head Seas', *Bulletin of the Faculty of Eng., Yokohama National Univ.,* Vol. 9 , pp.73-91, 1960.

27. FUJII, H. and TAKAHASHI, T., 'Practical calculation and application of added wave resistance in ship design' (In Japanese), *Journal of the Kansai Society of Naval Architects*, No.75, 1988.

9. AUTHOR BIOGRAPHY

Odd M. Faltinsen is Professor in marine hydrodynamics at CeSOS and Department of Marine Technology, NTNU. He was employed by DNV from 1968 to 1974 before becoming a professor at NTNU. Faltinsen has been member of 5 ITTC committees including 2 as chairman, and 3 ISSC Committees. He is a foreign member of the National Academy of Engineering, USA and the Chinese Academy of Engineering and has educated 50 PhD students.

SPT

As Agent for SPT Ltd.

AFRAMAX TANKERS • FULL SERVICE LIGHTERING • GLOBAL SUPPORT SERVICES • LNG

Ship to Ship Transfer
A Global Perspective
18 May, 2011

Cal Hayden
VP, Fleet Operations

Agenda

I Introduction to SPT
II History of the STS Industry
III Operating Perspective
IV Challenges and Opportunities
V Operating Limits
VI Risk Mitigation

SPT

SPT Organization

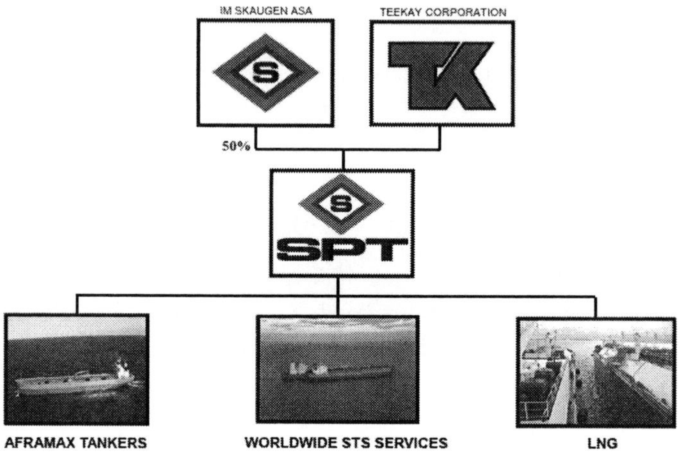

SPT

SPT Overview

Our History

- Founded by Jahre/Kosmos in 1981
- Acquired by I.M. Skaugen in 1989
- Teekay Shipping Canada, purchases 50% of SPT Oct 2003
- Acquired Melbourne Marine, Sunderland, May 2006 now known as SPT Marine Services.
- August 2007 – SPT Lightering Machines begin to arrive in the US Gulf.
- Over 14,000 operations completed - over 6.6 billion barrels lightered
- Approximately 8% of Total U.S. seaborne crude imports
- 830,000 average barrels lightered per day

SPT

SPT – Global STS Locations

- SPT operates in over 12 countries around the world with planned expansions in many other locations
- Current areas SPT can provide services:
  - US Gulf
  - US West Coast
  - Aruba
  - Panama
  - West Africa
  - Gibraltar
  - Malta
  - Cyprus
  - Northern Europe (multiple locations include, UK, Denmark, and The Netherlands)
  - Singapore
  - UAE

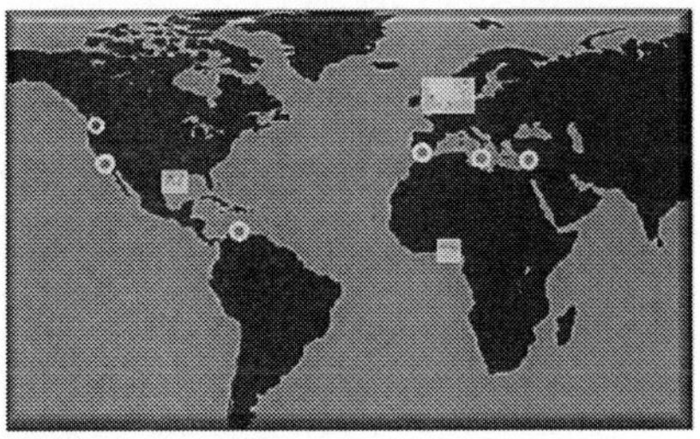

SPT

SPT's Safety and Environmental Protection Policy

SPT personnel are responsible for the prevention of accidents and pollution. SPT will ensure that approved safety standards and safe working practices are available to personnel.

Employees of SPT consider safety, health, and environmental protection as our highest priorities as we provide service that conforms to our customer's requirements.

Our targets are clear – *zero injuries, zero negative impact to the environment and zero equipment damage.*

We recognize the privilege of conducting business demands excellence in our execution of all aspects that affect safety and environmental performance. SPT will:

- Ensure the safety of our employees and third parties through a safety management system.
- Comply with appropriate laws, regulations, and company policies and procedures as well as utilizing accepted industry guidelines.
- Asses identified risks and establish appropriate safeguards or prohibit the practice.
- Monitor and openly report on our performance and seek continuous improvement.

SPT

SPT's Safety and Environmental Protection Policy

- Our targets are clear - *zero injuries, zero negative impact to the environment and zero equipment damage.*

SPT

ISO/ISM Certification

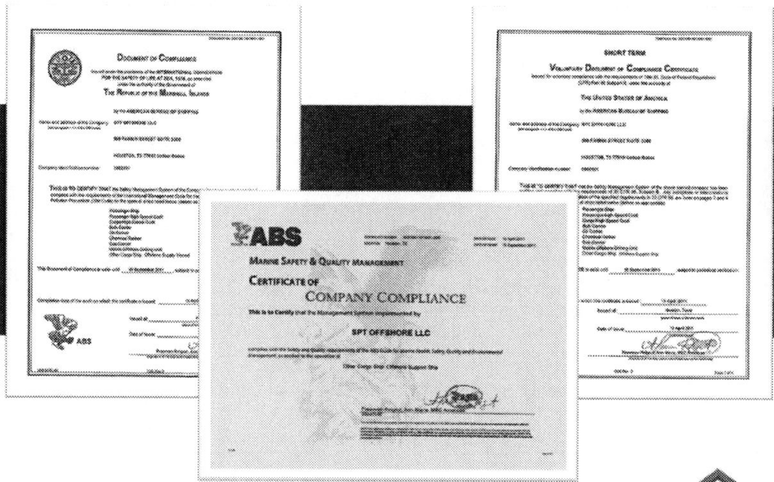

SPT

Agenda

I Introduction to SPT

II History of the STS Industry

III Operating Perspective

IV Challenges and Opportunities

V Operating Limits

VI Risk Mitigation

What is Lightering?

Transfers of Cargo From One Ship to Another

Products Lightered

Crude Oil

Liquefied Natural Gas

Refined Products

Containers

SPT

Reasons for Lightering

- Draft and Size Restrictions
- Port and Terminal Restrictions
- Cargo Blending
- Breaking Bulk Cargoes
- Consolidating Cargoes (Reverse Lightering)
- Strategic & Economic Purposes
- Emergency & Salvage

SPT

Agenda

I Introduction to SPT
II History of the STS Industry
III **Operating Perspective**
IV Challenges and Opportunities
V Operating Limits
VI Risk Mitigation

SPT

Lightering 101 Video

Click Here

SPT

Agenda

I Introduction to SPT
II History of the STS Industry
III Operating Perspective
IV Challenges and Opportunities
V Operating Limits
VI Risk Mitigation

SPT

Challenges

- **Regulation:**
  - MARPOL Emission Regulations
  - MARPOL Annex 1 Revision STS
  - ILO MLC 2006 Work Rules
- **Safety Performance:**
  - Hazardous Offshore Work Environment
- **Competitors:**
  - Low Barriers to Entry
  - Varying Degrees of Risk Aversion
- **Public Perception:**
  - Risky Operation
  - Substandard Tonnage
  - NIMBY (not in my back yard)

SPT

Opportunities

- **Regulation:**
  - SPT Tankers with low emission engines
  - SPT recognition for high operating standards
- **Safety Performance:**
  - SPT is a leader in offshore safety.
    - SPT has lightered over 7.3 billion barrels of oil Year To Date. Loss to the environment about 0.00000014% of barrels handled
    - 697 days since last Lost Time Injury
    - 504 days since last SS – STBL Touching
  - ISO and ISM Compliance
- **Competitors:**
  - Changing regulation will raise the bar for our competitors and level the playing field.
- **Public Perception:**
  - Industry outreach programs

SPT

Agenda

I Introduction to SPT

II History of the STS Industry

III Operating Perspective

IV Challenges and Opportunities

V Operating Limits

VI Risk Mitigation

SPT

Guidelines, Rules and Regulations

- OCIMF Ship-to-Ship Transfer Guide, 4th edition (Petroleum)
- ISGOTT
- USCG Code of Federal Regulations
  - Designated lightering zones
  - Operating conditions
  - Work hours
- Local Laws and Regulations
- SPT Lightering manual
- Oil company guidelines
- Owner specific procedures (BW, Tanker-Pacific)
- ITOL guidelines for cargo hose age and testing

SPT

Weather Restrictions

- Maximum conditions, shall not attempt to moor when:
    - Wind is 30 knots or higher
    - Wave height is 3 meters or more
    - Swell height exceeds 10 feet
  - Must also take into consideration
    - Size of vessels
    - Sea room available
    - Other factors
- Transfer operations to cease and hoses disconnected when
    - Wind exceeds 44 knots
    - Wave height exceeds 5 meters
- Safe distance –2 nautical miles of offshore structures
- Both vessels must maintain a bridge watch at all times
- Both vessels must be able to maneuver at all times

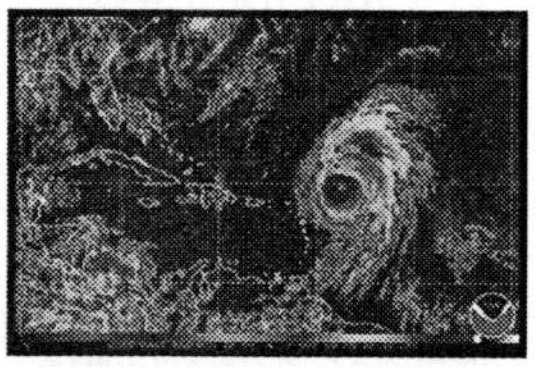

SPT

Work Hour Restrictions

Work hours

- Apply to both vessels and mooring master
- STCW or OPA 90 requirement
- SPT Policy
  - 15 hours in 24
  - 36 hours in 72

- ILO MLC 2006 coming into force in the near future
  - 14 hours in 24
  - 72 hours in 7 days (effectively 10 hour work day)
- Restrictions apply to all SPT operations

SPT

Agenda

I Introduction to SPT
II History of the STS Industry
III Operating Perspective
IV Challenges and Opportunities
V Operating Limits
VI Risk Mitigation

SPT

Risk Mitigation

We manage risk by focusing on:

- People
- Marine Assurance
- Equipment Selection and Maintenance

SPT

Mooring Masters

Selection Criteria:

- Class 1 Master's License
- Tankerman PIC Endorsement
- 3-years tanker command
- Prior lightering experience
- Prior Mooring Master experience
- Full-time Employees

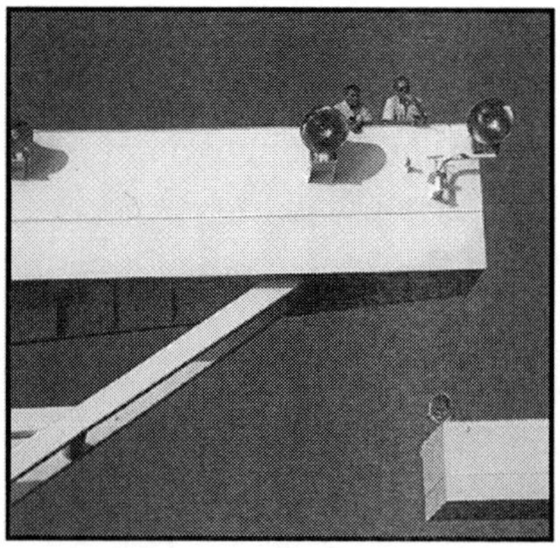

SPT

Mooring Masters

Training and Development Program:

- Office training on the business of lightering
- Orientation and training on STS Equipment construction, maintenance and repair at our offshore facility in Freeport, TX
- On the Job training in all aspects of lightering under supervision of approved SPT Mooring Masters
- Sign-Offs from three SPT Senior Mooring Masters
- Ship Handling Simulator Training:
  - Phase II within 12 months of hire
  - Phase III within 24 months of hire
  - Re-certification each 60 months
- Periodic Check Rides by Senior Mooring Masters
- Annual Performance Reviews

SPT

SPT Mooring Masters

Mooring Masters	Years of Experience	Years with SPT	Number of Operations
Jerry Barrett	6	4	83
Jerry Burgoyne	10	<1	0
Wolfgang Chidduck	10	1	76
Dick Berglund	8	6	200
Donald D'Souza	10	8	320
Nick Durant	6	2	51
Dave Edwards	10	3	84
Martin Eve	6	1	25
Giorgio Gerratana	5	4	9
Armstrong Hedgecock	21	5	1287
Nick Howard	6	<1	14
Kari Laine	9	5	138
Rajeev Khanna	15	10	521
David Madden	4	<1	9
Jari Makineste	<1	<1	0
Piotr Maziarksi	4	<1	0
Bogdan Mizstal	6	1	89
Anderson Scott	31	30	2110
Daniel Scroggie	11	6	278
David Yorke	6	<1	222

SPT

Marine Assurance

- All vessels are vetted for lightering suitability
- ITOL questionnaire
- Sire/USCG (PSIX) database
- Vessel performance input from Mooring Masters
- SPT Vessel Database
- Issues/concerns are forwarded to owner/operator for action/follow up

SPT

Marine Assurance

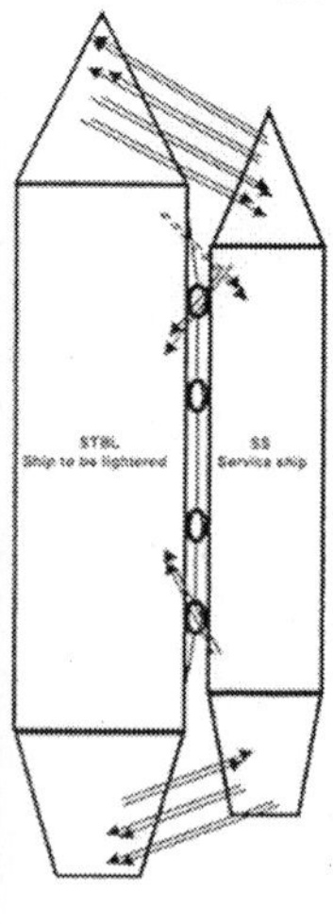

STBL MOORING PROCEDURES

STBL	SS
4 HEAD LINES	4 HEAD LINES
2 STERN LINES	2 FORWARD SPRING LINES
	2 AFT SPRING LINES
	4 STERN LINES

STBL - FORWARD GANG

NO. 1 RECEIVE 4 HEAD LINES FROM SS
NO. 2 GIVE 4 HEAD LINES FROM STBL
NO. 3 GIVE 2 FORWARD SPRING LINES FROM STBL (WHEN THESE LINES ARE REQUIRED)

STBL - AFT GANG

NO. 1 RECEIVE 2 FORWARD SPRING LINES FROM SS
NO. 2 RECEIVE 2 AFT SPRING LINES FROM SS
NO. 3 RECEIVE 4 STERN LINES FROM SS
NO. 4 GIVE 2 STERN LINES FROM STBL

TWO HEAD LINES AND 2 FORWARD SPRING LINES FROM SS WILL BE GIVEN AT THE SAME TIME. PLEASE ARRANGE FOR AFT GANG TO STAND-BY FORWARD TO RECEIVE SS SPRING LINES. WHEN SS SPRING LINES ARE ON BOLLARDS, AFT GANG PROCEED AFT TO RECEIVE SS BACK SPRINGS, THEN SS STERN LINES.

FORWARD GANG TO REMAIN FORWARD AND CONTINUE RECEIVING 4 HEAD LINES FROM SS, AND GIVE 4 HEAD LINES. (IF 2 FWD SPRING LINES FROM STBL BE USED, FWD GANG WILL GIVE THESE LINES, AFTER COMPLETING ALL HEAD LINES)

EACH SHIP MUST HAVE AT LEAST 2 HEAVING LINES AND 2 STRONG MESSENGER LINES (RIGGED AND READY, FORWARD AND AFT)

STBL MUST HAVE CHAIN STOPPERS RIGGED AND READY, IF SS HAS WIRE PENNANTS ON MOORING LINES

SUITABLE NEWLY SHARPENED FIRE AXES MUST BE PLACED AT ALL MOORING STATIONS

ALL MOORING ROPES/WIRES, AND FENDER WIRES MUST BE GREASED PROPERLY, AND REGULARLY

SPT

Equipment Selection and Maintenance

- Manufactured by Yokohama Rubber Company
- Size 3.3 x 6.5 and 4.5 x 9.0
- Weight 7300 kg
- Pressure 0.5 bar
- Max Deflection 60%
- Max Berthing Speed .8 Kn.
- Max Berthing Angle 10°
- Chain Net with Aircraft Tires
- All Fenders are Retired after 15 years

SPT

Fender Burst Tests

SPT

Cargo Hoses

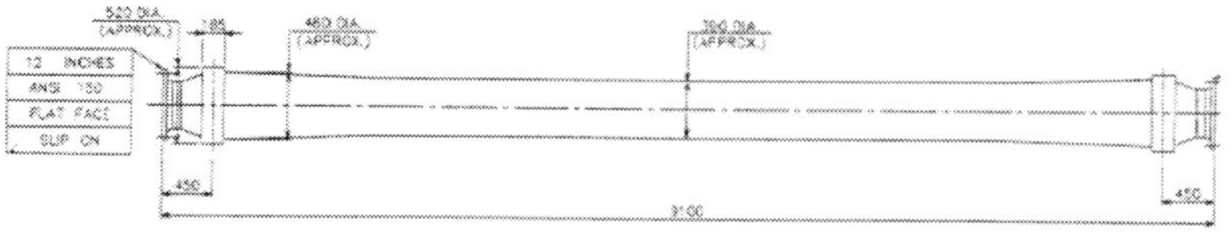

- Yokohama Rubber Company
- Fully Flexible
- 300 mm diameter (12")
- Weight each length: 720 kg
- Maximum working pressure: 21 Bars (300 PSI)
- Electrically Semi-Continuous
- Transfer rate (2 strings) 80,000 barrels /hour

SPT

Yokohama High Velocity STS Hose (Double Walled)

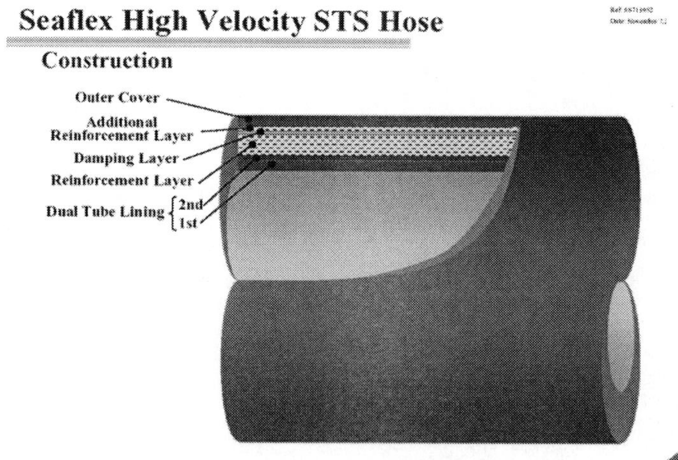

SPT

Emergency Release Coupling

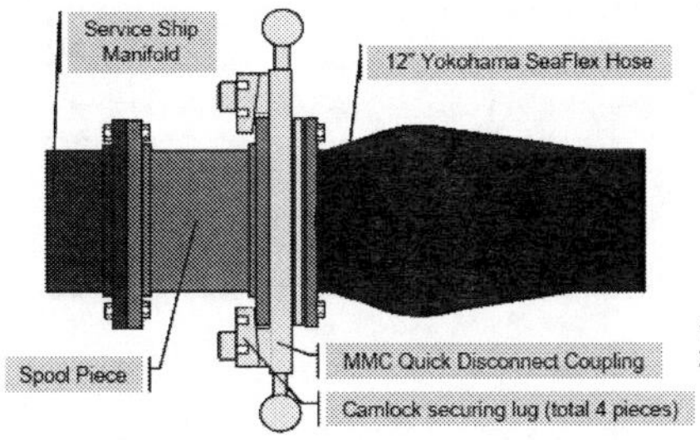

1. Support cargo hose using ship's crane.
2. Use camlock spanner to open 4 camlocks. Camlock spanner is attached with a chain to the outer ring.

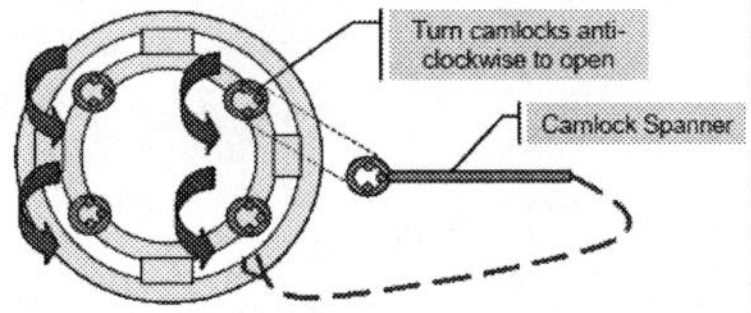

3. Once camlocks are open, place blind flange in position. Re-secure camlocks using camlock spanner.
4. Hose is now ready for transfer.

Revision 1; 3/15/02

SPT

Hose Maintenance

- **Testing**
  - USCG requires every 12 months
  - SPT policy, test every 6 months
  - Hoses Rotated
  - Stainless steel studs/nuts
  - Throughput Recorded

- **Age Policy**
  - 4 years or 4500 hours
  - Policy result of testing Yokohama (witnessed by ABS)

SPT

Hose Testing

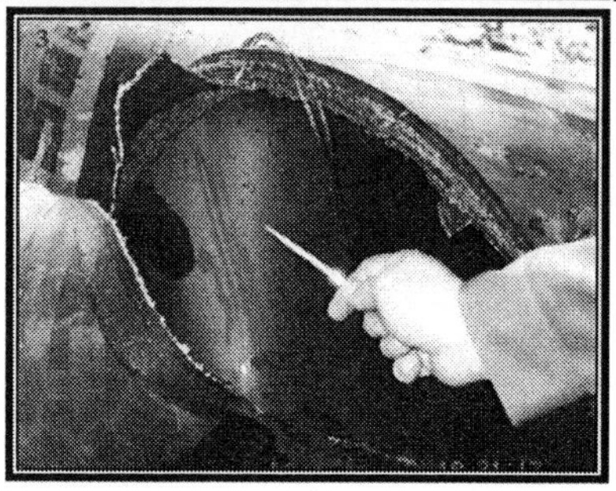

- Burst Test - 105 Bars
  - Kink Testing
  - 90° - 20,000 bends
  - 180° - 5,000 bends

SPT

A COMPARATIVE STUDY OF PIV EXPERIMENTS AND NUMERICAL SIMULATIONS OF FLOW FIELDS AROUND TWO INTERACTING SHIPS

T Arslan and **B Pettersen,** Dept of Marine Technology, Norwegian University of Science and Technology, Norway
J Visscher and **C Muthanna,** Norwegian Marine Technology Research Institute, MARINTEK, Norway
H I Andersson, Dept of Energy and Process Engineering, Norwegian University of Science and Technology, Norway

SUMMARY

The paper reviews the measurements and calculations of three dimensional (3D) unsteady flows past a pair of ship sections in close proximity. The ship sections have different breadth and draft and represent typical ship-to-ship marine operation in a cross current. The behaviour of the vortex-shedding around the ship hulls is investigated numerically for all cases by CFD methods. For the two ship sections, simulations for certain Reynolds numbers are done by using Smagorinsky LES turbulence model. Finally the cross flow past the ship sections in side-by-side position is simulated and vortex interaction between the two bluff bodies is reviewed by using the software Fluent. The numerical predictions are compared with PIV results taken in a circulation tank.

NOMENCLATURE

The following are the major symbols used:

3D	Three dimensional
AMG	Algebraic multi-grid
BARC	Benchmark on the Aerodynamics of a Rectangular Cylinder
CCD	Charged Coupling Device photosensor
CFD	Computational Fluid Dynamics
CWT	Circulating Water Tunnel
FOV	Field of view
KRISO	Korean Institute of Ships and Ocean Engineering
KVLCC	KRISO Very Large Crude Carrier
LDV	Laser Doppler Velocimetry
LES	Large Eddy Simulation
N-S	Navier-Stokes
PISO	Pressure Implicit with Splitting of Operators
PIV	Particle Image Velocimetry
RMS	Root mean square
SGS	Subgrid-scale
URANS	Unsteady Reynolds Averaged Navier-Stokes
U	Velocity (m s^{-1})
VOF	Volume of fluid
ν	Kinematic viscosity (m^2 s^{-1})

1. INTRODUCTION

The present work is motivated by phenomena occurring in marine operations with two ships loading or unloading in a side-by-side position in a cross current. The transfer operation of oil between a large tanker and a shuttle tanker is critical for oil companies. To understand the behaviour of the two ships near each other in a cross current is very important for safety of the operation. Knowing the detailed flow conditions around the two ships gives insight to people managing the operation, ship masters and pilots, in addition to better background for developing more realistic ship manoeuvring simulators for dedicated training on marine operations. Most of today's simulator models are based on data consisting of hydrodynamic forces and moments acting on the whole ship, either as interaction forces or corrections to existing deep water conditions.

Since 2008, MARINTEK has managed a 4-year long research project aiming for investigating hydrodynamic aspects and control strategies for ship-to-ship operations. The project focuses on traditional lightering operations and better understanding of the complex water flow close to and in the gap between two ships operating in close proximity.

Motivated by ongoing research activities in CFD [1] and experimental capabilities with PIV in our circulating water tunnel, we have studied ship-to-ship interaction in more detail. Both experiments and computations have been done with actual ship section geometries with the motivation to provide detailed field quantities (velocities, pressure).

In Figure 1, a KVLCC tanker and an Aframax shuttle tanker are seen in a transfer operation. Lataire et al. [2] reported experimental results of forces and moments acting on two ships advancing close to each other .

Figure 1: Two ships in load transfer operation.

Aero and hydrodynamics of bluff bodies have many industrial applications from civil engineering and marine related areas. The complexity of the flow structures brings forward some challenges in calculation of the flow field even at low Reynolds numbers. URANS models have been used for simulating such flows but today LES models are becoming capable and more popular to use for calculating the flow around bluff bodies.

Calculations of the flow around a square section at Reynolds number 22000 by using URANS and LES models are presented by Rodi *et al.* [3]. Experimental work by using PIV is reviewed by Lyn *et al.* [4]. The flow past a 5:1 rectangular cross section is simulated by Schewe [5] and Bruno *et al.* [6] at several Reynolds numbers. Liu *et al.* [7] investigated the flow past two square sections in tandem arrangements by doing experiments. Also Bhattacharyya *et al.* [8] performed simulations for the same geometry at low Reynolds numbers, close to a wall. These have acted as benchmark studies in this project.

2. NUMERICAL METHODS

2.1 COMPUTATIONAL DOMAIN

The effect of the bow and stern of the two ships are not considered in this work. Only the 3D parallel midship sections of both ship hulls are taken into account in the simulations. The 3D computational domain with the incoming flow direction can be seen as a grey zone in Figure 2. The entire computational domain is shown in Figure 3. The computational boundaries are far enough away from the ship sections in order to satisfy physical conditions. The distance between the ship sections and the inlet is $7B_1$. The depth of the domain is $2.2B_1$, and the spanwise length of the domain is B_1 where B_1 is the breadth of the larger ship. The outlet is $18B_1$ from the ship sections. The problem can now be considered as the flow around two interacting bluff bodies. The water depth is selected relatively small in order to see shallow water effects.

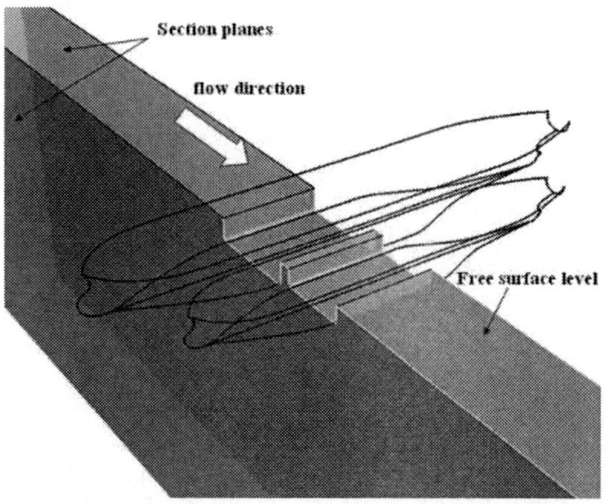

Figure 2: 3D computational domain with the two parallel midship sections.

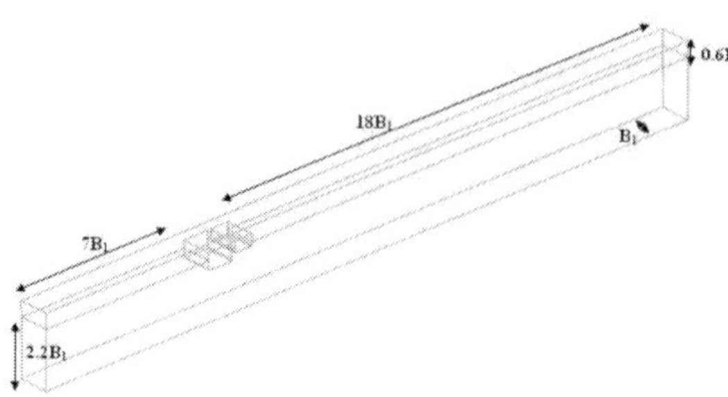

Figure 3: The dimensions of the entire computational domain. (B_1 is the breadth of the larger ship)

In Figure 4, the part of the computational domain close to the ship sections is shown. The distance from the free surface level to the top of the domain is A. The dimension of the ships sections, drafts and the gap between the ship sections are show in Table 1. The dimensions are in model scale 1:210. The problem can now be considered as flow around two bluff bodies and the interaction between them.

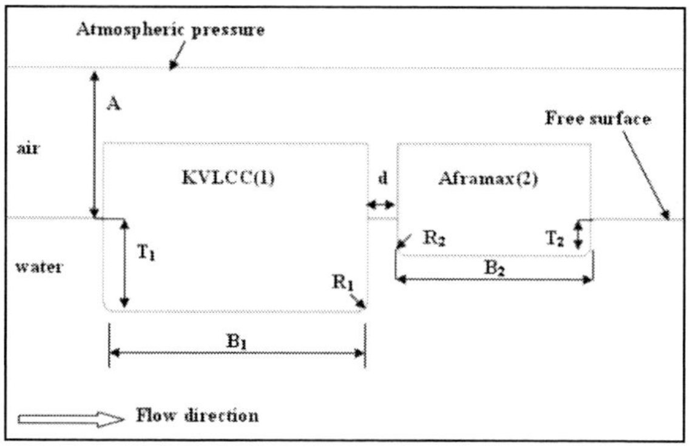

Figure 4: Details of the computational domain and ship dimensions.

	KVLCC (1) mm	Aframax (2) mm
Breadth (B)	275	200
Draft (T)	100	40
Bilge Radius (R)	10	20
Gap (d)	30 mm	
Water depth	610 mm	
A	160 mm	

Table 1: Dimensions of the computational domain in model scale. (1:210)

2.2 BOUNDARY CONDITIONS AND GRID RESOLUTION

In the numerical simulations, two incoming flow velocities are considered as inlet boundary conditions: 0.254 m/s and 0.0254 m/s. They correspond to Froude numbers 0.25 and 0.025 respectively. Froude numbers are calculated based on draft of the large ship section. Reynolds numbers based on breadth of the large ship are 68000 and 6800 respectively. Turbulence intensity in the inlet is taken as zero (ideal smooth flow). No-slip wall boundary condition is applied to the bottom of the computational domain in order to simulate shallow water conditions. The water depth is matching the value used in the experiments. At the top of the domain, the parts of the inlet and outlet which are open to the atmosphere, zero atm. pressure is used. The side walls are considered as periodical boundary conditions.

The computational domain consists of two parts: water and air. Total number of elements is about 1.4 million in the total (air+water) domain. An unstructured hexahedral grid is used in the domain except near the ship sections. In order to capture sufficient mesh resolution inside the boundary layer ($y^+ < 1$), high aspect ratio structured elements are used as layers expanding from ship surfaces normal to the direction of the wall. The number of elements in spanwise direction is 30.

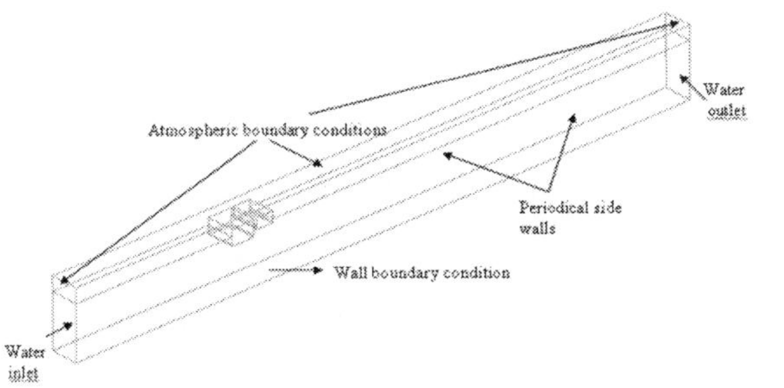

Figure 5: Boundary conditions at each surface of the computational domain.

2.3 CALCULATIONS

For calculations, the commercial software Fluent is used in this work. The code is using the Finite Volume Method and is able to solve unsteady, 3D, Navier-Stokes equations by using LES and/or RANS turbulence modeling. Also two phase flows including free-surface flows can be solved.

To calculate the multiphase flow, one of the surface-capturing methods, VOF, is used. The governing equations solved are a set of Navier-Stokes equations that are shared by gas (air) and liquid (water) phases or their mixture. The presence of two phases (air and water) in the computational domain is represented by a volume fraction [9].

In present work, the dynamic Smagorinsky model is used as LES turbulence model. The subgrid-scale (SGS) turbulence is modeled using the Smagorinsky's [10] eddy-viscosity model in combination with a dynamic procedure [11].

Fluent employs a cell-centered finite-volume method that permits use of polyhedral cells. Higher order methods are used for discretization of the N-S and VOF equations. For pressure-velocity coupling, the PISO algorithm is used. The discretized equation system is solved with AMG method by using second order time advancement schemes. The time increment is 0.0025 seconds.

In order to validate Fluent for flow past bluff bodies, vortex shedding behind a square section is simulated for Reynolds number 22000 based on the length of the square. As seen in Table 2, Fluent predicts the force coefficients well while the Strouhal number is underpredicted. The spanwise length of the 3D section is 4L, where L is the side length of the square. Strouhal number is calculated based on the frequency of the lift force f, freestream velocity U and L.

	St =fL/U	$C_{d,mean}$	$C_{d,rms}$	$C_{l,rms}$
CFD (num.)	0,121	2,17	0,2	1,29
Rodi [3] (exp.)	0,13	2,2	0,14	1,01
Lyn [4] (exp,)	0,132	2,1	-	-

Table 2: Numerical and experimental results for flow around the square (Reynolds number is 22000, Re=UL/ν)

A rectangular section which has 5:1 length to depth ratio is another ongoing benchmark case for calculation of the flow past bluff bodies (Bruno et al. [6]). In this study, simulations are done for Reynolds number 132000 based on the breadth by using the Smagorinsky LES model. The LES calculated Strouhal number, mean drag coefficient and RMS of lift coefficient are 0.121, 0.993 and 0.6 respectively compared to the reported experimental results by Schewe [5]: 0.111, 1.029 and 0.1. The drag coefficient is predicted well, while the lift coefficient is overpredicted. This may be because of the sensitivity of the pressure field and the separations at the top and bottom of the square body to the inlet turbulence conditions. In Figure 6, the instantaneous spanwise vorticity is illustrated. The spamwise length of the rectangular section is 5 times its depth.

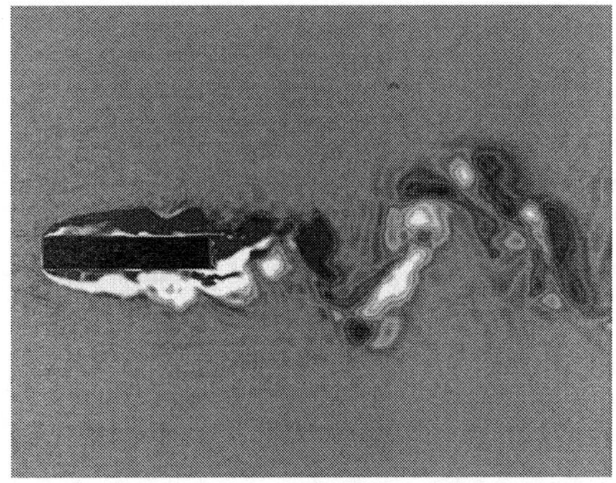

Figure 6: Instantaneous spanwise vorticity contours at the mid-section for 5:1 rectangular section (BARC benchmark case)

3. EXPERIMENTAL METHOD

3.1 TANK AND MODEL SETUP

The tests were carried out in the new Circulating Water Tunnel (CWT), a laboratory that is jointly operated by MARINTEK and NTNU's Dept. of Marine Technology since summer 2010. The tunnel has a transparent test section of 0.61 by 0.61 m in cross-section and 2.50 m in length and can be operated with or without a free surface. The flow speed can be varied between 0.01 and 1 m/s with a maximum turbulence intensity of 1%.

Within the experiments, flow velocities between 0.01 and 0.25 m/s were used. Results from 0.025 and 0.25 m/s are presented in Figures 11-14. They were defined through the average discharge over a cross-section normal to the inflow direction and the pump frequency adjusted accordingly. The actual pump rates were kept constant afterwards. Reference PIV measurements without any models show deviations in flow velocity of less than 1% from the theoretical mean value. The deviations visible in Figure 7 are due to optical aberrations and limited to the outer regions in the comparably large field-of-view.

The model arrangement with the CCD camera can be seen in Figure 8. Two ship hull sections were installed in the same configuration as used in the numerical study, as indicated in Figure 9. The models fill the entire cross-section of the tank, according to the two-dimensional character of the flow around the parallel midship sections. However, the confined water depth can not be ignored since it reduces the available cross-section from 0.36 m^2 to 0.3 m^2. Following the continuity equation, the average flow velocity increases by a factor of 1.2 in the cross-section below the larger hull section. This needs to be taken into account and adjusted for when compared to numerical results from infinite water depth. Within this study, calculations with an identical water depth are performed.

3.2 PIV SYSTEM AND VISUALIZATION

The PIV measuring system consists of a PCO Pixelfly CCD camera, a Quantel BSLT220 Nd:YAG laser and an ILA Net Synchronizer timing unit. The image processing was done with the ILA VidPIV 4.7 software. The typical sampling rate was 5 Hz.

In this study, one camera was used to measure instantaneous velocity vectors in a field of 0.3 by 0.2 m that covered the entire gap and parts of the two bodies up to the water surface (Figure 9). By using a single-camera PIV system, the flow could be studied inside the gap without optical occlusions that would appear in stereoscopic setups.

In addition to the PIV measurements, the experimental setup was used to visualize the flow around the larger hull by means of dye injection, as seen in Figure 10. The size and variability of the separation zone was of particular interest here.

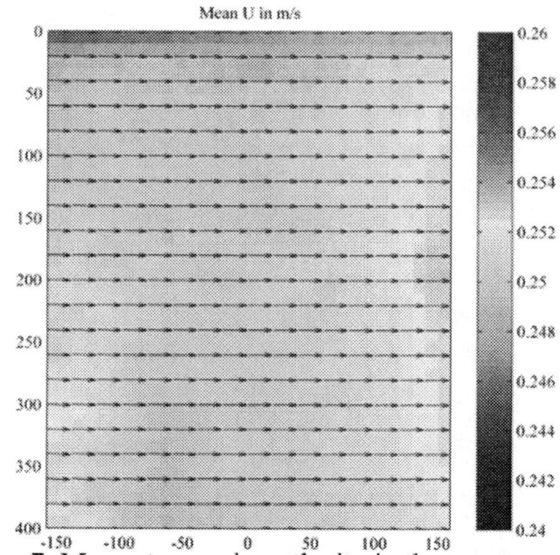

Figure 7: Mean streamwise velocity in the empty channel, measured in a 0.3 by 0.4 m field-of-view.

Figure 8: Model arrangement in the circulating water tunnel with PIV camera

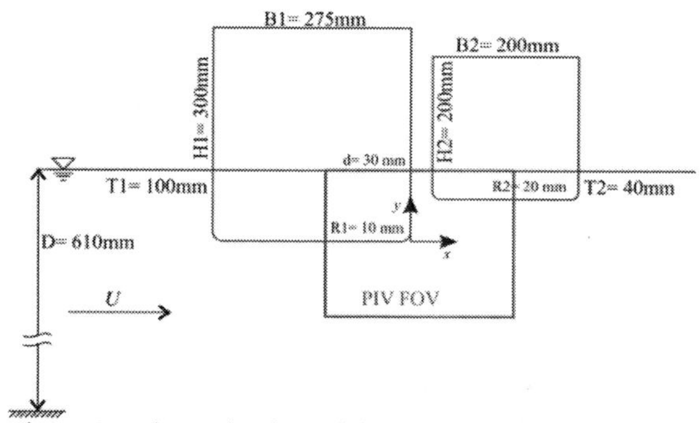

Figure 9: Schematic view of the hull sections in model scale with the PIV field-of-view (FOV) marked.

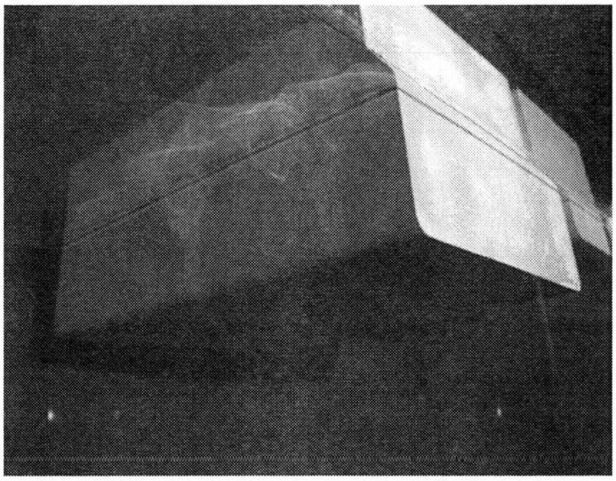

Figure 10: Dye visualization of the flow around the larger body at 0.25 m/s inflow velocity.

4. RESULTS

Figures 11 to 14 show the first results from numerical calculations on the left hand side versus experimental results on the right hand side, both on the same scale. The time-averaged fields of the velocity magnitude and of the streamwise component are normalized by the respective free stream velocity.

In Figure 11, the velocity magnitude contours in the midspan of the ship sections are shown for the free stream velocity 0.025 m/s. The shapes of the contours from calculations and experiments agree well, although the calculation slightly overpredicts close to magnitude at the bottom of the large ship section. The thickness and length of the separation bubble close to the large ship bottom is overpredicted as well. The velocity fields are matching well near the gap and the downstream side close to the small ship section. The same trend can be seen in the streamwise component of the velocity in the Figure 12, which indicates the position of the reattachment point through the turnover from negative to positive values. Some differences between numerical and experimental results regarding the separation bubble can be seen. The flow separation on the upstream edge can only be estimated because of the limited FOV of the PIV measurements.

In Figure 13, the contours of the mean velocity magnitude are shown for a free stream velocity of 0.254 m/s. The contour shapes are compared well in PIV and CFD results. The velocity field and the recirculation zone close to the bottom of the small ship section are matching well. As seen in Figure 14, the streamwise velocity component reveals that the size of the separation bubble is slightly underpredicted in the calculations. However, the locations of the reattachment points are matching reasonably well.

5. CONCLUSIONS

The present work is demonstrating comparative results between CFD calculations and PIV measurements. The present results and experience from other studies suggest that the flow around the first body being highly dependent on the inflow conditions and the detailed structure of the initial separation on the leading rounded bilge (R1). Experimental and numerical studies are therefore initiated to study variations in this particular part of the geometry. These would cover e.g.:

- Different configurations of the free surface. The experimental facility can be operated both with closed or free surface. Numerically, the surface can be treated through a free-slip boundary condition or with a two-phase flow.
- Limit the geometry to the first, larger hull only.
- Variation of the leading edge curvature, including a sharp edge.

It is expected that an even closer agreement of experimental and numerical results can be achieved when these aspects are known in detail.

With respect to the main research project it is also planned to extend the present ship hull configuration into different gap and loading conditions for both ships, including changing the flow direction. The experimental setup could be extended with more cameras in a side-by-side configuration to enlarge the field-of-view without losing spatial resolution, or by application of stereoscopic PIV to study three-dimensional flow features that might persist in the 2D geometry.

As seen in the Figure 12 and 14, there is a complex flow in the gap between the two ships. In order to understand the structure of the flow in this region, more detailed numerical studies will be performed. Additional simulations and experiments with only one ship section can give more information about the separation and reattachment of the flow on two ships.. This also gives more insight into flow inside the gap.

Instead of calculating multiphase flow with the VOF model, the usage of free slip boundary conditions without modelling the air side will be considered in the future. It is important not only for saving computing time but also to understand the free surface effects on the flow field and the exact location of flow separation. The reason of slight difference between experimental and numerical results can be investigated by using different sub-scale models for LES calculations with high resolution mesh in the simulations.

It is important to see the effect of the bow and stern to flow field between two ships. The full 3D study of the two ships, including the bow and stern, will be performed later.

6. ACKNOWLEDGEMENTS

The authors thank the Research Council of Norway (NFR), MAROFF Research and Development Program, for the financial support to this project. Computing time was supported by NFR and by NTNU.

7. REFERENCES

1. Marine-CFD project, http://www.ivt.ntnu.no/marinecfd

2. LATAIRE E., VANTORRE M., DELEFORTRIE G., 'Captive model testing for ship to ship operations', *Proceeding of the International Conference on Marine Simulation and Ship Maneuverability, Panama City*, 2009.

3. RODI, W., FERZIGER, J. H., BREUER, M., POURQUIE´, 'Status of Large Eddy Simulation: Results of a Workshop', *ASME J. Fluids Eng., Vol. 119 pp. 248-262*, 1997.

4. LYN, D. A., EINAV, S., RODI, W., AND PARK, J. H., 'A Laser-Doppler Velocimetry Study of the Ensemble-Averaged Characteristics of the Turbulent Near Wake of a Square Cylinder', *J. Fluid Mech., 304, pp. 285–319,* 1995.

5. SCHEWE, G., 'Influence of the Reynolds-number on flow-induced vibrations of generic bridge sections', *Proceeding of the Bridges International Conference, Dubrovnik, Croatia, 351-358.*, 2006

6. BRUNO L., FRANSOS, D., COSTE, N., BOSCO A., '3D flow around a rectangular cylinder: a computational study', *Proc. BBAA VI International Colloquium, Milan, Italy*, 2008.

7. LIU C-H., CHEN J. M., 'Observations of hysteresis in flow around two square cylinders in a tandem arrangement', *Journal of Wind Engineering and Industrial Aerodynamics, Vol. 90 pp. 1019-1050*, 2002.

8. BHATTACHARYYA S., DHINAKARAN S., 'Vortex shedding in shear flow past tandem square cylinders in the vicinity of a plane wall', *Journal of Fluid and Structures, Vol. 24 pp. 400-417.*, 2008.

9. HARLOW E. H., WELCH J. E., 'Numerical Calculation of the Time Dependent Viscous Incompressible Flow with Free Surface', *Physics of Fluids, Vol. 8, No. 12, pp. 2182-2189*, 1965.

10. SMAGORINSKY J., 'General circulation experiments with the primitive equations. I. The basic experiment', *Monthly Weather Review 91, 99–164.*, 1963.

11. GERMANO M., PIOMELLI U., MOIN P., CABOT W., 'A Dynamic Subgrid-Scale Eddy Viscosity Model', *Physics of Fluids A, Vol. 3, No. 7, pp. 1760-1765*, 1991.

8. AUTHORS BIOGRAPHY

Tufan Arslan holds the current position of a PhD scholarship at NTNU. He is responsible for CFD calculations in relevant projects. His previous experience includes numerical studies on marine hydrodynamics.

Jan Visscher holds currently the position of a Research Scientist at MARINTEK. His experience includes experimental methods and flow measurement techniques in marine hydrodynamics.

Bjørnar Pettersen holds a PhD in marine hydrodynamics. He is professor at Department of Marine Technology at NTNU. Main research activity is viscous flow simulations around bluff bodies.

Chittiappa Muthanna holds the current position of senior research scientist at MARINTEK. He is responsible for managing commercial projects and development of measurement instrumentation and techniques. His previous experience includes experimental methods, and flow measurement techniques and facilities in fluid dynamic studies.

Helge I. Andersson holds a PhD in fluid mechanics. He is a professor at Dept of Energy and Process Engineering. Main research activity is turbulence modelling and DNS.

APPENDIX

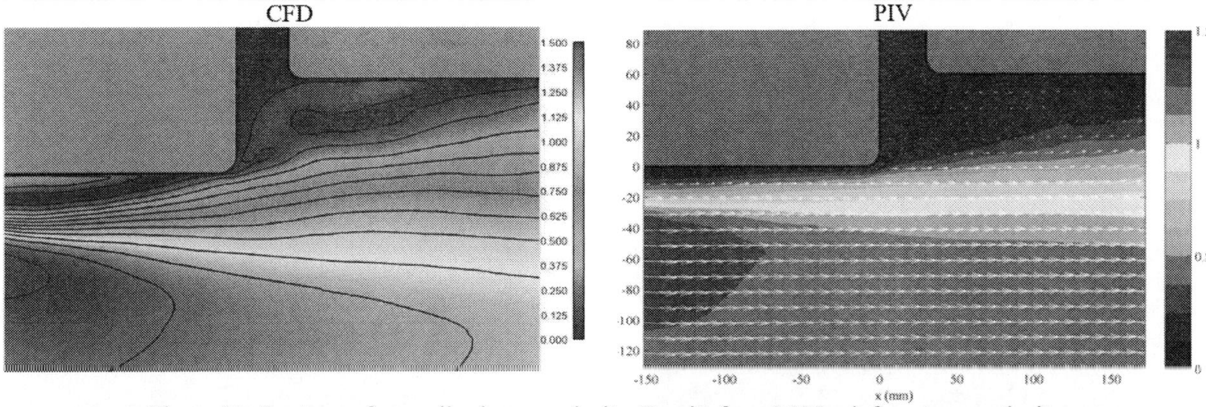

Figure 11: Contours of normalized mean velocity. Results from 0.025 m/s free stream velocity

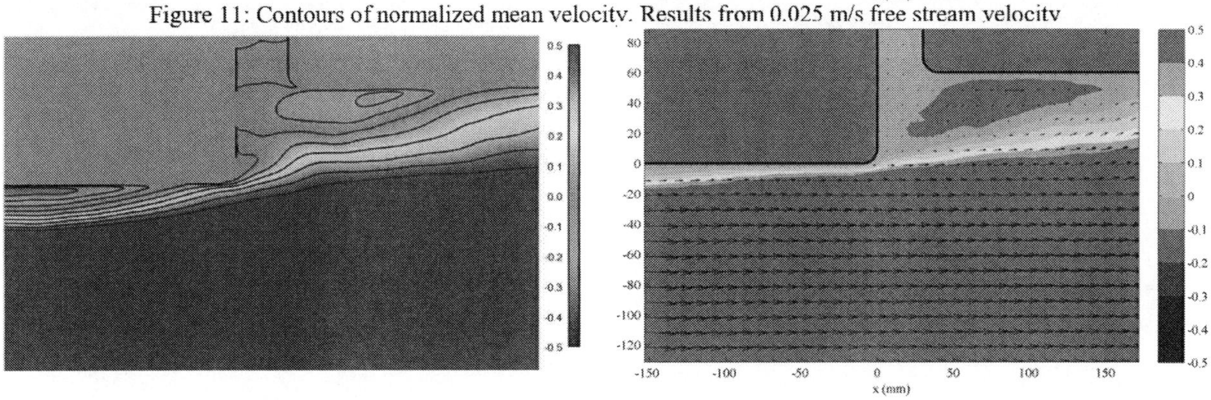

Figure 12: Contours of normalized mean streamwise velocity. Results from 0.025 m/s free stream velocity

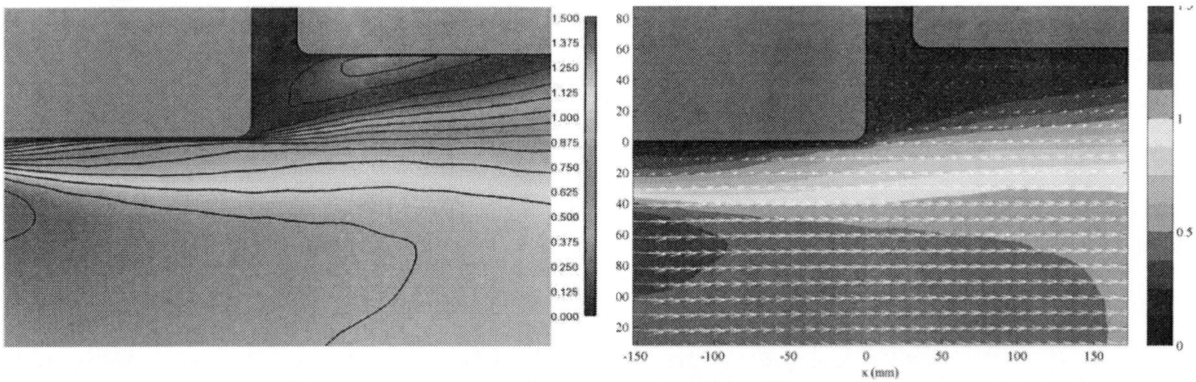

Figure 13: Contours of normalized mean velocity magnitude. Results from 0.254 m/s free stream velocity

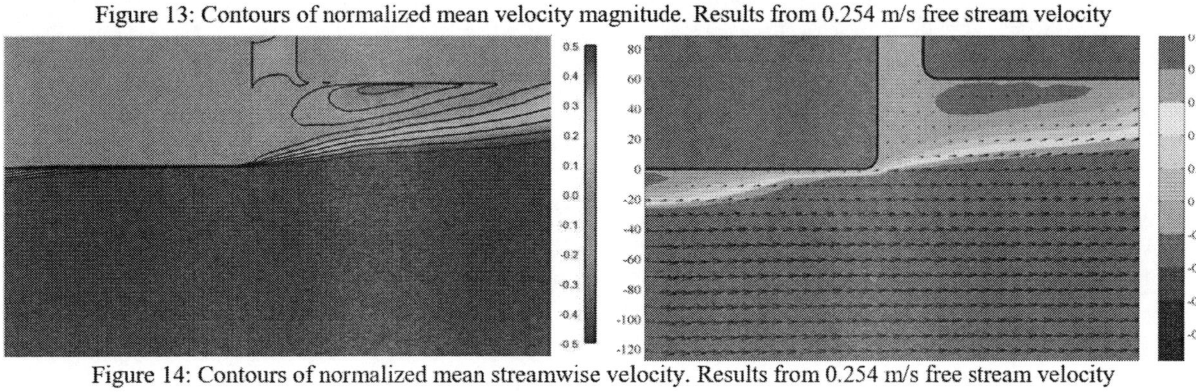

Figure 14: Contours of normalized mean streamwise velocity. Results from 0.254 m/s free stream velocity

IMPROVED SIMULATION MODEL AND METHODS FOR DETERMINATION OF THE REQUIRED PASSING DISTANCE BETWEEN SHIPS IN RESTRICTED FAIRWAYS

K Benedict and **W Köpnick,** Hochschule Wismar, University of Applied Sciences - Technology, Business and Design, Dept. of Maritime Studies Warnemuende, Maritime Simulation Centre Warnemuende / Germany
M Baldauf, World Maritime University WMU, Malmoe / Sweden
A Gronarz and **B Friedhoff,** Development Centre for Ship Technology and Transport Systems, Duisburg / Germany
Fernando de Mello Petey, Rheinmetall Defence Electronics Bremen / Germany

SUMMARY

Investigations of ship collisions of national and international Marine Investigation Institutions as well as practical experiences in ship handling show that there still is an urgent need for further quantification of the safe passing distance and the modelling of ship-ship interactions during overtaking situations. In restricted and narrow waters collisions occur especially during overtaking manoeuvres due to hydrodynamic interactions between the ships mainly. They often lead to damages of the ships and sometimes even the total closing of the entire fairway leading to massive economic losses. Within this paper methods and results regarding the improvement of the simulation of critical phases during overtaking and encounter situations are introduced. A test sample for training module development will be discussed using the higher quality of the mathematical models in ship handling simulator allowing for a more realistic training to avoid the risks in these situations. Furthermore a concept is described to derive information for a qualified assistance of ships' crews in order to estimate the safe passing distance for reducing the risk of collision in overtaking situations.

1. INTRODUCTION

Both the increasing maritime traffic accompanied by increasing ship dimensions and ship speed and the cumulative number of collisions in the last years support the urgency of investigations into the quantification of a safe passing distance for all types of encounter situations in shipping – meeting on opposite courses, overtaking situations and even encounters on crossing courses. On the other hand there is also a lack regarding potential steering strategies that could support the avoidance of a collision especially during an overtaking situation.

A case study of a recent collision scenario demonstrates the situation very well (Figure 1; [4]). Under good weather and visibility conditions a post-panmax container vessel (length 280 m; breadth 40 m) came from open sea sailing to the port in the river fairway over-taking a smaller container feeder vessel (length 101 m; breadth 18.2 m) which was keeping the right hand side of a fairway. The larger vessel had almost completely overtaken the smaller vessel. Its stern was just passing the fore ship area of the feeder vessel, when the latter suddenly lost her ability to steer due to suction effects caused by the hydrodynamic interactions between the passing vessels. The bow of the feeder very quickly turned to the aft ship of the big container vessel. Both vessels did touch only softly; but as a consequence of the impact there was temporarily a heavy list of up to 40° (see Figure 2), and apart from containers also one mariner fell overboard and lost his life. The final phase of the accident is given as a snapshot from the reconstructed scenario simulated in the full mission ship-handling simulator of Maritime Simulation Centre Warnemuende (MSCW) in the following figures.

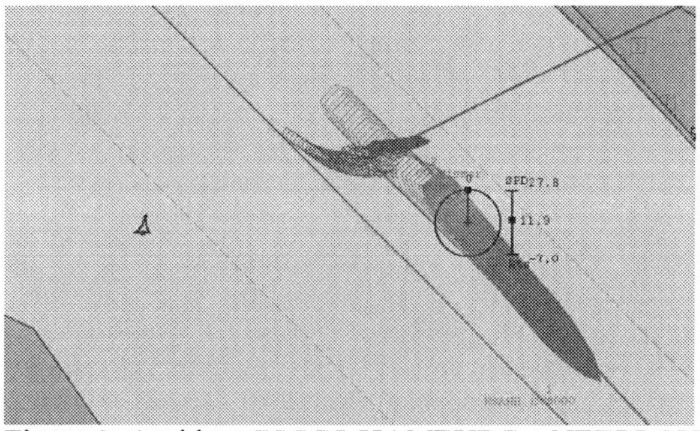

Figure 1: Accident COSCO HAMBURG – NEDLLOYD FINLAND (snapshot taken from replay of the reconstructed simulation scenario)

Figure 2: List of the overtaken vessel during final phase of the overtaking (photo taken from Accident Investigation Report BSU 345/04)

Obviously, neither the pilot nor any other members of the bridge team have been aware of the risk of collision. This illustrates very well the urgent need for the quantification of the hydrodynamic safe passing distance as a threshold for warnings on the one hand and on the other hand for an advisory tool for decision making to give suitable support to the navigators before dangerous hydrodynamic forces may lead to a collision.

As a result of this accident the collaborative research project SIPAS has been started for investigations into the following specific aspects:

- Operational methods for ships' command and application in ship-handling simulators. Parameterisation and analysis of accident cases as well as investigation of real passing situations for creating basic methods for decision support on board. Performed by Hochschule Wismar, University of Applied Sciences, institute ISSIMS Warnemuende.
- Hydrodynamic interactions during passing of seagoing vessels. Parameterisation and analysis of collision causes as well as investigations of passing situations by means of experimental and numerical methods. Performed by DST- Development Centre for Ship Technology and Transport Systems, Duisburg.
- Implementation of the improved mathematical model for hydrodynamic interactions during passing situations in ship-handling simulators, extension of existing wind and current models as well as 3-D visualisation of waves in a simulation environment by RDE- Rheinmetall Defence Electronics, Bremen.

2. INVESTIGATIONS FOR A NEW SHIP-SHIP INTERACTION MODEL

The realistic consideration of passing ship effects in ship handling simulators requires an appropriate representation of the complex hydromechanics. As direct real-time simulations are not feasible with today's computational power and numerical tools, the subproject SIPAS-MAP aimed at an efficient and detailed but still robust mathematical model of the characteristic quantities in unsteady passing situations.

DST has been investigating ship-ship interactions during the last couple of years within a consecutive series of research projects [7]. While former projects focused on special sub-domains of the wide field of unsteady interactions, the implementation into simulators requires a holistic model valid for the entire range of parameter combinations. Experimental and numerical fluid dynamics were used for the derivation of a new force model meeting these requirements.

2.1 PHYSICAL MODEL TESTS

To provide a basis for the new model and to improve the knowledge of the physical effects in overtaking situations extensive model test series have been conducted. In the large shallow water basin of almost 200 m length, a width of 10 m and a variable depth of up to 1.25 m two towing carriages are operated simultaneously. In most test runs both models are towed at constant speed on parallel tracks, but are free to sink and trim.

With a representative series of four sea-going ships reaching from a small feeder to a post panmax container vessel overtaking and encountering manoeuvres were simulated in scale 1:40. The combination of the 4 ship models with their corresponding form and arrangement parameters, the parameters of relative direction of motion, water depth, distance between the trajectories, and the dominant parameters speed of both vessels and speed difference are forming an extensive matrix of test conditions.

The typical character of time series of the forces and the yawing moment acting on the slower, passed ship during an overtaking manoeuvre can be seen in Figure 6 and Figure 8. The time is normally transformed to a non-dimensional relative position, -1 representing the beginning of the overlap, 0 the midship- midship position and 1 the resolution of the overlap. From the free surface elevation over the forces acting on the models and the rudders to the propulsion characteristics up to more than forty quantities are measured.

Further investigations dealt with details such as the influence of propulsion, the effect of laterally restricted waters or the manoeuvrability during a passing situation. Measuring the rudder forces on a model towed with constant rudder angle but varying speed and propeller rate helped to understand the steerability within a passing manoeuvre. It could be shown that the effect of changes in speed is small compared to the influence of the propeller rate on the rudder force for designs with the rudder mounted in the slip stream. A change of the rate of revolutions by 10% from the self-propulsion point results in a change of the rudder force by approximately 10%. Variations in speed of the same order of magnitude have a negligible influence when the propeller rate is kept constant. Figure 3 shows the axial velocity components in a horizontal plane cutting propeller and rudder. The distribution illustrates that the flow speed at the rudder is dominated by the slip stream rather than the uniform inflow, which supports the effects mentioned before.

The common recommendation to reduce the power of the slower vessel during overtaking situations has to be reviewed with this background. Even though it is helpful to reduce the passing time by means of an increased difference in speed it has to be ensured that the vessel remains able to keep its course. This is most important in the first phase of the sequence, when the yawing moment is turning the bow in. The second half with a bow out moment is not that critical, provided the water is free and sufficiently wide. The optimum would be a speed reduction long before the overlap begins and an increased power when the steerability is needed to keep the course followed by reduced power at the end of the manoeuvre if required.

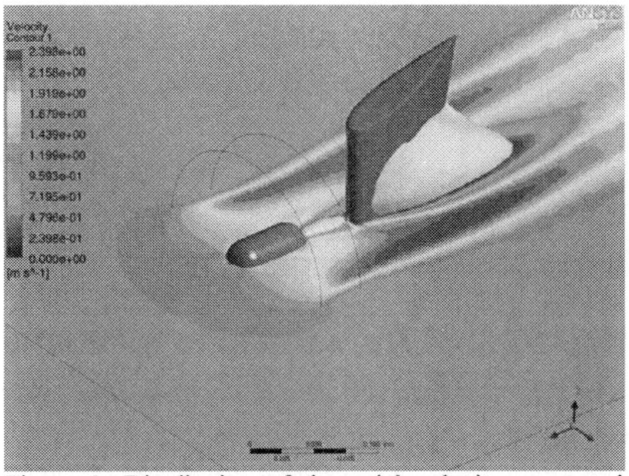

Figure 3: Distribution of the axial velocity computed with Ansys CFX in a horizontal plane cutting the propeller model and the rudder set to 20° rudder angle in uniform inflow corresponding to a typical advance ratio.

To validate the transformation of the force model into the ship motions in the ship handling simulator, selected situations were tested with the free running model of the feeder vessel. While the passing KCS is towed at constant speed on a straight track, the free running model is radio controlled by hand or by means of several autopilot strategies. In this constellation the trajectory of the model is measured instead of the forces. Figure 4 contains eight pictures taken during a characteristic overtaking situation without rudder activity. In this case no collision occurred but the position and heading of the free-running model visualises the typical sequence of effects.

Figure 4: Sequence of consecutive pictures taken during a characteristic passing manoeuvre without rudder activity.

To permit detailed analysis all time series of the tests are smoothed, non-dimensionalised and collected in a database. In a sophisticated approach using empirical characteristics and nonlinear regression analysis the required mathematical force model is derived. The forces computed for each passing situation occurring in a simulation scenario are transferred to surge, drift and yaw motions by the motion model implemented in the software of the ship handling simulator. Details about the mathematical methods and formulations can be found in a separate contribution to this conference prepared by Gronarz.

2.2 NUMERICAL INVESTIGATIONS

2.2.1 Simulation Methods and Scope

The mathematical model based on experimental data alone does not include the influence of different headings of the ships so far, as these situations cannot be reproduced with common experimental setups yet. Here numerical simulations were used after validation with measured situations to develop an empirical approach to include non-parallel passing situations into the model for ship handling simulators.

Due to arbitrary trajectories without topological restrictions, a very flexible method, able to deal with the relative motion of "objects" is required. Based on the software package MOUSE [3], a general overset method using overlapping grids has been developed. The underlying software package is an object oriented toolbox for data storage and manipulation with special focus on finite volume methods for field simulations using grids of arbitrary shapes. In past years, specific CFD-modules with various scopes have been developed. For the applications reported herein the following modules are chosen:

• Finite volume discretisation for incompressible flows via artificial compressibility.
• Transient time integration with dual time step iteration to ensure mass conservation.
• Free surface flows based on a level set approach.

More details on the basic scheme can be found in [3].

In the context of naval flow simulations, the applications for relative motion models are for example motions due to sea state, manoeuvres, ship to ship interactions or rotating propulsion systems and interactions with fixed objects such as hydraulic structures or varying restrictions of the waterway in confined waters. Since this list of applications accepts no restrictions concerning trajectories, most approaches for relative motions are not applicable. Grid distortion allows too little displacements. Combining the technique with remeshing tools, allows large displacements, however it is restricting the method to rather inefficient grid types. Sliding interfaces, in contrast, require knowledge of the motion patterns in advance, unacceptable for unknown and eventually crossing trajectories, like in the case of passing ships with unequal heading.

The overset method is thus the consequent choice and used for all current developments. In this approach, all moving objects carry their own grids, typically consisting of some element layers around their boundaries only. These objects are inserted into a basic frame of reference, which itself usually carries a grid for the surrounding domain. Figure 5 (left) shows a typical grid set up. Numerical coupling between overlapping grids of differing level is done via vice versa interpolation.

2.2.2 Numerical simulations of passing manoeuvres and comparison to experiments

After validation of the methods with test cases focusing on free surface elevation, drift forces or dynamic floating positions, several passing manoeuvres were simulated to validate the numerical methods presented in previous sections. In the simulations selected for this publication, the panmax KRISO container ship (KCS, length 230 m) is passing a smaller feeder (length 126 m). Draughts of the panmax and feeder vessels are 10.80 m and 6.50 m respectively. The water depth is 19.0 m. The panmax vessel is sailing at a constant velocity of 18 kn, while its degrees of freedom for secondary motions as sinkage and trim are neglected. Trim and sinkage are however taken into account for the feeder, as it encounters more critical situations.

In a first example, the non-moving feeder remains at a fixed position, while its degrees of freedom for trim and sinkage are activated, enabling the vessel to react on induced hydrodynamic loads (solution dependent motion). This set-up is chosen in accordance with the validating experimental set-up. The feeder is positioned with an angle of 90 degrees to the track of the panmax vessel. The clearance between both vessels corresponds to 32.0 m at full scale. Figure 5 shows the overset grid system used (left) and the free surface elevation at a point of time during the passing manoeuvre (right). Note that all plotted solutions inside the vessels are meaningless. These arise due to technical reasons only, as the inertial grid covers the space inside the vessel as well. A cosmetic plot filtering will soon be available to overcome this post processing problem.

The simulated horizontal forces (longitudinal and transversal), sinkage and trim are compared to experimental data, Figure 6. The results are in good agreement, except for the first part of the passing manoeuvre at relative ship positions in the range of -1.5 to -0.5. Due to an impulsive start of the panmax vessel in this simulation only about three ship lengths behind the feeder, an initial wave was generated that remained in the near vicinity in front of the bow, causing unphysical surface deflections and forces, still visible in the right part of Figure 5. Several approaches to overcome the problem exist. The simplest is a substantial increase of the computational domain, allowing the panmax vessel to start farther apart, giving all initial waves enough time to travel through the domain of interest before the vessels encounter. This, however at the cost of higher computational resources needed.

As a second example a parallel passing manoeuvre is simulated. Again, the velocity of the panmax vessel is 18 kn, while the feeder travels at only 2.5 kn. The clearance between both ships is 32.0 m. Figure 7 shows the surface deflection at different points of time during the passing manoeuvre. As in the previous example, the initial wave remaining from the impulsive start of the numerical simulation influences the results in the early stage of the passing manoeuvre.

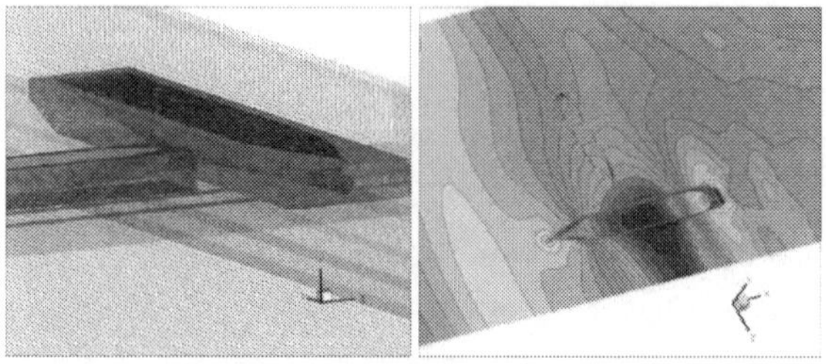

Figure 5: Overset grids (left) and the passing situation with surface deflection at a point of time during the passing manoeuvre (right).

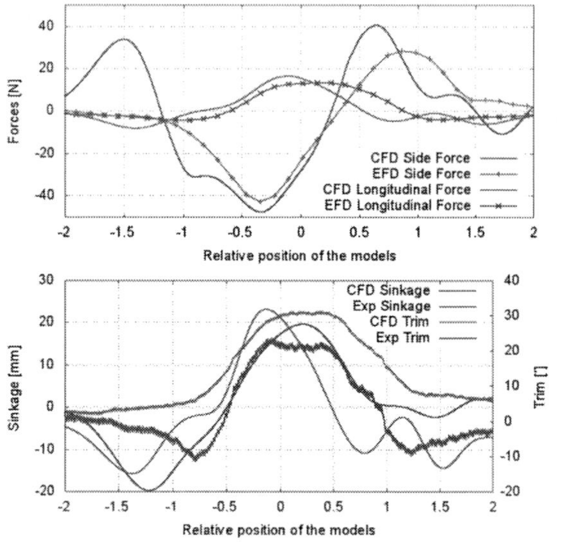

Figure 6: Comparison of numerical and experimental results for longitudinal and side forces (top), sinkage and trim (bottom).

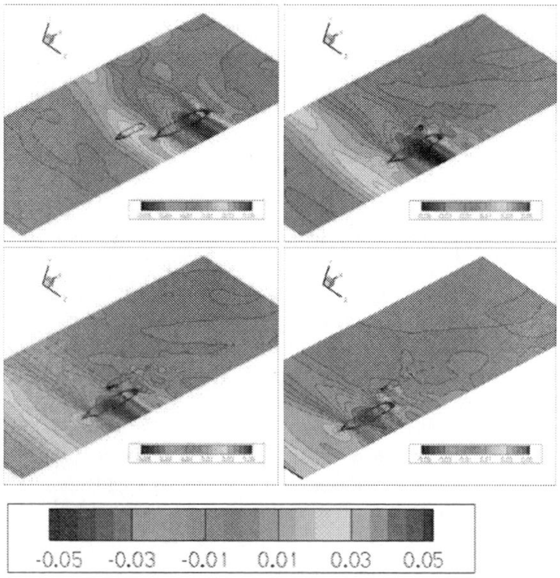

Figure 7: Deflection of the free surface during a parallel overtaking manoeuvre.

Figure 8 shows the induced loads for the feeder vessel in comparison to experimental results. Again, the differences due to the initial wave in the numerical simulation are mostly visible in the range of relative ship positions from -1.5 to -0.5.

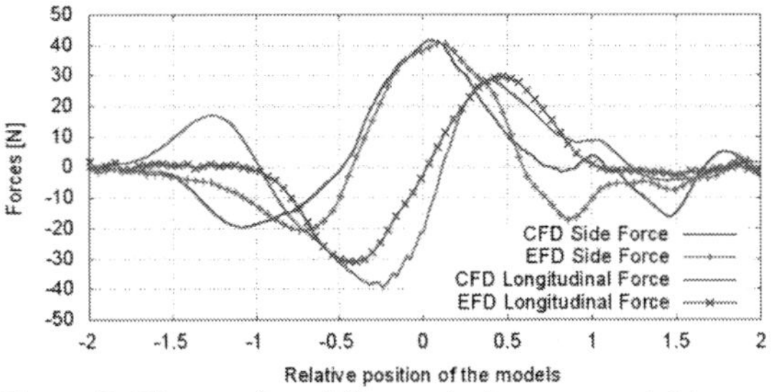

Figure 8: Time series of forces during an overtaking manoeuvre at model scale.

The last simulation case differs essentially from the previous two, as it involves the most important application of CFD in the current context, crossing trajectories. The physical set-up is sketched in Figure 9 (left). As in the previous example, the faster KCS sails at a speed of 18 kn, while the feeder vessel has a speed of 9 kn. The angle between the headings of both vessels is 30 degrees. In this example, the feeder crosses behind the panmax vessel.

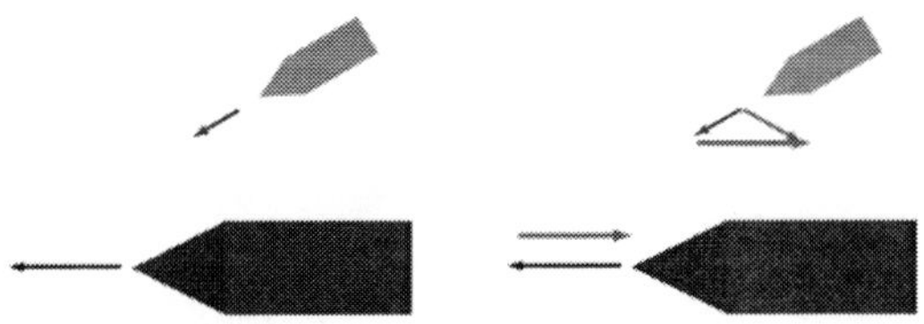

Figure 9: Sketch of the crossing situation (left) and its transformation (right).

Due to the problem with remaining initial waves after impulsive starts, as encountered in the previous cases, this computation is performed in transformed space. Speed transformation allows keeping the panmax vessel at a fixed position in the computational domain, while being exposed to an incoming parallel flow. The same transformation on the feeder vessel causes a relative motion direction to the lower right side, Figure 9 (right, red vector). This transformation allows starting the computation with a stationary flow solution around the panmax vessel, meaning that all of its initial disturbances are vanished before actually starting the transient manoeuvre simulations. After the initial, convergent result has been obtained for the panmax ship in the surroundings water, the feeder vessel is inserted as additional overset object. The method allows doing so with ease: In fact only the new grid has to be placed at an initial position somehow apart and its motion model is activated. Similar to the previous test cases, the trajectory of the feeder vessel is given, while sinkage and trim are subject to the dynamics of the transient solution process. As the width of the model basin and the capabilities of traversing mechanics are not sufficient in physical model tests, no experimental data for comparison are available. Anyhow, with the extensive validation of the methods the results are used for a simple approach to include interaction effects for passing situations with different headings of the vessels.

3. IMPROVED PASSING-SHIP MODULE FOR FULL MISSION SIMULATION AND ENHANCED VISUAL REPRESENTATION OF THE SHIP WAVE TRAIN

Over the past decades Rheinmetall Defence Electronics (RDE) has been building simulators for both commercial vessels and war ships. RDE simulation tools enable the customer to deal with ships of any size, fitted with almost any kind of propulsion/steering system.
The accuracy, performance, level of detail of the proposed modeling techniques have been validated in decades of simulation work involving different ship types such as tugs, container ships, oil tankers, mine hunters, frigates etc. Comparison between data recorded during ship full-scale tests and our simulated results show an excellent agreement.

Figure 10: RDE full mission ship handling simulator

Simulator. The Ship Handling Simulator ANS5000**professional** consists of a number of decentralized computers communicating via a network layer. The different simulator processes are distributed over these computers. The central processes (e.g. environment simulation) run on a server whereas the remaining ones (e.g. ownship model, ECDIS, radar, high-performance visual system with synchronized channels) run on the ship operating cubicle (SOC) computers. These processes consist of different modules designed according object-oriented principles (C++ language). The core of the simulator is its ownship model, which enables the customer to simulate realistically the motions of a vessel under the influence of wind, current, waves, anchor forces etc.

Ownship model. In the software the vessel is considered as a six-degree-of-freedom system. Its motions are calculated by numerical time-step integration of the six equations of motion (surge, sway, heave, roll, pitch and yaw) of the rigid body.
The three equations governing the vessel manoeuvring in the horizontal plane (i.e. surge, sway and yaw) are based on the well-known works by Norrbin and by Clarke et al. The remaining three equations (i.e. heave, pitch and roll) take into account all essential aspects of the physical phenomena: the roll equation e.g. includes roll inertia, roll damping, righting lever curve of the vessel and the coupling with the motion in the horizontal plane (centrifugal force).
The RDE ownship model takes into account:

- hull forces (vessel's resistance, added masses etc.),
- thruster forces (including tunnel thrusters, azimuth pod propulsion etc.),
- rudder forces
- wind forces,
- wave forces,
- current forces,
- bank effect
- squat
- line forces
- anchor forces etc.

All hydrodynamic maneuvering coefficients of the ship's hull depend on the water depth. The same applies to the rudder force, which increases in shallow water areas.
Included in the simulation is the high-frequency ship motion oscillation due to the waves, the magnitude of which is calculated using "response amplitude operators" determined according to the well-known strip theory.
Recently, the extended mathematical model developed by DST for the calculation of the hydrodynamic interactions between passing ships was implemented in the ship handling simulator ANS5000**professional** increasing substantially the accuracy of the simulation results.
Though the ANS5000**professional** ownship model is of high complexity, the amount of data needed for the ship modeling is kept deliberately to a reasonable volume, enabling even those customers, who are not closely familiar with the hydrodynamic ship theory, to model ships in a quick and user-friendly manner. These data are defined and stored in a ownship data base by using the so-called multimedia object class editor (MMOCE) developed by RDE.

Current. Magnitude and direction of a current may vary significantly along the ship's hull. The software therefore divides the vessel's hull longitudinally into a number of sections, determines the local current vector acting upon each section and calculates the current force for each section separately. Furthermore, the current (e.g. in a river) is affected by the presence of big objects. The simulated current shadowing has thus the effect of shielding vessels behind e.g. a big ship in their vicinity.

Wind. Wind gusts are modeled as the superposition of a large number of sinusoidal curves with different amplitudes, frequency and lengths. These curves are generated automatically from a so-called Davenport wind spectrum, which describes the physically correct distribution of the wind energy over the whole frequency range.
Since the wind speed and direction may vary significantly along the ship's hull, the software divides the vessel's hull longitudinally into a number of sections, determines the local relative wind vector acting upon each section and calculates the wind force for each section separately.
The leeward side of a big ship or a tall building is protected from the prevailing wind. The simulated wind shadowing has thus the effect of shielding vessels in the lee side (i.e. reducing the force exerted by the wind). In the simulator both wind shadowing

- by land mass (including buildings defined as three-dimensional models in the exercise data base) and
- by other ships (both traffic ships and ownships)

are considered.

High-performance visual system. The visual representation of the wave train of a moving ship in the simulator conveys to the helmsman important information about an impending passing situation. This visual perception (height of the characteristic bow wave peak, length of the transverse waves along the hull etc.) plays a decisive role in assessing the speed of the passing ship (awareness of danger).

Figure 11: High-performance visual system

A moving ship causes the surrounding water surface to deform, the most noticeable waves being a series of divergent waves starting with a large wave at the bow, followed by a complex system of transverse crests and troughs dying out along the hull. The shape of this wave train depends on: the shape of the hull, the speed of the vessel and the relative water depth (= depth / draught).
The customary representation of the wave train using a simple two-dimensional texture is by no means sufficient for a realistic visual perception. A new visualization

based on the three-dimensional deformation of the water surface has been therefore developed [6] and implemented in the simulator.

Figure 12: Three-dimensional visual representation (i.e. with deformation of the free water surface) of the wave train in simulator.

Finally, the new three-dimensional method must fulfill the general requirements imposed on a simulator:

- The computational effort must be compatible with the overall capacity of the visual system.
- In spite of the complexity of the visual system, the amount of data required for the wave train modelling in the ANS5000**professional** could be kept to a reasonable volume, enabling those customers, who are not closely familiar with the hydrodynamic wave theory, to model themselves ships in a quick and user-friendly manner.

4. OPERATIONAL METHODS TO SUPPORT THE BRIDGE TEAM AND APPLICATION IN SHIPHANDLING SIMULATORS

4.1 ANALYSIS OF PASSING SITUATIONS AND ACCIDENTS FOR DATA ACQUISITION AND IDENTIFICATION OF RELEVANT PARAMETERS

As a sound basis for the investigation a collection of data was started, both from observations and measurements of real time traffic and from accident data sources. Observations of passing situations in the research port Rostock were made to investigate data recorded on the ships, e.g. by a Voyage Data Recorder (VDR) and to gain detailed parameters of passing manoeuvres especially with regard to the lateral distance between the ships.

However, the focus was shifted towards the investigation of accidents. Some accidents were found in publications, the data were used for the analysis in the next chapter to create Figure 15. A real overtaking manoeuvre was analysed in detail by means of AIS recordings and additional information concerning a collision in the Kiel Canal [7], [8], see Figure 13. By means of special software for analysing the AIS-data developed at the Department of Maritime Studies Warnemuende the distance between the passing ships hulls could be determined for any defined moment of the manoeuvre. After one ship left the lock it starts overtaking another ship which had just left the berth. During the overtaking manoeuvre the overtaken ship was sucked to the stern of the bigger and faster ship, turned and ended up finally grounded.

During the analysis of the accidents it was very difficult to find out when the most critical situation was reached, i.e. when the accident could not be avoided anymore at a so called "point of no return". Therefore the data were measured for several situations at mostly three time points short before uncontrolled motions of the ship were taken from the time history and track to be used for Figure 15.

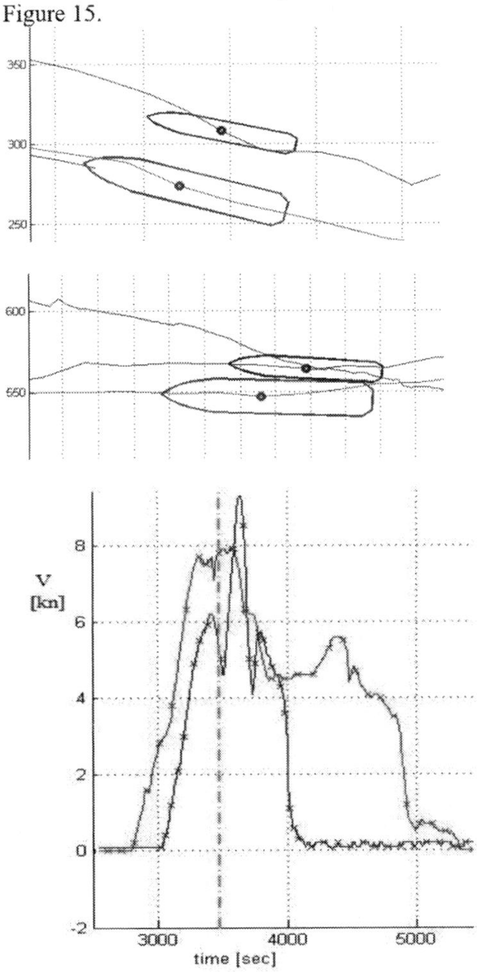

Figure 13: Collision between MT BIRTHE THERESA (blue, overtaken vessel) and MT VASI (red, overtaking vessel):

a) Ship shapes in two situations (track in x-y-coordinates [m]; top at 3500 and below at 3800 seconds)
b) speed history (bottom, speed [kn] versus time [s])

4.2 NEW APPROACH FOR ESTIMATION OF SAFE PASSING DISTANCE BASED ON REFERENCE DRIFT ANGLE FROM ACCIDENT ANALYSIS

Moving ships are surrounded by pressure and flow fields superposing each other during encountering or overtaking. Thereby changing of pressure distribution occurs and added unsteady forces and moments appear affecting the steering characteristics of the ships. During the passing manoeuvres the magnitude and the sign of the forces and moments alter permanently as indicated in Figure 6 and Figure 8; this leads to counter steering activities and course corrections by the navigators. A graphical demonstration of the transverse forces due to passing effect is given in Figure 14 – These forces are increasing with smaller distances and with the ratio of the different speed of the ships. With regard to risk assessment (specifically risk minimisation) it has to be investigated in principle whether the passing ships are able to compensate the existing maximum attraction and repulsion forces and moments by drift and rudder angles.

From investigations into overtaking collision cases performed so far it seems that the safety-critical passing distance in open sea area (unrestricted fairway) from hydrodynamic perspective seems to be in a range of 3.5 to 6.0×Bm (Bm is the mean breadth of the two ships involved). This range is indicating strong increase of the forces due to passing effect – and this is the same range where the accidents have occurred in Figure 15. Taking into account a safety margin it would be suggested to use a value greater than 3.5-6.0×Bm as an initial default value to determine the safe passing distance to avoid an accident during overtaking.

However, this approach results in relatively high values for suggestions for passing distances (e.g. as mentioned in [5]) and it is not flexible enough e.g. for considering the speed and different size of the ships.

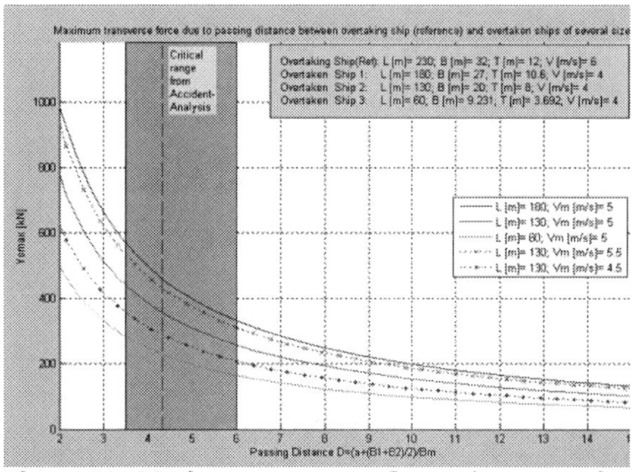

Figure 14: Maximum transverse forces due to passing effect as function of the passing distance and speed ratio related to the mean breadth of the ships.

The compensation of the yaw moment by corresponding rudder angles during a passing manoeuvre would be an analogue approach. Such analysis is very complex and would need data of rudder parameters from the ships which are normally not available (e.g. via AIS) for decision making on board.

Therefore, as a potential criterion, the determination of a "reference drift angle" is suggested derived from data of representative overtaking manoeuvres leading to accidents. During the overtaking manoeuvre both ships have to establish a drift angle to prevent a collision from the suction forces due to the passing effect. Assuming only minor effects due to the rudder forces the equilibrium of transverse forces can be focussed on the most important components: the transverse hull force $Y(\beta)$ and the force due to the passing effect YS.

$$Y(\beta) = YS \tag{1}$$

For the transverse force YS due to the passing effect a simplified formula can be used suggested by Brix [2] for the maximum of $YSmax$

$$YSmax = C_{YS}max \cdot \left(\frac{\rho}{2} V_m^2 \cdot L_m \cdot T_m\right) \cdot D_0/D \tag{2}$$

where $Vm=(V1*V2)/2$, $Lm=(L1+L2)/2$ and $Tm=(T1+T2)/2$ are the mean values of speed, length and draught of overtaking ship 1and overtaken ship 2. D is the passing distance measured between the ships centre lines with $D=a+(B1+B2)/2$ where a is the distance between the hulls. D_0 is a reference distance, which can be taken as $D_0=0.35*Lm$. $C_{YS}max$ is given as to $C_{YS}max=0.0275$ for passing procedures.

The hull forces due to drift can be represented by the linear approach from Clarke [9] for small drift angles

$$Y(\beta) = [\pi \cdot T^2/L^2 \cdot \left(1 + 0{,}40 \cdot c_B \cdot {}^{B}\!/_{T}\right)] \cdot \rho/2 \cdot V^2 \cdot L^2 \cdot \beta \tag{3}$$

The distance D of the ships centre lines can then be separated as to:

$$D = \frac{C_{YS}max \cdot \left(\frac{\rho}{2} V_m^2 \cdot L_m \cdot T_m\right) \cdot D_0}{[\pi \cdot T^2/L^2 \cdot \left(1 + 0{,}40 \cdot c_B \cdot {}^{B}\!/_{T}\right) \cdot \beta] \cdot \rho/2 \cdot V^2 \cdot L^2}$$

$$D = \frac{C_{YS}max \cdot 0.35}{\pi \cdot T^2/L^2 \cdot \left(1 + 0{,}40 \cdot c_B \cdot {}^{B}\!/_{T}\right) \cdot \beta} \cdot \frac{V_m^2 \cdot L_m^2 \cdot T_m}{V^2 \cdot L^2 \cdot T} \tag{4}$$

For a certain passing situation all parameters on the right side are given in order to calculate the necessary distance D for balancing the forces, except the drift angle β. The idea is now to estimate the critical drift angle β from analysis of accident cases during passing situations. The drift angle which might have occurred can be calculated as:

$$\beta = \frac{C_{YS}max \cdot (L_m \cdot T_m) \cdot D_0}{[\pi \cdot T^2/L^2 \cdot \left(1 + 0{,}40 \cdot c_B \cdot {}^{B}\!/_{T}\right)] \cdot L^2 \cdot D} \cdot V_m^2 / V^2 \tag{5}$$

In Figure 15 the drift angles were calculated from data collected from accidents of passing ships. For every accident there is a pair of drift angles: one for the overtaking ship indicated with 1 and one for the overtaken ship indicated with 2. The smaller and the slower a ship is the higher is the required drift angle. The blue line is representing the threshold for the bigger drift angles, normally occurring for the overtaken ship, with some exception, e.g. if a smaller ship runs in trouble when overtaking a bigger vessel (case 5 and 8).

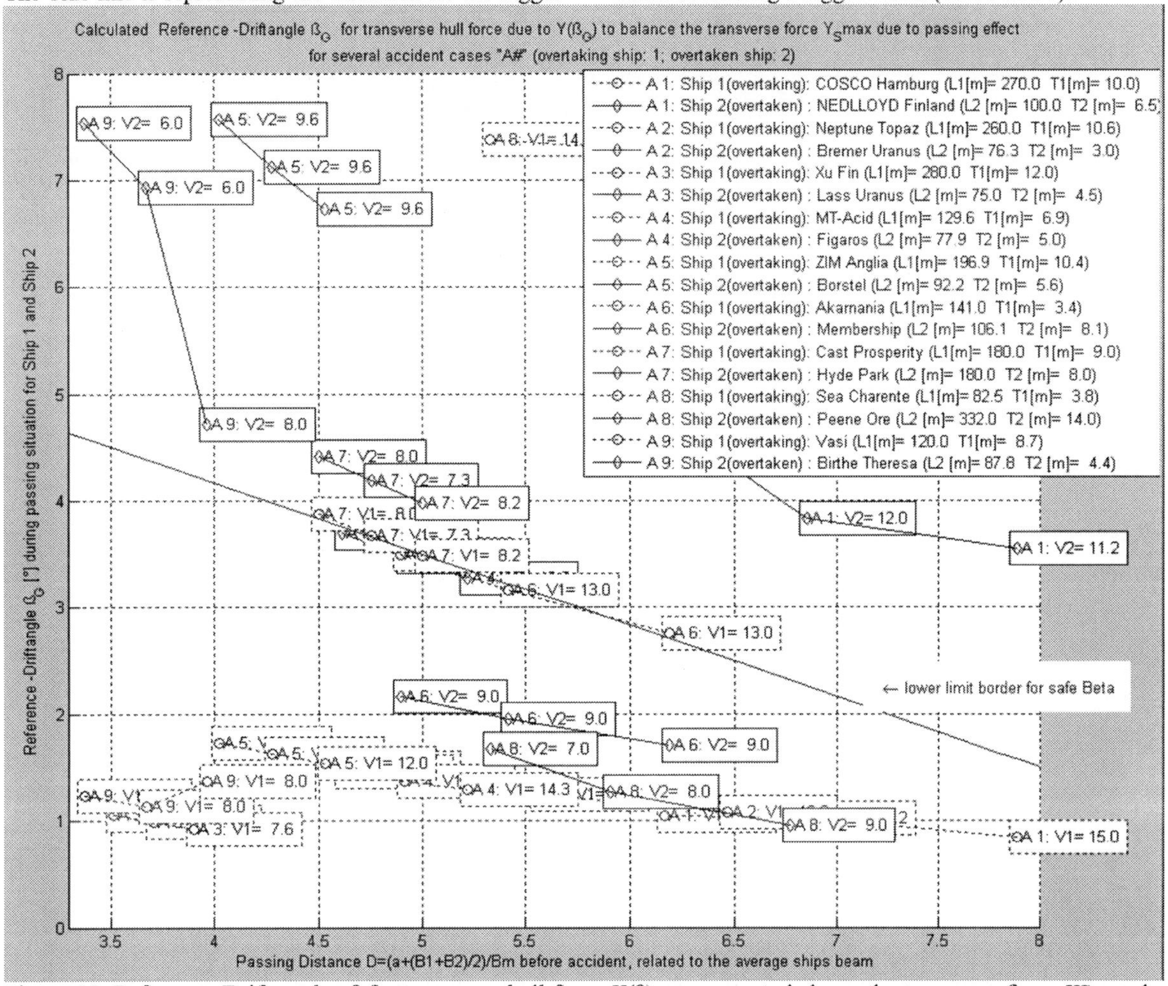

Figure 15: Reference–Drift angles β for transverse hull force $Y(\beta)$ necessary to balance the transverse force $YSmax$ due to passing effect for several accident cases "A#" (overtaking ship: 1; overtaken ship: 2) versus Passing Distance D related to average beam Bm

Such a threshold value for the drift angle can be taken as a reference for calculating the required safe distance. As a straightforward sample it is assumed that the reference drift angle is β=4.5° and taken as constant. Using this reference angle the safe passing distances can be calculated for both ships using the formula (4). It should be done for both ships' perspective in order to find out what maximum value is required. To demonstrate the results from this approach a parameter study was made for several ship sizes and passing conditions in Figure 16. For simplification some standard relations were used for ship dimensions (B=$L/7$, $T=B/2.5$, C_B=0.8).

The length of the overtaken ship is to be selected from the x-axis. The coloured curves are representing the result for passing distances due to the overtaking process: the curves with solid lines for the overtaking ship 1, the curves with broken lines for the overtaken ship 2. The basic curves are representing a speed ration Vm/V=1.0, this is not an overtaking process; both ships are running with the same speed and thereby need the required distances for the critical drift angles. For other speed ratios these values of the basic curves have to be multiplied with the square of the ratio Vm/V1 or $Vm/V2$ respectively to achieve the results for the interpolated curves of the sample ships with $L2$=230 m and $L1$=180 m in Figure 16.

This new approach is promising but needs to be further investigated. One astonishing result is that in the formula (4) only the ratio Vm/V is occurring – i.e. that the absolute speed of the ships has no impact and it would be the same whether the ships go with $V1$=12 kn and $V2$=8 kn or with $V1$=6 kn and $V2$=4 kn respectively: in both cases we get $Vm/V1$=5/6 or $Vm/V2$=5/4. It should be noted, that the lower the vessel speed the easier the interaction forces can be compensated by using engine and rudder reserve.

Furthermore it is planned to improve the accident data base: It should be mentioned that the available collision data have been often incomplete and not consistent so that the required input data have to be partly approximated. Generally the information concerning a maritime accident and the related parameters and details vary strongly and are commonly biased due to subjective opinion. Thereby a significant scattering of the calculated drift angles result.

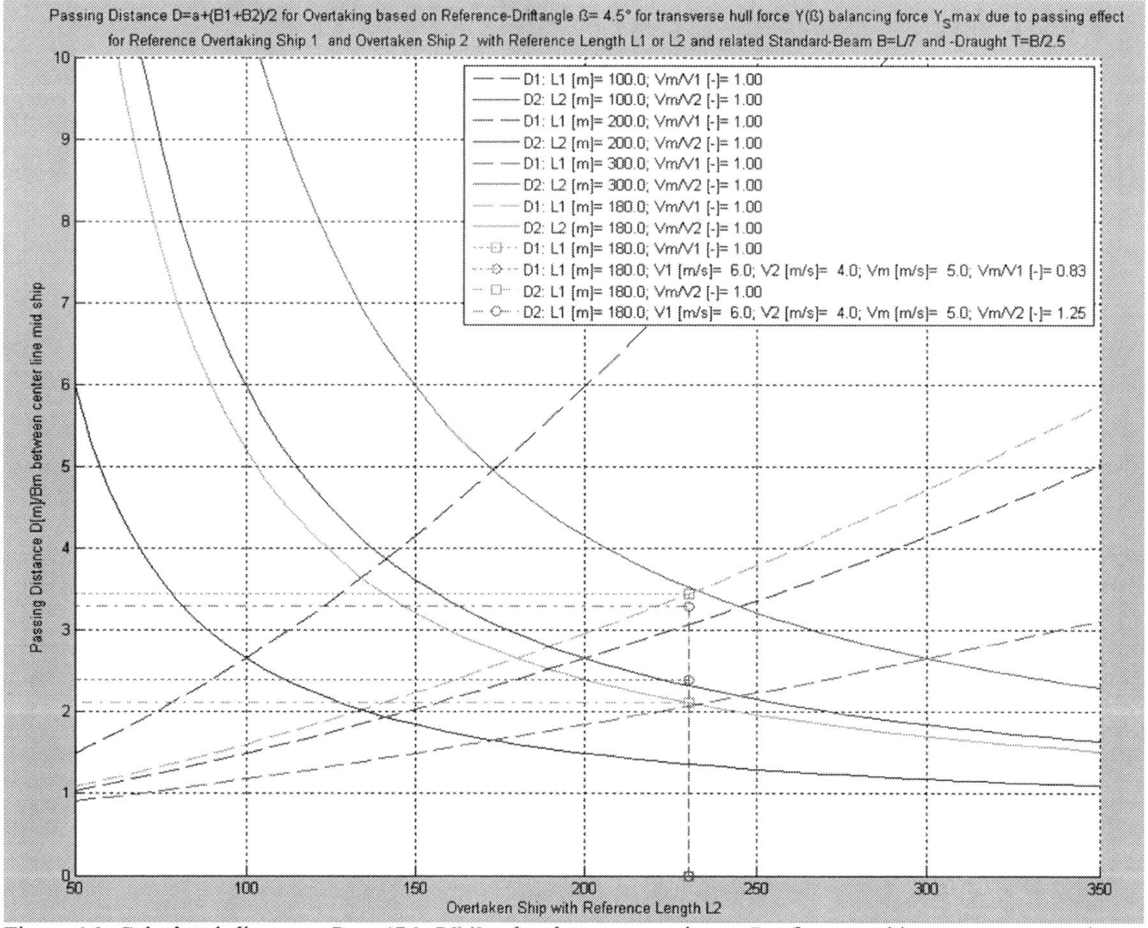

Figure 16: Calculated distances $D=a+(B1+B2)/2$ related to average beam Bm for overtaking manoeuvre to keep the balance of transverse forces $YSmax$ and $Y(\beta)$ by a constant drift angle β=4.5° and sample for overtaking a ship with $L2$=230 m by a ship $L1$=180 m

Present ongoing investigations are dealing with the further development and tuning of this methodology and will be followed by verification studies using other approaches for the forces due to overtaking, data of real collision cases and an enhanced ship handling simulator model.

4.3 SHIP MANOEUVRING SIMULATION FOR FUTURE INVESTIGATION OF PASSING MANOEUVRES AND NAVIGATION SUPPORT

4.3.1 Modelling of ships for simulation and analysis of passing modules used in the simulator

In order to test the enhanced passing effect module of the Ship Handling Simulator (SHS) four simulator ships were modelled. These simulator ships correspond to the dimensions and displacements of the original ships that are manufactured as ship models for the experimental investigations in the towing tank in Duisburg. With these models the old and the new module will be tested and compared.

In the passing ship effect module presently implemented in the SHS the interactions of passing ships are represented by induced flow fields. The induced flow speed components have effect on four ship current attack points uniformly distributed over the ship length and influence the course and speed of the Ownship via the equation of motion according to the distance, speed and course of the passing ship. The assessment of the current passing ship effect module in the SHS was made by means of selected overtaking manoeuvres with the simulator ships by variation of absolute and relative speed, lateral distance as well as water depth.

In order to evaluate the current passing ship effect module a data recording system and special software for analysing the data (SimDat) were developed at the Department of Maritime Studies in Warnemuende [1]. By means of this software all data of overtaking manoeuvres in the SHS can be logged and relevant manoeuvre parameters can be calculated and pictured graphically.
The results of the evaluation show some differences to the characteristics curves from existing calculation methods from literature and from the model tests in the towing tank. Therefore the enhanced module is implemented in the simulator now and will be tested and used for the extensive investigations of passing situations.

4.3.2 Future plans for investigations for enhanced decision support for navigating officers

The lack of sufficiently quantified hydrodynamic passing distances described in the first introductory chapter is one fundamental reason for the instigations in SIPAS project. After the validation tests with the new enhanced passing module the following activities are planned:

- Scenario design and simulation studies for testing the thresholds in the new approach for estimation of safe passing distance based on reference drift angle from accident analysis.
- Development of new warning functions for passing of ships: there is potential to provide more accurate information about passing distances related to the exact ship shapes taking into account the dimensions and other data (e.g. provided by AIS) of the involved ships: Assuming that reliable data about sea areas and fairways will be provided with the mandatory introduction of Electronic Chart Display and Information Systems (ECDIS), an automatic adaptation of the threshold for safe passing distance to the concrete situation parameters is possible and should be investigated in more detail. The presentation of specific features has to be studied regarding usability, practicability and user acceptance.
- Design of suitable manoeuvring strategies to avoid collisions or groundings due to passing effects: From the investigations regarding compensating of attraction / repulsion forces with countermeasures it is necessary to further investigate whether it is possible for the overtaken vessel to try to keep a safe distance during overtaking situation. Taking into account the outcome of earlier simulations (see [2]) it seemed to be successful that ordering rudder amidships in the final phase of the overtaking situation is potentially a better solution than trying to counteract with corresponding rudder angles to the suction motion. Also the results of the findings in chapter 2 about the impact of change of the rate of revolutions results in a significant change of the rudder force will be tested during manoeuvring scenarios.

5. CONCLUSIONS

A new and extended mathematical model for encountering and overtaking situations was developed based on comprehensive experimental and numerical investigations. Surge and sway forces as well as yaw moment can be calculated according to relevant parameters as ship's dimensions and speed, water depth, passing distance and angle of encounter. This enhanced mathematical model for calculating the hydrodynamic forces due to interactions between passing ships was implemented into the ship handling simulator ANS5000professional as improved 'Passing Ship Effect-Module'.

A new approach for estimation of safe passing distance based on reference drift angle from accident analysis was introduced. The data from nine accidents were analysed in order to find a method involving characteristic parameters as speed and size of the ships into the calculations of that distance.

This new approach will be tested during the projects in the simulator to validate the thresholds for safe passing manoeuvre, to develop new warning functions and also to find out suitable manoeuvring strategies to avoid collisions or groundings due to passing effects.

6. ACKNOWLEDGEMENTS

The research and results presented in this paper were partly carried out and achieved in the research projects "Safe passing distance between ships" (SIPAS) and "Identification of multi-variable parameter models for ship motion and control" (MULTIMAR) funded by the German Federal Ministry of Economics and Technology and the Ministry of Education and Research of Mecklenburg-Pomerania.
Further some parts of the research work were carried out within the Swedish-German RTD project ADOPTMAN which belongs to the European MARTEC program supported by the European Commission and are funded and supervised by the Swedish Governmental Agency for Innovation Systems (VINNOVA) and the German Research Centre Jülich (PTJ).
The authors like to thank Rainer Kaiser and Sven List (DST) and Roland Vilsmeier (numrax) for their substantial contributions for the numerical investigations reported herein.

7. REFERENCES

[1] BENEDICT, K. et al 2006 Combining Fast-Time Simulation and Automatic Assessment for Tuning of Simulator Ship Models. MARSIM - International Conference on Marine Simulation and Ship Manoeuvrability, Terschelling, Netherlands, Proceedings, M-Paper 19 pp. 1-9

[2] BRIX, J. et al: Manoeuvring Technical Manual. Seehafen Verlag Hamburg, 1993

[3] STUNTZ, N.: 'Numerische Simulation der Schiffsumströmung mit Berücksichtigung des Propellersogs', PhD thesis, University of Duisburg-Essen, 2006.

[4] BALDAUF, M., BENEDICT, K., FISCHER, S., HERBERG, S., KIRCHHOFF, M.: Situation Analysis of the Overtaking Manoeuvre between MV" COSCO Hamburg" and CMS "P&O Nedlloyd Finland" in the Elbe channel. Final Report /Accident analysis report for BSU (in German),. Warnemuende, 31. August 2004

[5] ELOOT, K.; VERWILLIGEN, J.; VANTORRE, M.: Safety Assessment of head-on Encounters and Overtaking Manoeuvres with Container Carriers in confined Channels through Simulation Tools. MARSIM09 – 36th AGA of IMSF and International Conference, Panama City, Panama; August 17-20, 2009. Conference Proceedings p. C-133 to C-144

[6] GRONARZ, A.: Ship-ship-interaction: overtaking and encountering of inland vessels on shallow water. MARSIM 2006 International Conference on Marine Simulation and Ship Manoeuvrability, Terschelling / Netherlands 25-30 June 2006

[7] -: Collision on the Kiel Canal between MT VASI and MT BIRTHE THERESA on 12 February 2009. Investigation Report of the German Federal Bureau of Maritime Casualty Investigation (BSU), 1. December 2010 (www.bsu-bund.de)

[8] Benedict, K.: Kollision auf dem NOK bei Kiel am 12.02.09 zwischen MT VASI und MT BIRTHE THERESA. Accident analysis report for BSU (in German), Rostock June 2010

[9] Clarke, D., Horn, J.R., "Estimation of Hydrodynamic Derivatives" Proceedings of the 11th Ship Control Systems Symposium, Southampton, U. K., Vol. 3, pp. 275-289, 1997

8. AUTHORS BIOGRAPHY

Knud Benedict is Professor for Ship's theory at Wismar University and Visiting Professor at World Maritime University Malmoe for Marine Technology. Since 1998 he is the Head of its Maritime Simulation Centre Warnemuende. He graduated from the Faculty of Naval Architecture of the Rostock University in 1972. He achieved his Doctorial Degree in Ship Hydrodynamics / Manoeuvrability (1978) and his Habilitation on Ship Operation Technology / Advisory Systems (1990).

Walter Köpnick is a graduate engineer for applied mechanics and is currently working as senior scientist on modelling ship dynamic at the Maritime Simulation Centre Warnemünde.

Michael Baldauf is presently employed at World Maritime University Malmö (Sweden). There he holds the position as Assistant Professor in Maritime Safety and Environmental Administration. He is researching and lecturing on maritime operational risk management and integrated maritime systems. He is project leader of several European RTD projects as e.g. TEAMSAFETY and ADOPTMAN. Michael Baldauf obtained his Ph.D. (Dr.-Ing.) in Safety Science from Bergische Universität Wuppertal, Germany and has a mariner's background.

Andreas Gronarz holds the current position of a senior research engineer at the DST (Development Centre for Ship Technology and Transport Research, Duisburg). He is responsible for all manoeuvring aspects of the work in the towing tank and for the hydrodynamic improvements on the inland waterway simulator SANDRA (Simulator for Advanced Navigation Duisburg, Research and Application) of the DST. Since 25 years he is working on the field of shallow water hydrodynamics including his PhD-thesis about the special aspects of the simulation of ship manoeuvres in shallow water.

Benjamin Friedhoff holds the position of a research engineer in DST's hydrodynamics department. He is the project manager of the subproject SIPAS-MAP at the DST. His previous experience includes shallow water effects and advanced methods of experimental fluid dynamics.

Fernando de Mello Petey is a naval architect and received his degree as Dr.-Ing. from the Institut für Schiffbau in Hamburg, Germany in 1987. He is a software engineer for navigation and nautical simulation with Rheinmetall since 2002.

OPTIMISATION OF TRIM WITH RESPECT TO SQUAT AND UNDER-KEEL-CLEARANCE

T Berndt and **J Reinking,** Jade University Oldenburg – Department of Civil Engineering and Geoinformation, GER
A Laupichler and **A Härting,** Jade University Elsfleth – Department of Maritime Studies, GER

SUMMARY

The paper deals with the relation between squat and trim and the corresponding implications to under-keel-clearance. A long-term study performed with one particular bulk carrier demonstrates that the dynamic change of trim is a good indicator of squat. Since trim can be measured in real-time, a method of real-time squat determination has been developed. On other types of vessels there is not always a significant correlation between squat and trim. There, the combination with different sensitive parameters is under investigation.

1. INTRODUCTION

In the past the squat of a moving ship was considered mostly with the assumption of an even static trim while, in practice, this postulate only holds in rare situations. In general the minimal under keel clearance (UKC) has to be determined by including the dynamic trim change in addition to the overall vertical lowering. The position of the minimal UKC can vary from bow to stern and might depend on the static trim.

The reduction of the static UKC (red_{UKC}) due to hydrodynamic squat (which is referred to as sinkage at LCF in this paper) and trim change of a moving ship was investigated in [4]. It has been shown that the practical handling of ship squat can substantially be improved if an uneven static trim and its dynamic change are properly considered.

The red_{UKC} is minimal if the ship is on even keel when underway. The authors of [4] found out from only three full-scale experiments on a bulk carrier (*MV Weser Stahl*) that an optimally chosen static trim decreases the dynamic reduction of UKC by up to 17 cm or 28%. Although the data base for the results of [4] is small the findings clearly show that reliable predictions of red_{UKC} are only possible on the basis of accurate and specific investigations for each individual ship. Rough rules-of-thumb, as they are presented by e.g. [1] or [3], are based on too restrictive assumptions on trim behaviour and may be misleading when attempting to optimize red_{UKC}.

The authors of [7] e.g. show that the dynamic trim change of a container ship (*MV Alianca Maua*) with a CB between 0.61 and 0.64 is strongly correlated with the draught of the ship. Whether the vessel trims dynamically to the bow or to the stern does not significantly depend on her static trim or CB. The results contradict the often repeated rule-of-thumb that ships with CB smaller than 0.7 trim by the stern when on even keel or always trim by the stern when statically trimmed by the stern.

Only on the basis of an extended data set can precise and detailed descriptions of the dynamic trim behaviour be developed. Therefore we initiated a long-term study on the *MV Weser Stahl* to confirm the findings of [4] and to precisely determine a reliable functional relation between squat and dynamic trim. This relation allows to optimise static trim with respect to minimised red_{UKC}.

2. LONG-TERM STUDY

The vessel under investigation (*MV Weser Stahl* Fig. 1) in the long-term study is a bulk carrier specially designed for the transport of iron ore from Rotterdam or Narvik to the steel works in Bremen [2]. The *Weser Stahl* was chosen because she frequently calls at the harbour of Bremen (on average once a week). The vessel always passes the Weser River at similar tide phases and under comparable loading conditions with a draft spectrum between 9.2 and 10.5 m. Therefore the processing and analysis of data is facilitated e. g. by using a simple, tide-related flow model for the calculation of the speed through water.

Figure 1: *MV Weser Stahl*

A GPS-based system (TrimSup, [2]) has been installed on the vessel to measure trim in real-time. The system consists of two GPS receivers with antennas at the fore and aft masts. The data of the receivers are sent by radio link to a computer on the bridge where the real-time solution is shown and the GPS raw-data are stored permanently.

During the last three years 74 data sets of the *Weser Stahl* were recorded. The data sets contain trim and squat of the loaded vessel over a distance of about 80 kilometers between the estuary and the berth at Bremen. The GPS-data were analyzed with respect to virtual reference stations along the track of the ship. The squat was derived as the difference between the estimated height of the longitudinal centre of floatation (LCF) and interpolated tide gauge readings.

Although the quality of the resulting squat data does not fully reach the level of data derived by the SHIPS method [8], the large number of data sets makes it possible to validate the results presented in [4] and to re-determine red_{UKC}. According to [4] a functional fit for squat and dynamic trim change with respect to speed through water was calculated using a total least squares (TLS) approach [9]. The combination of both functions for different static trim and velocities is shown in Fig. 2. The optimal static trim for the *Weser Stahl* leading to the smallest red_{UKC}, is significantly different from zero. Alike the findings in [4] the red_{UKC} decreases by up to 15 cm for a speed of 4.5 m/s if the static trim is chosen accordingly.

It is not surprising that squat and trim changes are correlated since they are both the result of the hydrodynamic change of pressure underneath the ship's hull. The trim change (green) and squat (blue) of one individual data set are plotted exemplarily as function of river kilometres in Fig. 3.

This data set shows a relatively constant ratio of squat and trim change which is a typical behaviour of this particular ship [2]. Thus the squat might be estimated by simply multiplying the observed trim change with a constant factor.

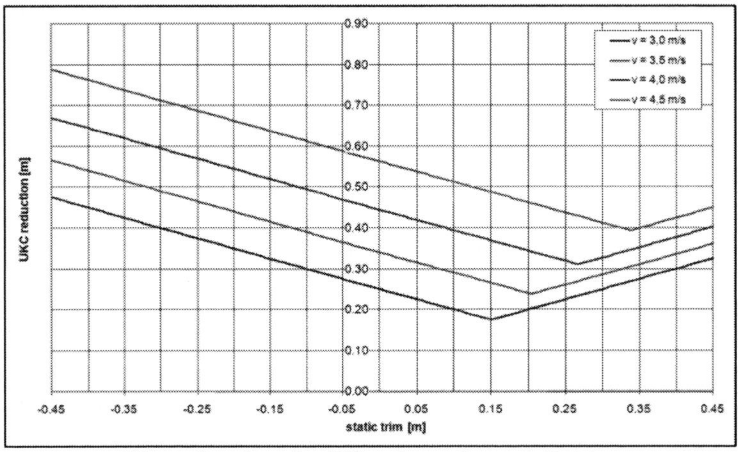

Figure 2: red_{UKC} for different speeds vs. static trim

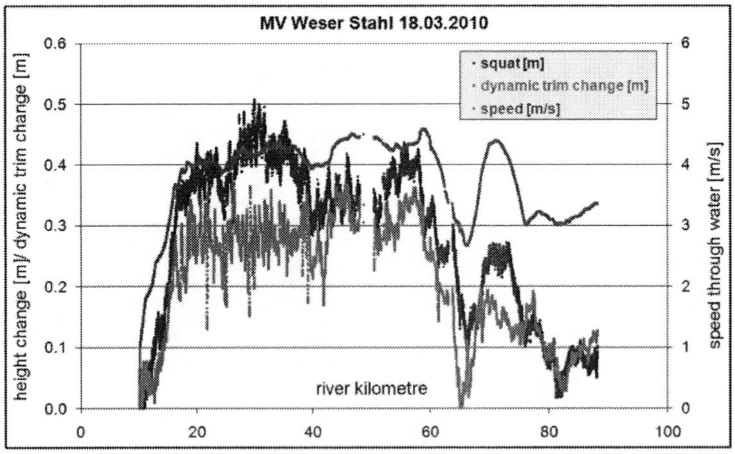

Figure 3: Data set *Weser Stahl* (2010-03-18). Trim and dynamic trim change is positive if bow is up. River kilometer starts at Bremen (0 km) and increases for outbound ships.

Fig. 4 presents the squat at LCF over the corresponding dynamic trim change for all data sets of the long-term study. To avoid misleading interpretation of a scattered point plot the data are presented as a histogram. The plot shows that most data points are concentrated close to a straight line while the number of outliers is reasonably small. Hence a linear function was fitted through these data using a TLS approach. This implies the functional relation for the *Weser Stahl*:

$$s = f \times dt \qquad \text{eq. 1}$$

with:

s ... squat in m,

dt ... dynamic trim change in m,

$f = 1.056 \pm 0.006$ ... constant factor.

To get an idea of the quality of the derived functional relation eq. 1 was used to model the squat from the observed dynamic trim changes. Fig. 5 presents the comparison of the modeled squat in blue and the observed squat in red for four data sets. Additionally, the difference is plotted in black. In Fig. 5a to Fig. 5c the two squat curves agree very well with few outliers more than 10 cm off. In Fig. 5b and Fig. 5c larger differences at river km 80 and onwards appear. There the observed squat may well be affected by inadequate salinity corrections or inaccurate interpolation of tide gauge readings. We assume that the squat modeled from dynamic trim changes by means of eq. 1 represents the actual squat much better because it is independent of uncertain corrections.

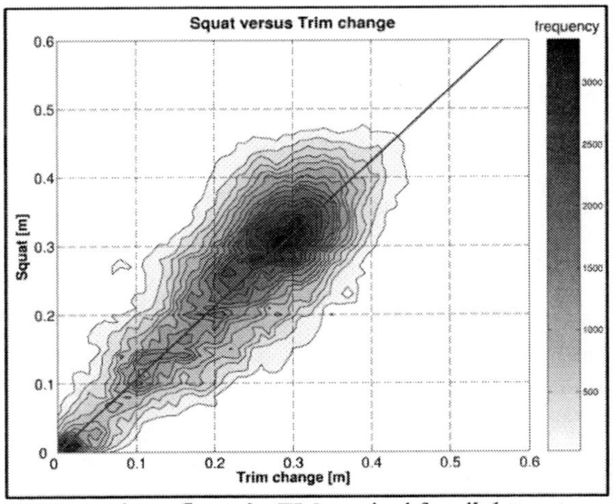

Figure 4: Linear fit on the TLS-method for all data sets

Fig. 5d shows an exceptional data set where the differences between observed and modeled squat reaches values of up to 25 cm. This data set was measured under the circumstances of a storm surge. It can be assumed that inevitable corrections are not longer valid under these conditions and the observed squat is not derived correctly. Hence, this data set was excluded from further analyses.

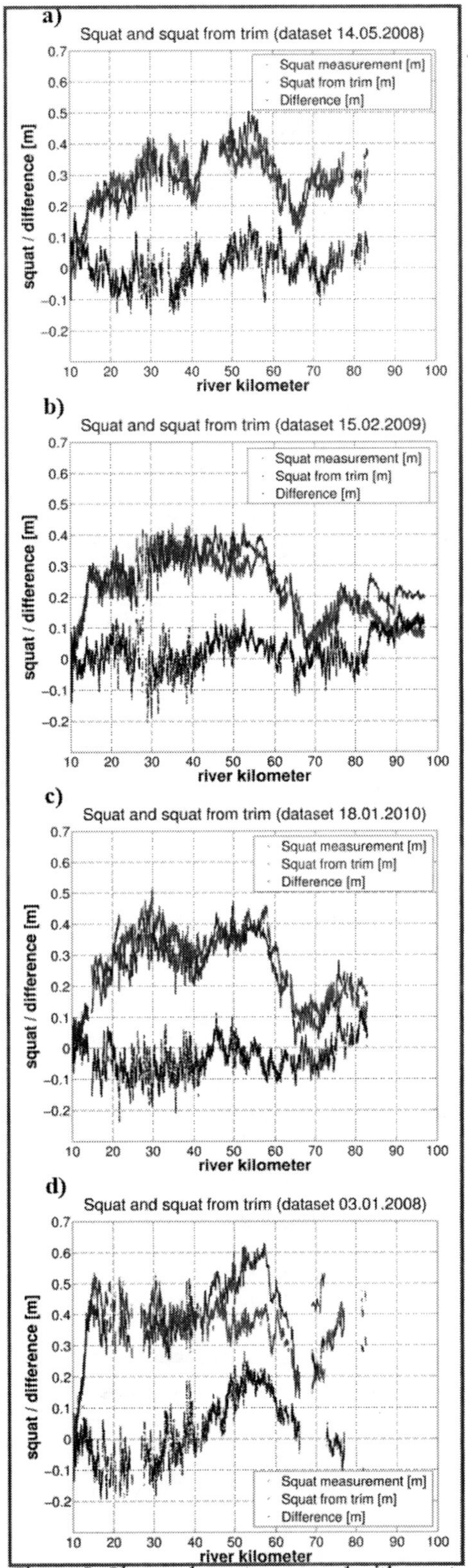

Figure 5: Observed squat compared with squat calculated from trim change using eq. 1

Fig. 6 shows the standard deviation calculated from the differences between observed and modeled squat for all data sets. Most of the data sets show a standard deviation between 4,5 and 6,5 cm. In two cases the standard deviation exceeds a value of 10 cm, but again, these data sets were observed under exceptional weather and sea conditions and have to be eliminated for further studies.

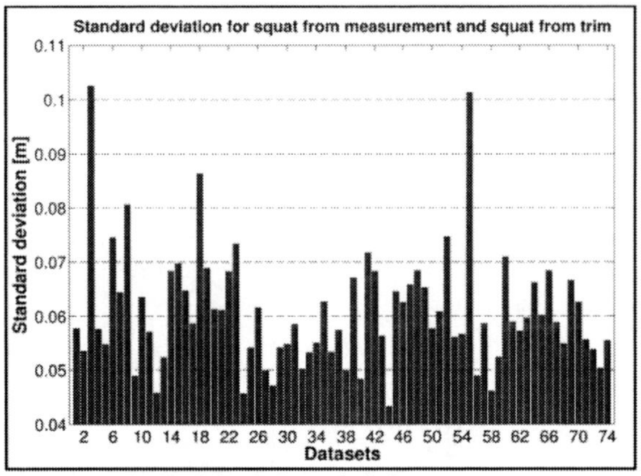

Figure 6: Standard deviation of all data sets

The results of the long-term study demonstrate explicitly the strong correlation between squat and dynamic trim change. The slope of the functional relation between squat and dynamic trim change is close to 1 for this particular ship under almost all conditions and it is possible to calculate the squat from observed dynamic trim changes. Precise observations of the dynamic trim in real-time are feasible if a GPS-based system like TrimSup is used. Hence it is worth to investigate the possibility of using such observations to derive squat information aboard a ship in real-time.

3. REAL-TIME-SQUAT ESTIMATION

In this chapter a method to estimate the squat of an inbound ship in real-time is described. The method is based on measuring the dynamic trim by means of, for instance, the TrimSup system and using the strong correlation between the dynamic trim change and squat established in Fig. 3 and 4. Of course, it is necessary to know the static trim, which, for a departing vessel would be no problem. For an inbound vessel, however, we must first address the question, how to determine the static trim on a ship underway with sufficient accuracy.

The static trim is estimated using the following equation:

$$st = ct - dt \qquad \text{eq. 2}$$

with: st... static trim in m,
ct... current trim in m,
dt... dynamic trim change in m.

As shown in the previous chapter, this trim change is a function of velocity, which can be determined for a particular ship from long term measurements.

In Fig. 7 the dynamic trim change of 23 measurements is shown together with a functional fit obtained by the total least squares method. From eq. 2 one gets the static trim for North Sea (outer Weser pilot position) water, where the average salinity is 30,98 ‰. In order to validate this function, trim values obtained from eq. 2 are compared to port readings.

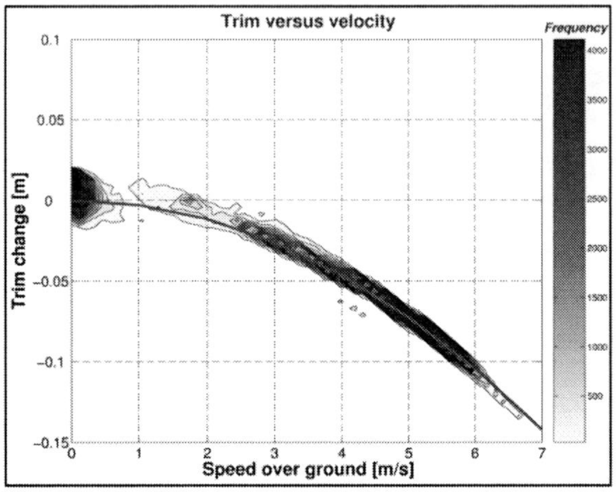

Figure 7: Fit of the dynamic trim change function

For a correct comparison the difference in salinities between the Weser pilot position and the port of Bremen must be taken into account. The lower salinity at Bremen causes an increase in draught, which moves the LCB towards the stern. This in turn changes the static trim by about 9,7 cm towards the bow, so that

$$st\_B = st\_NS - 0.097 \qquad \text{eq. 3}$$

with: st_B... static trim Bremen in m,
st_NS... static trim North Sea in m,

Now, we have two different static trim values, one for salt water and one for fresh water. Therefore, a function which describes the transition between these two limits is needed. Whereas the salinity values at both ends of our measurement trajectory are fairly constant, the behaviour in between depends strongly on recent meteorological events upstream. Here, it is assumed that at Weser km 115 the salinity starts to decrease linearly until fresh water level is reached at km 50. Any more elaborate model, applicable in real-time, would bear similar uncertainties [6].

Fig. 8 shows details of the analysis of one particular voyage. On top the abscissa has been magnified for small velocities up to 1 m/s.

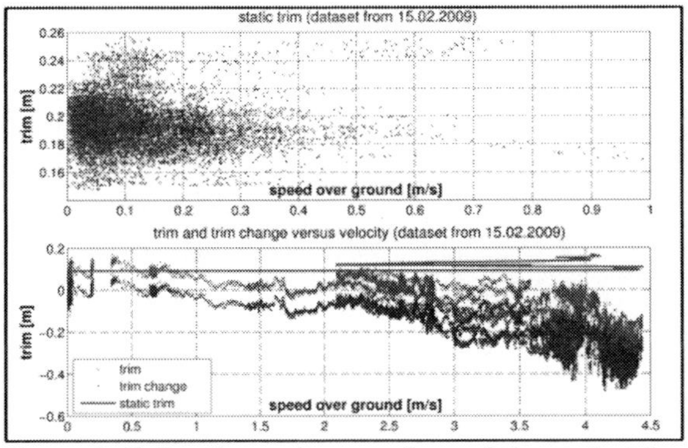

Figure 8: Estimation and comparison of the static trim evaluation for real-time use

This part, mostly encountered at anchor, was used for estimating the static trim by averaging. The red curve in the lower graph shows the trim and the blue one the dynamic trim change. The green curve indicates the static trim values.

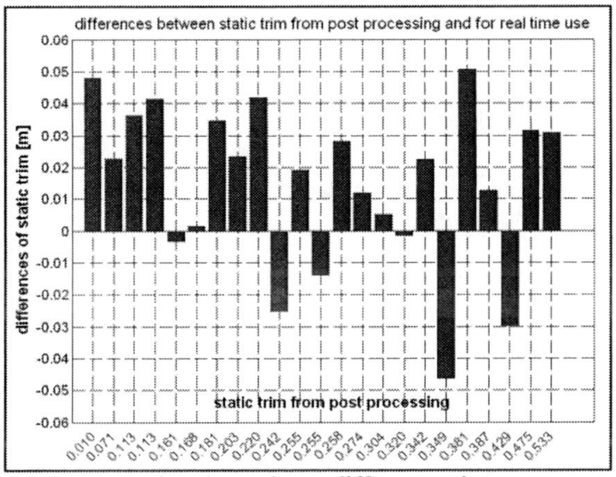

Figure 9: Static trim values: differences between post processing and real-time use

This procedure was followed for all 23 data sets. In Fig. 9 the static trim values for real-time use are compared to those estimated in post processing. It can be seen that there are differences between both ways of estimating, but the quality of the estimation of the static trim underway is sufficient, so that a real-time squat estimation using the dynamic trim change is possible. The positive mean of these values is probably caused by the simple salinity model. After the static trim has been obtained, the squat can be estimated as discussed in chapter two. Fig. 10 shows, for one typical data set, the measured squat (red) and the squat estimated from trim change with the real-time procedure (blue). This example was examined earlier (Fig. 5b), where the static trim had been input from arrival measurements. The black curves, showing the differences, agree closely.

A similar quality of the real-time estimation can be established for all data sets. Therefore, our results point out that, for this particular vessel, the dynamic trim change can be used to indicate squat in real-time.

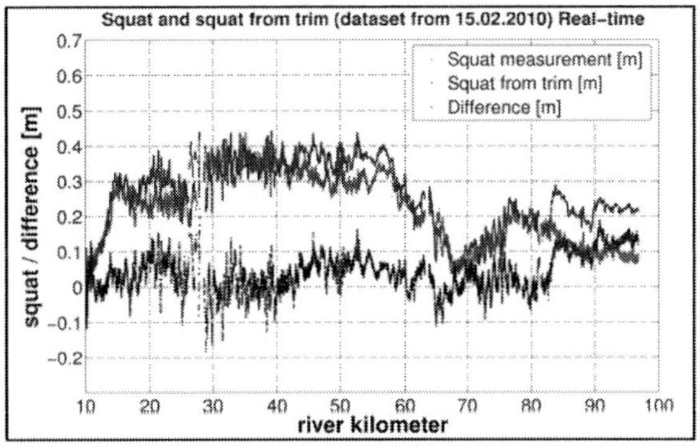

Figure 10: Observed and estimated squat from real-time

Naturally, one would expect that the same method can be applied to other kinds of ships. However, data show that matters are not necessarily straightforward and a unique correlation between trim and squat does not always exist. On modern container vessels, in particular, the trim behavior depends to a large extent on static trim and draft. Fig. 11 presents the trim of different vessels (all container carriers) over speed. The data are taken from squat measurements done with the SHIPS-method. The curves of the *Leverkusen Express* (red) and the *Hongkong Express* (yellow) do not show a significant trim change over the whole speed range, thus the speed dependent squat cannot be obtained from trim measurements.

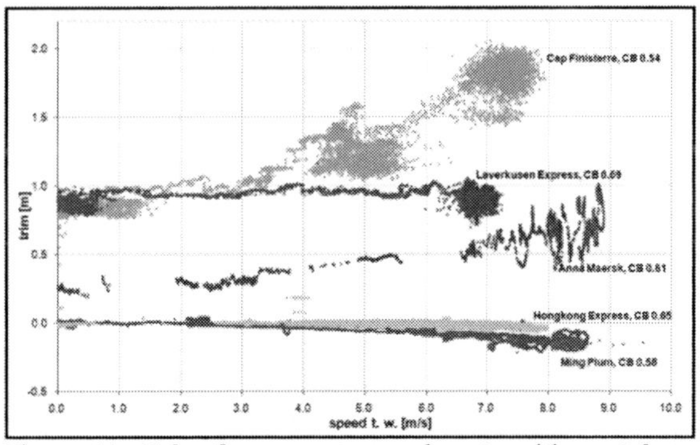

Figure 11: Trim from squat experiments with container vessels

4. SAGGING AND SQUAT

The dynamic change of trim is induced by the primary wave system generated by the moving vessel. As sketched in Fig. 12 the same wave system may create sagging due to the change of buoyancy distribution along the ship's length. So it might be possible that ships with no significant trim change over velocity instead show a correlation between sagging and speed. Depending on the position of GPS-antennas on board their distance changes with the deformation of the ship's hull.

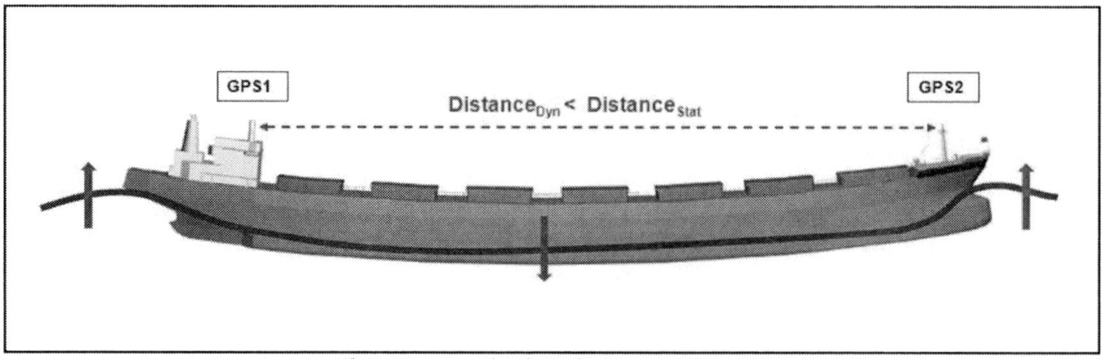

Figure 12: Sagging by primary wave system

If the antennas are mounted high enough and far enough apart, their changing distance can be sensitive to indicate sagging. On-going investigations [5] show that a system such as TrimSup, with antennas on both mast heads, can indeed provide real-time information on the dynamic increase of sagging (or decrease of hogging).

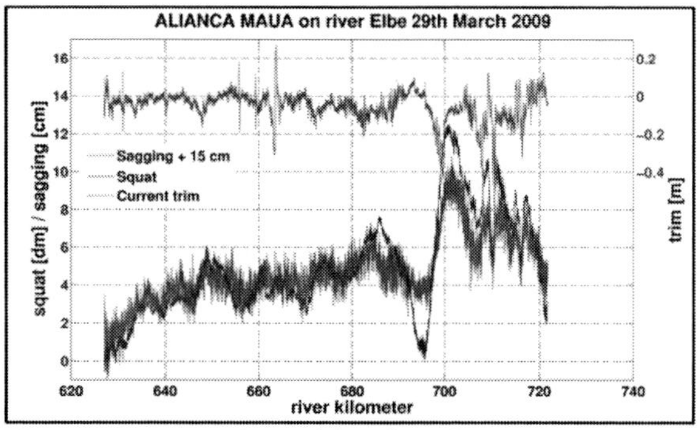

Figure 13: Squat, trim and bending observed on *Alianca Maua*

One experiment, suitable to test the above assumptions, was done with the container vessel *Alianca Maua* on the river Elbe in March 2009. The squat measurement was done with the SHIPS method using dedicated two-frequency receivers and TrimSup had been installed in parallel. Fig. 13 presents the trim (green) and the squat (blue) of the vessel on this day. It can be seen that there is no usable correlation between trim and squat. On the other hand the sagging curve (red) apparently correlates with the squat. In Fig. 14 squat is plotted over sagging to better visualize this existing correlation.

Further research is required to study the relationship between squat and sagging. In principle such a relationship could provide an alternative method to determine the squat of ships underway, which might be applicable where no significant change of trim with speed is present. With the possibility of measuring trim and sagging in real-time, a combination of both approaches seems promising to improve the consideration of squat in under-keel-clearance over existing methods.

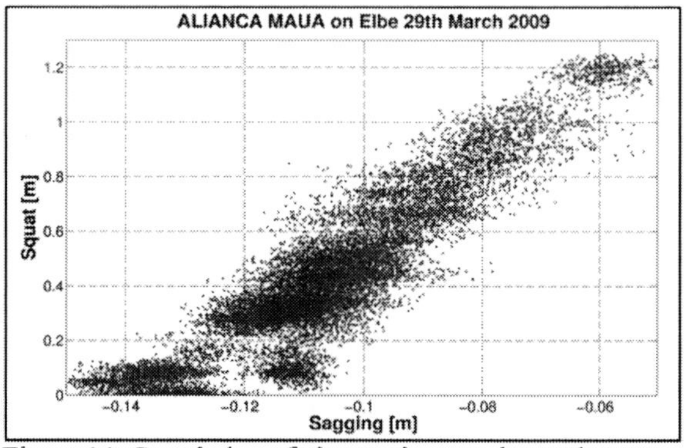

Figure 14: Correlation of observation sagging and squat on *Alianca Maua*

5. CONCLUSIONS

A method is proposed to determine the squat of a ship based on trim measurements. The method was tested during a long term study on the *Weser Stahl* giving good results compared to direct GPS-based squat measurements. By estimating the static trim as described in chapter 3 the new method is capable of real-time use. The method is probably applicable to ships with a similar hull form (bulkers). But it was found that other ships, in particular container vessels, might have no usable correlation between trim and squat, so the proposed method is not applicable there. For the *Alianca Maua* a correlation between squat and sagging was found if certain conditions regarding static trim and draught are met. Utilizing this correlation for a real-time squat prediction appears to be a promising approach for vessels with a weak trim-squat relationship and is subject of further investigation.

6. ACKNOWLEDGEMENTS

This work has been supported in part by Volkswagenstiftung, grant number ZN2172.

7. REFERENCES

[1] Barrass, C.B.: Maximum Squat of Victoria, The Naval Architect, The Royal Instution of Naval Architects, February 2009, 25-34.

[2] Berndt, T.; Reinking, R.; Härting, A.: Long-term Study of Squat and Trim on the River Weser, 3rd squat workshop "Nautical Aspects of Ship Dynamics", Bd.4 Elsflether Schriften zur Seeverkehrs- und Hafenwirtschaft, Härting, Reinking (Hrsg.), Elsfleth 2009.

[3] Briggs, M.J.;Vantorre, M.; Ulizcka, K.; Debaillon, P.: Prediction of Squat for Underkeel Clearance, in Handbook of Coastal and Ocean Engineering, Young C Kim (Ed.), World Scientific Publishing Company, Sept. 2009.

[4] Härting, A.; Laupichler, A.; Reinking, R.: Considerations on the squat of unevenly trimmed ships, in: Ocean Engineering 36 (2009) 193 - 201, Elsevier.

[5] Härting, A.; Reinking, R.: Bending of Ship Hulls Derived from GPS-Distances, 3rd squat workshop "Nautical Aspects of Ship Dynamics", Bd.4 Elsflether Schriften zur Seeverkehrs- und Hafenwirtschaft, Härting, Reinking (Hrsg.), Elsfleth 2009.

[6] Hilse, T.: Erstellung eines Modells zur Tiefgangsänderung eines tideabhängigen verkehrenden Schiffs unter Berücksichtigung der spezifischen Dichteänderung des Wassers. Elsfleth: University of Applied Sciences, Department of Maritime Studies, Diplomarbeit, 2008.

[7] Reinking, R.; Laupichler A.; Härting, A.: On the dynamic trim of container ships, 3rd squat workshop "Nautical Aspects of Ship Dynamics", Bd.4 Elsflether Schriften zur Seeverkehrs- und Hafenwirtschaft, Härting, Reinking (Hrsg.), Elsfleth 2009.

[8] Reinking, J.; Härting, A.: SHIPS: A New Method for Efficient Full-Scale Ship Squat Determination, Proceedings of the 30th PIANC Congress 2002. 1805-1813, Sydney 2002, Australia.

[9] Reinking, J: Total Least Squares?, in: ZfV 6/2008 133. Jg. 384 – 389.

[10] Reinking, J., Härting, A.: Geodetic Contributions to Ship Dynamics, PositionIT online Nov/Dec 2007, pp 18-23, EE Publishers , Muldersdrift, South Africa (http://www.eepublishers.co.za/images/upload/ServT - Geodetic.pdf)

8. AUTHORS' BIOGRAPHIES

Tobias Berndt is employed as a researcher and engineer at the Jade University of Applied Sciences in Oldenburg, Germany. He is also a master student at the named university. He is responsible for acquisition, processing and analysis of experimental data.

Jörg Reinking holds the current position of a Professor of geodesy, adjustment techniques and hydrographic surveying at the Jade University of Applied Sciences in Oldenburg, Germany. During the last decade he was engaged in the development of geodetic observation and analysis strategies for ship dynamic analysis (squat, trim and roll).

Anne Laupichler is employed as a researcher at the Jade University of Applied Sciences in Elsfleth, Germany. She is also a graduate student at the University of Oldenburg. She is responsible for the application of numerical models to ship dynamics and the analysis of experimental data.

Alexander Härting holds the current position of a Professor of Navigation and Physics at the Jade University of Applied Sciences in Elsfleth, Germany. Since more than ten years he has been active in the development of physical observation and analysis strategies for various aspects of ship dynamic (squat, trim and roll).

Ship Squat Comparison and Validation Using PIANC, Ankudinov and BNT Predictions

M J Briggs, Coastal and Hydraulics Laboratory, USA
P J Kopp, Naval Surface Warfare Center, USA
V Ankudinov, TRANSAS USA Inc., USA
A L Silver, Naval Surface Warfare Center, USA

SUMMARY

The PIANC (Permanent International Association of Navigation Congresses) has recommended several empirical and physics-based formulas for the prediction of ship squat. Some of the most widely used formulas include those of Barrass, Eryuzlu, Huuska, Römisch, and Yoshimura. The Ankudinov empirical formula is used by the Corp's ship simulator and has recently been updated. The BNT (Beck, Newman and Tuck) numerical predictions are used in the Corp's CADET (Channel Analysis and Design Evaluation Tool) model for predicting underkeel clearance due to ship motions and squat. The purpose of this paper is to compare Ankudinov and BNT predictions with these five PIANC empirical formulas for several ships with measured data from the Panama Canal. These comparisons demonstrate that both the Ankudinov and BNT predictions fall within the range of squat measurements and the PIANC predictions and can be used with confidence in deep draft channel design.

NOMENCLATURE

A_c	Wetted cross section area of canal (m^2)
A_s	Ship's underwater amidships cross section (m^2)
B	Ship's beam (m)
B_{Tr}	Stern transom width, Ankudinov
C_B	Ship's block coefficient
C_F	Correction factor for ship shape, Römisch
C_V	Correction factor for ship speed, Römisch
F_{nh}	Froude Depth Number
g	Gravitational acceleration (9.81 m s^{-2})
h	Water depth (m)
h_m	Mean water depth, Römisch (m)
h_{mT}	Restricted channel water depth, Römisch (m)
h_T	Height of dredged underwater trench (m)
K	Channel coefficient, Barrass
K_b	Correction factor for channel width, Eryuzlu
K_b^T	Bulbous bow factor, Ankudinov
K_C	Canal channel correction factor, Römisch
K_P^S	Propeller sinkage factor, Ankudinov
K_P^T	Propeller trim factor, Ankudinov
K_R	Restricted channel correction factor, Römisch
K_S	Channel width correction factor, Huuska
K_{Tr}	Trim coefficient, Ankudinov
K_{Tr}^T	Stern transom factor, Ankudinov
K_{T1}^T	Initial trim effect factor, Ankudinov
K_U	Unrestricted channel correction factor, Römisch
$K_{\Delta T}$	Squat at critical speed, Römisch
L_{pp}	Ship length between perpendiculars (m)
n	Inverse bank slope
n_{Tr}	Trim exponent, Ankudinov
P_{Ch1}	Channel effect parameter, Ankudinov
P_{Ch2}	Channel effect, trim correction parameter, Ankudinov
P_{Fnh}	Ship forward speed parameter, Ankudinov
P_{hT}	Propeller effect in shallow water on trim parameter, Ankudinov
$P_{+h/T}$	Water depth parameter, Ankudinov
P_{Hu}	Ship hull parameter for shallow water, Ankudinov
S	Blockage factor = A_s/A_c
S_s	Stern squat (m)
S_b	Bow squat (m)
S_h	Channel depth factor for R and C channels, Ankudinov
S_m	Midpoint ship sinkage (m m^{-1}), Ankudinov and BNT
s_1	Corrected blockage factor, Huuska
T	Ship draught (m)
T_{ap}	Draft at aft perpendicular
T_{fp}	Draft at forward perpendicular
Tr	Ship trim (m m^{-1}), Ankudinov and BNT
V_{cr}	Critical ship speed (m s^{-1})
V_e	Enhanced ship speed, Yoshimura (m s^{-1})
V_k	Ship speed (knots)
V_S	Ship speed (m s^{-1})
W	Channel width, measured at bottom (m)
W_{Eff}	Effective width of waterway (m)
W_{Top}	Channel width, measured at top (m)
∇	Ship volume of displacement (m^3)

1. INTRODUCTION

Historically, PIANC (Permanent International Association of Navigation Congresses) has recommended several empirical and physics-based formulas for the prediction of ship squat. These include those of Barrass [1], Eryuzlu et al. [2], Guliev [3], Hooft [4], Huuska [5], ICORELS [6], Millward [7], Norbin [8], Römisch [9], Tuck [10], and Yoshimura [11]. Most are functions of a limited number of ship and channel parameters in an effort to minimize the number of free parameters and increase the ease of use. Typical ship parameters include ship speed V_k, block coefficient C_B, and ship dimensions of length between perpendiculars L_{pp}, beam B, and draught T. Channel parameters include water depth h, type of channel cross-section A_c, side slope n, and bottom channel width W. Channel types are unrestricted (U) or open channels, restricted (R) or dredged with a trench, and canal (C) with sides that extend to the surface (PIANC 1997). However, many ship and channel parameters are not known with certainty. Channel cross-sections and dimensions can vary considerably along the length of the channel and are usually not as simple as the three idealized shapes.

The PIANC formulas are usually considered the standard for predicting ship squat if field or laboratory measurements are not available. Most of these PIANC formulas were developed over a decade ago for limited laboratory and field measurements, but are used for the newer generation of containerships, tankers, and bulk carriers. Some such as Tuck and Römisch are physics-based, while others like Barrass and Yoshimura are more empirical in nature. Although many pilots and channel designers have their "favorites," no one formula has demonstrated itself to be universally better for all ship types and channel shapes. Briggs [12] has recommended examining squat predictions with more than one formula and comparing the results based on the type of ship, channel, and formula constraints. Because there is no one formula that is universally accepted, an average, range, or maximum value might be considered in channel design.

The PIANC MarCom Working Group 49 (WG49) is updating the 1997 PIANC guidance [13] and expects to publish their report in 2011. The WG49 has reduced the number of squat prediction formulas to seven of the most popular formulas. These include the updated versions by Barrass, Eryuzlu, Huuska, ICORELS, Römisch, Tuck, and Yoshimura.

The Ankudinov ship squat formula is also an empirical formula that predicts maximum squat due to midpoint sinkage and vessel trim [14]. It has been used in the Coastal and Hydraulics Laboratory's Ship/Tow Simulator to account for underway sinkage and trim in the determination of instantaneous underkeel clearance. By comparison to other empirical formulas, it is somewhat more complicated as it includes additional input parameters to account for the effects of the ship's propeller, bulbous bow, stern transom, initial trim, and the channel's depth, blockage, and cross-section. Modifications have recently been incorporated to improve its accuracy.

The CADET (Channel Analysis and Design Evaluation Tool) is a computer program developed by the US Navy to determine the 'optimum' dredge depth for entrance channels [15]. It utilizes an external program called BNT (based on the work of Beck, Newman, and Tuck) to predict ship squat [16]. BNT is a potential flow program that predicts sinkage and trim from vertical force and pitching moment due to the dynamic pressure on the hull. Briggs et al. [17] found reasonable agreement between BNT and PIANC predictions for three containerships, tankers, and bulk carriers for two different channels.

The purpose of this paper is to compare the BNT predictions with the Ankudinov and PIANC formulas for four ships with measured data from the Panama Canal. This paper is an update of an earlier Coastal and Hydraulic Technical Note (CHETN) that compared the Ankudinov and PIANC predictions for these same ships [18]. These comparisons will demonstrate that both the BNT and Ankudinov predictions fall within the range of the PIANC predictions and squat measurements and can be used with confidence in deep draft channel design.

The second section in this paper describes the ship and channel parameters in the Panama Canal study. Section 3 describes the five PIANC empirical squat formulas used in this study. The Ankudinov ship squat formulas are presented in the next section. Section 5 describes the CADET/BNT ship squat program. Results are presented and discussed in Section 6 for the four ships. Finally, the last section gives a summary and conclusions.

2. SHIP AND CHANNEL PARAMETERS

This validation exercise involves field measurements [19, 20] made in the Gaillard Cut section of the Panama Canal (Figure 1) in December 1997 and April 1998 using Differential Global Positioning System (DGPS). The Gaillard Cut (Figure 2) is a typical "canal" cross-section and stretches from Culebra to Bas Obispo, a distance of approximately 9.1 km (5 nm) from station location 1,670 to 1,970 (in hundreds of feet). The channel width for all transits was 152 m. In the 1997 study, the minimum water depths in the center 91.4-m-section of the canal were 13.6 to 15 m, insuring a minimum underkeel clearance (UKC) for the deepest draft ships of 1.6 m. Similarly, in the 1998 study, the minimum water depths were 12.4 to 13.7 m, with corresponding UKC of 0.9 m. The DGPS measurements were made using dual frequency equipment mounted at three points on each ship (bow, and port and starboard bridge wings). The vertical accuracy levels were of the order of 1 cm.

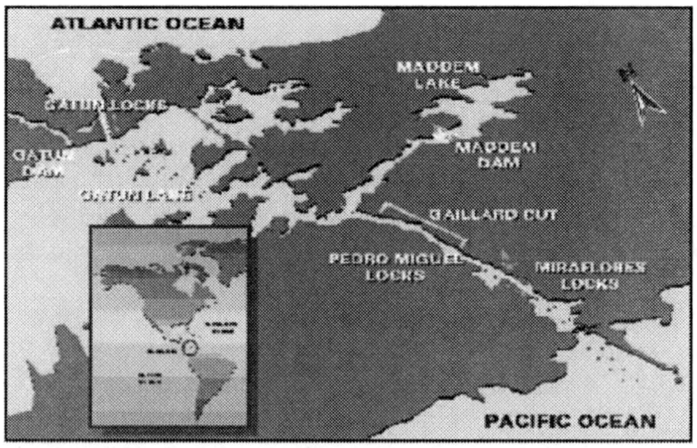

Figure 1: Gaillard Cut in Panama Canal.

Figure 2: Ship Transiting Gaillard Cut in Panama Canal.

Four of the ships from the 1997 and 1998 studies were selected for comparison. Table 1 lists the parameters for these four vessels that included a Panamax tanker (*Elbe*), Panamax bulk carrier (*Global Challenger*), and Panamax containership (*Majestic Maersk*), and one containership (*OOCL Fair*) shorter than Panamax length. The ships are grouped in Table 1 according to the year of the study.

Table 1: Ship Parameters For Panama Canal Study.

Ship Name	Type	L_{pp} (m)	B (m)	T_{fp} (m)	T_{ap} (m)	C_B	V_k (kts)
December 1997 – Minimum depth = 13.6 to 14.9 m							
Majestic Maersk	C	284.7	32.2	11.8	11.8	0.63	7-12
Global Challenger	B	216.0	32.3	11.7	11.8	0.83	9-10
April 1998 - Minimum depth =12.4 to 13.7 m							
Elbe	T	222.0	32.2	11.3	11.3	0.84	5-7
OOCL Fair	C	227.0	32.2	9.8	10.6	0.65	6-10

Notes:

1. Ankudinov: $K_P^S = K_P^T = 0.15$.

Figure 3 shows the ship speeds through the Gaillard Cut. All of the ships were traveling northward from the Pacific to the Atlantic Ocean, or from right to left in this figure. When calculating ship squat, one wants to avoid acceleration and deceleration. These transits obviously have some periods with non-steady ship speeds due to maneuvering concerns and bends in the channel (there are four in this section of the Panama Canal), but are included in the averages. The *Majestic Maersk* has the largest ship speeds and the most variation in speed. The *Elbe* had the smallest ship speeds since it was somewhat overloaded for the drought conditions in April 1998 and was required to go slower for the shallower depths and underkeel clearances.

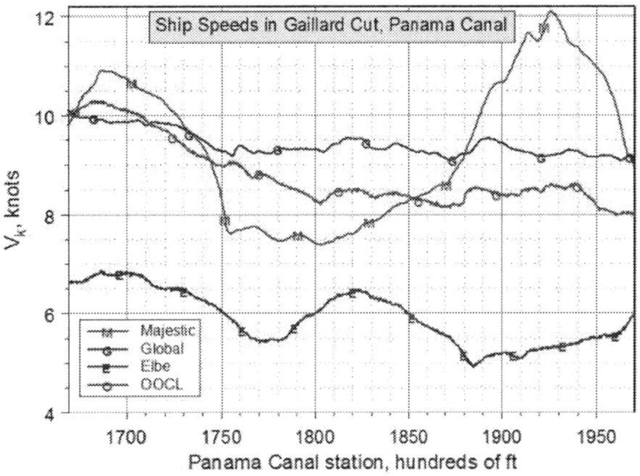

Figure 3: Ship Speeds during Panama Canal Transits. All ships were northbound, sailing from right to left.

3. PIANC SQUAT FORMULAS

In 1997 PIANC [13] included 11 empirical squat formulas in their design guidance for deep draft entrance channels. PIANC WG49 is in the process of updating this guidance and current thinking is to reduce these to 6 or 7 formulas that are the most appropriate and useful. Five of these squat formulas are evaluated in this paper, with the main emphasis on a canal (C) configuration. They include those of Barrass [1], Eryuzlu et al. [2], Huuska [5], Römisch [9], and Yoshimura [11]. Briggs [12] programmed these formulas in FORTRAN programs and Briggs et al. [21] summarized and illustrated them with examples. Even though some constraints and limitations for these formulas are exceeded, they are included in the results for this paper as this seems to be the accepted practice within the deep-draft navigation community.

All of these PIANC formulas give predictions of bow squat S_b, but only the Römisch method gives predictions for stern squat S_s for all channel types. Barrass gives S_s for unrestricted channels, and for canals and restricted channels depending on the value of C_B. According to Barrass [1], the value of C_B determines whether the maximum squat is at the bow or stern. Barrass assumes that full-form ships with C_B>0.7 tend to squat by the bow and fine-form ships with C_B<0.7 tend to squat by the stern. The C_B=0.7 is an "even keel" situation with maximum squat the same at both bow and stern. Of course, for channel design, one is mainly interested in the maximum squat and not necessarily whether it is at the bow or stern.

3.1 BARRASS (B3)

The Barrass ship squat formula has evolved and been revised at least four times. The one in this paper [1, 22] is considered the third version (i.e., B3) for both S_b and S_s.

It is a function of C_B, ship speed V_k in knots, and channel blockage coefficient K and is defined as

$$\frac{KC_BV_k^2}{100} = \begin{cases} S_b & C_B > 0.7 \\ S_S & C_B \leq 0.7 \end{cases} \tag{1}$$

If C_B >0.7, maximum squat occurs at the bow S_b. If C_B ≤0.7, it occurs at the stern and is equal to the stern squat S_s. Barrass's channel coefficient K is based on analysis of over 600 laboratory and prototype measurements for all three channel types [23]. It is defined as

$$K = 5.74S^{0.76} \qquad 1 \leq K \leq 2 \tag{2}$$

The limits on K are designed so that K=2 for C (also for R) channels. The blockage factor S is a measure of the relative cross-sectional area of the ship A_s to that of the channel A_c defined as

$$S = \frac{A_s}{A_c} = \frac{0.98BT}{Wh + nh^2} \tag{3}$$

The "0.98" factor is due to the radius on the corners of the hull. The A_c is a projection of the channel sides to the water surface. However, for the Panama Canal comparisons, the measured A_c was used. Finally, if S>0.25 for C (also for R) channels, the value of K is set to 2 to insure the limits required above for K.

3.2 ERYUZLU ET AL. (E2)

Eryuzlu et al. [2] developed a formula for S_b based on laboratory experiments. Although it is usually applied to only unrestricted (U) and restricted (R) channels, it is included in these comparisons since the Canadian Coast Guard uses it for ships in the St. Lawrence Seaway [24], a channel that is very similar to the Panama Canal. Therefore, it is included here even though the C_B constraint is technically exceeded. It is defined as

$$S_b = 0.298\frac{h^2}{T}\left(\frac{V_s}{\sqrt{gT}}\right)^{2.289}\left(\frac{h}{T}\right)^{-2.972}K_b \tag{4}$$

where the factor K_b is a correction factor accounting for relative channel width according to the ratio of channel width W to ship beam B.

$$K_b = \begin{cases} \dfrac{3.1}{\sqrt{W/B}} & \dfrac{W}{B} < 9.61 \\ 1 & \dfrac{W}{B} \geq 9.61 \end{cases} \tag{5}$$

3.3 HUUSKA/GULIEV (HG)

The next empirical squat formula was developed by Huuska [5] and Guliev [3] and is referred to as the "HG" formula. It is given by

$$S_b = \frac{2.4C_BBT}{L_{pp}}\frac{F_{nh}^2}{\sqrt{1-F_{nh}^2}}K_s \tag{6}$$

The channel width correction factor K_s is defined as

$$K_s = \begin{cases} 7.45s_1 + 0.76 & s_1 > 0.03 \\ 1.0 & s_1 \leq 0.03 \end{cases} \tag{7}$$

where the corrected blockage factor s_1=S for C channels. Huuska defined other values for s_1 for the other two channel types, but they are not presented here since we are only concerned with C channels in this paper.

3.4 RÖMISCH (R1)

Römisch [9] developed formulas for both S_b and S_s from physical model experiments for a C channel. The Römisch squat is defined as

$$S_b, S_s = C_VC_FK_{\Delta T}T \tag{8}$$

The factors in this equation are correction factors for ship speed C_V, ship shape C_F, and squat at critical speed $K_{\Delta T}$ defined as

$$C_V = 8\left(\frac{V_s}{V_{cr}}\right)^2\left[\left(\frac{V_s}{V_{cr}} - 0.5\right)^4 + 0.0625\right] \tag{9}$$

$$C_F = \begin{cases} \left(\dfrac{10BC_B}{L_{pp}}\right)^2 & \text{Bow} \\ 1.0 & \text{Stern} \end{cases} \tag{10}$$

$$K_{\Delta T} = 0.155\sqrt{h/T} \tag{11}$$

Critical ship speed V_{cr} for a canal is a function of wave celerity C and a channel shape correction factor K_C defined as

$$V_{cr} = CK_C = \sqrt{gh_m}\left[2\cos\left(\frac{\pi}{3} + \frac{Arc\cos(1-S)}{3}\right)\right]^{1.5} \tag{12}$$

where the mean water depth h_m is a function of the projected width at the top of the channel W_{Top} defined as

$$h_m = \frac{A_C}{W_{Top}} = \frac{A_C}{W + 2nh} \tag{13}$$

However, the h_m was provided for the Panama Canal data, so that value was used.

3.5 YOSHIMURA (Y2)

The last squat formula was developed by Yoshimura [11, 25] as part of Japan's Design Standard for Fairways in Japan. It was enhanced by Ohtsu [26] to include predictions for C (also for R) channels. It is defined as

$$S_b = \left[\left(0.7 + \frac{1.5T}{h}\right)\left(\frac{BC_B}{L_{pp}}\right) + \frac{15T}{h}\left(\frac{BC_B}{L_{pp}}\right)^3 \right] \frac{V_e^{\ 2}}{g} \quad (14)$$

where the enhanced ship speed term V_e is a function of ship speed V_s in m/s given by

$$V_e = \frac{V_s}{(1-S)} \quad (15)$$

4. ANKUDINOV SQUAT FORMULA

Ankudinov and Jakobsen [27] and Ankudinov et al. [14, 28] proposed the MARSIM 2000 formula for maximum squat based on a midpoint sinkage S_m and vessel trim Tr in shallow water. The Ankudinov method has undergone considerable revision as new data were collected and compared. The most recent modifications from a study of ship squat in the St. Lawrence Seaway by Stocks, et al. [29] and emails and telecons between Ankudinov and Briggs in April 2009 [30] are contained in this paper. These new revisions were programmed and documented in a technical note by Briggs [31]. The Ankudinov is the ship squat formula used in the ERDC Ship Tow Simulator.

The Ankudinov prediction is one of the most complicated formulas for predicting ship squat as it includes many empirical factors to account for the effects of ship and channel. The restriction on Depth Froude Number F_{nh} is values less than or equal to 0.6. The maximum ship squat S_{Max} is a function of two main components: the midpoint sinkage S_m and the vessel trim, Tr given by

$$S_{Max} = L_{pp}\left(S_m \mp 0.5Tr\right) \quad (16)$$

The S_{Max} can be at the bow or stern depending on the value of Trim. The negative sign in the equation above is used for bow squat S_b and the positive sign for stern squat S_s.

4.1 MIDPOINT SINKAGE S_m

The S_m is defined as

$$S_m = \left(1 + K_P^S\right) P_{Hu} P_{F_{nh}} P_{+h/T} P_{Ch1} \quad (17)$$

The ship, water depth, and channel parameters in this midpoint sinkage equation are described in the paragraphs below. The propeller parameter K_P^S is defined as

$$K_P^S = \begin{cases} 0.15 & \text{single propeller} \\ 0.13 & \text{twin propellers} \end{cases} \quad (18)$$

The ship hull parameter for shallow water P_{Hu} was recently modified by Ankudinov [30] as

$$P_{Hu} = 1.7C_B\left(\frac{BT}{L_{pp}^2}\right) + 0.004C_B^2 \quad (19)$$

The ship forward speed parameter P_{Fnh} is given by

$$P_{F_{nh}} = F_{nh}^{(1.8+0.4F_{nh})} \quad (20)$$

which is a numerical approximation to the term "$F_{nh}^2 / \sqrt{1-F_{nh}^2}$" that is in many of the PIANC empirical squat formulas. The F_{nh} is defined as

$$F_{nh} = \frac{V_s}{\sqrt{gh}} \quad (21)$$

The water depth effects parameter $P_{+h/T}$ is defined as

$$P_{+h/T} = 1.0 + \frac{0.35}{(h/T)^2} \quad (22)$$

The channel effects parameter P_{Ch1} for an R or C channel is given by

$$P_{Ch1} = 1.0 + 10S_h - 1.5(1.0 + S_h)\sqrt{S_h} \quad (23)$$

where the channel depth factor S_h is defined by

$$S_h = C_B\left(\frac{S}{h/T}\right)\left(\frac{h_T}{h}\right) \quad (24)$$

and h_T is the trench height measured from the bottom.

4.2 VESSEL TRIM

The second main component in the MARSIM squat equation is the vessel trim, Tr that was also recently modified by Ankudinov [30] as

$$Tr = -1.7P_{Hu}P_{F_{nh}}P_{h/T}K_{Tr}P_{Ch2} \quad (25)$$

In addition to the three parameters already described for the midpoint sinkage equation, the Tr also includes parameter $P_{h/T}$ and coefficient K_{Tr} to account for the effects of the ship propellers, bulbous bow, stern transom, and initial trim.

The vessel trim parameter $P_{h/T}$ accounts for the reduction in trim due to the propeller in shallow water and is defined as

$$P_{h/T} = 1 - e^{\left[\frac{2.5(1-h/T)}{F_{nh}}\right]} \quad (26)$$

The Trim coefficient K_{Tr} is a function of many factors and is given by

$$K_{Tr} = C_B^{nTr} - \left(0.15K_P^S + K_P^T\right) - \left(K_B^T + K_{Tr}^T + K_{T1}^T\right) \quad (27)$$

The first factor in this equation $C_B^{n_{Tr}}$ is the block coefficient C_B raised to the n_{Tr} power. The exponent is defined as

$$n_{Tr} = 2.0 + 0.8\frac{P_{Ch1}}{C_B} \tag{28}$$

The next two factors define the propeller effect on the vessel trim. The first factor K_P^S is the same as the propeller parameter for the midpoint sinkage and the second factor is the propeller trim parameter K_P^T

$$K_P^T = \begin{cases} 0.15 & \text{single propeller} \\ 0.20 & \text{twin propellers} \end{cases} \tag{29}$$

The last group of three factors define the effects of the bulbous bow K_b^T, stern transom K_{Tr}^T, and initial trim K_{T1}^T on the vessel trim. The K_b^T is given by

$$K_b^T = \begin{cases} 0.1 & \text{bulbous bow} \\ 0.0 & \text{no bulbous bow} \end{cases} \tag{30}$$

The K_{Tr}^T is defined by

$$K_{Tr}^T = \begin{cases} 0.1\left[\dfrac{B_{Tr}}{B}\right] = 0.04 & \text{stern transom} \\ 0.0 & \text{no stern transom} \end{cases} \tag{31}$$

where B_{Tr} is the stern transom width and is typically equal to 0.4B, although values as high as 0.7B have sometimes been used.

The K_{T1}^T is given by

$$K_{T1}^T = \frac{\left(T_{ap} - T_{fp}\right)}{\left(T_{ap} + T_{fp}\right)} \tag{32}$$

where T_{ap} is the static draft at the stern or aft perpendicular and T_{fp} is the static draft at the bow or forward perpendicular.

Finally, the channel effect trim correction parameter P_{Ch2} for both R and C channels is defined as

$$P_{Ch2} = 1.0 - 5S_h \tag{33}$$

5. CADET/BNT SINKAGE AND TRIM

Underway sinkage and trim may be provided externally by calculations or by model test data and imported into CADET. Alternatively, it can be calculated within CADET using the BNT (Beck, Newman, and Tuck) potential flow program by Beck et al. [16]. Although included with and loosely coupled to CADET, BNT is completely independent and standalone. Since channel geometry can vary from reach to reach, CADET supports the ability to define multiple sets of sinkage and trim data sets for the same ship and loading condition.

The BNT sinkage and trim prediction program is based on early work by Tuck [10, 32] investigating the dynamics of a slender ship in shallow water at various speeds for an infinitely wide channel and for a finite width channel such as a canal. This work was expanded to include a typically dredged channel with a finite-width inner channel of a certain depth and an infinitely-wide outside channel of shallower depth by Beck et al. [16].

Figure 4 is a schematic of the simplified channel cross-section used in BNT. In addition to the automatically-specified inside channel depth H, the user has the option to include the channel width W and outside channel depth H_{out} (i.e., similar to PIANC h_T trench height for restricted channels, but measured from the water surface to the top of the trench). The value of H_{out} remains the same for all H values. The walls of the channel are fixed as vertical since there is no input to specify the slope of the channel sides.

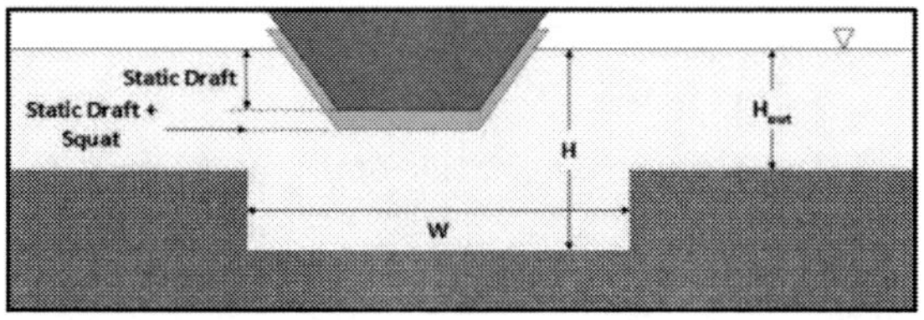

Figure 4: BNT channel geometry variables.

In his early work, Tuck [10] calculated the dynamic pressure of slender ships in finite-water depth and infinite and finite-water width by modeling the underwater area of the hull. This underwater area was defined by the 21 equally-spaced stations along the ship's length. Therefore, the ship's geometry file, draft, speeds, and water depths are used in the BNT squat calculations. Within this analysis, the fluid is assumed to be inviscid and irrotational, and the hull long and slender. Input hull definition is provided in terms of the waterline beam and sectional area at 21 equally-spaced stations between the forward and aft perpendiculars (Figure 5). Typically, generic ship lines from a ship database are used and adapted for a particular ship as ship lines are proprietary and not readily available for newer ships

The dynamic pressure is obtained for each F_{nh} by differentiating the velocity potential along the length of the hull. The sinkage and trim predictions are obtained from the dynamic pressure by calculating the vertical force and pitching moment which are translated to vertical sinkage and trim angle. The proper use of this BNT program requires that channel depths be of the same order as the draft of the ship, therefore satisfying the shallow-water approximations assumed in Tuck [10].

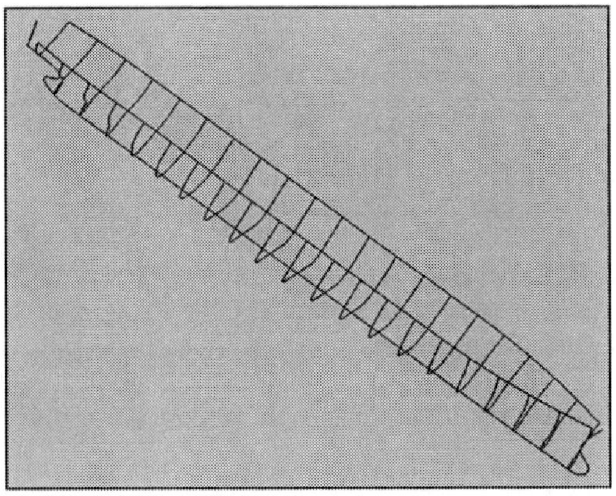

Figure 5: CADET and BNT Ship Lines Hull Geometry.

The BNT program numerically calculates midship sinkage S_m and trim Tr as a function of F_{nh}. Sinkage is measured in ft, positive for downward movement. Trim in ft is the difference between sinkage at the bow and stern, positive for bow down. The equivalent bow S_b and stern S_s squat are given by

$$\begin{aligned} S_b &= S_m + 0.5(Tr) \\ S_s &= S_m - 0.5(Tr) \end{aligned} \tag{34}$$

This is a simplistic representation of the squat at the bow and stern as it assumes they are equidistant, fore and aft, from the midpoint of the ship. An Excel spreadsheet was created from the BNT output to iterate between water depths and ship speeds at each measurement location.

6. PANAMA CANAL RESULTS

Although FORTRAN programs had been written for the Ankudinov [31] and PIANC [12] formulas, an Excel version was created for this application due to the speed, depth, and channel area variations at each measurement location along the Panama Canal. They were used to "spot check" the accuracy of the squat predictions at several locations for each ship. The output was imported into an Excel file for post-processing and iteration for the exact ship speed and water depth combinations.

As mentioned previously, only five of the PIANC formulas were included in this study. Although all five can predict bow squat, only the Barrass and Römisch formula were appropriate for stern squat predictions for canal channels. Therefore, only these two predictions were used to calculate the PIANC average for stern squat. Depending on the value of C_B, only bow or stern squat predictions were calculated as dictated by the constraints of each formulation. Again, the PIANC values were used to calculate an average bow or stern squat prediction at each location for each ship to compare with the measured DGPS values.

6.1. *Majestic Maersk* Containership

Figures 6 to 9 compare Ankudinov, BNT, and PIANC average bow or stern squat predictions to the measured DGPS values for the four ships along the Gaillard Cut. Figure 6 illustrates the stern squat for the *Majestic Maersk* Panamax containership since this ship has a $C_B < 0.7$. The right-side axis shows the ratio of Ankudinov, BNT, and PIANC predictions to measured squat at each location. Values above 1.0 indicate overprediction, with underprediction for smaller values.

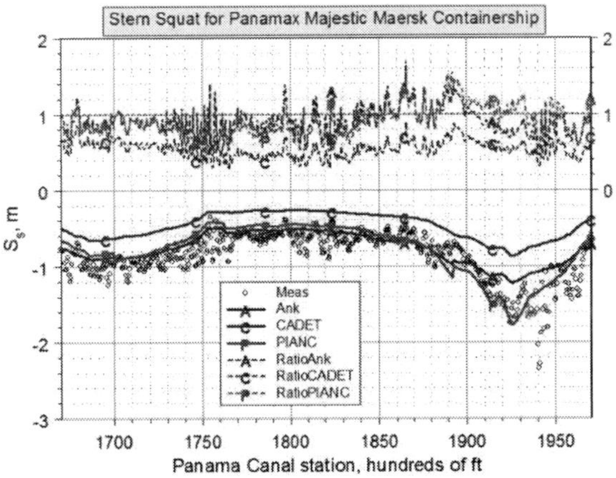

Figure 6: Measured and predicted stern squat for *Majestic Maersk* Panamax containership. Ship northbound, sailing from right to left.

Table 2 lists the minimum, average, and maximum values of the Ankudinov, BNT, and PIANC ratios for each ship for the 9.1 km length of the Gaillard Cut. A ratio of 1.0 is a perfect match, whereas values greater than one indicate overprediction and less than one signify underprediction. Table 3 lists minimum, average, and maximum differences between measured and predicted bow and stern squat for each ship. Again, negative values indicate underprediction and positive values imply overprediction.

Table 2: Predicted to Measured Squat Ratios: Ankudinov, BNT, and PIANC Formulas.

Ankudinov			BNT			PIANC		
Min	Ave	Max	Min	Ave	Max	Min	Ave	Max
Majestic Maersk Containership - Stern								
0.46	0.91	1.70	0.30	0.54	0.89	0.45	0.89	1.62
Global Challenger Bulk Carrier - Bow								
0.96	1.21	1.61	0.53	0.69	0.90	0.72	0.92	1.21
Elbe Tanker - Bow								
0.96	1.29	1.78	0.45	0.64	0.84	0.60	0.83	1.11
OOCL Fair Containership - Stern								
0.78	1.04	1.48	0.54	0.78	1.16	0.70	0.96	1.33

Notes:
1. Ratio = 1.0 is prefect match
$<$ 1.0 is underprediction
$>$ 1.0 is overprediction.

Table 3: Differences Between Measured and Predicted Squat: Ankudinov, BNT, and PIANC Formulas.

Ankudinov (m)			BNT (m)			PIANC (m)		
Min	Ave	Max	Min	Ave	Max	Min	Ave	Max
Majestic Maersk Containership - Stern								
-1.28	-0.1	0.29	-1.62	-0.39	-0.05	-1.01	-0.10	0.41
Global Challenger Bulk Carrier - Bow								
-0.05	0.22	0.48	-0.65	-0.35	-0.08	-0.38	-0.10	0.17
Elbe Tanker - Bow								
-0.02	0.11	0.22	-0.28	-0.16	-0.04	-0.21	-0.08	0.03
OOCL Fair Containership - Stern								
-0.17	0.01	0.19	-0.35	-0.13	0.07	-0.21	-0.04	0.14
Notes: 1. Negative value is underprediction. 2. Positive value is overprediction.								

Another measure of the goodness of fit of the data is the Mean Square Error (MSE) defined as

$$MSE = \frac{\sum_{j=1}^{n}\left(S_{Pred,j} - S_{Meas,j}\right)^2}{n} \tag{35}$$

where $S_{Pred,j}$ is the predicted squat at the bow or stern at location j, $S_{Meas,j}$ is the measured squat at location j, and n is the number of measurement points. Table 4 lists the MSE for each of the four ships and average for all four ships.

Table 4: MSE (m^2) Between Measured and Predicted Squat: Ankudinov, BNT, and PIANC Formulas.

Ship	n	Ankudinov	BNT	PIANC
Majestic Maersk Containership - Stern	649	0.043	0.190	0.042
Global Challenger Bulk Carrier - Bow	633	0.062	0.138	0.025
Elbe Tanker - Bow	1015	0.016	0.027	0.009
OOCL Fair Containership - Stern	681	0.005	0.023	0.006
Average:	2,978	0.029	0.085	0.019

For the *Majestic Maersk*, the Ankudinov ratios ranged from 0.5 to 1.7 times the measured stern squat, with an average underprediction of 0.9. The Ankudinov differences ranged from a worst underprediction of 1.3 m (station location 1,940) to an overprediction of 29 cm, with an average underprediction of 10 cm. The MSE was 0.043 m^2.

The BNT ratios ranged from 0.3 to 0.9, with an average underprediction slightly larger than half of the measured squat. The BNT differences ranged from an underprediction of 5 cm to 1.6 m (Station 1940.73), with an average underprediction of 39 cm. The extreme underprediction at Station 1940.73 is probably an error in the measured data as they all appear to look unusually large in this section of the canal from 1940 to 1950. The MSE was 0.19 m^2.

The PIANC ratios ranged from 0.4 to 1.6, with an average underprediction of 0.9 times the measured squat. The PIANC differences ranged from an underprediction of 1.0 m to overprediction of 41 cm, with an average underprediction of 10 cm. The MSE was 0.042 m^2.

While the BNT model underpredicted the measured values, it showed the same trends as the measured data and the other predictors. From station 1,670 to 1,850, both Ankudinov and PIANC tended to underpredict stern squat, with Ankudinov predictions slightly better. Around station 1,880 to 1,960, the PIANC formula overpredicted stern squat. In general, both Ankudinov and PIANC tended to underpredict stern squat by 10 percent, with minimum and maximum predictions about the same.

6.2. *Global Challenger* Bulk Carrier

Figure 7 shows the bow squat for the *Global Challenger* Panamax bulk carrier. This ship was trimmed 12 cm by the stern (i.e., deeper draft at the stern). In general, the Ankudinov formula overpredicted and BNT and PIANC underpredicted bow squat. Table 2 shows that the Ankudinov ratios varied from 1.0 to 1.6 times the measured bow squat, with an average overprediction of 1.2. The differences between measured and predicted squat ranged from underpredictions of 5 cm to overpredictions of 48 cm, with an average underprediction of 22 cm for the Ankudinov formula. The MSE was 0.062 m^2.

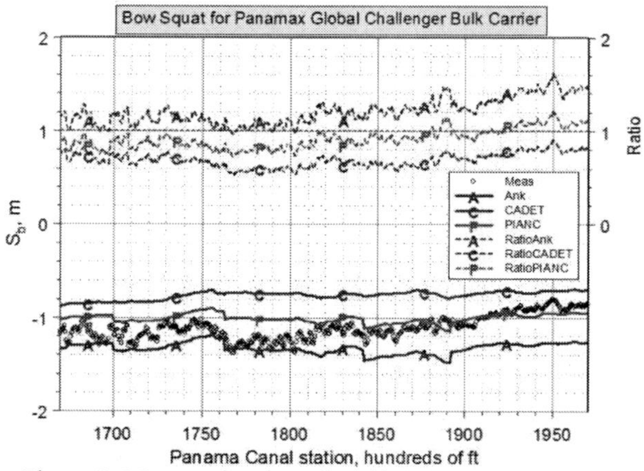

Figure 7: Measured and predicted bow squat for *Global Challenger* Panamax bulk carrier. Ship northbound, sailing from right to left.

The BNT ratios ranged from 0.5 to 0.9, with an average underprediction of 0.7 times the measured squat. The BNT differences ranged from an underprediction of 65 cm to an overprediction of 8 cm, with an average underprediction of 35 cm. The MSE was 0.138 m^2.

The PIANC ratios with five formulas in the calculations were closer to the measured bow squat, especially above the location at station 1,850. They ranged from 0.7 to 1.2 times the measured squat, with an average underprediction of 0.9. The PIANC differences ranged from an underprediction of 38 cm to an overprediction of 17 cm, with an average underprediction of 10 cm. The MSE was 0.025 m^2.

Again, BNT underpredicted the measured squat, but followed the same trends as the two other predictors. The PIANC formulas overpredicted the measured squat for a short section from location 1920 to 1970. Thus, the Ankudinov formula averaged 20 percent and 21 cm overprediction and the PIANC 10 percent and 9 cm underprediction. For design purposes, overprediction is more conservative and potentially safer.

6.3. *Elbe* Tanker

Figure 8 shows the bow squat for the *Elbe* Panamax tanker. Table 2 lists a range of Ankudinov ratios from 1.0 to overpredictions up to 1.8, with an average of 1.3 times the measured bow squat. The Ankudinov differences ranged from underprediction of 2 cm to overpredictions of 22 cm, with an average overprediction of 11 cm. The MSE was 0.016 m^2.

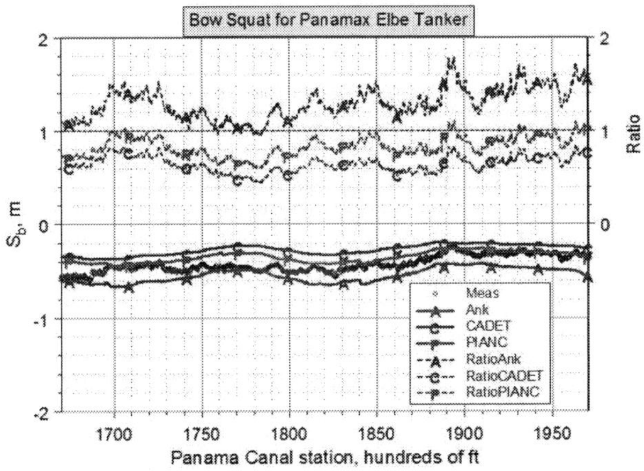

Figure 8: Measured and predicted bow squat for *Elbe* Panamax tanker. Ship northbound, sailing from right to left.

The BNT ratios ranged from 0.5 to 0.8, with an average underprediction of 0.6 times the measured squat. The BNT differences ranged from underpredictions of 28 to 4 cm, with an average underprediction of 16 cm. The MSE was 0.027 m^2.

The PIANC averages ranged from 0.6 to 1.1 times the measured bow squat, with an average underprediction of 0.8. The PIANC differences ranged from an underprediction of 21 cm to an overprediction of 3 cm, with an average underprediction of 8 cm. The MSE was 0.009 m^2.

In general, the BNT predictions were smaller than measured values, but again followed the same trends and were closer to the PIANC predictions. The Ankudinov formula overpredicts by 30 percent and 12 cm while the PIANC underpredicts by 20 percent and 9 cm. Again, overpredictions would be more conservative design.

6.4. *OOCL Fair* Containership

Finally, Figure 9 shows the stern squat for the *OOCL Fair* containership. This ship had the most trim with a value of 0.8 m by the stern. Table 2 lists a range of Ankudinov ratios from 0.8 to 1.5 times the measured stern squat, with an average ratio of 1.0 (near exact match with measured data). The Ankudinov differences ranged from underprediction of 17 cm and overpredictions of 19 cm, with an average of 1 cm (near exact match). The MSE was 0.005 m^2.

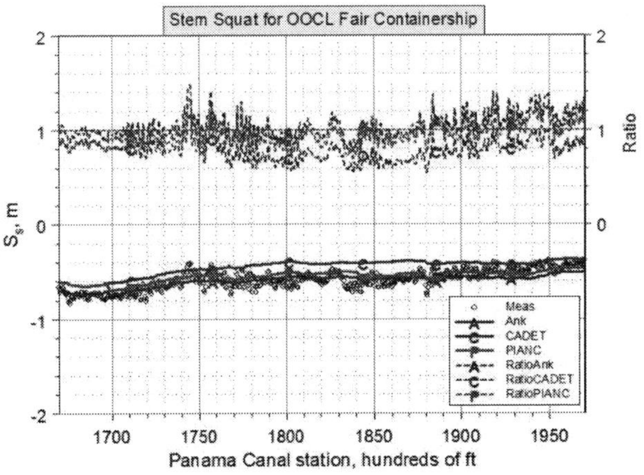

Figure 9: Measured and predicted stern squat for *OOCL Fair* containership. Ship northbound, sailing from right to left.

The BNT ratios ranged from 0.5 to 1.2, with an average underprediction of 0.8 times the measured squat. The BNT differences ranged from underpredictions of 35 to 7 cm, with an average underprediction of 13 cm. The MSE was 0.023 m^2.

The PIANC ratios varied from underpredictions of 0.7 to overpredictions of 1.3, with an average of 1.0 (near exact match) times the measured stern squat. The PIANC differences ranged from an underprediction of 21 cm to an overprediction of 14 cm, with an average underprediction of 4 cm. The MSE was 0.006 m^2.

Again, BNT underpredicted the measured data, but not by as much as some of the other ships. The Ankudinov predictions were about the same as the PIANC predictions in this case, and both very good.

6.5. Discussion

On average for the ships in this Panama Canal dataset, the Ankudinov formula overpredicts squat by 25 percent for the bow and underpredicts by 5 percent for the stern. The average MSE for all four ships based on a total of 2,978 measurements for the Ankudinov predictions is 0.029 (Table 4). BNT underpredicts squat by 30 to 35 percent for the bow and 20 to 45 percent for stern squat.

The average MSE was 0.085. The PIANC underpredicts squat by 15 percent for both bow and stern. The PIANC has the smallest average MSE of 0.019. All of the 2,978 data points were used in the comparisons even though there are a lot of turns or bends in this section of canal. Ships experience acceleration, deceleration, and roll while turning which effects squat, but is not accounted for in these squat predictors. Also, portions of the *Majestic Maersk* measurements look to be inordinately large from Station 1940 to 1950.

These comparisons indicate that the Ankudinov formulas are conservative in most instances as they tend to overpredict ship squat. The BNT predictions are generally lower than the measured values. Possible reasons for the smaller BNT predictions might be that the actual ship lines were not used due to proprietary issues. The use of generic ship lines, while appropriate, can misrepresent the water line beam and sectional area curves of the ships during transit. A sensitivity study by Kopp [33] showed that there can be a 3 to 8 percent variation in predicted squat due to 10 percent variations in fore and aft sectional areas. BNT has shown better agreement with US Navy projects at NSWCCD when actual ship lines and measured model-scale data are available for comparison. As previously mentioned, Briggs found good agreement between BNT and PIANC predictions for a range of ship and channel types. Additionally, the BNT results have been found to be comparable to those produced by more expensive higher-order CFD predictions. Therefore, variations in the generic ship lines that are most representative of the ships in this study could have a significant effect on the predicted squat. It should be noted that CADET is not restricted to using the BNT model results as the user can always import other specific or model test squat data. The PIANC predictions are based on averages (sometimes only one or two formulas for stern squat). In some instances, one or more of the PIANC formulas might match measured data much better than the averages.

7. SUMMARY AND CONCLUSIONS

This paper has compared Ankudinov, BNT, and PIANC ship squat predictions with DGPS measurements of four ships in the Gaillard Cut section of the Panama Canal. These ships included a Panamax containership, Panamax bulk carrier, Panamax tanker, and a containership. In general, the Ankudinov formulas overpredicted measured bow squat by a factor of 1.25 and underpredicted stern squat by a factor of 0.98. BNT underpredicted bow and stern squat by factors of 0.67 and 0.66, respectively. PIANC underpredicted bow and stern squat by 0.93 and 0.88, respectively. Thus, the Ankudinov predictions are slightly larger than the PIANC predictions, although the Ankudinov predictions match canal channel types like the Panama Canal better than PIANC. The BNT predictions were generally smaller than the measurements, but showed the same trends as the other predictors. Thus, all three predictors appear to give reasonable predictions of ship squat and can be used with confidence in deep draft channel design.

8. ACKNOWLEDGEMENTS

The authors wish to acknowledge the Headquarters, U.S. Army Corps of Engineers, the Naval Surface Warfare Center, Carderock Division, and TENSAS for authorizing publication of this paper. It was prepared as part of the Improved Ship Simulation work unit in the Navigation Systems Research Program.

9. REFERENCES

[1] BARRASS, C.B., 'Ship Squat and Interaction', *Witherby Seamanship International Ltd*, Bell & Bain Ltd, Glasgow, 2009.

[2] ERYUZLU, N.E., CAO, Y.L., D'AGNOLO, F., 'Underkeel Requirements for Large Vessels in Shallow Waterways', *Proceedings of the 28th International Navigation Congress*, PIANC, Paper S II-2, Sevilla, Spain, 17-25, 1994.

[3] GULIEV, U.M., 'On Squat Calculations for Vessels Going in Shallow Water and Through Channels', *PIANC Bulletin 1971*, Vol. 1, No. 7, 17-20, 1971.

[4] HOOFT, J. P. 'The Behavior of a Ship in Head Waves at Restricted Water Depth', *International Shipbuilding Progress*, No. 244, Vol 21, 367-378, 1974.

[5] HUUSKA, O., 'On the Evaluation of Underkeel Clearances in Finnish Waterways', *Helsinki University of Technology*, Ship Hydrodynamics Laboratory, Otaniemi, Report No. 9, 1976.

[6] ICORELS (International Commission for the Reception of Large Ships), 'Report of Working Group IV', *PIANC Bulletin No. 35, Supplement*, 1980.

[7] MILLWARD, A. 'A Comparison of the Theoretical and Empirical Prediction of Squat in Shallow Water', *International Shipbuilding Progress*, Vol. 39, No. 417, 69-78, 1992.

[8] NORRBIN, N.H. 'Fairway Design with Respect to Ship Dynamics and Operational Requirements', *SSPA Research Report No. 102*, SSPA Maritime Consulting, Gothenburg, Sweden, 1986.

[9] RÖMISCH, K. 'Empfehlungen zur Bemessung von Hafeneinfahrten', *Wasserbauliche Mitteilungen der Technischen Universität Dresden*, Heft 1, 39-63, 1989.

[10] TUCK, E. O. 'Shallow-Water Flows Past Slender Bodies', *JFM*, Vol. 26, Part 1, 81-95, 1966.

[11] YOSHIMURA, Y., 'Mathematical Model for the Manoeuvring Ship Motion in Shallow Water' *Journal of*

the Kansai Society of Naval Architects, Japan, No. 200, 1986.

[12] BRIGGS, M.J., 'Ship Squat Predictions for Ship/Tow Simulator', *Coastal and Hydraulics Engineering Technical Note CHETN-I-72*, U.S. Army Engineer Research and Development Center, Vicksburg, MS, 2006.

[13] PIANC, 'Approach Channels: A Guide for Design', *Final Report of the Joint PIANC-IAPH Working Group II-30 in cooperation with IMPA and IALA*, Supplement to Bulletin No. 95, June 1997.

[14] ANKUDINOV, V., DAGGETT, L.L., HEWLETT, J.C., AND JAKOBSEN, B. K. 'Prototype Measurement of Ship Sinkage in Confined Water', *Proceedings of the International Conference on Marine Simulation and Ship Maneuverability (MARSIM 2000)*, Orlando, FL, May 8-12, 2000.

[15] KOPP, P.J. and SILVER, A.L. 'Program Documentation for the Channel Analysis And Design Evaluation Tool (CADET)', *David Taylor Model Basin, Carderock Division*, Naval Surface Warfare Center NSWCCD-50-TR-2005/004, May 2005.

[16] BECK, R.F., NEWMAN, J.N. and TUCK, E.O. 'Hydrodynamic Forces on Ships in Dredged Channels', *Journal of Ship Research*, Vol 9, No 3, Sep 1975.

[17] BRIGGS, M.J., KOPP, P.J. SILVER, A.L. and MATHIS, I. 'Comparison of PIANC and CADET Ship Squat Predictions', *PIANC 125th Congress*, Liverpool, UK, May 11-14, pp 18, 2010.

[18] BRIGGS, M.J. and DAGGETT, L. 'Ankudinov Ship Squat Predictions – Part II: Laboratory and Field Comparisons and Validations', *Coastal and Hydraulics Engineering Technical Note ERDC/CHL CHETN-IX-20*, Vicksburg, MS: U.S. Army Engineer Research and Development Center, 2009.

[19] DAGGETT, L. and HEWLETT, C. 'Study of Ship Squat in the Panama Canal', *Waterway Simulation Technology, Inc.*, March, 1998a.

[20] DAGGETT, L. and HEWLETT, C. 'Study of Ship Sinkage in the Panama Canal, Phase II – Low Water Conditions', *Waterway Simulation Technology, Inc.*, June, 1998b.

[21] BRIGGS, M.J., VANTORRE, M., ULICZKA, K., and DEBAILLON, P., 'Chapter 26: Prediction of Squat for Underkeel Clearance', *Handbook of Coastal and Ocean Engineering*, World Scientific Publishers, Singapore, 2009.

[22] BARRASS, C.B. 'Ship Squat – A Guide for Masters', *Private report*, www.ship-squat.com, 2002.

[23] BARRASS, C.B., 'Ship Squat and Interaction for Masters', *Private Report*, www.ship-squat.com, 2007.

[24] CANADIAN COAST GUARD, 'Safe Waterways (A Users Guide to the Design, Maintenance and Safe Use of Waterways), Part 1(a) Guidelines for the Safe Design of Commercial Shipping Channels, Software User Manual Version 3.0', *Waterways Development Division*, Fisheries and Oceans Canada, December 2001.

[25] OVERSEAS COASTAL AREA DEVELOPMENT INSTITUTE OF JAPAN, 'Technical Standards and Commentaries for Port and Harbour Facilities in Japan', 2002.

[26] OHTSU, K., YOSHIMURA, Y., HIRANO, M., TSUGANE, M. AND TAKAHASHI, H. 'Design Standard for Fairway in Next Generation', *Asia Navigation Conference*, No. 26, 2006.

[27] ANKUDINOV, V.K., and JAKOBSEN, B.K. 'Squat Predictions at an Early Stage of Design,' *Workshop on Ship Squat in Restricted Waters* (October 4, Washington), SNAME, Jersey City, NJ, pp. 48-69, 1996.

[28] ANKUDINOV, V., DAGGETT, L. HUVAL, C., and HEWLETT, C. 'Squat Predictions for Maneuvering Applications', *International Conference on Marine Simulation and Ship Maneuverability, MARSIM' 96*, Balkema, Rotterdam, The Netherlands, pp. 467-495, Copenhagen, Denmark, September 9-13, 1996.

[29] STOCKS, D.T., DAGGETT, L.L., and PAGE, Y. 'Maximization of Ship Draft in the St. Lawrence Seaway Volume I: Squat Study', Prepared for Transportation Development Centre, *Transport Canada*, June, 2002.

[30] ANKUDINOV, V. and BRIGGS, M.J. Email and telecons with updates to equations 2, 4, 5, 10, and 18 in *CHETN-IX-19*, April 22 and 29, 2009.

[31] BRIGGS, M.J. 'Ankudinov Ship Squat Predictions – Part I: Theory and FORTRAN Programs', *Coastal and Hydraulics Engineering Technical Note ERDC/CHL CHETN-IX-19*, Vicksburg, MS: U.S. Army Engineer Research and Development Center, 2009.

[32] TUCK, E.O. 1967. 'Sinkage and Trim in Shallow water of Finite Width', *Schiffstechnik*, Vol. 14, No. 73, pp. 92-94.

[33] KOPP, P.J. 'BNT Sensitivity Study for the Tanker Elbe', (Unpublished Memorandum), *David Taylor Model Basin, Carderock Division*, Naval Surface Warfare Center, January 2011.

10. AUTHORS BIOGRAPHY

Dr. Michael J. Briggs is a Research Hydraulic Engineer at the Coastal and Hydraulics Laboratory, U.S. Army Engineer Research and Development Center, Vicksburg, MS. His research interests include deep draft navigation, laboratory simulation and analysis of multidirectional waves, and tsunami wave modeling. Dr. Briggs is currently a member of PIANC's WG49 that is revising the horizontal and vertical design guidance for approach channels. He is a registered Professional Engineer, Fellow in ASCE and COPRI, Coastal and Ocean Engineering Diplomate in ACOPNE, and has authored over 300 technical reports and papers.

Paul J. Kopp is a Naval Architect at the Naval Surface Warfare Center, Carderock Division, in the Numerical Simulation Branch of the Seakeeping Division. He is an experienced engineer and software developer and is responsible for the maintenance and development of the Channel Analysis and Design Evaluation Tool (CADET). Mr. Kopp received B.S.E. and M.S.E. degrees in Naval Architecture and Marine Engineering from the University of Michigan and is a member of the SNAME Ship Controllability Panel (H-10).

Dr. V. K. Ankudinov, TRANSAS Group Director of Hydrodynamics and Research, has over 35 years experience in senior management; in development of advanced computer and simulation systems; and in application of theoretical, experimental, and applied ship hydrodynamics for ship and port designs and advanced simulators. He is a recognized expert in analysis of full scale and model test data, hydrodynamic software development and realistic, simulation ship-model predictions. These have been used for designing numerous commercial and military vessels, training of ship operators, and various R&D studies for port and navigational channel development. He has worked in various European R&D labs (St.-Petersburg, Delft, and Trondheim), U. of Michigan, Hydronautics, and BMT in the US, and authored over 250 technical papers. He is a member of the SNAME Ship Controllability Committee (H-10 Panel).

Andrew L. Silver is a member of the Modeling and Simulation Branch of the Seakeeping Division at the Naval Surface Warfare Center, Carderock Division. Through his career, beginning in 1980, Mr. Silver has developed operator guidance software tools for predicting ship motions in shallow water. These tools have been employed by the Navy to predict underkeel clearance of deep draft ships in shallow channels, and by the Corps of Engineers to determine the required dredge depths for the entrance channels to ports of the United States. Mr. Silver is an acknowledged expert in this area with over 40 reports and refereed journal papers. Mr. Silver has also led three studies that compared model test data to the output of simulation codes that predict the motions of multiple ships forming a base at sea to transfer military equipment and personnel.

SIMULATION OF THE LAUNCH/RECOVERY OF A SUBMERSIBLE FROM A HOST SHIP

D. Cassidy, Naval Surface Warfare Center-Panama City, U.S.
P. Carrica, IIHR, University of Iowa, U.S.
M. Hyman, Naval Surface Warfare Center-Panama City, U.S.

SUMMARY

Candidate concepts for ship-board autonomous vehicle operations often involve deployment and recovery over the stern of the host ship. The flow in this region, even at slow speeds, can be violent at the length scale of the autonomous service craft, potentially leading to unacceptable motions. We investigate motions of a small, deployed craft as it approaches the host ship under a range of scenarios, including sea states and wave incidence. Both vehicles respond to ambient conditions and are controlled by autopilots. The CFD code CFDShip-Iowa is used for the computations, which is a semi-coupled two-phase level set RANS/DES solver with extensive motion and multi-ship interaction capabilities. The objective of the study is to obtain conditions under which safe operation is achieved throughout the time history of the approach. Results include trajectories, time histories of motions and velocity fields, as well as analysis of the physical processes that lead to dangerous conditions.

1. INTRODUCTION

Since the advent of ships large enough to carry other craft, it has been commonplace for host (or mother) ships to deploy and recover smaller craft. These craft, usually manned surface vessels, are typically launched and recovered via cranes over the side or stern of the host ship. Relatively calm seas have generally been a prerequisite for these operations; in less than ideal conditions, launch and recovery (L&R) is dangerous to both crew and vehicles and can consume a considerable amount of time, waiting until both craft are in a favorable position for the operation to take place. The operation places a strong reliance on highly trained personnel, both on the host ship and the deployed craft for its success.

With the introduction of autonomous or semi-autonomous vehicles in maritime operations, L & R is often desired in sea-conditions that are far from ideal and with demanding time constraints. Obviously, the advantage of having an experienced pilot onboard the deployed craft is lost. Generally, deployed craft are designed to carry out a specific task, not for the L&R operation. Therefore the vehicle is not likely to be ideal from an L&R perspective. All these issues suggest that L&R is no longer an afterthought and must be considered well before any L&R evaluation of a new system is initiated. While physical simulation at the model scale is very helpful, conditions under which those simulations can be carried out are limited. As a result, computational simulation is an attractive means of augmenting an understanding of how L&R under a wide range of conditions might proceed, and can help define a safe operational envelope for a given set of craft (host ship and deployed vehicle). Physics-based computational tools, applied in a marine environment, are now able to provide the needed accuracy and fidelity to be useful for engineering work.

In the present paper, we describe a means by which physics-based simulation can be used to provide an accurate and meaningful understanding of a portion of the recovery process under realistic conditions. We show a set of four simulations comprising of two craft, a waterjet propelled host ship and a semi-autonomous, deployed, semi-submersible. Both craft are controlled by autopilots. In the case of the host ship, the autopilot is configured to emulate the actions of a helmsman in two modes and in the case of the deployed craft, the autopilot is configured to emulate the actions of a ship-based pilot relaying commands to the deployed craft. There is no communication between the two autopilots. We will consider cases in which the host ship is controlled by moving nozzles at the waterjet outlet as well as cases in which it is controlled by keeping the nozzles fixed but varying waterjet thrust. Carrying out L&R operations with waterjet nozzles splayed out (away from the ship centerline) is an obvious means of reducing flow unsteadiness near the ship centerline.

It is helpful to review the fluid-dynamic environment in which a physics-based simulation will be performed. Generally, both craft are operating in a sea-way. In addition, the host ship (and often the deployed craft) must have a small forward velocity in order to maintain directional control. In our simulations, this velocity is on the order of 7 knots full scale. Because of these (sea-way and forward speed), the host ship is generally experiencing some amount of motion in all degrees of freedom and it's propulsors are moving in space both because of ship motion and because of steering inputs from the helmsman (cases in which the waterjet exit stream is steerable in the lateral plane). The deployed craft is also undergoing sea-state induced motion and, as it approaches the host ship, increasingly encounters an unsteady flow environment due to flow around the ship transom and due to unsteady waterjet flow. Both of these types of motion can be appreciable. As a result, the flow simulation must accurately resolve flow features on each hull for force/moment assessment and accurately resolve flow structures in the host ship wake.

Finally, there are time scale issues. Time scales for autopilots are often not the same as those of the flow environment (a design parameter). In the far field, this is not a problem, body motion time scales are on the order of the autopilot time scales (indeed, autopilot time scales are selected in this basis). In the near field however, body motion time scales can be shorter than autopilot time scales. This can lead to problems with the L&R operation. Naturally, the shortest resolved flow structures drive the choice of simulation time scale but it is essential to realize the potentially large variation of time scales that are encountered during a simulation and be aware of the effect of each.

2. METHODOLOGY

In recent years, the scale and complexity of physics-based simulation for practical marine problems has advanced enormously. The problems treated have evolved from flow around bare hulls to fully appended hulls to appended hulls with moving propellers to hulls free to sink and trim. At the SIMMAN 2008 workshop, physics-based simulations of ships undergoing maneuvers, both turning and zigzag, due to moving appendages were performed with surprisingly small errors in both macro-scale metrics such as trajectory as well as more local metrics like roll, pitch and yaw [1]. More recently, simulations of ships in close proximity have been undertaken with the goal of understanding interaction forces. In the present work, we are presenting simulations in which both craft are free to move independently of each other and be controlled using realistic controllers. This is only possible by the use of very large numerical grids that can undergo varying degrees of independent movement.

Broadly speaking, the simulations to be presented were accomplished by first, solving for the flow field at a given time step around each body in a unified grid system, ie., simultaneously. Next, the forces and moments on each body are computed and used to move the bodies (the grids that define the bodies in a CFD context). Next, the response of each body's controller to the movement is determined and used to actuate various control mechanisms on each body – this can entail further grid motion. Finally, interpolation coefficients, central to this simulation approach, are updated to correspond to the new grid locations and used to generate a new overall grid on which the flow solution at the next time step can be obtained. By means of repeating these steps, the trajectory of the two craft can be computed as a function of time and ambient sea conditions.

The fundamental flow solution is obtained on a set of structured volume grids with the CFDSHIP-Iowa solver. Large grids are decomposed into smaller grids to allow efficient solution via parallel computation. Geometric features and moving appendages are also resolved with smaller grids. Bodies are defined as a collection of surfaces in non-orthogonal, body fitted volume grids. The flow solution is obtained with a semi-coupled two-phase flow solver [2] – both liquid and air domains are included. This allows each body to undergo large amplitude motion and also allows future inclusion of wind effects on the craft superstructure. The critical air-water interface is resolved with a level-set function in which large free-surface deformations can be captured without undue interface smearing [3]. In the simulations to be presented, high Reynolds number flow solutions were obtained using a two equation turbulence model – the blended kw/ke model of Menter [4], with Detached Eddy Simulation (DES) [5]. The overall solution was computed with fourth-order accurate finite difference stencils. DES is an important aspect of the liquid solution. It allows temporal and spatial resolution of flow structures that are small relative to the host ship but large relative to the deployed craft. These structures are very often destroyed by a combination of overly dissipative UnRaNS turbulence modes and grids that are too coarse to support the smaller length scales associated with the structures.

The flow solution is obtained on a set of grids that are allowed to move relative to one another. These overlapping grids communicate with their neighbors via interpolation coefficients at or near the grid surface. Identification and determination of the needed interpolation data is provided by the SUGGAR algorithm [6]. SUGGAR is called at the end of each time step, once the final position and orientation of the grid blocks have been determined, in a multi-process approach involving several iterations [7]. SUGGAR is used to compute second-order accurate interpolation coefficients implying that at grid block boundaries, the flow solution can be no higher than second order irrespective of the stencil accuracy used in the block interior.

Once a flow solution at a time step is obtained, forces and moments on solid surfaces can be computed. These are obtained by summation of viscous nd pressure components on each surface grid element and then an integration over the grid elements on a given body. Grid movement is computed by using these forces and an explicit 6DOF solver to determine where each grid will move during a computational time step [8].

The effect of propulsors is modeled with an actuator disk model to reduce overall grid size. Using this model, an axial and circumferential momentum source term is included in volume grid cells in the volume occupied by each propulsor. While the model assumes that the axial and circumferential distribution of momentum is constant, it radially distributes the additional momentum sources in these cells such that the source is zero at the centerline and at the tip, but has a maximum at approximately 70 % of the radius. It matches the integrated axial and circumferential momentum in the propulsor volume to input values. Of particular note, the model is not able to incorporate inflow conditions into its' radial distribution.

Waves are implemented based on initial and boundary conditions for linear waves. A variety of wave models are available in the code, ranging from single wavelength, unidirectional regular waves to multi-wavelength, multi-directional spectal models like JONSWAP, Pierson-Moscowitz, and Bretschnieder [9]. Care must be taken to ensure that the grid on which the waves are traveling has a resolution which does not damp an influential wavelength.

3. APPLICATION

The cases chosen for presentation purposes utilize the same two craft. Simulation is intended to help answer questions such as: are there preferred orientations of propulsors during L&R operations, what are the influences of sea state on L&R operation, what are expected motions of the deployed craft as it either departs from or returns to the host ship in calm conditions? Answers to these questions are attempted by simulation of the approach of the deployed craft to the host ship. At the beginning of each simulation, the host ship and deployed craft are stopped with the deployed craft in a position of approximately one length (host ship length – L) astern the host ship. The two craft rapidly achieve a set velocity, the host ship slower than the deployed craft and the deployed craft begins its' approach to the host ship.

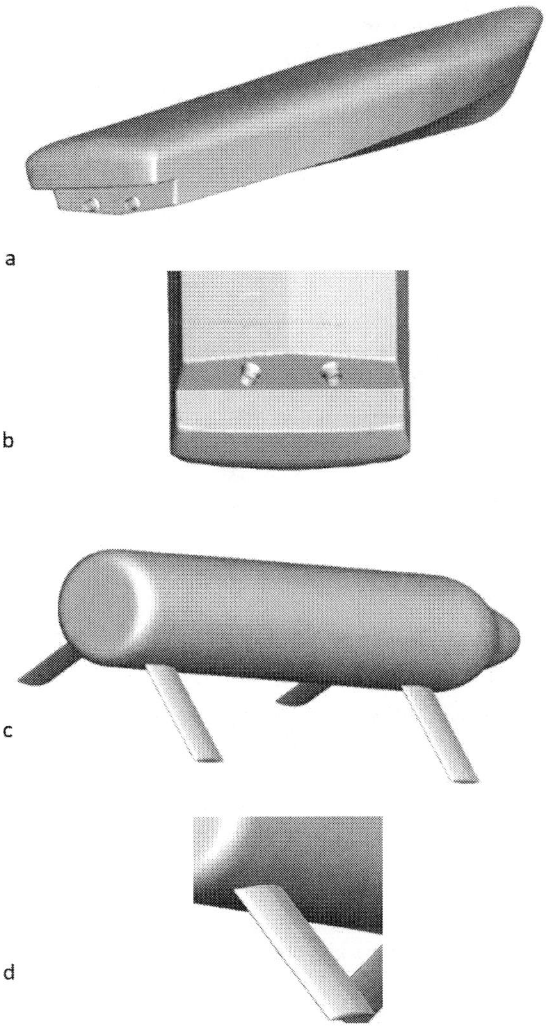

Figure 1: a. Host ship, b. Transom of host ship – waterjets splayed out, c. Deployed craft, d. Deployed craft control surface (deflected)

3.1 HOST SHIP

The host ship is a waterjet propelled utility craft. This ship, shown in figure 1a, is a hard-chine workboat hull form with flat buttock lines at the stern and no skeg. Thus its' directional stability is minimal. It is propelled by two waterjets, reasonably close to the ship centerline with steerable nozzles. For purposes of this simulation, the nozzle shape was transformed from a rectangular to a circular cross-section. In addition, the nozzles are allowed to move only in the lateral plane – figure 1b shows the nozzles oriented in a splayed position. The host ship speed is controlled by varying the thrust of both waterjets. The host ship direction can be controlled in one of two modes. In the first mode, the waterjet nozzle orientations are fixed and waterjet thrust is varied differentially, ie, to turn the ship to starboard if the nozzles are straight back, the port waterjet thrust is increased. Note, that the orientation of the nozzles is critical to determination of thrust distribution to the jets; if they are splayed out, the distribution is the reverse to that required for an inward splay. The second mode of directional control is to move the steering nozzles retaining equal thrust on the two jets (if both are used, the single jet if just one is used). Use of both modes is explored in the test cases.

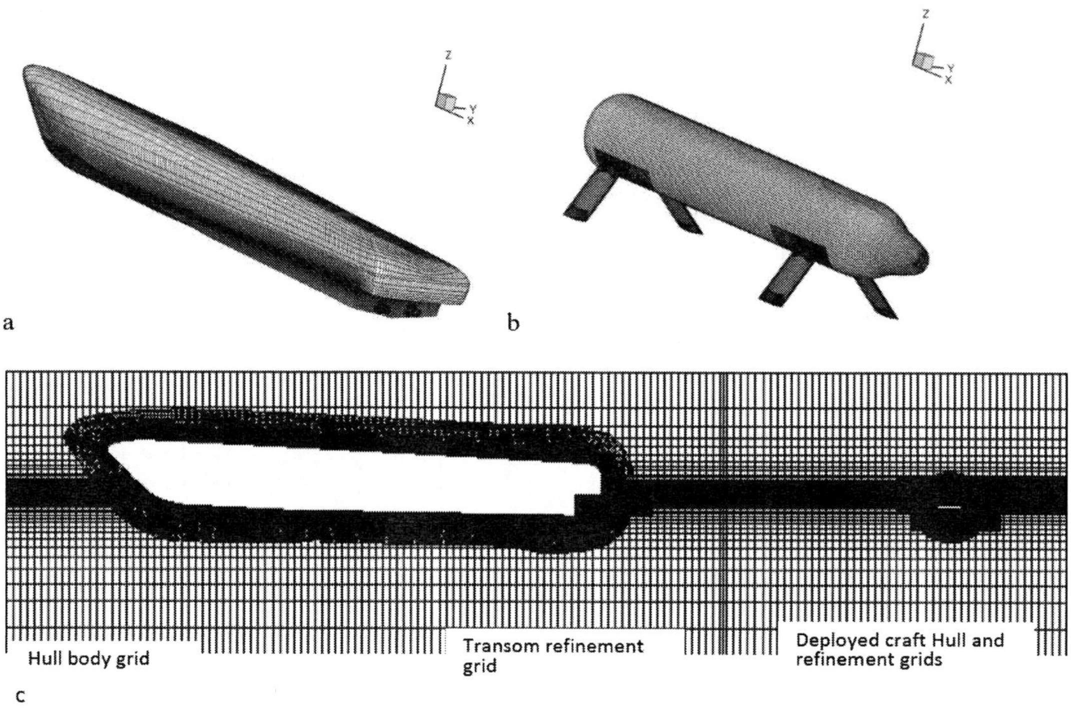

Figure 2: a. Host ship surface grids, b. Deployed craft surface grids, c. Overall grid system

3.2 DEPLOYED CRAFT

The deployed craft is a small notional submersible, 1/20 L where L is the length of the host ship, with circular cross-section, controlled by a set of four moveable lifting surfaces and propelled by an in-line multi-bladed propeller. Shown in Figure 2a with control surfaces not deflected and in figure 2b deflected to 15 degrees, the craft has control, via these surfaces, of its' movement in sway, yaw, pitch, heave and roll. Its' control system accomplishes this by a combination of coordinated and/or differential control surface deflections. Like the waterjets on the host ship, the deployed craft propeller is modeled by the actuator disc model. The craft uses a control algorithm is based on that developed for the Remote Minehunting Vehicle RMV [10, 11] that provides control of all 6 degrees of freedom via the control surfaces and thruster. In the present simulation, it is assumed that as the deployed craft approaches the host ship, craft control is split between a simulated shipboard operator and the resident control algorithm. Heave, pitch and roll remain controlled by the resident controller while thrust, sway and yaw are controlled by the operator. Very close to the transom, the vehicle rises to the surface for recovery to begin. The operator is modeled by controlling the deployed craft to a point just astern the host ship transom (the target location). The deployed craft's speed (in excess of the host ship) is proportional to its' distance from the target location – far from the host ship, the deployed craft speed is several knots faster than the host ship, at the target location, the two craft controllers are trying to impose the same forward speed.

3.3 OPERATING ENVIRONMENT

The environment in which the two ships are functioning varies with case. In three of the cases, the sea state is calm and in another two cases, a JONSWAP spectrum is used with a "sea-state" level of 2. This corresponds to the wave height of the highest 1/3 waves equal to 0.004 L and a wavelength of 0.81 L. In each case with nonzero sea-state, the incoming waves have a peak that is 4 degrees off the port bow and have a directional spread of 90 degrees to either side of the peak with $\cos^2$ distribution.

3.4 GRID LAYOUT

Because the grid plays such an important role in these simulations, some effort should be devoted to describing it. Figure 4 shows the overall grid system which includes a set of grids around the host ship, another around the deployed craft and finally a background grid in which both sets move. The host ship hull grids including those associated with the waterjets are shown in figure 5. These grids are designed to both resolve the hull shape with sufficient fidelity that accurate force and moment calculations can be achieved and allow the boundary layer to be sufficiently well represented (for the same purpose). The hull grid near the free surface must be well resolved to allow free-surface deformations to be captured, although, since the hull moves (quite a lot in high SS), this is difficult to accomplish. The refinement grid at the stern is used to allow DES to resolve flow structures of a size that are on the order of the deployed craft length, ie, those length scales that will most influence its' motion as it comes in close proximity to the host ship. These will involve both submerged flow structures and their free-surface manifestations. The hull, waterjet and refinement grids move together in the simulation in yaw, sway, surge, roll, pitch and heave. The nozzle grids just astern the waterjets are constrained to move with the hull system and are allowed to rotate laterally about fixed hinge points relative to the hull system.

The deployed craft grid system is similarly complex. It is required, like the hull grid, to resolve the hull geometry and boundary layer for force and moment calculation and to allow the control surface grids to move relative to the body grid in a manner similar to the waterjet nozzles. The deployed craft grids are immersed in a refinement grid. This system, like the host ship system, moves inside the background grid.

The background grid has two functions. The first is to provide a common region for the host ship and deployed craft to move within, transporting any flow disturbances from one to the other. In addition, it must be sufficiently fine that the most influential waves in the (nonzero) sea-state spectrum are allowed to propagate through it and not be damped by grid-induced dissipation. The background grid moves longitudinally and laterally with the host ship, guaranteeing that the host ship will always remain inside the background grid. No such guarantee exists for the deployed craft implying that it can move out of the computational domain (the background grid), in which case, the computation comes to a halt.

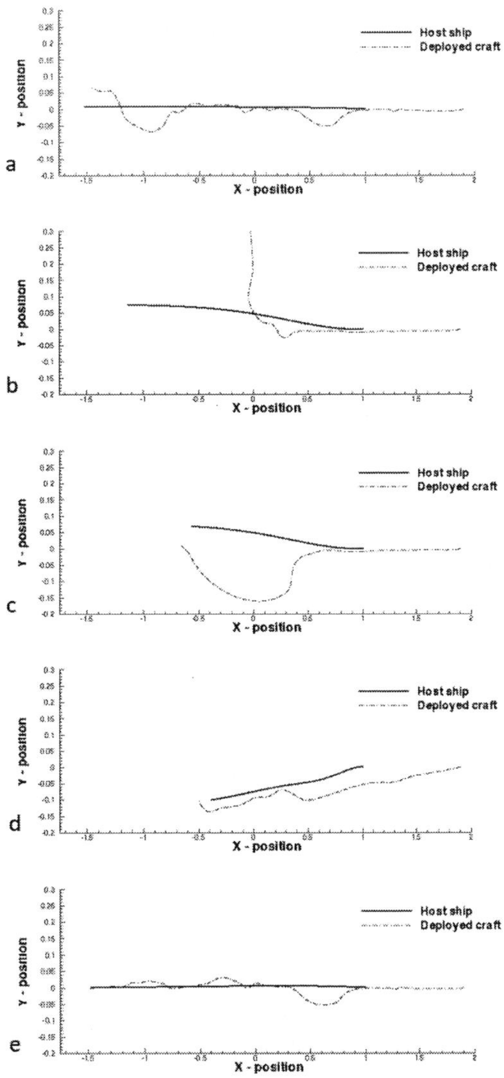

Figure 3: Plot of lateral and longitudinal trajectory of both craft, a. Calm water, nozzle control, b. Calm Water, Thrust control, c. Sea State 2, nozzle control, d. Calm water, single nozzle, thrust control, e. Sea State 2, thrust control

4. CASES

As alluded to earlier, five cases are presented, three in a calm sea and two in sea-state 2. In each sea-state, two simulations are performed, one with the waterjet nozzle direction used to steer the host ship and one with the nozzles splayed outboard and host ship steerage maintained by thrust differential. It was initially hypothesized that splaying the waterjets outboard would create a calm region just astern the ship along its centerline for L&R operation. These simulations were intended to explore the hypothesis. A final case in calm water is included to explore the effect of shutting down one waterjet and commanding the deployed vehicle to approach the host ship just astern the idle propulsor. In this case, the active propulsor is used for both speed and streerage control. The case set is summarized in table 1. Figures 3a-4e show trajectories of the two craft as viewed from above (lateral and longitudinal displacement).

Table 1. Simulation cases

Case	Sea state	Control	Remarks
1	0	Differential thrust	0.005 (nondimensional time) Re = 4.593×10^8 20 degree splay (outboard)
2	0	Moving nozzle	0.005 (nondimensional time) Re = 4.593×10^8 Max nozzle deflection less than 1 degree
3	2	Differential thrust	
4	2	Moving nozzle	Max nozzle deflection less than 2 degrees
5	0	Single moving Nozzle	Max. nozzle deflection less than 1 degree, ship sideslip 4 degrees

5. RESULTS

Results of simulations corresponding to the cases above are illustrated in the next figures. Of primary interest are the trajectory plots in figures 3a-3e. These show the lateral and longitudinal (sea gull's view) of the movement of both craft from beginning to end of each simulation. In these plots, the track of the host ship is given as a heavier, solid line and that of the deployed craft as a dashed line. As an aside, it is clear that the course controller using nozzle control is superior to that using thrust control as evidence by the nearly zero deviation in course-made-good in both calm sea and sea-state 2 (case 1 and 3) in contrast to the small but nonzero deviation seen in cases with thrust control. In cases 1, 2 and 3, the deployed vehicle trajectory is sufficiently perturbed by the wake of the host ship that it is forced to go into a re-acquire mode of operation, ie, it must return to reattempt the recovery. Interestingly, case 4 (in sea-state 2) with nozzle control proceeds with the deployed craft approaching the host ship relatively smoothly and the results indicate sufficient time for the host ship to attach recovery cables. This must be regarded as a less common stochastic event given the behavior seen in case 1 in calm water. In that case, the deployed craft initial trajectory is similar to case 4 but is perturbed by something in the host ships' near wake. The end result is the deployed craft abruptly turning to port, the control system overcorrecting and causing the craft to cross the stern of the host ship.

Clearly, splaying the waterjets (cases 2 and 4) does not improve the probability of L&R success. In both cases, the deployed craft suddenly rolls, turns (to starboard in one case, to port in the other) and exits the host ship transom wake. The reasons for this will be looked at more closely below. Only in case 5, with the host ship operating with a single waterjet, does L&R appear likely to be successful. The trajectory in figure 3e shows only small deviations of the deployed craft track as it approaches the host ship, remaining near the stern for a relatively long interval. Note that the track of the host ship, steadily to port, is a result of use of a single waterjet. It is the course-made-good while the ships' heading is 0. Even in this case, reverse flow at the transom will be seen to limit the time available for L&R.

Figures 4, 5 and 6 provide a better understanding of the flow environment that the deployed craft encounters as it approaches the host ship. In figure 4a -4i, a sequence of time steps is given showing the deployed craft approaching the host ship between the two operating waterjets. Since these jets are always moving laterally (though not by much) it is almost inevitable that the deployed craft will be caught by one jet or the other and be deflected away from the target. Clearly, while the deployed craft has sufficient control authority to enable it to maintain course in the open sea (it was designed for that operation) that authority is a little too weak (and possibly too slow) to maintain the craft's course under the influence of a waterjet.

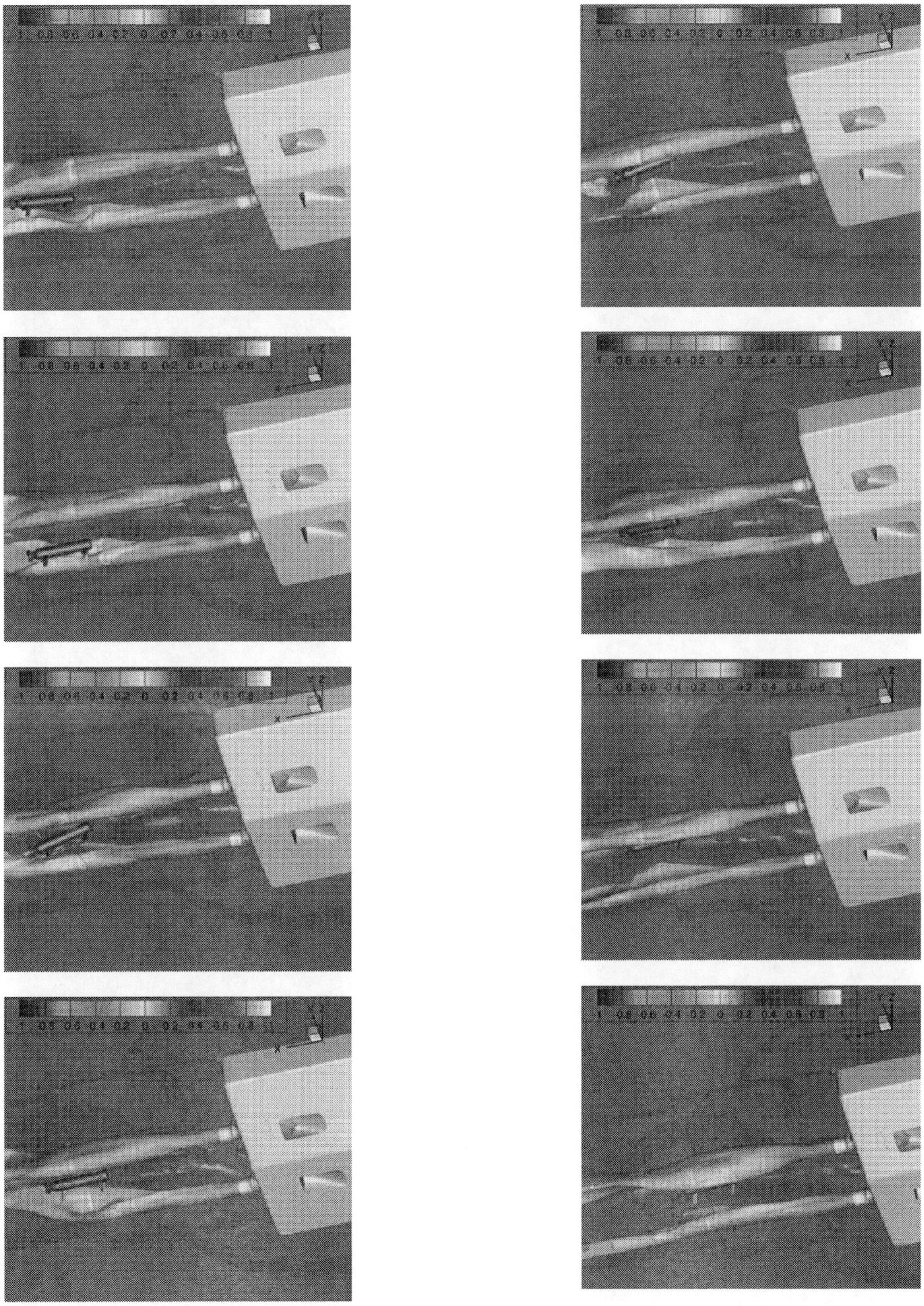

Figure 4: Time history calm water with Nozzle control - dimensionless velocity isosurfaces, 0.5, at 11.7 sec per step and 93 sec total duration

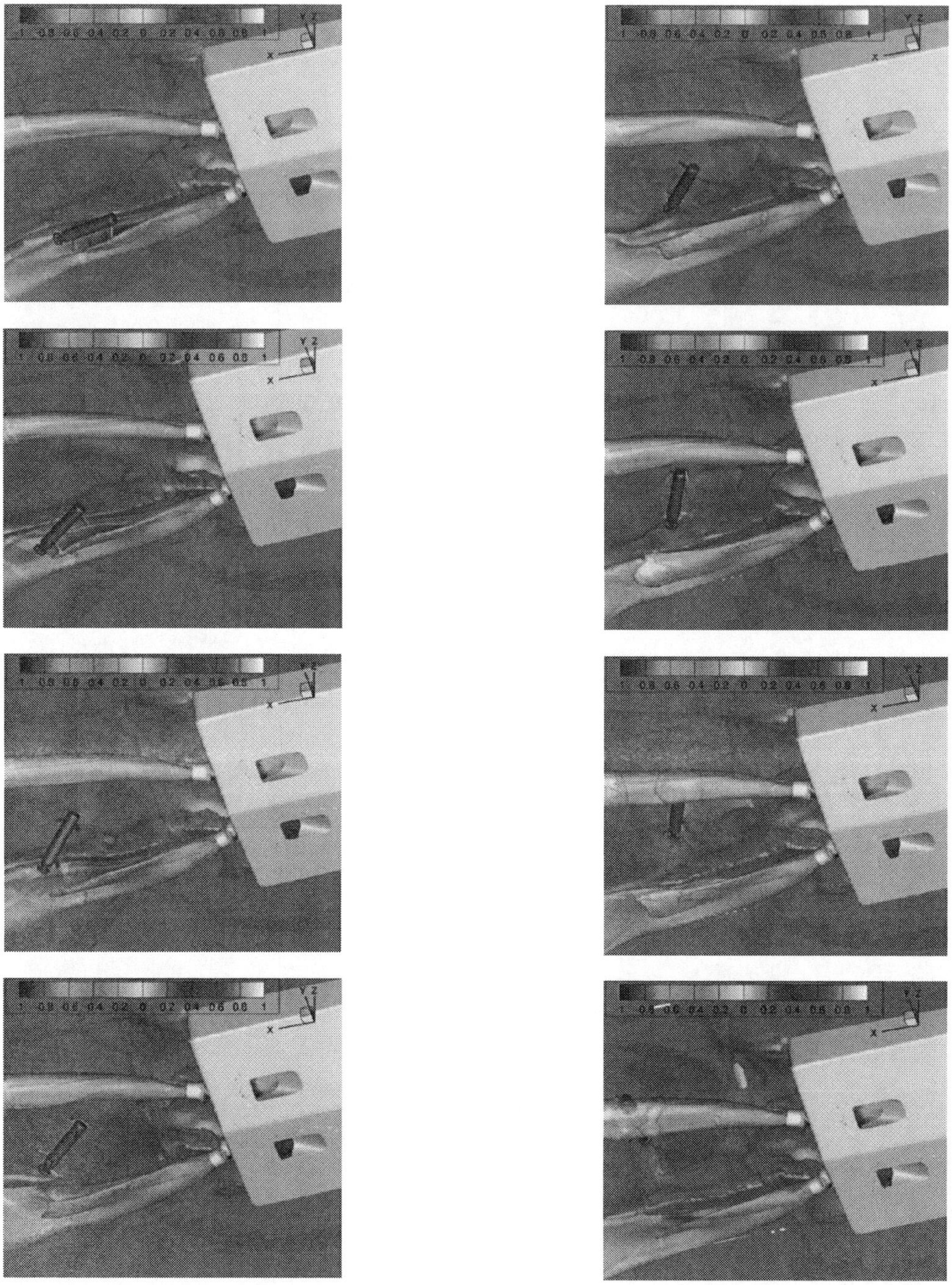

Figure 5: Time history Calm Water With Thrust Control - Dimensionless Velocity Isosurfaces, 0.5, at 11.7 sec per step

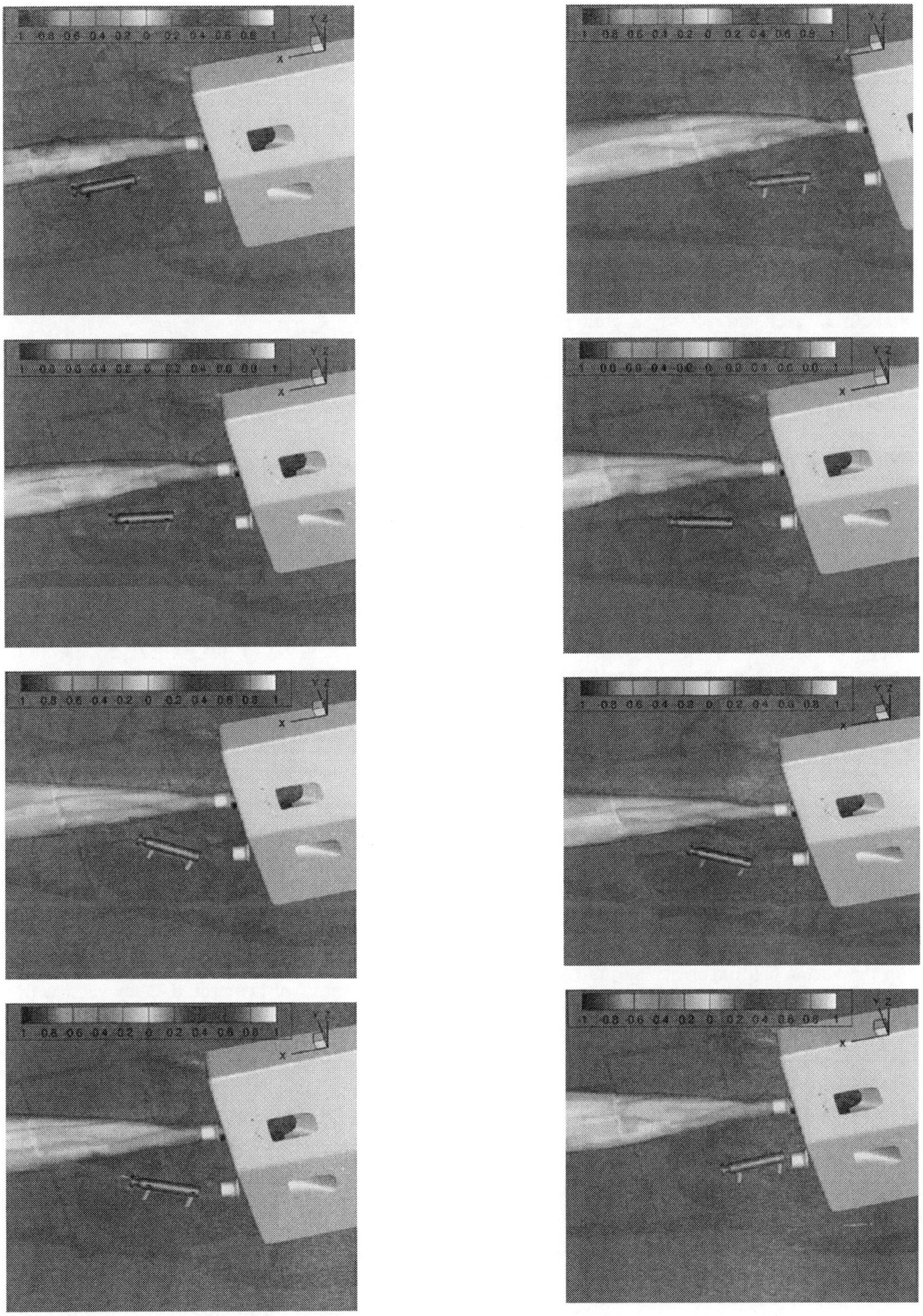

Figure 6. Time history, calm water with (single) nozzle control - dimensionless velocity Isosurfaces, at 11.7 Sec per step

A similar sequence of time steps from case 2 is given in figure 5a – 5i. In these, it can been seen that the deployed craft encounters a violent perturbation in the host ship and undergoes a large roll and a large yaw angle – both of which are more than 90 degrees. In addition, the deployed craft undergoes extreme pitch and partially exits the water. This occurs just after the last frame in figure 5

Finally, figure 6e-6g shows the deployed craft approaching the host ship with a single waterjet operating. While the trajectory is not perfect, no large amplitude course deviations are seen and recovery is quite possible. As will be seen, this occurs because use of only one waterjet creates a fairly large zone on one side of the transom that is reasonably free of intense vortex shedding.

In the above frame sequences the motion of the deployed craft is easily seen to be strongly influenced by the waterjet race but the reason for the craft to be forced into that race is unclear. Extracting flow solutions along the host ship centerline shows that a series of vortices roll off the transom with a wavelength that varies from 0.5 to 2 deployed craft lengths (l). The cross-dimension of these structures is about 1/3 l. These are exactly the range of forcing dimensions that will most strongly influence deployed craft motion. These structures of course emanate from the transom and decay with increasing distance astern. In the simulations, they tend to lose a substantial fraction of their energy by 4-5 l astern the host ship but it is unclear whether this observation is real or a reflection of the extent of the transom refinement grid.

To further investigate this issue, cross-flow plots of the host ship wake near the transom are presented. The first of these, in figure 7a, shows the deployed craft (visible as a shadow in the vector field) approaching the host ship. The vorticity field in the immediate vicinity of the craft is not strong. The craft is in a region of reverse flow (towards the transom) and some vertical shear is evident. This apparently is sufficient to cause the craft to roll and suffer the large course deviation noted earlier.

In contrast, the cross-flow plot in figure 7b (corresponding to case 3) does not exhibit regions with reverse flow. Instead, the velocity excess due to the waterjets is clearly evident as are a series of strong axial vortex structures near the surface. It is almost certain that the craft will encounter one of these structures and experience a large impulse leading to a course perturbation.

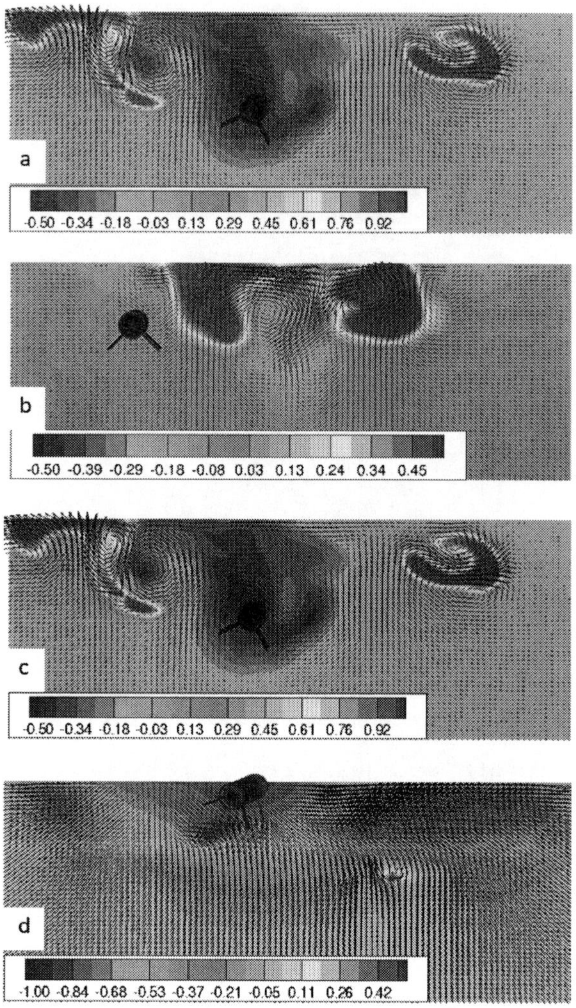

Figure 7: Velocity fields astern host ship a. Calm water with thrust control, b. Sea state 2 with nozzle control, c. Sea state 2 with thrust control, d. Calm water with single nozzle control

The plot in figure 7c corresponding to case 4 makes clear the danger associated with weak directional control. In this figure, the deployed craft encounters a particularly difficult set of forcing flow structures – both a region of reverse flow and, imbedded inside, a vortex. This shows up as a classic "hook" feature. The structure is likely due to a mild but significant host ship sideslip caused by a nonrobust mode of directional control (through differential thrust).

Finally, the plot in figure 7d, corresponding to case 5 with a single waterjet operating shows a relatively benign environment for L&R operations to be undertaken. Even in this case, the figure shows the deployed craft broaching the surface, albeit not violently enough to cause lateral course deviation.

6. CONCLUSIONS

A series of simulations including two craft of greatly contrasting size in close proximity to each other have been carried out using a physics-based computational model. In the simulations, craft motion was allowed to occur in response to both the craft environment as well as craft control mechanism movement. The simulations were performed to attempt an understanding of the launch and recovery operation with a view of learning, via simulation rather than physical model or full scale testing, what modes of recovery are more likely to be successful and which to avoid. The simulations have made clear that characteristics of both host ship and deployed craft must be carefully considered: the wake environment created by the host ship is critical to L&R success as is the control algorithm and control authority of the deployed craft. Use of high resolution simulation tools like that presented allows the designer a means of gaining insight into the physical mechanisms that dominate the L&R operation and how to either overcome those mechanisms or ensure that they are avoided.

7. ACKNOWLEDGEMENTS

The support of Dr. Patrick Purtell, ONR Code 33, in particular, his approval for modification of CFDShip-Iowa, is gratefully acknowledged as is the support of the DOD supercomputing center at Vicksburg, MS. where all computations were performed.

8. REFERENCES

1. Carrica PM, Ismail F, Hyman M, Bhushan S, Stern F, 'Turn and zigzag maneuvers of a surface combatant using a URANS approach with dynamic overset grids,' SIMMAN 2008, Copenhagen, Denmark, 2008.

2. Huang J, Carrica PM, Stern F, 'Semi-coupled air/water immersed boundary approach for curvilinear dynamic overset grids with application to ship hydrodynamics,' Int. J. Num. Meth. Fluids 58, 591-624, 2008.

3. Carrica PM, Wilson RV, Stern F, 'An Unsteady Single-Phase Level Set Method for Viscous Free Surface Flows,' Int. J. Num. Meth. Fluids 53, 229-256, 2007.

4. Menter FR, 'Two-equation eddy viscosity turbulence model for engineering applications,' AIAA J. 32, 1598-1605, 2004.

5. Travin A, Shur M, Strelets M, 'Physical and numerical upgrades in the detached eddy simulation of complex turbulent flows,' R. Friedrich and W. Rodi, eds, Advances in LES of Complex Flows, 239-254, 2002.

6. Noack R, 'SUGGAR: a general capability for moving body overset grid assembly,' AIAA paper 2005-5117, 17th AIAA Comp. Fluid Dyn. Conf., Toronto, Ontario Canada, 2005.

7. Carrica PM, Huang J, Noack R, Kaushik D, Smith B, Stern F, 'Large-scale DES computations of the forward speed diffraction, pitch and heave problem for a surface combatant,' Comput. Fluids 39, 1095-1111, 2010.

8. Carrica PM, Wilson RV, Noack R, Stern F, 'Ship motions using single-phase level set with dynamic overset grids,' Comput. Fluids 36, 1415-1433, 2007.

9. Carrica PM, Wilson RV, Noack R, Xing T, Kandasamy M, Shao J, Sakamoto N, Stern F, 'A Dynamic Overset, Single-Phase Level Set Approach for Viscous Ship Flows and Large Amplitude Motions and Maneuvering', 26th Symp. Naval Ship Hydrodynamics, Rome, Italy, 2006.

10. Larkin, L., Thomsen, C. and Feldman, J., 'Dynamics of a High Speed Autonomous Semi-submersible Vehicle,' Oceans 2002.

11. Larkin, L.J, Thomsen, C.J., ' Simulation validation for a unmanned semi-submersible vehicle, Oceans 2003 Proceedings, pp 1431-1436, 2003.

9. AUTHOR BIOGRAPHIES

Dan Cassidy is a Research Engineer at the Naval Surface Warfare Center - Panama City where he is engaged in CFD research and application on a range on naval problems including ship motions and cavitating flows.

Pablo Carrica is an Associate Professor in Mechanical Engineering at the University of Iowa, where his research is focused in two-phase flow and ship hydrodynamics. He is the main developer of the code CFDShip-Iowa V4.5.

Mark Hyman holds the position of Research Engineer at the Naval Surface Warfare Center - Panama City where he is engaged in CFD research and application on a range on naval problems including ship motions, vehicle dynamics and control and bubbly wakes.

NUMERICAL INVESTIGATION OF BANK INFLUENCE ON A SHIP MOTION

A.M. Chetvertakov, M.P. Lebedeva and **D.V. Nikushchenko,** Kongsberg Maritime, Russia

SUMMARY

The paper contains results of numerical simulation of maneuvering of a ship in restricted waters. The main goals of the research presented were the development of a method of forces and moments prediction and the elaboration of criteria of maneuverability of a ship. Results of numerical computations are presented and compared with results of experimental measurements. A regression analysis was made to find dependences taking into consideration additional forces and moments which appeared on the ship's hull while maneuvering nearby banks. With the help of these formulas additional forces and moments for several ships were calculated. There were several simulations in the training simulator "Polaris" to validate the formulas.

NOMENCLATURE

υ	Velocity (m/s)
ρ	Density of water (kg m^{-3})
P	Pressure (N m^{-2})
Y	Sway force (N)
N	Yaw moment (N m)
L	Length of the ship (m)
Lpp	Length between perpendiculars (m)
B	Breadth of the ship (m)
T	Draught of the ship (m)
β	Drift angle (deg)
Fn	Froude number
$Y0$	Distance between lateral bank and side of the ship (m)
$\overline{Y}$	$\overline{Y} = \frac{Y0}{B}$ relative distance to the lateral bank.
Cy	Sway force coefficient
Mz	Yaw moment coefficient
H	Depth of the water
HT	HT=T/H relative depth

1. INTRODUCTION

One of the most important problems of ship hydrodynamics is prediction of forces and moments acting on the ship hull during she's motion. The main reason is complex character of 3D viscous flow around the ship hull.

Usually maneuvering of a ship takes place in conditions of different external effects (like waves, wind, speed of vessel, heading) and depends on hull geometry (length, beam, draught, etc.) on the one hand, and external parameters on the other hand. One of the most important problems of ship maneuverability in restricted waters is influence of banks. In this case it is important to take into account that influence of bank changes the coefficients of forces and moments on the ship hull. Therefore it is necessary to make respective corrections of the coefficients.

There are several experimental investigations of this effect [1, 2, 3, 4].

2. NUMERICAL MODEL

All results were obtained with help of computational fluid dynamics (CFD) approach. Motion of incompressible viscous fluid describes by Navier-Stokes equations [5]

$$\rho\left(\frac{\partial \overline{v}}{\partial t} + \overline{v}\cdot\nabla\overline{v}\right) = -\nabla p + \mu\nabla^2\overline{v} + \overline{f}$$

$$\nabla\cdot\overline{v} = 0$$

This set of equations is not closed and it is necessary to use additional equations - models of turbulence. The two-parametric $k-\varepsilon$ turbulence model was used in present work. Transport equations for k and ε are [6]

$$\frac{\partial k}{\partial t} + \overline{u}_j\frac{\partial k}{\partial x_j} = \frac{\partial}{\partial x_j}(\nu + \nu_t/\mathrm{Pr}_k)\frac{\partial k}{\partial x_j} + P - \varepsilon$$

$$\frac{\partial \varepsilon}{\partial t} + \overline{u}_j\frac{\partial \varepsilon}{\partial x_j} = \frac{\partial}{\partial x_j}(\nu + \nu_t/\mathrm{Pr}_\varepsilon)\frac{\partial \varepsilon}{\partial x_j} + 1.44\cdot P\cdot\frac{\varepsilon}{k} - 1.92\cdot\frac{\varepsilon^2}{k}$$

We used ANSYS FLUENT® for the investigation.

3. BANK EFFECT MODEL IN POLARIS

In the "Polaris" training simulator , the bank effect is taken into consideration by including additional force and moments like below

$$Y = mgFn^2C_{BY},$$

$$N = mgLFn^2C_{BN},$$

Where non-dimensional force and moment coefficients are

$$C_{BY} = C_{BY}^{(0)}C_{BY}^{(1)}\left(y_B/B, H/T\right)C_{BY}^{(2)}(\beta)C_{BY}^{(3)}\left((H-H_B)/H\right)$$

$$C_{BN} = C_{BN}^{(0)}C_{BN}^{(1)}\left(y_B/B, H/T\right)C_{BN}^{(2)}(\beta)C_{BN}^{(3)}\left((H-H_B)/H\right)$$

Here the coefficient $C_{BY}^{(2)}$ determines by next dependence:

$$C_{BY}^{(2)} = C_{BN}^{(2)} = \sin\beta.$$

The other coefficients are determined based on published results of experimental data analyses and need to be corrected for each ship.
In this paper we offer a new approach to determine the bank effect based on another representation of the force and moment coefficients:

$$C_{BY} = C_{\substack{\overline{Y}\to\infty \\ H/T\to\infty}} + C_Y(Fn) + C_Y(\beta) + C_Y(\overline{Y}) + C_Y(H/T)$$

$$C_{BN} = C_{\substack{\overline{Y}\to\infty \\ H/T\to\infty}} + C_N(Fn) + C_N(\beta) + C_N(\overline{Y}) + C_N(H/T)$$

4. TWO-DIMENSIONAL APPROACH

In a first stage of the investigation, 2D flow around a ship hull was simulated. Sway force and yaw moment were taken as functions of several parameters of motion and geometrical parameters. As main parameters Froude number Fn, drift angle β and distance to the lateral bank $\overline{Y}$ were used. Influence of each of the parameters was supposed to be independent of other parameters. The research was carried out in several phases. For each only one parameter was variable while the others were kept constant.

4.1 PROBLEM DIFINITION

The ship model had a length of 1.9 m and a breadth of 0.095 m. The origin was located in centre of the ship. Motion of the fluid was considered in a relative coordinate system. There was no motion between ship and bank. Top view of the ship and rule of signs are indicated in Figure 1.

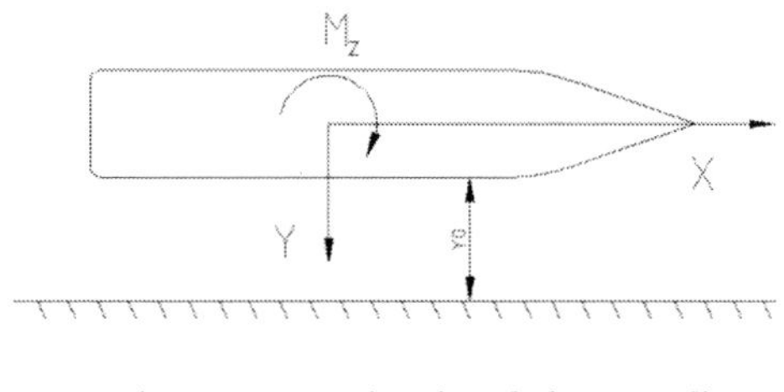

Figure 1. Top view in relative coordinates

4.2 DRIFT ANGLE INFLUENCE

The goal of this part of the investigation was to study the experimental measurements, obtained by planar mechanism tests – where the model of the ship were moving with the bank around the point of rotation. The distance between the model and the vertical bank was about two and a half times the breadth. This distance allows to realize any drift angles and in general it was possible to rotate the model near the bank.
The same parameters of motion were used in numerical research. The drift angle ranged between 0 and 180 degrees with a step of about 15 degrees. The Froude number assumed as Fr=0.2. The influence of the drift angle on the sway force and yaw moment is shown in Figures 2 and 3, along with the experimental data measured. It can be seen that the two-dimensional problem definition gives overrated values of forces and moments. However, the qualitative behavior is correct. At a drift angle of 90° there is a maximum of lateral force and a rather small vertical moment.

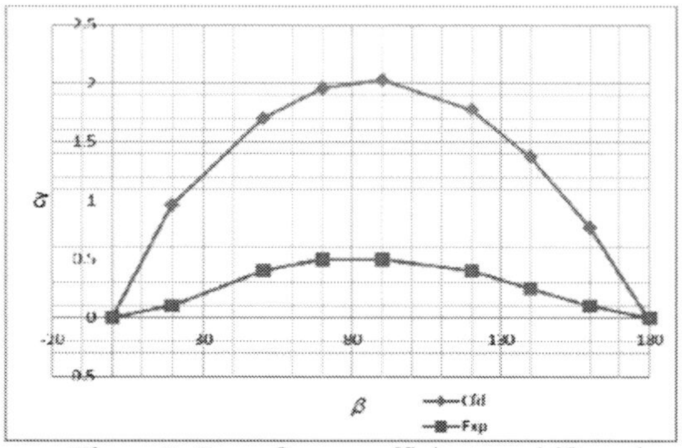

Figure 2. Sway force coefficient vs. drift angle

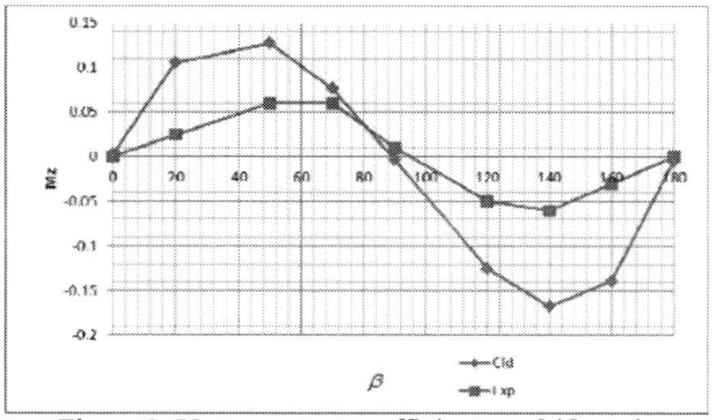

Figure 3. Yaw moment coefficient vs. drift angle

4.3 DISTANCE AND FROUDE NUMBER INFLUENCE

The influence of distance and relative speed of the ship were investigated next. In the beginning, the distance from the bank was fixed. The drift angle was equal to 0°. Three numerical calculations for different velocities (Fr=0.1, Fr=0.2, Fr=0.3) were made. The distance from the bank was then changed and the calculations repeated.

The results of the numerical calculations are presented in Figures 4-7. It can be seen that the influence of the speed is rather small. At the same time, the influence of the distance on the hydrodynamic characteristics is rather large.

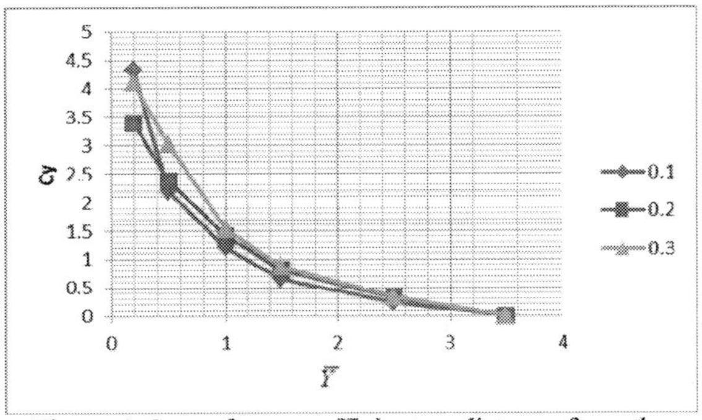

Figure 4. Sway force coefficient vs.distance from the bank for three different Froude numbers

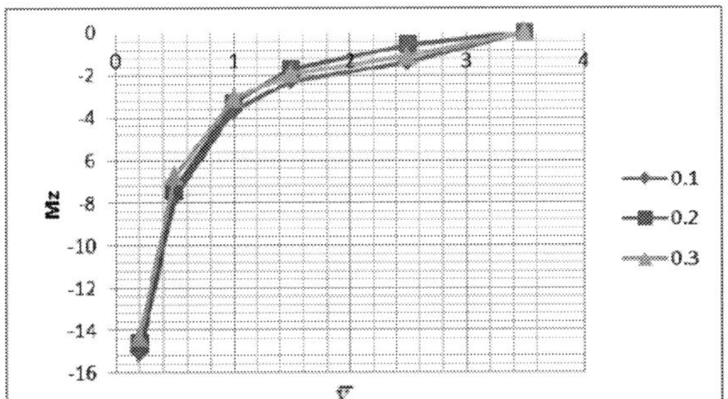

Figure 5. Yaw moment coefficient vs. distance from the bank for three different Froude numbers

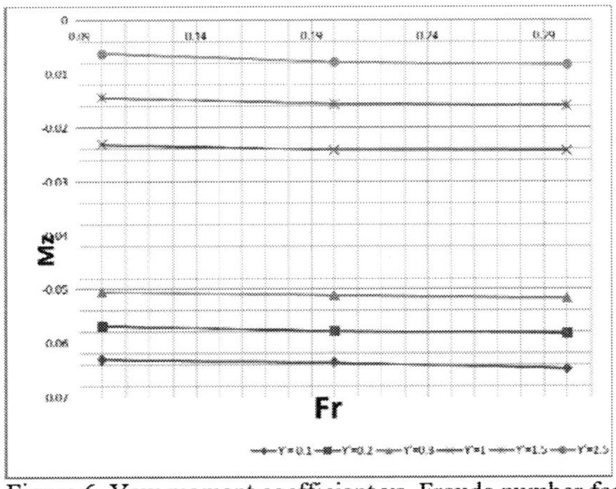

Figure 6. Yaw moment coefficient vs. Froude number for six different distances from the bank

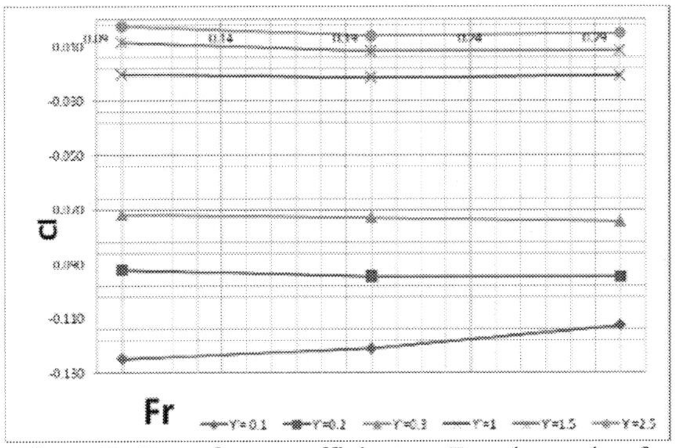

Figure 7. .Sway force coefficient vs. Froude number for six different distances from the bank

4.4 REGRESSION ANALYSIS

Regression analysis was used to determine additional forces and moments during motion along a bank, and a solution with square-law dependence was found. A general overview of formulas represented below. For the sway force, the following solution is obtained:

$$C_L = ((-212{,}5 + 12{,}9\beta)Fr^2 + (29{,}35 - 1{,}86\beta)Fr + (-3{,}64 + 0{,}08\beta))Y_0^2 + ((36{,}3 - 16{,}9\beta)Fr^2 + (21{,}9 + 1{,}96\beta)Fr - (0{,}716 + 0{,}01\beta))Y_0 + ((-26{,}9 + 7{,}16\beta)Fr^2 - (5{,}77 + 0{,}18\beta)Fr + (0{,}312 + 0{,}026\beta))$$

Formula 1

And for the yaw moment :

$$C_m = ((79{,}5 - 16{,}7\beta)Fr^2 + (-121{,}8 + 16{,}47\beta)Fr + (14{,}3 - 1{,}53\beta))Y_0^2 + ((60 + 34{,}2\beta)Fr^2 + (25{,}7 - 20{,}25\beta)Fr + (-3{,}51 + 1{,}72\beta))Y_0 + ((-66 - 12{,}9\beta)Fr^2 + (0{,}2 + 8{,}16\beta)Fr + (0{,}11 - 0{,}66\beta))$$

Formula 2

These formulas are corrected in scope of $Fr=0.1..0.3$; $\beta=0^{\circ}..15^{\circ}$; $\bar{Y}=0.1..2.5$.

The dependences obtained are in accordance with formulas above:

$$C_{\mathrm{mod}}(i) = \frac{\rho\, Fn^2}{2} LT \frac{1}{mg\, Fn^2} C_L(i)$$

$$M_{Z\mathrm{mod}}(i) = \frac{\rho\, Fn^2}{2} LTT \frac{1}{mgL\, Fn^2} C_M(i)$$

Finally, we have:

$$C_{BY} = C_Y \frac{L^2 T}{V}$$

$$C_{BN} = C_N \frac{L^2 T}{V}$$

5. 2D SIMULATION

To validate the formulas obtained, several exercises were made in a training simulator. We have used the tanker «Axel Spirit» as test object, which has Lpp=239 m, B=43.8 m, T=14.9 m. With help of the offered method it is possible to calculate the additional forces and moments for maneuvering near a bank, and, finally, to predict the ship behavior in motion along a bank.

5.1 MOTION NEAR A VERTICAL BANK

In the simulation, the ship moves along a vertical bank at a distance one breadth from the bank with speed Fr=0.1.

Tracks of original ship and ship with renewed hydrodynamics characteristics calculated with the help of CFD are presented in Figures 8 and 9.

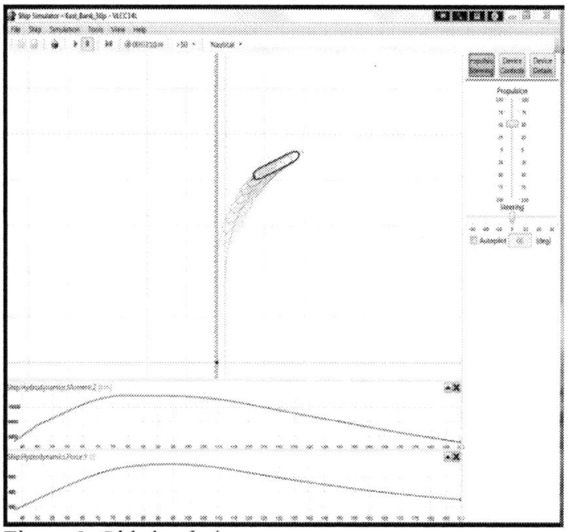

Figure 8. Old simulations

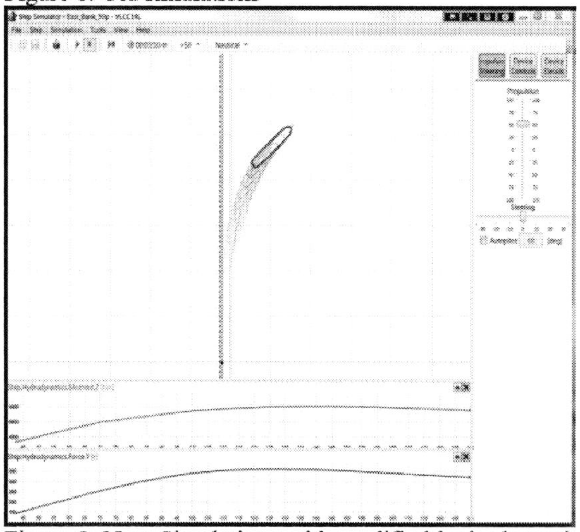

Figure 9. New Simulations with modified hydrodynamic parameters

The original and modified behavior of the ship are quite similar, but in the modified case the influence is much stronger which is in accordance with measurements.

5.2 CONCLUSIONS OF 2D INVESTIGATION

Bank influence on ship's motion has an effect in the region of three and a half breadths of ship from the bank. 2D definition can be used to evaluate bank effect in case of motion along a vertical bank.

6. THREE-DIMENSIONAL APPROACH

With goal of increasing of accuracy and taking into account of shallow water influence, 3D computations of the described problem were also made. The hull geometry, water region and grid were made using Gambit® mesh generator, as shown in Figure 9..

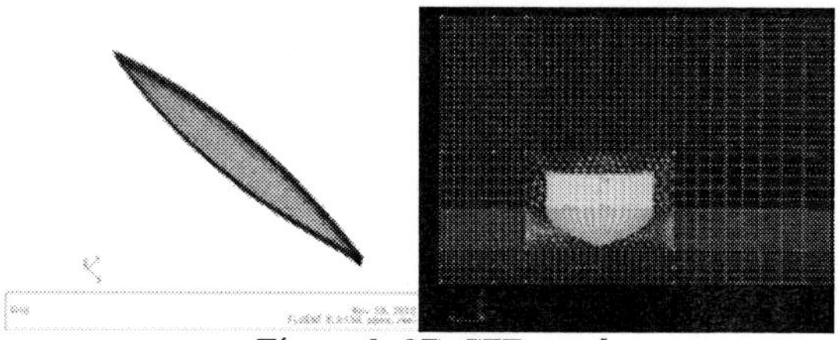

Figure 9. 3D CFD mesh

6.1 DEPTH INFLUENCE

The main geometrical parameters of the problem are shown on Figure 10. The results of the numerical research are presented in Figures 11 and Figure 12.

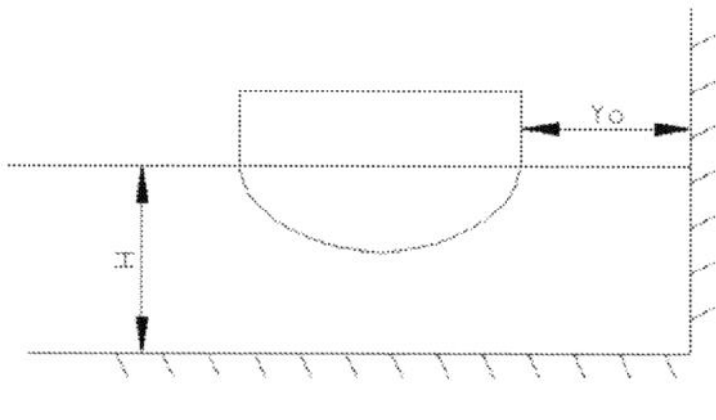

Figure 10.

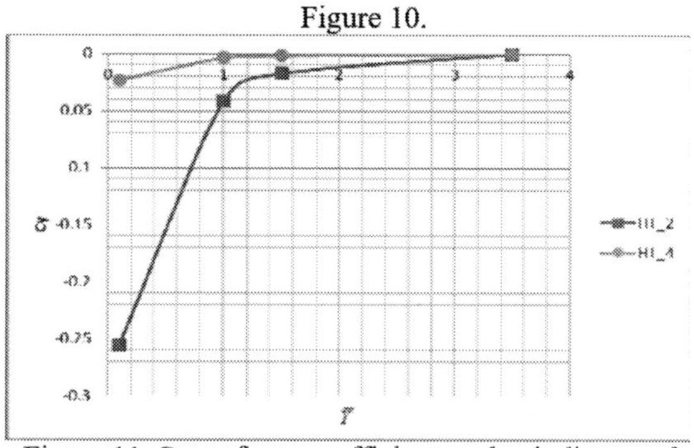

Figure 11. Sway force coefficient vs. bank distance for different relative depths

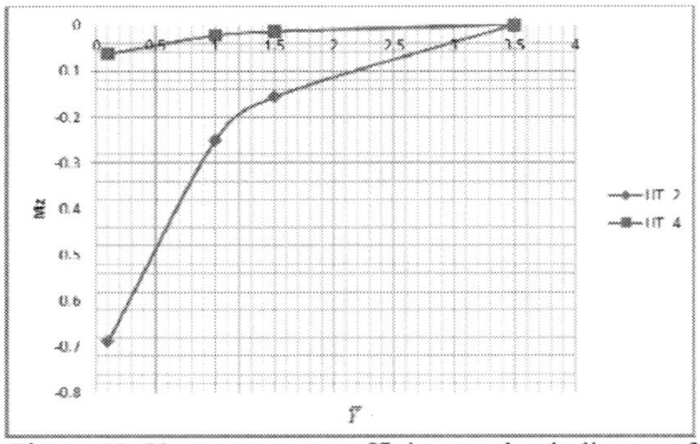

Figure 12. Yaw moment coefficient vs. bank distance for different relative depths

6.2 FINAL FORMULAS

The new results of 3D computations were added into dependencies above (cf. Section 4.4). The general structure of the formulas for sway force and yaw moment coefficients is as follows:

$$C_Y = (f_{C_Y}(Fr,\beta)(\frac{H}{T})^2 + f_{C_Y}(Fr,\beta)(\frac{H}{T}) + f_{C_Y}(Fr,\beta))(Y_0)^2 +$$
$$+(f_{C_Y}(Fr,\beta)(\frac{H}{T})^2 + f_{C_Y}(Fr,\beta)(\frac{H}{T}) + f_{C_Y}(Fr,\beta))(Y_0) +$$
$$(f_{C_Y}(Fr,\beta)(\frac{H}{T})^2 + f_{C_Y}(Fr,\beta)(\frac{H}{T}) + f_{C_Y}(Fr,\beta))$$

$$f_{C_Y}(Fr,\beta) = (a_1 + b_1 Fr)\beta^2 + (a_1 + b_1 Fr)\beta + (a_1 + b_1 Fr)$$

$$M_z = (f_{M_z}(Fr,\beta)(\frac{H}{T})^2 + f_{M_z}(Fr,\beta)(\frac{H}{T}) + f_{M_z}(Fr,\beta))(Y_0)^2 +$$
$$+(f_{M_z}(Fr,\beta)(\frac{H}{T})^2 + f_{M_z}(Fr,\beta)(\frac{H}{T}) + f_{M_z}(Fr,\beta))(Y_0) +$$
$$(f_{M_z}(Fr,\beta)(\frac{H}{T})^2 + f_{M_z}(Fr,\beta)(\frac{H}{T}) + f_{M_z}(Fr,\beta))$$

$$f_{M_z}(Fr,\beta) = (a_2 + b_2 Fr)\beta^2 + (a_2 + b_2 Fr)\beta + (a_2 + b_2 Fr)$$

7. 3D SIMULATIONS

Several simulations of ship motions in conditions of bank effect were made with the newly modified formulae. The tanker «Axel Spirit» again was used again in the simulations.

7.1 MOTION IN SUBMERGED CHANNEL

At first, a simulation in a channel was carried out where the left border was a vertical wall, and the right bank was submerged. The track of original simulation model is shown on Figure 13, and behavior of the modified simulation model based on CFD computations is shown in Figure 14.

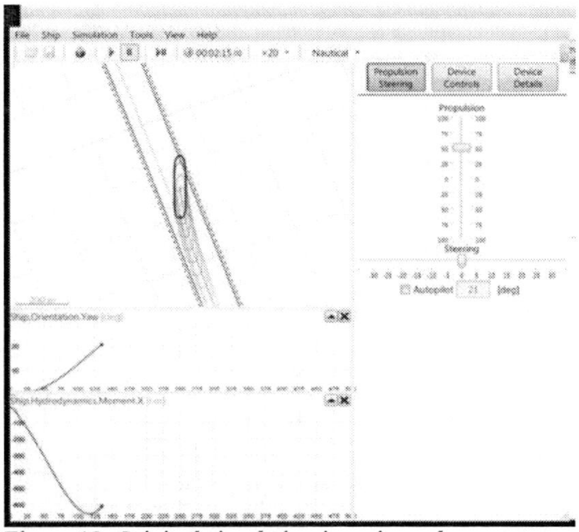

Figure 13. Original simulation in a channel

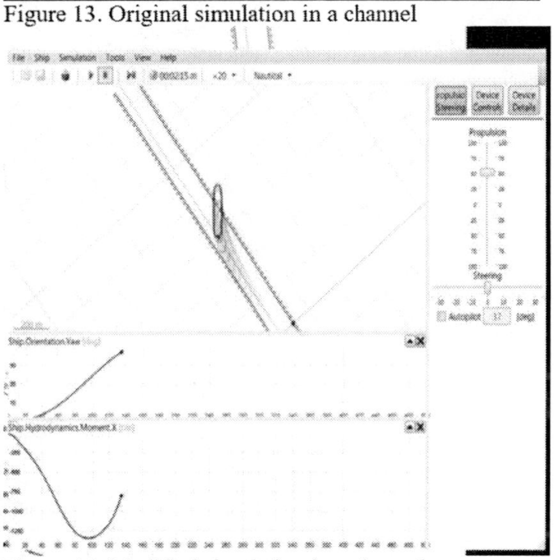

Figure 14. Modified simulation in a channel

In our opinion the second, modified, case is much more probable then the first one.

Results of simulation of the ship motion on a base of Polaris simulator provided by Mr. Garland Hardy are presented in Figures 15 and 16.

The difference in behavior of the old and modified model is obvious during comparison of Figures 15 and 16 respectively. It's possible to see that in the second, modified, case the ship has much more bank effect which is in accordance with measurements.

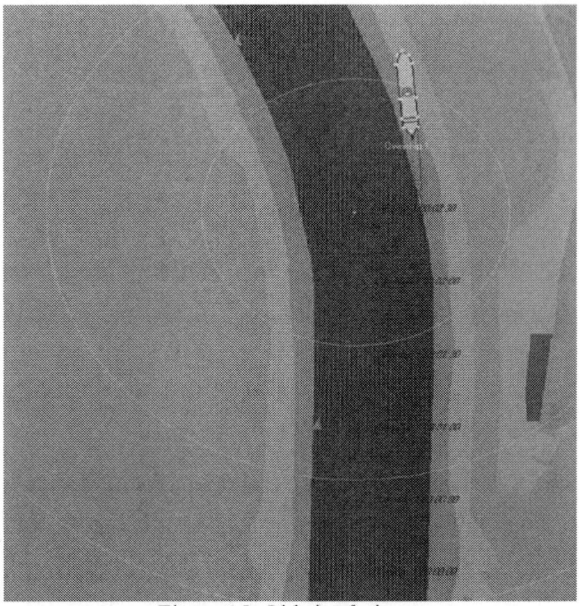

Figure 15. Old simulation

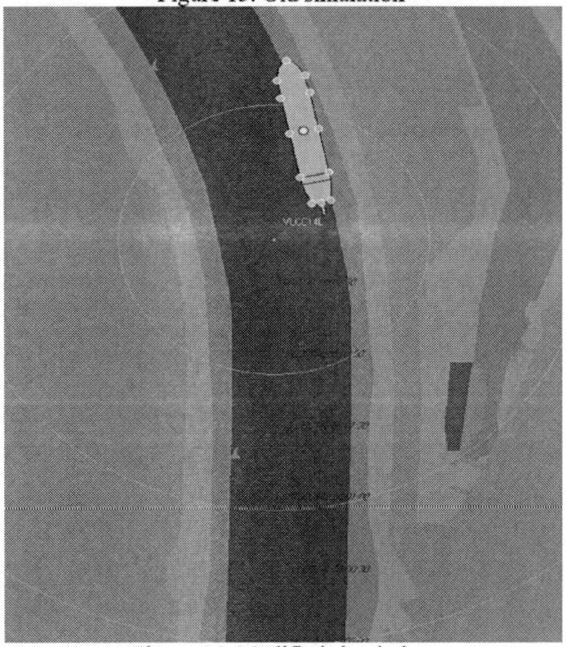

Figure 16. Modified simulation

8 CONCLUSIONS

On a base of CFD approach new formulae for sway force and yaw moment coefficients acting taking bank effects into account were obtained

The new formulas allow to predict bank effect with minimum effort but practically acceptable accuracy.

9. ACKNOWLEDGEMENTS

Special thanks to Garland Hardy (LANTEC Marine Inc., Canada) for testing models in "Polaris" and Sergey Zaykov (Kongsberg Maritime AS, Norway) for help during work.

10 REFERENCES

1. Dand, I.W., 'Some measurements of Interaction induced by Surface – Piercing and Flooded Banks', *NMIR 110*, 1981.

2. Vantorre, M., Delefortie, G., Eloot, K., Laforce, E., 'Experimental investigation of ship-bank interaction forces', *MARSIM'03*, 2003.

3. Basin, A.M., Velednatskyi, I.O., Lahovickyi, A.G., 'Ship's hydrodynamics in shallow water', *Shipbyilding*, 1976.

4. Mastushkin, Yu.M., 'Controllability of Catcher Boats', *Light and Food Industry*, 1981.

5. Milne-Thomson, L.M., 'Theoretical hydrodynamics', *Cambridge University Press*, 1932.

6. ANSYS. Fluent 6.3 Documentation.

11. AUTHORS BIOGRAPHY

Dmitry Nikushchenko is an assistant professor of State Marine Technical University and Manager of Department of Hydrodynamic Modelling of Kongsberg Maritime AS. He is author of 56 publications in area of Computational Fluid Dynamics, Ship Manoeuvrability, Ship propulsion.

Marina Lebedeva is an assistant professor of State University for Waterway Communications and senior researcher in Department of Hydrodynamic Modelling of Kongsberg. She is an specialist in Ship Manoeuvrability and Ship propulsion

Alexey Chetvertakov is an engineer-researcher in Department of Hydrodynamic Modelling. He is author of 4 publications in area of Computational Fluid Dynamics, Ship Manoeuvrability.

POST WIDENING SHIP/SHIP INTERACTION AND OPERATIONAL DECISIONS IN THE PANAMA CANAL

L. L. Daggett, Waterway Simulation Technology, Inc., Vicksburg, MS, USA
J. C. Hewlett, Waterway Simulation Technology, Inc., Columbia, SC, USA
R. S. Gray, Retired Panama Canal Pilot, Whidbey Island, WA, USA

SUMMARY

In 1999 The Panama Canal Commission (PCC) widened two reaches of the Gaillard Cut from 152.4m to 192.0m in straight reaches and 222.5m in bends to meet growing demands for service by increasing the capacity of the Panama Canal. This widening was to allow two-way traffic for Panamax and Supers in the Gaillard Cut and thus eliminate convoying operations of a rapidly increasing percentage of the canal traffic. To test the plan and develop operating guidelines, the PCC conducted a set of meetings of these ships for the newly widened canal sections of Bas Obispo and Las Cascadas with a select set of pilots. Some of the ship meeting tests were measured using Dual-frequency Differential Global Positioning System (DDGPS) gages mounted on the ship's bow and both bridge wings. The meetings test program focused on ships with combined beams greater than operationally allowed in the canal at that time. Measurements and operational guidelines are presented in this report.

1. INTRODUCTION

In 1991, the construction project to widen the Gaillard Cut (Figure 1) was approved. In 1999 widening of two reaches in the northern end of Gaillard Cut near Gamboa, Bas Obispo and Las Cascadas, was completed (Figure 2). The primary purpose of the present study was to provide data and information to be used in establishing the operating procedures for meeting Panamax ships in the widened Gaillard Cut. Issues of particular interest included added sinkage during passing in a restricted channel, strength and controllability of ship/ship interactions during meeting

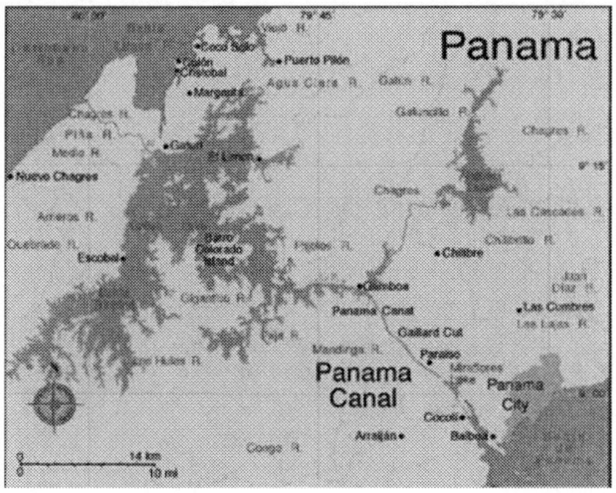

Figure 1: Study Location

operations, control during recovery following a meeting, and the combined effects of ship/ship interactions and bank effects during the setup, meeting and recovery. The study consisted first of pre-testing with increasingly larger vessels without instrumentation, and, secondly, DDGPS testing with position instrumentation installed aboard meeting vessels. The participating Panama Canal pilots for the pre-testing phase predominantly were the same during the latter phase.

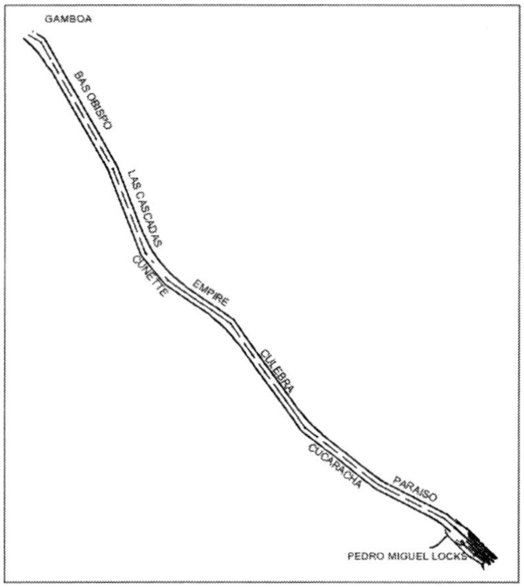

Figure 2: Reaches of Gaillard Cut

2. STUDY APPROACH AND RESULTS

2.1 PRE-TESTING

The PCC assigned a group of senior pilots to special duty aboard selected ships. They began meeting ships in Bas Obispo and Las Cascadas Reaches. Two senior pilots were assigned to each ship; one as the conning pilot and the second as an observer and assistant to the former. The plan for pretesting was to begin with the smaller ships that were just over the restriction beam of 91 ft and progress to larger ships as experience was gained. The purpose was to determine concerns related to the meeting of two Panamax ships in the cut.

A secondary purpose was for the pilots to obtain experience in performing these meeting operations and to identify if specific ship types might be more critical to measure in the later instrumentation phase. These tests began on May 3, 1999, and continued as opportunity permitted, ending with the DDGPS study. Table 1 shows combined beams of meeting ships during the pre-testing phase; if more than one meeting was experienced druing a single transit, these are listed in the 2nd, 3rd, and 4th columns. The average daily combined beam of the observed vessel meeting, is shown graphically below the table; this indicated a general trend of increasing pilot confidence during the period.

Table 1: Combined Beam during Pre-testing (m)

Date (1999)	Meetings Encountered During a Single Transit			
	1st Meeting	2nd Meeting	3rd Meeting	4th Meeting
May 3	54.0	32.6	-	-
May 4	43.3	50.3	-	-
May 5	50.0	43.6	-	-
May 6	54.0	52.4	52.7	51.2
May 10	55.2	58.2	-	-
May 11	50.3	44.8	-	-
May 12	50.6	64.3	45.7	59.4
May 14	57.6	-	-	-
May 17	60.7	-	-	-
May 18	53.0	42.4	-	-
May 19	55.2	53.9	-	-
May 20	63.1	64.6	-	-
May 21	60.7	62.5	-	-
May 24	52.7	62.2	62.2	-
May 25	64.6	64.6	64.6	-
May 26	62.5	64.6	-	-
May 30	52.1	52.4	-	-

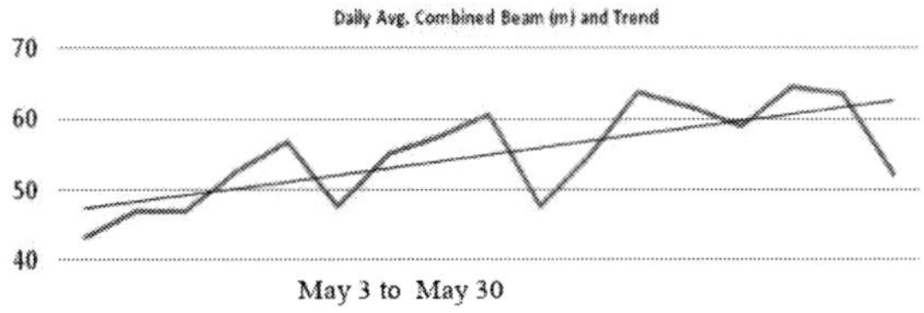

A pilot evaluation form on which key parameters and observations were recorded was completed for each meeting situation. In most cases a special GPS-based navigational aid identified as a CTAN (Communication Traffic management And Navigation system which was a specially designed ECDIS unit) unit was carried aboard that provided pilots with additional position and traffic information. Table 2 shows abridged results from some of the pre-instrumentation meetings.

Generally this pre-testing identified no significant problems during meeting of Panamax ships. Only one of the meetings resulted in pilot comments that described the tested maneuver as ill-advised - most likely because of the meeting location in or near a channel bend. Several of the ships had centerline cranes present which restricted visibility of meeting ranges. Some concern was expressed about the sensitivity of the mid-channel and meeting ranges to indicate position across the channel. This lead to the decision to have the CTAN on board in addition to the GPS measurement equipment to assist the pilots in setting up the meeting situations at the desired location and to be in the proper lateral position in the channel. It was decided based on these documented pilot assesments that combinations of Panamax containerships and bulk carrier/tankers would be the most significant ship types to test during the DDGPS study phase.

Table 2: Meeting Assessments during Pre-testing

Date (1999)	Combined Beam (m)	Meeting* Location	Comb. Speed (kts)	Pilot Assesment (abridged)
May 17	60.7	Mid B.O.	-	Good, CL Cranes, little interaction
May 18	53.0	Mid B.O.	-	Good, plenty of room
May 19	55.2	Mid B.O.	10.9	Good, little interaction
May 20	64.6	Mid B.O.	10.5	Good, weak interaction
May 21	60.7	South L.C.	-	Bad, too close to bend CL Cranes
May 21	62.5	South B.O.	10.9	Good, slight interaction
May 24	62.2	North B.O.	9.0	Good, medium interaction
May 24	52.7	South L.C.	9.5	Good (1st ship) Good (2nd ship)
May 24	62.2	North L.C.	10.2	Good (1st ship) Good, CL Cranes (2nd ship)
May 25	64.6	Mid B.O.	11.7	Good (1st ship) Difficult (2nd ship)
May 25	64.6	B.O.	11.7	Good, little interaction
May 25	64.6	North B.O.	11.9	Good, little interaction
May 25	64.6	Mid L.C.	9.5	Good, little interaction
May 26	64.6	South B.O.	-	Good, little interaction
May 30	52.1	South L.C.	17.9	Good, no interaction

* B.O. - Bas Obispo
L.C. - Las Cascadas

2.2 DDGPS

2.2.1 Study Description

The pre-testing phase was conducted in order to pinpoint what combinations of meeting ships should be instrumented during the GPS phase of the study. DDGPS measurements were made using dual-frequency receivers, capable of accuracy on the order of one centimeter, mounted at three points on each monitored ship. GPS receivers were positioned on the bow and on the port and starboard bridge wings. Use of the two receivers on the bridge wings was of particular importance in measuring ship roll, which is usually significant for containerships. Both ships in each of twelve meetings between predominantly Panamax size vessels were instrumented (Table 3). These meetings all took place in either Bas Obispo or Las Cascadas Reach, both of which had been recently widened to about 192m. In addition, the widened Las Cascadas reach was constructed with a different alignment than for the original canal requiring repositioning of channel centerline navigational range markers (discussed later in this paper). Of the twelve meetings, six were containership/bulk combinations, four were bulk/bulk combinations and two were containership/containership combinations. The measurements defined the vertical (sinkage) and horizontal motions of the ships during transits through the Gaillard Cut and Miraflores Lake and entering and exiting the Pedro Miguel and Miraflores Locks. In Table 3 the sinkage increase factor refers to how much sinkage increased during the meeting compared to that which was occurring before and after the meeting. For the determination of sinkage on ships while underway, recordings of the static elevations of the DDGPS equipment mounted on the ship were made while the ship was stationary in the locks. These static elevations were referenced to the vertical datum used by the canal and ship data was used to compute the bow and stern keel elevations.

Meeting Ships c - cont. b - bulker t - tanker	**Draft (m)**	**Length (m)**	**Beam (m)**	**Meeting Speed (knots)**	**Meeting Distance (m) Center-to-Center**	**Sinkage Increase Factor**
Mette Maersk (c)	11.6	294.2	32.3	5.3	84	1.2
Neapolis (c)	10.9	228.7	32.3	5.2		1.0
Protank Orinoco (t)	8.0	228.7	32.3	6.7	89	1.5
Maren Maersk (c)	11.9	294.2	32.3	6.7		1.25
Michele Iuliano (b)	12.0	225.0	32.3	4.7	66	2.0
Ming Asia (c)	11.1	275.9	32.3	6.2		1.5
Marit Maersk (c)	11.5	294.2	32.3	6.5	72	1.5
Tokio Express (c)	11.4	287.8	32.3	6.6		1.25
Ever Excellent (b)	11.9	225.0	32.3	5.5	88	1.5
Sea Clipper (b)	12.0	225.0	32.3	5.5		1.5
Zim Israel (c)	10.7	234.0	32.3	6.6	83	1.75
Mehmet Bey (b)	12.0	225.0	32.3	5.8		1.25
Hanjin Marseilles (c)	10.6	289.6	32.3	7.6	86	1.35
Harmony (b)	12.1	225.0	32.3	6.0		1.5
Yick Zao (b)	11.4	177.7	29.0	5.5	77	2.0
Ever Dainty (c)	11.5	294.2	32.3	8.9		1.5
CCNI Atacama (b)	11.4	185.1	32.3	6.0	91	1.5
Beauty Sea (b)	11.4	225.0	32.3	5.6		2.0
Jin Qiang (b)	11.6	185.7	31.1	6.3	67	1.6
St. Joseph (b)	11.8	220.1	32.3	5.8		2.0
Ever Right (c)	10.5	293.9	32.3	6.9	64	2.0
Manzanillo (c)	9.2	242.7	32.3	6.7		2.0
Auk Arrow (b)	10.6	187.5	29.0	6.3	59	2.0
CO-OP Phoenix (b)	12.0	225.0	32.3	6.7		1.2

Table 3: Meeting Maneuvers during DDGPS Measurements

As during the pre-testing, five senior pilots were assigned to con the ships during the meeting situations and to observe the conditions, strategies, and results and to assist the conning pilot in accomplishing his meeting strategy. Four junior pilots were also assigned to assist in carrying aboard, setting up and taking down the GPS equipment, setting up and operating the CTAN units, and recording observations about the meeting situations in addition to the senior pilot.

For each transit, team members recorded the engine and rudder commands at the time the helmsman or first mate responded to the pilot, particularly during the ship approach and meeting and the lock entrance and exits. Other items of interest, e.g. the meeting of another ship, were recorded with the time synchronized to the GPS time. This provided a log of events useful in interpreting study results and for other types of analysis of the measured data. For example, an analysis of the required control and the horizontal response of the ship could be performed using this log and the coordinate position of the ship. This type of analysis could be used to investigate the control response of a ship in a highly restricted waterway. It should be noted that during data logging if the pilot commanded the helmsman to hold a steady course, only the commanded-course was recorded and not every rudder movement.

When available, ship geometry data were gathered from the captain during the transit. Sources of these data included copies of ship hydrostatic tables, ship particulars listings and ship bridge maneuvering cards along with squat tables and ship geometry drawings.

Video recordings were made of the rudder and engine rpm indicators located on the ships' bridge wings. These recordings were time synchronized with the GPS measurements. The data-logs of pilot orders and other pertinent information backed up these videotaped data. The videotapes and data-logs provided information for analysis of the meeting situations recorded during the study. Videotapes of the bridge-wing indicators were used when available and for those ships for which the videotapes were not available the data logger information was used. Engine rpm and rudder postion time records were manually developed from the videos.

The pilot evaluation forms that were used during the pre-testing continued to be used during the testing with instrumented ships. In addition, junior pilots were assigned to assist in recording observations and comments made during the tests. The information from these forms provided a subjective evaluation of the meeting conditions and documented key observations for use by other pilots.

One of the critical aspects for conducting safe and controlled meetings in the widened Gaillard Cut with Panamax ships was knowing where the ship was in the channel, i.e., how far off the channel centerline and bank the ship was as it approached and met the traffic ship. The primary guidance for this information came from two sets of ranges placed at the ends of the channel reaches. For these tests the channel centerline was marked with temporary ranges with two small rectangles elongated vertically. The new meeting ranges, positioned 30.1m starboard of the channel ceterline, were square with vertical black lines. The center ranges were international orange and the meeting ranges were white.

Figure 3 shows the channel range setup for the instrumented meeting tests with a northbound ship completing the turn into Bas Obispo and the bridge lined up on the Las Cascadas meeting ranges. In the figure the two sets of temporary ranges are seen on the outside of the four visible range sets. The temporary channel centerline ranges are the ranges farthest to the left and the temporary passing lane ranges are the ranges farthest to the right. There was a change in heading of the widened Las Cascadas reach and, therefore, the temporary ranges are offset from the normal range sets. Note the difference in separation between the front and rear stations of the regular centerline ranges compared to that of the temporary ranges. The short distance between the forward and rear temporary ranges made it difficult to identify how far to the right of the range line the ship was located. Also noticeable is the minimal clearing done around the ranges which left in place an extensive green background that made it difficult to see the ranges at a distance. The orange temporary centerline ranges were nearly invisible against the green background at this distance. The pilots noted that these ranges were not easily seen during low visibility, e.g. afternoon rainstorms. The cranes positioned down the centerline of the ship also made it difficult to see the ranges when the pilot was at the center of the bridge and the ship was lined up on the ranges.

Figure 3. Northbound Ship Approaching Temporary Ranges Used for Meeting Tests

2.2.2 Example of GPS-Instrumented Ship Meeting

Results for the twelve meetings are recorded and presented in a report to the Panama Canal Commission[1]; however, in the present paper the meeting between the *Marit Maersk* and *Tokio Express* was chosen to serve as an example. These two ships were containerships with lengths of 294.1m and 287.7m, respectively, the sum of which resulted in the highest combined LOA meeting measured. Figure 4 shows the trackplots of the two ships during the passing maneuver in Bas Obispo Reach. Figures 5 & 6 show control measure plots for the two ships during the meeting. (The reader should note that English units are used on these plots.) In the control measure plots the following directional conventions pertain:

- positive roll denotes starboard side down
- positive drift angle denotes ship motion vector directed clockwise (to starboard) from ship heading,
- positive lateral speed denotes ship movement to starboard
- positive rate of rotation (ROR) denotes clockwise angular speed (bow to starboard)
- positive yaw denotes clockwise angular rotation of ship heading off the channel heading
- added width indicates additional cross-channel width of ship due to yaw angle.

The *Marit Maersk* experienced a fairly large repulsion (positive lateral speed) and inward rotation (negative ROR) when abeam the passing ship. Prior to the meeting, starboard rudder was used as needed to counteract this bank suction and moment on the ship to maintain a course parallel to the bank during the meeting. No port rudder was used during the entire meeting maneuver with the submerged channel banks providing countering moment to the ship's starboard rudder. A slightly increased roll was initiated during the meeting. Increase in sinkage was estimated to be about 50% during the meeting when compared to that which was experienced before and after the meeting. The senior pilot aboard reported good visibility and no problem recovering. He also said that the rudder was at midship during the meeting; although, the indicator showed hard starboard was used just prior to when the ships were beam-to-beam. The added width went to zero when the ships came abreast of each other and then rose to 15.9m as the ship recovered from the meeting.

The *Tokio Express* also experienced a slight increase in roll which started during the meeting. Sinkage increased about 25% compared to that before the meeting. A five- to fifteen-degree starboard rudder was used prior to the meeting and ten degrees starboard rudder was used during half of the meeting

[1] *Panamax Ships meeting in the Gaillard Cut, Panama Canal,* Waterway Simulation Technology, Inc., June 1999

maneuver. Starboard rudder was used almost exclusively throughout the meeting with only a very short application of port rudder after the ships had passed each other. The *Tokio Express* experienced very little change in heading during the meeting with the starboard rudder resisting inward (toward other

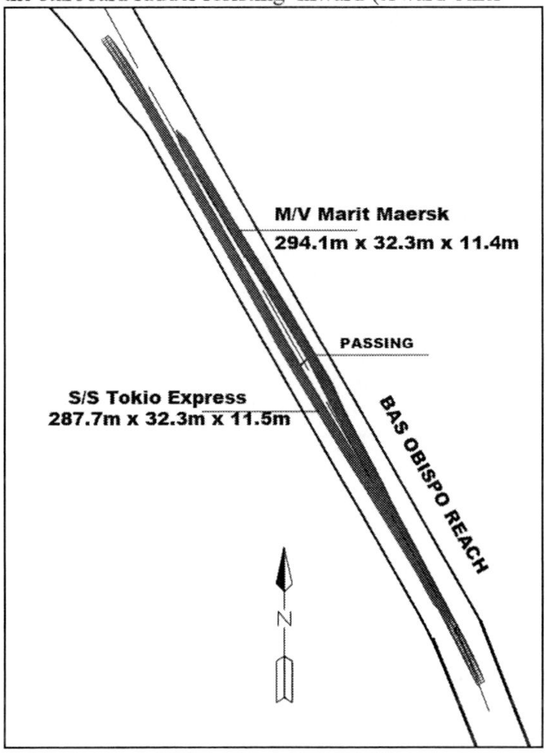

Figure 4. Meeting Track Plots of *M/V Marit Maersk* and *S/S Tokio Express*

ship) rotation of the ship. The senior pilot aboard reported good visibility, a slight motion to port as the bows met and a roll to starboard. The added projected beam width was 11.5m following the meeting. Clearances were measured for all the recorded meetings and the values for the subject scenario are shown below. For the last value, "toe" indicates the submerged bottom of the side slope from the bank (i.e., the channel edge) and the clearance represents the minimum value experienced by either ship.

Min Clear. (hull-hull)	Centerline Dist. Between Ships			Min Clear. (to toe)
	@ Bow-Bow	@ Abeam	@ Stern-Stern	
35.4m	66.1m	71.6m	71.3m	39.0m

The pilot(s) aboard the two ships provided the following journal entries for documentation of the meeting maneuver. One of the pilots noted difficulty in determining cross-channel position of his ship using the channel ranges.

Marit Maersk meeting *Tokio Express*

Senior Pilot:

Visibility: Good; Engine Response: Good; Rudder Speed: Good; Rudder Effect: Good

- Have pilots eye video for this meeting, no problem recovering, other vessel felt crowded, passed with rudder MS.

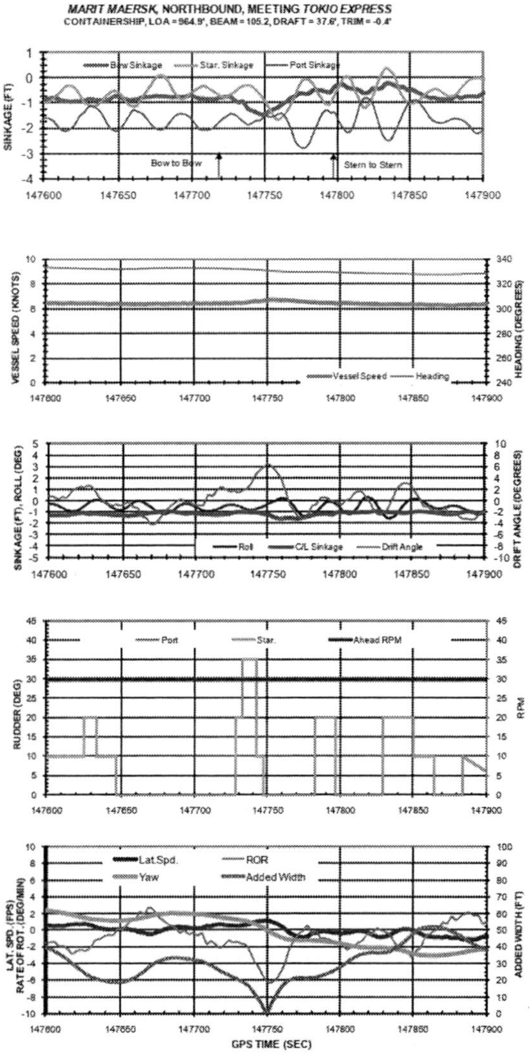

Figure 5. Measured Ship Responses of *M/V Marit Maersk* Meeting *S/S Tokio Express*

Junior Pilot:

11:08 – Hard port, suction acting on port quarter on the way out, after clearing the arrow rudder midship

11:23 – While [the senior pilot] on the vessel's con made a suggestion of ck [..] distance of the bank with side positioned containers.

11:29:41 – Port 20 deg due to suction of the port quarter and bow cushion off Gold Hill turn.

11:39:38 – Turning into Empire, c/hoist turn vessel listed about 2 ½ deg.

11:58:32 – Left of the old CL ranges trying to correct vessel's position.

12:02:20 – ([junior pilot's] View)Vessels abeam, 1deg roll to port, hard over rudder to stbd due to stbd quarter suction, problem with the ranges position giving a false way of your position, meeting speed for NB 6.5 kts, SB 6.4 kts, ([senior pilot's] View) Came into reach with an angle towards the bank to prevent the vessel to move parallel bodily towards the bank using later the bank suction in the stbd quarter to keep in position at the time of the meeting.

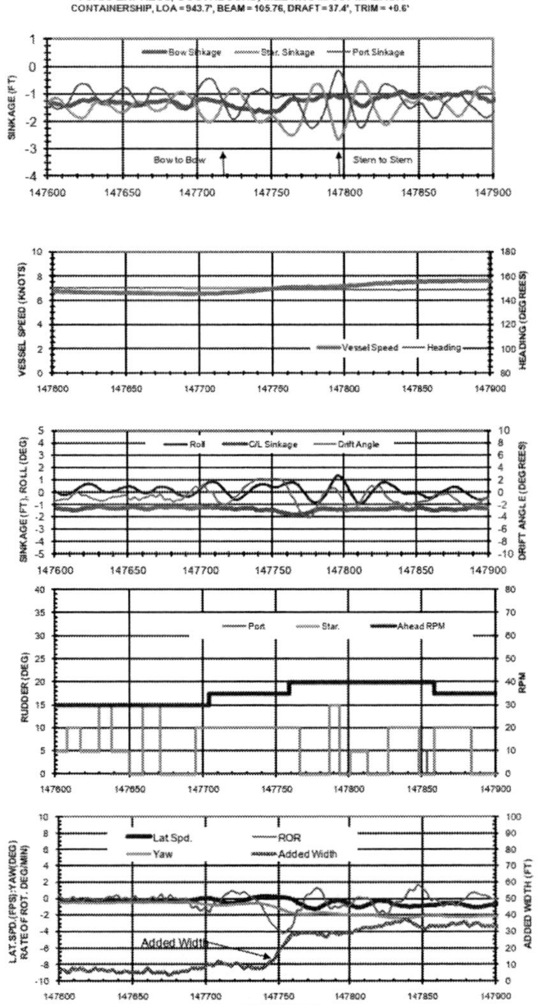

Figure 6. Measured Ship Responses of *S/S Tokio Express* Meeting *M/V Marit Maersk*

Tokio Express* meeting *Marit Maersk

Senior Pilot:

Visibility: Good; Engine Response: Good; Rudder Speed: Good; Rudder Effect: Good

Weather: Overcast, good visibility, no wind

- Just before meeting rudder midship, slight motion to port as bows met, counteracted motion, then port 10 deg then steadied, as ships met I felt roll to stbd first then continued after ships met.

3.0 CONCLUSIONS

The study assisted in determining critical factors related to incipient two-way Panamax vessel traffic in the widened canal and there was a noticeable increase in pilot confidence and resolve during the time period to establish procedures and move forward with such maneuvers.

Based on the tests and measurements, the following conclusions related to the behavior of ships during meeting and passing situations in the Gaillard Cut are presented. These conclusions are based on the analysis of all the meeting scenarios observed during the study.

- Ship meetings in the study reaches required good visibility. During the study good visibility was sometimes affected by visual obstructions on the deck (cranes, containers, ship bridge layout, etc.). In addition, it was critical that the pilot be able to determine his position in the channel which would require channel ranges that are easily visible and are sensitive to cross-channel position. Adequate size and separation distance between front and rear ranges are critical factors related to determining ship position. Placement of a third set of ranges near the water surface assists in maintaining visibility of the ranges during foggy conditions.
- The results show (corroborated by pilot comments) that with good visibility and an approach speed of six to seven knots there was adequate channel width and reserve maneuvering power to safely meet and pass all ships in the Bas Obispo and Las Cascadas Reaches, up to and including Panamax size.
- The characteristic sequence of horizontal motion of the vessels during meetings was as follows:
  - After the bows met a strong repulsion force and bow-in moment with the ship rotating toward the other ship,
  - A reversal before the sterns met of varying magnitude, and
  - Frequently, another reversal of drift and rotation before final drift back toward the center of the channel after completely past the other ship.

  In some cases there was evidence of a bow-out rotation as the bows came together as was expected. The lack of this motion in many cases may be due to anticipation of this motion by the pilot and his strategy to counter the effect.
- Horizontal control measures such as drift angle, lateral speed and rotation rate

experience the greatest increase during meetings for the bulk carrier/tanker type vessels. These measures also increased for containerships, but less so. This was the rule probably because the containerships were always bigger and more powerful than the bulk carrier/tanker ships.

- A characteristic ten to twenty degree starboard rudder was required to keep the ship against the bank in the passing lane before meeting the other ship. Pilots expressed a concern about the ship shearing off the bank in front of the approaching ship.
- Normally the application of ten degrees rudder was enough to control the rotation towards the passing ship after the bows meet. In some cases twenty degrees or hard-over rudder was used for short periods of time.
- Meetings increased sinkage on the order of 50% generally but could be as much as 100%. The sinkage increased more for bulk carrier/tanker type vessels than for containerships.
- The primary effect in the vertical direction for containerships in meeting situations is increased roll.
- There was a temporary, small increase in speed of all ships during the meetings.

4.0 Application of Study Results and Conclusions

Despite subsequent widening of the remainder of the Gaillard Cut reaches, two-way traffic for Panamax vessels was not implemented until 2009. There were many reasons for this delay. It took several years to complete the channel widening and "straightening" or widening the bends in the Cut. During this time blasting and additional dredging equipment was in the waterway. This was also the time when the Panama Canal was transferred from the United States to Panama and the Panama Canal Commission yielded authority to the Panama Canal Authority (ACP). Although training of Panamanian pilots had been ongoing to build up the pilot force in preparation for the transfer, there was a departure of some of the senior pilots, particularly some of the pilots that participated in these tests and development of recommended procedures based on the tests. Also during this time, many other modernization and improvement projects were underway throughout the canal by the ACP.

While this was the background for negotiations between the pilot union and ACP, there were specific requirements being demanded by the pilots that resulted from these tests. A primary position of the pilots involved the location and sensitivity of the meeting ranges. A previous simulation study conducted to assist in the design of the widened channel recommended a channel width of 192 m in the straight reaches and 222.5 m wide in the bends and that the ranges used to position the ships for meeting, i.e., meeting ranges, should be placed 38 m from the centerline; thus the location of the temporary ranges. However, due to the pilots concern about the bank effects and the results of the ship behavior during the meetings, the pilots strongly recommended placing the ranges 32 m from the centerline so that a separation distance of 30.5 m could be maintained between the meeting ships. This allowed reduced bank effects during the approach and more room for the ship's stern as the ship rotated back towards the center of the channel.

In addition, they strongly insisted on meeting ranges that would allow the pilot to detect a lateral displacement of 10 % of the ship's beam, i.e., about 3 m for a Panamax ship. This evolved into a request for a minimum of 20% cross track factor based on United States Coast Guard design criteria. Also, because of the topography, they requested that a third set of ranges be placed near the water level so that it would be visible in fog.

ACP replaced the original Single Point Ranges with much more accurate Vertical Array type ranges with better lateral sensitivity. On top of the higher range tower, they also installed PEL sector lights throughout the Cut. PEL lights will be installed throughout the rest of the Canal as part of the canal expansion.

The experience in using the CTAN unit was disappointing to the pilots. The units were unreliable and were not considered by the pilots to be usable during a meeting situation. However, they were considered to be helpful in setting up the location and timing of meeting another ship. It was recommended that these units and the supporting system be improved for use in introducing two-way traffic.

There was also concern about the mantainance of the navigable part of the Cut channel. This concern was increased with the increase in ship squat observed during the ship meetings. The pilots requested that hydrographic survey data be provided in a form of charts and that dredging of the channel toes and shoaling be undertaken in a timely manner. ACP management requested that buoys be placed along the channel toeline at 152.4 m intervals and fitted with the appropriate lights to assist the pilots in identifying the channel lateral limits.

Because bank lights that lined the water's edge along the cut to assist the pilots at night in positioning the ships in the channel had been removed during channel construction, the pilots also requested that they be replaced prior to starting two-way traffic of Panamax vessels in the Cut. These bank lights were replaced with new lights designed to reflect off the water surface and to be more easily serviced from shore.

Because of the pilot's concern over the relatively narrow channel, determination of the ship's position across the channel was critical to the safety of the meetings. Visibility and accuracy of the ranges were considered critical and, therefore, obstructions on the ship (cranes, masts, etc.) restricting visibility from the conning position were considered to be too limiting, Under these conditions it was the pilot's opinion that the ship should be classed as "clear cut" and meetings between Panamax vessels should not occur. Similar concern was expressed about ships known to be poor handling and these ships were recommended to be classed as "clear cut".

Again, concern about being able to clearly see the ranges and correctly position the ship within the channel brought about a recommendation to not schedule two-way traffic of Panamax ships after 1200 noon during the rainy season.

As the channel improvements have been completed, the ranges improved, the channel buoyed and bank lights reinstalled, the Panama Canal pilots have begun a program of meeting incrementally increasing ship sizes in the Gaillard Cut. Consequently, two-way traffic of Panamax and Supers has begun to occur as additional improvements, including dredging to widen and deepen the Cut and Gatun Lake, has begun for the expansion project. The latest negotiated contract with the pilot's union includes a new table of restrictions and ship size combinations that can be allowed to meet, including allowing some of the largest ships to transit the Cut and Lake at night with some meeting restrictions in the Cut. This comes at a time when the traffic mix has shifted dramatically from 40% supers and 60% regular size ships to 70% supers and 30% regular size ships.

5.0 **References:**

Panamax Ships meeting in the Gaillard Cut, Panama Canal, Waterway Simulation Technology, Inc., June 1999.

6.0 **Authors Biographies**

Dr. Larry L. Daggett is a registered engineer and diplomate of the Academy of Coastal, Ocean, Port, and Navigation Engineers, American Society of Civil Engineers, Society of Naval Architects and Marine Engineers, and a past member of the Marine Board and Transportation Research Board of the National Academy of Science and Engineering. Dr. Daggett retired as Chief of the Navigation Division of the U.S. Army Engineer Waterways Research Station and is the Chairman of the Maritime Navigation Advisory Board for the Panama Canal Authority. He has over 40 years of experience in the design of navigation channels, ship and tow simulation, and application to navigation.

Mr. J. Christopher Hewlett is a registered professional engineer and was founder of Waterway Simulation Technology (WST). He was in charge of developments and applications of ship and tow simulation to many Corps of Engineers projects while employed at the U.S. Army Engineer Waterways Experiment Station and is now the President of WST. He has over 30 years of experience and continues to be active in design and testing of navigation and marine terminals with ship and tow simulation and prototype ship maneuvering measurements along with Dr. Daggett.

Captain Robert S Gray is a retired professional pilot with the highest rating as a Panama Canal Senior Pilot. Capt. Gray served as the Panama Canal Pilot's member of the Technical Committee which was an organization of Panama Canal Commission management and pilots that worked together to address key issues involving the safe handling of ships through the Panama Canal. Capt. Gray also has an endorsement as First Class Pilot, Cook Inlet from Nikiski to Anchorage including the ports of Nikiski, Chisik Island and the Port of Anchorage.

SQUAT OF SHIPS SAILING IN A NON CENTERED TRACK IN A CANAL

H J de Koning Gans, Delft University of Technology, The Netherlands

SUMMARY

Squat is a serious problem when very large container ships are entering harbours with small underkeel margins. The squat effects are that the ship acquires sinkage and trim. Due to this sinkage and trim the keel clearance decreases drastically. The distance between bottom and ship becomes very small and measures must be taken to avoid contact between the ship and the bottom.

Also when the ships are not sailing in the centre of the canal, the squat effects become more serious. The flow around the ship is not symmetrical anymore, and also the gap between the wall and the ship becomes smaller and higher return flows are expected. This will cause extra forces to act on the ship and as a result, the ship will suffer extra squat-effects.

This study looks at several sailing tracks at the centre and parallel of the centre line of the port. The flow calculations are carried out with a three-dimensional potential flow model. For this model the wetted hull geometry has only to be determined of the underwater body (wetted hull) in still water conditions. From these calculations the pressures on the wetted hull are determined and by integrating the pressures, the forces are determined. From each track the forces are determined and the trim and sinkage are calculated. From these results, it appears that, when the ship sails more closely to the wall the forces will increase drastically.

NOMENCLATURE

B	Breadth of the ship. (m)
F_j	$1 \leq j \leq 3$ Force in j direction (N)
F_j	$4 \leq j \leq 6$ Moment in j direction (Nm)
F_3	Sinkage force amidships (N)
F_5	Trimming moment amidships (Nm)
g	Acceleration of the gravitation (ms^{-2})
h	Initial water depth (m)
K	Keel clearance or minimum distance between ship and bottom (m)
L	Length of the ship (m)
n	Normal vector ( )
p	Pressure (on ship surface) in water (N)
s	Sinkage in amidships (positive to beneath) (m)
T	Draught of the ship (m)
t	Trim angle (positive ahead) (rad)
U	Ship Speed (ms^{-1})
u	Longitudinal velocity component of a water particle (ms^{-1})
v	Cross velocity component of a water particle (ms^{-1})
w	Vertical velocity component of a water particle (ms^{-1})
r	Distance between two points (m)
x	Local coordinate in longitudinal direction (m)
y	Local coordinate in cross direction (m)
z	Local coordinate in vertical direction (m)
φ	Potential function (m^2/s)
μ	Dipole strength (m^2/s)
σ	Source strength (m/s)
ζ	Change of local water level (positive above) (m)
ρ	Specific weight of water (kgm^{-3})
Ω	Surface area of wetted hull (m^2)

1 INTRODUCTION

Recently, the Rijkswaterstaat[1] got a message that a big inland vessel was sailing close and parallel to the quay and this ship was not able to sail to the centre of the canal. The only solution was to stop the main engines and to manoeuvre the ship with low speed to the centre of the canal. When it arrived there, the full speed manoeuvre could be continued.

Apparently side forces occur, when ships are sailing close to quays. These forces are so big that a ship is not able to react for a safe sailing through the canals. What happens is the following. The ship is controlled with a rate of turn feedback indicator. This steering controller is very well able to keep a ship on course, but it is not able to control in cross direction (and it time derivatives). When the ship has an initial offset, the ship experiences a force in the direction of the quay. Due to this force, the ship will accelerate and move to the quay. And when the ship becomes closer to the quay the (side) forces become higher. The controller will cause that the ship will not get a drift angle, so it is not possible to generate a force on the ship in the centre direction of the canal. The result is, that the ship will sail parallel and closer to the quay. And when the captain does not pay attention to this effect, the ship will touch the quay almost parallel.

From the above phrase, it appears that side forces occur when ships are not sailing in the centre of a canal.

Another point of view is that the most formulae's, to predict the squat (sinkage and/or trim) (see [1], [17] and [19]), are only valid in the centre of the canal. It is expected that the squat will increase when the ship is not sailing in the centre of the canal. To get an impression to predict squat effects, this study has been

[1] Dutch Federal Authorities of inland navigation

set up. In previous research it is already proven that panel methods are a suitable tool to predict the sinkage and trim (see [9], [10] and [11]). To make a proper investigation in this research, a panel method is used.

2 CALCULATION METHOD

2.1 PANEL METHODS, THE THREE DIMENSIONAL METHOD

In the last part of the previous century numerical methods have been developed to calculate the potential flow, with or without wave pattern around ship hulls. The advantage of numerical methods is that the velocities and pressures can be calculated at each point of the wetted surface of the ship hull as well as at the free surface. Also the forces and moments can be determined by integrating the pressure with respect to the wetted hull. So more and new insights concerning the behavior of the flow can be analyzed. The tests can be carried out with or without free surfaces effects. When the geometry or flow direction is changed a quick calculation can be made in order to gain more insight regarding the characteristics of the flow. At the TUDelft a numerical method based program was developed. This program is based on a panel method using a Dirichlet boundary condition. The panels are placed on the wetted hull surface and on the free surface around the ship. The program can predict the double body flow when panels are only placed on the wetted hull. This program can also be used for wave pattern calculations, but as it will be mentioned in section 2.6, but at low Froude (based on the ship length) wave pattern calculations are not important for problems concerning sinkage and trim. The model is based on a potential flow. Some important properties and characteristics of the potential theory are discussed below.

For the simulation of the flow around a floating object a simplified flow model is used. One of them is the potential flow model. This model is a theoretical flow model and it has some important characteristics, which can be used for calculations of the flow. One of these characteristics is, that an integral method can be used for simulating the flow. Now some of the characteristics thereof will be discussed. The potential flow is assumed to simulate an incompressible and inviscid flow around a profile or body. This means that only mass inertia forces are simulated and no drag can occur. And also this means that there will be no vorticity in the flow. Due to the fact that the flow is incompressible, the flow will be divergence free. The means that in the flow no material will be created nor destroyed and is called the equation of conservation of mass. Now the potential flow can be described as a solution of the Laplace equation:

$$\Delta\Phi = \frac{\partial^2\Phi}{\partial x^2} + \frac{\partial^2\Phi}{\partial y^2} + \frac{\partial^2\Phi}{\partial z^2} = 0 \qquad (1)$$

This is a homogeneous and linear differential equation and so the superimposing property can be used. With this property, it is possible to transform the Laplace equation into the Fredholm's integral equation with help of Green's theorem.

2.2 INTEGRAL EQUATIONS

For solving the potential, the Fredholm integral equations of a second kind are used. On the surface of the body a source and a dipole layer will be placed. Thus the following integral equation is used, with the Greens function, which represents physical sources and dipoles:

$$\varphi(\bar{x}) = \iint_{\Omega} \{\sigma(\frac{-1}{4\pi r}) + \mu(\frac{-<\vec{n}\cdot\vec{r}>}{4\pi r^3}\} d\Omega \qquad (2)$$

In which r is the distance between an arbitrary point and a point at the surface Ω. The surface Ω is divided into the wetted surface of the double body Ω_B and the canal wall and bottom Ω_C. To solve the integral equation the inner Dirichlet boundary condition is used. The source strength σ at the fixed wetted surface Ω_B and canal wall and bottom Ω_C then is:

$$\sigma = \frac{\partial\varphi}{\partial n} = -<\vec{U}_\infty \cdot \vec{n}> \qquad (3)$$

For simulating the flow with a potential flow solution a number of boundary conditions have to be obeyed. The boundary conditions have to be obtained at the wetted surface of the ship body, at the free surface and far away in front of the ship the wave height and the derivatives thereof will be zero. When no free surface flow is obtained, all the boundary conditions can be applied as no leakage condition or kinematic boundary condition. The flow at the solid body Ω_B and canal wall and bottom Ω_C will be tangential to the surface. This means that there is no velocity component in normal direction. This can be written as an inner product:

$$<\nabla\Phi \cdot \vec{n}_b>=0 \qquad (4)$$

By making use of the potential jump relation of a dipole layer and of the choice that the inner potential $\varphi=0$, the perturbation potential can be determined at the outer surface according to the so-called jump relation:

$$\varphi(P \in \Omega) = -\mu(P) \qquad (5)$$

According to this equation the potential at Ω_B is known. The first (partial) derivative of the potential in any direction delivers the velocity in that direction. So, by taking the first spatial partial derivatives of the potential along the wetted hull, the velocities are obtained on the surface of the body. The fluid velocity at a control point is:

$$\vec{v} = \overrightarrow{U_\infty} + \sigma\vec{n} + \nabla\varphi \qquad (6)$$

2.3 PRESSURES AND FORCES

The pressure at any particular point can be determined with the Bernouilli equation:

$$p=-\tfrac{1}{2}\rho<\vec{v}\cdot\vec{v}>+C=-\tfrac{1}{2}\rho<\vec{v}\cdot\vec{v}>+\tfrac{1}{2}\rho U_\infty^2 \quad (7)$$

The constant is determined at 'infinity' far in front of the ship, where the water is in rest and so the perturbation velocities and perturbation potentials are 0. ($C=\frac{1}{2}\rho U_\infty^2$) The velocities in this equation (7) are determined from equation (6) by taking the gradient of the potential. By integrating the pressures from equation (7) over the ship surface the forces F_1 to F_3 are determined:

$$\vec{F}=\iint_\Omega -p\vec{n}d\Omega \quad \text{For the first 3 components (8)}$$

The moments on the ship $M_1..M_3 \equiv F_4..F_6$ are determined by using the cross product of the normal vectors and the distance vector from the ship centre and these results are integrated over the ship surface Ω:

$$\vec{M}=\iint_\Omega -p\vec{n}\times\vec{r}d\Omega \quad \text{For the last 3 components (9)}$$

To find the sinkage and trim, the equations (12) and (13) can be used by filling in the values of F_3 and F_5 respectively.

2.4 SINKAGE AND TRIM

The sinkage and trim can be obtained by using Archimedes' law. When the sinkage is defined as s (positive downwards) and the trim is defined as t (positive ahead), the equilibrium according to Archimedes' law for a trim and sinkage ship is:

$$\begin{aligned}\rho g\int_{-\frac{L}{2}}^{\frac{L}{2}}(s+xt)B(x)dx &= s\rho g\int_{-\frac{L}{2}}^{\frac{L}{2}}B(x)dx+t\rho g\int_{-\frac{L}{2}}^{\frac{L}{2}}xB(x)dx\\ &=\rho g(A_w s+S_y t)=-F_3\\ \rho g\int_{-\frac{L}{2}}^{\frac{L}{2}}(s+xt)xB(x)dx &= s\rho g\int_{-\frac{L}{2}}^{\frac{L}{2}}xB(x)dx+t\rho g\int_{-\frac{L}{2}}^{\frac{L}{2}}x^2B(x)dx\\ &=\rho g(S_y s+I_y t)=F_5\end{aligned} \quad (10)$$

In this equation the following known quantities appear: A_w: the water area or zero order area, S_y: the static moment or first order area and I_y: the moment of inertia or second order area of the water plane defined from amidships. Substituting these parameters in equation (10), the following matrix equation is obtained:

$$\rho g\begin{pmatrix}A_w & S_y\\ S_y & I_y\end{pmatrix}\begin{pmatrix}s\\ t\end{pmatrix}=\begin{pmatrix}-F_3\\ F_5\end{pmatrix} \quad (11)$$

This matrix equation can be solved easily, the results are:

$$s=\frac{-F_3I_y-F_5S_y}{\rho g(A_wI_y-S_y^2)} \quad \text{For the sinkage} \quad (12)$$

And

$$t=\frac{F_5A_w-F_3S_y}{\rho g(A_wI_y-S_y^2)} \quad \text{For the trim} \quad (13)$$

The forces F_3 and F_5 are already obtained by equations (8) and (9). The keel clearance is obtained by the subtracting the draught T of the ship and the influence of the sinkage and trim from the water depth h: (See also Figure 1):

$$K=h-T-s-\frac{|t|}{2}L \quad (14)$$

For the keel clearance it does not matter if the trim is by head or by stern, when the origin of the coordinate system of the ship is defined amidships. By the same amount of trim (ahead or astern) one part of the keel reaches the same distance closer to the bottom.

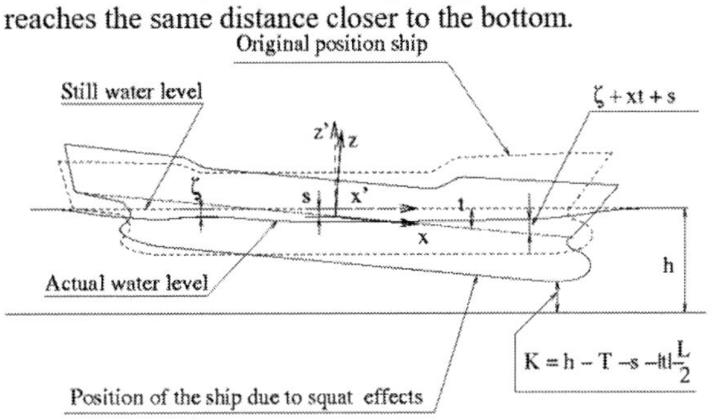

Figure 1 New situation when a ship sails in squat conditions

When a heel angle is taken into account, then this heel angle can be calculated as follows:

$$\varphi=\frac{F_4}{\rho g\nabla GM} \quad (15)$$

So, when a heel angle is present, the minimum value between the ship bottom and canal bottom has to be found. It is more practical to introduce a frame box around the under water ship and this frame box has the dimension of the length, breadth and draft. Now the virtual sinkage at one of the corner points of the frame can be determined. In this case the maximum total sinkage (s_{max}) is:

$$s_{max}=s+\frac{|t|}{2}L+\frac{|\varphi|}{2}B \quad (16)$$

2.5 PANEL METHODS FOR SQUAT CALCULATIONS

For free surface calculations, including the wave pattern, the speed of the ship has to be considerable. The most wave pattern prediction programs require a panel density of 12 panels per wavelength and for an accurate calculation the programs require even a density of 20 panels. When a ship sails with a speed of 10 *knots* = 5.144*m/s* the generated wavelength is:

$$\lambda = \frac{2\pi U^2}{g} \approx 17.0m$$

This means, that the panel length at the free surface should have a length between 1.0m to 2.0m. For a stable and a realistic wave pattern calculation the free surface has to be panelised almost a half of the ship length in front of the ship and a ship length aft of the ship. For a ship of a length of 300m or more at least 500 panels has to be generated in longitudinal direction. For a more accurate calculation, the number of panels should be even higher than 1000 panels. Also in cross direction the panel width should be the same dimension as the panel length. So, a lot of panels in width direction have to be generated too. The panel distribution on the ship hull should have almost the same measurements as the free surface panels. This means, that an enormous number of panels has to be generated ($\pm 10^6$). Further the matrix is a full matrix and it has to be stored inside the computer. This requires a lot of computer memory, which is not available. Also, due to the large amount of numbers the inversion process will deliver poor results, due to a bad conditioned matrix (, the von Neumann condition of a matrix is also linear with the number of rows.) From this point of view, it is expected that no reliable or unstable results may occur and it is not recommendable to use a free surface calculation.

Fortunately with this very low speed of the ship the wave effects on squat effects are negligible. The wave influence in vertical direction reduces very rapidly and at one wave length under the free surface the influence is approximately 0.2% of the effect at the free surface. The bottom of the ship does not experience much of the disturbed pressure of the generated Kelvin waves. Also the amplitude of the wave is very low. So, at low speeds of the container ships in a harbour, the free surface waves or Kelvin waves do not influence the squat. This means that when a panel method is used, the free surfaces do not have to be modelled.

2.6 ADAPTATION OF THE PANEL METHOD FOR SHALLOW WATER

The above methods describe, that the sinkage is consists of a Froude depth number dependent part and a geometrical or shape of the ship hull. The squat can be calculated according to:

$$s = f(F_h)g(Geom) \qquad (17)$$

For the linear squat calculations, when a panel method is used, the same principle is used. When only the panel method in the original case is used, the surfaces of the bodies are rigid. Furthermore the free surface is removed and the calculations are carried out by a mirrored body model, where the hull and the bottom are mirrored in the free surface. The calculated pressures are proportional to the square of the velocity. From this, the forces, moments, sinkage and trim are also proportional to the square of the velocity:

$$p, F, M, s, t \sim v^2 \sim F_h^2 \qquad (18)$$

However, by using the panel method, the shallow water equation is not built in. Tuck (see [19]) has derived, that for shallow water problems, the potential equation has to fulfill the shallow water hydrodynamics. From this shallow water hydrodynamics equation the calculated pressures, forces and moments are proportional to:

$$p, F, M, s, t \sim \frac{F_h^2}{\sqrt{(1 - F_h^2)}} \qquad (19)$$

3 THE SQUAT RESULTS

3.1 THE DIMENSION OF THE CONTAINER VESSELS

The calculations have been carried out for one type of container ship, one canal depths and four different offsets of the center of the canal. The type of containers ship is the so called 'MEGA JUMBO'. The canal depth is 19.65m and the initial keel clearance (this is the keel clearance when the ship is in rest) is 2.65m. The dimensions and the specific ship parameters are shown in table 1.

Ship	Dim.	MEGA JUMBO
Container Cap.	[*TEU*]	12500
Length o.a. L_{oa}	[*m*]	382.0
Length p.p. L_{pp}	[*m*]	364.9
Breadth B	[*m*]	57.0
Depth H	[*m*]	30.0
Draught T	[*m*]	17.0
Displacement ∇	[m^3]	242911
Block coef. c_b	[ ]	0.6870
Slenderness coef. $\lambda = \frac{L}{\sqrt[3]{\nabla}}$	[ ]	5.848
Loc. Of center of buoyancy L_{cb}	[*m*] [%]	-5.65 (=-1.55)
Waterline area A_w	[m^2]	16902
Waterline area coef. c_{WL}	[ ]	0.8126
Loc. of waterl. cent. x_{WL}	[*m*] [%]	-16.44 (=-4.51)
Static moment of W.L.[1] S_y	[m^3]	-277919
Mom. of inertia of W.L.[2] I_y	[m^4]	13818744

Table 1 Dimensions and specific parameters of the tested container ship.

The dimensions of the canal are 600*m* (width) and 19.65*m* (depth). The simulations are carried out at the center of the canal and at three different offsets. The offsets are respectively 25%, 50% and 75% of the half

[2] defined from amidships

width of the canal. The corresponding offsets in meters are respectively 0, 75, 150 and 225 *m*. All the simulations are carried out with a unit speed of 1 ms^{-1} and without a drift angle.

To obtain the real forces and sinkage and trim can be calculated as follows. The real values are obtained by using the results at the unity speed. These values have to be multiplied with the square of the real velocity. According to these multiplications the linear results are obtained. When the shallow water equation (*swe*) correction is included, the linear (*lin*) results have to be multiplied again with the so called shallow water correction:

$$s_{swe} = \frac{s_{lin}}{\sqrt{1-F_h^2}} \qquad (20)$$

3.2 THE SQUAT RESULTS OF THE PANEL METHODS

3.2.(a) The pressure distribution

The results of the panel method are presented in figure 2 to figure 9. In the first four figures (2 to 5) the velocities along the hull and canal walls are shown as well as the pressure. In the last four figures (6 to 9) only the pressures at the wetted hull of ship are shown. Looking at the pressures in these figures already some remarks can be made and insight can be obtained. From these figures, it is obvious that the pressures become less when the ship sails more in the vicinity of the quay. Due to these lower pressures, it is expected that a larger sinkage force and trim moment will occur and so, the ship will have a sinkage and trim. This sinkage and trim will be larger when the ship is sailing closer to the quay.

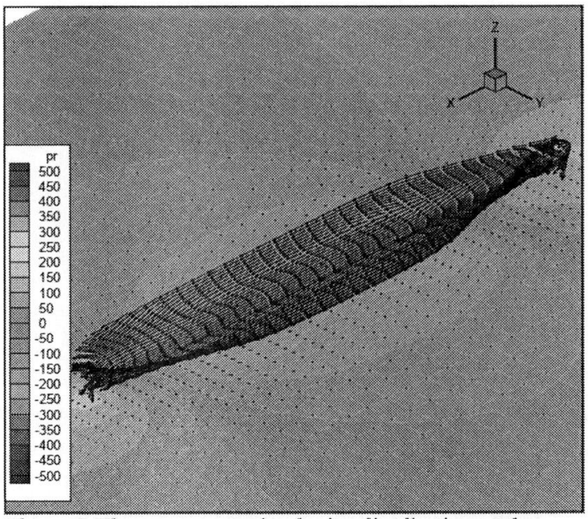

Figure 2 The pressure and velocity distribution at the ship and bottom at a offset of $y = 0m$.

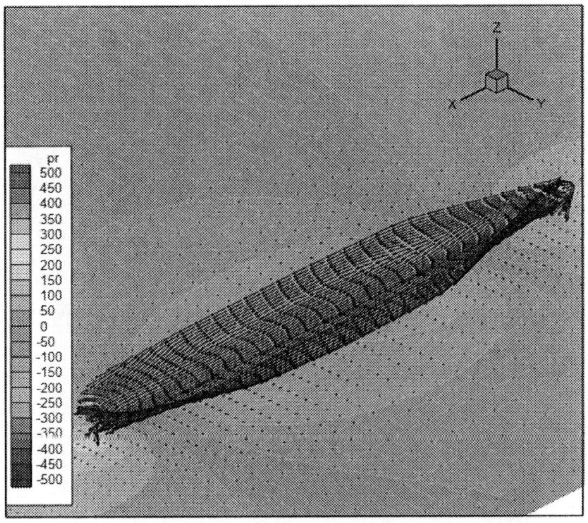

Figure 3 The pressure and velocity distribution at the ship and bottom at a offset of $y = 75m$.

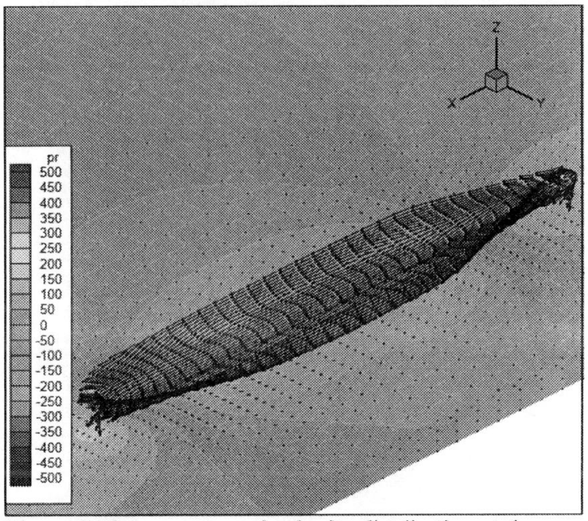

Figure 4 The pressure and velocity distribution at the ship and bottom at a offset of $y = 150m$.

The most important calculated variables are presented in Table2. In this table the results are obtained at the unity speed.

y (m)	0	75	150	225
F_2 (kN)	0.00	8.571	30.82	135.5
F_3 (MN)	-2.246	-2.280	-2.438	-3.111
F_4 (MNm)	0.00	-0.1189	-0.4224	-1.853
F_5 (MNm)	21.06	21.39	22.67	25.93
s (10^{-3}m)	13.49	13.69	14.64	18.72
t (10^{-6} rad)	11.88	12.06	13.04	18.89
φ (10^{-6} rad)	-0.02	-9.793	-34.77	-152.6
s_f (10^{-3}m)	15.65	15.89	17.02	22.16
s_s (10^{-3}m)	11.32	11.49	12.26	15.27
s_φ (10^{-3}m)	0.00	0.2791	0.991	4.348
s_{max} (10^{-3}m)	15.65	16.17	18.01	26.51

Table2 Squat results at unity velocity

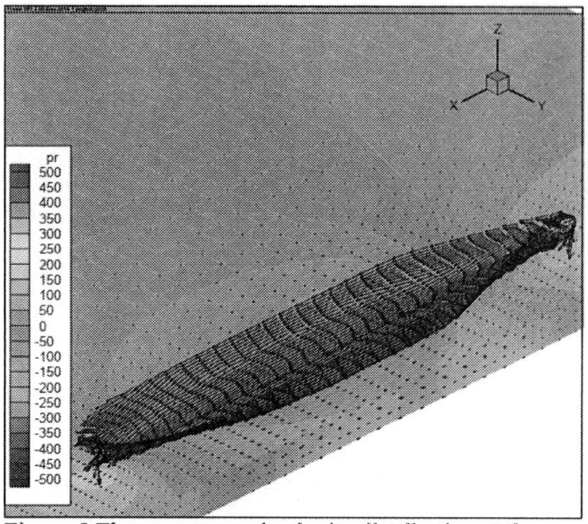

Figure 5 The pressure and velocity distribution at the ship and bottom at a offset of $y = 225m$.

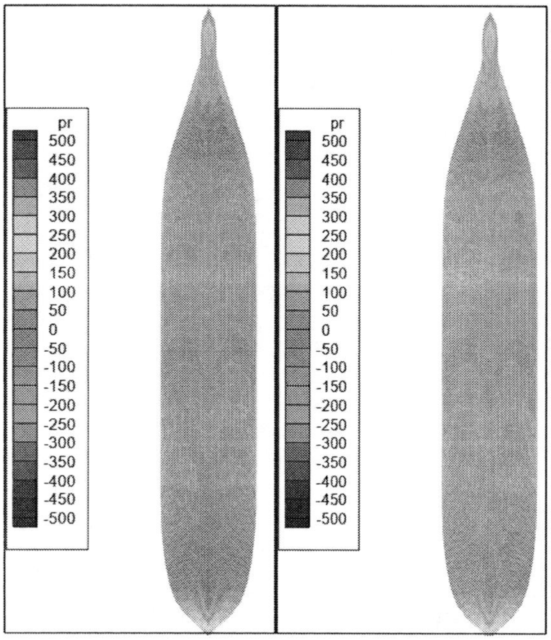

Figure 6 and 7 The pressure distribution at the wetted hull of the ship at a offset of $y = 0m$ (left) and $y = 75m$ (right) respectively.

From the results presented in Table2, it is obvious, that when the ship is sailing more in the vicinity of the quay that all the forces and moments increases. Not only the vertical force is increasing, but also a side force (F_2) occurs. This force becomes greater, when the ship sails in the vicinity of the quay. Also the moments are increasing and when the ship sails closer to quay a heel moment occurs as well.

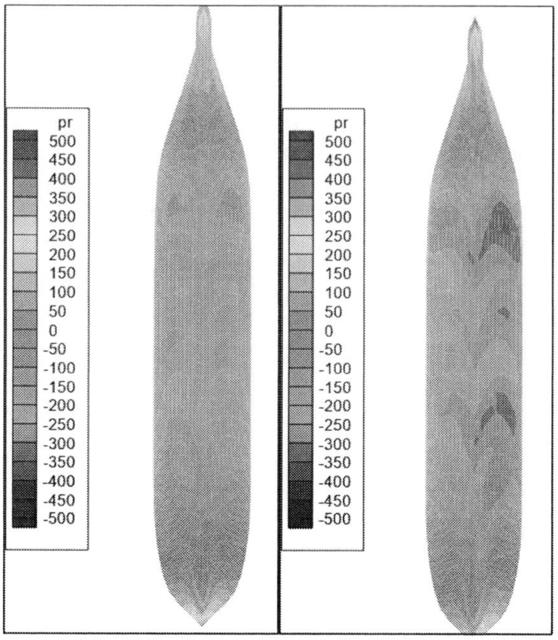

Figure 8 and 9 The pressure distribution at the wetted hull of the ship at a offset of $y = 150m$ (left) and $y = 225m$ (right) respectively.

From the results, shown in Table2, it is possible to present the sinkage as function of the ship speed. The presented sinkage is calculated amidships and at the place where the frame box[3] of the underwater ship has it highest sinkage. The sinkage, presented in figure 10 and figure 12, is proportional to the square of the velocity. The sinkage, presented in figure 11 and 13

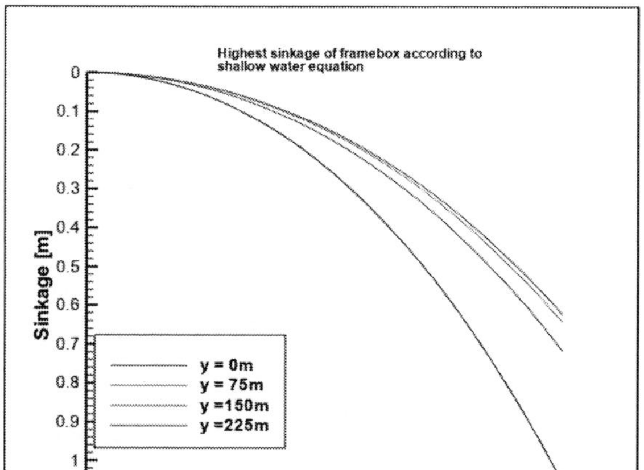

Figure 13, is according to the shallow water equation.

[3] The frame box is a fictive rectangular box around the ship and has the dimension of the length *L*, breadth *B* and draft *T*

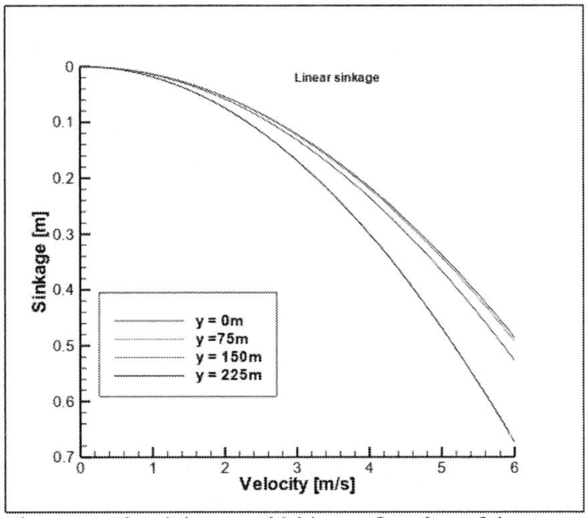

Figure 10 The sinkage amidships as function of the velocity at several offsets according to the linear equation.

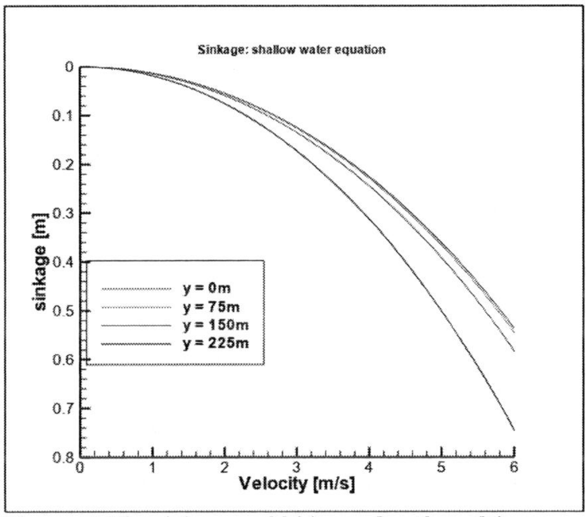

Figure 11 The sinkage amidships as function of the velocity at several offsets according to the shallow water equation.

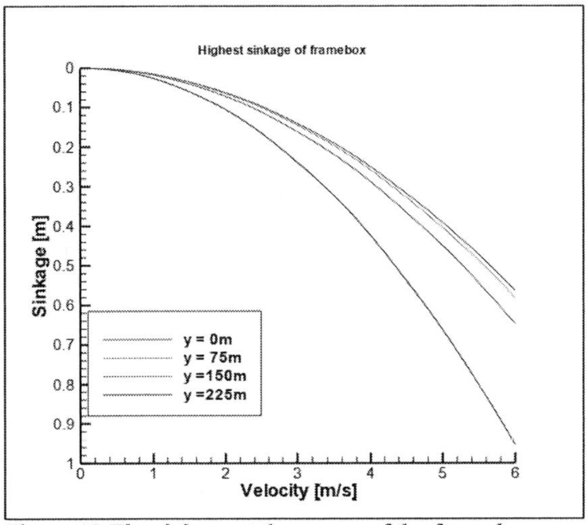

Figure 12 The sinkage at the corner of the frame box as function of the velocity at several offsets according to the linear equation.

From these figures (10 to 13), it appears that the sinkage according to the shallow water equation is a slightly more than according to the square velocity.

In this case the Froude depth number is not very high as well (The maximum Froude number is $F_h = 0.428$ at 6 ms^{-1} = 11.66 *knots*) and the correction of the shallow water equation is a factor 1.1 ($=1/\sqrt{(1-F_h^2)}$).

The figures 14 (amidships) and 15 (frame box) present the sinkage as function of the offset of the track from the center of the canal. The ship speed in the presented figures is 5 ms^{-1} = 9.72 *knots*, which is a realistic velocity in a port environment. From these last two figures, it appears that the squat increases when the ship in more close to the quay. However the sinkage increases drastically when the ship is in the proximity of the quay.

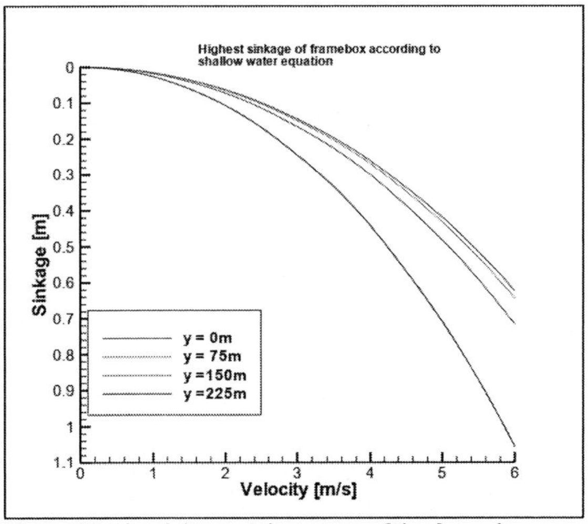

Figure 13 The sinkage at the corner of the frame box as function of the velocity at several offsets according to the shallow water equation.

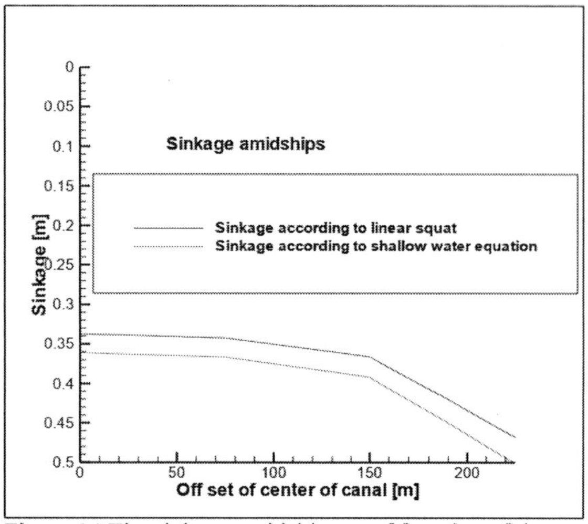

Figure 14 The sinkage amidships as of function of the offset at a velocity of $5m/s$

Not only an extra sinkage, trim and heeling occur when the ship is not sailing along the center track, but also a side force (F_2) and a yawing moment appear. The side force is presented in figure (16). The yaw moment is very dependent of the ship geometry and this is not presented here. The side force (F_2) as function of the offset of center of the canal is presented in figure. In this figure the ship is 5 ms^{-1} and this is a realistic situation.

From this figure (16), it is obvious, that the side force increases drastically, when the ship sails more in the vicinity of the quay. Due to this force, the ship will accelerate to the quay and when no opposite forces are introduced, the ship will really move to the quay. To remain on a stable course, opposite forces are needed. Opposite forces can be obtained by wind forces, tugs or by introducing a drift angle, such that the bow points to the center of the canal. How closer the ship is in the vicinity of the quay, how larger the drift angle should be.

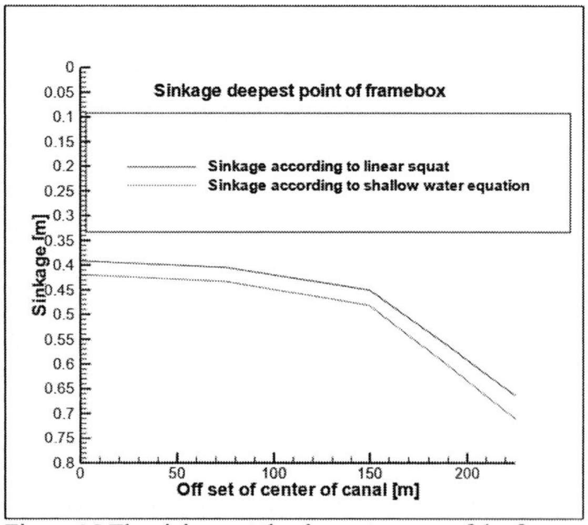

Figure 15 The sinkage at the deepest corner of the frame box as of function of the offset at a velocity of $5m/s$

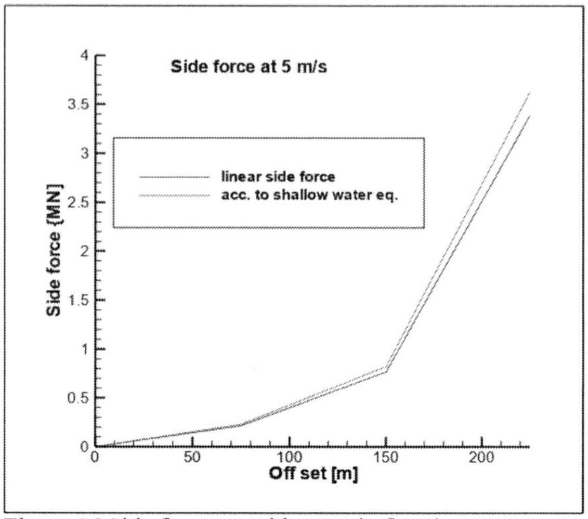

Figure 16 Side force at a ship speed of $5m/s$

During the simulation tests no drift is taken into account. From previous research (see [11]), it appears, that for ships sailing under a drift angle, the squat effects increases as well. From this research, it appears that the extra sinkage is of the same order as the sinkage at the center of the canal. However, it is unknown, when the ship sails in the vicinity of the quay, how large the (extra) sinkage will be. when the simulation are carried with a drift angle.

4 SOME REMARKS AND DISCUSSIONS

The simulated tests are carried out in a potential flow model. This model simulates only the inertia forces of water, while the viscous forces are not taken into account. How important the viscous influences are, is not researched, but comparing squat results from tank tests, the viscous forces are not quite important (see [7]).

In this research no propeller interaction is taken into account. When a propeller is used, this propeller will suck the water under the aft ship and lower pressure will occur. The result of this is, that an extra trim moment will occur and that the ship will be trimmed more to the stern.

5 RECOMMENDATIONS

- Equations, which predict the squat, have to be adapted for ships, which are not sailing at the centre of the canal.
- New calculations have to be carried out with drift angles. These calculations can also be done with a panel method (or boundary element method). Two types of calculations can be distinguished, with or without lift effects. When lift effects are applied a vortex sheet behind the ship has to be created as well and a Kutta

condition has to be applied at the end of the centre plane of the ship.

- More simulations with other type of ships have to be carried out to give more insights of squat effects, when ships are not moving in the centre of the canal.
- To validate these squat results, towing tests have to be carried out.

6 CONCLUSIONS

- Panel Methods give an indication to predict squat, even when ships sails in the vicinity of the quay. However the results should have to validate with experiments.
- From the calculations, it appears that forces increase when ships sail closer to the quay. The forces grow drastically when the ship sails in the vicinity of the quay.
- Also the squat effects (sinkage and trim) become larger. This means, that equations used to predict the squat, are only valid for ship sailing at the centre of the canal.
- Beside the trim and sinkage, also a heel angle occurs when the ship sails close to the quay.

7 REFERENCES

[1] Barrass, C.B. "Ship Squat - A Reply", The Naval Architect, November 1981

[2] Beck, R.F., "Forces and Moments on a Ship Moving in a Shallow Channel", Journal of Ship research, Volume 21, No 21 June 1977, pp 107-119

[3] Breit, S.R., "The Potential of a Rankine Source between Parallel Planes and in a Rectangular Cylinder", Journal of Engineering Mathematics 25 pp.151-163, Kluwer Academic Publishers, Printed in the Netherlands, 1991

[4] Flügge G. & Uliczka, K., "Fahrverhalten grosser Containerschiffe in extrem flachem Wasser, Das dynamische Fahrverhalten und die Wechselwirkungen mit der Fahrrinennsohle von sehr grossen Containershiffen unter extremen Flachwasserbedingungen.", Hansa-Schiffahrt-Schiffbau-Hafen-138.Jahrgang-2001-Nr 12

[5] Hermans, A.J., "Ship Manoeuvring and Hydrodynamic Forces Acting on Ships in Confined Waters", Proceeding of Symposium on Aspects of Navigability, Delft, 1978

[6] Katz, J. & Plotkin, A., "Low-speed aerodynamics from wing theory to panel method", ISBN 0-07-100876-4, McGraw-Hill, Inc., Singapore, 1991

[7] Koning Gans, H.J. de, "Squat effects of very large container ships sailing in a harbor environment", Report No.1407-O,Laboratory of Ship Hydromechanics, TUDelft (Delft University of Technology) and Port Research centre Rotterdam-Delft Publications , November 2004, Delft, ISBN 90-5638-134-2

[8] Koning Gans, H.J. de, "Squat Results from Calculations of Panel Methods", Report No.1408-O, Laboratory of Ship Hydromechanics, TU Delft (Delft University of Technology) and Port Research centre Rotterdam-Delft Publications, November 2004, Delft, ISBN 90-5638-135-0

[9] Koning Gans, H.J. de, "Squat Effects of very Large Container Ships Sailing in a Harbor environment", 'International Maritime-Port Technology and development Conference 2005, Rotterdam', September 2005, Rotterdam, ISBN 90-80989-21-5 (old) ISBN 978-90-80989-21-4 (new)

[10] Koning Gans, H.J. de, "A Method to Predict Forces on Passing Ships under Drift", 9th International Conference in Numerical Ship Hydrodynamics, August 2007, Ann Harbor

[11] Koning Gans, H.J. de, "Squat Effects of Very Large Container Ships with Drift in a Harbor Environment", International Maritime-Port Technology and Development Conference, September 2007, MTEC 2007, Singapore, ISBN:978-981-05-9102-1

[12] ICORELS, "International Commission for the Reception of Large Ships, ICORELS Expression in Seehafenzufahrten - Ein Leitfaden zur Bemessung (Übersetzung) PIANC/IAPH AG II-30, Bulletin Nr. 95, Bonn 1997.

[13] Lap, A.J.W., "Fundamentals of Ship Resistance and Propulsion Part A: Resistance", Publicati-on No.129a of the N.S.M.B., Reprinted from International Shipbuilding Progress (I.S.P.)Rotterdam

[14] Newman, J.N., "Marine Hydrodynamics", ISBN 0-262-14026-8, The MIT Press, Cambridge, Massachusetts and London, England, 1977

[15] Newman, J.N., "The Green Function for Potential Flow in a Rectangular Channel",Journal of Engineering Mathematics 26 pp.51-59, Kluwer Academic Publ., Printed in the Netherlands, 1992

[16] Schuster, S., "Untersuchingen über Strömungs- und Widerstand-verhältnisse bei der Fahrt van Schiffen in beschränktem Wasser"

[17] Schmiechen, M.,"Squat-Formeln, Das Schiff" in begrenzten Gewässern",18 Duisburger Kolloquium Schiffstechnik / Meerestechnik, Duisburg, 1997

[18] Timman, R., Hermans, A.J. \& Hsiao, G.C.,"Water Waves and Ship Hydrodynamics: An Introduction", ISBN 90-247-3218-2, Martinus Nijhoff Publishers and Delft University Press, Delft, the Netherlands, 1985

[19] Tuck, E.O. \& Taylor P.J., "Shallow wave problems in ship hydrodynamics",8th On Navel Hydrodynamics, pages 627-659, Rome 1970

[20] Uliczka, K., "Das Schiff in Wechselwirkung mit der Wasserstrasse (The ship in interaction with the waterway)", Verkehrswasserbau an Seeschifffahrtsstrassen (Hydraulic engineering applied to approach channels, 13-15 June 2002

[21] Yeung, R.W., "Application of a Slender-Body Theory to Ships Moving in Restricted Shallow Waters", Proceeding of Symposium on Aspects of Navigability, Delft, 1977

8 AUTHORS BIOGRAPHY

Henk de Koning Gans holds the current position of assistant professor at Delft University of Technology. He is responsible for the education and research of mathematical and numerical ship hydromechanics. His previous experience includes shallow water problems, panel methods. He applied the panel methods for propeller calculations, free surface (wave pattern) calculation and of passing of ships.

THE IMPORTANCE OF THE FORM OF TIME DOMAIN FORCES AND MOMENTS ON BERTHED SHIP INTERACTION

J T Duffy & M R Renilson, Australian Maritime College, Australia

SUMMARY

An investigation has been undertaken into the interaction forces and moments experienced by berthed ships due to passing ships. The aims were:

- To quantify the influence of waterway geometry on the magnitude of the peaks and the form of the time domain interaction forces and moments;
- To quantify the influence of the form of the interaction forces and moments on berthed ship motions and mooring line loads.

The second aim was considered important, as in the past the peak values of the forces and moments were used, together with the assumption of an idealised form of the interaction forces, to predict the motions, and hence loads on the mooring lines. It was felt necessary to verify this assumption, particularly for a range of different waterway geometries.

Results were obtained from model scale experiments conducted in the Towing Tank at the Australian Maritime College to measure the time domain interaction forces and moments on a berthed ship due to a passing ship for different waterway geometries. The measured time domain interaction forces and moments are presented for three different channel geometries. The influence of the different channel geometries on the magnitude of the peaks and the form of the interaction forces and moments has been quantified.

The interaction forces and moments were used as input to a time domain numerical simulation model to predict berthed ship motions and mooring line loads. The influence of the form of the interaction forces and moments on the maximum berthed ship motions and maximum mooring line loads has been quantified for a typical mooring arrangement. From this the key features of the form of the forces and moments in the time domain was obtained, and the applicability of using an open water idealised form for cases with different channel widths and near bank geometries has been assessed.

NOMENCLATURE

B	Ship beam (m)
C_B	Block coefficient
L_{PP}	Length between perpendiculars (m)
N	Yaw moment (Nm)
N''	Non-dimensional yaw moment $[N/(0.5\rho U^2 L_{PP}^2 T)]$
T	Ship draught (m)
U	Passing ship speed (m s^{-1})
X	Surge force
X''	Non-dimensional surge force $[X/(0.5\rho U^2 BT)]$
Y	Sway force (N)
Y''	Non-dimensional sway force $[Y/(0.5\rho U^2 L_{PP} T)]$
ρ	Density of water (kg m^{-3})

1. INTRODUCTION

When a moving ship passes a berthed ship, interaction forces are generated causing the berthed ship to move on its moorings. These motions should be minimised to prevent excessive mooring line loads and to reduce the likelihood of interruptions to loading/unloading procedures and damage to mooring equipment.

Model scale experiments are an effective method to quantify the interaction forces and moments experienced by a berthed ship due to a passing ship. Such techniques have been used in past studies to measure such interaction forces and moments [1,2,3,4].

Remery [1] used model scale experiments to investigate interaction forces and moments on tanker models and found that plotting the measured forces and moments to a base of the position of the passing ship relative to the berthed ship showed that the character of the curves was similar for the various sizes and passing distances of the passing ship.

Dand [2] conducted physical model scale experiments to measure the interaction forces and moments on a berthed ship due to a passing ship and developed formulae to predict the forces and moments.

Theoretical techniques have also been developed in previous work to predict the interaction forces and moments on a berthed ship due to a passing ship [5,6,7,8].

Muga and Fang [5] used their theoretical method to calculate the interaction forces and moments on a berthed ship due to a passing ship for a range of cases and observed that the time pattern (or phase relations) of the occurrence of the peaks and the sequences of the loadings were very similar for different cases. Hence, they suggested that each loading function can be described by a single value of the ordinate, perhaps the maximum. They also observed that the time pattern of the loading functions as predicted by theory were all very similar to each other and corresponding experimental cases.

Using the results from Remery [1], Flory [9] found that if the interaction surge force, sway force and yaw moment were non-dimensionalised the various plots for different tanker sizes and separation distances were essentially the same, allowing one set of graphs to represent all cases. Flory [9] developed empirical formulae to predict the peaks of the time domain interaction forces and moments for cases where the passing ship path is parallel to the longitudinal axis of the berthed ship and the waterway width is infinite (open water).

Once the interaction forces and moments on the berthed ship are obtained, numerical simulation techniques can be used to predict berthed ship motions and mooring loads. This information is important to ship operators and port authorities to ensure that the berthed ship motions and mooring line loads satisfy applicable operational criteria. Berthed ship motions due to external forces have been predicted using numerical simulations in previous studies [6,4,10,11].

In the present study results were obtained from model scale experiments conducted in the Towing Tank at the Australian Maritime College to measure the interaction forces and moments on a berthed ship due to a passing ship for different channel geometries with varying near bank slopes and channel widths. The results were used to quantify the influence of the different channel geometries on the magnitude of the peaks and the form of the time domain interaction forces and moments.

The interaction forces and moments were then used as input to a time domain numerical simulation model to predict berthed ship motions and mooring line loads. The influence of the form of the interaction forces and moments on the maximum berthed ship motions and maximum mooring line loads was quantified for a typical mooring arrangement. From this the applicability of using the generic form for open water cases proposed by Flory [9] was assessed when applied to cases with different near bank slopes and channel widths.

2. SCALE MODEL EXPERIMENTS-MEASUREMENT OF INTERACTION FORCES AND MOMENTS

2.1 GENERAL

Knowledge of how channel geometry influences the forces and moments acting on a berthed ship due to a passing ship is important to enable realistic real-time simulation of the behaviour of a berthed ship. An effective method for determining such knowledge is through the use of model scale experiments [1,2,3,4].

In the present study results have been obtained from physical model scale experiments conducted at the Towing Tank at the Australian Maritime College to measure the interaction forces and moments on a berthed ship due to a passing ship. The tests were conducted with stationary models to simulate the berthed ships, and a towed model to simulate the passing ship. Interaction forces and moments were measured on the berthed ship model, which was constrained in surge, sway and yaw and free to heave, pitch and roll. Interaction forces and moments were not measured on the passing ship model.

2.2 SHIP MODEL DETAILS

All tests were carried out using two models of typical 130,000dwt bulk carriers. Both models were MarAd L series bulk carriers [12]. The passing ship model was fitted with the standard cylindrical bow, whilst the berthed ship model was fitted with a bulbous bow. Principal particulars of each of the models are given in Table 1.

Model	L_{PP}/B	B/T	C_B
Cylindrical bow bulk carrier	5.00	4.05	0.85
Bulbous bow bulk carrier	5.00	4.05	0.85

Table 1: Principal particulars of the models

2.3 TEST PROGRAM

Forces and moments on the berthed ship due to the passing ship were measured for three different channel geometries. The channel cross section along the Towing Tank was constant for each of the conditions. The channel width for Case A represented the full width of the Towing Tank and is shown in Figure 1. Case B had a sloping near bank on the starboard side of the berthed model and is shown in Figure 2. For Case C the starboard bank was the same as Case B, but the distance between the passing ship model and the port bank was reduced, as illustrated in Figure 3. The water depth to draught ratio was 1.07 for the passing and berthed ship model. The passing ship path was parallel to the longitudinal axis of the berthed ship. The direction of travel of the passing ship is shown in Figure 4. The results in the present paper are for a passing ship speed corresponding to a Froude length number of 0.074. The speed of the passing ship model was constant for the duration of the passing manoeuvre.

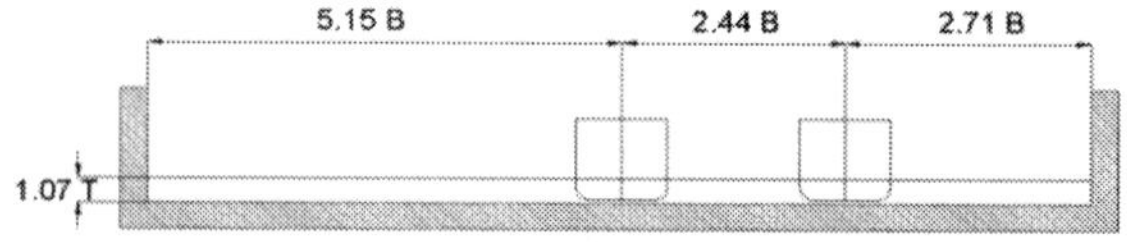

Figure 1: Channel cross section for Case A

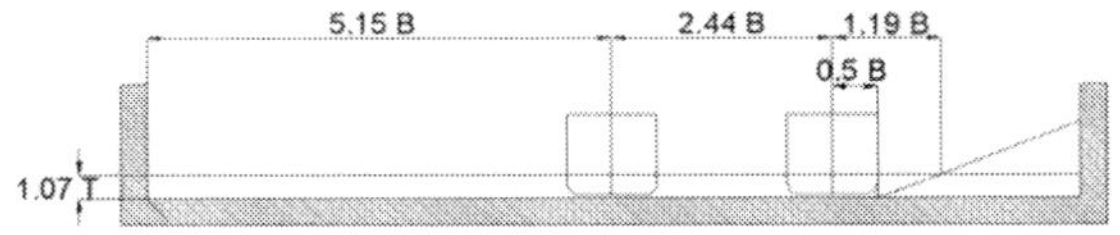

Figure 2: Channel cross section for Case B

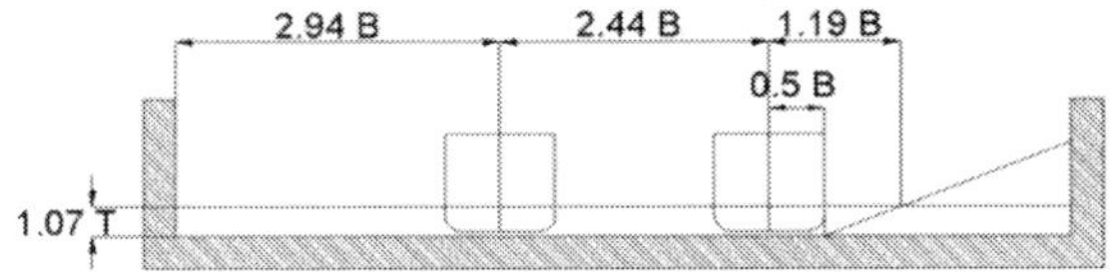

Figure 3: Channel cross section for Case C

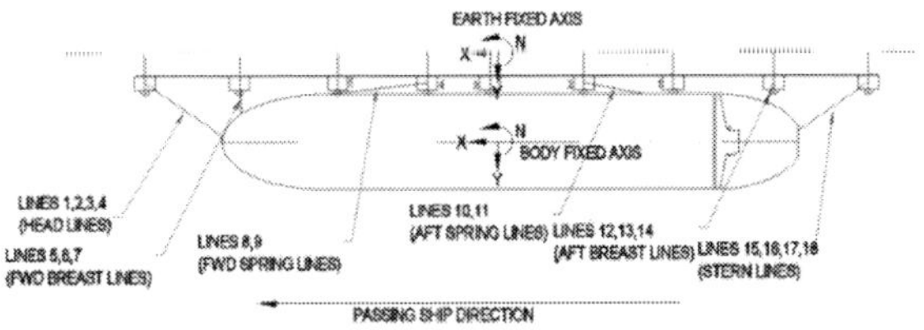

Figure 4: Schematic of the mooring arrangement and sign convention

2.4 TEST PROCEDURE

The passing ship model was towed along the centreline of the Towing Tank. The passing ship model reached the desired constant speed well before it reached the berthed ship model. To maintain consistency in the tank test section and to synchronise the beginning of each measurement, the data acquisition was commenced at a common location for each test run. Sufficient time was allowed between consecutive runs to achieve calm water conditions.

The experiments yielded a time domain trace of interaction surge force, sway force and yaw moment. Examples of these traces are given in Section 4.

3. NUMERICAL SIMULATION OF BERTHED SHIP MOTIONS AND MOORING LINE LOADS

3.1 GENERAL

The time domain interaction forces and moments obtained from physical model experiments were used as input to numerical simulations to predict the motions of the berthed ship and associated mooring line loads. The berthed ship motions and mooring line loads were predicted using Termsim II, which is a time domain numerical simulation program suite developed by MARIN (Maritime Research Institute Netherlands) [11].

3.2 MOORING LINES

A generalised mooring arrangement was used for the present study, as shown in Figure 4. The properties of each mooring line were based on Nylon double braided lines, 88.9mm in diameter with breaking strength of 1638kN. The pre-tension on each mooring line was set at 147kN.

3.3 FENDERS

The fender reaction characteristics were modelled in the numerical simulations. Bridgestone C3000H cell fenders were modelled.

3.4 BERTHED SHIP MOTIONS AND MOORING LINE LOADS

Time domain traces of the motions of the berthed ship and the associated mooring loads were obtained from the numerical simulations. The surge, sway and yaw motions and mooring line loads are considered in the present paper.

Each numerical simulation was run for 200 seconds (full scale) before the passing ship force was applied to allow the berthed ship to obtain an equilibrium position. Examples of time domain berthed ship motions are presented in Section 4.

4. RESULTS

4.1 GENERAL

In this section the measured interaction forces and moments on the berthed ship are presented for each case and compared to the idealised force and moment traces from Flory [9]. In each case the magnitude of the largest peak of the idealised force trace has been matched to the corresponding peak for the measured force trace. The results are presented such that the maximum negative surge force occurs at the same time for both the measured and idealised force traces.

The influence of channel geometry on the magnitude of the peaks and the form of the force and moment traces has been identified.

The sign convention is given in Figure 4 with the origin at midships. The midship location of the berthed and passing ships coincided at approximately 175 seconds.

4.2 TIME DOMAIN INTERACTION FORCES AND MOMENTS

4.2 (a) Case A

The non-dimensional surge force, sway force and yaw moment on the berthed ship for Case A are presented in Figures 5a, 5b and 5c, respectively.

From Figure 5a it can be seen that the measured surge force was initially in the opposite direction to the passing ship direction (negative) followed by a positive surge force after the midship locations of the ships coincided until the end of the pass. Dand [2] measured interaction surge forces with similar general form. The idealised surge force trace deviated from the measured surge force trace in the early stages of the pass, as it predicted a positive surge force. The magnitude of the negative peak of the idealised surge force has been matched to the measured value. The magnitude of the positive peak at approximately 200 seconds was overestimated by the idealised curve. The measured surge force was positive in the latter stages of the pass; however the idealised trace predicted a negative surge force after approximately 250 seconds.

As can be seen from Figure 5b the measured sway force was towards the fenders as the passing ship approached, then as the passing ship was abeam the berthed ship the sway force was away from the fenders. The berthed ship experienced a sway force towards the fenders in the latter stages of the pass. The magnitude of the idealised positive sway force peak was matched to the measured value. The magnitude of the idealised first negative peak correlated well the measured value, whereas the magnitude of the second negative peak was underestimated by the idealised curve.

The yaw moment was generally bow away from the wharf as the passing ship approached, as illustrated in Figure 5c. Between approximately 120 seconds and 170 seconds the yaw moment was bow in to the wharf. The berthed ship experienced a yaw moment bow away from the wharf as the passing ship moved away, followed by a small bow in yaw moment. In general, the magnitude of the idealised peak yaw moments correlated well with the measured values, as illustrated by Figure 5c.

From Figures 5a, 5b and 5c it can be seen that, in general, the times at which the peak forces and moments occurred for the idealised traces correlated well with the measured traces.

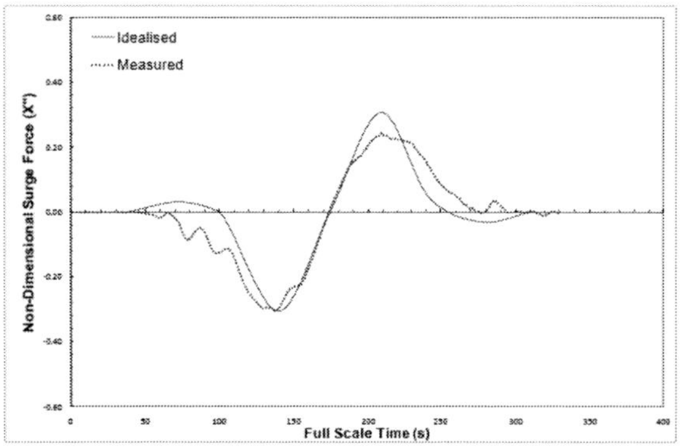

Figure 5a: Non-dimensional interaction surge force on the berthed ship for Case A

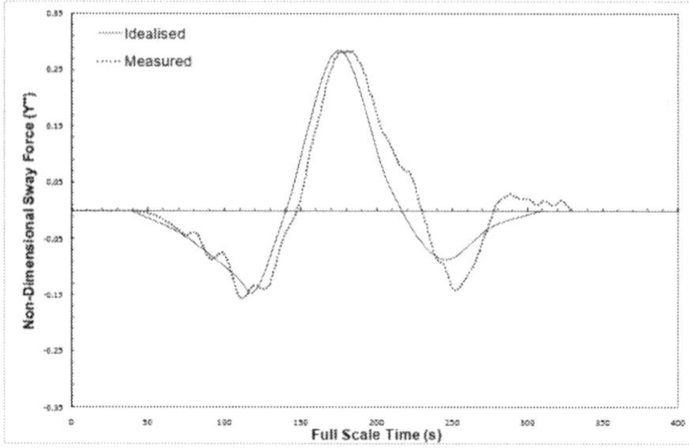

Figure 5b: Non-dimensional interaction sway force on the berthed ship for Case A

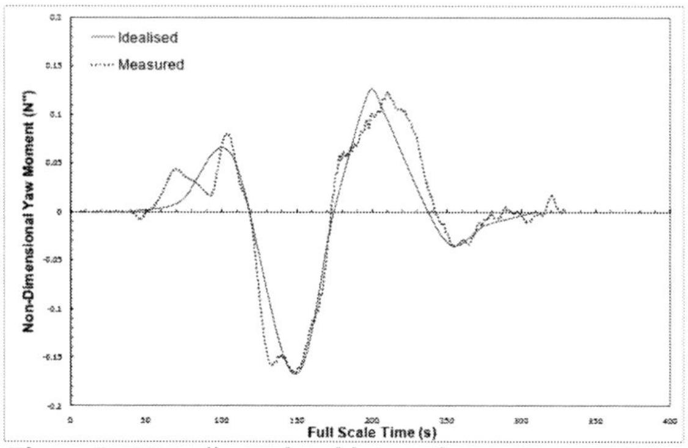

Figure 5c: Non-dimensional interaction yaw moment on the berthed ship for Case A

4.2 (b) Case B

The non-dimensional interaction surge force, sway force and yaw moment on the berthed ship for Case B are presented in Figures 6a, 6b and 6c, respectively.

The general form of the force and moment time domain traces is similar to those for Case A.

Comparing Figures 5a and 6a it can be seen that when the sloping bank was alongside the berthed ship (Case B), the magnitude of the peak surge force was greater than for the wider channel case (Case A). Comparing Figures 5b and 6b it can be seen that when the sloping bank was alongside the berthed ship, the magnitude of the peak sway force was less than that for the wider channel case. Pinkster and Ruijiter [8] found a similar trend when comparing forces for a vertical quay and open water.

From Figure 6a it can be seen that the idealised surge force trace deviated from the measured surge force trace in the early stages of the pass, as it predicted a positive surge force compared to the negative measured surge force. The magnitude of the idealised peak surge force at approximately 200 seconds overestimated the measured peak, as can be seen in Figure 6a. The magnitude of the negative sway force peaks was underestimated by the idealised force trace. Similarly, the magnitude of the positive yaw moment peaks was underestimated by the idealised trace, as illustrated in Figure 6c.

From Figures 6a, 6b and 6c it can be seen that, in general, the times at which the peak forces and moments occurred for the idealised traces correlated well with the measured traces.

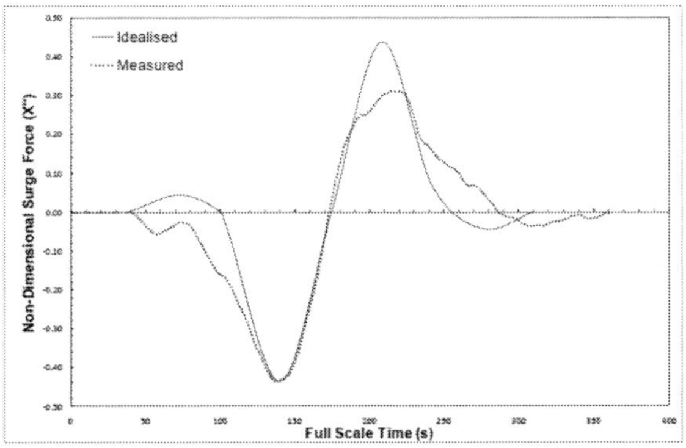

Figure 6a: Non-dimensional interaction surge force on the berthed ship for Case B

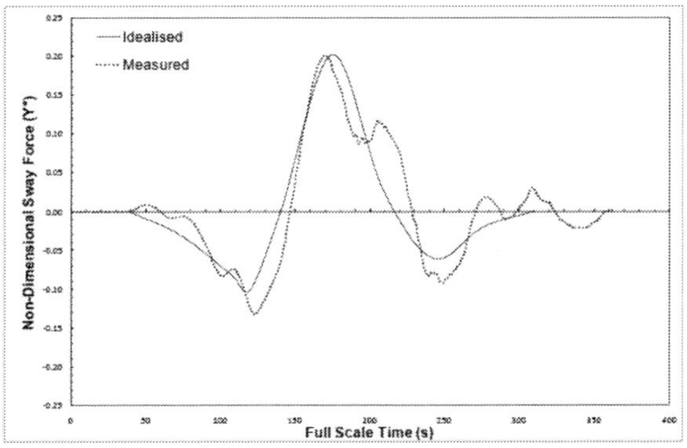

Figure 6b: Non-dimensional interaction sway force on the berthed ship for Case B

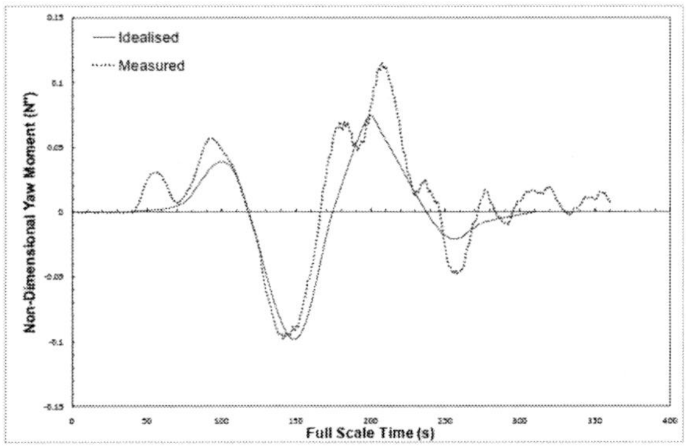

Figure 6c: Non-dimensional interaction yaw moment on the berthed ship for Case B

4.2 (c) Case C

The non-dimensional interaction surge force, sway force and yaw moment on the berthed ship for Case C are presented in Figures 7a, 7b and 7c, respectively.

The general form of the force and moment time domain traces is similar to those for Cases A and B.

On comparing Figures 5a, 6a and 7a it can be seen that the maximum interaction surge force was greatest for cases with a sloping near bank and narrow channel width. From Figures 5b, 6b and 7b it can be seen that the smallest maximum interaction sway force occurred for the case with a sloping near bank and narrow channel width.

From Figure 7a it can be seen that the idealised surge force trace deviated from the measured surge force trace in the early stages of the pass, as it predicted a positive surge force compared to the negative measured surge force. The magnitude of the idealised peak surge force at approximately 200 seconds overestimated the measured peak, as can be seen in Figure 7a. The magnitude of the positive yaw moment peaks was slightly underestimated by the idealised trace, as can be seen in Figure 7c.

From Figures 7a, 7b and 7c it can be seen that, in general, the times at which the peak forces and moments occurred for the idealised traces correlated well with the measured traces.

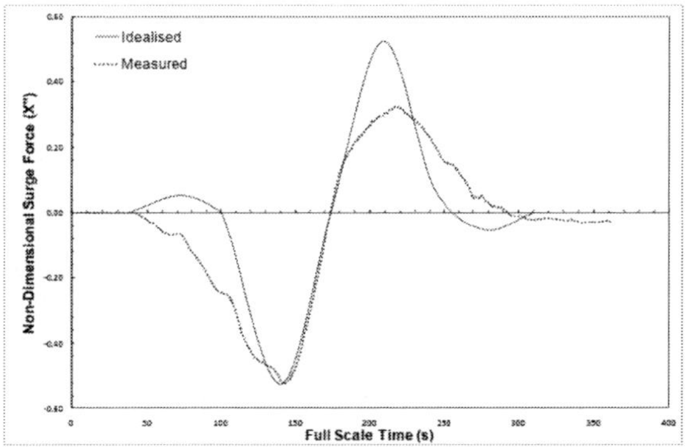

Figure 7a: Non-dimensional interaction surge force for Case C

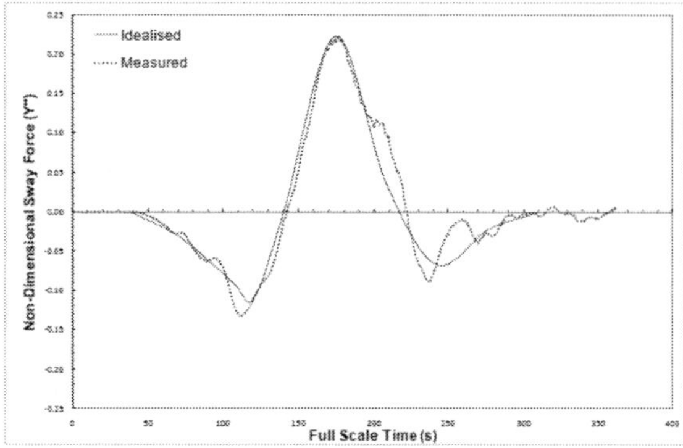

Figure 7b: Non-dimensional interaction sway force for Case C

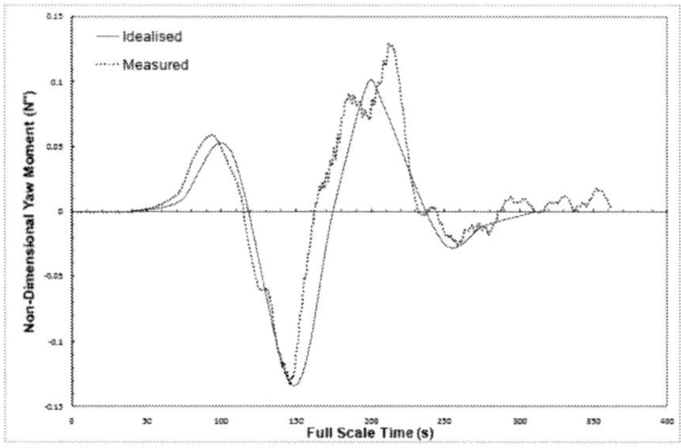

Figure 7c: Non-dimensional interaction yaw moment for Case C

4.3 PREDICTED TIME DOMAIN BERTHED SHIP MOTIONS

The measured and idealised time domain interaction forces and moments on the berthed ship were used as input to Termsim II to perform numerical simulations of the berthed ship behaviour. Predicted berthed ship surge, sway and yaw motions for Case A are presented in Figures 8a, 8b and 8c, respectively. The non zero motion at zero time represents the equilibrium position of the berthed ship. The midships of the passing and berthed ships coincided at 175 seconds.

As can be seen from Figure 8a the berthed ship was initially surged towards the passing ship on the approach, followed by a movement forward of the start position, followed by a movement aft of the start position in the latter stages of the pass.

The berthed ship was pushed into the fenders as the passing ship approached, moved away from the fenders as the passing ship was adjacent the berthed ship and was pushed in to the fenders in the latter stages of the pass, as illustrated in Figure 8b.

From Figure 8c it can be seen that the bow of the berthed ship was rotated away from the fenders as the passing ship approached, followed by the bow being rotated towards the fenders. The bow was then rotated away from the fenders, followed by a rotation in the opposite direction in the latter stages of the pass.

The general form of the predicted berthed ship motions were found to be similar for Cases B and C and hence are not presented in the present paper. Instead, the maximum berthed ship motions and mooring line loads for each case are presented in the next section.

From Figures 8a, 8b and 8c it can be seen that the general form of predicted motions was similar when using the measured force and moment input and the idealised force and moment input. However, there were some minor discrepancies in the magnitude of the motion peaks when using the different force and moment inputs, as follows:

- The second negative surge peak motion, the second negative sway peak motion and the second positive yaw motion were overestimated;
- The first negative sway peak was underestimated.

The above differences were only minor and may be within experimental uncertainty.

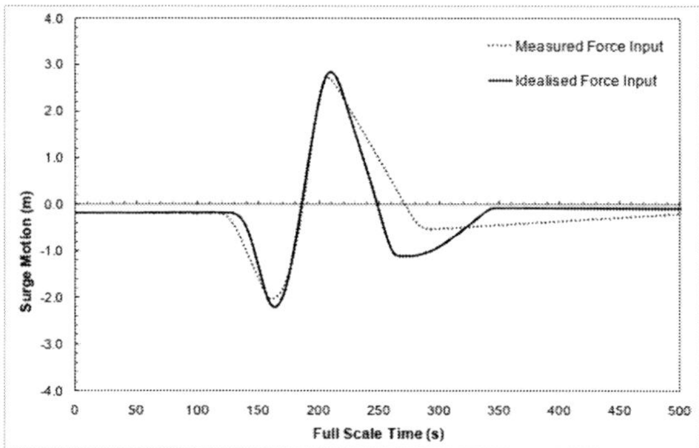

Figure 8a: Predicted berthed ship surge motion for Case A

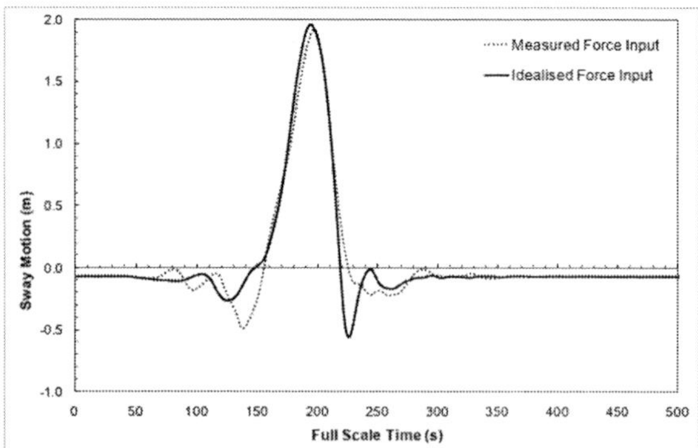

Figure 8b: Predicted berthed ship sway motion for Case A

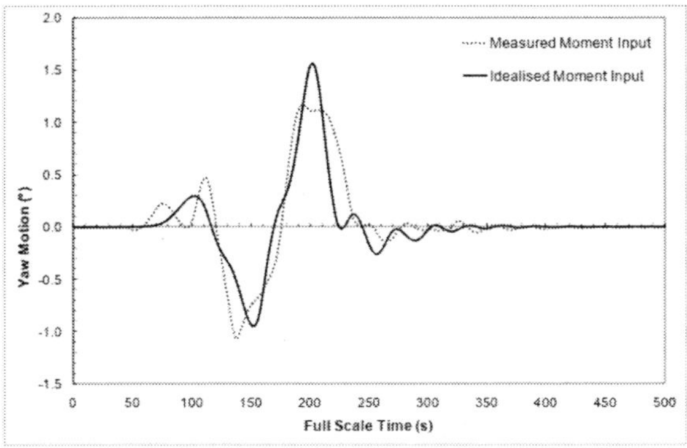

Figure 8c: Predicted berthed ship yaw motion for Case A

4.4 PREDICTED MAXIMUM BERTHED SHIP MOTIONS AND MOORING LINE LOADS

Predicted maximum berthed ship motions and maximum mooring line loads are presented for each case using the measured and idealised interaction force and moment traces. The magnitude of the predictions using the different force and moment inputs is assessed to determine the applicability of the idealised interaction force and moment traces for the cases tested.

4.4 (a) Maximum Berthed Ship Motions

The peak to peak surge and yaw motions and zero to peak sway motion for Cases A, B and C are presented in Figures 9, 10 and 11, respectively.

As expected, the surge motion was greatest for the cases with the sloping near bank and narrow channel. The largest sway and yaw motion was experienced for the wide channel case (Case A).

The predicted surge motion was marginally larger when using the idealised force and moment traces for all cases. This could be partly attributed to the steeper slope on the idealised surge force traces leading to the peak at approximately 150 seconds, as can be seen in Figures 5a, 6a and 7a.

The difference between the predicted berthed ship motions using the measured force and moment input and the idealised force and moment input for each case is given in Table 2. As can be seen, the maximum difference between the predictions using the idealised force and moment and the measured force and moment was 13% for the yaw motion in Case A. The maximum difference between the predicted motions for the idealised and the measured force and moment inputs was 12% for Case B and 10% for Case C. This indicates that, for the cases tested, there was only a marginal difference in the predicted maximum berthed ship motions when using the idealised force and moment trace that is based on open water data. The differences in the predicted values may be within experimental uncertainty.

From Table 2 it can be seen that, in general, the difference in berthed ship motions did not increase for the cases with narrow channel width, despite the idealised forces and moments being based on open water.

From the present analysis the maximum predicted berthed ship motions using the idealised forces and moments were similar to those predicted with the measured forces and moments. Hence, within experimental uncertainty, the idealised force and moment curves appear to be suitable to predict the maximum berthed ship motions for early design work for the cases tested, in the absence of measured data. Analysis of further cases with varying degrees of lateral channel restriction is being undertaken to assess the applicability of using the idealised interaction force and moment traces to predict the berthed ship motions.

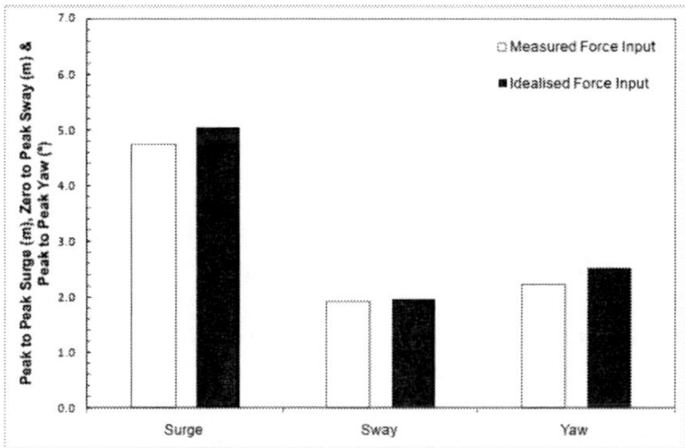

Figure 9: Predicted maximum berthed ship motions for Case A

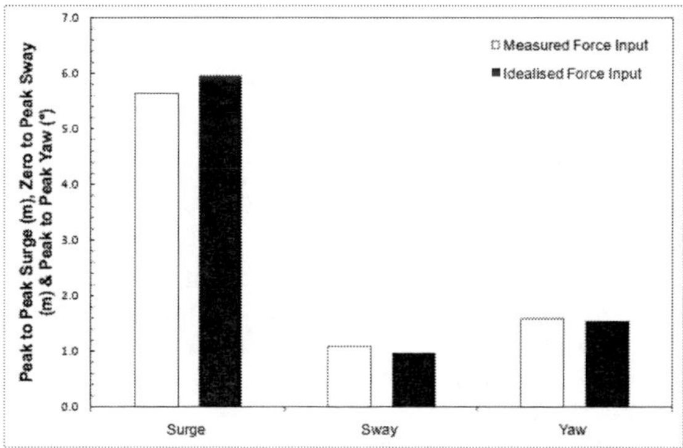

Figure 10: Predicted maximum berthed ship motions for Case B

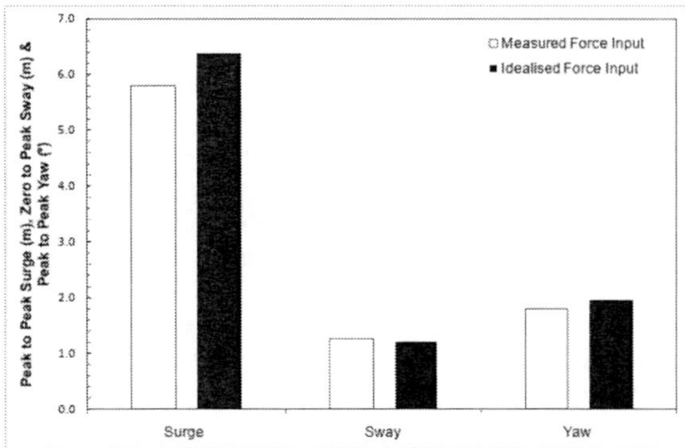

Figure 11: Predicted maximum berthed ship motions for Case C

Case	Surge	Sway	Yaw
A	6	2	13
B	5	12	3
C	10	4	9

Table 2 Percentage difference in berthed ship motions predicted with measured and idealised forces and moments

4.4 (b) Maximum Mooring Line Loads

The predicted maximum mooring line loads from the numerical simulations for each case are given in Figures 12, 13 and 14. It can be seen that the largest loads were experienced by the relatively short forward breast lines for each case. The largest load in the forward and aft breast lines occurred for Case A, where the large sway force was acting. The load in the spring lines was largest for Case C, due to the large surge force acting with the sloping near bank and the smallest channel width.

In general, the majority of the mooring line loads were slightly over predicted when using the idealised interaction force and moment traces.

The difference in the predicted maximum mooring line loads when using the idealised and the measured force and moment traces is shown in Table 3. For Case A (widest channel) the largest difference was 16% for the forward breast lines. The maximum difference in mooring line load for Case B was 24% on the aft spring lines and for Case C it was 19% on the forward spring lines. It can be seen that the difference between the maximum predicted mooring line loads was slightly higher than the difference in predicted maximum berthed ship motions. However, the differences may be within experimental uncertainty when measuring the interaction forces and moments.

From the present analysis the maximum predicted mooring line loads using the idealised forces and moments were generally similar to those predicted with the measured forces and moments. Hence, within experimental uncertainty, the idealised force and moment curves appear to be suitable to predict the maximum mooring line loads in the early stages of berth design for the cases tested, in the absence of measured data. Analysis of further cases is being undertaken to assess the applicability of using the idealised interaction force and moment traces to predict berthed ship motions and mooring loads for cases where there is varying degrees of lateral channel restriction.

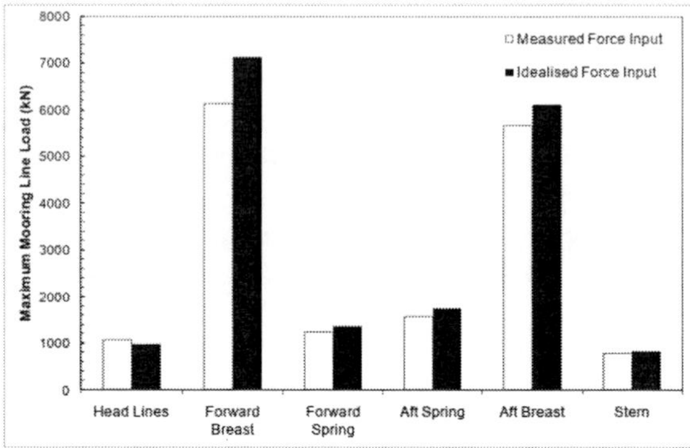

Figure 12: Predicted maximum mooring line loads for Case A

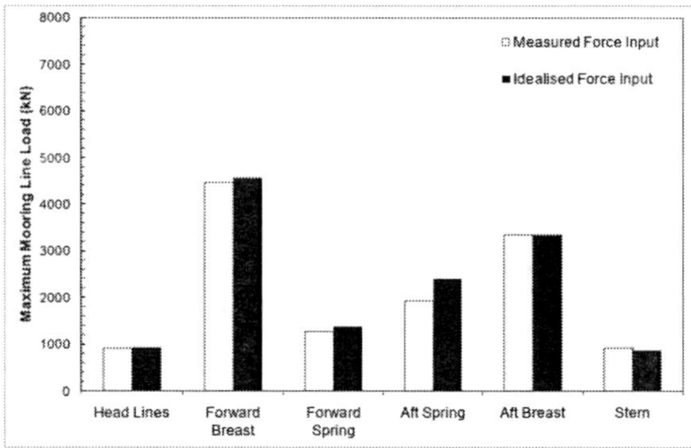

Figure 13: Predicted maximum mooring line loads for Case B

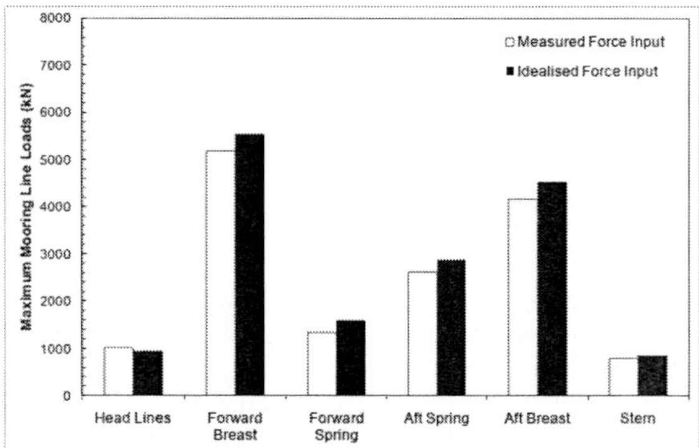

Figure 14: Predicted maximum mooring line loads for Case C

Case	Head	Forward Breast	Forward Spring	Aft Spring	Aft Breast	Stern
A	11	16	11	10	8	6
B	1	2	9	24	1	5
C	7	7	19	10	9	10

Table 3 Percentage difference in maximum mooring line load predicted with measured and idealised forces and moments

5. CONCLUDING REMARKS

An investigation has been undertaken into the interaction forces and moments experienced by berthed ships due to passing ships. The influence of waterway geometry on the magnitude of the peaks and the form of the time domain interaction forces and moments has been quantified for a selection of test cases. Berthed ship motions and mooring line loads have been predicted for cases with varying lateral channel restriction using both interaction force and moment data from physical model experiments and idealised forces and moments based on open water data. The influence of the form of the interaction forces and moments on maximum berthed ship motions and maximum mooring line loads has been quantified for the test cases.

From the results it was found that:

- The general form of the interaction forces and moments on the berthed ship were similar for the test cases with varying degrees of lateral channel restriction;
- The magnitude of the peak interaction surge force was largest for the narrow channel configurations;
- The greatest peak interaction sway force was experienced with the widest channel configurations;
- The maximum berthed ship surge motion was marginally overestimated for each test case when using the idealised force and moment traces;
- The maximum mooring line loads were marginally overestimated for the majority of the lines for each test case when using the idealised force and moment traces.

From the present analysis the maximum berthed ship motions and maximum mooring line loads predicted using the idealised forces and moments were generally similar to those predicted with the measured forces and moments. Hence, within experimental uncertainty the idealised force and moment curves appear to be suitable to predict the maximum berthed ship motions and maximum mooring line loads in the early stages of berth design for the cases tested, in the absence of measured data.

Further analysis is being undertaken to assess the applicability of the idealised interaction force and moment trace for different channel geometries.

6. REFERENCES

1. REMERY, G. F. M., 'Mooring Forces Induced by Passing Ships', OTC 2066, *Proceedings of 1974 Offshore Technology Conference,* Richardson, Texas, 1974.

2. DAND, I.W., 'Some Measurements of Interaction between Ship Models Passing on Parallel Courses, *National Maritime Institute*, Report NMI R108, 1981.

3. SIMMONS, T., ISAACS, L.T., RENILSON, M.R., 'Bank Effects on Vessel Interaction Forces', *Proceedings of the 11th Australasian Fluid Mechanics Conference*, University of Tasmania, Hobart, Australia, 14-18December, 1992.

4. DUFFY, J.T., WEBB, G., 'Berthed Ship-Passing Ship Interaction: A Case Study for the Port of Newcastle', *Coasts and Ports 2003,* Auckland, New Zealand, 9-12 September, 2003.

5. MUGA, B.J. & FANG, S.T., 'Passing Ship Effects from Theory and Experiment", OTC 2368, *Proceedings of 1975 Offshore Technology Conference*, Offshore Technology Conference, Richardson, Texas, 1975.

6. SPENCER, J.M.A., McBRIDE, M.W., BERESFORD, P.J., GOLDBERG, D.G., 'Modelling the Effects of Passing Ships', *International Colloquium on Computer Application in Coastal and Offshore Engineering*, Kuala Lumpa, June 1993.

7. PINKSTER, J.A., NAAIJEN, P., 'Predicting the Effect of Passing Ships', *Proceedings of the 18th International Workshop on Water Waves and Floating Bodies*, Le Croisic, France, 2003.

8. PINKSTER, J.A., RUIJTER, M., 2004, 'The Influence of Passing Ships on Ships Moored in Restricted Waters', *Offshore Technology Conference*, Houston, Texas, U.S.A., 3 - 6 May 2004.

9. FLORY, J.F., 'The Effect of Passing Ships on Moored Ships', *Proceedings of Prevention First 2002 Symposium, California State Lands Commission*, Long Beach, California, 10-11September, 2002.

10. DAND, I.W., 'Simulation of the Behaviour of a Moored Ship when Passed by Other Ships, *National Maritime Institute*, Report NMI R111, 1981.

11. VAN DER MOLEN, W., LIGTERINGEN, H., VAN DER LEM, J.C., de WAAL, J.C.M., 'Behavior of a Moored LNG Ship in Swell Waves', *Journal of Waterway, Port, Coastal and Ocean Engineering*, January/February, 2003.

12. ROSEMAN, D.P. (ed.), 'The MarAd Systematic Series of Full Form Ship Models', *Society of Naval Architects and Marine Engineers Publications*, 1987.

7. AUTHORS' BIOGRAPHIES

Jonathan Duffy holds the current position of Research Engineer/Lecturer at the Australian Maritime College. He is responsible for undertaking commercial consultancy projects, conducting research and lecturing Naval Architecture subjects. His previous experience includes work in the field of ship hydrodynamics; including prediction of ship behaviour in shallow and restricted water and prediction of berthed ship motions and mooring loads due to passing ships, waves, wind and current.

Martin Renilson holds the current position of Chair of Hydrodynamics at the Australian Maritime College. He established the Ship Hydrodynamics Centre at the Australian Maritime College in the mid 1980s, was director of the Australian Maritime Engineering Cooperative Research Centre, and then Head of Department of Naval Architecture and Ocean Engineering. He also spent six years as Technical Manager for Hydrodynamics at DERA/QinetiQ in the UK. He has a significant research interest in ship manoeuvring, particularly in restricted waters.

SQUAT DURING SHIP-TO-SHIP INTERACTIONS IN SHALLOW WATER

K Eloot, Ghent University, Flanders Hydraulics Research, Belgium
M Vantorre, Ghent University, Belgium
J Verwilligen, Flanders Hydraulics Research, Belgium
H Prins, Rijkswaterstaat, Netherlands
T W F Hasselaar, Maritime Research Institute MARIN, Netherlands
M Mesuere, Shipping Assistance Division, Belgium

SUMMARY

Ships navigating in a channel for two way traffic will suffer hydrodynamic effects due to the presence of other shipping traffic. As squat is an important issue for ships navigating with limited under keel clearance the sinkage and trim induced by ship-to-ship interactions can be of high importance. In order to investigate the influence of two ships sailing on parallel courses at different speeds a comprehensive captive model test program was carried out. A large number of parameters including ship type, ship dimensions, draft, water depth, lateral distance and speed were studied.

Apart from model tests two other methods were used to study squat induced by ship-to-ship interactions. The first method implies a RANSE-calculation of two container ships meeting in a channel with a lateral distance equal to a ship beam. The second method concerns a RTK measurement of ultra large container ships on the Western Scheldt meeting each other.

NOMENCLATURE

ξ	Non-dimensional relative longitudinal ships' position (-)
AP	Aft Perpendicular (-)
B	Beam (m)
B_O	Beam of own ship (m)
B_T	Beam of target ship (m)
C_B	Block coefficient (-)
FP	Fore Perpendicular (-)
DoF	Degrees of Freedom (-)
h	Water depth (m)
L_{OA}	Length over all (m)
L, L_{PP}	Length between perpendiculars (m)
L_O	Length of own ship (m)
L_T	Length of target ship (m)
n	Propeller revolutions (rps)
SOG	Speed Over Ground (kn)
STW	Speed Through Water (kn)
STS	Ship to Ship (-)
T	Draft (m)
T_O	Draft of own ship (m)
T_{ref}	Reference draft for STS (m)
T_T	Draft of target ship (m)
y_{BB}	Clearance between ships (m)
y'_{BB}	Non-dimensional distance between ships' boards: $y'_{BB} = \dfrac{y_{BB}}{\min(B_O; B_T)}$ (-)
z_{VF}, z_{VA}	Dynamic sinkage due to ship motion at FP and AP (m)

1. INTRODUCTION

Squat is an important issue for ships navigating with limited under keel clearance. In channels for two-way traffic, a ship will not only be affected by hydrodynamic effects due to the vicinity of the bottom and the banks of the waterway, but also by the presence of other shipping traffic. Meeting or overtaking manoeuvres may, among other effects, result in an additional sinkage and trim which temporarily reduce the net under keel clearance and, hence, increase the probability of bottom touch. Squat effects being very sensitive to the ship's speed, they can be diminished by speed reduction, but such a measure may lead to a significant decrease of the operational capacity of the waterway.

Figure 1: Passing of two Ultra Large Container Ships on the Western Scheldt

All over the world waterways authorities are challenged by the scale enlargement in the shipping world to adapt existing navigation channels to present and future requirement. As a famous example, the expansion program for the Panama Canal can be mentioned [1]: the Third Set of Locks will allow access of larger vessels to the Canal, which will result in a completely different mix of ships that may interact with each other in the confined waters of the Gaillard Cut and the Gatún Lake. The Western Scheldt (Figure 1), the access channel to the Port of Antwerp, is a second example: recently ship length restrictions have been released drastically after model test based simulation studies [2], while extensive dredging works have led to an increase of the allowable draft.

In the scope of these challenges, the present paper intends to provide more background on squat during ship-to-ship interactions by combining different research methods: i.e. model tests, RANSE-calculations and full scale measurements.

Squat is the sinkage and trim of a ship while sailing at a non-zero speed. Due to the hydrodynamic pressure (and water level) variations around the moving hull the static draft of the ship at fore and aft perpendicular, resulting from the ship's loading condition and referred to an undisturbed waterline, will be increased with dynamic sinkages z_{VF} and z_{VA} (see Figure 2). These sinkages are influenced by many parameters as summarised in [3] whereas ship geometry and type, ship's speed and water depth to draft ratio will be discussed in this paper in relation to other parameters characterising a ship-to-ship interaction:

- Type of manoeuvre: meeting, overtaking or passing a moored ship
- Passing distance
- Relative longitudinal position of the ships

Figure 2: Definition of ship squat as combination of sinkage and trim

The passing of moored ships can be considered as a special case where the speed of the moored ship is zero but during passing dynamic sinkages will occur. Another special case is the sinkage involved during lightering operations where the relative speed between the ships is approximately zero.

2. MODEL TEST BASED MEASUREMENTS

In [4] a summary is given of the extensive model test program for ship-to-ship interaction tests executed in 2000 in the Towing Tank for Manoeuvres in Shallow Water, co-operation Flanders Hydraulics Research (FHR) and Ghent University (UGent). Ship models and test parameters are fully described while the measurements and modelled results focus on forces and yawing moment executed on the own ship while passing a target ship.

2.1 MODEL TEST INFORMATION

Four ship models have been tested with model characteristics as shown in Table 1. Ship C is a bulk carrier, ship E and H are tankers and ship D a container ship. Ship C, D and E have approximately same lengths but varying breadth while the dimensions of model H are overall smaller. The reference draft T_{ref} is used in the mathematical models described in [4] but more drafts are examined to investigate the influence of the loading condition. All tests were executed at the model self propulsion point. The clearance between both ship models during the tests varied between 0.25 and 2 times the smallest of both ship beams. The ships' speeds varied from 0 to 0.95 m/s on model scale. The under keel clearance was varied between 10 and 70% of draft.

Table 1: Ship model characteristics

Ship		C	D	E	H
L_{pp}	m	3.984	3.864	3.824	2.210
B	m	0.504	0.550	0.624	0.296
T_{ref}	m	0.180	0.180	0.207	0.178
C_B (at T_{ref})	-	0.843	0.588	0.816	0.830
T	m	0.155 – 0.200	0.155 – 0.200	0.136 – 0.256	0.125 – 0.178

For most waterways the interaction is not restricted to equally sized ships, as sea-going vessels of all tonnages and even inland ships are operating close to each other. The length/draft ratio for each ship is balanced so that taking into account the required minimum gross under keel clearance for the larger ship (e.g. 12.5% of the ship draft on the Western Scheldt) the sinkage of the smaller ship during a passing will be of lower importance thanks to the larger UKC.

2.2 SQUAT MEASUREMENTS FOR SHIPS WITH SIMILAR LENGTH

For ships of approximately equal lengths the relative increase of sinkage at aft and fore perpendicular is examined during meeting and overtaking manoeuvres. The relative sinkage is referred to the maximum sinkage (fore or aft) measured in the steady phase before or after the interaction took place which means a ξ value of respectively -3 or +3. ξ is a non-dimensional parameter representing the longitudinal distance between the own and target midships sections divided by the mean ship length $(L_O+L_T)/2$; the sign of ξ is selected in such a way that the value increases in time during any manoeuvre. Table 2 summarises the selected test series for this chapter.

Table 2: Ship to ship model parameters for ships with similar length

Code	Own	Target	T_O (m)	T_T (m)	h (m)
EII	E	D	0.207	0.180	0.248
EKI	E	D	0.207	0.180	0.228
ERI	E	C	0.207	0.180	0.248
ETI	E	C	0.207	0.180	0.228
DKI	D	E	0.180	0.207	0.228

2.2 (a) Meeting manoeuvre

In Figure 3 sinkages at AP and FP are shown for ship E for different non-dimensional clearances y'_{bb}, defined as the clearance between the ship boards divided by the minimal beam B_{min} ($min(B_O,B_T)$) of the passing ships.

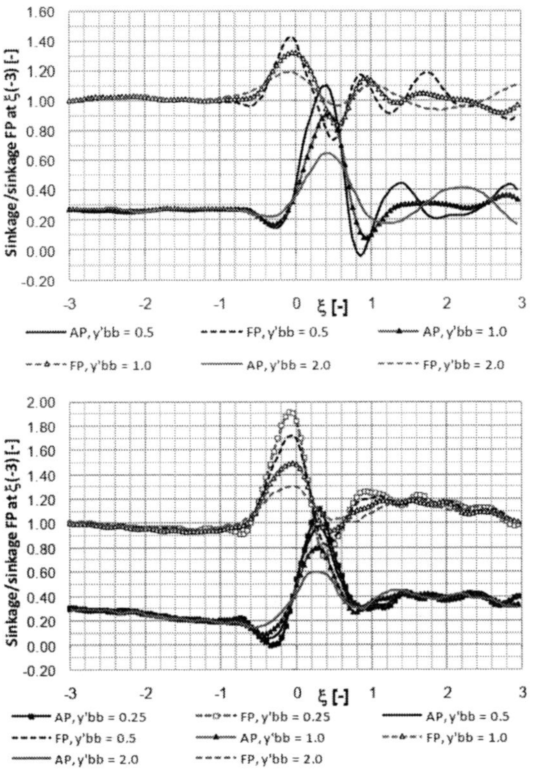

Figure 3: Relative sinkage for ship E due to meeting with ship D, code EII, both ships at 12 kn (upper) or 8 kn (lower figure), 20% UKC for ship E

For normal ship passages the distance y'_{bb} will be at least greater than 1 as summarised in [5] where passing distances were examined according to different sources such as concept design methods and real (simulated) passages. Nevertheless, most model tests have been executed with y'_{bb} values varying between 0.25 and 2.0.

For ship E at 20% UKC the sinkage at the bow and stern increases to a maximum value respectively before and after the midships sections are alongside each other (ξ = 0). The trim bow down for this tanker increases before the midships are meeting but becomes smaller or changes sign when the ships' sterns are coming alongside (ξ = 0.5). Despite some fluctuations in squat measurements after the meeting (especially at higher ship speeds) the same bow down trim is found.
Generally, for the risk of bottom touch the maximum values both at AP and FP are most important with an additional risk for damages at the propeller and the rudder at the AP.

With decreasing water depth the variation of sinkages at AP and FP during the meeting differ considerably. While at 20% UKC the maximum sinkage during the complete manoeuvre is still measured at the fore peak, in Figure 4 a maximum relative sinkage aft of 1.25 to 1.6 times the initial sinkage fore is found for passing distances of 1.0 to 0.25. The gross UKC of 10% of the ship's draft for ship E blocks the water flow below the keel so that the relative sinkage at the bow reduces during the meeting while the sinkage aft still increases considerably due to the improved flow evacuation caused by the propeller action.

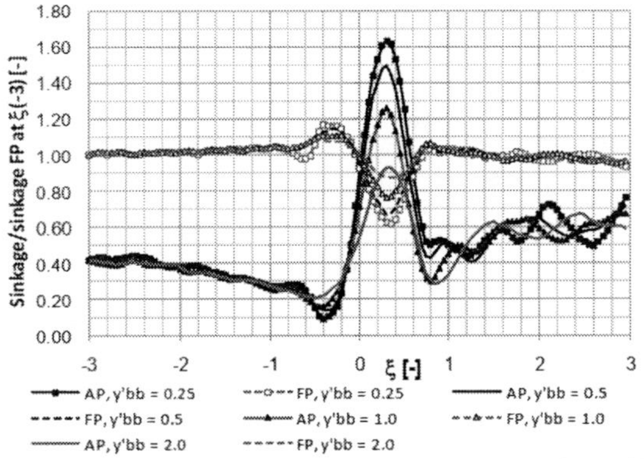

Figure 4: Relative sinkage for ship E due to meeting with ship D, code EKI, both ships at 8 kn, 10% UKC for ship E

Relative sinkages measured fore and aft on container carrier D for the same meeting manoeuvre are shown in Figure 5. The gross UKC is much larger and corresponds to 27% of the ship's draft. The sinkages roughly change in time as for ship E at 20% UKC although some differences could be recognised:

- A trim bow up is measured before the meeting, which is mostly typical for slender ships.
- The increase of sinkage at AP is much higher than at FP so that for a passing distance y'_{bb} between 1.0 and 0.25 the maximum sinkage is occurring at the stern and grows to respectively 3.5 to 4.6 times the sinkage aft found before the meeting.

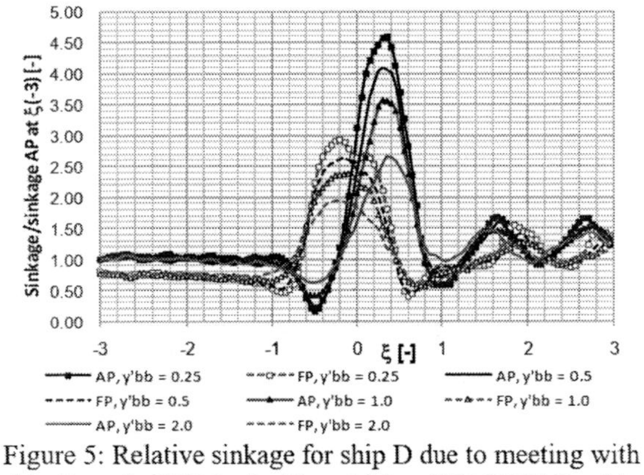

Figure 5: Relative sinkage for ship D due to meeting with ship E, code DKI, both ships at 8 kn, 27% UKC for ship D

Figures 3 to 5 represent squat measured during a meeting between a tanker and a container carrier. How the squat of tanker E is changing if two full ships are passing with target ship C loaded at the same draft as ship D, is illustrated in Figure 6. The gross UKC is identical and equals 10% but the beam of ship C is smaller than the beam of ship D.

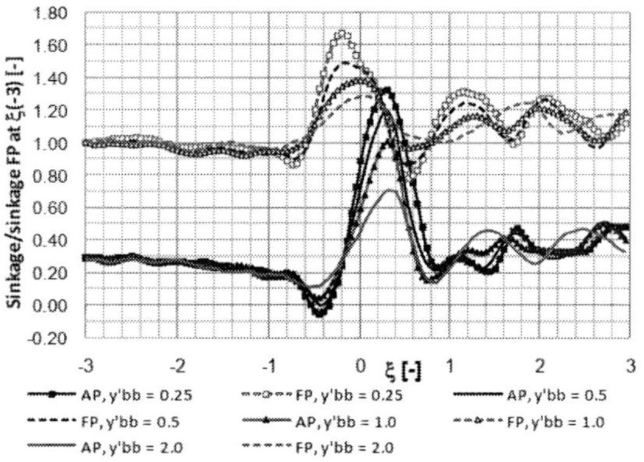

Figure 6: Relative sinkage for ship E due to meeting with ship C, code ETI, both ships at 8 kn, 10% UKC for ship E

The variation of sinkages fore and aft are probably in between the variations found in Figures 3 (lower) and 4 as the passing distances and the blocking of the water flow around the ships are smaller due to the smaller minimal beam of ship C. Additionally, the long parallel midship body of ship C compared to the slender container carrier D will probably induce an extended sinkage of the water level around ship C giving a clear de-phasing of the extrema of sinkages fore and aft of tanker E.

2.2 (b) Overtaking manoeuvre

In Figure 7 ship E is meeting with (upper) or overtaken by ship D (lower figure) giving relative sinkages fore and aft which are much higher during being overtaken than the values found during a meeting. Bow down trim before the ships interact, changes in bow up trim before the midships are alongside followed by a larger bow down trim in the second phase of the overtaking manoeuvre.

As forces and moment act much longer on both ships during an overtaking manoeuvre compared to a meeting and squat is also increasing considerably, in confined two-way traffic waterways the overtaking manoeuvre between two large ships sailing at higher speeds is often not admitted. At smaller UKC's (i.e. 10% UKC) bottom touch and thus damage at the ship model had to be prevented during the model tests by the mechanical limits so that model test results are scarce.

Ship E overtaken by bulk carrier C is affected by maximum sinkages which act during a longer period than the overtaking manoeuvre by container carrier D (Figure 8). Ship E overtaking other ships of approximately similar length is not included in the cases described in this paper but the passing of a moored ship can be evaluated as a special passing manoeuvre.

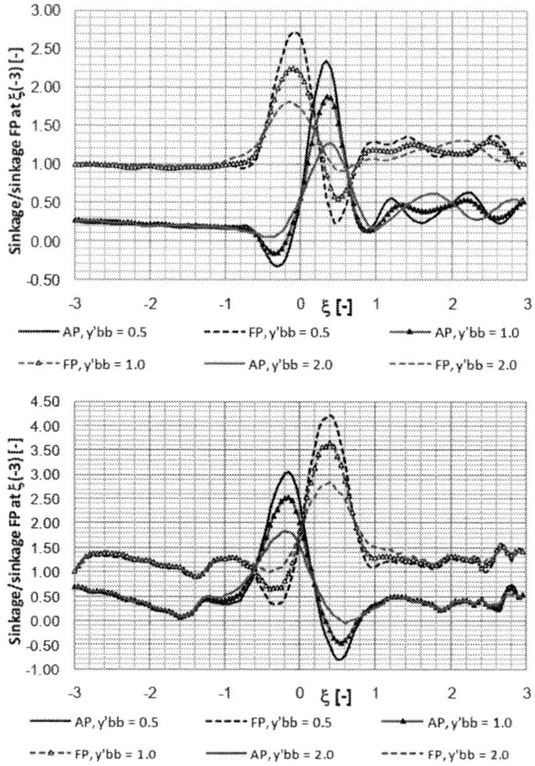

Figure 7: Relative sinkage for ship E due to a meeting with ship D (upper) or due to being overtaken by ship D (lower), code EII, ship E at 8 kn and ship D at 12 kn, 20% UKC for ship E

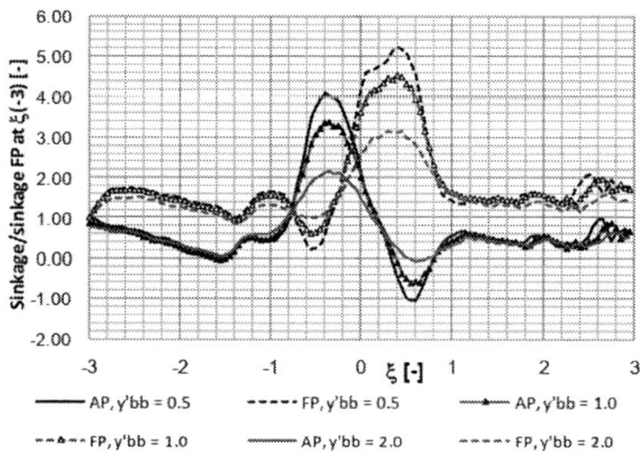

Figure 8: Relative sinkage for ship E due to being overtaken by ship C, code ERI, ship E at 8 kn and ship C at 12 kn, 20% UKC for ship E

2.2 (c) Passing of a moored ship

The effect of passing a moored ship is especially important for the forces in the mooring lines and the maximum sinkage of the moored ship. The effect on the mooring lines is not discussed in this paper and only an example is given for the completeness of the squat measurements due to ship-to-ship interaction in this chapter and 2.3 (c). In Figure 9 the relative sinkage aft does not change remarkably during the passing while the sinkage fore increases significantly once the clearance between the ships y'_{bb} becomes smaller than 1.

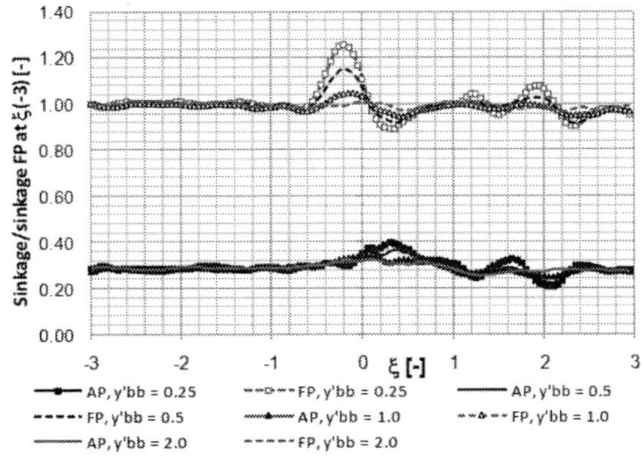

Figure 9: Relative sinkage for ship E due to passing moored ship D, code EII, ship E at 12 kn, 20% UKC for ship E

2.3 SQUAT MEASUREMENTS FOR SHIPS WITH SIGNIFICANTLY DIFFERENT LENGTH

The smaller tanker H has been tested with all other ship models with larger length. Taking into account the results of the previous chapter where sinkages become critical for an UKC below 20% of the ship's draft, the sinkages of the smaller tanker H will only be affected considerably if the draft of the interacting target ship is comparable and the water depth is restricted. The selected test series parameters are summarised in Table 3.

Table 3: Ship to ship model parameters for ships with significantly different length

Code	Own	Target	T_O (m)	T_T (m)	h (m)
HCI	H	C	0.178	0.180	0.198
HDI	H	D	0.178	0.180	0.248
HEI	H	D	0.178	0.180	0.228
DQI	D	H	0.180	0.125	0.228
DRI	D	H	0.180	0.125	0.198

2.3 (a) Meeting manoeuvre

Tanker H has a bow down trim at a gross UKC of 11% (Figure 10). Due to the meeting with bulk carrier C sinkages fore and aft are increasing to values which are almost 3 times the reference sinkage fore before the meeting. The maximum sinkages are measured just after the midships meet and give a net UKC of approximately half the gross UKC for the smallest lateral clearance. Increasing this clearance to at least one ship beam of the tanker H reduces the maximum sinkages to about one third of the gross UKC. In addition, the variation of the sinkages in time for ship H differ from the variation seen for ships of similar length.

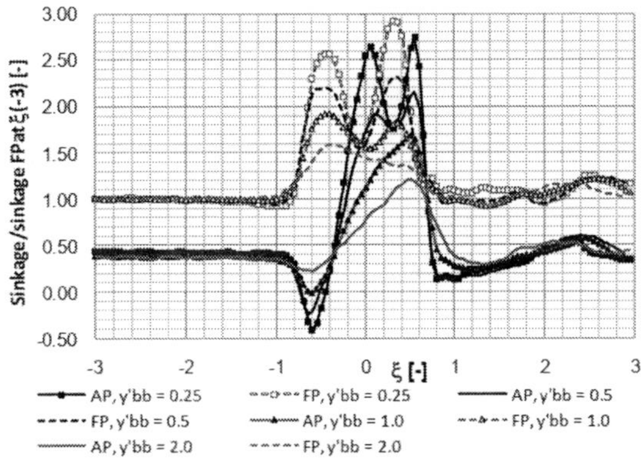

Figure 10: Relative sinkage for ship H due to meeting with ship C, code HCI, both ships at 8 kn, 11% UKC for ship H

2.3 (b) Overtaking manoeuvre

To discuss some tendencies measured during overtaken and overtaking, some figures will be generated for interacting ships with significantly different length.

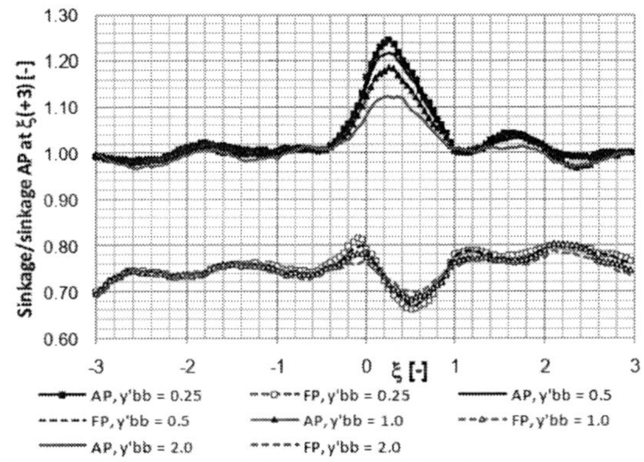

Figure 11: Relative sinkage for ship D due to overtaking ship H, code DQI, own D at 16 kn and target H at 6 kn, 27% UKC for ship D

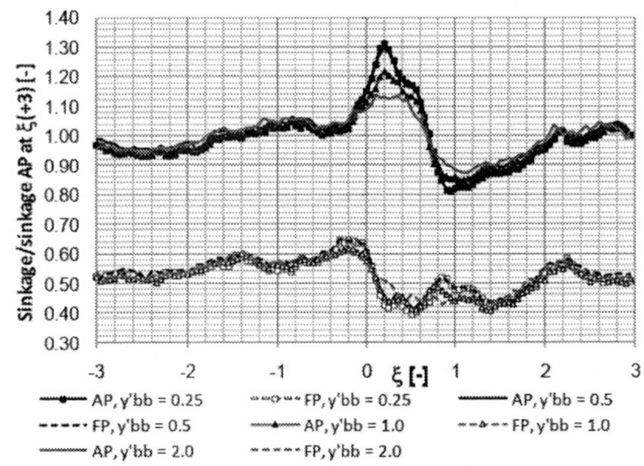

Figure 12: Relative sinkage for ship D due to overtaking ship H, code DRI, own D at 8 kn and target H at 4 kn, 10% UKC for ship D

In Figure 11 ship D is overtaking ship H at a speed of 16 knots in 27% UKC. The maximum sinkage aft (bow up trim) increases with 18 to 25% for clearances y'_{bb} of 1.0 to 0.25 while the trim is increasing as well.

Reducing the water depth and consequently the ship speed gives even higher relative sinkages with 20 to 30% increase at the AP as shown in Figure 12. Due to the very small UKC the measured sinkages fore and aft are rather unstable. The results for ship H will not be discussed as the UKC is much larger than for ship D thanks to the loading condition (see Table 3).

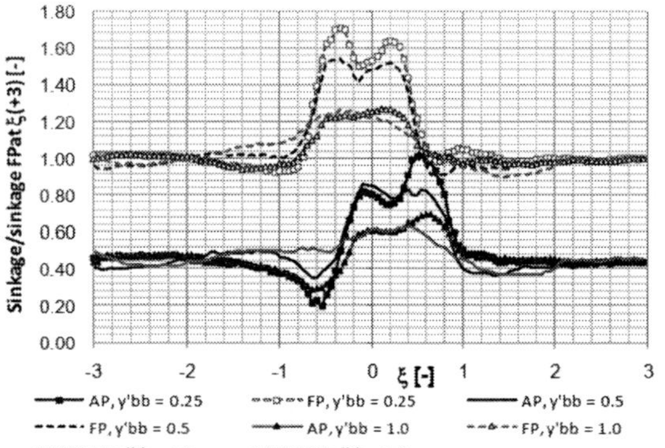

Figure 13: Relative sinkage for ship H due to overtaking ship C, code HCI, own H at 8 kn and target C at 4 kn, 11% UKC for ship H

In Figure 13 ship H is overtaking ship C in very shallow water so that the maximum sinkages remain smaller than the sinkages measured during a meeting manoeuvre where both ships are travelling at the same speed. The reduction of speed of target ship C helps in limiting the additional sinkages due to ship-to-ship interaction.

2.3 (c) Passing of a moored ship

The effect on sinkage for a smaller tanker H passing a moored larger ship is estimated to be of minor importance taking into account the small effect seen on Figure 9 for ships of similar length. Therefore, although it does not belong to the scope of this paper the absolute sinkages are shown in Figure 14 for moored ship H passed by ship D at a reduced speed of 8 knots and varying clearances. Thanks to the gross UKC which is greater than 20% the absolute sinkages are negligible compared to the available water depth.

Generally, as the water depth at the berth is sometimes less than in the approach channel, bottom touch of moored ships should be taken into consideration if the UKC for the moored ship is restricted. Moreover, the vertical motions – which cannot be affected by the mooring lines – cause an additional loading for the mooring lines.

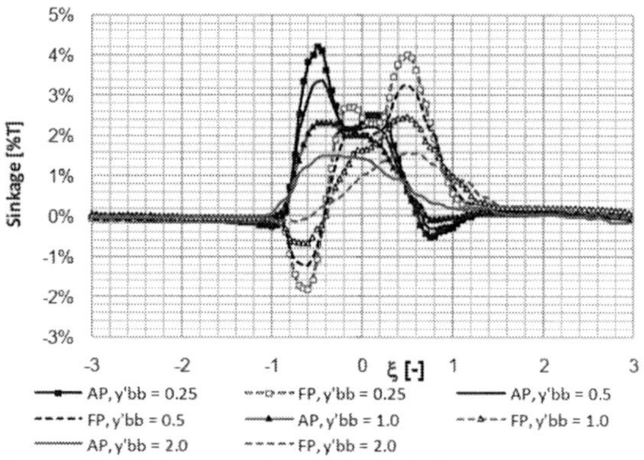

Figure 14: Sinkage as % of draft T for moored ship H due to passing of ship D, code HEI, own H at 0 kn and target D at 8 kn, 28% UKC for ship H

2.4 SQUAT MEASUREMENTS DURING LIGHTERING

The squat measured during lightering operations can be considered as a special case of the aforementioned overtaking/overtaken manoeuvre. The speed ratio of interacting ships is approximately one as ship's speeds are almost identical. As illustration sinkages fore and aft measured on the service ship during multi-modal tests with constant forward speed and varying lateral distance (and speed) are shown in Figure 15. The midship sections are alongside each other giving $\xi = 0$. The lateral distance between the ship boards y_{bb} is varying at full scale between 100 m - 4 m – 100 m during the test or y'_{bb} values between 2.4 and 0.1.

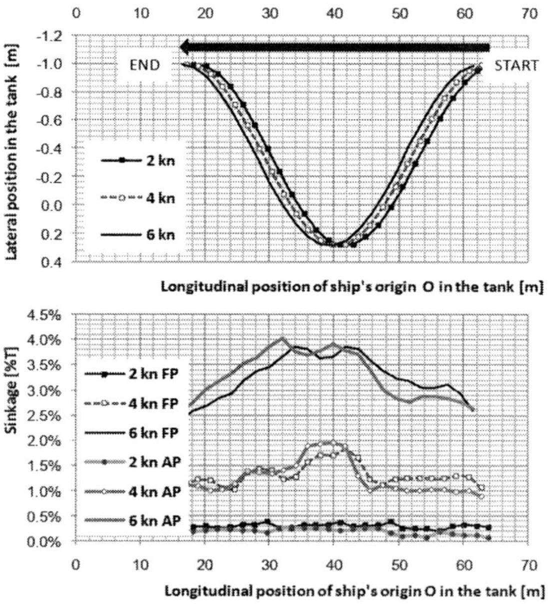

Figure 15: Sinkage as % of draft T (7.5 m) for the Aframax service ship during lightering at 130% UKC

3. RANSE-CALCULATIONS

In order to evaluate the accuracy of RANSE calculations of ship-to-ship interactions, FHR ordered the calculation of a head-on encounter of two container vessels to Cd-Adapco. This company with major CFD expertise performed the dynamic calculation using the commercial code STAR-CCM+ V4.05. The calculated case applied to a head-on encounter of two identical models of 8400 TEU container vessels meeting each other at different speeds (Table 4).

The unsteady calculation was performed with a volume of fluid model applied to an unstructured hexahedral mesh with 1.7 million cells. The calculation supplied horizontal forces, yawing moment and vertical motions for Ship 1 (4 DoF fixed and 2 DoF released).
A water current speed equal to the ship speed of Ship 1 was specified. Ship 2 was added to the environment as a moving body with fixed longitudinal speed.

Table 4: Conditions for RANSE calculation of a head-on encounter

	Ship 1	Ship 2
Capacity Full scale	8400 TEU	8400 TEU
Scale (-)	1/80.8	1/80.8
L_{OA} (m)	4.36	4.36
B (m)	0.53	0.53
T (m)	0.18	0.18
rudder	Fixed 0°	Fixed 0°
propeller	Actuator disk	Actuator disk
n (1/s)	6.7	10.0
SOG (kn)	0.89	1.34
STW (kn)	0.89	1.34
h (m)	0.228	
y_{bb} (m)	0.5	

The sinkages at fore and aft resulting from the RANSE calculations were compared to those obtained from a towing test at comparable conditions (Figure 16). Although the RANSE calculations seem to underestimate the sinkage at the fore peak before and after the meeting, during the meeting the results are in good agreement with those measured in the towing tank. Furthermore the results from the towing test show a very symmetrical evolution of sinkage fore and sinkage aft during the meeting while RANSE calculations predict a higher sinkage at the aft.

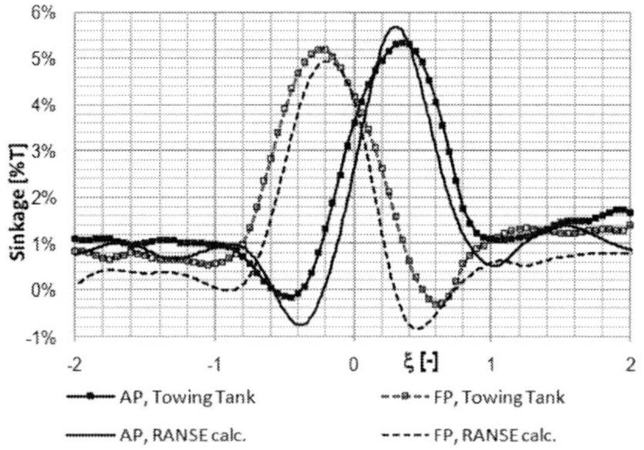

Figure 16: Sinkage as a percentage of the draft obtained from RANSE calculation and towing test for a meeting of two scale models of an ULCS

4. FULL SCALE MEASUREMENTS

Since April 2009 ultra large container ships (ULCS) with length over 360 m are allowed to sail to the port of Antwerp. In order to validate the accessibility studies that are summarised in [2], the ship behaviour during the voyage with the MSC Daniela (366 m x 51.3 m) on May 27th 2009 was monitored by Maritime Research Institute Netherlands (MARIN). The ship behaviour in terms of sinkage was measured using a RTK GPS on the bridge with cm accuracy (positioned 147 m from the fore peak). The pitch and roll angle were determined using a PHINS fiber-optic gyroscope and motion sensor. During her voyage on the Western Scheldt to Antwerp the ship performed a meeting with an outbound sailing container vessel with a length of 350 m (Figure. 17). From the RTK measurements performed by MARIN, it could be observed that this meeting had a major effect on the sinkage and trim of MSC Daniela. The meeting between MSC Daniela and COSCO Guangzhou was performed at conditions listed in Table 5.

FHR and UGent developed a mathematical model for squat prediction of ULCS [3]. For ship to ship interactions the model calculates the maximum supplementary sinkage without taking into account the evolution of sinkage during the meeting.

In Figure 18 the registration of squat during the meeting of MSC Daniela and COSCO Guangzhou is compared to the results of the mathematical model mentioned above. Although the mathematical model predicts maximum sinkage at the aft peak at UKC 50%, the full scale measurements reveal maximum sinkage at the fore peak (bow down trim). Figure 18 illustrates that the predicted maximum sinkage increases 60% compared to the sinkage before the meeting ($\xi = -3$), while the maximum sinkage during the full scale measurement increases 80% during the meeting.

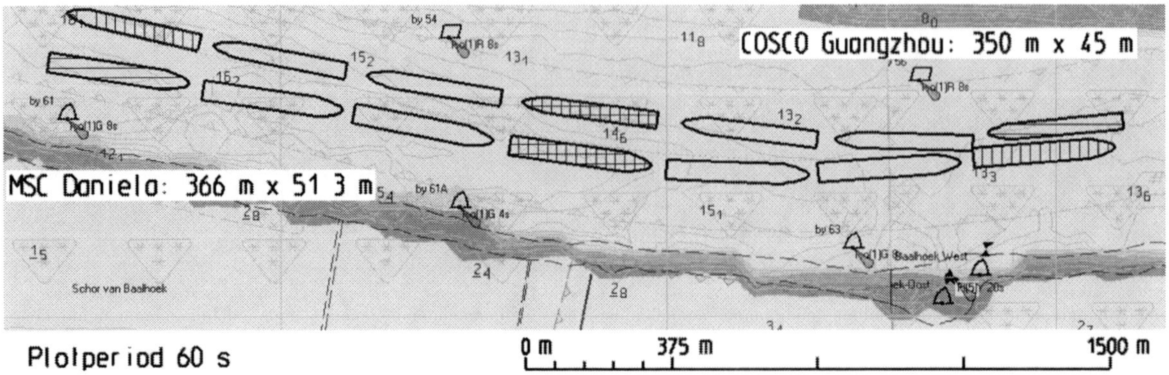

Figure 17: Meeting of two ULCS on the Western Scheldt

Table 5: Conditions during meeting of two ULCS on the Western Scheldt

	MSC Daniela	COSCO Guangzhou
Capacity	14000 TEU	9500 TEU
L_{OA} (m)	366	350
B (m)	51.3	45
T (m)	11	Not available
SOG (kn)	13.2	13.1
STW (kn)	11.6	14.6
Data source	MARIN, Trial and monitoring department	Beheer- en Exploitatieteam Schelderadarketen
h (m)	16.5	
y_{bb} (m)	70	

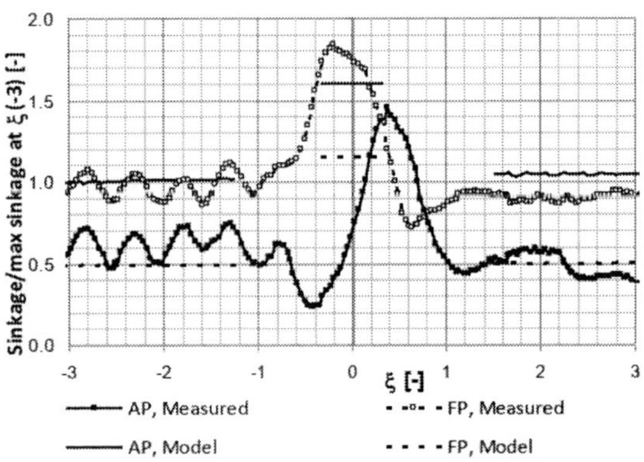

Figure 18: Meeting of two ULCS on the Western Scheldt compared to results of mathematical model

The meeting of the MSC Daniela and the COSCO Guangzhou was – except for the lateral distance y_{bb} - performed at similar conditions as the towing tank test presented in Figure 16. Comparison of both meetings reveals that the initial trim measured in the towing tank was bow up while the full scale measurement indicates an initial bow down trim. During the meeting the trim evolution in both towing tank and full scale measurement change from bow down to bow up. Both towing tank test and full scale measurements reveal a similar increase in sinkage during the meeting at the aft peak and the fore peak.

5. CONCLUSIONS

Due to the increasing ship sizes not only in the container traffic but also for many other ship types, port and waterway authorities are tackling the key problem of combining the demands for expansion with the demands of reducing the development costs. If two-way traffic has to be allowed not only the horizontal limits of a waterway must be considered but also the additional vertical space necessary for the increasing squat of passing ships.

This paper collects research on the influence of ship-to-ship interaction from three different methodologies: experimental and computational fluid dynamics and full scale measurements.

The influence of ships' passages on squat cannot be denied. Taken from the model test results the effect is of minor importance as long as the gross UKC for both ships is larger than 20% of the ship's draft. Once a waterway is accessible for ships with loading conditions resulting into smaller gross UKC the additional sinkages due to ship-to-ship interaction must be controlled by a revision of the operational conditions. Especially for restricted waterways in vertical ánd horizontal dimensions the combination of increased squat due to traffic and bank clearance may cause risks of bottom touch.

The maximum sinkages fore and aft are of utmost importance as the real time dependence of the sinkages is less important for freely sailing ships (contrary to moored ships). The type of manoeuvre, the ship speed, the clearance between the ships and the water depth to draft ratio have been modified during model tests to detect their influence. Many examples are summarised in this paper to illustrate them but validation based on other methodologies such as RANSE calculations and full scale measurements helps in translating these lab conditions to real life. CFD is still under development and considering Figure 16 the results are promising but must be extended to other conditions.

6. RECOMMENDATIONS AND FURTHER RESEARCH

The design of two-way navigation channels not only requires the determination of the horizontal dimensions, but attention has to be paid to the depth as well. The latter should be sufficient to account for the additional squat caused by ship-to-ship interaction during meeting manoeuvres. Although typically, width and depth design are, especially during a concept design stage, only weakly linked, horizontal and vertical channel dimensions cannot be selected independently from each other. This is particularly the case for the aspect highlighted in this paper: squat, including the additional sinkage due to interaction, clearly depends on the lateral dimensions, such as width, blockage, hull clearance, bank clearance. Underestimation of this effect may lead to an inefficient use of the channel, as speeds may have to be reduced during meeting manoeuvres. Further parametric study involving all relevant parameters is therefore recommended.

This paper considers squat during ship-to-ship interaction of sea-going ships. In inland waterways, clearances to both other ships and banks are often restricted to values less than one ship's beam inducing a superposition of these different sources to the maximum sinkages (Figure 19). Concept design guidelines for inland waterways often recommend gross UKC's of approximately 30% of the ship's draft which reduces the loading condition of the vessels considerably. Further research as described in this paper could help in defining the operational limits or the design criteria for vessel or waterway.

Figure 19: Meeting of two class Va (CEMT) inland vessels during simulation

7. ACKNOWLEDGEMENTS

The authors want to acknowledge all representatives of Flemish and Dutch authorities involved in any part of the nautical chain for their contribution in the validation of mathematical models through the execution of full-scale measurements.

8. REFERENCES

1. QUIJANO, J.L., 'The Panama Canal Expansion', *International Conference on Marine Simulation and Ship Maneuverability, Panama*, 2009.

2. ELOOT, K., VERWILLIGEN, J., VANTORRE, M., 'Synergy between theory and practice for Ultra Large Container ships', *32nd PIANC Congress, 125th anniversary PIANC, Liverpool, UK*, 2010.

3. ELOOT, K., VERWILLIGEN, J., VANTORRE M., 'An Overview of Squat Measurements for Container Ships in Restricted Water', *International Conference on Safety and Operations in Canals and Waterways, Glasgow, UK. pp. 106-116*, 2008.

4. VANTORRE, M. et al., 'Model Test Based Formulations of Ship-Ship Interaction Forces', *Ship Technology Research, Vol. 49*, 2002.

5. ELOOT, K., VERWILLIGEN, J., VANTORRE, M., 'Safety Assessment of Head on Encounters and Overtaking Manoeuvres with Container Carriers in Confined Channels through Simulation Tools', *International Conference on Marine Simulation and Ship Maneuverability, Panama*, 2009.

9. AUTHORS BIOGRAPHY

Katrien Eloot is co-ordinator of the nautical research group at Flanders Hydraulics Research and professor in ship hydrodynamics, specialism of ship manoeuvrability, at Ghent University, Belgium. This cooperation between FHR and UGent led in 2008 to the formation of the Knowledge Centre for Manoeuvring in Shallow and Confined water (www.shallowwater.be). Her experience includes modelling of ship behaviour based on model tests, fast- and real-time simulation and training through ship manoeuvring simulators.

Marc Vantorre, naval architect, is full senior professor of marine hydrodynamics and head of the Maritime Technology Division at Ghent University, Belgium. His research focuses on ship behaviour in shallow and confined waters, mainly in close co-operation with Flanders Hydraulics Research in Antwerp. He is member of PIANC Working Groups and former member of the ITTC Manoeuvring Committee.

Jeroen Verwilligen holds the current position of nautical researcher at Flanders Hydraulics Research. He is experienced with simulation studies and full-scale measurements on several vessels to Flemish Harbours. He is responsible for initiating CFD research at FHR.

Hank Prins became a Master Mariner in 1973 and obtained his Licence in International Public Law in 1995. He obtained his PhD in 2003 at the Department of Political and Social Sciences at Ghent University. At present, he is senior expert at Rijkswaterstaat Netherlands and he is responsible for decision making and implementation of nautical safety related projects for the Dutch authorities.

Thijs Hasselaar holds the current position of project manager at the trails and monitoring department at MARIN. He is responsible for ship performance monitoring projects and is specialised in ship efficiency and position measurements.

Martin Mesuere, Master Mariner, holds the current position of Head Nautical Management at Vessel Traffic Services VTS River Scheldt. He is responsible for tactical and strategically planning of shipping movements. His previous experience includes ship's master, pilot and head of Flemish coastal pilots.

SLACK TOWLINE ON TOW-TOWED SHIPS INTERACTION DURING MANEUVERING

A. Fitriadhy and H. Yasukawa, Hiroshima University, Japan
Y. Masaki, Mitsui Engineering and Shipbuilding Co., Ltd, Japan

SUMMARY

Concern for safety, phenomenon of a slack towline on tow-towed ships interaction during maneuvering is important subject to be studied. This paper is aimed at clarifying a mechanism of slack towline motion, which comes into derivation of slack towline criteria. Two components i.e., towline length and deflection-rudder angle were treated in numerical analysis to capture this phenomenon. Coupled maneuvering motion equations of tow-towed ships associated with the towline motion was derived. 2D lumped mass method was applied to express a dynamic towline motion. The results revealed that when a towline has slackened-off, tow ship will be vulnerable to mishandle or loose restriction of towing to towed ship. The large towline tension at some time after the slackening may introduce successive sea accidents such as breakage of the towline, which then followed by loss towing of a towed ship and often causes collision between tow ship and other vessel especially in the confined waters with high sea-traffic mode.

NOMENCLATURE

B_i	Breadth of ship i (m)
β_i	Drift angle of ship i (deg)
C_i	Course stability index of ship i
d_i	Draft of ship i (m)
δ_1	Deflection-rudder angle (deg)
G_i	Centre of gravity for ship i
γ, σ	Angle between towline and $x-$direction of ship i (deg)
F_N	Normal rudder force (tonne)
$F_x^{(i)}, F_y^{(i)}$	Surge and sway forces around ship i
F_{Tx}, F_{Tx}	Surge and sway forces due to towline tension
$I_z^{(i)}$	virtual moment of inertia coef. for ship i
J'_{zi}	Added moment of inertia coef. for ship i
k_{zzi}	Radius of gyration of ship i.
L_i	Length of ship i (m)
ℓ'	Ratio of towline length with respect to length of towed ship
ℓ'_B, ℓ'_T	Ratio of tow point distance with respect to each COG to length of towed ship
m'_{xi}, m'_{yi}	Added mass coef. of ship i for surge and sway motions
M_{Tz}	Yaw moment due to the towline tension
N	Sum of lumped mass for segmented towline
N'_{vi}, N'_{ri}	Yaw moment derivatives of ship i
ψ_i	Heading angle of ship i (deg)
r'_i	Yaw rate of ship i
S_L, S_d	Ratio of length and draft of tow ship to length and draft of towed ship, respectively
$\dot{\theta}_i$	Angular velocity of ship i
T_C	Tension of towline at tow point (tonne)
u_i, v_i	Advance and lateral velocity ship i (m/s)
U_i	Instantaneous speed of ship i
θ_i	Angle between $X-$axis and i-th segmented towline
$\dot{v}_i, \ddot{\psi}_i, \ddot{\theta}_i$	Acceleration of ship i
X'_{uui}	Resistance's coef. ship i
Y'_{vi}, Y'_{ri}	Sway force derivatives of ship i

1. INTRODUCTION

Improper use of towline length and set of deflection-rudder angle in turning motion of a ship towing may introduce a slack towline motion. This condition causes a tow ship will loose restriction of towing to towed ship. Once a towline has been slackened, a sudden increase of towline tension occurs and often cause the towline to eventually fracture, with resulting loss of a towed ship. This large tension also is undoubtedly dangerous since apart from damaging structure especially on tow points both at tow and towed ships. Furthermore, when tow and towed ships felt into irregular motions once a towline slackened-off, it may enable tow ship to collide with towed ship or another vessel especially in the confined waters with high sea traffic mode. Concern for safety, an extensive study of slack towline on tow-towed ships during maneuvering is therefore required.

In order to capture a slack towline, Shigehiro et al. carried out turning model tests of tow-towed ships [1]. Four conditions of ship towing were involved in this experiment; two conditions of the towed ship were selected i.e., stable and unstable towed ships in course stability and two conditions of towline length were employed i.e., shorter and longer towline than turning radius of the tow ship. In case of the towed ship was unstable and the towline was longer turning diameter of the tow ship, the results revealed that the towed ship stopped on the way of the motion while the tow ship continued to turn. Finally, the tow ship collided with the towed ship.

Free-running test can offer new insights towards enhancing understanding of the slack towline phenomenon during maneuvering motions. This has proven to be very costly and time-consuming. Considerable thought has been given to achieving fast and cheap, and this can flexibly adapt to various conditions where computer simulation has proved very useful to analyse the slack towline problems on tow-towed ships

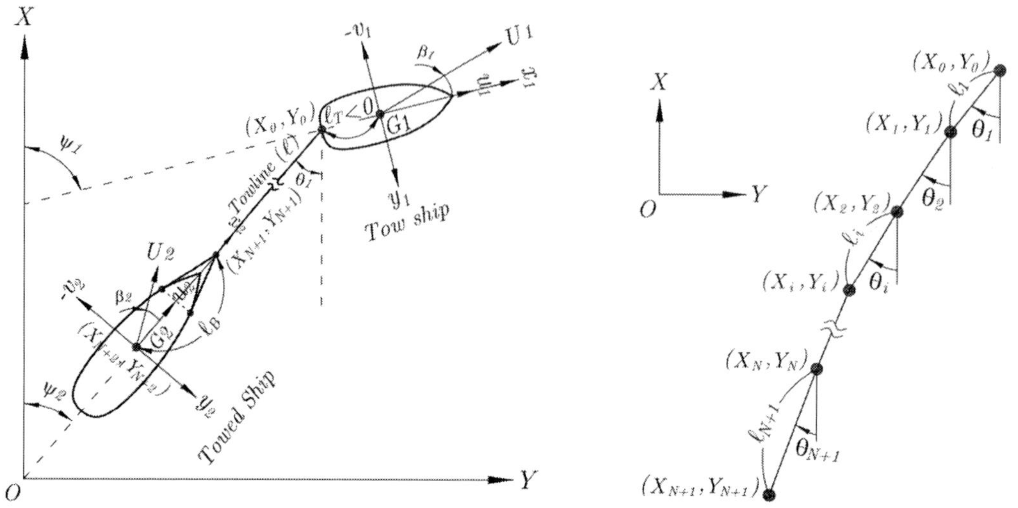

Figure 1: Coordinate systems of tow and towed ships (left) and lumped mass model for towline (right)

during maneuvering.

On the basis of recent studies, a numerical approach developed for maneuvering motion of the tow and towed ships, which involves phenomenon of slack towline, is so far still insufficient and often unreliable assessment. In this paper, authors developed a numerical tool, which is capable of predicting slack towline on the way of turning of a ship towing. The realistic model of a ship towing system, where a towed vessel coupled to a tow ship through towline is performed to achieve aforementioned objective. The 2D lumped mass method was then adopted for modeling the towline motion. Furthermore, the linearised motion equations was derived to gain knowledge in identifying certain circumstance on slack towline problem in which a boundary limit of turning ability of tow-towed ship was determined.

2. DEVELOPMENT OF MATHEMATICAL MODELING

The development of mathematical model describes the tow and towed ship motions incorporated with the towline motion in the calm water, which are expressed in three-degrees-of-freedom and defined in horizontal plane i.e., surge, sway and yaw motions.

2.1 COORDINATE SYSTEM

Three coordinate systems are used, Figure 1 (left). One set of axes is fixed to the earth coordinate system and denoted as $O - XY$, and two set of axes $G_1 - x_1y_1$ and $G_2 - x_2y_2$ are fixed relative to each ship's moving coordinate system aligned with its origin at centre of gravity. In the moving reference, x_i-axis points forward and y_i-axis to starboard. $i = 1$ and 2 designate as tow ship and towed ship, respectively. The heading angle ψ_i refers to the direction of the ship's local longitudinal axis x_i with respect to the fixed $X-$axis. The instantaneous speed of ship U_i can be decomposed into an advance velocity u_i and a transverse velocity v_i. The angle between U_i and the x_i-axis is the drift angle $\beta_i \equiv -\tan^{-1}(v_i/u_i)$.

The towline is represented as the sum of lumped mass of finite number (N). Each mass is connected by a segmented element over the entire truss element. The lumped mass particulars describe the towline characteristics, such as the mass, the density and the drag. The coordinates of i-th lumped mass is labeled by (X_i, Y_i), where $(i = 1, 2, 3, \ldots, N+1)$. The angle between $X-$axis and length of the i-th segmented towline ℓ_i is denoted as θ_i. Their connection points through a towline in the earth fixed coordinate systems are denoted as (X_0, Y_0) and (X_{N+1}, Y_{N+1}). With respect to each centre of gravity of the local ship coordinate systems, the distance of these tow points are defined as $(\ell_T, 0)$ and $(\ell_B, 0)$. The coordinate system (X_i, Y_i), which defines θ_i and ℓ_i can be expressed as follow:

$$X_i = X_0 - \sum_{j=1}^{i} \ell_j \cos\theta_j, \quad Y_i = Y_0 - \sum_{j=1}^{i} \ell_j \sin\theta_j \quad (1)$$

where $\theta_{N+2} = \psi_2$ and $\ell_{N+1} = \ell_B$.

2.2 MOTION EQUATIONS OF TOW-TOWED SHIPS AND TOWLINE

The motion equations of the towed ship are written in Eqs.(2) and (3) [2]. $M_x^{(2)} = (m_2 + m_{x2})$ and $M_y^{(2)} = (m_2 + m_{y2})$ represent the virtual mass components in x_2 and y_2, respectively. $I_z^{(2)} = (I_2 + J_2)$ is the virtual moment of inertia. These are expressed as sum of mass (moment of inertia) and added mass (added moment of inertia) components. $F_x^{(2)}$, $F_y^{(2)}$ and $M_z^{(2)}$ are the surge force, the sway force, and the yaw moment acting on the towed ship, respectively. The superscripts (1) and (2) in the notations mean the tow and the towed ship, respectively.

$$- \sum_{j=1}^{N+2} \ell_j \left(M_{x1} \sin\theta_j - M_{y1} \cos\theta_j\right) \ddot{\theta}_j$$

$$= \sum_{j=1}^{N+2} \ell_j \left(M_{x1}\cos\theta_j + M_{y1}\sin\theta_j\right)\dot{\theta}_j^2$$
$$+ T_{V1} + M_{x1}\ddot{X}_0 + M_{y1}\ddot{Y}_0 \quad (2)$$

$$I_z^{(2)}\ddot{\theta}_{N+2} + \sum_{j=1}^{N+2} \ell_j\ell_B \sin\phi \left(M_{y2}\cos\theta_j - M_{x2}\sin\theta_j\right)\ddot{\theta}_j$$
$$= \sum_{j=1}^{N+2} \ell_j\ell_B \sin\phi \left(M_{y2}\sin\theta_j + M_{x2}\cos\theta_j\right)\dot{\theta}_j^2$$
$$- \ell_B \sin\phi(T_{V2} - M_{x2}\ddot{X}_0 - M_{y2}\ddot{Y}_0) + M_z^{(2)} \quad (3)$$

where

$$\begin{aligned}
M_{x1} &= M_x^{(2)}\sin\phi\cos\psi_2 + M_y^{(2)}\cos\phi\sin\psi_2 \\
M_{y1} &= M_x^{(2)}\sin\phi\sin\psi_2 - M_y^{(2)}\cos\phi\cos\psi_2 \\
M_{x2} &= M_x^{(2)}\cos\phi\cos\psi_2 - M_y^{(2)}\sin\phi\sin\psi_2 \\
M_{y2} &= M_x^{(2)}\cos\phi\sin\psi_2 + M_y^{(2)}\sin\phi\cos\psi_2 \\
T_{V1} &= (M_x^{(2)}v_2\sin\phi + M_y^{(2)}u_2\cos\phi)\dot{\psi}_2 \\
&\quad -(F_x^{(2)} - M_y^{(2)}v_2\dot{\psi}_2)\sin\phi \\
&\quad +(F_y^{(2)} - M_x^{(2)}u_2\dot{\psi}_2)\cos\gamma \\
T_{V2} &= (-M_x^{(2)}v_2\cos\phi + M_y^{(2)}u_2\sin\phi)\dot{\psi}_2 \\
&\quad +(F_x^{(2)} + M_y^{(2)}v_2\dot{\psi}_2)\cos\phi \\
&\quad +(F_y^{(2)} - M_x^{(2)}u_2\dot{\psi}_2)\sin\phi \\
\phi &= \theta_{N+1} - \psi_2
\end{aligned}$$

The Lagrange's motion equations were used to describe the dynamic motion of the towline as derived in Eq.(4). m_i and k_{Fi} are the mass and the added mass coefficients of the i-th lumped masses, respectively.

$$\sum_{i=k}^{N}\left\{\sum_{j=1}^{i}\left(m_{si}\sin\theta_k\sin\theta_j + m_{ci}\cos\theta_k\cos\theta_j\right)\ell_k\ell_j\ddot{\theta}_j\right\}$$
$$+ \ \ell_k\sin\left(\theta_k - \theta_{N+1}\right)\sum_{j=1}^{N+2}\ell_j(M_{x2}\sin\theta_j - M_{y2}\cos\theta_j)\ddot{\theta}_j$$
$$= \ Q_{0k} - Q_{1k} - \ell_k T_{V3}\sin\left(\theta_k - \theta_{N+1}\right)$$
$$(k = 1, 2, 3, ..., N) \quad (4)$$

where

$$\begin{aligned}
m_{si} &= m_i\left(1 + k_{Fi}\sin^2\theta_i\right) \\
m_{ci} &= m_i\left(1 + k_{Fi}\cos^2\theta_i\right) \\
T_{V3} &= \sum_{j=1}^{N+2}\ell_j\left(M_{x2}\cos\theta_j + M_{y2}\sin\theta_j\right)\dot{\theta}_j^2 + M_{x2}\ddot{X}_0 \\
&\quad + M_{y2}\ddot{Y}_0 - T_{V1} - T_{V2} \\
Q_{0k} &= -\ell_k\sin\theta_k\sum_{i=k}^{N}\left(R_{Ci}\sin\theta_i - F_{Ci}\cos\theta_i\right) \\
&\quad - \ell_k\cos\theta_k\sum_{i=k}^{N}\left(R_{Ci}\cos\theta_i + F_{Ci}\sin\theta_i\right)
\end{aligned}$$

$$\begin{aligned}
Q_{1k} &= \ell_k\sin\theta_k\sum_{i=k}^{N}\left(\ddot{X}_0 + \sum_{j=1}^{i}\dot{\theta}_j^2\ell_j\cos\theta_j\right)m_{si} \\
&\quad - \ell_k\cos\theta_k\sum_{i=k}^{N}\left(\ddot{Y}_0 + \sum_{j=1}^{i}\dot{\theta}_j^2\ell_j\sin\theta_j\right)m_{ci} \\
&\quad + 2\ell_k\sum_{i=k}^{N}\left(\dot{X}_i\sin\theta_k + \dot{Y}_i\cos\theta_k\right)m_i k_{Fi}\dot{\theta}_i\sin\theta_i\cos\theta_i \\
&\quad - \left(\dot{X}_k^2 - \dot{Y}_k^2\right)m_k k_{Fk}\cos\theta_k\sin\theta_k
\end{aligned}$$

2.3 MOTION EQUATION OF TOW SHIP

The equation of the tow ship motion was derived adequately as follows:

$$\begin{aligned}
M_x^{(1)}\dot{u}_1 - M_y^{(1)}v_1\dot{\psi}_1 &= F_x^{(1)} + F_{Tx} \quad (5) \\
M_y^{(1)}\dot{v}_1 + M_x^{(1)}u_1\dot{\psi}_1 &= F_y^{(1)} + F_{Ty} \quad (6) \\
I_z^{(1)}\ddot{\psi}_1 &= M_z^{(1)} + M_{Tz} \quad (7)
\end{aligned}$$

$M_x^{(1)} = (m_1 + m_{x1})$ and $M_y^{(1)} = (m_1 + m_{y1})$ represent the virtual mass components in the direction x_1 and y_1, respectively, $I_z^{(1)} = (I_1 + J_1)$ the virtual moment of inertia. $F_x^{(1)}$, $F_y^{(1)}$ and $M_z^{(1)}$ denote surge force, sway force and yaw moment around the centre of gravity of the tow ship, respectively. F_{Tx}, F_{Ty} and M_{Tz} denote surge force, sway force, and yaw moment due to the towline tension acting at the connection point of the tow ship:

$$\begin{aligned}
F_{Tx} &= -T_X\cos\psi_1 - T_Y\sin\psi_1 \quad (8) \\
F_{Ty} &= T_X\sin\psi_1 - T_Y\cos\psi_1 \quad (9) \\
M_{Tz} &= \ell_T(T_X\sin\psi_1 - T_Y\cos\psi_1) \quad (10)
\end{aligned}$$

The towline tension components T_X and T_Y are expressed as, Fitriadhy and Yasukawa [3]:

$$T_X = \sum_{i=1}^{N} m_{si}\left(\sum_{j=1}^{i}\ell_j\sin\theta_j\ddot{\theta}_j\right) + \sum_{j=1}^{N+2}\ell_j\Big(M_{x2}\sin\theta_j$$
$$- \ M_{y2}\cos\theta_j\Big)\ddot{\theta}_j\cos\theta_{N+1} + T_X^{(1)} \quad (11)$$

$$T_Y = -\sum_{i=1}^{N} m_{ci}\left(\sum_{j=1}^{i}\ell_j\cos\theta_j\ddot{\theta}_j\right) + \sum_{j=1}^{N+2}\ell_j\Big(M_{x2}\sin\theta_j$$
$$- \ M_{y2}\cos\theta_j\Big)\ddot{\theta}_j\sin\theta_{N+1} + T_Y^{(1)} \quad (12)$$

where

$$\begin{aligned}
T_X^{(1)} &= \sum_{i=1}^{N}\Bigg[\left(\ddot{X}_0 + \sum_{j=1}^{i}\ell_j\cos\theta_j\dot{\theta}_j^{\,2}\right)m_{si} \\
&\quad + 2m_i\dot{X}_i k_{Fi}\sin\theta_i\cos\theta_i\dot{\theta}_i + R_{Ci}\sin\theta_i \\
&\quad - F_{Ci}\cos\theta_i\Bigg] + T_{V3}\cos\theta_{N+1} \\
T_Y^{(1)} &= -\sum_{i=1}^{N}\Bigg[\left(\ddot{Y}_0 + \sum_{j=1}^{i}\ell_j\sin\theta_j\dot{\theta}_j^{\,2}\right)m_{ci}
\end{aligned}$$

$$- 2m_i \dot{Y}_i k_{Fi} \sin\theta_i \cos\theta_i \dot{\theta}_i + K_{Ci}\cos\theta_i$$
$$+ F_{Ci}\sin\theta_i\Big] + T_{V3}\sin\theta_{N+1}$$

The resultant towline tension at the tow point of the tow ship can be expressed as $T_C = \sqrt{T_X^2 + T_Y^2}$.

3. A LINEAR THEORY ON STEADY TURNING OF TOW-TOWED SHIPS

Since the coupled motion equations of tow-towed ships associated with the dynamic towline motion introduces so complex problems with highly nonlinear motions, a linear approach has been applied to give insight into basic mechanisms describing the dynamic interactions between tow and towed ships during maneuvering.

3.1 LINEARISATION OF MOTION EQUATIONS

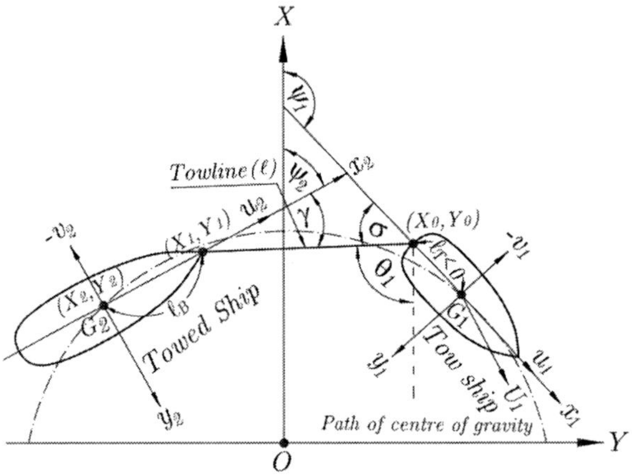

Figure 2: Steady turning coordinate system of tow and towed ships

In accordance with Figure 2, the linearized motion equations of the steady turning of the tow and the towed ships has been derived. In the linearisation, the towline is treated as non-extensible catenary model. The connection or towing points at tow and towed ships through towline are denoted as (X_0, Y_0) and (X_1, Y_1), respectively. Equation of the towline motion is then simplified from Eq.(1) as

$$X_i = X_0 - \sum_{j=1}^{i}\ell_j, \qquad Y_i = Y_0 - \sum_{j=1}^{i}\ell_j\theta_j \quad (13)$$

Refer to Eqs.(2),(3),(6) and (7), respectively, the linearised motion equations are expressed as follows:

$$(m_2 + m_{y2})\left(-\ell\ddot{\theta}_1 - \ell_B\ddot{\psi}_2 + \ell_T\ddot{\psi}_1 + \dot{v}_1\right) = -Y_{v2}\ell\dot{\theta}_1$$
$$+\left[Y_{r2} - (m_2 + m_{x2})U - Y_{v2}\ell_B\right]\dot{\psi}_2$$
$$+\left[Y_{v2}\ell_T - (m_2 + m_{y2})U\right]\dot{\psi}_1$$
$$-X_{uu2}U^2(\theta_1 - \psi_2) + Y_{v2}\left[U(\psi_1 - \psi_2) + v_1\right] \quad (14)$$

$$(I_2 + I_{z2})\ddot{\psi}_2 = -N_{v2}\ell\dot{\theta}_1 + (N_{r2} - N_{v2}\ell_B)\dot{\psi}_2 + N_{v2}\ell_T\dot{\psi}_1$$
$$-\ell_B X_{uu2}U^2(\theta_1 - \psi_2) + N_{v2}\left[U(\psi_1 - \psi_2) + v_1\right] \quad (15)$$

$$(m_1 + m_{y1})\dot{v}_1 = \left[Y_{r1} - (m_1 + m_{x1})U\right]\dot{\psi}_1 + Y_{v1}v_1$$
$$+X_{uu2}U^2(\theta_1 - \psi_1) + Y_{\delta 1}\delta_1 \quad (16)$$

$$(I_1 + I_{z1})\ddot{\psi}_1 = N_{r1}\dot{\psi}_1 + X_{uu2}\ell_T U^2(\theta_1 - \psi_1)$$
$$+N_{v1}v_1 + N_{\delta 1}\delta_1 \quad (17)$$

Once the equilibrium condition of the steady turning is achieved, the magnitude of towing angles (σ) and (γ) at the tow and towed ships are remain constant, where $\sigma = \psi_1 - \theta_1$ and $\gamma = \theta_1 - \psi_2$. At the same time, the angular velocities of tow and towed ship ($\dot{\theta}_1 = \dot{\theta}_2 = r$) are also constant and their acceleration equals to zero ($\dot{v}_1 = \ddot{\psi}_1 = \ddot{\theta}_1 = \ddot{\theta}_2 = 0$). Thus, Eqs.(14)~(17) are re-written as

$$0 = Y_{v1}v_1 + \left[Y_{r1} - (m_1 + m_{x1})U\right]r$$
$$-X_{uu2}U^2\sigma + Y_{\delta 1}\delta_1 \quad (18)$$

$$0 = N_{v1}v_1 + N_{r1}r - X_{uu2}\ell_T U^2\sigma + N_{\delta 1}\delta_1 \quad (19)$$

$$0 = Y_{v2}v_1 + \left[Y_{v2}(\ell_T - \ell_1 - \ell_B) + Y_{r2} - (m_2 + m_{x2})U\right]r$$
$$+U(Y_{v2} - X_{uu2}U_1)\gamma + Y_{v2}U\sigma \quad (20)$$

$$0 = N_{v2}v_1 + \left[N_{v2}(\ell_T - \ell_1 - \ell_B) + N_{r2}\right]r$$
$$+U\left(N_{v2} - \ell_B X_{uu2}\gamma\right) + N_{v2}U\sigma \quad (21)$$

Furthermore, the simplified motion equations were applied to obtain v_1 and r by eliminating the value of γ from Eqs.(20) and (21). From this, the value of σ is substituted into Eqs.(18) and (19), which yields to simple expression as

$$0 = (Y_{v1} + X_{uu2}U)v_1 + \left[Y_{r1} - (m_1 + m_{x1})U\right.$$
$$\left.X_{uu2}U(\ell_T - K_{\sigma 2})\right]r + Y_{\delta 1}\delta_1 \quad (22)$$

$$0 = (N_{v1} + X_{uu2}U\ell_T)v_1$$
$$+\left[N_{r1} + X_{uu2}U(\ell_T - K_{\sigma 2})\ell_T\right]r + N_{\delta 1}\delta_1 \quad (23)$$

where

$$K_{\sigma 2} = \ell + \ell_B + \frac{C_2}{X_{uu2}U(N_{v2} - \ell_B Y_{v2})} - \frac{N_{r2} - \ell_B\left[Y_{r2} - (m_2 + m_{x2})U\right]}{N_{v2} - \ell_B Y_{v2}} \quad (24)$$

Eq.(24) describes effect of the towline (ℓ) and towed ship on steady turning performance. Stability criterion of towed ship is noted as $C_2 \equiv N_{r2}Y_{v2} - N_{v2}\left[Y_{r2} - (m_2 + m_{x2})U\right]$. Solved using elimination method, v_1 and r can be obtained from Eqs.(22) and (23) as

$$v_1 = \delta_1 \frac{N_{\delta_1}\left[Y_{r1}^* - (m_1 + m_{x1})U - Y_{\delta_1}N_{r1}^*\right]}{C_1^*} \quad (25)$$

$$r = \delta_1 \frac{Y_{\delta_1}N_{v1}^* - N_{\delta_1}Y_{v1}^*}{C_1^*} \quad (26)$$

where

$$Y_{v1}^* = Y_{v1} + X_{uu2}U \quad (27)$$

$$Y^*_{r1} = Y_{r1} - X_{uu2}U\left(K_{\sigma 2} - \ell_T\right) \quad (28)$$
$$N^*_{v1} = N_{v1} + X_{uu2}U\ell_T \quad (29)$$
$$N^*_{r1} = N_{r1} - X_{uu2}U\left(K_{\sigma 2} - \ell_T\right)\ell_T \quad (30)$$
$$C^*_1 = N^*_{r1}Y^*_{v1} - N^*_{v1}\left[Y^*_{r1} - (m_1 + m_{x1})U\right] \quad (31)$$

Several towing parameter, which affected on the turning performance of tow ship can be described from Eqs.(27) to (31). The effect of towed ship and tow point ℓ_T are assigned on second parts of the right side Eqs.(27) to (30). The values of $K_{\sigma 2}$ and X_{uu2} associated with tow point ℓ_T will have significant influences on steady turning characteristics of tow ship. Similar to what obtain v_1 and r, the equation of σ and γ are obtained as

$$\gamma = -\frac{r}{U}\frac{C_2}{X_{uu2}U\left(N_{v2} - Y_{v2}\ell_B\right)} \quad (32)$$
$$\sigma = \frac{r}{U}K_{\sigma 2} \quad (33)$$

3.2 NON-DIMENSIONALISATION

Refer to Eqs.(25) and (26), these non-dimensional equations can be obtained through dividing added mass and hydrodynamic derivatives of tow and towed ships by L_i, d_i and U. The notations of L_i and d_i are the length and the draft of tow ship ($i = 1$) and towed ship ($i = 2$); U = is the tow speed. Here, ℓ' and ℓ'_B denote the ratios of towline length and tow point to length of the towed ship, respectively, ($\ell' = \ell/L_2$) and ($\ell'_B = \ell_B/L_2$). In addition, ℓ'_T denotes the ratio of tow point to length of the tow ship ($\ell'_T = \ell_T/L_1$). Through non-dimensionalising of v_1 ($v'_1 = v_1/U$) and r_1 ($r' = rL_1/U$), this leads to a relationship between r' and δ_1 as follows:

$$\frac{v'_1}{\delta_1} = \frac{N'_{\delta_1}\left(Y^{*\prime}_{r1} - m_1 - m_{x1}\right) - Y'_{\delta}N^{*\prime}_{r1}}{C^{*\prime}_1} \quad (34)$$
$$\frac{r'_1}{\delta_1} = \frac{Y'_{\delta_1}N^{*\prime}_{v1} - N'_{\delta_1}Y^{*\prime}_{v1}}{C^{*\prime}_1} \quad (35)$$

where

$$Y^{*\prime}_{v1} = Y'_{v1} + \frac{X'_{uu2}}{S_LS_d} \quad (36)$$
$$Y^{*\prime}_{r1} = Y'_{r1} - \frac{X'_{uu2}}{S_LS_d}\left(K'_{\sigma 2}/S_L - \ell'_T\right) \quad (37)$$
$$N^{*\prime}_{v1} = N'_{v1} + \frac{X'_{uu2}}{S_LS_d}\ell'_T \quad (38)$$
$$N^{*\prime}_{r1} = N'_{r1} - \frac{X'_{uu2}}{S_LS_d}\left(K'_{\sigma 2}/S_L - \ell'_T\right)\ell'_T \quad (39)$$
$$C^{*\prime}_1 = N^{*\prime}_{r1}Y^{*\prime}_{v1} - N^{*\prime}_{v1}\left(Y^{*\prime}_{r1} - m'_1 - m'_{x1}\right) \quad (40)$$

$$K'_{\sigma 2} = \ell' + \ell'_B + \frac{C'_2 - X'_{uu2}\left\{N'_{r2} - \ell'_B\left(Y'_{r2} - m'_2 - m'_{x2}\right)\right\}}{X'_{uu2}\left(N'_{v2} - \ell'_BY'_{v2}\right)} \quad (41)$$

where $C'_2 = N'_{r2}Y'_{v2} - N'_{v2}\left(Y'_{r2} - m'_2 - m'_{x2}\right)$. The notations of $S_L = L_1/L_2$ and $S_d = d_1/d_2$.

This non-dimensionalisation yields also to the values of γ and σ.

$$\gamma = -\frac{r'_1}{S_L}\frac{C'_2}{X'_{uu2}\left(N'_{v2} - \ell'_BY'_{v2}\right)} \quad (42)$$
$$\sigma = \frac{r'_1}{S_L}\left(K'_{\sigma 2} - \ell'_TS_L\right) - v'_1 \quad (43)$$

3.3 CRITERIA OF SLACK TOWLINE

The derivation of a representative equation to describe slack towline phenomenon becomes an important step to improve better insight of slack towline mechanism analytically before nonlinear numerical simulation is to be employed. In general, if tow and towed ships remain in steady turning condition, of course a slack towline does not occurs. However, the slack towline may happen due to improper use of towing parameters and introduce instabilities for ship towing system. Pertaining to operational safety of ship towing system during maneuvering, the authors propose a linearised equation to justify several variables of towing parameters that lead to slack towline condition. By substituting Eq.(41) into Eq.(43), we obtain as follows:

$$\frac{r'_1}{S_L}\left[\ell' + \ell'_B - \ell'_TS_L + \frac{C'_2 + X'_{uu2}\left\{\ell'_B\left(Y'_{r2} - m'_2 - m'_{x2}\right) - N'_{r2}\right\}}{X'_{uu2}\left(\ell'_BY'_{v2} - N'_{v2}\right)}\right] - v'_1 < \frac{\pi}{2} \quad (44)$$

Eq.(44) above is then called as slack towline's formula, where the value of $\sigma < \pi/2$ becomes a criterion as maneuverability requirement of tow-towed ships to prevent from slackening of towline. However, the condition of slack towline occurs if $\sigma > \pi/2$. According to Eq.(44), several considerable points of towing components that indicates a towline to be slackening, as follow:

- An increase of $\ell' + \ell'_B - \ell'_TS_L$ causes longitudinal separation distance between centre of gravities of tow and towed ships is getting far. Merely, employing the longer towline will have significant effect to bring the magnitude of σ exceeds $\pi/2$. It should be noted that when $\sigma > \pi/2$ is vulnerable to loose restriction to the barge. This is quite reasonable since the towline is not tight, while the tow ship continues to turn and possibly fall into irregular motion.
- When a yaw rate of tow ship (r'_1) raised, it enables the towline to be slackened-off and implies tow ship to have small diameter of turning.
- Shortening of tow ship's dimension (S_L) allows considerably the towline to be easier slackening than longer S_L.
- Increment of C'_2 indicates that towed ship motion tends to be stable. This implies the towline to suffer from slackening.

From a practical viewpoint as illustrated in Figure 2, in case of $\sigma = 90°$, it inherently will bring a towline to be slackened-off in the way of their turning motions and successively causes the towline tension tends to zero due to loss of towing object. Whilst tow ship continues to turn, an instability of ship towing system in form of an excessive towline tension occurs at some after slackening and may lead to breakage of towline.

Table 1: Principal dimensions of tug and barge

Description	Tug	Barge
S_L	0.66	-
LBP (m)	40.0	60.96
Breadth (m)	9.0	21.34
draft (m)	2.2	2.74
Volume (m^3)	494.7	3292.4
LCB (m)	2.23	-1.04
Block coef.	0.63	0.92
k_{zz}/L	0.25	0.252
L/B	4.44	2.86

Table 2: Resistance coefficient, linear hydrodynamic derivatives on maneuvering and added mass coefficients

Symbol	Tug	Barge
X'_{uu}	-0.033	-0.0641
Y'_v	-0.3579	-0.4373
Y'_r	0.127	0.1355
N'_v	-0.0698	-0.0491
N'_r	-0.0435	-0.0742
m'_x	0.0187	0.0391
m'_y	0.1554	0.2180
J'_z	0.0056	0.0124
Y'_δ	-0.05	-
N'_δ	0.025	-

4. SIMULATION CONDITION

4.1 SHIPS

Tow and towed ships associated with set of towing parameters were employed to achieve the objective of this paper. The principal dimension of tug and barge as tow and towed ships are presented in Table 1. The length of the tug and the barge are denoted as L_1 and L_2, respectively. The notation of k_{zz} is radius of gyration for yaw. Twin-propellers in which coupled with twin-rudder were equipped on the tug. The data of CPP propellers with 1.8m in diameter, the propeller revolution of 300rpm and the total engine power of 1430ps were used to achieve 7.0 knots of advance speed. Refer to Pfennigstorf's article [4], a rudder is assumed as rectangular shape with 2.0m × 2.0m for the span and cord length, respectively. The steering speed of the rudder was set to 2.0°/s. The simulation of a solely tug turning motion was examined, as shown in Figure 3. Based IMO criteria, the maneuverability of the tug was achieved satisfactorily by measuring the magnitude of tactical diameter at $\delta_1 = 15°$, $\delta_1 = 20°$ and $\delta_1 = 35°$ are nearly about $4.5L_1$, $3.0L_1$ and $1.7L_1$, respectively.

4.2 HYDRODYNAMIC DERIVATIVES ON MANEUVERING

The hydrodynamic derivatives both of the tug and the barge including resistance coefficients in full scale (X'_{uu}) were obtained by captive model test in the towing tank and completely summarized in Table 2. Added mass coefficients (m'_x, m'_y, J'_z) were calculated using singular distribution method under the rigid free-surface condition.

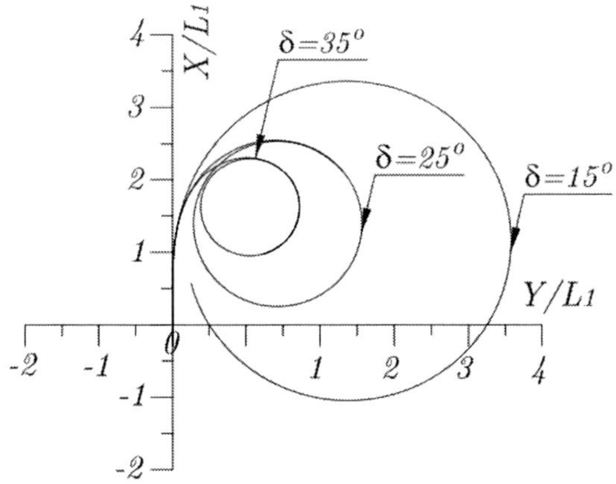

Figure 3: Turning trajectories of solely tug

4.3 NUMERICAL PROGRAM

The nonlinear simulation program of turning motions for the tow and towed ships was developed. The reliability of this program was previously examined to analyse a slewing motion performance of towed ship, which revealed good agreement with experimental model test in the towing tank [2].

5. RESULTS

This section presented the numerical analysis of slack towline condition on the tug and the barge during maneuvering.

5.1 SLACK TOWLINE: LIMIT OF TURNING MOTION OF TOW-TOWED SHIPS

Trajectories of time histories for turning motion of tow-towed ship due to effect of towline length are shown in Figure 4. Due to subsequent lengthening of the towline ($\ell' = 1.0, 2.0$ and 3.0), it was proportional with increasing their diameter of turns as displayed in Figures 4a, 4b and 4c, respectively. Based these results, we obtained that the advance diameters of the barge were 141m, 145m and 91m, respectively, while the tactical diameters of the barge were 324m, 438m and 471m, respectively. In the turning trajectories, the tug was liable to turn outwardly with respect to the turning motion of the barge, while the turning offset between the tug and the barge was getting large as the towline

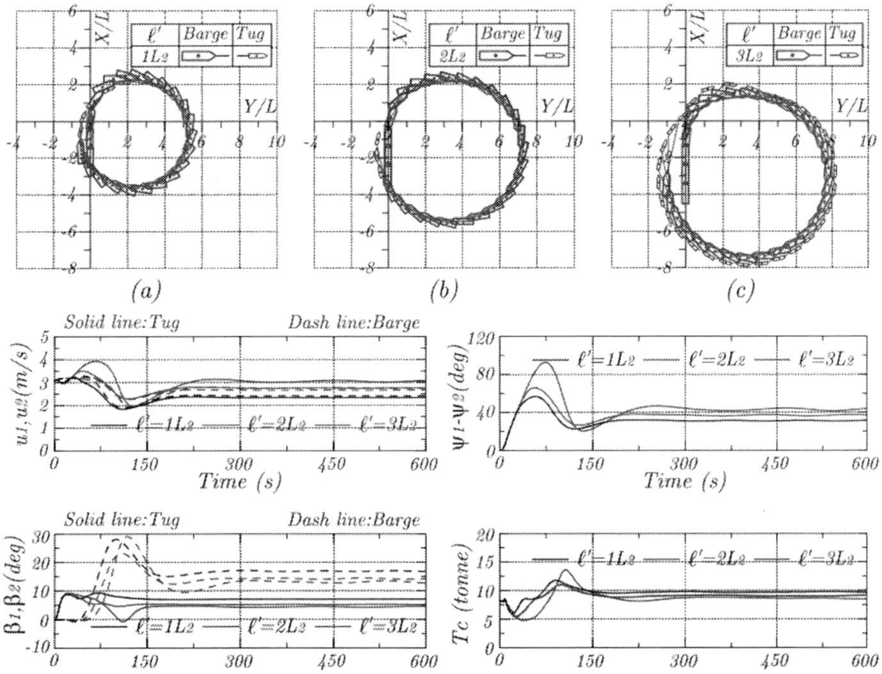

Figure 4: Trajectories and time histories of turning motions of tow-towed ships due to increase of towline ($\delta_1 = 15°$, $S_L = 0.66$ and $\ell'_T = -0.5$)

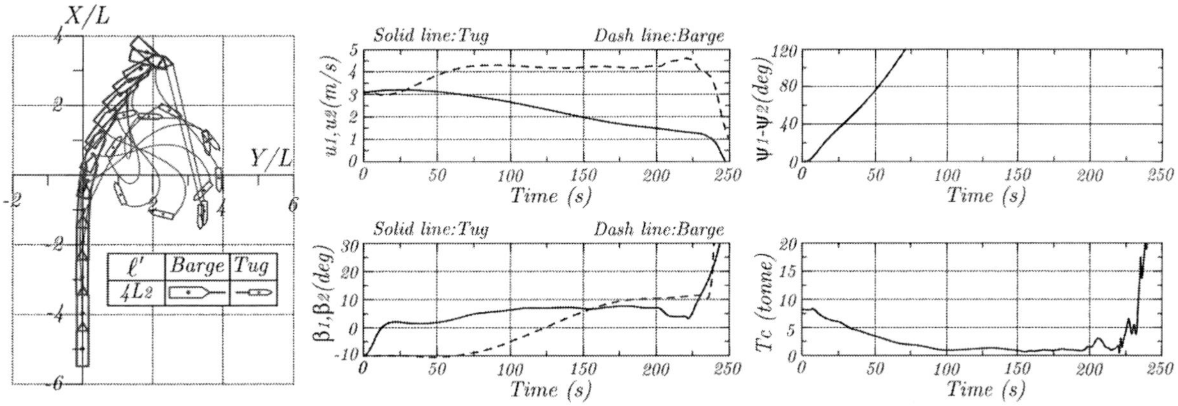

Figure 5: Slack towline condition due to increase of towline ($\delta_1 = 15°$, $S_L = 0.66$ and $\ell'_T = -0.5$)

increased. For this reason the behavior of the tug trajectory was related to its turning speed performance, where u_1 was generally somewhat faster than u_2 especially in the case of $\ell' = 2.0$ and 3.0. Furthermore, the time histories of β_1 and β_2 were decreased and inversely the deviation of $\psi_1 - \psi_2$ was increased due to the increment of the towline from $\ell' = 1.0$ to 3.0. By taking advantage of this lengthening towline, the towline tension was reduced from 9.95t to 8.82t, while the mean advance velocity of the tug (u_1) was raised from 1.7m/s up to 2.6m/s.

The further increase of towline ($\ell' = 4.0$), slack towline on the tug and the barge interaction during maneuvering occurred as demonstrated in Figures 5. It was seen in the time histories, when the tow and the barge commenced their turns, the towline tension gradually appeared. At t = 98s, the towline tension decreased and tended to be zero since the barge voluntarily retarded its speed and finally ceased. This condition led to slacken the towline, which caused the tug was vulnerable to loose restriction of towing to the barge. At t = 71s, deviation of $\psi_1 - \psi_2$ was increased excessively, while the tug kept turning until reached its steady speed by t = 202s. Furthermore, β_1 and β_2 unduly increased and the tug turned into small radius of turn. Once the tug completed its first turning motion as illustrated in the slack towline trajectory, it promptly created the large tension of towline at t = 238s. This could be explained by the fact that when the barge halted, it would be burdensome and caused the large tension forces, which imposed on the tow point ℓ_T after slakening of the towline. Therefore, the tug decreased

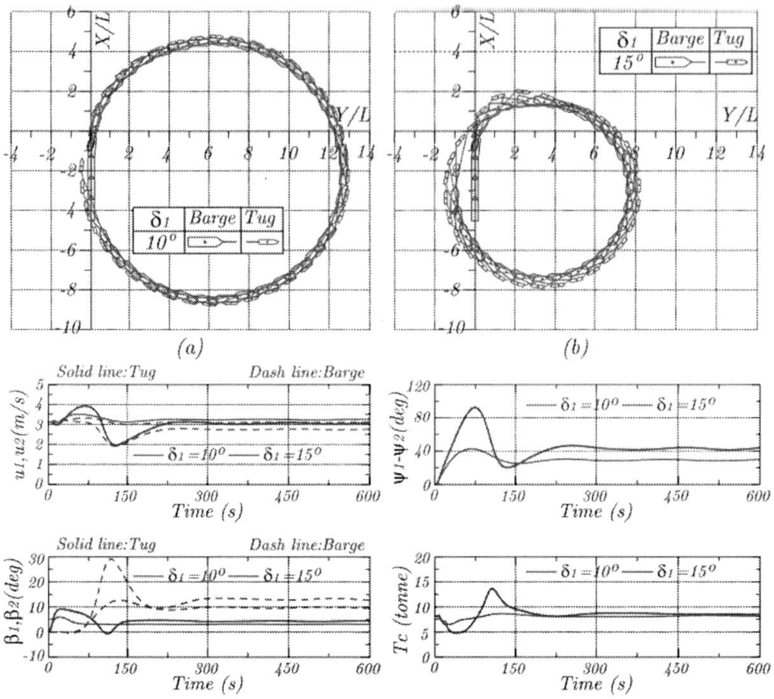

Figure 6: Trajectories and time histories of turning motions of tow-towed ships due to increase of towline ($\delta_1 = 15°$, $S_L = 0.66$ and $\ell'_T = -0.5$)

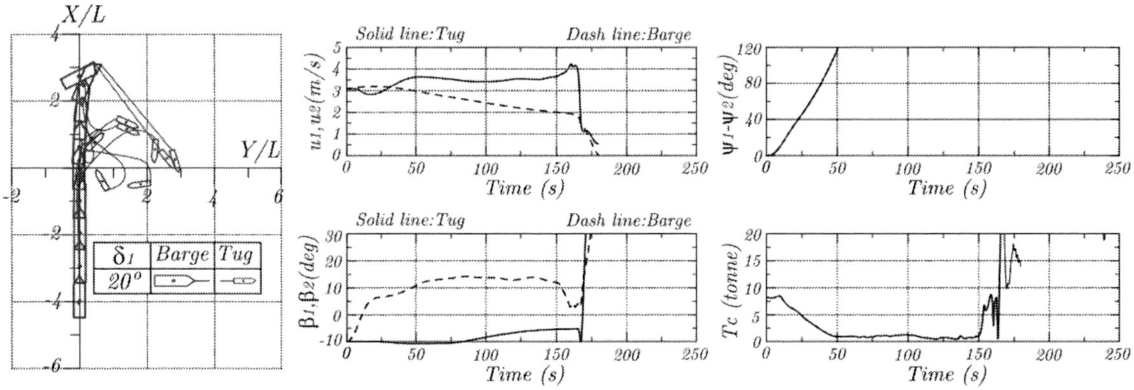

Fig.7: Slack towline condition due to increase of deflection-rudder angle ($\ell' = 3.0$, $S_L = 0.66$ and $\ell'_T = -0.5$)

its turning speed, which was liable to zero at $t = 250$s. It should be noted here that a solution to prevent the slackening of the towline was proposed by Shigehiro et al i.e., the employed towline length should be shorter than the turning diameter of the tow ship [1].

Trajectories of time histories for turning motion of tow-towed ship due to effect of deflection-rudder angle are shown in Figure 6. In the case of $\delta_1 = 10°$, the advance and tactical diameters of the barge were 265m and 775m, respectively. According to this result, the decrement in δ_1 from 15° to 10° was able to increase significantly the advance and tactical diameters of the barge about 124m and 304m, respectively. This was due to the fact that the deviation of $\psi_1 - \psi_2$ and the discrepancy between β_1 and β_2 were decreased considerably and implied the tug to have large turning diameter, which followed inherently the barge's turning diameter. On the other hand, the increase of δ_1 was directly correlated well with the turning speed of the tow and towed ships. By changing δ_1 from 10° to 15°, it had proven that u_1 and u_2 reduced. The reason for this was possibly that in larger deflection-rudder angle the rudder normal force (F_N) increased, which produced more additional resistance to the tug. However, the influence of the increase of δ_1 on the towline tension was insignificant.

In the case of δ_1 was angled further to 20°, however, the slack towline was happened as presented in Figure 7. Although the towline was shortened to be $\ell' = 3.0$ as compared to the previous slack towline at $\ell' = 4.0$, the tug and the barge were unable to maintain their turning once the deflection-rudder angle was set to $= 20°$.

Similar analysis to what was explained above, the barge slowed down its speed (u_2) and the remarkable increase of $\psi_1 - \psi_2$ associated with their drift angles caused the towline slackened-off at t = 180s. Coincidently, the sudden high tension of the towline appeared at some after the slackening.

Typical characteristics of slack towline condition during maneuvering were revealed by nonlinear and linear analysis. A series of simulation for the slack towline on tow-towed ships interaction during maneuvering due to effect of δ_1 with respect to the towline have been carried out as plotted in Figure 8. Based on this slack towline diagram, the computed results using the linear analysis was agree qualitatively well with the non-linear numerical simulations. Hence, this slack towline assessment was practically meaningful for the ship operators to prevent the slack towline on the tow-towed ships interaction during maneuvering. Merely, it could be seen that the longer towline incorporated with the greater deflection-rudder angle would lead to the slack towline condition.

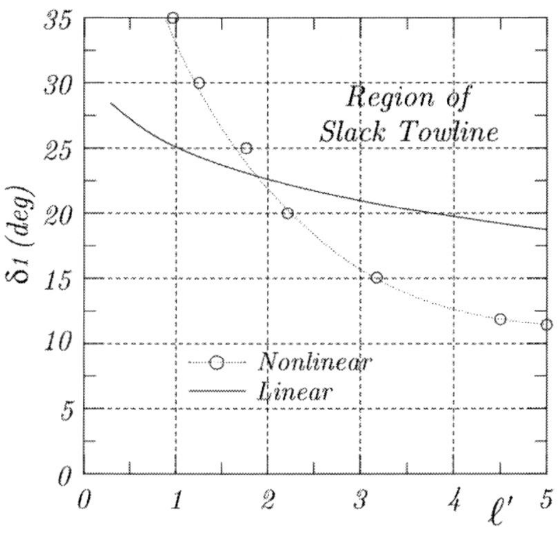

Figure 8: Slack towline diagram due to effect of deflection-rudder angle with respect to towline ($S_L = 0.66$ and $\ell'_T = -0.5$)

6. CONCLUSIONS

Based on linear analysis, the mechanism of the relations between several towing variables to the slack towline condition were clearly clarified. The lengthening of towline ℓ, replacing of the tow point ℓ_B (more far away from COG of the barge), reducing the length of the tow ship S_L and increment in C_2), were proportional to the magnitude of σ. In the case of $\sigma > \pi/2$, the slack towline on the tow-towed ships interaction during maneuvering occurred. Using the non-linear numerical analysis, the slack towline on the tug and the barge interaction during maneuvering was captured successfully due to effect of the towline (ℓ) and the deflection-rudder angle (δ_1). When the towline was slack, it tended to bring the towline tension to be zero, while the barge voluntarily retarded its speed. This condition caused the tug was vulnerable to loose restriction to the barge. Coincidently, it implied the tug to turn into small diameter of turning. The sudden increase of the towline tension was then appeared once the tug completed its first turning motion. This might threaten safety of the ship towing system, which could introduce successive towing accidents such as breakage of the towline and it enabled the uncontrolled tug to collide with the barge or another vessel especially in the confined waters with high sea traffic mode. Concluding the slack towline condition by comparing the analytical results, the computed results using the linear analysis was agree qualitatively well with the non-linear numerical simulations.

References

[1] SHIGEHIRO, R., UEDA, K., NAKAYAMA, H., 'Experimental Studies For Stopping and Hard Turning Maneuver of Tow And Towed Vessels' (in Japanese), *Journal Kansai Society. Naval Architects*, Japan, No. 230, pp.165-175, 1998.

[2] YASUKAWA, H., HIRATA, N., NAKAMURA, N., MATSUMOTO, Y., 'Simulations of Slewing Motion of a Towed Ship' (in Japanese), *Journal of The Japan Society of Naval Architects and Ocean Engineers*, Vol. 4, pp.137-146, 2006.

[3] FITRIADHY, A., and YASUKAWA, H.,'Course Stability of a Ship Towing System', *Journal of Ship Technology Research Shiffstechnik*, Vol. 58, pp.4-24, 2011.

[4] PFENNIGSTORF, J., 'D. Schubschiffahrt, Handbuch der Werften, X. Band, bearbeitet von Wendel, K., Schiffahrts-Verlag HANSA, C. Schroedter and Co., Hamburg 11'

Authors Biography

Ahmad FITRIADHY holds the current position of Ph.D student in Hiroshima University. He is responsible for a development of course stability and maneuverability of tow and towed ships.

Hironori YASUKAWA holds the current position of Professor in Hiroshima University. He is responsible for research on the ship dynamics and hydrodynamics. Now, he is maneuvering committee member of ITTC.

Yusaku MASAKI holds the current position as naval architect engineer at Mitsui Engineering and Shipbuilding Co., Ltd, Japan.

NUMERICAL STUDY OF SHIP-TO-SHIP INTERACTION FORCES ON THE BASIS OF VARIOUS FLOW MODELS

J M A Fonfach, CENTEC, Instituto Superior Técnico, Technical University of Lisbon, Portugal
S Sutulo, CENTEC, Instituto Superior Técnico, Technical University of Lisbon, Portugal
C Guedes Soares, CENTEC, Instituto Superior Técnico, Technical University of Lisbon, Portugal

SUMMARY

A comparative CFD study of the interaction between a smaller tug boat and a large tanker sailing parallel at low speed was undertaken on the basis of four flow models: (1) double-body inviscid flow; (2) double-body turbulent viscous flow; (3) inviscid free-surface flow, and (4) viscous free-surface flow. Two commercial CFD codes: *Star CCM+* and *ANSYS CFX* were used for computations. Special attention was given to generation of the computational mesh inside the gap between the both vessels. Computations were carried out at various lateral distances and partly compared with available experimental data. Analysis of numerical results revealed a relatively weak influence of viscosity while free-surface effects are changing dramatically the character of the interaction at small lateral distances even at very low values of the Froude number based on the ship length while the Froude number based on the lateral distance becomes more informative in the case of side-by-side motion.

NOMENCLATURE

CFD	Computational Fluid Dynamics
RANS(E)	Reynolds-Averaged Navier–Stokes (Equations)
H	Water depth (m)
N	Yaw moment (Nm)
p	Pressure (Pa)
u_i	Fluid velocity component (m/s)
V	Ship speed (m/s)
x, y, z	Coordinates in body axes (m)
X	Surge force (N)
Y	Sway force (N)
η'_s	Dimensionless lateral displacement of the own ship relatively to the target
ν	Fluid viscosity (m^2/s)
ν_T	Eddy viscosity (m^2/s)
ξ'_s	Dimensionless longitudinal shift of the own ship with respect to the target ship
ρ	Density of water (kg m^{-3})
τ_{ij}	Stresses in the fluid (N/m^2)
∇	Ship displacement (m^3)

1. INTRODUCTION

Modern bridge simulators have reached a high degree of versatility and permit to model a large variety of situations when certain interaction forces are of major importance and cannot be neglected in the manoeuvring models. For instance, rather typical [1] is the simulation exercise involving one or more assisting tugs navigating near an assisted vessel. This interaction involves various aspects, such the direct desired interaction through the towline or through a direct mechanical contact in pushing mode, the propeller wash both from the tug to vessel and—if the ship's engine is not stopped and the tug enters the ship's wake—from the vessel to tug. There are also other components of the ship wake also modifying manoeuvring forces applied to the tug. However, last but not least is the effect of direct hydrodynamic interaction between the hulls which is mostly of potential nature although the influence of ship-generated waves is sometimes obvious. This kind of interaction is also observed in overtaking and encountering manoeuvres of vessels of more or less equal size. Only interaction forces of this last category will be discussed here.

Literature on the hydrodynamic interaction of ship hulls with other ships and other obstacles and boundaries is abundant. Different methods and techniques proposed to estimate the interaction loads can be classified according to various properties and features:

- Experimental or theoretical/numerical
- Dealing with actual hull form or exploiting simplifying assumptions (e.g. slender hulls)
- Fully nonlinear or linearised in some sense
- Including or excluding free surface effects
- Including or excluding viscous effects
- Covering or not all possible locations and motion of interacting ships (in particular: is the method unsteady or quasi-steady)
- Applicable to the deep or shallow water case or to the both
- Allowing or not fast online computations.

One of first fundamental theoretical studies of the ship-to-ship was carried out by Abkowitz et al. [2]. The formulation was based on the Havelock hypothesis which states that the potential flow model without free-surface effects is sufficient to explain the interaction phenomena. This hypothesis looked very reasonable at low Froude numbers when most interactions occur and was exploited in many studies. Abkowitz et al. [2] introduced the additional slenderness assumption and the hulls were approximated with bodies of revolution simulated with axial source and dipole distributions. Horizontal dipoles originated from lateral motions of the own ship and from other ship's induction, and vertical dipole were added in shallow water. The dipoles' strength was determined as result of a fast iteration process and the forces and moments were computed basing on the Lagally–Cummins formulae. It was found that unsteadiness of the flow affects much more the yawing moment, than the sway force.

Beck [3] applied the matched asymptotic expansions method to the problem of ship-to-bank interaction in a shallow channel. In the outer region, the shallow water theory was applied, so the final results depended on the depth Froude number. Interesting theoretical studies were undertaken by Yeung [4] who applied the slender-body theory and the method of matched asymptotic expansions. The method results in relatively fast algorithms and accounts for the Kutta condition on the ship's hull but it heavily underestimates sway forces and yaw moments in the most interesting case of small separation between the hulls.

Tuck and Newman [5] performed one of the best asymptotic studies of the ship-to-ship interaction problem. Two situations were investigated, both at zero Froude number. In the first case, the water depth was assumed to be infinite or finite but relatively large. The so-called first-order theory was constructed i.e. each ship was just sailing in the flow field of its partner. Then, the lateral loads on each of the slender hulls were computed using the generalized Taylor theorem for a body moving in an accelerated uniform flow. The Kutta condition was accounted for sterns of finite height. However, the theory predicted no interaction with the target (not own) ship is at rest. The second studied case related to the pair of ships moving with the same speed in extremely shallow water. Commenting on the missing flow factors it was supposed that viscous effects would be more significant than those related to the wavemaking.

One of the first applications of a numerical method belongs likely to Ivanov [6] who successfully applied the potential flow algorithm by Hess and Smith [7] to the zero-Froude-number interaction problem. A more advanced but still offline application of another version of the 3D panel potential code was undertaken by Korsmeyer et al. [8] who studied body-to-body interaction in a rectangular canal. Only parallel motion was studied. A direct 3D potential-flow formulation was also exploited by Yasukawa [9] who used an unspecified implementation of the panel method for determining the flow potential and the interaction forces which were expressed through a set of generalized added masses. Finally, a RANSE code was applied by Chen et al. [10] for tackling the problem of ship-ship interaction in shallow water. Good agreement with experimental data was demonstrated except for the case of relatively small distance between the boards when insufficient finesse of the grid was suspected.

A large group of studies was based on experimental methods. An approximate empiric method for predicting forces in overtaking manoeuvres was developed by Brix [11]. Extensive automated tests with scaled models of 4 different shapes were carried out by Vantorre et al. [12] covering only parallel motion i.e. overtaking, passing by and encountering situations in shallow water. On the basis of the huge amount of assembled data a simple approximate empiric method for estimating peak values of the interaction forces was proposed. Ch'ng et al. [13] carried out bank-interaction experiments with *MarAd* and *S-175* models equipped with propellers but without rudders. Linear regressions were devised and it was demonstrated that the propeller thrust may influence substantially the interaction sway force and yaw moment.

A more specialized experimental study focusing on inland vessels was performed by Gronarz [14]. Interesting details about this study are shown original measurement records showing uncertainty levels, investigation of the propeller's influence and convenient approximations for the forces.

Varyani et al. [15] studied mooring-lines loads originating from the hydrodynamic interaction between the moored vessel and the passing by ship. The zero-Froude-number theory was used to estimate the flow field around the moving ship and the pressure integration resulted in estimates for the surge and sway forces and the yaw moment acting on the moored ship. Although the final formulae do not look obvious and the back interaction from the moored ship was completely neglected, comparisons with experimental results showed fair agreement. Later this and other similar studies resulted in a generic set of approximations of interaction sway forces and yaw moments proposed by Varyani [16].

Apparently, approximations based on physical experimentation currently prevail in proprietary mathematical models implemented on bridge simulators. They, however, suffer from an irreparable drawback: it is impossible to carry out experiments covering all possible situations of mutual positions and motions of the interactive ships let alone necessity to perform new tests for ship forms substantially different from those in the database.

That is why, assuming that the Havelock hypothesis mentioned above was valid and bearing in mind computing power of modern systems it was judged that a numerical double-body potential code could be suitable for an adequate online modelling hydrodynamic interaction in simulation systems. Sutulo and Guedes Soares [17] have developed a code based on the classic Hess and Smith [7] method and suitable for online computations of interaction forces for arbitrary number of ship hulls with arbitrary relative positions and motion. This algorithm, primarily developed for only deep water was later fused with the manoeuvring simulation code and an uncontrolled and controlled motion of two vessels in overtaking maneuver [18] was simulated. Later, the code was extended to cover also the flat bed shallow water case and even was generalized to treat the bottom and banks of arbitrary shape [19].

In [20], the interaction code was validated against experimental results obtained in the towing tank for a specific tug boat operating near a typical large vessel. Only results related to parallel motion with equal speeds were considered there. The comparisons showed that in most cases the potential code provides reliable predictions for the sway force and yaw moment but gave worse estimate of the surge interaction force. Large quantitative and even qualitative discrepancies were also obtained for the sway force and yaw moment at very small gaps between the sidewalls. It was suspected that this happened due to neglected wavemaking and viscous effects but it was not quite clear which of these two was really essential.

In the present study it was decided to use the power of modern commercial CFD codes for exploring contribution of various factors not accounted for by the panel method. Two CFD codes (*ANSYS CFX* [21] and *Star CCM+* [22]) were applied for modelling some of the cases explored in [20]. For each case, computations were carried out in 4 modes:

- Perfect fluid; free surface as rigid wall.
- As above but free surface with exact boundary conditions.
- Viscous fluid, turbulent flow; free surface as rigid wall.
- As above but free surface with exact boundary conditions.

It is clear that viscous effects were only accounted for in the last two cases while the wavemaking effects—in the second and fourth case. Due to historical and organizational reasons, the code *CFX* was used for waveless computations and computations with the deformable free surface were performed with *Star CCM+*. In general, using various codes for such comparative computations is not desirable but in this particular case, as show the comparisons, this did not lead to fatal devaluation of the results.

2. PROBLEM FORMULATION

2.1 FRAMES AND PARAMETERS

The Earth-fixed and body axes generally used in hydrodynamic interaction problems are shown in Figure 1. However, in the present study, the both ships will travel parallel to the axis $O\xi$ without drift at equal speed V. The "own" ship (tug, numbered 1) is always located from the portside of the "target" ship (tanker, numbered 2) with various longitudinal stagger ξ and lateral distance η. Also, corresponding non-dimensional parameters ξ_s' and η_s' will be defined as: $\xi_s' = 2\xi / L_2$ and $\eta_s' = 2\eta / B_2$. The hydrodynamic forces of surge X, sway Y and the yaw moment N will be only computed for the tug and their dimensionless analogues are: $X' = 2X/(\rho AV^2)$, dimensionless sway forces $Y' = 2Y/(\rho AV^2)$, and the dimensionless yaw moment $N' = 2N/(\rho ALV^2)$, where $A = \sqrt[3]{\nabla_1 \nabla_2}$ is the reference area, and $\nabla_{1,2}$ are displacements of the both interacting ships.

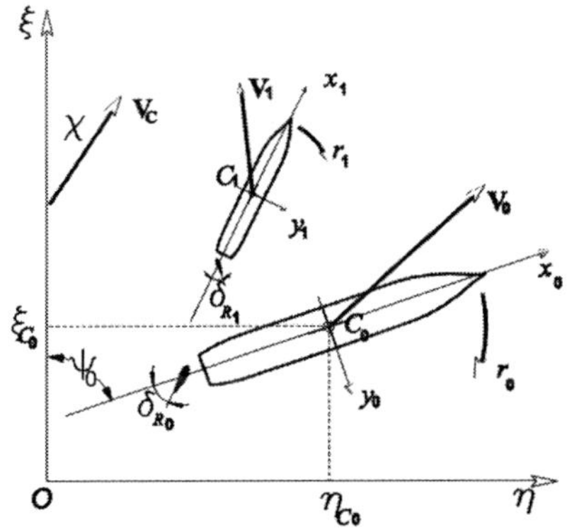

Figure 1: Frames of reference and kinematic parameters

All the cases studied in this contribution refer to the shallow water of constant depth H with the absolute and relative depth parameters shown in Table 1.

Table 1: Depth conditions

Absolute depth in full scale	Depth relative to:	
	Tug's draft T_1	Tanker's draft T_2
$H_1 = 12.25$m	2.66	1.19
$H_2 = 15.57$m	3.38	1.51

Two speed regimes (Table 2) were considered, both characterized by the Froude number value which is usually considered in ship manoeuvring and even in ship resistance as small, with negligible free surface effects.

Table 2: Speed regimes data

Speed value in full scale		Froude number based on:			
kn	m/s	Tug's length L_1	Tanker's length L_2	H_1	H_2
4	2.06	0.13	0.048	0.188	0.167
6	3.09	0.195	0.072	0.282	0.25

2.2 SHIP PARTICULARS AND HULL FORMS

Main particulars of both interacting ships are given in Table 3 and their hull forms are shown in Figure 2. As can be seen, the difference in the dimensions is substantial which explains why forces on the "target" ship are of smaller interest.

Table 3: Main particulars of tug and tanker model

Main Particulars	Unit	Tug (Ship 1)		Tanker (Ship 2)	
		Full scale	Model	Full scale	Model
Length overall, L_{OA}	m	29.5	1.18	189.5	7.58
Length between perpendiculars, L	m	25.6	1.024	186.2	7.44
Breadth, B	m	11.0	0.44	31.6	1.26
Draft, T	m	4.6	0.184	10.3	0.41
Displacement, ∇	m^3	649	0.0416	49197	3.15

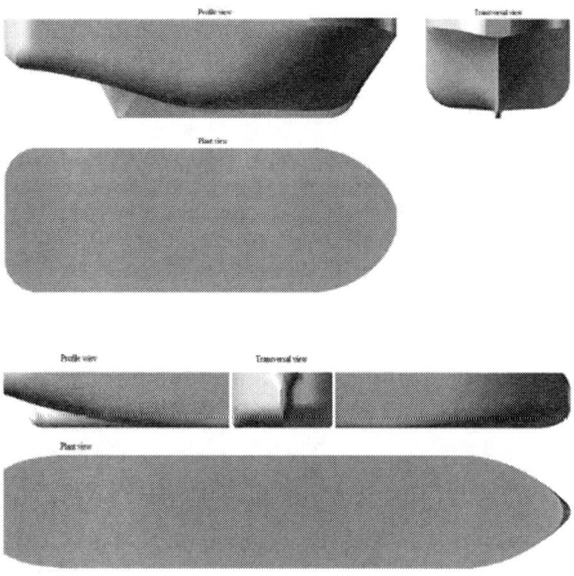

Figure 2: Hull forms of the tug (top) and tanker (bottom)—in different scales.

2.3 HYDRODYNAMIC EQUATIONS AND BOUNDARY CONDITIONS

2.3 (a) Equations

The Reynolds-averaged Navier–Stokes equations [23] form the spine chord of any commercial CFD code. The corresponding momentum and continuity equations written in the component form with the Einstein summation rule applied are:

$$\frac{\partial \bar{u}_i}{\partial t} + \frac{\partial (\bar{u}_i \bar{u}_j)}{\partial x_j} = F_i - \frac{1}{\rho}\left(\frac{\partial \bar{p}}{\partial x_i} + \frac{\partial \bar{\tau}_{ij}}{\partial x_j}\right),$$
$$\frac{\partial \bar{u}_j}{\partial x_j} = 0, \tag{1}$$

where F_i is the volumetric force' component (the gravity force in this application), the bar over a symbol means the averaged value and $\bar{\tau}_{ij}$ are the turbulent stresses which disappear when the perfect fluid model is considered and the momentum equation becomes the Euler equation.

2.3 (b) Boundary conditions

As all the considered cases were steady, the flow inversion principle could be applied and then the no-slip boundary condition $\bar{u}_i = 0$ was applied on the wetted surface of each hull. In the non-viscous case the no-penetration boundary condition is used instead: $u_i n_i = 0$, where n_i is the external unity normal's component. The latter condition is also applied on the free surface treated as rigid wall.

In the general case, the free-surface conditions include the kinematic and the dynamic condition. As in *Star CCM+* free surface formulations are treated as two-phase flows, and the upper phase is obviously the air, these conditions take the form:

$$u_{i\,\text{water}} = u_{i\,\text{air}} \quad \text{and} \quad p_{\text{water}} = p_{\text{air}} \tag{2}$$

at every point on the free surface.

The boundary conditions presented above can be called natural. Besides, additional boundary conditions are necessary to form a suitable numerical towing tank. Namely, the inlet boundary condition takes the form $u_1 = V,\ u_2 = u_3 = 0$ and the outlet boundary condition is specified for the pressure: $p = p_{\text{hydrostatic}}$.

2.3 (c) Turbulence models

A suitable turbulence model is used for closing the set of fluid dynamics equations. The turbulent stresses are expressed as

$$\bar{\tau}_{ij} = \rho \nu_T \left(\frac{\partial \bar{u}_i}{\partial x_j} + \frac{\partial \bar{u}_j}{\partial x_i}\right), \tag{3}$$

where ν_T is the turbulent (or eddy) viscosity and the molecular viscosity ν can be neglected for flows at high Reynolds numbers.

In the computations with the wavemaking i.e. performed with *Star CCM+* was used the standard Spalart–Allmaras turbulence model with the governing equation:

$$\frac{\partial \tilde{\nu}}{\partial t}+\bar{u}_j\frac{\partial \tilde{\nu}}{\partial x_j}=C_{b1}\xi\tilde{\nu}-C_{\omega 1}f_{\omega}\left(\frac{\tilde{\nu}}{d}\right)^2 + \frac{1}{\sigma}\frac{\partial}{\partial x_k}\left[(\nu+\tilde{\nu})\frac{\partial \tilde{\nu}}{\partial x_k}\right]+\frac{C_{b2}}{\sigma}\frac{\partial \tilde{\nu}}{\partial x_k}\frac{\partial \tilde{\nu}}{\partial x_k}, \tag{4}$$

where $\tilde{\nu}=\nu_T/f_{\nu 1}$ is the modified eddy viscosity; $f_{\nu 1}=(\nu\tilde{\nu})^3/[(\nu\tilde{\nu})^3+C_{\nu 1}^3]$, and the remaining symbols $C_{b1},\ldots,C_{b2},C_{\nu 1}$ represent various auxiliary constants, variables and functions.

In waveless computations performed with *CFX* the Shear Stress Transport $k-\omega$ model was activated. According to this model, $\nu_T=\rho k/\omega$, where k is the turbulence' kinetic energy and ω is its specific dissipation rate. The transport equations for the last two variables are:

$$\begin{aligned}
&\rho\bar{u}_i\frac{\partial(\bar{u}_j k)}{\partial x_j}\\
&=\frac{\partial}{\partial x_j}\left(\frac{\nu_t}{\sigma_\varepsilon}\frac{\partial k}{\partial x_j}\right)+\nu_T\frac{\partial \bar{u}_i}{\partial x_j}\left(\frac{\partial \bar{u}_i}{\partial x_j}+\frac{\partial \bar{u}_j}{\partial x_j}\right)\frac{\partial \bar{u}_i}{\partial x_j}-\rho k\omega,\\
&\rho\bar{u}_i\frac{\partial(\bar{u}_j \omega)}{\partial x_j}\\
&=\frac{\partial}{\partial x_j}\left(\frac{\nu_t}{\sigma_\varepsilon}\frac{\partial \omega}{\partial x_j}\right)+\alpha\frac{\omega}{k}\nu_T\frac{\partial \bar{u}_i}{\partial x_j}\left(\frac{\partial \bar{u}_i}{\partial x_j}+\frac{\partial \bar{u}_j}{\partial x_j}\right)-\beta\rho\omega^2,
\end{aligned} \tag{5}$$

where α and β are constants.

3. NUMERICAL SOLUTION

3.1 GENERAL REMARKS

The used commercial codes are both based on the Volume of Fluid method [23]. The maximum number of iterations in the CFX code was taken equal to 500 although in most cases convergence was reached at not more than 300 iterations. Simulation in the Star CCM+ was treated as unsteady with duration of 30s and time step 0.05s with 10 iterations for each step. Typical convergence condition was 1.0×10^{-4}. For stability reasons, the air in the two-phase modelling was assumed compressible. The hydrostatic forces caused by the density's difference were also modelled in this case. The surface tension was not applied. Computations with both codes were carried out for the full-scale value of the Reynolds number (5.3×10^7 and 7.9×10^7 for the tug) and computations with the CFX code were also repeated for the model values of the Reynolds number (4.2×10^5 and 6.3×10^5 respectively).

3.2 COMPUTATION DOMAIN AND GRIDS

The used control volume had a rectangle box shape with the inlet boundary at a distance of $1.5L_2$ from the stem, where L_2 is the length of the larger ship (tanker). The outlet boundary was located $2.6L_2$ downstream from the stern and the width was $1.5B_2$. In free-surface computations the height of the air phase was $0.25L_2$.

General view of computational domains (numerical towing tank) is shown in Figure 3 for one of configurations (relative position of the tug with respect to the tanker). It is seen that it was assumed *a priori* that the flow around the tanker's hull remained symmetric in spite of the presence of the interacting tug. This assumption permitted to decrease the required number of cells and while in the general interaction case it should not be used, here it was possible due to large difference in the ships' dimensions.

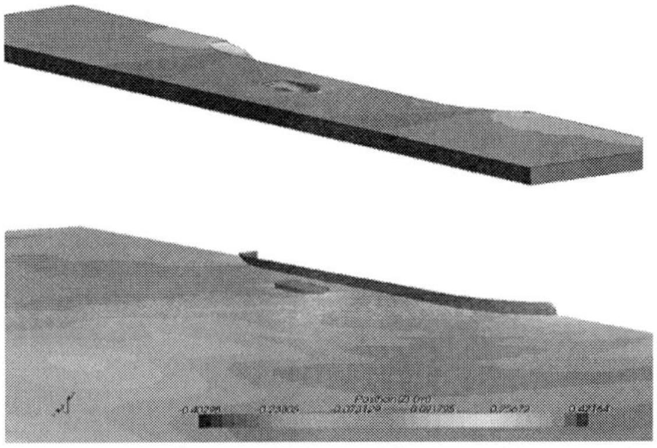

Figure 3: Computational domains: top—waveless computations with CFX; bottom—computations with free surface.

Due to resource limitations, it was not possible in this case to carry out special mesh convergence study: only one mesh was generated for each of explored cases. However, the previous experience obtained in another problem [24] turned out useful in the present work and apparently the discretisation errors remained acceptable.

In CFX computations, an unstructured tetrahedral grid was generated. Example of this mesh is shown in Figure 4. The target cell size was $0.0025L_1$ near the tug and $0.01L_2$ near the tanker. In Star CCM+ computations quite a different hexahedral mesh was created (Figure 5). That mesh was covering not only the underwater (below the design waterline) part but the whole hulls and it was refined near the free-surface zone.

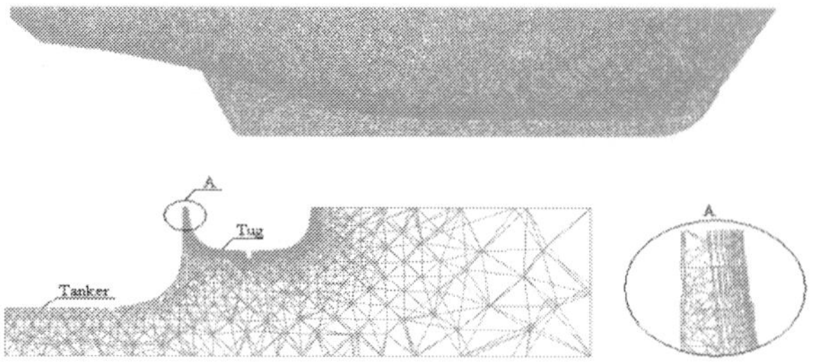

Figure 4: Tetrahedral mesh: top—over the tug's hull, bottom—transverse section at $x/L_1 = -0.15$.

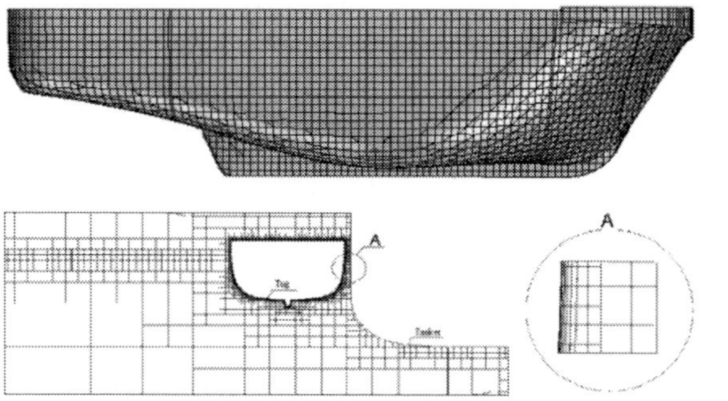

Figure 5: Hexahedral mesh: top—over the tug's hull, bottom—transverse section at $x/L_1 = -0.15$.

4. RESULTS OF COMPUTATIONS

4.1 INTERACTION FORCES AND MOMENTS

Only pure interaction forces and moments will be presented.

The interaction force/moment components are defined as

$$X_I = X_t - X_e,\quad Y_I = Y_t - Y_e,\quad N_I = N_t - N_e, \tag{6}$$

where the subscript I stands for "interaction", t— for "total", and e—for "eigen" that is proper loads on an isolated object. The latter category of loads was estimated from separate computations.
The computed forces and moments and also the results obtained with the panel method and experimental data [20] are shown, as functions of the tug's dimensionless lateral displacement ξ_s' from the tanker on Figure 6 for the case with the smaller water depth H_1 and on Figure 7—for the depth H_2. Comparison and analysis of the displayed data permits to formulate the following observations:

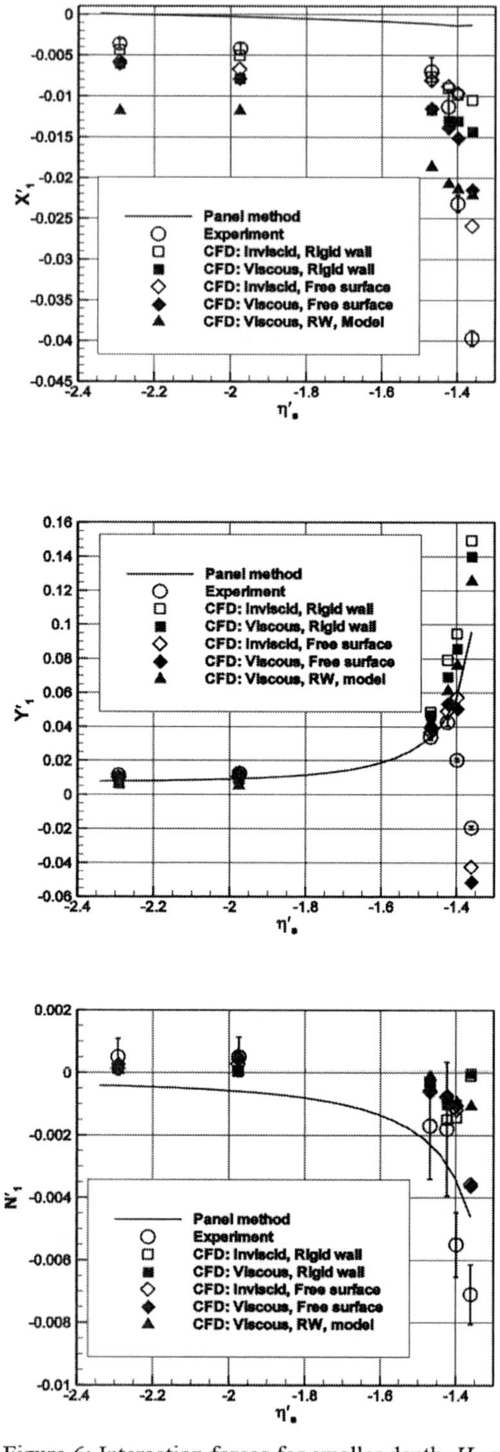

Figure 6: Interaction forces for smaller depth H_1 and the depth Froude number 0.188: top—surge force, centre—sway force, bottom—yaw moment.

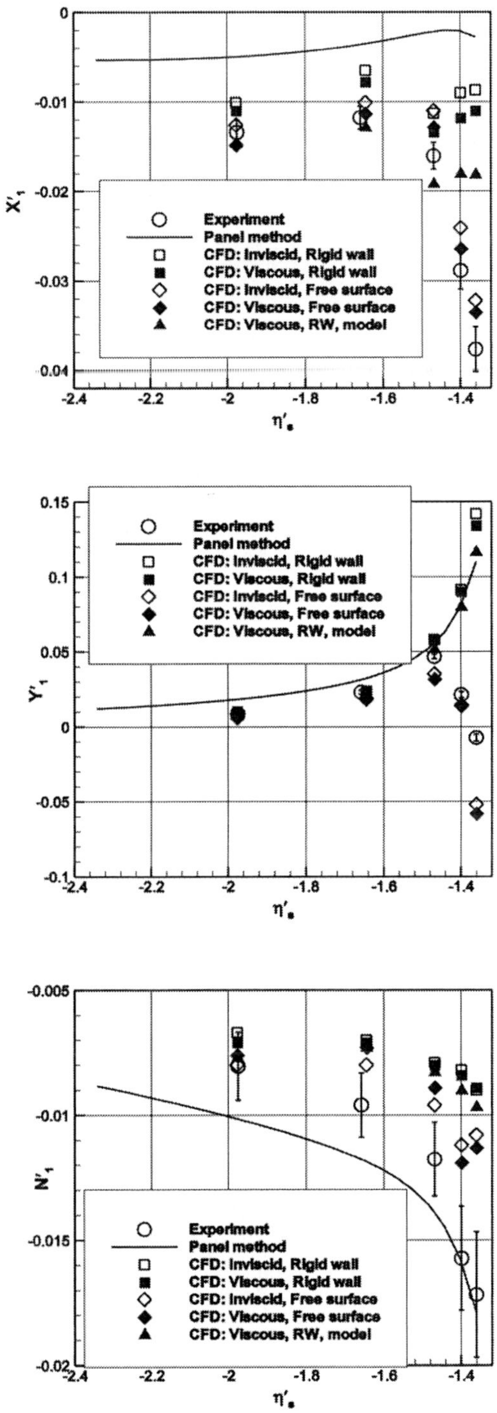

Figure 7: Interaction forces for larger depth H_2 and depth Froude number 0.25: top—surge force, centre—sway force, bottom—yaw moment.

- There is, in general, a certain overall qualitative agreement between results of computations according to various methods and options and the experimental data except for the sway force at very close distances between the ships.
- The worst overall agreement is observed for the surge force.
- Although inviscid rigid-wall CFD results should remain close to those obtained by the panel method, this was not clearly demonstrated by the present data: the difference is tangible in almost all the cases except for the sway force in smaller water depth and at larger lateral distances. Causes of such discrepancy still remain unclear, especially when it goes about the depth H_1 when the panel method was used with 3000 panels and signs of convergence were obvious [20].
- Rigid-wall CFD usually showed better agreement with the experiment than the panel method except for the yaw moment and the sway force at the depth H_1.
- In all explored cases influence of the fluid viscosity was not very pronounced but in most cases accounting for the viscosity somewhat improved agreement with experimental data. However, it can be expected that the influence of viscosity will become greater when the non-parallel motion with significant drift angles is considered and or when one of the ships is located in the another ship's wake.
- As expected, influence of viscosity was greater for the model Reynolds numbers and, as a rule, these results showed better agreement with experimental data.
- While the rigid-wall CFD results showed at small lateral distances the same trend of increasing suction as did the panel method, free-surface computations, both viscous and inviscid, have captured the experimentally observed effect of the suction reduction followed by transition to repulsion at very small lateral distances.
- CFD computations in general, and especially those with the wavemaking, are much better than the panel method predicting the surge force' dependence on the lateral distance.

4.2 PRESSURE AND VELOCITY DISTRIBUTIONS

Graphical representation of the pressure distribution over the tug's hull is shown in Figure 8 for the case when the computations took into account both viscosity and wavemaking effect for two relative lateral distances. The pressure values are given for the full scale, include the hydrostatic pressure of water but exclude the base atmospheric pressure $p_a = 1.013 \times 10^5$ Pa . The views are from the port and from the bow and the interacting tanker is located from the port side. Due to presence of the hydrostatic component, the differences in the pressure distribution are not very pronounced at small speed but some growth of the pressure responsible for the repulsion effect is slightly traceable.

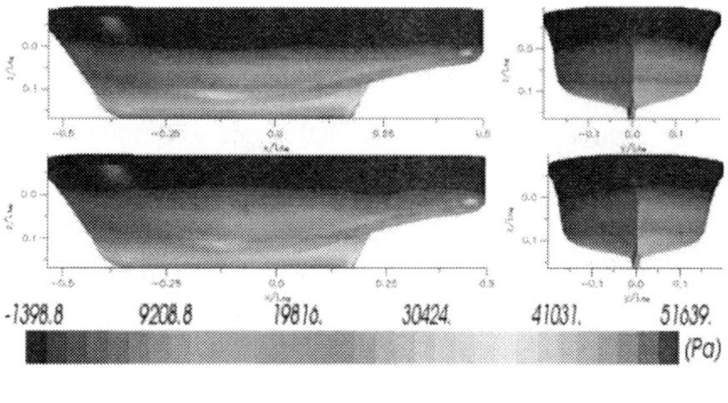

Figure 8: Pressure distribution on the hull of the tug for the depth H_1 , depth Froude number 0.188 and two lateral separations: top—$|\eta_s'| = 1.95$, bottom—$|\eta_s'| = 1.34$.

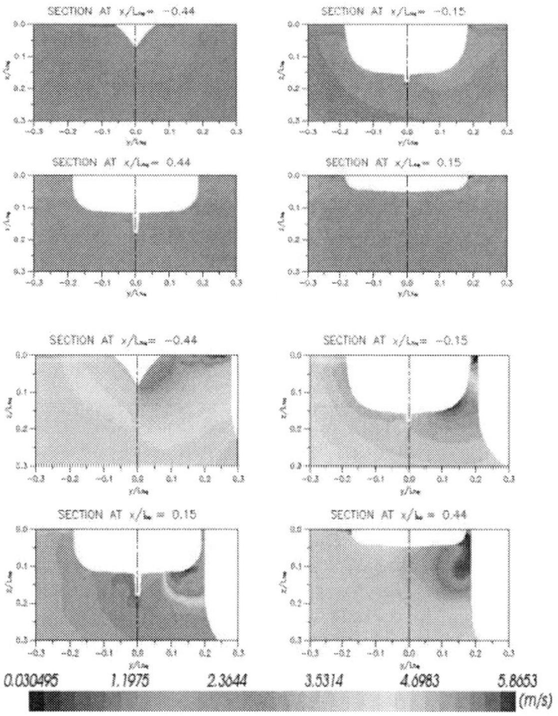

Figure 9: Velocity distribution on the hull of the tug for the depth H_1 , depth Froude number 0.188 and two lateral separations: top 4 plots—$|\eta_s'| = 1.95$, bottom 4 plots—$|\eta_s'| = 1.34$.

Velocity fields for the same two cases are presented in Figure 9. Negative abscissae correspond there to fore sections and the data are taken from perfect fluid computations with free surface. The velocity colour scale corresponds to full-scale ships advancing at the velocity 2.87m/s. At a larger lateral separation the velocity field looks very homogenous and areas of the flow's deceleration clearly appear. This deceleration is caused by wavemaking effects and results in the increased pressure and repulsion force. When the wavemaking is neglected, the flow is accelerated in the same areas resulting, on the contrary, in strong suction.

Finally, free surface elevations near the hull of the tug are shown in Figure 10 for the same conditions and also in the full scale in the form of contour plot and cuts along the sidewalls from the both sides of the centerplane.

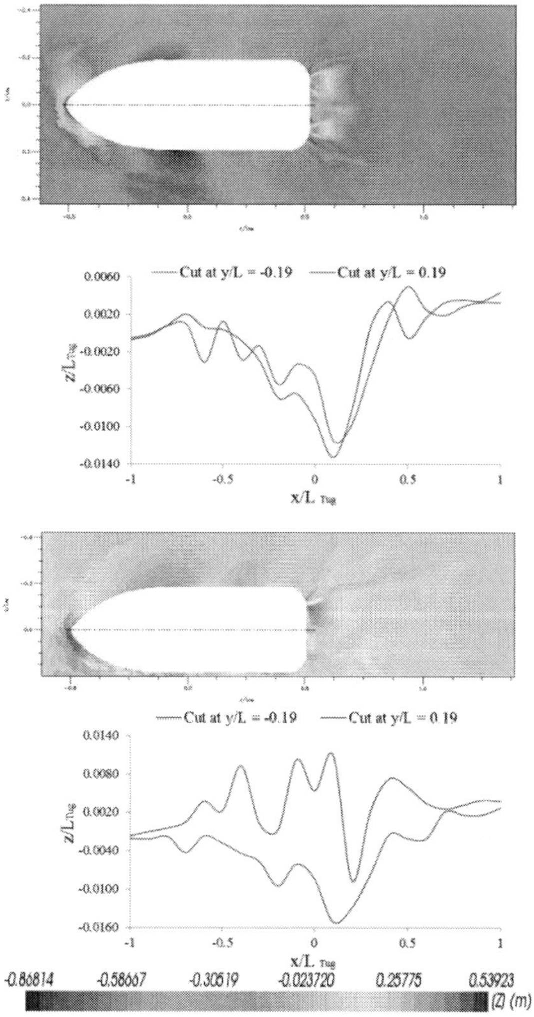

Figure 10: Free surface elevation near the hull of the tug for the depth H_1 , depth Froude number 0.188 and two lateral separations: top—$|\eta_s'| = 1.95$, bottom—$|\eta_s'| = 1.34$.

It is evident that presence of the interacting vessel changes considerably the wave pattern around the smaller ship. Also, strong wavemaking effect at relatively small ship Froude numbers is not very surprising: the Froude number based on the distance between the sidewalls even at small ship speeds reaches values from 0.5 to 1.0 and this explains why local wave making effects can become significant in interaction problems.

5. CONCLUSIONS

The numerical study whose results are presented in this paper helped to get a deeper insight into hydrodynamics of the ship-to-ship interaction. In particular, comparative computations with and without wavemaking for both perfect and viscous fluid models permitted to establish greater importance of free-surface effects compared to those of viscosity at least in the case of parallel ship motion. At the same time, some uncertainties still remain unresolved. One of them is the faced difference between the results obtained with the *CFX* code and the well-validated panel method with the double-body approximation. Apparently, the focus in the future work must be given to the CFD codes convergence issues although closer inspection of the panel code and its possible residual errors also deserves attention.

However, the phenomenon of the sway force inversion usually happens at very small distances between the sidewalls and, meaning impossibility of running a RANSE code online, application of the panel code in simulation systems may become justified. Unfortunately, at present no other method can be expected to be run in real time and from this viewpoint the problem of reliable and fast prediction of interaction forces and moments in arbitrary motion and mutual positions of the ships including the case of the non-parallel motion with large difference in heading angles remains unsolved. In this more general case a numerical study on the relative significance of various neglected factors must be also performed: the conclusion about relatively small influence of viscosity may be disavowed in the cases which, say, include crabbing motion of the tug near the assisted ship.

As the free-surface computations were carried out with exact free-surface condition it remains unclear whether the linear wave theory would capture here the largest part of free-surface effects or they are substantially nonlinear. In the latter case, perspective of their online account in manoeuvring simulators becomes even less clear.

6. ACKNOWLEDGEMENTS

The study was carried out within the framework of the research project PTDC/ECM/65806/2006 ''Dynamics and Hydrodynamics of Ships in Approaching Fairways'' financed by Fundação para a Ciência e a Tecnologia (FCT), Portugal.

7. REFERENCES

1. HENSEN, H., 'Ship Bridge Simulators: a Project Handbook', *The Nautical Institute*, London, 1999.

2. ABKOWITZ, M.A., ASHE, G.M., FORTSON, R.M., 'Interaction effects of ships operating in proximity in deep and shallow water', *Proceedings of the 11th Symposium on Naval Hydrodynamics, London, UK, March 28–April 02 1976, London–New York,* 1977, pp. 671–691.

3. BECK, R. F., 'Forces and Moments on a Ship Moving in a Shallow Channel', *J. Ship Res., Vol. 21, No. 2,* 1977, pp. 107–119.

4. YEUNG, R.W., 'On the Interaction of Slender Ships in Shallow Water', *J. Fluid Mech.,* Vol. 85, part 1, 1978, pp. 143–159.

5. TUCK, E.O., NEWMAN, J.N., 'Hydrodynamic Interactions between Ships', *Proc. 10th Symposium on Naval Hydrodynamics,* Cambridge, Mass., USA, 1974, pp. 35–69.

6. IVANOV I.V. 'Interaction between Two Manoeuvring Vessels (in Russian)', PhD Thesis, *Leningrad Shipbuilding Institute*, 1981.

7. HESS, J.L., SMITH, A.M.O. 'Calculation of Nonlifting Potential Flow About Arbitrary Three-Dimensional Bodies', *J. Ship Res. Vol. 8,* 1964, pp. 22–44.

8. KORSMEYER, F.T., LEE, C.-H, NEWMAN, J.N., 'Computation of ship interaction forces in restricted waters', *J. Ship Res.*, Vol. 37, 1993, pp. 298–306.

9. YASUKAWA, H., 'Simulation of Ship Collision Caused by Hydrodynamic Interaction between Ships', *Proceedings International Conference on Marine Simulation and Ship Maneuverability (MARSIM'03),* 25–28 August 2003, Kanazawa, Japan, Vol. 3, 2003, pp. RC-31-1–RC-31-9.

10. CHEN, H.-CH., LIN, W.-M., LIUT, D.A., HWANG, W.-Y., 'An Advanced Viscous Flow Computation Method for Ship-Ship Dynamic Interactions in Shallow and Restricted Waterway', *Proceedings International Conference on Marine Simulation and Ship Maneuverability (MARSIM'03),* 25–28 August 2003, Kanazawa, Japan, Vol. 3, 2003, pp. RC-35-1–RC-35-10.

11. BRIX, J., Ed., 'Manoeuvring Technical Manual', Seehaven Verlag, Hamburg, 1993.

12. VANTORRE, M., VERZHBITSKAYA, E., LAFORCE, E., 'Model test based formulations of ship-ship interaction forces', *Ship Technology Research,* Vol. 49, 2002, pp. 124–141.

13. CH'NG, P.W., DOCTORS, L.J., RENILSON, M.R., 'A Method of Calculating the Ship-Bank Interaction Forces and Moments in Restricted Water', *Int. Shipbuild. Progr.*, Vol. 40, No. 421, 1993, pp. 7–23.

14. GRONARZ, A., 'Manövrierverhalten völliger Schiffe auf begrenzter Wassertiefe', *Versuchsanstalt für Binnenschiffen e.V.*, Duisburg, Bericht 1226, 1988.

15. VARYANI, K.S., KRISHNANKUTTY, P., VANTORRE, M., 'Prediction of Load on Mooring Ropes of a Container Ship Due to the Forces Induced by a Passing Bulk Carrier', *Proceedings International Conference on Marine Simulation and Ship Maneuverability (MARSIM'03),* 25–28 August 2003, Kanazawa, Japan, Vol. 3, 2003, pp. RC-33-1–RC-33-9.

16. VARYANI, K.S., 'Guide to New Generic Equations for Interaction Forces/Moments between Moored-Passing Ship and between Encounter-Overtaking-Overtaken Ships', *Proceedings International Conference on Marine Simulation and Ship Manoeuvrability (MARSIM 2006),* 25–30 June 2006, Terschelling, The Netherlands, 2006, pp. M-23-1–M-23-10.

17. SUTULO, S., GUEDES SOARES, C., 'Simulation of the Hydrodynamic Interaction Forces in Close-Proximity Manoeuvring', *Proceedings of the 27th Annual International Conference on Offshore Mechanics and Arctic Engineering (OMAE 2008), Estoril. Portugal, 15–19 June 2008,* Paper OMAE2008-57938, 10p., 2008.

18. SUTULO, S., GUEDES SOARES, C., 'Simulation of Close-Proximity Maneuvers Using an Online 3D Potential Flow Method', *Proceedings of International Conference on Marine Simulation and Ship Manoeuvrability MARSIM 2009, Panama City, Panama, 17-20 August 2009,* pp. M-9-1–M-9-10, 2009.

19. ZHOU, X. Q., SUTULO, S., GUEDES SOARES, C., 'Computation of Ship-to-Ship Interaction Forces by a 3D Potential Flow Panel Method in Finite Water Depth', *Proceedings of the 29th Annual International Conference on Offshore Mechanics and Arctic Engineering (OMAE 2010), Shanghai, People's Republic of China, 6–11 June 2010, Paper OMAE2010-20497,* 2010

20. SUTULO, S., GUEDES SOARES, C., OTZEN, J.F., 'Validation of Potential Flow Estimation of Interaction Forces Acting upon Ship Hulls in Side-to-Side Motion', Submitted for publication, 2010.

21. 'ANSYS CFX-Solver: Manager User's Guide, Release 12.0', ANSYS, Inc., 2009.

22. 'STAR CCM+ Version 5.04.006: User Guide, *CD-adapco*, 2010.

23. FERZIGER, J.H., PERIĆ, M., 'Computational Methods for Fluid Dynamics', Springer-Verlag, 2002.

24. FONFACH, J., GUEDES SOARES, C., 'Improving the Resistance of a Series 60 Vessel with a CFD Code', *Proceedings of the 5th European Conference on Computational Fluid Dynamics (CFD2010), ECCOMAS,* 2010.

8. AUTHORS BIOGRAPHY

José M. Fonfach is Research Assistant at the Centre for Marine Technology and Engineering (CENTEC) at Instituto Superior Técnico (IST). He is specializing in application of CFD methods to Ship Hydrodynamics. His previous experience included carrying out CFD computations with applications to Ship Resistance.

Serge Sutulo holds the current position of Researcher at CENTEC and of Adjunct Associate Professor at IST. He is responsible for research and teaching in the field of Ship Manoeuvrability. His previous experience includes various theoretical and experimental studies in the mentioned area.

C. Guedes Soares is President of CENTEC and holds position of Full Professor at IST. He is supervising different research projects in various areas and has worked in some fields of Naval Architecture, mainly in the interaction between ship hydrodynamics and ship structures including uncertainty modelling.

PREDICTION OF LINEAR HYDRODYNAMIC DERIVATIVES IN SHALLOW WATER

Y. Furukawa and Y. Nakiri, Kyushu University, Fukuoka, Japan
K. Kijima, Nagasaki Institute of Applied Science, Nagasaki, Japan

SUMMARY

The authors propose practical prediction method for linear derivatives for lateral force and yaw moment in shallow water. The linear derivatives in shallow water can be calculated by multiplying those in deep water by correction factor accounting for shallow water effect. The correction factor is a function of the ratio of ship's draught to water depth. It was obtained based on the hydrodynamic forces measured by captive model tests. Parameters contained in the correction factor can be estimated with ship's principal particulars and a few parameters representing stern form. It is shown that calculated hydrodynamic forces using the linear derivatives estimated with the correction factor agree well with measured results. It can be said that the correction factor is useful to estimate linear hydrodynamic derivatives in shallow water from the practical points of view.

NOMENCLATURE

a, a_1, a_2, a_3	Coefficients for $F(h)$ (-)
B	Ship's breadth (m)
C_b	Block coefficient (-)
C_{pa}	Prismatic coefficient of the aft hull between A.P. and midship (-)
C_{wa}	Waterplane area coefficient of the aft hull between A.P. and midship (-)
$D(\infty)$	Hydrodynamic derivatives in deep water
$D(h)$	Hydrodynamic derivatives in shallow water
d	Ship's draught (m)
e_a, e'_a	Coefficients representing the fullness of aft run (-)
$f(h)$	Added correction factor for hydrodynamic derivatives in shallow water (-)
$F(h)$	Correction factor for hydrodynamic derivatives in shallow water (-)
h	Ratio of ship's draught to water depth (-)
H	Water depth
K	Form factor (-)
L	Ship's length (m)
m	Ship's mass (kg)
m_x, m_y	Longitudinal and lateral components of added mass of ship (kg)
m', m'_x, m'_y	Nondimensional ship's mass and added masses of ship (-)
N_H	Yaw moment acting on ship's hull measured at midship (Nm)
N'_H	Nondimensional yaw moment acting on ship's hull measured at midship (-)
$N'_\beta, N'_r, N'_{\beta\beta}, N'_{rr}, N'_{\beta\beta r}, N'_{\beta rr}$	Linear and nonlinear derivatives for yaw moment (-)
r	Yaw rate (rad s^{-1})
r'	Nondimensional yaw rate (-)
Y_H	Lateral force acting on ship's hull measured at midship (N)
Y'_H	Nondimensional lateral force acting on ship's hull measured at midship (-)
$Y'_\beta, Y'_r, Y'_{\beta\beta}, Y'_{rr}, Y'_{\beta\beta r}, Y'_{\beta rr}$	Linear and nonlinear derivatives for lateral force (-)
U	Ship's speed (m s^{-1})
β	Drift angle (rad)
Δ	Manoeuvring criteria (-)
σ_a	Coefficient representing the fullness of the aft hull between A.P. and midship (-)
ρ	Density of water (kg m^{-3})

1. INTRODUCTION

It is important to know the hydrodynamic forces acting on ship precisely to evaluate ship manoeuvrability with high accuracy by numerical simulations. Furthermore shallow water effect should be taken into account for the simulation of ship manoeuvring in confined water. There are several ways to examine the shallow water effect. Eloot et al. conducted captive model tests for 8000 TEU container ship and investigated the influence of combination of draught and under keel clearance in detail [1]. Simonsen et al. carried out CFD simulation of flow around the bare hull in deep and shallow water [2].

The authors had proposed approximate formulae for estimating hydrodynamic derivatives in deep and shallow water [3][4]. The approximate formulae are available at initial design stage because they consist of principal particulars of ship and some parameters relating stern form. However there exists troublesomeness to calculate many parameters to obtain correction factor.

In this paper, the authors reanalyzed the measured hydrodynamic forces of seven model ships and propose new practical approximate formulae for estimating linear derivatives for lateral force and yaw moment in shallow water. The linear derivatives in shallow water can be calculated by multiplying those in deep water by correction factor accounting for the shallow water effect. The correction factor is a function of the ratio of ship's draught to water depth. It was obtained based on the hydrodynamic forces measured by captive model tests. Parameters contained in the correction factor can be estimated with ship's principal particulars and a few parameters representing stern form. It is shown that calculated hydrodynamic forces using the linear derivatives estimated with the correction factor agree well with measured results. It can be said that the correction factor is useful to estimate linear hydrodynamic derivatives in shallow water from the practical points of view.

2. HYDRODYNAMIC DERIVATIVES IN DEEP AND SHALLOW WATER

The authors have the data of hydrodynamic forces in both deep and shallow water conditions for seven ships shown in Table 1. Several kinds of ships are contained in the data and the range of block coefficients is about 0.57 to 0.83. Table 2 shows the conditions of water depth on the model experiment for each ship. All the hydrodynamic forces were measured using bare hull model of 2.5 m long (except for Ship C of 3.0 m long) conducted at the Seakeeping and Manoeuvring Basin in Kyushu University.

Lateral force Y_H and yaw moment N_H measured at midship are analyzed to derive hydrodynamic derivatives based on the following mathematical models which contain second order terms about drift angle β and nondimensional yaw rate r' ;

$$\left.\begin{aligned} Y'_H &= Y'_\beta \beta + Y'_r r' + Y'_{\beta\beta}\beta|\beta| + Y'_{rr} r'|r'| \\ &\quad + (Y'_{\beta\beta r}\beta + Y'_{\beta rr} r')\beta r', \\ N'_H &= N'_\beta \beta + N'_r r' + N'_{\beta\beta}\beta|\beta| + N'_{rr} r'|r'| \\ &\quad + (N'_{\beta\beta r}\beta + N'_{\beta rr} r')\beta r'. \end{aligned}\right\} \quad (1)$$

The superscript " ′ " refers to the nondimensionalized quantities as follows;

$$Y'_H = \frac{Y_H}{\frac{1}{2}\rho L d U^2}, \quad N'_H = \frac{N_H}{\frac{1}{2}\rho L^2 d U^2}, \quad r' = \frac{rL}{U}. \quad (2)$$

Table 3 displays the ratio of the values of hydrodynamic derivatives in shallow water to those in deep water ($H/d = 6.0$) for the seven ships shown in Table 1 and Fig. 1 shows the tendency of change of hydrodynamic derivatives to water depth. It is observed from Fig. 1(a) and (c) that the values of Y'_β and N'_β increase as water depth becomes shallow except for Y'_β of Ship A. The tendencies of change for Ship D to Ship G are well alike especially. The values of their C_b are larger than 0.8. As for the $Y'_r-(m'+m'_x)$ shown in Fig. 1(b), change of the value by the difference in water depth is smaller as compared with other derivatives. However, the tendencies of change of the value differ for every ship. It seems that the tendencies of change of N'_r for Ship D to Ship G shown in Fig. 1(d) are similar as same as Y'_β. It can be said that the change of hydrodynamic derivatives depending on water depth could be classified into two

Table 1: Principal particulars of ships

Ship	Ship type	L/B	B/d	C_b
Ship A	Container Ship	6.9	2.67	0.572
Ship B	Cargo Ship	5.96	2.99	0.698
Ship C	Chemical Tanker	5.96	2.74	0.751
Ship D	Coal Carrier	5.58	3.94	0.812
Ship E	Tanker	5.0	2.73	0.820
Ship F	Bulk Carrier	5.0	4.76	0.823
Ship G	Tanker	6.13	2.4	0.831

Table 2: Water depth conditions

Ship	H/d				
Ship A	6.0,			1.5,	1.2
Ship B	6.0,			1.5,	1.2
Ship C	6.0,			1.5,	1.2
Ship D	6.0,			1.5,	1.2
Ship E	6.0,			1.5,	1.2
Ship F	6.0,		2.0,	1.5,	1.2
Ship G	6.0,	3.0,	2.0,	1.5,	1.2

Table 3: Change of linear derivatives by the difference in water depth

Y'_β

H/d	6.0	3.0	2.0	1.5	1.2
Ship A	1.000	–	–	0.606	2.299
Ship B	1.000	–	–	1.811	4.883
Ship C	1.000	–	–	1.558	3.381
Ship D	1.000	–	–	2.420	5.751
Ship E	1.000	–	–	2.577	4.614
Ship F	1.000	–	1.357	1.885	4.050
Ship G	1.000	1.100	1.256	2.191	5.311

$Y'_r-(m'+m'_x)$

H/d	6.0	3.0	2.0	1.5	1.2
Ship A	1.000	–	–	1.128	–0.350
Ship B	1.000	–	–	1.088	1.290
Ship C	1.000	–	–	1.322	1.175
Ship D	1.000	–	–	1.333	1.404
Ship E	1.000	–	–	1.620	1.598
Ship F	1.000	–	1.342	1.345	1.551
Ship G	1.000	0.668	0.653	1.331	1.683

N'_β

H/d	6.0	3.0	2.0	1.5	1.2
Ship A	1.000	–	–	2.389	3.543
Ship B	1.000	–	–	2.100	4.291
Ship C	1.000	–	–	2.170	3.217
Ship D	1.000	–	–	3.056	5.171
Ship E	1.000	–	–	1.944	3.020
Ship F	1.000	–	2.021	2.959	5.115
Ship G	1.000	1.241	1.516	1.947	4.084

N'_r

H/d	6.0	3.0	2.0	1.5	1.2
Ship A	1.000	–	–	1.048	4.802
Ship B	1.000	–	–	0.733	1.254
Ship C	1.000	–	–	1.073	0.671
Ship D	1.000	–	–	1.387	2.276
Ship E	1.000	–	–	1.955	2.555
Ship F	1.000	–	1.280	1.392	1.915
Ship G	1.000	1.154	1.142	1.298	1.380

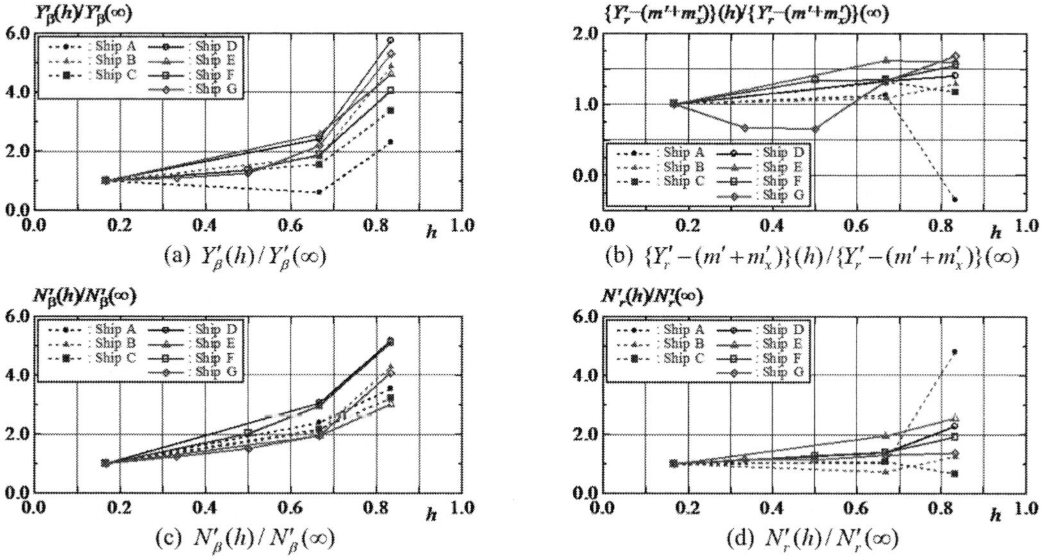

Fig. 1: The ratio of hydrodynamic derivatives in shallow water to those in deep water (H/d = 6.0)

groups, such as full ships of which C_b is larger than 0.8 and the others.

3. CORRECTION FACTOR FOR LINEAR DERIVATIVES IN SHALLOW WATER

In order to express change of hydrodynamic derivatives by the difference in water depth, the authors proposed correction factor $F(h)$ as function of h (= d/H) which represents the ratio of ship's draught to water depth [4]. Multiplying hydrodynamic derivatives in deep water by the correction factor, hydrodynamic derivatives in shallow water with arbitrary water depth are calculated as follows;

$$D(h) = F(h) \times D(\infty)\,. \tag{3}$$

$D(h)$ and $D(\infty)$ indicate hydrodynamic derivatives in shallow and deep water respectively. The author assumed that $F(h)$ had following forms [4];

$$F(h) = \left\{ \frac{1}{(1-h)^a} - h \right\} \times f(h) \quad \text{for } Y'_\beta \text{ and } N'_\beta, \tag{4}$$

$$F(h) = (a_3 h^3 + a_2 h^2 + a_1 h + 1) \times f(h) \quad \text{for } Y'_r-(m'+m'_x) \text{ and } N'_r. \tag{5}$$

a in Eq. (4) and a_1, a_2, a_3 in Eq. (5) are constants which have a inherent value for every derivative and $f(h)$ is added correction factor which is defined for every derivative. Although the constants contained in Eqs. (4) and (5) are calculated by approximate formulae consist of principal particulars and a few parameters which represents the characteristics of stern form, it is very troublesome because many formulae are required for every classification of ship based on C_b or other parameters.

This difficulty originates in having dealt with all the ships shown in Table 1 simultaneously. As mentioned above, the tendency of change of derivatives depending on water depth is classified into two groups. In this paper, the authors deal with only Ship D to Ship G of which the values of C_b are larger than 0.8, because fine ships have the complicated characteristics such as $Y'_r-(m'+m'_x)$ and N'_r of Ship A needing further investigation or experiment to understand the phenomena.

In order to reduce the number of parameters contained in the correction factor, the authors introduce new correction factor which has the following form;

$$F(h) = \frac{1}{\{1-(h-1/6)\}^p} - q \times (h - 1/6). \tag{6}$$

p and q are constants which have a inherent value for every ship and derivative. The value of $F(h)$ calculated by Eq. (6) is equal to 1.0 at the conditions of H/d = 6.0 (h = 1/6) which is regarded as deep water. By applying Eq. (6) for the experimental results shown in Table 3, the values of p and q were obtained. Table 4 shows the values of p and q for Ship D to Ship G.

4. ESTIMATION OF THE PRAMETERS CONTAINED IN THE CORRECTION FACTOR

Table 4: The values of p and q for Ship D to Ship G

		Ship D	Ship E	Ship F	Ship G
Y'_β	p	1.788	1.415	1.495	1.746
	q	2.066	0.179	1.703	2.262
$Y'_r-(m'+m'_x)$	p	−0.427	−0.426	−0.152	1.031
	q	−1.171	−1.564	−1.026	2.018
N'_β	p	1.385	1.062	1.398	1.429
	q	−0.888	0.289	−0.696	1.146
N'_r	p	1.062	0.647	0.668	−0.009
	q	1.401	−0.779	0.274	−0.585

It would be useful if the values of p and q could be obtained for arbitrary ships using a few parameters such as principal particulars. Then correlation among p, q and several nondimensional parameters are investigated. The nondimensional parameters consist of combination of principal dimensions and the following parameters;

$$\left.\begin{aligned} K &= \left(\frac{1}{e_a'} + \frac{1.5}{L/B} - 0.33\right)(0.95\sigma_a + 0.40), \\ e_a &= \frac{L}{B}(1 - C_{pa}), \quad e_a' = \frac{e_a}{\sqrt{1/4 + 1/(B/d)^2}}, \\ \sigma_a &= \frac{1 - C_{wa}}{1 - C_{pa}}. \end{aligned}\right\} \tag{7}$$

Equation (7) was proposed by Mori for the estimation of form factor K considering the characteristics of stern form [5]. These parameters are also used in the approximate formulae for hydrodynamic derivatives in deep water proposed by the authors [6].

From the above mentioned investigation on correlation among p, q and nondimensional parameters, adequate parameters were obtained for every derivative. Then fittings of p and q were carried out by the secondary function of the nondimensional parameter to each derivative as shown in Fig. 2. The solid lines are expressed by the following equations.

(a) For Y_β';

$$\left.\begin{aligned} p &= 166.0 \times \{(C_b B/L)K\}^2 \\ &\quad - 29.19 \times (C_b B/L)K + 2.630, \\ q &= -2486.0 \times \{(C_b B/L)K\}^2 \\ &\quad + 202.4 \times (C_b B/L)K - 1.928. \end{aligned}\right\} \tag{8}$$

(b) For $Y_r' - (m' + m_x')$;

$$\left.\begin{aligned} p &= 5381.0 \times (C_b B/L)^2 - 1659.0 \times C_b B/L + 127.1, \\ q &= 10970.0 \times (C_b B/L)^2 - 3405.0 \times C_b B/L + 262.0. \end{aligned}\right\} \tag{9}$$

(c) For N_β';

$$\left.\begin{aligned} p &= -4236.0 \times [\{d(1 - C_b)/B\}K]^2 \\ &\quad + 147.2 \times [\{d(1 - C_b)/B\}K] + 0.2478, \\ q &= -20710.0 \times [\{d(1 - C_b)/B\}K]^2 \\ &\quad + 887.0 \times \{d(1 - C_b)/B\}K - 8.349 \end{aligned}\right\} \tag{10}$$

(d) For N_r';

$$\left.\begin{aligned} p &= 12990.0 \times [\{d(1 - C_b)/B\}K]^2 \\ &\quad - 527.6 \times \{d(1 - C_b)/B\}K + 5.275, \\ q &= 8554.0 \times [\{d(1 - C_b)/B\}K]^2 \\ &\quad - 438.2 \times \{d(1 - C_b)/B\}K + 4.833. \end{aligned}\right\} \tag{11}$$

Figure 3 shows comparison between measured results and calculated results by Eq. (6) using the values of p and q calculated by Eqs. (8) to (11). Most of calculated results are estimated well for all ships except for $Y_r' - (m' + m_x')$ shown in Fig. 2(c). It seems that this deviation in $Y_r' - (m' + m_x')$ originates in the peculiar characteristics of Ship G differing from other ships.

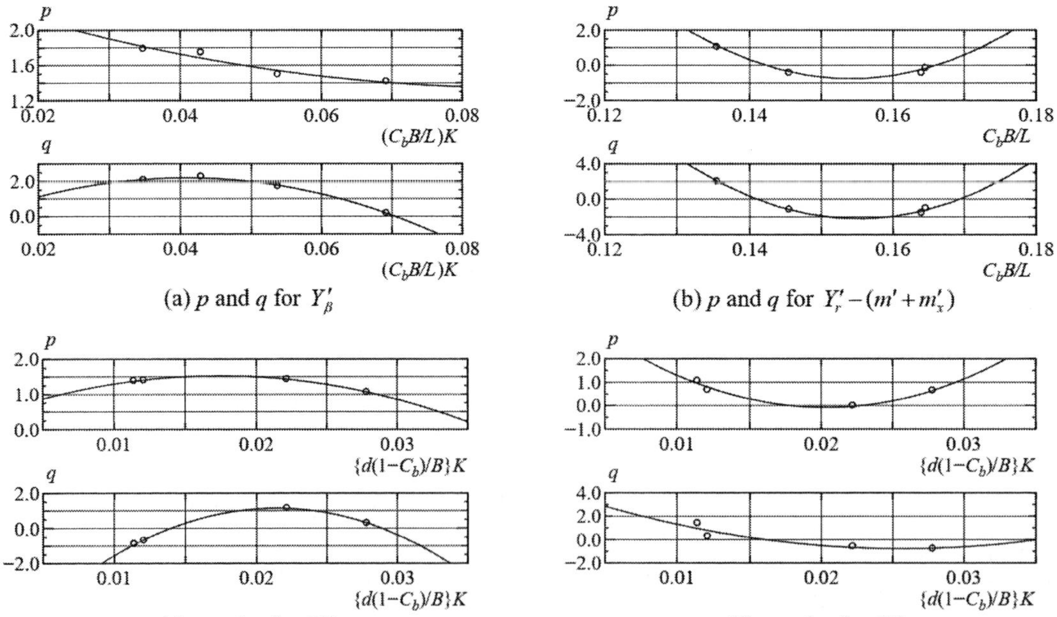

(a) p and q for Y_β'

(b) p and q for $Y_r' - (m' + m_x')$

(c) p and q for N_β'

(d) p and q for N_r'

Fig. 2: Parameter p and q as function of nondimensional parameters consist of principal particulars and stern form parameters

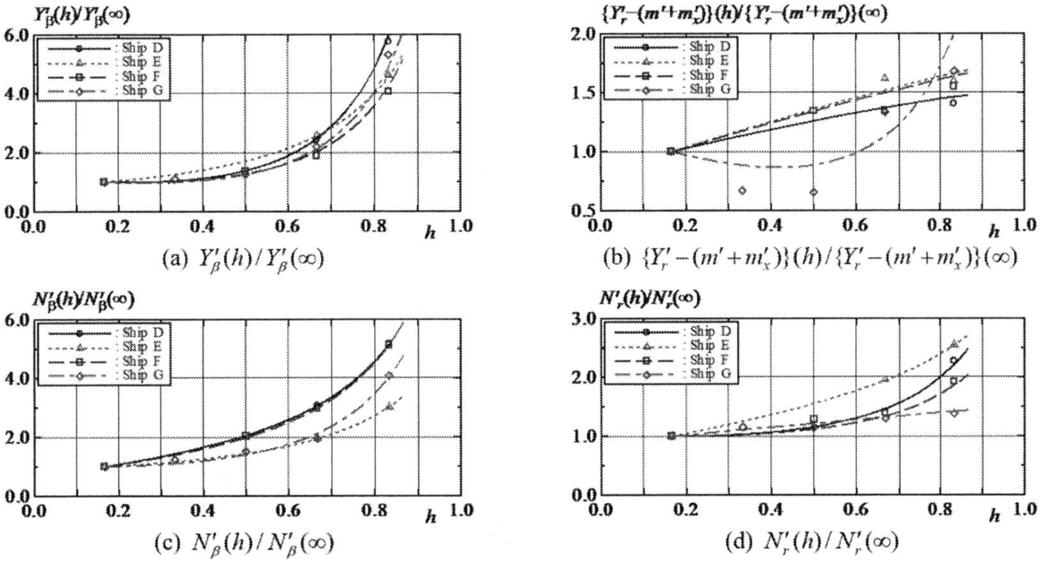

Fig. 3: Comparison between calculated correction factor and experimental results

Although the other ships' correction factor $F(h)$ for $Y'_r-(m'+m'_x)$ increases monotonically to h, the values of $F(h)$ at $h = 1/3$ and $1/2$ for Ship G once decrease and the values at $h = 1/1.5$ and $1/1.2$ are not so different from those of other ships. It means that the accuracy of fitting by Eq. (6) is inadequate for Ship G. Furthermore, signs of p and q for Ship G differ from those of other ships as presented in Table 4. On the other hand, the numbers of data at $h = 1/3$ and $1/2$ are insufficient for other ships.

5. COMPARISON BETWEEN CALCULATED AND MEASURED HYDRODYNAMIC FORCES

Figures 4 and 5 show comparison between calculated lateral forces and yaw moments by Eq. (1) and measured results for the conditions of $H/d = 1.2$ and 1.5 respectively. The linear derivatives used in Eq. (1) are estimated multiplying hydrodynamic derivatives in deep condition by the correction factor given by Eq. (6) with the parameters p and q calculated by Eqs. (8) to (11). The experimental values are used for nonlinear derivatives. There is a discrepancy between calculated and experimental results exist when values of drift angle and nondimensional yaw rate are big, but it is not so large.

Comparison between manoeuvring criteria calculated with experimental or calculated linear derivatives is shown in Table 5. The manoeuvring criteria is calculated by the following equation;

$$\Delta = Y'_\beta N'_r - \{Y'_r - (m' + m'_x)\} N'_\beta. \qquad (12)$$

As expected from the results shown in Fig. 3, difference between experimental and calculated values for Ship G is not good. However it would have enough accuracy to evaluate yaw checking and course keeping abilities at initial design stage.

Table 5: Comparison of manoeuvring criteria

	$H/d = 1.2$		$H/d = 1.5$	
	Exp.	Cal.	Exp.	Cal.
Ship D	0.0991	0.1058	0.0877	0.0891
Ship E	−0.0920	−0.0803	0.0032	−0.0031
Ship F	0.1519	0.1641	0.0887	0.1015
Ship G	0.0851	0.1283	0.0307	0.0252

6. CONCLUSIONS

The authors proposed correction factor for estimating linear hydrodynamic derivatives in shallow water. The correction factor is a function of the ratio of ship's draught to water depth and the parameters contained in the correction factor can be estimated with ship's principal particulars and a few parameters representing stern form. It can be said that the correction factor is useful to estimate linear hydrodynamic derivatives in shallow water from the practical points of view.

However the correction factor is derived from the data of only four model ships. It is required to collect more measured hydrodynamic forces both in deep and shallow water for the improvement in accuracy of estimation.

7. REFERENCES

1. ELOOT, K., VANTORRE, M., DELEFORTRIE, G., 'Prediction of Ship Manoeuvrability of an 8000 TEU Container Ship in Deep and Shallow Water: Mathematical Modelling and Captive Model Testing', *Proceedings of International Conference on Marine Simulation and Ship Manoeuvrability (MARSIM 2006, Netherlands), pp.M-4-1-M-4-11*, 2006.

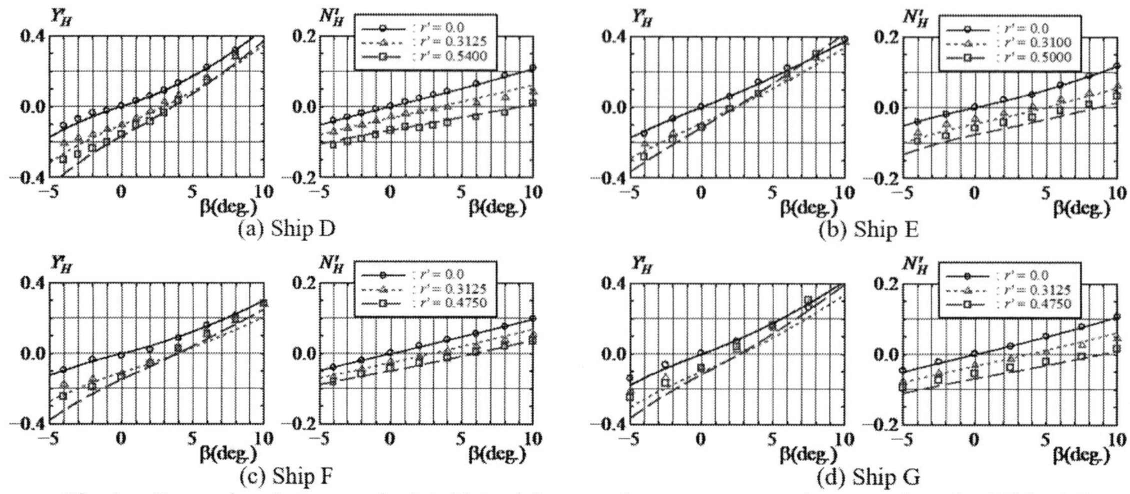

Fig. 4: Comparison between calculated lateral forces and yaw moments and measured results (H/d = 1.2)

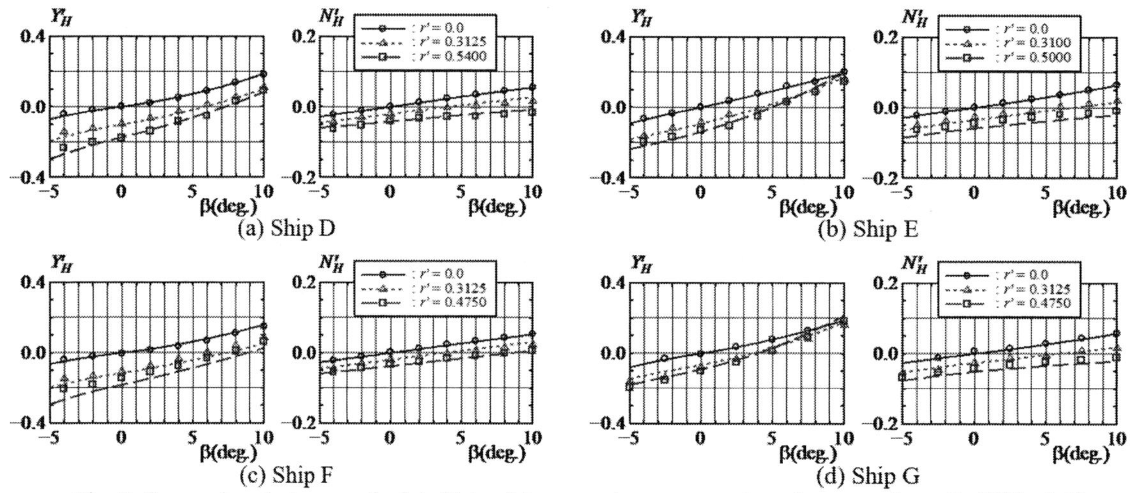

Fig. 5: Comparison between calculated lateral forces and yaw moments and measured results (H/d = 1.5)

2. SIMONSEN, C.D., STERN, F., AGDRUP, K., 'CFD with PMM Test Validation for Manoeuvring VLCC2 Tanker in Deep and Shallow Water', *Proceedings of International Conference on Marine Simulation and Ship Manoeuvrability (MARSIM 2006, Netherlands), pp.M-4-1-M-4-11*, 2006.

3. KIJIMA, K., NAKIRI, Y., TSUTSUI, Y., MATSUNAGA, M., 'Prediction Method of Ship Manoeuvrability in Deep and Shallow Waters', *Proceedings of Joint International Conference on Marine Simulation and Ship Manoeuvrability (MARSIM & ICSM 90, Tokyo), pp.311-318*, 1990.

4. KIJIMA, K., NAKIRI, Y., 'On the Practical Method for Prediction of Ship Manoeuvring Characteristics', *Proceedings of the Fourth Conference for New Ship and Marine Technology (New S-Tech 2004, Shanghai), pp.153-160*, 2004.

5. MORI, M., 'A Note on Hull Form Design (Part 24)', *Ship Science, Vol.48, pp.40-49*, 1995 (written in Japanese).

6. KIJIMA, K., NAKIRI, Y., 'On the Practical Prediction Method for Ship Manoeuvrability in Restricted Water', *Transactions of the West-Japan Society of Naval Architects, No.107, pp.37-54*, 2003 (written in Japanese).

8. AUTHORS BIOGRAPHY

Yoshitaka Furukawa is a professor of Kyushu University. He is responsible for education and research in Department of Marine Systems Engineering.

Yasuaki Nakiri is a technical staff in Kyushu University. He is responsible for model experiment and data analysis in the Seakeeping and Manoeuvring Basin.

Katsuro Kijima is the president of Nagasaki Institute of Applied Science and emeritus professor of Kyushu University. His previous experience includes the member of 19th an 20th ITTC Manoeuvrability Committee and the chairman of 21st ITTC Manoeuvrability Committee.

INTERACTION FORCES IN TUG OPERATIONS

S Geerts, Delft University of Technology, The Netherlands
M Vantorre, Ghent University, Belgium
K Eloot, Flanders Hydraulics Research, Belgium
R Huijsmans, Delft University of Technology, The Netherlands
N Fierens, Unie van Redding- en Sleepdienst, Belgium

SUMMARY

An assessment is made of the hydrodynamic interaction forces acting on a tug sailing freely in the vicinity of the bow of a large container vessel. For this purpose, a series of towing tank test have been performed during which the model of an azimuth stern drive tug was towed together with the model of a container carrier. Forces and moments on the tug model were measured for a range of relative positions and drift angles at multiple forward speeds. The model test results were incorporated in a fast-time simulation program to assess the steering action required to keep the tug at a fixed station by controlling the thrust and azimuth angle of both thrusters. Through simulations at different positions at which the towline is passed, an assessment is made of the most suitable position to perform this action.

NOMENCLATURE

B	Breadth of tug (m)
COG	Centre of gravity (-)
I_{xx}	Mass moment of inertia of tug about longitudinal axis (kg m^2)
I_{zz}	Mass moment of inertia of tug about vertical axis (kg m^2)
K	Roll moment on tug (Nm)
L	Length of tug (m)
m	Mass of tug (kg)
N	Yawing moment on tug (Nm)
N'	Non-dimensional yaw moment on tug (-) as described in (1)
N_i	Yawing moment on tug due to interaction with container vessel, positive when bow-out (Nm)
$N_{\dot{r}}$	Added moment of inertia for yaw (kg m^2)
$N_{\dot{v}}$	Hydrodynamic sway acceleration derivative of the yawing moment (kg m)
p	Roll velocity of tug (rad/s)
$\dot{p}$	Roll acceleration of tug (rad/s^2)
r	Rate of turn of tug (rad/s)
$\dot{r}$	Yaw acceleration of tug (rad/s^2)
T	Draft of tug (m)
U	Velocity component is tug's forward direction (m/s)
$\dot{u}$	Acceleration of tug in longitudinal direction (m/s^2)
v	Velocity component is tug's lateral direction (m/s)
$\dot{v}$	Acceleration of tug in lateral direction (m/s^2)
V_0	Forward speed of carriage and thus of both models (m/s)
X	Longitudinal force on tug (N)
X'	Non-dimensional longitudinal force on tug (-), as described in (1)
x_0	Longitudinal position of carriage (m)
x_G	Longitudinal coordinate of COG of tug (m)
X_i	Longitudinal force on tug due to interaction with container vessel, positive in forward direction (N)
x_{rel}	Relative longitudinal position of the tug with respect to the container vessel (m)
x'_{rel}	Relative longitudinal position of the tug with respect to the container vessel, non-dimensionalized with the tug's length. x'_{rel} = 0 when the tug's COG coincides with the midship's section of the container vessel; the tug's COG is at the longitudinal position of the bow of the container vessel when x'_{rel} = 3.3 (-)
$X_{\dot{u}}$	Added mass for surge (kg)
Y	Lateral force on tug (N)
Y'	Lateral force on tug, non-dimensionalized as described in (1)
y_0	Transverse position of planar motion mechanism with respect to centre line of tank (m)
Y_i	Lateral force on tug due to interaction with container vessel, positive when repulsive (N)
$Y_{\dot{r}}$	Hydrodynamic yaw acceleration derivative of the lateral force (kg m)
y_{rel}	Transverse position of both models (m)
y'_{rel}	Relative transverse position of both models, non-dimensionalized with the tug's breadth. y_{rel} = 0 when the centrelines coincide; the sides touch each other when y_{rel} = 1.9 (-)
$Y_{\dot{v}}$	Added mass for sway (kg)
z_G	Vertical coordinate of COG of tug (m)
$z_{\dot{v}}$	Vertical coordinate of centre of effort of hydrodynamic inertia force due to sway acceleration (m)
ρ	Density of water (kg/m^3)
ϕ	Roll angle (rad)
ψ	Heading angle (rad)

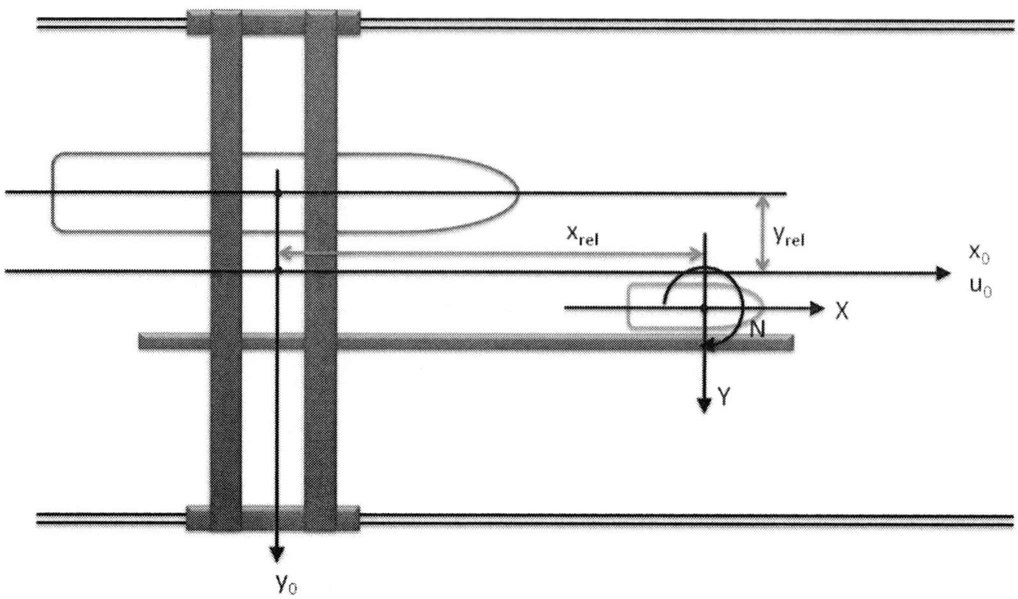

Figure 1: Overview of model topology in the towing tank

1 INTRODUCTION

Waterways and harbour entrances have not grown significantly in the last decades and navigational infrastructure such as bridges and locks is updated very slowly. Meanwhile, the carrying capacity of cargo vessels has not ceased to grow in latest years. Especially the size of container vessels has increased significantly. Together with this increase in deadweight and carrying capacity, the installed power has gone up which results in a higher dead slow speed.

This higher dead slow speed has a great influence on the way these bigger ships enter port and perform harbour manoeuvres. For the cargo vessel to stay manoeuvrable, especially on a river with multiple turns, such as the river Scheldt, a higher forward speed is maintained throughout the passage.

Figure 2: "Union Ruby" assisting at the rear on the river Scheldt. (Photo: Stefan Geerts)

Modern tugs are more powerful than their predecessors but have not grown in size at the same rate. Due to the higher manoeuvring speed of the cargo vessel they have to handle, the tug will have to sail at a higher speed through water and, therefore, at a higher Froude number at the moment the towline has to be passed. This has implications on the hydrodynamic forces acting on the tug and may even create a potentially dangerous situation for both tug and crew.

In this paper, the focus will lie on the hydrodynamic forces acting on a tug assisting at the bow of a large cargo vessel. To pass the towline successfully, a tug has to spend a considerable time at a fixed station along the hull of the assisted vessel. To pass a bowline, the tug will pass a considerable amount of time at a fixed station near the forward shoulder of the cargo vessel to pass the line before moving forward of the bow to assist the vessel. The flow in the vicinity of the bow is highly two-dimensional due to the high forward speed and the restricted water depth that often occur when performing this kind of operation. This generally causes an amplification of the interaction effects (sway force, yaw moment) experienced by the tug.

When sailing in the vicinity of another vessel, interaction effects occur. The close proximity of a large vessel near the free-sailing tug will greatly influence the hydrodynamic forces acting on the tug. In particular, a sway force and a yawing moment will be induced due to the asymmetry of the flow, but also the resistance will be affected. These interaction forces are expected to be intensified due to the two-dimensional character of the flow mentioned above.
In order to assess these hydrodynamic interaction forces and moments, a series of model tests have been

carried out in the *Towing tank for manoeuvres in shallow water (co-operation Flanders Hydraulics Research – Ghent University)* in Antwerp (Belgium), on the initiative of URS (Unie van Redding- en Sleepdienst, Antwerp), the leading towage and salvage company for all Belgian and Dutch ports on the river Scheldt, with support of Delft University of Technology. During these tests, the model of an azimuth stern drive tug was towed together with the model of a container carrier, see Figure 1. Forces and moments on the tug model were measured for a range of relative positions and drift angles at multiple forward speeds.

The interaction forces registered during the model tests were incorporated in a fast-time simulation program to assess the steering action required to keep the tug at a fixed station by controlling the thrust and azimuth angle of both thrusters. Through simulations at different positions at which the towline is passed, an assessment is made of the most suitable position to perform this action.

This paper intends to describe the model test setup and discuss the results of the tests and of the simulations. Firstly, an overview of the use of an azimuth stern drive tug for harbour (approach) manoeuvres will be given.

2 CASE DESCRIPTION

The need to assess the interaction forces acting on the tug's hull while sailing in close proximity of a large cargo vessel has arisen from a number of incidents and accidents that have occurred over time, some of them with serious consequences. In this research the use of an azimuth stern drive (ASD) tug as a fore-boat (Figure 3) has been considered since the tug fleet of URS consists for a large part of ASD-tugs.

Figure 3: ASD-tug "Union Emerald" assisting bulk carrier "Aizu" stern-to-bow as seen from the wheelhouse. (Photo: Stefan Geerts)

Different assisting methods exist for an ASD-tug assisting at the bow, such as stern-to-bow and bow-to-bow. When assisting bow-to-bow, the ASD-tug sails in reverse with the thrusters pulling the tug through the water, the tug is assisting with the towline attached at the bow of the tug. It effectively acts as a tractor tug. When assisting stern-to-bow, the tug sails in the forward direction with the towline attached at the aft towing post which is for the most part of the ASD-fleet situated a little aft of amidships.

The direction of travel of the tug, forward or reverse, will influence the interaction forces, since these depend on the underwater shape of the tug's hull. In this research, only the stern-to-bow method is considered.

A very critical part of harbour tug operations occurs while passing the lines to the assisted vessel. The tug spends some considerable time very close to the assisted vessel to pass a series of lines. First a small diameter messenger line is thrown from the cargo vessel to the tug; the speedy execution of this manoeuvre depends largely on the experience and capabilities of the crew on board the cargo vessel. Secondly, a feeder line is sent up from the deck of the tug via the messenger line, which will feed the towline to the ship where it is clasped on a bollard.

To facilitate throwing, often the tug will have keep station near the forward shoulder of the vessel, since the freeboard there is less than on the bow. Due to time restrictions, and increasingly inexperienced deckhands, the tug will have to approach much closer to the bow.

This does not pose problems per se, but for an ASD-tug to manoeuvre away from a cargo vessel without bow thruster, it will first have to swing the aft around closer to the hull of the assisted vessel before being able to thrust forward and move away from the assisted vessel because the tow point is located closer to amidships than the propulsive force. A tractor type tug, for example a Voith-Schneider (VS) tug, will pull away directly from the hull of a cargo vessel since the tow point is located more aft than the propulsive force (Figure 4).

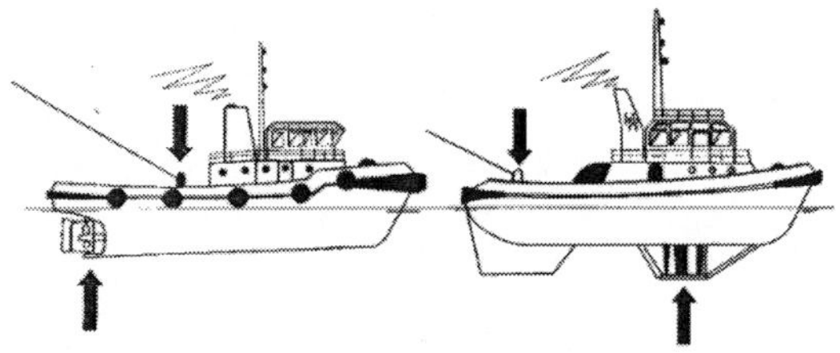
Figure 4: Difference between a conventional type tug and a tractor tug. Left: conventional tug, identical to ASD-tug. Right: tractor type tug, Voith-Schneider type.

An ASD-tug is equipped with often two azimuth stern drives, hence the name. An azimuth stern drive can rotate through 360 degrees and so provide thrust in all directions. Both drives can be steered independently or coupled from the wheelhouse by the tug master. In normal transit conditions, the thrusters will be steered coupled via an autopilot and only small thrusters angles will be needed to manoeuvre the tug.

An experienced ASD-tug master is able to use both thrusters independently and so vector thrust in any given direction, comparable to a VS tug. Very complicated manoeuvres, like crabbing sideways, can be accomplished by vectoring thrust from both thrusters independently. This versatility of the ASD-tug explains its popularity among fleet operators.

3 TOWING TANK EXPERIMENTS

3.1 FACILITIES

To measure interaction forces, towing tank experiments were performed in the towing tank for manoeuvres in shallow water (Figure 5) at Flanders Hydraulics Research in Antwerp, Belgium, a research institution of the Flemish Government specialised not only in nautical but also in hydraulic and hydrological research. The tank is dedicated to manoeuvring and seakeeping experiments in very shallow water. It is equipped with a planar motion mechanism, a wave generator and an auxiliary carriage for ship-ship interaction tests. The towing carriage is fully computerized and automated making clock round testing possible during every day of the week. All three motion modes of the planar motion mechanism can be controlled independently, placing every motion in the horizontal plane within grasp. The main tank dimensions are:

- Length over all: 88.0 m
- Useful length: 67.0 m
- Width: 7.0 m
- Maximum water depth: 0.5 m
- Typical ship model length: 3.5 – 4.5 m

Figure 5: Towing tank for manoeuvres in shallow water (co-operation Flanders Hydraulics Research – Ghent University), Antwerp (photo: Flanders Hydraulics Research)

3.2 EXPERIMENTAL SETUP

To test two ships simultaneously in the tank, one model is attached to the planar motion mechanism of the main carriage and a second model is attached to an auxiliary frame. As both models must have the same forward speed, this frame is also attached to the main carriage.

To identify the influence of longitudinal and transverse separation of the two models, it should be possible to vary the position of the tug relative to the container vessel. It is therefore decided to attach the model of the container vessel to the planar motion mechanism of the main carriage which will then be used to create transverse separation between the models. The ASD-tug model is attached to a secondary carriage which can be positioned along the length of the auxiliary frame attached to the main carriage as can be seen on Figure 6.

Figure 6: Tug-Container vessel interaction testing at Flanders Hydraulics Research (photo: Flanders Hydraulics Research)

The linkage between tug model and secondary carriage allows the tug model to be set at a specific drift angle relative to the direction of motion. The position of the tug model is incorporated in the automated system of the towing tank; in this way, no manual manipulation has to be done to set both longitudinal and transverse separation and testing can be continued unsupervised.

The ASD-tug considered is the "Union Emerald" of URS:

- Length: 33.00 m
- Breadth: 11.00 m
- Draught: 5.60 m
- Gross tonnage: 497
- Bollard pull (ahead): 65 tons
- Bollard pull (astern): 60 tons

The cargo vessel considered is a container vessel with following main dimensions:

- Length: 229.00 m
- Breadth: 30.50 m
- Draught: 8.20 m
- Displacement: 33725 m^3

The scale used for the models and the test program is 1/50. The somewhat limited size of the assisted vessel is due to the restrictions on model size in the towing tank for the interaction experiments. Test results are expected to be subject to scale effects on viscous forces.

3.3 TEST PROGRAM

As already mentioned, the longitudinal and transverse separation of tug and container vessel is varied, together with the drift angle of the tug relative to the forward direction of travel of the container vessel.

Only stationary conditions have been tested due to time restrictions, so the longitudinal and transverse separation is kept constant during the whole run. The following variations are done:

Forward speed [knots], full scale	5, 6, 7, 8, 9
Longitudinal separation x_{rel} [-], from amidships of container vessel to centre of gravity of tug.	0.73, 1.25, 1.77, 2.29, 2.82, 3.33, 3.88, 4.41, 4.93
Transverse separation y_{rel} [-], from centreline to centreline	2.6, 3.3, 4.0, 4.4, 5.3
Drift angle [°], negative angle bow-in since tug model is situated on starboard in the towing tank.	-10, -5, 0, 5, 10

Table 1: Test matrix for tug-container vessel interaction

Additionally, a limited number of tests were performed with only the tug model, so without interaction. These are done at a full scale velocity of 5, 7 and 9 knots and the same drift angles mentioned in Table 1.

3.4 RESULTS

The conventions and topology shown in Figure 1 dictate that a bow-out yaw moment and a repulsive sway force are positive.

The results of the towing tank tests are non-dimensionalized as follows::

$$X' = \frac{X}{\frac{1}{2}\rho V^2 LT}$$

$$Y' = \frac{Y}{\frac{1}{2}\rho V^2 LT} \quad (1)$$

$$N' = \frac{N}{\frac{1}{2}\rho V^2 L^2 T}$$

3.4 (a) Side force and turning moment due to interaction

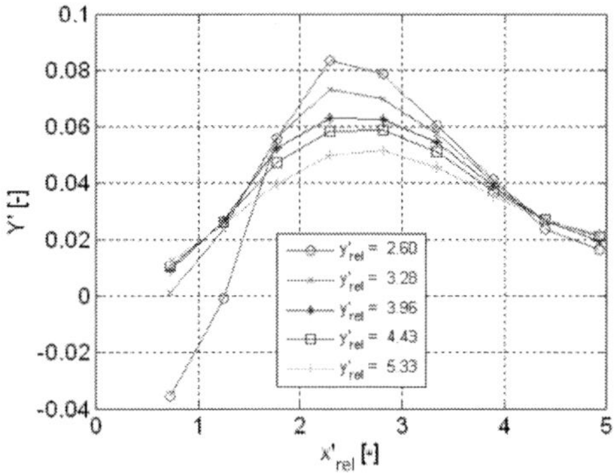

Figure 7: Interaction sway force as a function of relative longitudinal position for different relative transverse separations; Fn = 0.20.

The change in interaction side force along the length of a container vessel sailing at eight knots full scale speed is shown in Figure 7. An interaction side force is clearly present and changing when sailing along the hull of a container vessel.

The highest side forces appear when the tug's centre of gravity is at a longitudinal position of about 2.5. This puts the bow of the tug a little behind the bow of the container vessel. Compared to literature 3, 4 the maximum side force occurs further aft in the more recent experiments, this is probably due to the smoother underwater shape of the container vessel. The cargo vessel models used in 3 and 4 are much fuller and will create a more significant pressure field at the fore shoulders. Meanwhile a modern container vessel, designed for higher cruising speed, will have a smoother entrance in the water. Neither does the container vessel considered have a pronounced parallel middle body, which might explain why the "wall effect" discussed by Dand is not observed here.

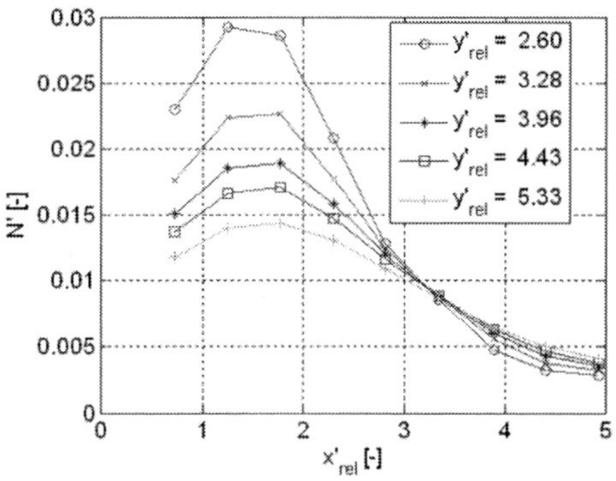

Figure 8: Interaction yaw moment as a function of relative longitudinal position for different relative transverse separations; Fn = 0.20, sides touch at y_{rel} = 1.9, both ships are bow on bow at x_{rel} = 3.3.

The change in interaction yaw moment acting on a tug sailing along the length of a container vessel can be found in Figure 8. A large bow-out moment can be observed between 1 and 2. The magnitude of the interaction yaw moment clearly depends on longitudinal and transverse position.

The largest yaw moment occurs when the tug is abeam of the shoulder of the container vessel. A large bow-out moment is possibly created by the cross-flow of the container vessel catching the rather full fore-body of the tug. In literature, 3, 4, a large bow-in moment is found when the tug is forward of the bow of the container vessel. In the present experiments, this was not found. This is possibly due to the absence of propulsive devices or lift generating surfaces at the rear of the tug during the tank experiments.

3.4 (b) Effect of separation distance

As expected, both interaction sway force and yawing moment decrease with increasing lateral separation distance between cargo ship and tug, as is clearly illustrated in Figure 9 and Figure 10, respectively.

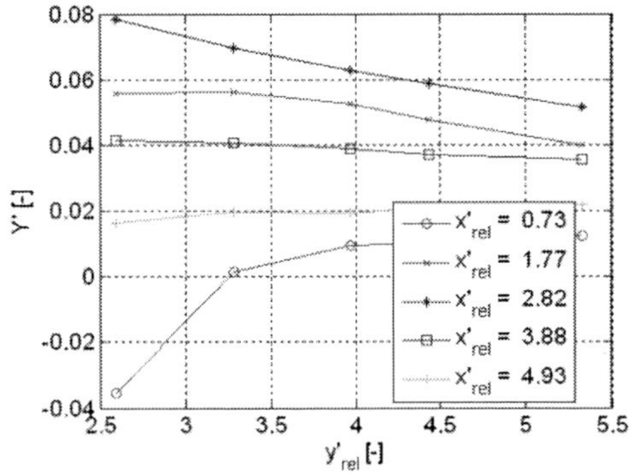

Figure 9: Interaction sway force as a function of relative transverse separation for different relative longitudinal positions; Fn = 0.20.

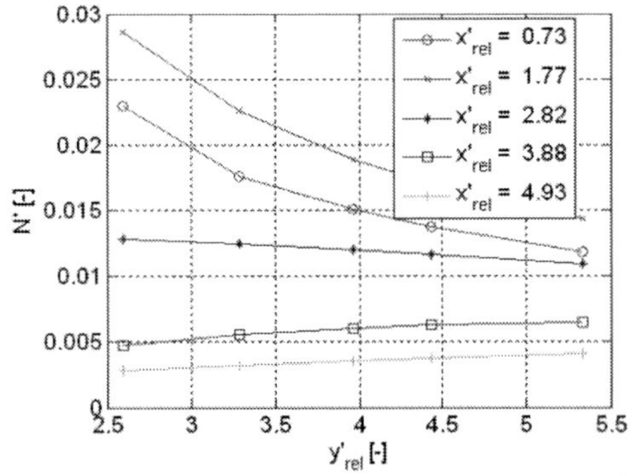

Figure 10: Interaction yaw moment as a function of relative transverse separation for different relative longitudinal positions; Fn = 0.20.

3.4 (c) Effect of speed

In Figure 11, the sway force acting on the tug at a lateral distance of 3.3 with zero drift angle is displayed, for forward speed values of 5 to 9 knots. These speeds correspond with Froude numbers 0.14 to 0.26, respectively, determined with the tug's length. The interaction sway force, made non-dimensional with the square of the forward speed, is found to be without large variation in a significant zone along the container vessel's hull. The maxima found are between 5% and 15%.

The variation of the non-dimensional interaction yaw moment on the tug has more variation along the container vessel's hull. One can see that far enough forward of the bow of the container vessel, the interaction yaw moment will reduce to zero. At higher velocities, the peak situated between 1 and 2 – i.e. in the vicinity of the fore shoulder of the assisted ship – appears to be more pronounced, about 30%.

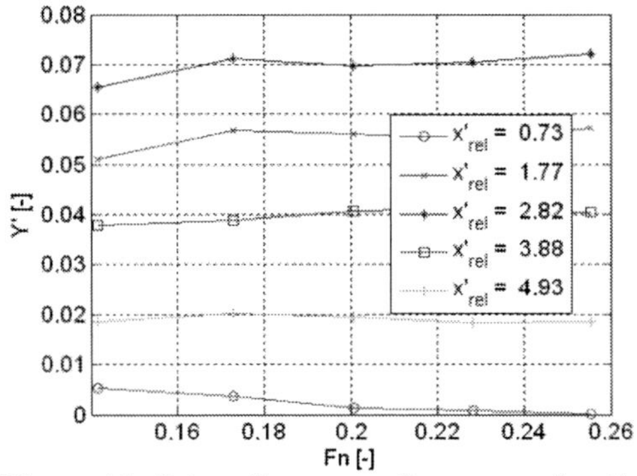

Figure 11: Interaction sway force as a function of Froude number for different relative longitudinal positions; $y_{rel} = 3.28$.

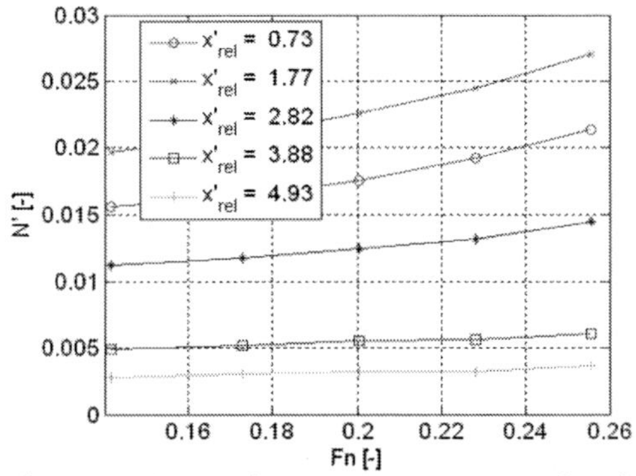

Figure 12: Interaction yaw moment as a function of Froude number for different relative longitudinal positions; $y_{rel} = 3.28$.

3.4 (d) Effect of drift angle

The total sway force acting on the tug, consisting of a component due to the oblique course and an interaction component, is presented in Figure 13 as a function of the drift angle of the tug. The tug and container vessel sail at 7 knots and the transverse separation of both vessels is of magnitude 4. Drift angles between -5 and 10 degrees are considered in steps of 5 degrees; a negative drift angle indicates a bow-in position.

Comparing the oblique paths with a straight course path in Figure 13, the same trend can be observed in sway force along the hull of a container vessel. The magnitude of the sway force is naturally largely dependent on the drift angle, therefore a more deconstructed analysis is shown in Figure 15.

The total yaw moment shows a similar trend for bow-in values, only the magnitude changes significantly with drift angle (Figure 14). For bow-out drift angles on the other hand, a change in trend for different positions along the hull of the container vessel is observed. To consider this change a more detailed analysis is presented in Figure 16.

The interaction sway force is obtained by subtracting the sway force measured with no assisted ship present from the total measured sway force.

In Figure 15 the total sway force is normalised with the nominal sway force to obtain the interaction sway force. The sway force measured at the same drift angle with no assisted ship present, the nominal sway force, is subtracted from the measured sway force with interaction to obtain the interaction sway force. A similar approach is used to determine the interaction yaw moment.

One now observes that for bow-in drift angles only very little variation in normalized interaction side force can be observed. Bow-out drift angles on the other hand produce large added interaction sway force components depending on the drift angle. A reason for this phenomenon can be contributed to the presence of the large skeg at the stern of the ASD-tug close to the hull of the container vessel. The strong two-dimensional accelerated flow close to the hull of the container vessel will intensify the lift produced by this skeg.

The interaction yaw moment is obtained by subtracting the yaw moment measured with no assisted ship present from the total measured yaw moment.

The total interaction yaw moment of Figure 14 is normalised with the nominal yaw moment measured at the same drift angle with no interaction present and represented in Figure 16. One can observe the same trend as in Figure 15, although less clear. A similar added interaction yaw moment can be observed at bow-in angle; meanwhile bow-out angles produce a much larger variation in added yaw moment.

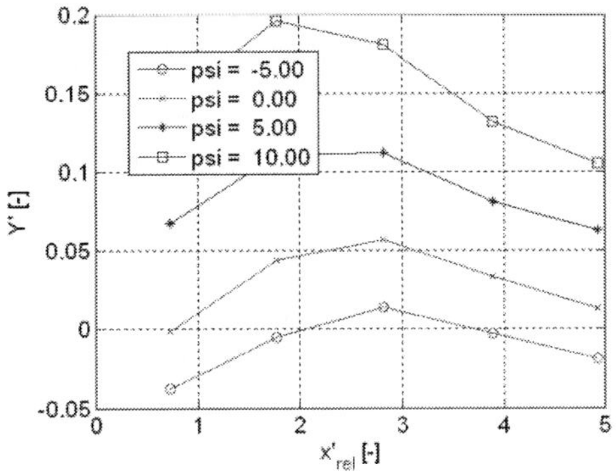

Figure 13: Total sway force as a function of relative longitudinal position for different drift angles, Fn = 0.20 and y_{rel} = 3.96.

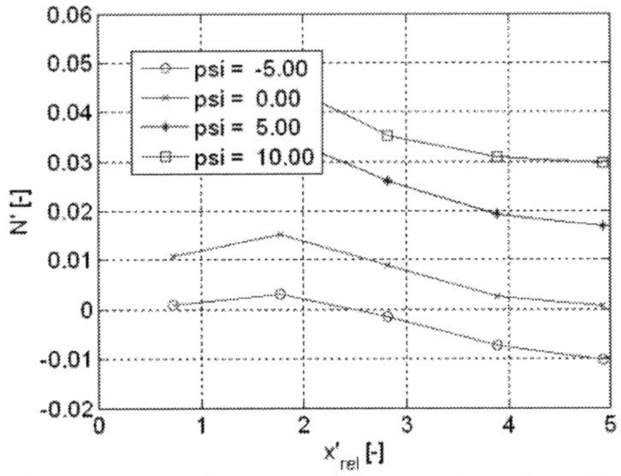

Figure 14: Total yaw moment as a function of relative longitudinal position for different drift angles, Fn = 0.20 and y_{rel} = 3.96.

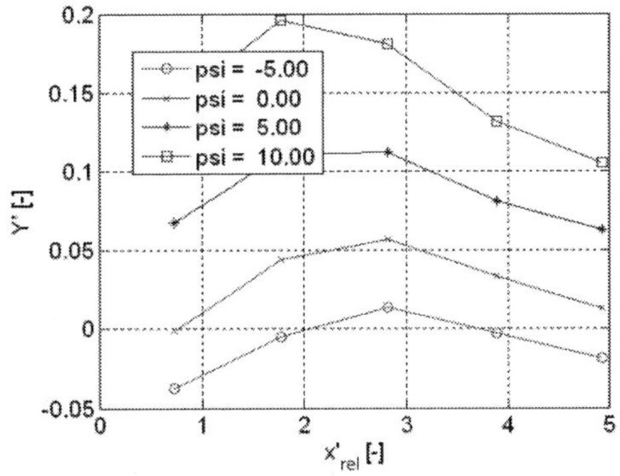

Figure 15: Interaction sway force as a function of relative longitudinal position for different drift angles, Fn = 0.20 and y_{rel} = 3.96.

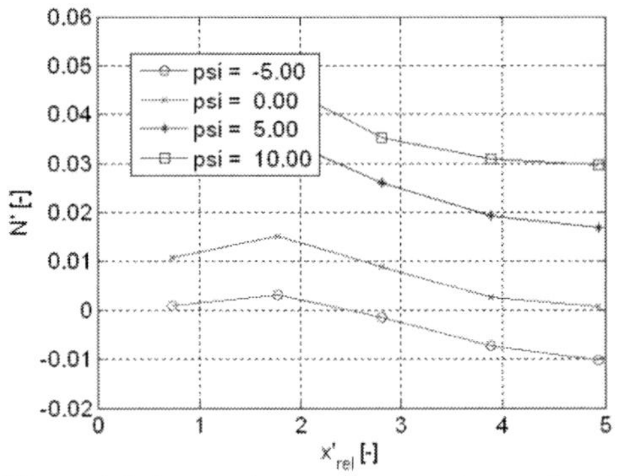

Figure 16: Interaction yaw moment as a function of relative longitudinal position for different drift angles, Fn = 0.20 and $y_{rel} = 3.96$.

3.4 (e) Effect on resistance

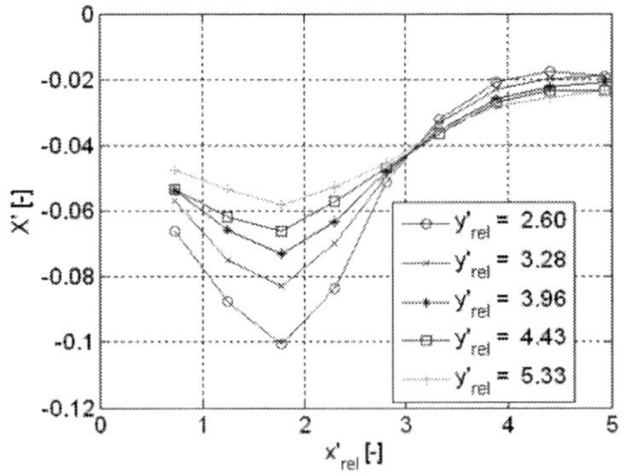

Figure 17: Total longitudinal force as a function of relative longitudinal position for different relative transverse separation values; Fn = 0.20,.

As known from literature, 3, 4, 0, 0, a change in resistance is experienced when sailing along the hull of a larger vessel. In Figure 17 is shown how this change in resistance intensifies aft of the forward shoulder of the container vessel, especially between 1.5 and 2. Here, the complete hull of the tug is encased in the divergent wave system generated by the fore shoulder of the container vessel.

One notices that forward of the bow of the container vessel the total resistance is diminishing; due to the high pressure field forward of the bow of the container vessel, the tug will experience a forward thrust by the container vessel, reducing resistance.

4 REAL-TIME SIMULATION

To properly investigate the behaviour of an ASD-tug in the vicinity of a container ship's bow, dynamic behaviour has to be introduced. Since mainly interaction forces are determined during towing tank tests, the full behaviour of the ASD-tug cannot be described with the results of these tests alone.

4.1 EQUATIONS OF MOTION

The equations of motion used to model the behaviour of the tug are based on those for motions in the horizontal plane with the addition of a fourth equation for the roll motion. These are given by:

$$\begin{aligned}
&(m - X_{\dot{u}})\dot{u} = X + mvr + mx_G r^2 - mz_G pr \\
&(m - Y_{\dot{v}})\dot{v} + (mx_G - Y_{\dot{r}})\dot{r} - mz_G \dot{p} = Y - mur \\
&(mx_G - N_{\dot{v}})\dot{v} + (I_{zz} - N_{\dot{r}})\dot{r} = N - mx_G ur \\
&(-mz_G - Y_{\dot{v}} z_{\dot{v}})\dot{v} - Y_{\dot{r}} z_{\dot{v}} \dot{r} + I_{xx}\dot{p} = K + mz_G ur \qquad (2)
\end{aligned}$$

From the first equation, $\dot{u}$ can be calculated independently from the other equations. The equations for sway, heel and yaw are coupled and will be solved using the method of Cramer. The body forces (X, Y, K and N) consist of a propulsion component, a hydrodynamic force component and an interaction component describing the hydrodynamic forces between container vessel and tug.

4.1 (a) Hydrodynamic force components

- Resistance
  A full scale investigation was done at URS to determine the speed/power ration of this type of ASD-tug. The results from this investigation were used to determine a speed dependant resistance curve. The added mass in X-direction is calculated with the program SEAWAY. By calculating the response of the tug at very low circular wave frequencies and determining the limit value at zero frequency, an approximation of added mass is attained.

- Sway force and yaw moment
  From existing literature, 0 and a limited number of towing tank tests carried out with only the tug model, force coefficients are determined as a function of forward speed and drift angle. The sway added mass and yaw added moment of inertia are determined using an analog method as described for resistance. The damping is approximated using a semi-empirical representation from 2 and 0.

- Heeling moment
  Roll added inertia and damping are determined using the methods described for resistance and sway incorporating non-lineair effects due to the skeg. The stability moment $mgGM \sin \phi$ is also included in this formulation.

4.1 (b) Propulsion forces

The tug is propelled by two azimuth stern drives. The propellers fitted are Kaplan 4-70 propellers in a 19A duct. For this Ka 4-70/19A, a four quadrant thrust coefficient diagram can be found in [14]. The thrust depends on ship velocity, inflow velocity at the thruster and azimuth angle. As a rudimentary approximation, the inflow velocity is determined to be the component of the local velocity vector $\vec{v}_{local} = \vec{v}_{global} + \vec{\Omega} \times \vec{r}$ with $\vec{d}$ the position vector of the azimuth drive, normal to the thruster axis, thus dependent on the azimuth angle.

4.1 (c) Interaction forces

The interaction forces are determined with the stationary towing tank tests. Depending on the lateral and longitudinal separation, a zero-heading interaction force component is determined by interpolation on the test-results. To this, a component is added that depends on separation and drift angle, i.e. a cross-term between drift and separation. This method is analogous to the decomposition in 0.

4.2 STATIONARY EVALUATION

At the time of writing of this paper, only stationary situations of the station keeping manoeuvre have been simulated. Only limited evaluation will be reported in this paper, the complete evaluation can be found in 0.

Interaction forces have been determined by towing tank tests only in a limited area around the bow of the container vessel. One therefore has to take care not to introduce a step function when starting a dynamic simulation. It is thus necessary to determine a whole range of initial values across the simulation domain. A stationary initial condititon will be determined by a drift angle, a propulsion rate and an azimuth angle, all depending on the longitudinal and transverse separation and the forward velocity of the container vessel.

To reach a stationary intial state, a cost function combining the forces acting on the tug is minimized to compute the optimal combination of drift angle, azimuth angle and propeller rate for the tug to remain on station. A stationary condition cannot be achieved at all combinations of relative longitudinal position and transverse separation, as is depicted in Figure 18.

At low velocity, the influence of the interaction forces will be greater, resulting in highly unstable situations. Figure 19 shows the stability of calculated locations at a much higher Froude number, Fn = 0.23. It can be seen that only a limited number of unstable combinations of relative longitudinal position and relative transverse separation can be found.

In Figure 20, one can find the stationary results as a function of relative longitudinal position and relative transverse separation for Froude number 0.19. Figure 21 and Figure 22 show the corresponding azimuth angle and propeller rate needed to keep station.

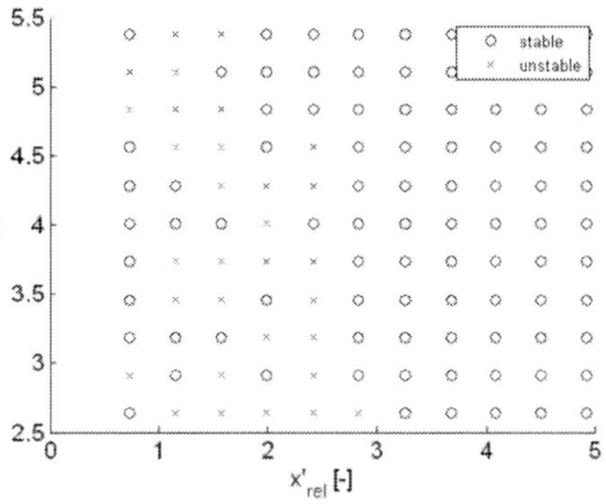

Figure 18: Stable and unstable stationary conditions referred to the relative longitudinal position x_{rel} and the relative transverse separation y_{rel}; Fn = 0.14, sides touch at y_{rel} = 1.9, both ships are bow on bow at x_{rel} = 3.3.

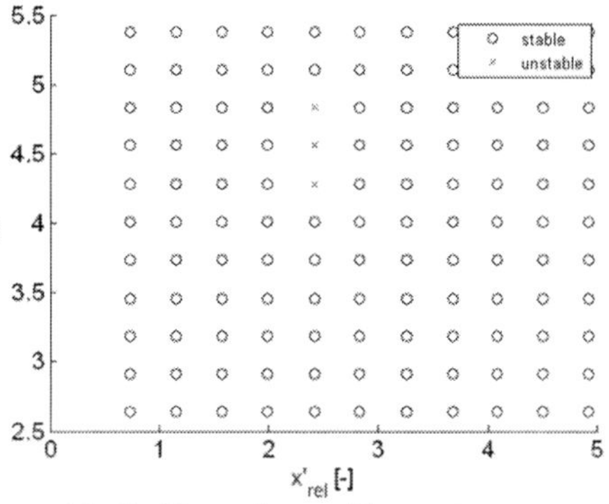

Figure 19: Stable and unstable stationary conditions referred to the relative longitudinal position x_{rel} and the relative transverse separation y_{rel}; Fn = 0.23, sides touch at y_{rel} = 1.9, both ships are bow on bow at x_{rel} = 3.3.

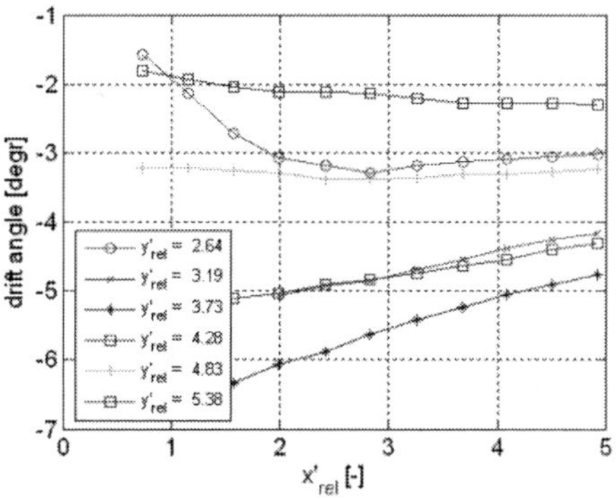

Figure 20: Stationary drift angle as a function of relative longitudinal position for different values of relative transverse separation; Fn = 0.19

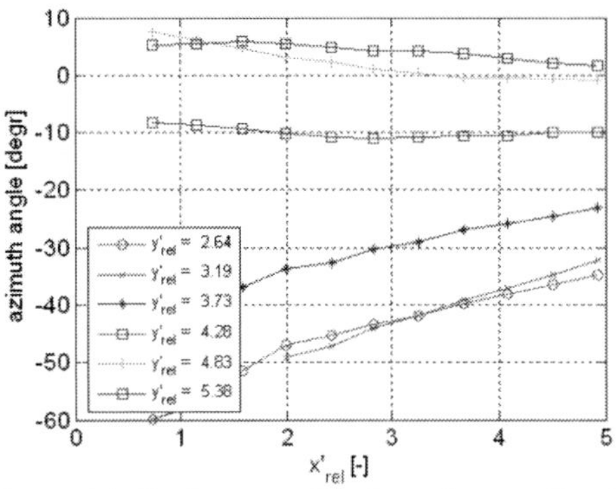

Figure 21: Stationary azimuth angle as a function of relative longitudinal position for different values of relative transverse separation; Fn = 0.19

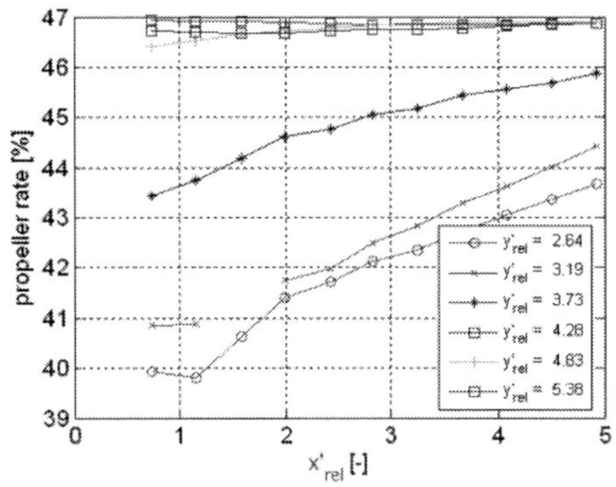

Figure 22: Stationary propeller rate as a function of relative longitudinal position for different values of relative transverse separation; Fn = 0.19

One can observe in Figure 20 that the stationary drift angle varies significantly with both longitudinal position and transverse separation. The maximum drift angle is reached at about 3.7 times the tug's breadth of separation between the centrelines and will generally diminish when moving forward along the container vessel's hull.

The corresponding azimuth angle and propeller rate show a different evolution; generally the azimuth angle will diminish when moving further away transversely from the container vessel. The propeller rate will reach a limit value when moving further to the side, which will logically be the propeller rate in transit with no assisted vessel present.

5 CONCLUSIONS

An ASD-tug sailing in close vicinity of the bow of a large container vessel will experience strong interaction forces. A changing interaction sway force, yaw moment and resistance alteration will be experienced along the length of the hull of the container vessel. These depend on forward velocity and horizontal and transverse separation of the two hulls.

A tug master should be aware of these interaction phenomena since these interaction forces are of significant magnitude to introduce unwanted tug handling behaviour. The tug will experience either a repulsive or attractive interaction sway force or a bow-in or bow-out interaction yaw moment. A combination of these forces will necessitate a steering force to enable station keeping.

For the tug to stay on station, equilibrium is needed in resistance – propulsion, and the combination sway force, yaw moment – azimuth angle, drift angle. To keep station, the tug will generally need to sail with a constant drift angle applying a constant azimuth angle to counterbalance the yaw moment and sway force. This phenomenon will diminish when moving further forward and to the side away from the assisted vessel. Presently, not in all relative longitudinal positions and relative transverse separations, a stationary solution is found. Further investigation will determine if this instability is of a numerical or physical nature.

6 ACKNOWLEDGEMENTS

This research into interaction forces between tug and container vessel has been funded by the Towage and Salvage Union (Unie van Redding- en Sleepdienst) and Flanders Hydraulics Research, with the support of Delft University of Technology and Ghent University, and has been performed in the scope of a master thesis on the same subject by the first author of this paper.

7 REFERENCES

1. BERTRAM, V.: 'Practical Ship Hydrodynamics', *Elsevier Science*, 2000

2. BRIX, J.E.: 'Manoeuvring Technical Manual', *Seehafen Verlag*, 1993

3. DAND, I.W.: 'Some aspects of tug-ship interaction', *Proceedings of the Fourth International Tug Convention, New Orleans, pp. 61-80*, 1975

4. DAND, I.W.: 'Small craft interaction in shallow water', *RINA Occasional Publication No. 4: Collected Papers from Conference on Small Ships in Shallow Water, pp. 1-37*, 1982

5. GEERTS, S.: 'Master thesis on Ship-Tug interaction in confined waters', *Delft University of Technology*, 2011, in progress

6. GERRITSMA, J.: 'Scheepsbewegingen, sturen en manoeuvreren 3', *Delft University of Technology*, 2003

7. GRONARZ A.: 'Investigation of the influence of a vertical wall on a ship moving with drift angle', *International Conference on Ship Manoeuvring in Shallow and Confined Water: Bank Effects, Antwerp, Belgium, pp. 77-83*, 2009

8. HENSEN, H.: 'Tugs use in Port', *The Nautical Institute*, 1997

9. HENSEN, H.: 'Bow tug operations with azimuth stern drive tugs', *The Nautical Institute*, 2006

10. SMITH, J.R., BIRMINGHAM, R.: 'Creating the 'Virtual Tug'', *17th International Tug & Salvage Convention*, 2002, Day 1, Paper No. 3

11. VANTORRE, M.: 'Experimental study of bank effect on full form ship models', *Mini Symposium on Ship Manoeuvrability: Proceedings of Mini Symposium on Ship Manoeuvrability;* Fukuoka, Japan, pp. 85-101, 1995

12. VANTORRE, M.: 'Manoeuvreer- en Zeegangsgedrag van Maritieme Constructies', *Ghent University*, 2005

13. VANTORRE, M.: 'Model Test Based Formulations of Ship-Ship Interaction Forces for Simulation Purposes', *Ghent University*

14. VAN MANEN, J.D.; VAN OOSSANEN, P. Propulsion. In: Principles of Naval Architecture, Volume II – Resistance, Propulsion and Vibration (editor: LEWIS, E.V.). The Society of Naval Architects and Marine Engineers, Jersey City, NJ, 1988.

A NEW APPROACH IN MODELLING THE INTERACTION FORCES

A Gronarz, DST, Duisburg, DE

SUMMARY

Usually the additional hydrodynamic forces and moments acting on the own ship passing a target ship are modelled by a damped oscillation with factors describing the dependency of the situation parameters. This approach is practicable for vessels of comparable size and speed, but some configurations cannot be covered by it.

- Variation of length

Short vessels overtaking long ones and long vessels overtaking short ones show force developments which are far from harmonic oscillations.

- Variation of speed

Changing the velocity of the own and the target ship from +Vmax through zero to –Vmax covers all situations between of overtaking and encountering manoeuvres and can also not be modelled easily by sine or cosine functions.

A pragmatic way of describing the characteristic curves of the forces and moments for these special cases is introduced. This method is complemented by factorial influence functions with variable exponents, which are using a nonlinear regression analysis of model test results.

This so called "Hybrid Regression" is not a single formula any more but an algorithm formulated as a subroutine. For more convenient use a structural framework can be set up with a formula parser and a special input data set in order to compute the interaction forces with a high level of flexibility depending on the source of information.

NOMENCLATURE

B	Breadth	[m]
g	Gravity	[m/s²]
h, H	Water depth	[m]
L	Ship length	[m]
m	Mass	[kg]
N	Yaw momentum	[Nm]
t	Time	[s]
T, D	Draught	[m]
V, U	Velocity	[m/s]
X	Longitudinal force	[N]
Y, S_P	Lateral clearance	[m]
Y	Lateral force	[N]
Indices:	o	Own ship
	t	Traffic ship
	m	Mean value o & t

1. INTRODUCTION

Since decades the hydrodynamic institutes have carried out model tests (captive and free running) to investigate the reaction of ships during overtaking and encountering manoeuvres. While in the 60's [1 and 2] the focus was on understanding and describing the effects, later the introduction of the numerical motion simulation led to systematic model tests [3], fig. 1 and mathematical formulations of the forces and moments [4], fig. 2.

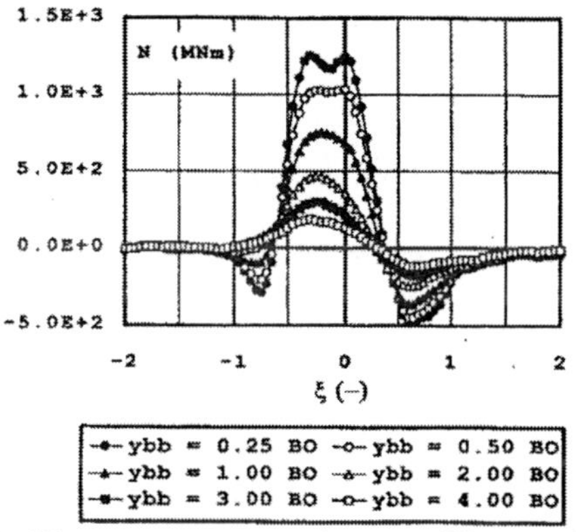

Fig. 1 - Yaw momentum vs. clearance (Vantorre 2002)

$$C_M = 0.34\sin\left(-0.65\pi(t-0.05)\right)e^{(-1.5t^2)}(1-0.18t)A(t)$$
$$\times\left[\frac{H/D}{1.5}\right]^{-2.2}\left[2\frac{S_P}{L}\right]^{-1.3}\left[\frac{L_1}{L_2}\right]^{-0.35}\left[0.5\left(1.5\frac{U_1}{U_2}-1\right)\right]$$

Fig. 2 - Formula for yaw momentum (Varyani 2004)

Within the last 6 years the DST has carried out several projects to carry out a lot of captive model tests with inland and seagoing ships to measure forces and moments during overtaking and encounter manoeuvres [5, 6 and 7]. The results lead to different modelling strategies, starting with the attempt to find a straight mathematical formulation, fig. 3.

$$F = \hat{y}^{Ay} \cdot \hat{h}^{Ah} \cdot d\hat{V}^{AdV} \cdot \hat{V}_t^{AV_t} \cdot \hat{m}_B^{AmB} \cdot \hat{V}_{mx}^{AV_{mx}} \cdot \hat{m}_{ot}^{Am_{ot}} \cdot A \cdot e^{\left(-1.5 \cdot |x_r|^{Ee}\right)} \cdot \sin\left(\pi \cdot \omega \cdot \hat{y}^{Wy} \cdot x_r - \varphi \cdot \hat{y}^{Py} \cdot \hat{m}_B^{Pm_B} \cdot \hat{V}_{mx}^{PVmx} \cdot \hat{m}_{ot}^{Pm_{ot}}\right)$$

Fig. 3 - Initial mathematical formulation

It consisted of a harmonic function which was damped to zero outside the vicinity of the own ship. The amplitude of the function was corrected using power functions of non dimensional parameters describing the variations during the test series.

Some variations carried out led to graphs with a behaviour that made it impossible to describe it with the conservative approach. This problem triggered the new modelling approach discussed in this paper.

2. MODEL TESTS

The complete test programme of all projects covered both inland and seagoing vessels of different size in overtaking and encounter situations. The main variation parameters were speed, lateral clearance and the water depth. Additionally, some special variations have been carried out described and discussed in this chapter.

Forces and moments have been measured during the passing manoeuvre. For comparison purpose the measurement time scale has been transpired to the relative passing time 'tr', which is defined as

- tr = -1: Begin of the manoeuvre when the perpendiculars meet at first (bow to stern when overtaking, bow to bow when encountering).
- tr = 0 : mean time when the **midship** sections meet.
- tr = +1 : End of the manoeuvre when the perpendiculars meet the last time.

This definition results in a further evaluation independent of length and speed of both ships involved in the passing situation.

2.1 LENGTH VARIATION

The DST has a stock of inland vessel in the scale 1:16 with a breadth of 11.4 m comprising of several bow, stern and midship parts. Using this construction kit a series of fictive inland vessels could be assembled starting with a length of 50 m up to 250 m in steps of about 25 m.

The test series carried out with this model family was with constant water depth and lateral clearance, speeds of 10 km/h and 15 km/h; as special feature a variation of the value Lo/Lt from 0.2 up to 5.0 was included. This means that a short model overtakes a long one, models of same length overtake each other and a long model overtakes a short one (with several length steps in between).

Fig. 4a presents the results for the longitudinal force X for the model overtaken measured in Newton. The curves in this diagram are quite different from a harmonic one. While the Lo/Lt-values near 1 (bold black) could be represented by a sine curve with reasonable accuracy this was not possible for the extreme values of the variation parameter. In order to understand the form of the curves some explanations are needed.

- The case is: a long vessel overtakes a short one (Lo/Lt > 4, dashed blue and magenta)
- When the bow of the overtaking traffic ship 't' meets the stern of the own ship 'o' this falls back into the trough due to the surface deformation alongside the traffic ship. This causes a decelerating negative longitudinal force (A).
- Then the flow alongside the long vessel influences the model in the trough, which results in an additional resistance and a nearly constant negative X-force (B).

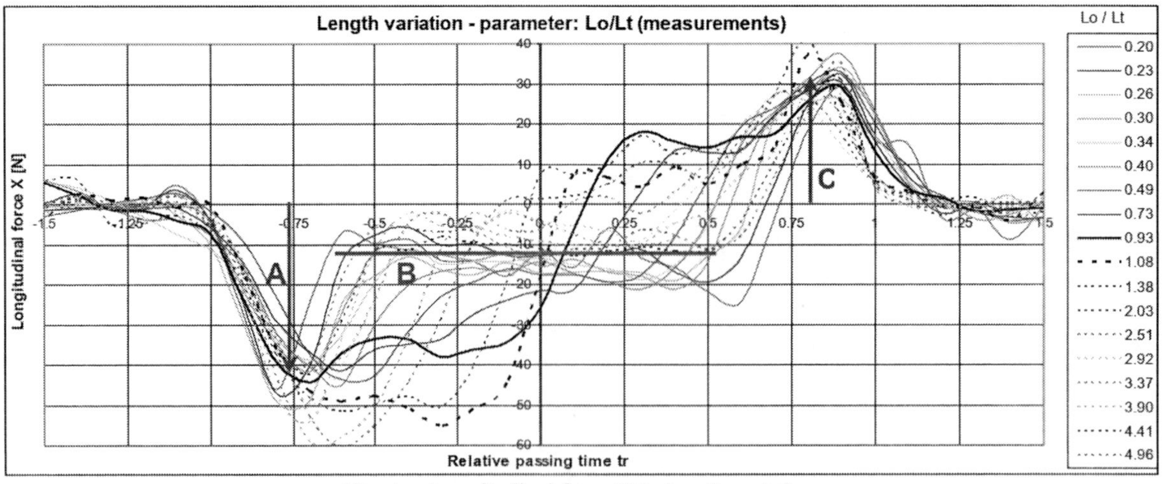

Fig. 4a - Longitudinal force X in length variation

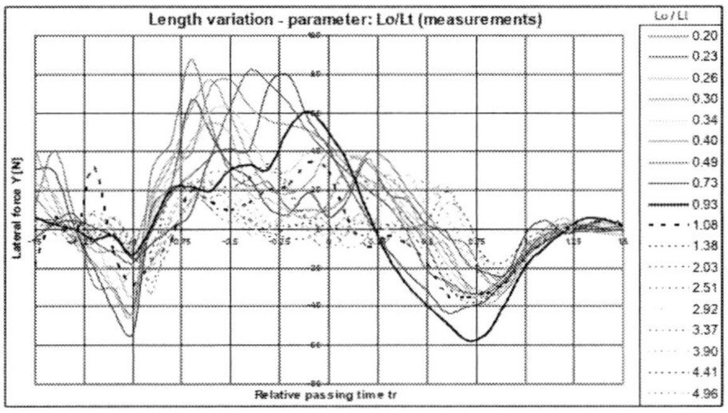

Fig. 4b - Lateral force Y in length variation

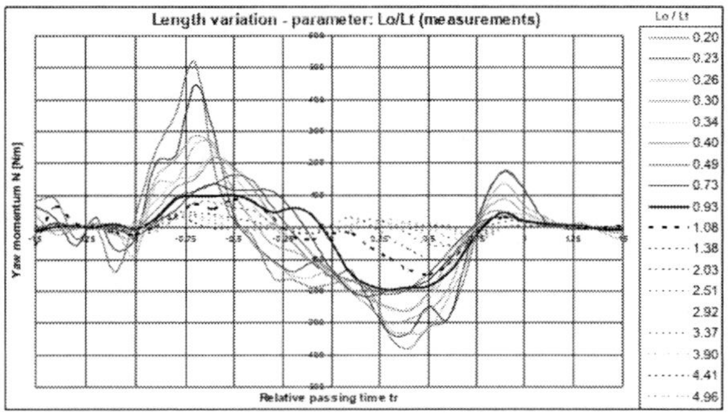

Fig. 4c – Yaw momentrum in length variation

- The small oscillations in (B) are from the ship waves of the overtaking ship and should be neglected in a modelling strategy.
- At the end of the manoeuvre the stern is lifted to the former level and an accelerating downhill-slope force can be measured (C).

This special behaviour needs an alternative approach to be modelled in a satisfactory way.

For the sake of completeness the measured data of the lateral force Y and the yaw momentum N are shown in fig. 4b and 4c. Although the dependency of the length is also obvious in this data, the drawback to the main parameter Lo/Lt is cannot so easily be performed.

2.2 SPEED VARIATION

Within a test series with seagoing vessels a variation of the speed of the own ship has been carried out. The traffic ship running always 18 kn overtook the own ship which was moving slower each run down to a speed of zero. Then it was turned and was running faster each test in the encountering tests. The speed values were made positive for the overtaking and negative for the encounter situation. The measured values for the yaw momentum are shown in fig. 5a.

The bold lines indicating the overtaking situation have an acceptable harmonic behaviour while the measurements of the momentum during the encounter tests diverge significantly. There is an additional sub-oscillation deriving from the wave system of the passing ship. This phenomenon will be neglected in the further evaluation because the ship waves are of secondary order in comparison with the primary wave system.

The black lines indicate the situation when a traffic ship is passing a moored own ship. A comparison between the directions of the own ship - traffic ship approaching first from bow or from stern - shows that there is nearly no difference in the yaw momentum affecting the own ship.

Although the significance of the speed parameter is not very obvious for the longitudinal force X and the lateral force Y they are also presented in fig. 5b and 5c.

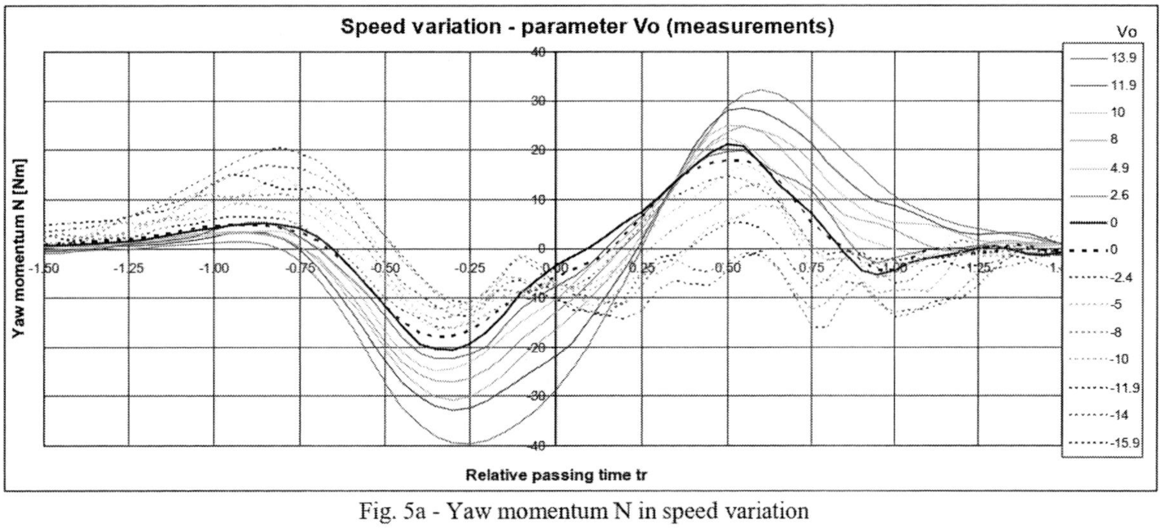

Fig. 5a - Yaw momentum N in speed variation

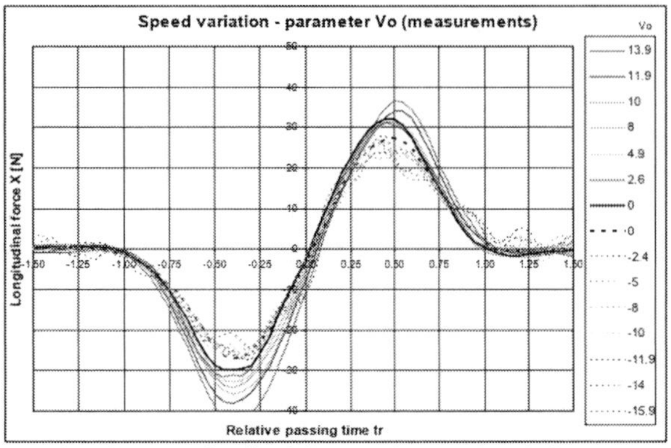

Fig. 5b - Longitudinal force X in speed variation

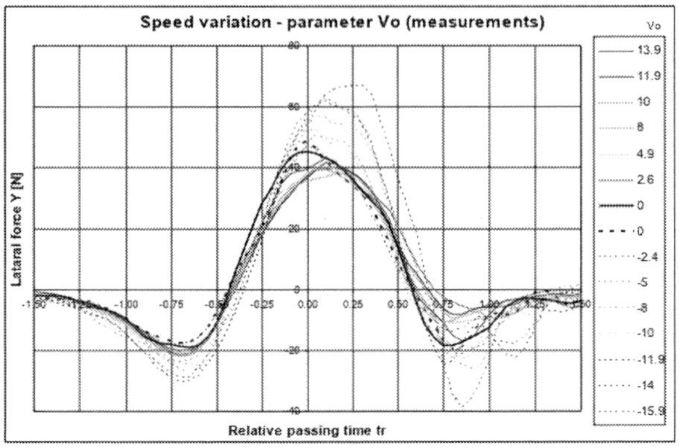

Fig. 5c - Lateral force Y in speed variation

3. NONDIMENSIONAL REPRESENTATION

In the further elaboration the forces and moments but also the influencing parameters are treated in non dimensional way. Due to the fact, that the origin of the effects of the ship-to-ship interaction is the traffic ship, the mass 'mt' is used for the normalization. Because the yaw moments occur at the own ship, the length Lo is used for the non dimensional representation of N.

$$X'=X/(m_t \cdot g)$$

$$Y'=Y/(m_t \cdot g)$$

$$N'=N/(m_t \cdot g \cdot L_o)$$

The variation parameters are also used in non dimensional way by means of relevant data of the actual situation.

$B_m/(y+B_m)$	Lateral clearance
T_m/h	Water depth
$(V_o^2+V_t^2)/V_{max}^2$	Speed of both ships
Fnh_t	Speed of traffic ship
$(B_o \cdot T_o)/(B_t \cdot T_t)$	Ratio of main section
L_t/L_o	Length ratio

The value Vmax is a theoretical maximum velocity. In shallow water (h/T<3) it is defined as Vmax=√g·h, in deep water it is fixed to Vmax=√g·3·Tm.

4. HYBRID REGRESSION

Because of the difficulties to use harmonic functions for the basic representation of the forces and moments according to the two special cases above a new empirical approach has been developed.

4.1 BASIC CURVES

4.1 (a) Length variation

The curves have been analysed by detecting maxima and minima (separately for the positive and the negative half) of the abscissa 'tr' and special features as begin and end of parallel intervals. The positions and the values are plotted over the variation parameter using the example of the length variation. For convenience not Lo/Lt is used, but the logarithm of it.

In fig. 6 the change of the values is shown using symbols. In light colours the positive half of the tr-axis is analyzed, in dark colours the negative half. The green symbols show the data at the mean position, indicating the height of the parallel region.

For the further processing a cubic polynomial was calculated fairing the values over the abscissa and delivering the coefficients for a numerical representation. In the legend they are marked as 'Apx'.

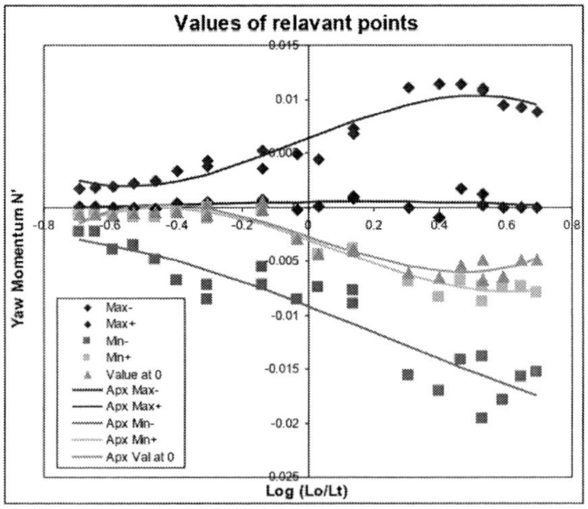

Fig. 6 – Values of relevant points

Fig. 7 shows the position on the abscissa for these points. For the parallel region (Value at 0) begin and end is plotted as 'Parallel from' and 'Parallel to'.

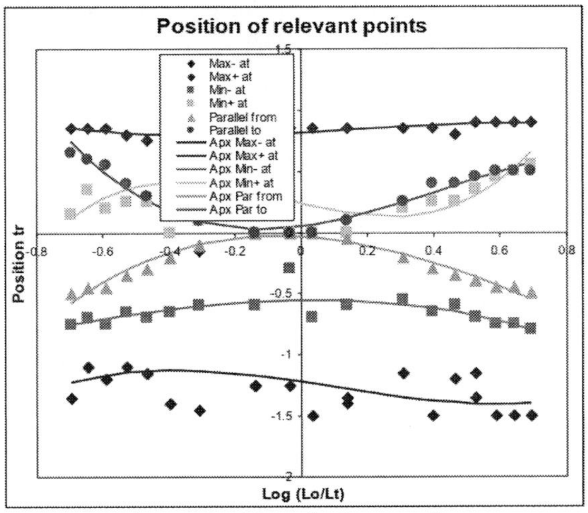

Fig. 7 – Position of relevant points

For the reconstruction of the original diagram some functions have been programmed taking into account special boundary conditions, see fig. 8:

- LIN : Linear interpolation between two points, used for the parallel interval
- QUAD1 : Quadratic interpolation with horizontal tangent left
- QUAD2 : Quadratic interpolation with horizontal tangent right
- CUB : Cubic interpolation comprising horizontal tangents left and right.

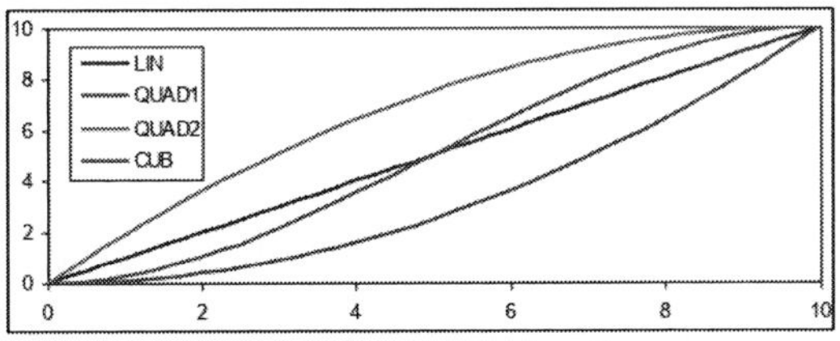

Fig. 8 – Interpolation functions

With these functions, the values and positions of the relevant points, the longitudinal force at the length variation can be generated over the whole range of 'tr':

1. – infinite to -1.2 : Zero
2. -1.2 to Min- : Cubic between (-1.2|0) and (xMin-|Min-)
3. Min- to ParFrom : Cubic between (xMin-|Min-) and (ParFrom|ValueAt0)
4. ParFrom to ParTo : Linear between (ParFrom |ValueAt0) and (ParTo|ValueAt0)
5. ParTo to Max+ : Cubic between (ParTo|ValueAt0) and (xMax+|Max+)
6. Max+ to +1.2 : Cubic between (xMax+|Max+) and (1.2|0)
7. 1.2 to +infinite : Zero

A special feature is used for ships of equal length without any significant parallel interval. In this case QUAD1 is used for the interval between (xMin-|Min-) and (0|ValueAt0) and QUAD2 for (0|ValueAt0) and (xMax+|Max+). Within the range of Log(Lo/Lt) of ±0.2 this type is smoothed to the type using a parallel interval.

Using the generation scheme 1. - 7. and the coefficients for the polynomials the basic curve for the length variation can be calculated. It is shown in fig. 9.

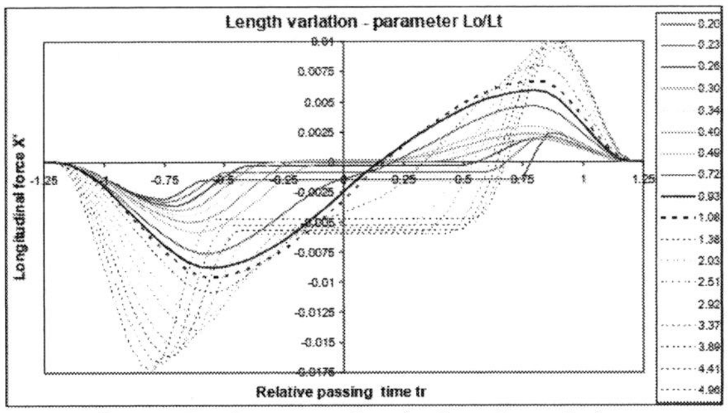

Fig. 9 – Basic curves for the length variation

4.1 (b) Speed variation

Using the same approach the other special case 'speed variation' can be reconstructed. As there is no parallel part in these curves the construction scheme is different:

1. – infinite to -1.3 : Zero
2. -1.3 to Min- : Cubic between (-1.3|0) and (xMin-|Min-)
3. Min- to Max+ : Cubic between (xMin-|Min-) and (xMax+|Max+)
4. Max+ to Min+ : Cubic between (xMax+|Max+) and (xMmin+|Mmin+)
5. Min+ to +1.3 : Cubic between (xMmin+|Mmin+) and (1.3|0)
6. 1.3 to +infinite : Zero

The parameter in this non dimensional representation is the Froude depth number Fnh. The range is extrapolated up to a value of 0.9, corresponding to a speed higher than reasonable on shallow water. The result is shown in fig. 10. It can be seen, that the secondary ship waves of the traffic ship during the encountering are not reproduced in the diagram but the tendency of a negative yaw

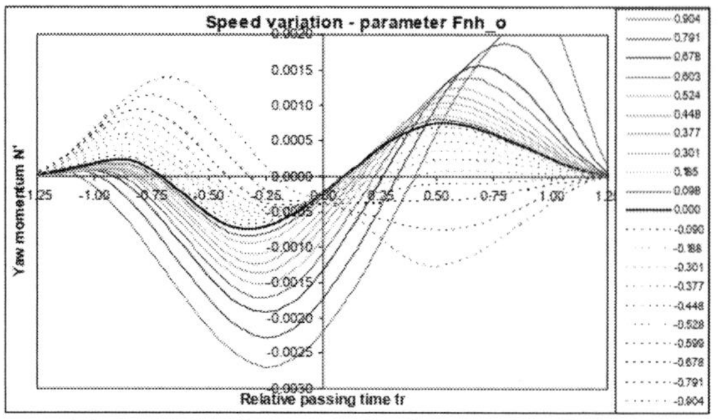

Fig. 10 – Basic curves for the speed variation

momentum at the and of the encounter manoeuvre can clearly be seen.

4.2 ADDITIONAL INFLUENCES

When modelling further influences multiplicative correction functions are used. They were either determined by a special analysis of certain test series covering a special variation or by using the tool "nonlinear regression' for the computation of the coefficients.

The basic curves have been modelled for a certain setting of the influencing variables e.g. Tm/h=0.4789 for the speed variation. To allow the use of correction functions for the calculation of changes from the settings for the basic curves, they must have the value 1 for the basic curve. This is achieved by dividing the function by the reference value in each case.

4.2 (a) Water depth

The evaluation of a series with 4 different water depths leads to a very smooth distribution of the peak values of the forces and moments. The nonlinear regression using the formula

$$F = C_o + C_n \cdot (T_m / h)^n$$

gives curves describing the steady rise of the forces with decreasing water depth. They start with constant values for infinite depth as shown in fig. 11.

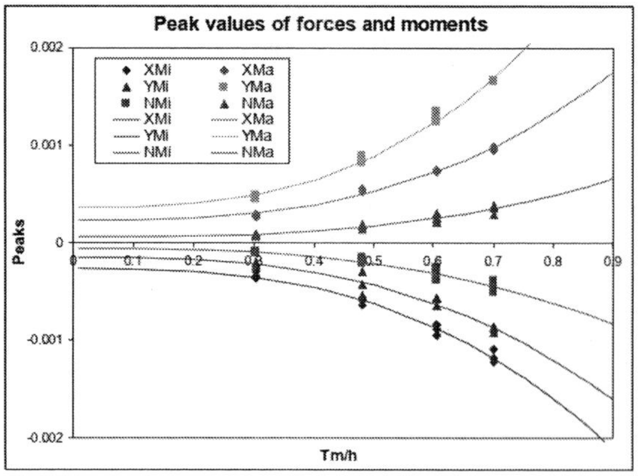

Fig. 11 – Peak values vs. relative water depth

By dividing all data by the values of the reference water depth all curves came to a rather similar shape. Thereby a single correction function for all components X, Y and N could be used. In Fig 12 the final correction function is given, showing, that its value at the reference water depth is 1.

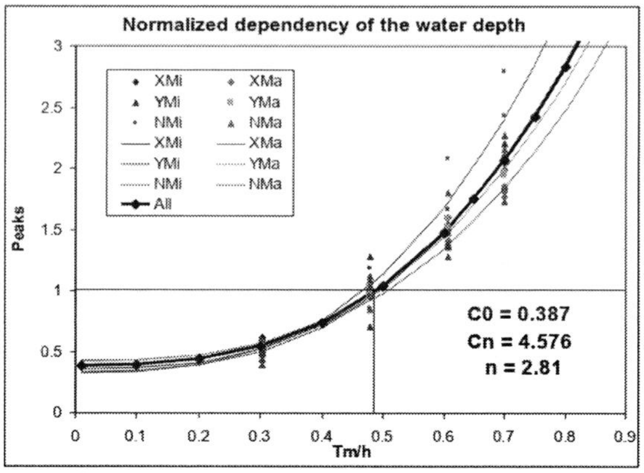

Fig. 12 – Determination of correction function

4.2 (b) Lateral clearance and other influences

For the lateral clearance and the other influences power functions are chosen. The general form is

$$F = Coefficient \cdot (Func / Refval)^{Exponent}.$$

The function 'Func' is one of those mentioned in chapter 3, except that for the length variation Lo/Lt is not used. Due to the different reference values for the parameters in the two main test series described above, for both cases different coefficients and exponents have been derived using the nonlinear regression.

5. DYNAMIC SOFTWARE

One method of application is programming the complete algorithm for the hybrid regression including the coefficients, operational steps and formulae for the additional influences in a subprogram. The disadvantage of this procedure is the work needed to alter code and recompile the software whenever a coefficient or a formula changes. A better way is, to program only the principle of the structure of the hybrid regression and use external data to control the process and input the coefficients.

5.1 Basic curves

For the construction of the basic curves a traceback of the operations carried out delivers the following facts:

- A basic curve consists of N segments, defined by N+1 points.

- The coordinates of these N+1 points are defined by pairs of cubic polynomials with 4 coefficients for x and y.
- For each of the N intervals it is recorded, which of the 4 functions shall be used for the interpolation.

This information is listed in a table which is read and processed by the software.

5.2 Formula parser

The abscissa of the basic curve and the additional influences are described in a formula. The calculation of the formula is done by a formula parser which carries out consecutive operational steps of one or two operators and puts the interim results onto a stack. The last content of the stack is the final result of the formula.

Operations using 2 operators		
1	+	Addition
2	-	Subtraction
3	*	Multiplication
4	/	Division
5	^	Powering
Functions using 1 operator		
-1	Exp( )	Exponential function
-2	Sin( )	Sine
-3	Cos( )	Cosine
-4	Sqr( )	Square root
-5	Abs( )	Absolute value
-6	Log10( )	Common logarithm

The operands can be

a) Constants – These can have a value between -1000 and +1000
b) Variables – They are taken from a list and coded as '1000 + ListNo'
c) Stacks – They are coded as '-(1000 + StckNo)'

5.3 Local variables

The variables used in the computation are listed in a table and can be addressed using their list number:

1	y	2	h
3	Stb_Pt	4	Lm
5	Bm	6	Tm
7	Vmax	8	FnMax
9	Type		

Own ship			
11	Lo	12	Bo
13	To	14	CBo
15	mo	16	Psio
17	Vo	18	Fno
19	Fnho		

Traffic ship			
21	Lt	22	Bt
23	Tt	24	CBt
25	mt	26	Psit
27	Vt	28	Fnt
29	Fnht		

'Stb_Pt' is 1, if the traffic ship is on starboard of the own ship and -1 if it is on port.
'Type' is 1 for overtaking and -1 for encountering.

5.4 Additional influences

The correction functions for the results of the basic curve are fragmented and their operational sequence is listed in tables. As example the function for the change of the amplitude with in relation to the water depth is noted and broken up into single processing steps using the rules described above:

$$Func(h) = 0.387 + 4.576 \cdot (T_m / h)^{2.81}$$

Op1	Op2	Operation / Function	
4			Number of steps
1006	**1002**	**4**	Tm / h = stack(1)
-1001	**2.81**	**5**	Stack(1) ^ 2.81 = stack(2)
-1002	**4.576**	**3**	4.576 * stack(2) = stack(3)
0.387	**-1004**	**1**	0.387 + stack(4) = result

The Froude depth number of the own ship would be

$$Fnh_o = V_o / \sqrt{g \cdot h}$$

Op1	Op2	Operation / Function	
3			Number of steps
9.81	**1002**	**3**	g * h = stack(1)
-1001	**0**	**-4**	Sqr(stack(1)) = stack(2)
1017	**-1002**	**4**	Vo / stack(2) = result

6. RESULTS

Using basic curves, which are already shown in the previous chapter, and correction functions a software has been programmed to analyse the behaviour of the new modelling approach.

Fig. 13 shows the behaviour of the lateral force with increasing clearance. While with close lateral clearance significant values can be obtained it diminishes with greater distances. It has to be noted, that the value 'y' is not the distance of the centrelines of the ships, but the clearance between the sides, which comes to zero in case of contact.

Another example in fig. 14 shows the behaviour of the water depth function. Contrary to the clearance the peaks never come to zero but reach constant small values for infinite water depth.

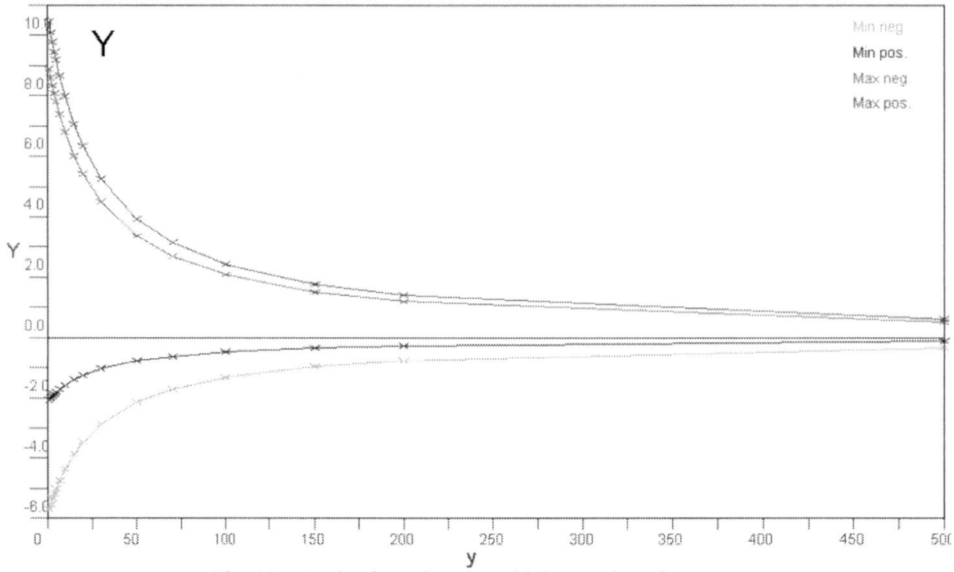

Fig. 13 – Reduction of peaks with increasing clearance

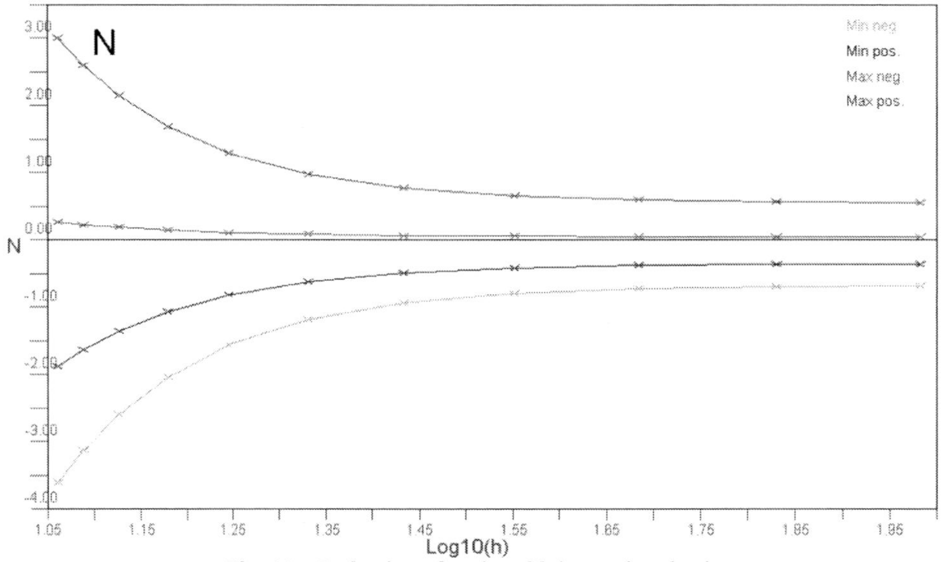

Fig. 14 – Reduction of peaks with increasing depth

7. CONCLUSIONS

- The 'Hybrid regression' is a combination of an empirical basic function for complicated characteristics of time dependant interaction forces with the tool of nonlinear regression.
- The generation of the basic function is flexible and can be adapted to any parameter using specific points of a curve family.
- Additional corrections are made by multiplication with power functions.
- Using a formula parser the algorithm can be coded externally for simple changes without modifying and recompiling a software.
- Unfortunately it was up to now not successful to combine the two basic functions to a more global approach because they are based on different reference parameters.

8. ACKNOWLEDGEMENTS

The research work and the extensive model tests presented here have been carried out at the DST within the three research projects 'Sicherbin', Simubin' and 'Sipas' which have thankworthy been funded by the German Federal Ministry of Economics and Technology.

9. REFERENCES

1. MÜLLER, E., 'Investigation of the forces and moments occurring during overtaking and encounter manoeuvres on shallow water' (in German), VBD-Report 269, 1962.

2. MÜLLER, E., 'Investigations of the mutual interference of the course on inland waterways' (in German), VBD-Report 413, 1967.

3. VANTORRE, M., VERZHBITSKAYA, E., LAFORCE, E , 'Model Test Based Formulations of Ship-Ship Interaction Forces', Ship Technology Research Vol. 49, 2002

4. VARYANI, K.S. THAVALINGAM, A., KRISHNANKUTTY, P., 'New generic mathematical model to predict hydrodynamic interaction effects for overtaking manoeuvres in simulators', Journal of Marine Science Technology, Volume 9, 2004

5. GRONARZ, A., 'Ship-ship-interaction: overtaking and encountering of inland vessels on shallow water', MARSIM 2006

6. GRONARZ, A., 'SIMUBIN – Modelling and simulation of realistic ship motions on inland waterways' (in German), DST-Report 1939, 2009

7. BENEDIKT, K. et al., 'SIPAS – Safe passage clearance of seagoing vessels during overtaking and encounter manoeuvres on restricted waterways' (in German), BmWi-Statustagung, Warnemünde, 2010

10. AUTHORS BIOGRAPHY

Andreas Gronarz holds the current position of a senior research engineer at the DST (Development Centre for Ship Technology and Transport Research, Duisburg). He is responsible for all manoeuvring aspects of the work in the towing tank and for the hydrodynamic improvements on the inland waterway simulator SANDRA (Simulator for Advanced Navigation Duisburg, Research and Application) of the DST. Since 25 years he is working on the field of shallow water hydrodynamics including his PhD-thesis about the special aspects of the simulation of ship manoeuvres in shallow water.

CRITICAL ELEMENTS OF A LIGHTERING OPERATION

Capt. G. Gudmundseth, Ship Manoeuvring Simulator Centre AS, Norway

SUMMARY

The Ship Manoeuvring Simulator Centre in Trondheim, Norway, has delivered different levels of training courses for Pilots and Deck Officers performing in Sea Lightering Operations.

Course participants have been professional Mooring Masters, each with individual experience of more than 1000 STS-operations, experienced Master Mariners with many years' of service, where STS-operations have been a frequent method used during discharging of cargo, Harbour Pilots conducting STS from time to time and Master Mariners/Ship Officers with little or no experience in this way of berthing method.

Observations and discussions obtained, during many years of arranging training courses, are the main source for this paper, together with my own experience in lightering operations, in command of both the vessel to be lightered and the service ship or moving vessel.

Based on the observations done during such training courses, a wide range of key elements pertaining to STS-operations have come to light, the nature of some being of great value relating to the safety aspect of such advanced operations.

This paper will discuss normal, as well as critical elements in the final approach phase of such operations.

The major topics, but not limited to, will highlight following elements:

- Procedures and checklists
- Observations and monitoring during approach
- Mooring Master condition
- Communication
- Forces to be handled during approach
- Time frames available during final approach
- Handling of non-conformities
- Action priorities
- Analysing non-conformities
- Conclusions

The OCIMF Ship to Ship Transfer Guides [1] [2] is the main operation guide used by most of the operators who are selecting STS-operations as berthing/unloading method.

Elements from said guide are referred to in this paper, in order to adhere to the content of same.

NOMENCLATURE

STBL	Ship to be lightered
SS	Service ship (Moving vessel)
Mooring Master	In charge of the operation
STS	Ship to ship operation
STS area	Designated operation area
LSV	Lightering support vessel
Operation commence	Checklist 2 completed
Run in	Checklist 3 completed
VLCC	Very Large Crude Carrier
ULCC	Ultra Large Crude Carrie
LPG-C	Liquid Petroleum Gas Carrier
LNG-C	Liquid Natural Gas Carrier
OBO	Oil/Bulk/Ore Carrier
ROT	Rate of Turn
OCIMF	Oil Companies International Marine Forum

1. INTRODUCTION

Guidelines and other sources pertaining to STS-operations do not often give too much details relating to how and when things are done during conducting of this type of operation, with vessels berthing alongside each other, underway or at anchor, both often in open sea area.

Nor is it easy to obtain any specific information from individuals performing such operations for different reasons.

The latter may be as easy as they just do not know how to explain or give understandable arguments to questions like; "What, when and why did you do what you just did?"

The most common answers to such questions are; "I felt it was suitable to do it like that now!"

This we can easily imagine, like when the Master approach the anchorage after the sea-passage, evaluate the space available for anchoring, moving traffic, external forces affecting his vessels and other items, hence, based on his evaluation, commence reducing the vessel's speed by ordering half ahead in the engine.

If you, in such an event, ask the Master for an explanation, either by mathematical, hydrodynamic calculations, or by known factors relating to sea-trials, his answer will most probably be that; based on my experience with this type of vessel, it was "suitable" to start reduce the speed now, and it was also "suitable" to reduce to half ahead in the engine at this time.

Above said, there are no doubt that the Masters experience, his knowledge, understanding and skill will bring his vessel safe to the selected anchorage position and he will anchor as "planned", but a documented detailed plan, with description of his performed step-by-step actions, for the events that took place, can be found.

Incidents occurring in operations at sea are nowadays occasionally dictating the need for operation procedures. In order to avoid unwanted situations caused by Masters "not so experienced", detailed procedures for the operations should be developed and implemented.

The way of handling a vessel based on what is "suitable", is not in all situations taking care of an acceptable safety standard.

Today, regulating organizations, guideline publicises, oil majors and other bodies involved in STS-operations are requiring STS-operation Manuals on board vessels involved is this type of operations, and some of them contain procedures for the approach part of the manoeuvring.

However, important details relating to the approach manoeuvre are not included, and thus the way of doing the handling may contain a broad range of limits, rather than reasonable, safe limits within the capability of the tool available, i.e. the vessels involved.

STS-operations, especially during the final approach stage, are very limited relating to available time-frame.

A lot of elements need to be observed/monitored; communication must be short and accurate in order to avoid occupation of radio equipment in case of any urgency occurring, orders for rudder and engine must be given and confirmed, crew handling the moorings must be monitored and any observed non-conformities must be dealt with immediately.

The Masters involved and the crew need to have the necessary knowledge, understanding and skill to conduct a safe STS-operation.

2. LIGHTERING OPERATIONS

2.1 PROCEDURES AND CHECK LISTS

2.1 OCIMF Ship to Ship Transfer Guides [1] [2] contains the STS-procedures and check-lists which most of the companies involved in lightering operations have included in their STS-manuals.

The approach procedures is describing a "no go area" in the vicinity of the STBL, being an area bordered by limits of no closer than ½ nautical astern of the STBL and 1 cable off starboard side of the STBL.

This is valid for operations done when both vessels are underway, often with an operation speed of 4, 5 to 5, 5 knots.

The Check Lists contain elements covering the key manoeuvring equipment, navigation equipment, mooring equipment, fender equipment, safety and communication.

Checklist no. 1 is the suitability of the vessels to conduct STS-operation. Same is completed between the STBL Chartering Company and the SS provider.

Checklist no. 2 is completed between the STBL Master and the SS Master, before operation commence.

Checklist no. 3 is completed between the STBL Master and the SS Master, before final run in/mooring.

Use of the existing OCIMF Ship to Ship Transfer Guides [1] [2] procedures and check lists will be referred to under the other chapters in this paper.

2.2 OBSERVATIONS AND MONITORING DURING APPROACH

2.2 Normal procedures for final approach dictate the moving vessel to come to a position 1 cable (180 meters) on starboard side of the STBL, on a parallel course, and maintain this distance during adjustment of the vessels speed in order to maintain also the relative longitudinal position, where the two vessel's manifolds are about in line with each other.

Only when the two vessels have equal speed, i.e. the relative longitudinal movement between the vessels is close to zero, they are maintaining the same course and the distance apart is not changing, then the SS commence the final approach by altering course to port, converging towards the STBL.

There are arguments that this approach starting distance, being 180 meters, is far too much, and it should be safe for the SS to commence converging towards the STBL from a less distance than the one recommended by the OCIMF Ship to Ship Transfer Guides [1] [2] guidelines.

However, from this position (180 meters apart), and inwards until the vessels are alongside each other (both separated by the fenders), a lot of observations, monitoring and decisions should be done.

Further, any risk for potential non-conformities may be observed and acted upon before they disturb the operation, or in worst case, ends up in an incident/accident.

There are 3 key elements which needs attention and must be continuously evaluated and acted upon. Those are:

a. The manoeuvring and movement of the vessels
b. The behaviour of the personnel
c. The communication between the involved personnel

The Mooring Master must continuously monitor the SS's movement, the STBL's movement, and adjust both vessel's course and speed in order to obtain a safe landing on the fenders, protect the vessels hull, as well as provide a safe condition for the mooring teams to commence and complete the mooring of the vessels.

a. In this respect, he will need to continuous monitor, but not limited to, following equipment and conditions:

On the SS:

Speed
ROT
Engine RPM
Rudder Position
Course
Fender position and movement
Distance apart
Approach speed
Mooring team readiness
Wind direction and speed
Wave direction and speed
Swell direction and speed
Current direction and speed
Drift angle
Interaction forces
Rolling, yawing, sway

On the STBL:
Propeller wash water
Use of rudder
Funnel smoke
Rolling, yawing, gearing
Mooring team readiness
Mooring equipment, including machinery

b. In this respect, he will need to continuous monitor, but not limited to, following equipment and conditions:

On the SS:
Performance of the helmsman
Performance and behaviour of the Master
Communication equipment
Response to orders given
Bridge team work and attitude
Mooring teams' skill

On the STBL:
Performance of the helmsman
Performance and behaviour of the Master
Communication equipment
Response to orders given
Bridge team work and attitude
Mooring teams' skill

c. In this respect, he will need to continuous monitor, but not limited to, following:
Response and reply to orders given
Use of radio equipment relating to short and clear communication

Above listed 36 items, but not limited to same in an normal STS-operation, should keep the Mooring Master pretty busy during the time he use to cover the 180 meters.

But in addition to above, the normal parameters to be monitored during approach, variable items always occur, which have to be observed and dealt with as found necessary.

Those may be a few only, but when one occurs, other parameters tend to be affected, and often a lot of variables, disturbing the stability of the operation, have to be handled correctly in order to regain stability/balance of the on-going operation.

The importance and usefulness of above parameters should not be neglected. If they are carefully monitored, and acted upon if unwanted deviation occurs, the Mooring Master is more comfortable with the situation, and as such, his capacity to handle critical variables is better.

2.3 MOORING MASTER CONDITION

2.3 Commanding a STS-operation require that the Mooring Master is in a state of being comfortable with what he is observing, monitoring and doing.

Due to an occasionally very limited time-frame to act on non-conformities, it is very important that he does not permit any deviations from a well-balanced and smooth run in.

Should deviations occur, he has to act as soon as possible to adjust the unwanted situation.

During such an event, he will experience that his state of condition is shifted from being comfortable, to a state where he is uneasy, worried, scared or even close to panic.

The latter depends on several factors, like how dangerous the situation is, his experience, knowledge, understanding and skill.

If we describe the unwanted conditions that may occur during the STS-operation in a table, it could look like given in table 1.

Condition	1.Comfortable
	2.Uneasy
	3.Worried
	4.Scared
	5.Panic

Table 1

Important to note also, is the possible different conditions that be lurking above the "comfortable condition".

Those conditions could look like given in table 2.

Condition	5.Negligent
	4.Poor focus on key elements
	3.Focus on non-relevant talk
	2.Excessive self-confidence
	1.Comfortable

Table 2

The main point is to regain control of the situation, and thus he will also get himself back to a condition as close to comfortable as possible.

By succeeding this, he will also have more of his capacity available to handle other unwanted items which may occur during the operation.

2.4 COMMUNICATION

2.4 The quality of equipment for communication and how it is used, is a very important key factor in STS-operations.

Loss of communication between the vessels involved, i.e. occupying/blocking the selected working radio channel by unnecessary talk, qualifies for an abort operation action.

Being equipment not only needed to relay orders and report back on same, but also to report immediately any non-conformity observed.

The time-frame available is occasionally very limited in situations where information/orders have to be exchanged, and an established good radio discipline has to be in place.

Long essays to report vessels present course back on radio to the Mooring Master should be avoided.

The same is valid for any other reply on parameters requested by the Mooring Master, or others involved in the operation, using the selected working channel.

A plan describing who, how and when necessary communication shall take place, should be known and understood by all involved using the equipment.

2.5 FORCES TO BE HANDLED DURING APPROACH

2.5 The knowledge and understanding of externalforces acting in an STS-operation is important.

The Mooring Master normally selects the operation speed and course.

Based on his experience, he aim to minimize unwanted movement (See figure 1; Rolling, yawing and sway) of the vessels, but occasionally he is restricted by limited permitted operation area, and have to find the most favourable operation course within a limited sector.

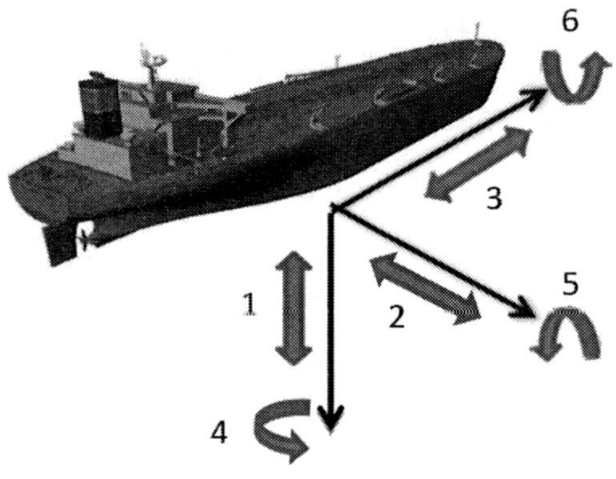

Figure 1. Vessels move along the three axes: 1. heave, 2. sway, 3. surge, 4. yaw, 5. pitch, 6. roll

External forces like wind, waves, swell and current are normally known, and can be rather easy be considered in relation to find the best operation course and speed.

Not so with unexpected variables like current layers, black squalls/strong winds, severe interaction and sudden change of waves/swell due to current variables.

Unexpected variables will occur from time to time, and only a focused, well trained and experienced Mooring

Master may observe and identify them in time to prevent unwanted incidents to occur.

The first indication of such unexpected variables may be sudden increasing deviation from course and/or drop of speed, this due to an unknown source, as no such movement is verified caused by any rudder/engine movement, nor by any other visual observations within the operation.

Those forces may be very strong and unpredictable, and due to same, abort operation may be the best choice if the vessels distance apart permit.

The action to be taken will be further discussed under Chap. 2.7 in this paper

2.6 TIME FRAMES AVAILABLE DURING FINAL APPROACH

2.6 Mariners may not always be the best source to give you time needed to reach a specific position, without calculating same or obtain the answer from navigation equipment, but they can almost at any time tell you how far away the same specific position is.

STS-operations, especially the final approach, is controlled and based on what the Mooring Master observes from the port bridge wing, together with the other information obtained by radio communication from participating personnel.

The Mooring Master can tell you pretty accurate, at any time, the vessels distance apart, but the "time apart" with present approach speed may be a different matter.

Normal approach speed is dictated by the capacity of selected fenders and the moving masses (vessels) involved.

The approach speed is observed and evaluated at any time by the Mooring Master to be "suitable".

However, due to lack of measuring equipment for this parameter, what is 0,15 knots for one Moring Master, may not considered to be the same for another one.

This is due to our differences of what is felt as "safe" speed, i.e. 0,3 knots approach speed may be comfortable for one Mooring Master, whilst it can far too high, or low for that sake, to another Mooring Master.

It depends on what the individual is comfortable with.

Thus it is understood that an error range may be present, and to some, what is believed to be 0,15 knots may actually be 0.3 knots.

The quality of measuring the approach speed is improving as the distance apart is decreasing, and approach speed adjustment is occasionally required.

The Mooring Master may be very comfortable with all parameters from 180 meters and inwards, but from about 30 meters and beyond, his options to abort operation commence to be more limited, and his focus and stress level will increase.

As an example, shown in table 3, is the time needed to cover remaining distance apart at different approach speed levels.

The table shows the vessels abeam relative approach speed in the first column, the vessels distance apart in the second column, and the time needed to cover this distance apart is given in column three.

Approach speed (knots)	Distance to go (meters)	Time needed (min)
0.10	30	9,68
0.15	30	6,48
0.20	30	4,86
0.30	30	3,24
0.40	30	2,44
0.50	30	1,95

Table 3

The main point with this chapter is not to blame Mariners for their eventually lack of ability to predict the time aspect correct, but to show the limited time actually available to handle any occurring non-conformities from distances apart of 30 meters and beyond.

Needless to say, the question should be what is most important when non-conformity occur, a "suitable" approach speed or maximum time available to handle the unwanted situation, and also remember the maximum permitted approach speed for the selected fenders.

2.7 HANDLING OF NON-CONFORMITIES

2.7 In STS-operations unwanted events occur from time to time, and there is a wide range of the nature of them.

Common is excessive use of rudder, misunderstood orders for helm and engine, loss of power, wrong speed setting and so on.

When the vessels are more than 50 meters apart, most of those unwanted events are relative easy to rectify.

There is both time and space (distance apart) to re-establish balance and control of the vessel.

If satisfying control is not established, the operation is aborted and not commenced before the non-conformities are rectified.

As we are closing up to the distance 30-35 meters apart, an additional element occurs. The question of whether it is possible to abort operation or not.

The unwanted event in most cases takes the vessel out of balance relating to a steady course and speed. Thus rudder and engine has to be used to regain balance, acceptable approach angle and so on.

If the disturbing element is observed early, this is possible, but if the nature of the unwanted event actually dictate an abort operation action, same may be very difficult to do safe when the vessels are only 30-35 meters apart.

To abort, the SS needs to regain balance, alter course to starboard to stop the approach, maintain longitudinal position in order to have all fenders in in place if landing must be done due to unsuccessful attempt to abort the operation.

Abort operation often result in excessive angles between the vessels, and failed attempt to abort often results in steel to steel damages.

When the vessels are less than 30 meters apart, it is often the best practice to complete the approach and land as parallel as possible on all the fenders at the same time.

Bearing in mind the time-frame available to handle unwanted events that occurs beyond 30 meters, it is often seen that the SS is almost on the fenders after the major part of available time to go is used for observing, identifying and selection the right measures to regain control and acceptable approach angle.

With the limited time available to handle unwanted situations when they occur, potential situations should be identified, considered and measures to handle same should be planned.

2.8 ACTION PRIORITIES

2.8 In the aftermath of incidents and accidents, there are always questions about what was done to minimize the damages.

In this respect, it is of major importance to select correct action priorities.

Again, within the very limited time-frame available, risk for wrong action priorities will be present.

Often, if potential situations is not pre-evaluated and counter-measures planned, wrong priority of orders and actions occurs.

This can be seen by an example, were priority of orders/actions may be as described.

Event: The vessels are 20 meters apart, the STBL suddenly get her rudder stuck hard to port.

Tools still available to the Mooring Master are following:
SS rudder and engine
STBL engine

Mooring Master's Action Priority orders:

1. STBL stop your engine
2. SS Hard a Port, half ahead in engine
3. Observe result/development of orders given
4. Adjust SS rudder to port and engine setting based on conclusion of item 3.
5. STBL amidships rudder once same can be moved
6. Follow STBL in a turn to port, maintain parallel vessels and longitudinal position
7. Maintain ROT to port less than ROT of the STBL in order to increase the distance between the vessels
8. Abort operation once sufficient distance between the vessels is obtained.

During planning of potential incidents, it is important to be able to justify priorities given, by explaining each of them, i.e. why should that order be the first one and the next number two and so on?

2.9 ANALYSING NON-CONFORMITIES

2.9 Analysing and planning of measures to handle unwanted events should be done by Mooring Masters and other personnel in charge of STS-operations.

Potential situations should be described and analysed.

One method is to first identify which unwanted parameters have occurred, hence consider which one of them are developing fastest, will cause most damage and have to be reduced and stopped first.

In STS-operations there are also some wanted parameters.

Those are, among other things, following:
STBL to maintain a steady course
STBL to maintain a steady speed
SS to maintain a steady course and angle of approach
SS to maintain a steady speed
SS to maintain the relative longitudinal position

Any deviation from above parameters is unwanted, and such deviations may, as earlier discussed in this paper, be caused by a wide range of sources.

The direct or indirect causes for unwanted events to occur, may be due to technical malfunctions, human errors and external forces.

3. CONCLUSIONS

Mooring Masters, Master Mariners and others involved in the operational/manoeuvring part of STS-operations, should obtain an acceptable level of knowledge, understanding and skill to handle the key elements both during normal and critical conditions of STS-operations.

Detailed procedures have to be developed and implemented in the Company's STS-manual.

The procedures should contain detailed suggestions for what is best practice relating to elements such as angle of approach versus distance apart and guidelines relating to best longitudinal position.

Further, a section with detailed contingency plans should also be developed and implemented.

The latter should give details relating to potential incidents, analysis of same and methods of manoeuvring, in order to prevent or minimize injury and damages.

4. REFERENCES

1. OCIMF: Ship to Ship Transfer Guide, Petroleum, 2005 edition
2. OCIMF: Ship to Ship Transfer Guide, Liquefied Gases, 1995 edition

5. AUTHORS BIOGRAPHY

Capt. Gunnar Gudmundseth holds the current position of Operation Manager at Ship Manoeuvring Simulator Centre AS.

He is responsible for managing the Course department, the Technical support department and the Sales department.

His previous experience includes 28 years of service with previous Bergesen D.Y. AS (Now BW-Gas), during which time he served on board several VLCC, ULCC and OBO-vessels, frequently taking part in STS-operations as Master and Mooring Master.

During 5 years of service as Marine Superintendent, he worked several times as Mooring Master on board LPG tankers, conducting STS-operations in the Gulf of Oman.

He has 6 years of experience as Course Instructor on the Ship Manoeuvring Simulator, being main responsible for the STS-operation courses, LNG-cargo operation courses and Emergency Towing courses.

INTEGRATION AND TESTING OF AN STS DECISION SUPPORT SYSTEM IN A FULL-MISSION SHIP MANEUVERING SIMULATOR.

D. Husjord, Norwegian University of Science and Technology, Norway.
T. A. Øritsland, Norwegian University of Science and Technology, Norway.
E. Pedersen, Norwegian University of Science and Technology, Norway.

SUMMARY

Interviews of and a questionnaire survey among ship officers in charge of lightering operations with crude oil tankers have revealed what kind of information that is required to improve the decision-making process during the approach between two vessels that aim to get alongside with forward speed in open waters. This is a part of the ongoing development of a navigation support and guidance system for ship-to-ship operations. Graphical user interfaces (GUIs) tailored for the typical approach phases in ship-to-ship operations have been proposed, developed and tested on a full-mission ship-handling simulator to assess the potential for enhancing operational safety. The structure of a real-time monitoring system for ship-to-ship operations is presented and discussed.

1. INTRODUCTION

A ship-to-ship, STS, lightering operation typically involves two tankers maneuvering in close proximity, at a speed range of 4-6 knots, in order to come alongside and commence cargo transfer. This is a challenging task for the officer in charge of the operation, the Mooring Master, whose decision-making process during the approach is commonly based on radar measurements until the vessels are approx 0.1 n.mil apart. The final approach phase is carried out using visual observations only. In an STS lightering operation, the Captains are, as always, in charge of their respective ships. However, it is industry standard to have a Mooring Master onboard the maneuvering ship, acting as a Pilot and advising the Captain and his crew on how to navigate and maneuver. The Mooring Master is normally authorized to terminate the operation if safety is at risk. Experienced and trained officers are needed to minimize the risk of miscommunication between the ships, [1].

An STS lightering operation can either be carried out while two ships are under power making way through water, or when one ship is already positioned at anchor. In the case of two ships under power and sailing, one ship is required to maintain speed and course, and is referred to as the Constant Heading Ship, or Ship to be Lightered (STBL). The Maneuvering Ship, or Service Ship (SS), will approach until it is parallel with the STBL before commencing the final approach phase, which is to maneuver until the ships are side by side and mooring can take place, Figure 1, [2].

A user survey among European and US Mooring Masters has revealed potential improvement to the overall safety in such operations. Real-time information about distances between the two vessels and their relative movement, when presented graphically in an understandable way and relayed to the decision-maker on the bridge, could speed up the decision-making process, [3].

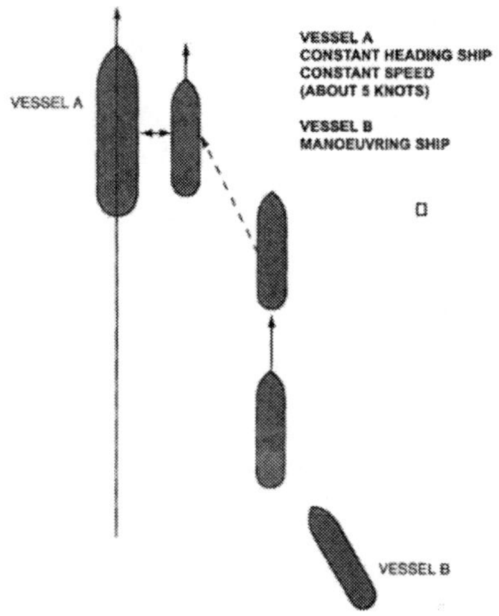

Figure 1: The principle of an approach maneuver in STS lightering operations, [2].

2. SIMULTANEOUS OBSERVATION OF RELATIVE MOVEMENT

Simultaneous observation of the relative movements of two dynamic points is a challenge to the human eye. In STS operations, the decision-maker (mooring master, pilot or Captain) must maintain control during the landing phase and keep a close eye on the relative movements between the two vessels. In the interviews, mooring masters especially emphasized the transverse movement of the bow and the angle between the two ships. How the bow-to-bow distance changes over time and indicates changes in the relative yaw-movement was pointed out as significant.

In a lightering operation, the SS is normally approaching the STBL with a relative angle of up to 5°. This is normally done by observing the course of one's own ship and taking for granted that the STBL maintains steady on its course, something for which there is no guarantee.

Therefore, in some situations, such as in heavy swells, the Mooring Masters require an assistant Mooring Master to be shipped onboard to the STBL, to monitor course and speed and communicate with the SS.

Consider an approach phase where the SS yaws starboard, away from the STBL, and the angle between the vessels is widening. This yaw movement can be the result of hydrodynamic interaction forces or other external forces acting on the SS vessel. Imagine now that the decision maker gives the helm order: port 10. When does the unwanted relative starboard movement stop? This stopping-point is important. It tells you how much rudder force, over a period of time, is needed to stop the vessel's movement. Observations of the yaw movement or changes in the rate of turn (ROT) will determine how long the existing order can stand before it has to be changed, for example by a reduction as slacking the rudder to midships. If the vessel's yaw movement is not stopped "in time", the force must be increased by a larger rudder order to port, or a kick forward in the engine.

However, an experienced decision-maker knows that excessive rudder force may lead to an uncontrollable situation if one's own ship's rate of turn stops too quickly and builds too rapidly in the opposite direction. For example, look at normal coastal navigation, where the decision-maker will hold his view steady forward, and observe the relative movements of the ship's bow in relation to fixed objects on land. This is normally performed by observing the movement of the vessel's mast in the bow relative to a background, like a tongue of land or a rock. The decision-maker can observe this, without a change in his own point of view.

The situation is different in the landing phase of an STS operation. From the point when the two vessels come into sight of each other and during the whole approach, the decision-maker has to observe the bow of his own ship, or port anchor bed, but at the same time, the STBL's hull and bow post must also be observed. According to normal procedures, the final phase will start when the two vessels are at same speed and moving almost parallel to each other, around one cable off. The decision-maker must simultaneously observe the aiming line from the port bridge wing on the SS to the port anchor bed and observe the aiming line from the decision-maker to the STBL's hull or bow post. On parallel courses one cable off, we find the angle, α, between the two aiming lines to be approximately 43°, with a distance between the bridge and bow of $L_{BB} = 200$ (m), Figure 2. Visually focusing on two different spots simultaneously is difficult to accomplish with the human eye. The human eye has limited depth perception, which makes it hard for the decision-maker to observe the angle between the two ships and verify relative yaw. Depth perception is mainly based on comparisons and experience, or experienced pictures, [4].

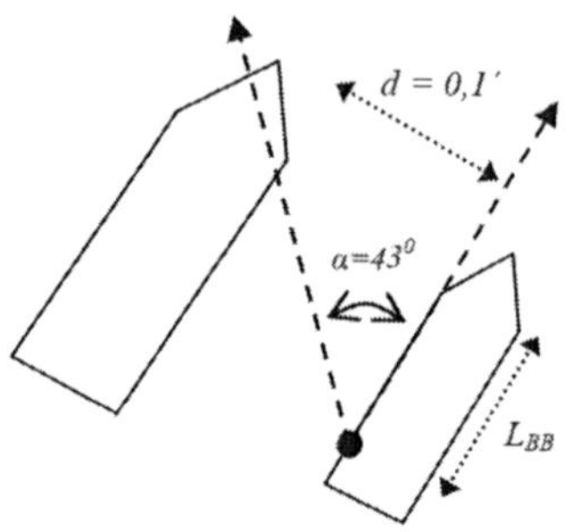

Figure 2: Visual aiming lines from SS port bridge wing.

During interviews with some experienced Mooring Masters, they claimed to have full control using only visual observations. These experienced Mooring Masters did not see the limited depth perception capabilities of the human eye as a problem in their decision-making process. They trusted their observations and did not feel a great need for more information in their decision-making process. Practical experience accumulated over several years of conducting lightering operations is partly conscious and partly not. Studies conducted into these natural decision-making processes find the experience and knowledge of decision-makers to be subconsciously primed. And this occurs without any conscious optimization of the solution, [5]. In some literature, this human characteristic is called tacit knowledge or implied knowledge. The original tacit knowledge held by individuals is unique to them, a product of their whole experience, and not a direct source of generalizable knowledge, [6].

An experienced and well-trained mooring master relies on similar experiences he has experienced in the past to make a decision, [1]. Ergo, his decisions are not the result of observations by the eye alone, but of his processing these observations in a well-known situation for which he is trained. Knowledge about the vessels, their crews, and local environmental forces acting on the ships is also part of the Mooring Masters decision-making process.
An example of the difficulty of observing the angle between two vessels is the practice of performing port of calls along the coast of Norway. Norwegian Pilots have always written down the orientations of the quays, so the required course to steer during the approach phase can be deduced. This is because the angle between the ship and the quay cannot easily be deduced visually.

3. USER-CENTERED DESIGN APPROACH

A Mooring Master on a bridge in a lightering operation is a part of a cognitive system, together with the crew on both vessels. Lintern (2005) explains that the cognitive work expands the view of what is cognitive beyond the individual mind to encompass coordination between people and their use of resources and materials, [7]. This view is also in accordance with the theory of distributed cognition enunciated by Hutchins (1995), [8].

A Mooring Masters cognitive work consists of knowing, understanding, planning, observing, comparing, finding alternative solutions, predicting, and making decisions, Figure 3, while he is fully integrated with perceiving and acting. In operations like lightering, experience and training play a major role in the decision-making process, [1].

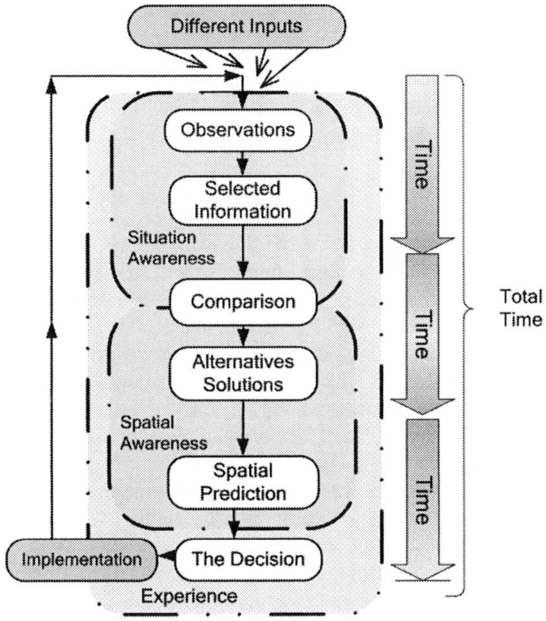

Figure 3: A model of a decision-making cycle in STS lightering. Situational awareness is important in the first phase, while spatial awareness is important in the last phase of the process. Experience comes into play throughout the whole process.

A user-centered study, as a background for the development of a decision-making support system, commenced in the autumn of 2008. Questionnaires were handed out to some experienced Mooring Masters and interviews were conducted in groups directly connected to the development of graphical user interfaces (GUIs).

The interviews were conducted at the Ship Manoeuvring Simulator Center (SMSC) during lightering courses for Mooring Masters in the spring and autumn of 2009, [9]. Different types of information and graphical presentations were discussed, focusing on the decision-making process. The interviews included open- and closed-type questions regarding how the users considered the observations they received and how they made a decision based on this information.

Finally, how the perceptual situation could be improved by presenting different types of information – parameters and naturalistic graphics – on the GUI in a decision support system, were performed.

3.1 HAND-HELD DECISION SUPPORT SYSTEM, DSS

One of the earliest findings in the user-centered study was the obvious request for a hand-held display. The Mooring Master is on the bridge wing in the final decisive phase, Figure 4. An immobile device inside the bridge will be inaccessible, but a hand-held device can communicate the information needed, while being within reach for the Mooring Master at all times.

Handheld displays are rarely used in navigation at sea, in contrast to personal urban navigation, which navigation features are integrated in mobile phones and other Personal Desktop Accessories (PDAs). A handheld display in STS operations would have to be both static and dynamic: static because the two ships must be on the screen all the time, and dynamic in the way that the graphics change from smaller to larger ships when the final phase commences and the two ships are getting closer to the landing.

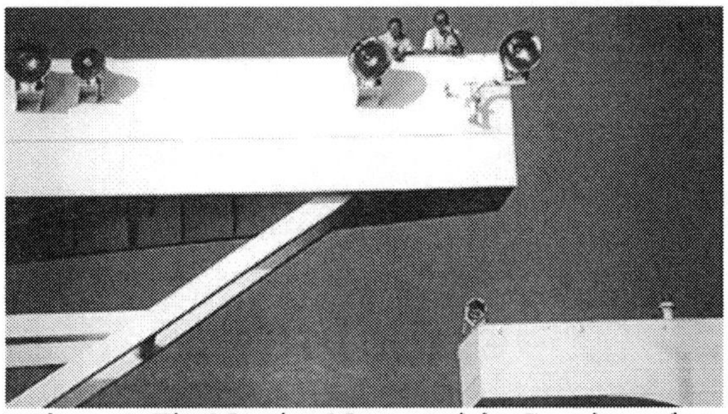

Figure 4: The Mooring Master and the Captain on the bridge wing. (Courtesy of SPT Ltd.)

In response to the question of whether the same information should be presented inside in the bridge, one group of mooring masters' answers were a bit surprising: If this information is made available to the Captain on board, it could create conflicts and uncertainties. These Mooring Masters believed that by presenting the captain with more detailed information, they would see more interference with the operation. For example, the Captain could get the impression from the DSS that his vessel is coming in too hard during the landing phase.

In this context, it may be added that many oil tankers perform lightering operations on an occasional basis only, which means some Captains have little experience with such operations. Any arguments as to the benefit of an open flow of information on the bridge and the importance of having the same information available to all persons involved did not make any impression on the Mooring Masters in this group.

3.2 USER-REQUESTED INFORMATION

In the user-centered study, figures and sketches on the blackboard, and various proposals for information parameters were discussed. Figure 5 shows a sketch of a standard STS lightering operation just prior to the SS entering the final phase and closing in on the STBL. In front of the vessels, the 6 degrees of freedom are outlined. Several figures like this were used during the discussion with the Mooring Masters to identify the information needed to improve the decision-making process.

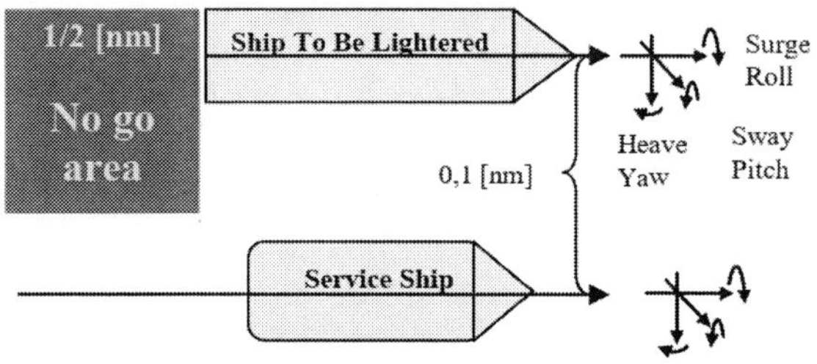

Figure 5: The sketch used for discussions with the Mooring Masters.

Relative speeds, distances and the angle between the two vessels are parameters Mooring Masters use as references in their decision-making process. This information is considered to be key parameters in the process. Additional real-time information, from both ships, could be: course, speed, rate-of-turn (ROT), and rudder angle. During STS operations in waters subjected to heavy swells, (e.g. off the coast of California), information about roll and heave movements would be beneficial.

In order to gain more information from the Mooring Masters, different paper prototype displays were used in the study. Later on, different mock-ups of various sizes were handed out during the interviews, Figure 6. By holding the mock-ups and pointing at a particular part of the GUI, the Mooring Masters easily shared their opinions and discussed this with the author and the others in the group. The GUI has to present key information in a simple way. This is necessary, both from navigational and safety points of view, and because the system has to be capable of operating in any sea/weather conditions.

In terms of size, the Mooring Masters preferred the cell phone size (left), which measured 10.5x7x2 cm. However, the size of the hand-held DSS that is developed in the prototype system is quite a bit larger, a tablet-PC, as this is an easier and more flexible platform on which to program the GUIs and set up the communication with the simulator system.

Figure 6: Proposed mock-ups of hand-held devices.

In addition, several contextual observations were carried out during lightering courses at SMSC. During 4 simulator exercises, each 45 minutes long, the Mooring Masters were observed and debriefed.

All these activities took place in 2009 and resulted in proposals of guidance and decision-support displays to be further integrated and tested in a simulator setting. One of these can be seen in Figure 7, GUI-Alternative 1, which shows the rudder angles displayed in red to port and green to starboard. This makes it easier to perceive the information and it is also in agreement with maritime rules and traditions. The relative movements are all labeled with an arrow and digits. The severity is also indicated by the thickness of the arrow and the color surrounding the arrow. Relative movements in opposite directions have different colors. The information above the vessels is linked to the vessels by having the same color and by being on the same side of the display. When the vessels are closing in, Figure 7 (right), the information about the relative movement can be placed on top, one layer over the ship illustrations. If the background is transparent, the ships' graphical movement can still be observed.

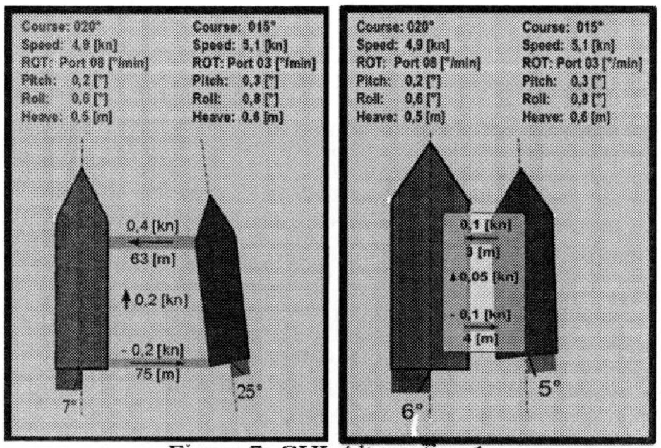

Figure 7: GUI-Alternative 1.

The second iteration was carried out in 2010. Figure 8, GUI-Alternative 2, is primarily the result of the work with mock-ups. This GUI has the same "bird's eye view" as Alternative 1, but with a different graphical presentation. A grid gives the distance between the vessels, while a longer arrow, representing the course, is proposed in order to easily identify the relative angle between the vessels.

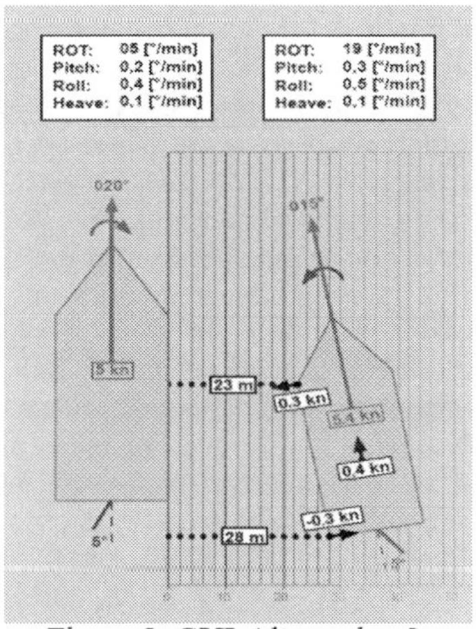

Figure 8: GUI-Alternative 2.

4. THE DECISION SUPPORT SYSTEM, DSS

4.1 PHYSICAL STRUCTURE

The Master PC collects data from sensors, or in this case from the simulator, and processes this data, calculating positions and movements. Results are presented on the screen. The size of both vessels and the positions of their GPS antennas, must be entered into the Master PC before startup. The portable Tablet-PC receives data wirelessly from the Master PC and presents the result as GUIs, Figure 9.

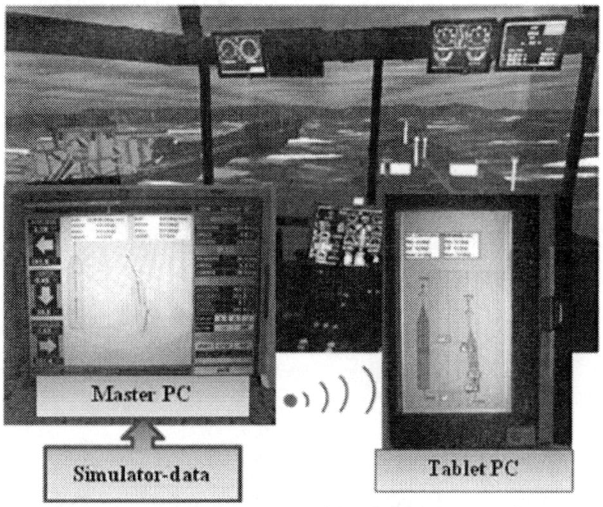

Figure 9: System structure with wireless connection. (Courtesy of SMSC AS.)

Different sensors can be used as input to the decision support system. Data can be received from various systems, such as AIS, GPS, Galileo, or short-distance-position-systems, e.g. microwave systems. Velocity Information-GPS (VI-GPS) and AIS-data have been used in this project. But, an integrated system, consisting of normal pseudo-range GPS and Inertial Sensors, [10], will make the decision support system more likely to become implemented in a larger part of the worldwide tanker fleet, [9].

4.2 GRAPHICAL USER INTERFACE, GUI

Screenshots from the Tablet-PC after the second iteration are seen in Figures 10 and 11. In the initial phase, Figure 10, less information is needed.

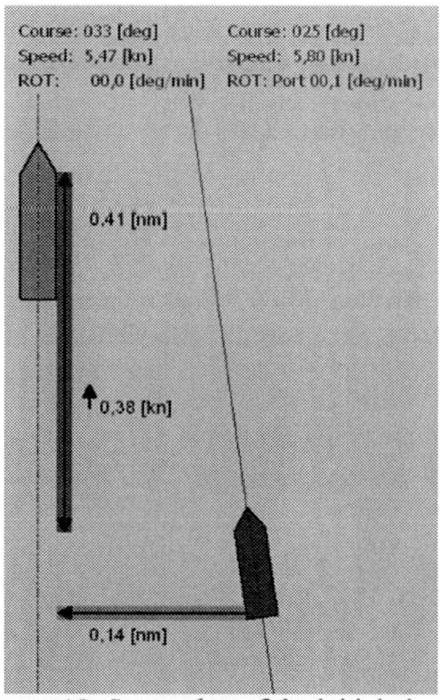

Figure 10: Screenshot of the initial phase.

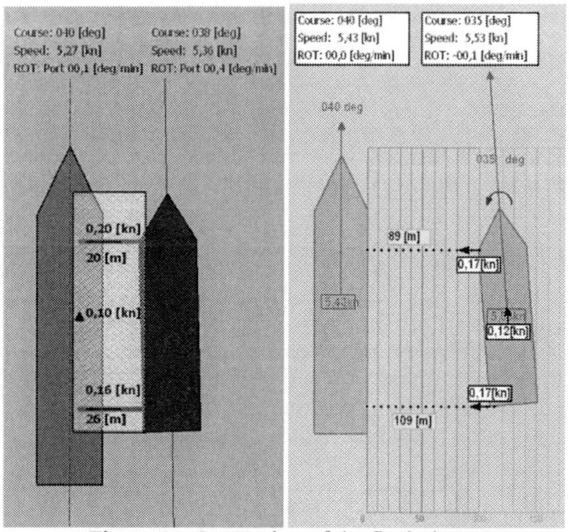

Figure 11: Screenshot of the final phase.

Alternative 1 on the left and Alternative 2 on the right. The two different alternatives, 1 or 2 in Figure 11, is selected on the Tablet-PC. This second iteration is a further development of GUI Alternative 1 in Figure 7, and can be seen in Figure 11, on the left. Information about rudder and some vessel movements are removed because this data is not provided by the simulator.

The process of programming the GUIs introduced some challenges. For example, the limitations of the library in the program used, (Borland C++), and the fact that the DSS is built for real operations, complicated the development of the GUIs.

5. TESTING THE DECISION-SUPPORT SYSTEM

Two field observation and testing programmes onboard tankers performing lightering operations were carried out in the autumns of 2007 and 2009. Data was recorded from the Voyage Data Recorders onboard the SS and video recordings were also made to document the process.

A test with AIS-data from the STBL and phase measuring GPS-data from the SS, Figure 12, took place during lightering operations in the Gulf of Mexico, just outside Freeport, in November 2009. The results from this test showed sufficient accuracy, [11]. Standard deviations in longitudinal and transverse velocities of the bow were approx. 2 cm/sec as measured in real-time by VI-GPS, [12].

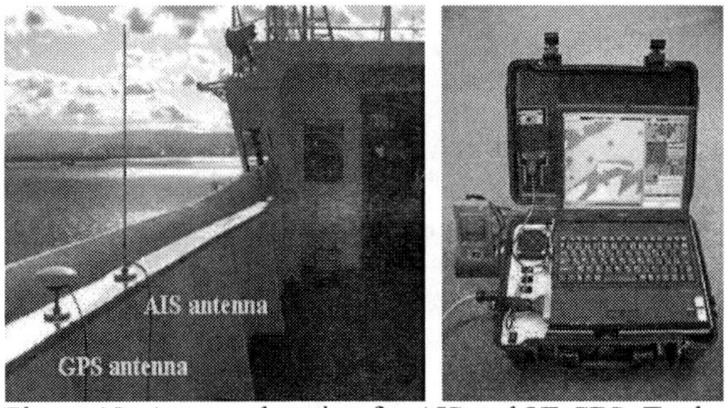

Figure 12: Antenna location for AIS and VI-GPS. To the left is the technology platform for the DSS by Akishima Laboratories, [11].

In January 2011, pre-tests of the graphical DSS were carried out at SMSC in Trondheim. This included technical tests of the system connected to the simulator, comparing physical parameters on the simulator with the parameters presented on the DSS. Real-time measurements were performed to identify any time-delays in the calculation of the GUI presentations.

5.1 PROTOTYPE TESTING

The prototype test of the graphical user interfaces was performed at SMSC during a lightering course for Mooring Masters in March 2011. The test was an interactive process, focusing on the context of using the DSS during the operational setting within the simulator environment. The Mooring Masters did not use any electronic instrument other than the DSS during these prototype tests.

At first, the intention was to use a test like a NASA-TLX (Task Load Index), [13]. This is a workload test with multi-dimensional rating procedures that provide a subjective overall workload score based on a weighted average of ratings on six subscales: mental demands, physical demands, temporal demands, own performance, effort, and frustration. These parameters are not easy to detect and scale in a simulator environment, especially with experienced operators. Therefore, the GUIs were tested using different observations and analysis techniques, including questionnaires, interviews, and contextual observations. The questionnaires used Likert scales, which are psychometric scales subjected to the user experience and commonly used in questionnaires, [14]. This method is open to neutral answer, but may also be irrelevant.

After each session, the Mooring Master's opinions were documented on multi-item scaled questionnaires. Each questionnaire had one parameter to be tested, specified by one question with three or four different descriptive alternatives, to be rated by the Mooring Masters on a scale of 1 to 9. (1 was the lowest score and 9 was the highest score.) The descriptive alternative differed in the questionnaires relative to the parameter being tested. The scales in this test can only be defined with reference to a particular context, [15]. This means that there are no absolute measures of the test. The first questionnaire is included in Table 1.

Table 1: Example of a Likert-scaled questionnaire, [14].

PART 1: OVERALL USER REACTIONS.											
Please circle the numbers that best reflect your impression of this decision-support system.											
Overall reaction to the system:											
1.1	terrible							excellent			NA
	1	2	3	4	5	6	7	8	9		
1.2	frustrating							satisfying			NA
	1	2	3	4	5	6	7	8	9		
1.3	difficult							easy			NA
	1	2	3	4	5	6	7	8	9		
1.4	not in real time							in real time			NA
	1	2	3	4	5	6	7	8	9		

Semi-structured interviews were performed after the Likert tests, and revealed more feedback on the operational level and a deeper understanding of the Mooring Masters' thoughts and ideas concerning this system.

The test persons in this test were two mooring masters, one more experienced than the other. An experienced Captain with a background from various types of tankers was also participating. All three persons were tested while using both GUI alternatives.

The Mooring Master watched the Tablet PC and used this information to adjust the propulsion of the Service Ship in the decision to set the steering course. Course orders were given to the helmsman, who applied the rudder force needed by watching the bridge course and ROT indicators to maintain the required course.

The test scenario was a normal lightering approach, where the service ship started about 3-4 cables astern of the STBL's starboard side. The duration of the scenario was about one and a half hours. Contextual observations from the bridge were applied, including how the GUI was used in the decision-making process. The Mooring Masters occasionally pointed out what parameters they perceived from the hand-held GUI and how this information was used to correct the ship's course and speed.

5.2 PARAMETERS TESTED

Table 2 shows an example of a questionnaire where one particular parameter was tested. In total, seven different parameters were tested for each GUI.

- Part 1 & 8: Overall user reactions.
- Part 2 & 9: The decision-making.
- Part 3 & 10: The vessel's yaw movement.
- Part 4 & 11: The angle between the vessels.
- Part 5 & 12: Decision about course commands.
- Part 6 & 13: The distance between the vessels.
- Part 7 & 14: The degree of accuracy in the presentations on the DSS.

Table 2: Example of questionnaire for one parameter.

PART 3: THE VESSEL'S YAW-MOVEMENT.										
Please circle the numbers that best reflect your impression of this decision-support system.										
How well does the DSS describe yaw movement?										
3.1 incorrect								correct		NA
	1	2	3	4	5	6	7	8	9	
3.2 unreadable								readable		NA
	1	2	3	4	5	6	7	8	9	
3.3 not operational								operational		NA
	1	2	3	4	5	6	7	8	9	
3.4 not in real time								in real time		NA
	1	2	3	4	5	6	7	8	9	

The semi-structured interviews were carried out after the questionnaires, and focused on the same parameters, but revealed more detailed information about the Mooring Masters' reactions to the DSS, and how the two different GUIs were able to provide more detailed information about the operation. This also included particular parameters like: graphics, colors, size, and fonts, and was strongly connected to the Mooring Masters' operational experience. The last part of the interview focused on the Mooring Masters' final suggestions for improvements.

6. ANALYSIS

6.1 EXPECTATIONS

The ideal outcome of the introduction of the DSS is that the decision-making process speeds up. A good GUI will provide more exact information, especially through situation awareness, and less time is spent selecting and comparing information with internalized experiences of the Mooring Masters, Figure 13. The decision-making process is made faster, and commands from the Mooring Master could potentially be given earlier with lower degrees of uncertainty. After some time using the DSS, Mooring Masters are expected to adapt the DSS using their own experience. The graphical prototypes presented to the Moring Masters have potential for improvements. These tests were carried out to set the course for future development.

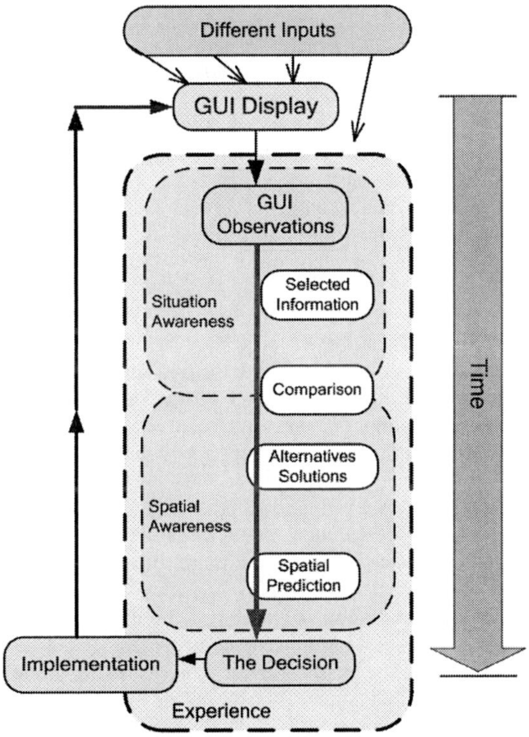

Figure 13: With information from the GUI, the decision-making process can move faster.

6.2 RESULTS

Table 3 shows the different scores provided by three Mooring Masters to the seven different questions for each of the two GUI alternatives. The figures highlighted in light blue are the results from the second test-run, and the numbers highlighted in yellow are the scores from the operational impressions, which may indicate usability in real lightering operations.

The results of the test, especially Parts 5 and 12, were influenced by the fact that the Mooring Masters did not use rudder commands, but course commands.

Table 3: The scores of the scaled questionnaires.

	GUI Alternative 1							GUI Alternative 2						
USER / PART	1	2	3	4	5	6	7	8	9	10	11	12	13	14
I	8	8	7	8	7	8	6	7	7	7	7	5	7	6
Experianced	8	8	8	8	7	8	8	7	7	8	7	6	7	8
	9	8	8	8	7	8	8	7	7	8	8	5	7	8
	9		8	8				7		7	7			
II	8	8	7	7	8	8	8	8	8	7	7	7	8	8
Less experianced	8	8	7	7	8	8	8	7	7	6	7	7	8	8
	8	8	7	7	8	8	8	8	7	7	7	7	8	8
	8		7	7				8		6	7			
III	7	7	5	8	5	8	8	6	6	5	8	5	7	8
Captain / Instructor	7	7	5	8	5	8	8	6	6	5	7	5	8	8
	8	6	7	8	5	7	8	7	5	7	8	5	7	8
	7		7	7				6		7	7			
Average score	7,9	7,6	6,9	7,6	6,7	7,9	7,8	7,0	6,7	6,7	7,3	5,8	7,4	7,8
Average Total	**7,5**							**6,9**						
Operational score	**7,4**							**7,2**						

All participants had a very positive response to the system. They felt the graphics and the data presented on the GUIs were easy to understand. The contextual observations made on the bridge also gave the impression that the information gained from the GUI was easily and quickly perceived.

The Mooring Masters looked at the GUI frequently during the tests, up to 20 to 30 percent of the time, but they did so less frequently when the two ships were close to the landing, about 10 meters apart. The information they received about distances and relative speeds resulted in adjustments in the vessel's propulsion and course.

The Mooring Masters found Alternative 1 of the GUIs to have the best presentation, as this was more intuitive to both read and perceive than the more directly integrated information from the SS (Alt. 2, Figure 11, left). The amount of information connected to the graphics of the SS made this information harder to perceive because it changed when the vessel moved. The results in Table 3 give the same indications.

7. CONCLUDING REMARKS

7.1 CONCLUSIONS

The development of the decision-support system for STS operations was based on a user-centered design approach. Mooring Masters actively participated in the design of the user interface. This was achieved by user-centered techniques and the application of artifacts like paper prototyping, mock-ups, and programmed graphical displays. The Mooring Masters interacted with these different prototypes, providing specific feedback through interviews and questionnaires focusing on the operational safety aspect.

7.2 RECOMMENDATIONS FOR FURTHER WORK

Suggestions for improving the unit include maximizing the information on the GUI screen, especially using larger fonts. The data could be hard to read when exposed to the environment on the bridge wing. The shift from nautical miles to meters can be mentally challenging. In the initial approach phase, it is most convenient to present nautical miles, but the change to meters can be delayed until the vessels are closer together, approx. 0.02 nautical miles apart.

Another suggestion is to implement thermology speed over ground, SOG, and course over ground, COG, and include the Service Ship's heading on the GUI. This would clear up possible misunderstandings. A rudder indicator and engine settings, e.g. RPM, for the SS could be presented in a corner of the GUI. And information about the connection-point, the manifold, could be included in the graphics.

In the development further, Alternative 1 combined with the underlying grid-division and the ROT arrows in Alternative 2 is expected to offer the best presentation of the DSS.

This test group of Mooring Masters argued for one fixed display inside on the bridge and one handheld display which could also be strapped to the bridge wing.

8. REFERENCES

1. Husjord D. and Pedersen E., 'Operational Aspects on Decision-making in STS Lightering,' *Proc. of the 19th International Offshore and Polar Engineering Conference and Exhibition (ISOPE 2009)*, Osaka, Japan, 21 – 26 June 2009.

2. OCIMF & ICS, 'Ship To Ship Transfer Guide,' *Petroleum 4th Ed.* 2005, pp 3.50.

3. Husjord, D. and Pedersen, E., 'On Guidance and Decision-Support Displays in Ship-to-Ship Lightering Operations,' *eNewsletter of IEEE Systems, Man and Cybernetics Society*, June 2010.

4. Sand, O., Sjaastad, Ø.V., Haug, E., og Bjålie J.G., 'Menneskekroppen – Fysiologi og anatomi,' *Gyldendal Forlag*, 2006.

5. Klein, G. A., 'Sources of Power,' *MIT Press, Cambridge*, 1998.

6. Rust, C., 'Design Enquiry: Tacit Knowledge and Invention in Science,' *Design Issues: Volume 20, Number 4, MIT*, pp 76- 85, 2004

7. Lintern, G. 'What is a Cognitive System?' *Proceedings of the Fourteenth International Symposium*

on Aviation Psychology, (pp. 398-402). April 18-21, 2005, Dayton, Ohio.

8. Hutchins, E. 'Cognition in the wild,' *MIT Press, Cambridge.* 1995, pp 70 247.

9. Husjord, D., Pedersen, E. and Øritsland, T. A., 'Integration of Decision-support System in STS Lightering Operations,' *Proc. of Techno-Ocean 2010,* Kobe, Japan, 14 – 16 October 2010.

10. Farrell J. L., 'GNSS Aided. Navigation & Tracking,' *American Literary Press*, Baltimore, Maryland. 2007

11. Pedersen, E., Husjord, D., Yoo, Y. and Shimizu, E., 'Field Testing of GPS Applications in Ship-to-Ship Lightering Operations,' *Proc. of the ION International Technical Meeting 2010*, San Diego, California, USA, 25 – 27 January 2010.

12. Yoo Y., Pedersen E., Tatsumi K., Kouguchi N. and Arai Y., 'Application of 3-D Velocity Measurement of Vessel by VI-GPS for STS Lightering,' *Proc. of the 8th International Symposium on Marine Navigation and Safety at Sea Transportation (TransNav 2009)*, Gdynia, Poland, 17 – 19 June 2009.

13. Hart, S. G. & Staveland, L. E., 'Development of NASA-TLX (Task Load Index): Results of empirical and theoretical research. In: Human Mental Workload,' *(P. A. Hancock and N. Meshkati (Eds.)), 139-183. North-Holland: Elsevier Science.* 1988.

14. Moser & Kalton and Rugg & Cantril cited by Converse, J. M. & Presser, S. 'Survey Questions — Handcrafting the Standardized Questionnaire' *Sage Publications, London* 1986.

15. Sharp H., Rogers Y. and Preece J., 'Interaction Design - Beyond Human- Computer Interaction,' *2nd Ed, Wiley,* 2007.

9. AUTHOR BIOGRAPHIES

Dagfinn Husjord is a PhD candidate in the field of nautical science at the Norwegian University of Science and Technology. Husjord has several years experience as a senior navigation officer onboard merchant ships and is currently also a senior lecturer in navigation courses at the University of Tromsø in Norway.

Trond Are Øritsland is an Associate Professor with the Department of Product Design at NTNU. Dr. Øritsland received his doctorate in 1999. His courses and research projects focus on: interaction design, graphic user interface design, ergonomics, and man-machine interaction.

Egil Pedersen is a Professor of Marine Technology at the Norwegian University of Science and Technology. Dr. Pedersen received his doctorate in nautical science from the university in 1997 and prior to his current assignment he has worked in the international offshore oil service industry and as a post-doc researcher at Marine Technical College, Kobe University, and the National Maritime Research Institute in Japan.

A STUDY ON ESTIMATION OF HYDRODYNAMIC FORCES ACTING ON A SHIP HULL IN SHALLOW WATER BY CFD

Y Kimura, Oshima College of Maritime Technology, Institute of National Colleges of Technology, Japan
E Kobayashi, Graduate School of Maritime Sciences, Kobe University, Japan
Y Tahara, Center for CFD Research, National Maritime Research Institute, Japan
S Koshimura, Disaster Control Research Center, Tohoku University, Japan

SUMMARY

Computational fluid dynamics (CFD) has recently undergone remarkable progress in the field of naval architecture. Much research has been devoted to CFD evaluation particularly in the early design stage. This study explored the application of CFD to the estimation of the hydrodynamic forces acting upon a ship hull manoeuvring in shallow water. First, CFD code developed by one of the authors, which has proven effective in flow field analysis, was adapted to the study of manoeuvring hydrodynamic forces in shallow water by introducing requisite improvements in capability and accuracy. Next, this code was applied to a model of a very large crude carrier which has previously been subject to detailed analysis concerning manoeuvring motion. The obtained computational results were found to be in good agreement with experimental data in deepwater, suggesting that this code could also accommodate more complex conditions in the future.

NOMENCLATURE

d	Ship draft
Fn	Froude number
H	Water depth
I_{zz}	Mass moment of inertia of ship about z axis
J_{zz}	Added mass moment of inertia of ship about z axis
L	Ship length
m	Mass of ship
m_x, m_y	Added mass of ship in x, y direction
N_H	Yaw moment acting upon a ship hull
N_0	Yaw moment in space-fixed coordinate system
p	Piezometric pressure normalized by ρU_0^2
r	Rate of turn
Rn	Reynolds number
S	Wetted surface area
t	Time
U_0	Reference velocity
u, v	Velocity components in x, y direction
X_H, Y_H	Longitudinal and Lateral forces acting upon a ship hull
X_0, Y_0	Drag force and Lateral force in space-fixed coordinate system
x^i, x^j	Dimensionless co-ordinates normalized by characteristic length
β	Drift angle
k	Turbulent kinematic energy
ν	Kinematic viscosity
ν_t	Isotropic eddy viscosity
ρ	Density of water
ψ	Heading angle of ship

1. INTRODUCTION

The estimation of ship manoeuvring motion is an important issue for safe navigation in the open sea. The most reliable method to estimate such ship manoeuvring motion is computer simulation based upon reliable mathematical models. Of these models, the modular type MMG model—comprising ship components such as the hull, propeller, rudder, and thruster—is currently the most commonly used in Japanese shipyards.

In a mathematical model, it is very important to understand the hydrodynamic forces acting upon a ship hull, which depend upon flow conditions such as turning and/or oblique running.

In addition, the present authors have recently investigated the behaviour of the ships in a port area during a tsunami. A tsunami has a strong and horizontal flow and the water level changes gradually. To evaluate the ship's behaviour in a tsunami, it is crucial to examine the hydrodynamic forces acting upon the hull [1]. Moreover, the hull force may differ in shallow water and under restricted conditions such as near a quay.

One reliable method to gain an understanding of the hydrodynamic forces acting upon the hull is to carry out tank tests such as oblique towing, circular motion and PMM tests. However, this is very expensive and takes a great deal of time, and it is rare for such experiments to be carried out in shallow water or under restricted conditions.

On the other hand, remarkable progress in Computational fluid dynamics (CFD), as applied to naval architecture, has been made in recent years. Several studies have reported that the computational results of the hydrodynamic forces were in good agreement with experimental results [2-7]. Many studies have validated their CFD codes without carrying out any experiments in a towing tank, especially in the estimation of ship propulsive performance in the early design stage.

As reviewed so far, a variety of investigations have contributed to the estimation of ship's manoeuvring force. However, compared to the amount of research on ship hydrodynamic forces using CFD technique, few studies have been done to analyze this under restricted conditions such as in shallow water [8]. Therefore, the purpose of this study is to examine the hydrodynamic forces acting upon a hull in shallow water by CFD.

In this paper, CFD code that was developed by one of the authors and has shown many good results in flow field analysis [9-11] is improved to have enough capability and accuracy to understand manoeuvring hydrodynamic forces. The improved CFD code is applied to a model of a very large crude carrier (VLCC) which has been previously subject detailed analysis concerning manoeuvring motion. The present work covers estimation of manoeuvring forces acting upon a bare hull without rudder and propeller. These computational results have been compared with experimental data published in literature [12,13] to verify the numerical method and to examine that this CFD code has enough practicality to estimate hydrodynamic forces acting on a ship hull in shallow water.

2. HYDRODYNAMIC FORCES OF MATHEMATICAL MODEL

Hydrodynamic forces acting upon a ship hull are expressed by the following equations:

$$\left.\begin{aligned}(m+m_x)\dot{u}-(m+m_y)vr&=X_H\\(m+m_y)\dot{v}+(m+m_x)ur&=Y_H\\(I_{zz}+J_{zz})\dot{r}&=N_H\end{aligned}\right\}\quad(1)$$

where

m, m_x, m_y : mass of ship, added mass of ship in x, y directions,

I_{zz} : mass moment of inertia of ship about z axis,

J_{zz} : added mass moment of inertia of ship about z axis,

u, v, r : velocity components in x, y direction and rate of turn,

X_H, Y_H, N_H : longitudinal and lateral force and yaw moment acting upon ship hull.

The coordinate systems and definition of positive direction of forces and moments are shown in Fig. 1.

Lift, drag and yaw moment in a space-fixed coordinate system are obtained from CFD computation. Therefore, hydrodynamic forces of manoeuvring motion in a body-fixed coordinate system are calculated by the following relational equations:

$$\left.\begin{aligned}X_H&=X_0\cos\psi+Y_0\sin\psi\\Y_H&=-X_0\sin\psi+Y_0\cos\psi\\N_H&=N_0-Y_0\cdot x_{OG}\end{aligned}\right\}\quad(2)$$

where

$\psi=\beta$ under this diagonal condition,

x_{OG} : distance from z_0 axis in x_0 direction.

3. CFD CALCULATION METHOD

CFD code developed by one of the authors which has proven effective in flow field analysis has been improved in terms of capability and accuracy to study the hydrodynamic forces under manoeuvring.

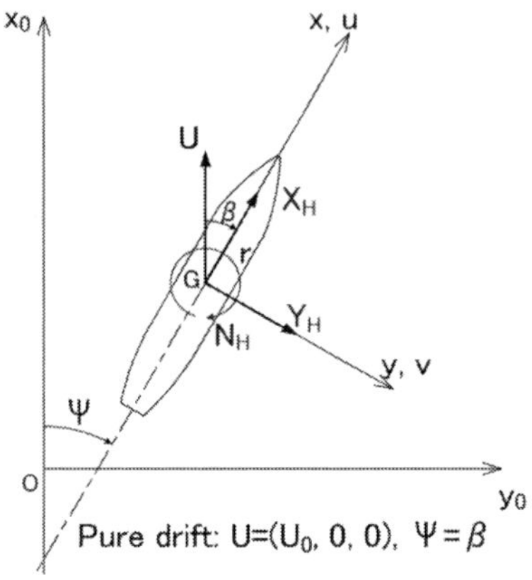

Figure 1: Coordinate system

3.1 GOVERNING EQUATIONS

The non-dimensional Reynolds-averaged Navier-Stokes (RANS) equations for unsteady, three-dimensional incompressible flow can be written in Cartesian tensor notation as,

$$\frac{\partial U_i}{\partial t}+U_j\frac{\partial U_i}{\partial x^j}+\frac{\partial\overline{u_iu_j}}{\partial x^j}+\frac{\partial p}{\partial x^i}-\frac{1}{Rn}\nabla^2U_i=0\quad(3)$$

$$\frac{\partial U_i}{\partial x^i}=0\quad(4)$$

where

$U_i(i=1,2,3)=(U,V,W)$, and $u_i(i=1,2,3)=(u,v,w)$: Cartesian components of mean and fluctuating velocities, respectively, normalized by the reference velocity U_0,

$x^i(i=1,2,3)=(X,Y,Z)$: dimensionless co-ordinates normalized by characteristic length L,

$Rn=U_0L/\nu$: Reynolds number, ν : kinematic viscosity,

$-\overline{u_iu_j}$: barred quantities of Reynolds stresses normalized by U_0^2,

$p=\overline{p}+Z/Fn^2$: piezometric pressure normalized by ρU_0^2,

$\overline{p}$ is static pressure,

$Fn=U_0/\sqrt{gL}$: Froude number. For zero-Fn case (i.e., double-model flow simulation), $p=\overline{p}$.

If $-\overline{u_iu_j}$ are related to the corresponding mean rate of strain through an isotropic eddy viscosity ν_t, i.e.,

$$-\overline{u_iu_j}=\nu_t\left(\frac{\partial U_i}{\partial x^j}+\frac{\partial U_j}{\partial x^i}\right)-\frac{2}{3}\delta_{ij}k\quad(5)$$

where $k=\left(\overline{uu}+\overline{vv}+\overline{ww}\right)/2$ is the turbulent kinematic energy, Eq. (3) becomes

$$\frac{\partial U_i}{\partial t}+\left(U_j-\frac{\partial \nu_t}{\partial x^j}\right)\frac{\partial U_i}{\partial x^j}-\frac{\partial \nu_t}{\partial x^j}\frac{\partial U_j}{\partial x^i}$$
$$+\frac{\partial p}{\partial x^i}\left(p+\frac{2}{3}k\right)-\frac{1}{R_\varphi}\nabla^2 U_i=0 \qquad (6)$$

where $1/R_\varphi = 1/Rn + \nu_t$, and $\varphi = U_i(i=1,2,3)$. Equations (4) and (6) can be solved for U_i and p when a suitable turbulence model is employed to calculate the eddy-viscosity distribution.

3.2 COMPUTATIONAL METHOD

In this study, free surface effects were not considered, thus the boundary condition imposed upon the water plane is undisturbed.

The conditions for computation were as follows: L=1, U_0=1, Rn=4,000,000, without rudder and propeller.

The governing equations of the flow field, as mentioned above, were the unsteady RANS and continuity equations for viscous incompressible flow written in the physical domain using Cartesian coordinates. A Baldwin-Lomax (BL) model was used for the turbulence model. The equations were transformed from Cartesian coordinates in the physical domain to numerically generated, non-orthogonal, curvilinear coordinates in the computational domain.

The transport equations were discretized using the finite analytic method. Pressure-velocity coupling was accomplished using the PISO type/One step algorithm.

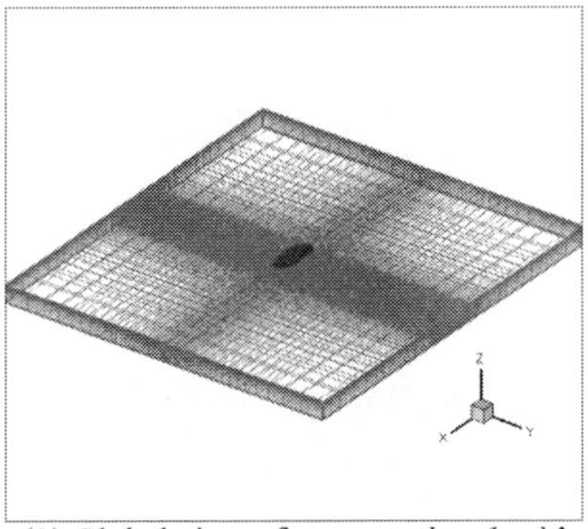

(1) Global view of computational grid

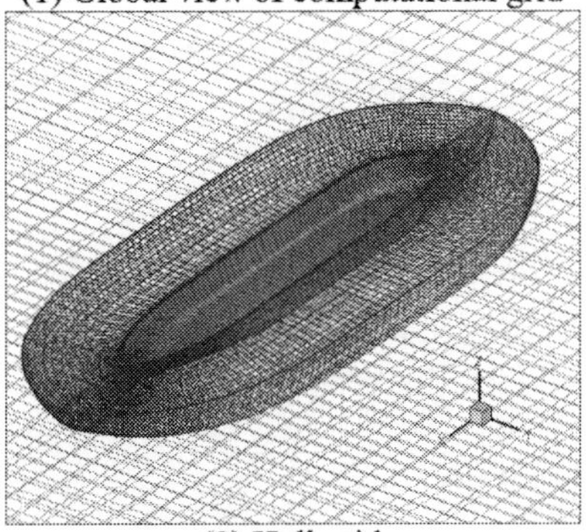

(2) Hull grid

Figure 2: Computational grid

To improve the CFD code from the calculation code used by the former report stability [14], and to expand the setting condition, the following improvements were carried out:

- Expansion of calculating area and an increase of depth direction lattice
- Re-examination of calculation parameter
- Improvement of calculation output file

The forms of the computational grids were two blocks and the over-set multi block grids. A global view of the computational grid and a hull grid are shown in Fig. 2(1) and (2).

The number of hull grid points (O-O type) was 121 × 39 × 79 (longitudinal direction × normal direction × direction of the girth). The number of background grid points (H-H type) was 88 × 59 × 21 (main flow direction × horizontal direction × direction of depth). The number of grid points in the former report was 121 × 40 × 79 and 88 × 59 × 15, respectively.

The section forms of the grids used in the former report and this study are shown in Fig. 3(1) and (2). They are the forms in the vicinity of the midship seen from stern. Some depth settings are different. The grid arrangement in the bottom part of a ship was shoaled and the grid form of the bilge was changed so that the calculation under a shallower condition may become possible. Moreover, the number of background grid points has been increased in the direction of depth.

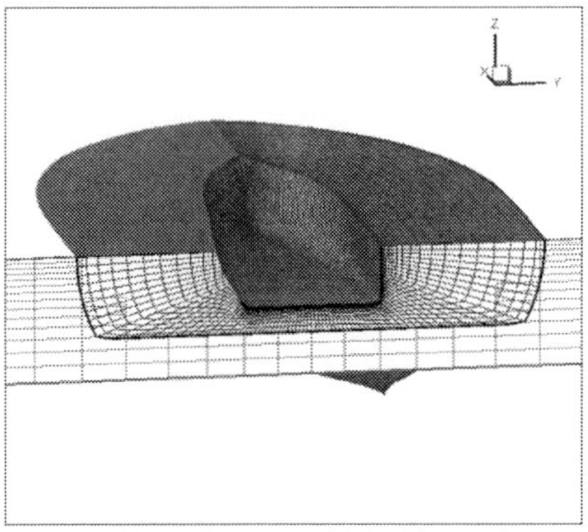

(1) A former grid (H/d=2.25)

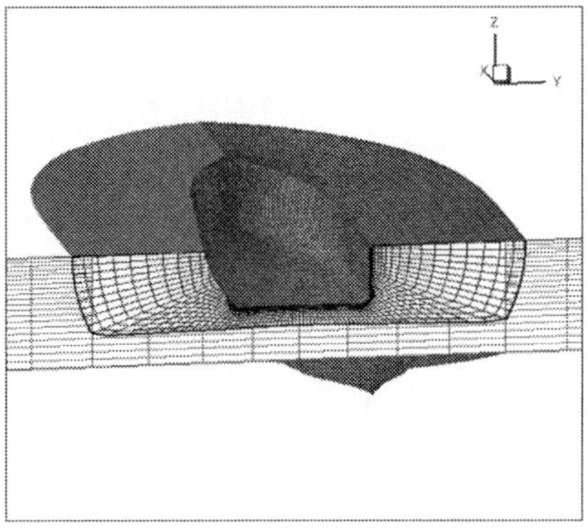

(2) A modified grid (H/d=2)

Figure 3: Section forms of grid

3.3 RESULTS OF CFD COMPUTATION

Hydrodynamic forces acting upon a hull were calculated from the results of CFD computations of flow fields. Typical examples of computational results at small and large drift angles are shown in Fig. 4(1) and (2). Comparisons of the pressure contour on a free surface in a typical deep water case are shown in these figures. It is clear that the positive and negative pressure regions are larger than those in small angle.

Next, comparisons of the axial velocity contour and streamline on a free surface in a typical shallow water case and a deep water case are shown in Fig. 5(1) and (2). It is observed that vortical flows are created in the negative axial velocity region. These flows in shallow water are more complicated downstream from the bow and stern part than in deep water.

Hydrodynamic forces and moment concerning each axis were calculated from these analytical results.

Fig. 6 provides the convergence history of the computations for the hydrodynamic forces and a moment. Hydrodynamic forces X_0', Y_0' and Z_0' are shown in the upper left graph, the upper right graph and the lower left graph. Yaw moment N_0' is shown in the lower right graph. X_0', Y_0' and N_0' are drag, lift and yaw moment values that are non-dimensionalized using the combinations of fluid density ρ, the reference velocity U_0, wetted surface area S and characteristic length L. They are defined by the following equations:

$$X_0', Y_0' = X_0, Y_0 \Big/ \frac{\rho}{2} S U_0^2, \quad N_0' = N_0 \Big/ \frac{\rho}{2} S U_0^2 L \qquad (7)$$

According to the Fig. 6(1), it tends to shift the value in the former code gradually. On the other hand, it tends to converge in the modified code by about iteration number (it) 2000, although there is a difference in drift angle.

Thus, excellent convergence became possible with the new code even under the condition that made convergence difficult for the code used in the former report.

From the CFD results, hydrodynamic forces acting upon a hull were mainly pressure components. Their values were obtained from the averaged values after convergence. The non-dimensional values of hydrodynamic forces are shown in Fig. 7 with drift angles. The trends of lift, drag and yaw moment in a space-fixed coordinate system are obtained from CFD computation.

As previously described, these CFD computational values were converted to hydrodynamic forces of manoeuvring motion X_H, Y_H and N_H.

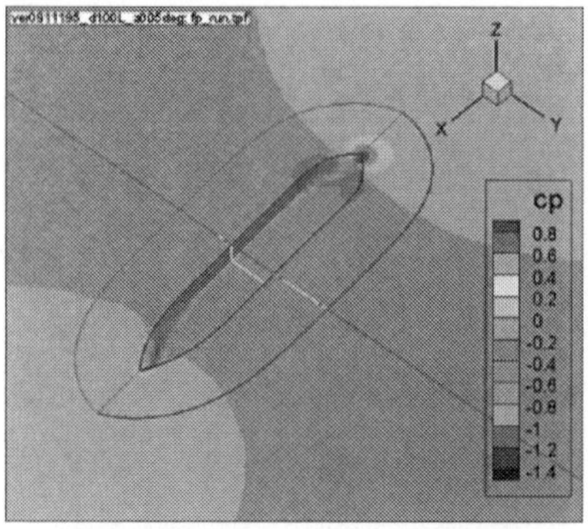

(1) Small drift angle (β =5[deg.])

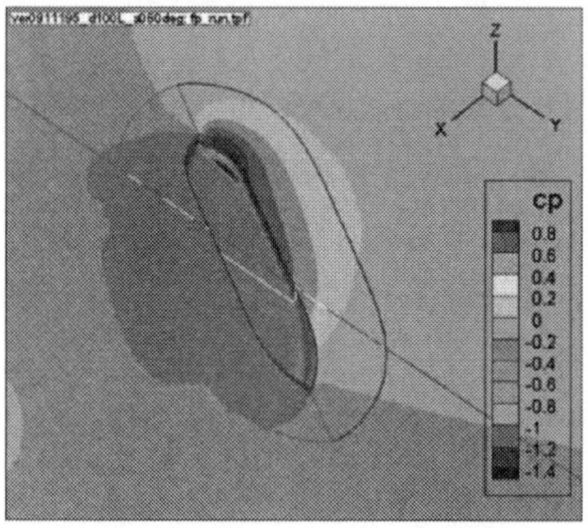

(2) Large drift angle (β =60[deg.])

Figure 4: Pressure contour

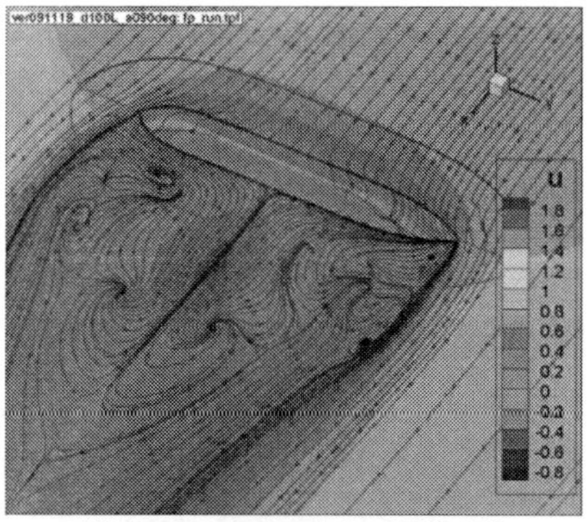

(1) In deepwater (β =90[deg.])

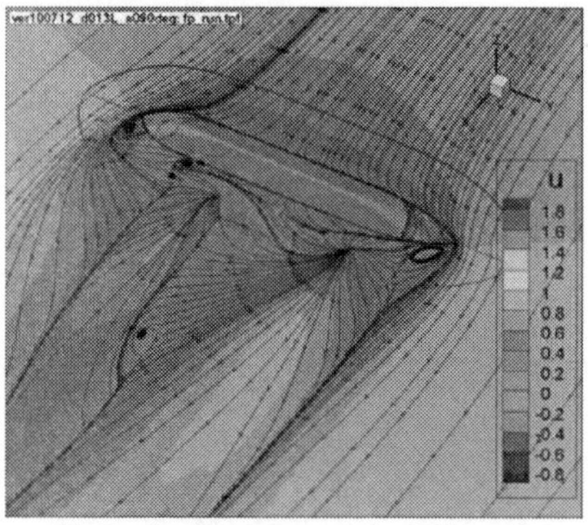

(2) In shallow water (β =90[deg.])

Figure 5: Axial velocity contour and streamline

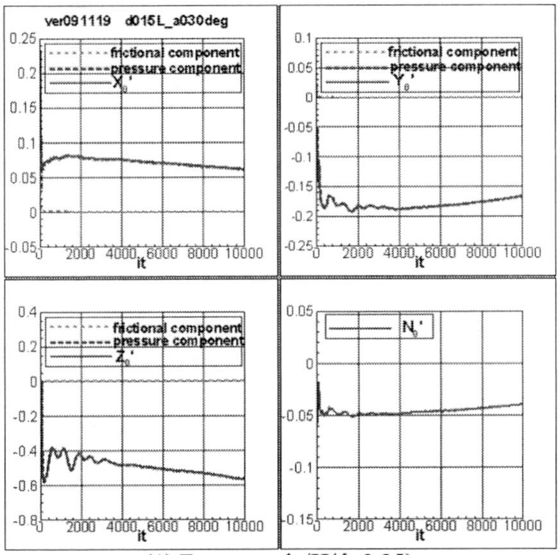

(1) Former code(H/d=2.25)

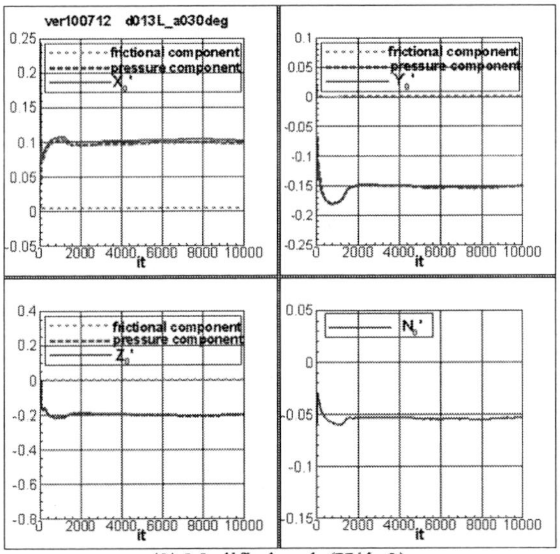

(2) Modified code(H/d=2)

Figure 6: Convergence history of computation

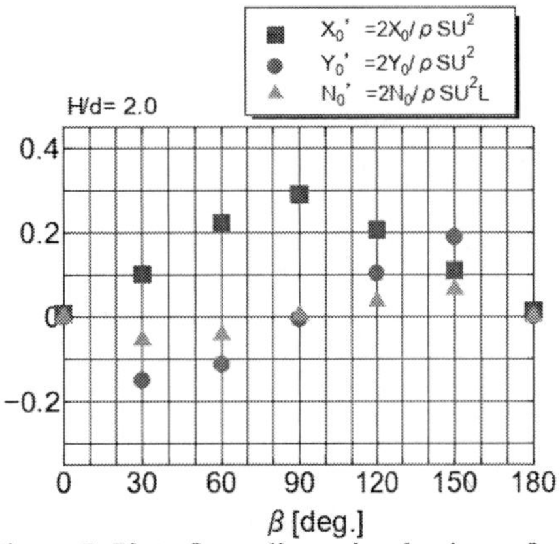

Figure 7: Plot of non-dimensional values of hydrodynamic forces against drift angle

4. RESULTS AND DISCUSSION

4.1 NORMAL DRIFT ANGLES IN DEEP WATER (β < 25°)

A VLCC was selected to confirm the validity of this CFD calculation for manoeuvring hydrodynamic forces and moment [12].

The principal dimensions of the VLCC are shown in Table 1.

The variables computed by CFD were non-dimensionalized using combinations of ship length L, draft d, reference speed U_0 and fluid density ρ. In non-dimensional form, the hydrodynamic forces acting upon a ship hull are as follows:

$$X_H', Y_H' = X_H, Y_H \Big/ \frac{\rho}{2} L d U_0^2 \quad , N_H' = N_H \Big/ \frac{\rho}{2} L^2 d U_0^2 \qquad (8)$$

The hydrodynamic forces acting on a ship hull were investigated in deep water. Figure 8 shows the comparison of the values between computations and experiments at the normal drift angles (β < 25°).

The non-dimensional values of hydrodynamic forces are shown with drift angles. It can be seen that longitudinal force X_H' shows a similar trend, although the computational data are slightly larger than the experimental data on the whole. As for lateral force Y_H', computational values are larger than the experimental values at a drift angle of 10° and above. As for yaw

Table 1: Principal dimensions

	SHIP	2.5m model	4.0m model	4.6m model
Lpp(m)	325.00	2.50	4.00	4.60
B(m)	53.00	0.4077	0.6523	0.7509
D(m)	31.00	-	-	-
d(m)	21.73	0.1672	0.2682	0.3080
Wetted surface area(m^2)	27680	1.6350	4.1929	5.5660

moment N_H' , it corresponds well, although the computational values are a slightly larger than the values of the experimental data on the whole.

From the results of the comparison of non-dimensional values of computations and experiments, it was shown that there were similar trends for the hydrodynamic forces acting upon a ship hull as a whole. The computation was carried out as a number of about 500,000 grids. The computation could be executed with practicable accuracy in a drift angle range of 0-25°, and it was possible to estimate the hydrodynamic forces using the CFD code with enough accuracy.

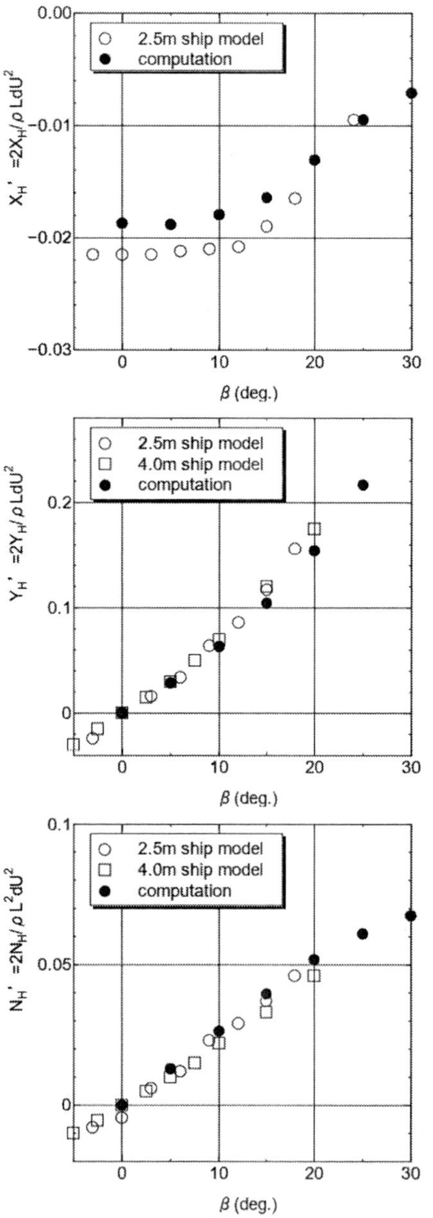

Figure 8: Comparison of hydrodynamic forces between computations and experiments (normal drift angles)

4.2 VARIATIONS WITH DIFFERENT WATER DEPTH

Lift force Y_0, drag force X_0 and yaw moment N_0 in a space-fixed coordinate system are obtained from CFD computation. The variations of non-dimensionalized hydrodynamic forces with different water depth are shown in Fig. 9a (1), Fig. 9b (2) and (3) in four drift angles, respectively. First, the non-dimensionalized values almost remain constant by the ratio of water depth to draft around 6.7. Next, The values increase with the decrease of the ratio of water depth to draft, although Y_0' and N_0' at 90° do not increase.

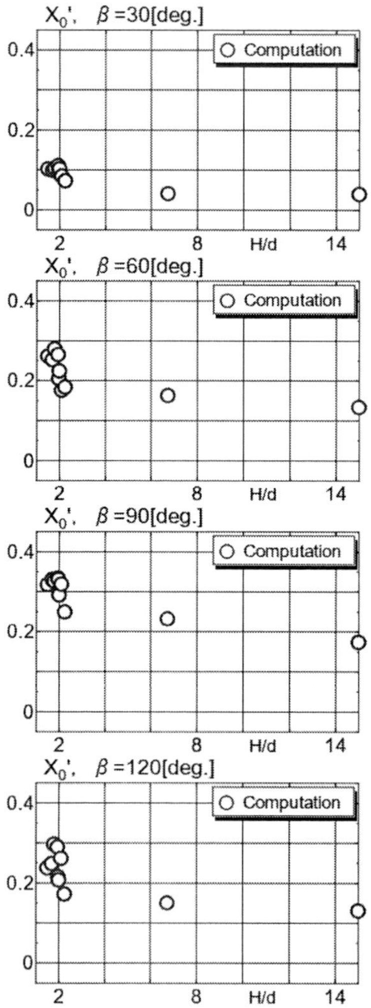

(1) Drag force

Figure 9a: Variations in hydrodynamic forces with different water depth at four drift angles

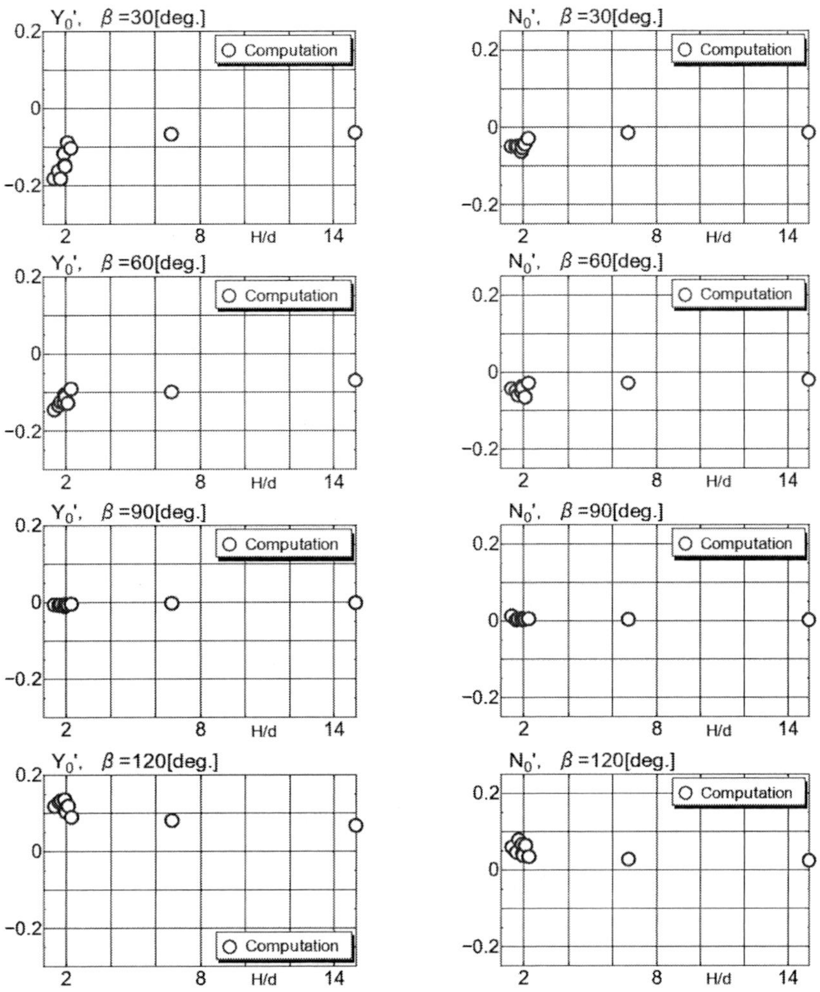

(2) Lift force (3) Yaw moment

Figure 9b: Variations in hydrodynamic forces with different water depth at four drift angles

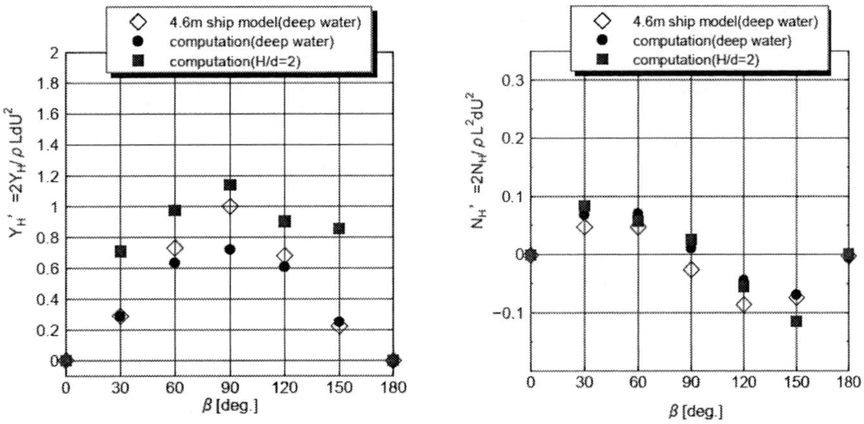

(1) Lift force (2) Yaw moment

Figure 10: Comparison of hydrodynamic forces between computations and experiments (large drift angles)

4.3 LARGE DRIFT ANGLES ($\beta \geq 25°$)

In this case, the experimental data Y_H' and N_H' of a 4.6 m scale model published in literature [13] were used for comparison.

Fig. 10 (1) and (2) shows the comparison of the values between computations and experiments for a deepwater case at large drift angles. In addition, computational data for a shallow water case is shown in these figures. First, according to the figure, the CFD computations in deep water do not completely agree with the measurement data, although they show a certain trend in the comparison result. A further study of verifying the experimental accuracy should be conducted. Next, concerning non-dimensional lift force, there is a dramatic increase in the computations in shallow water. However, non-dimensional yaw moment shows that there is no particular increase between deepwater and shallow water in this case, although maximum and minimum values show a slight change, respectively. From the CFD results, convergence history showed unstable yaw moment in several shallow water cases. However, it does not appear clearly in these analytical results using averaged hydrodynamic force values.

5. CONCLUSIONS

In this study, flow field analysis for a VLCC was performed to verify the computation results using experiment data published in literature, and to obtain basic CFD data under oblique running conditions. The following were the main findings: (a) the CFD code has enough practicality to estimate hydro-dynamic forces acting upon a ship hull under oblique running conditions of normal drift angles in deepwater and (b) the computational results show that this CFD code expresses the shallow water effect in hydrodynamic forces.

It is expected that a CFD calculation that considers turning motion can be executed with further study. The results confirmed that this CFD code can adequately estimate the hydrodynamic forces acting upon a ship hull subject to a tidal wave that assumes uniform flow. This suggested that this code can accommodate more complex conditions in the future.

On the other hand, the computational results in shallow water have not been validated by comparing the experimental results. It will be very beneficial to consider comparison of theoretical results for shallow water with the experimental one. Further empirical studies for shallow water and large drift angles are needed.

Furthermore, the researchers confined CFD calculation to analyze hydrodynamic forces in deep and shallow water. Additional studies, which could provide some detailed results from the CFD calculations showing the effect of viscous terms (not available in potential flow calculations) close to the ship and sea bottom (pressure, velocities), and where CFD can help in understanding flow physics not available in integrated EFD force/moment results, are suggested.

6. ACKNOWLEDGEMENTS

This work was partly supported by a Grant-in-Aid for Scientific Research (B) from the Japanese Ministry of Education, Sports, Culture, Science, and Technology (No. 21360430). The authors also wish to thank members of the maritime safety system laboratory at Kobe University for their assistance.

7. REFERENCES

1. KOBAYASHI, E., KOSHIMURA, S. and MIZUNOE, T., 'Study on the Evacuation Maneuver from a Tsunami Attack, *Conference Proceedings of the MARSIM'09, C-18*, 2009

2. MIYAZAKI, H., NONAKA, K., HINO, T., HIRATA, N., NIMURA, T. and UENO, M., 'Computation of Hydrodynamic Forces acting on a Ship in Manoeuvring Motion', *Journal of the Society of Naval Architects of Japan (187), pp.121-130.*, 2000

3. YAMAZAKI, K., MATSUDA, S. and KATSUI, T., Application of CFD to the Study of Ship Manoeuvrability on Initial Stage of Hull Design, *Journal of the Society of Naval Architects of Japan(189), pp.55-62.*, 2001

4. SHUTO, C., KARASUNO, K., MAEKAWA, K. and MIYOSHI, J., 'Computation of Hydrodynamic Forces acting on a Ship with Large Drift Angles in Oblique Towing Tests', *Journal of the Kansai Society of Naval Architects,No.236, pp.137-143.*, 2001

5. TODA, Y., LEE, Y. and SADAKANE, H., 'Numerical Investigations of Hydrodynamic Forces Acting on a Ship Hull under Lateral Low Speed Motion', *Journal of the Kansai Society of Naval architects,No.238, pp.77-83.*, 2002

6. SIMONSEN, C. D., STERN, F., 'Verification and Validation of RANS Maneuvering Simulation of Esso Osaka: Effects of Drift and Rudder Angle on Forces and Moments', *Computers & Fluids*, Volume *32*(10), *pp.1325-1356.*, 2003

7. SIMONSEN, C. D., STERN, F., 'RANS Maneuvering Simulation of Esso Osaka With Rudder and a Body-Force Propeller', *Journal of Ship Research, Volume 49, No.2, pp.98-120.*, 2005

8. WANG, H-M., ZOU, Z-J. and YAO, J-X., 'RANS Simulation of the Viscous Flow around a Turning Ship in Shallow Water', *Conference Proceedings of the MARSIM'09, M-10*, 2009

9. TAHARA, Y., 'Computation of Viscous Flow around Series 60 Model and Comparison with Experiments', *Journal of the Kansai Society of Naval Architects,No.220, pp.29-47.*, 1993

10. TAHARA, Y., TOHYAMA, S., KATSUI, T., 'CFD-Based Multi-Objective Optimization Method for Ship Design,', *International Journal for Numerical Methods in Fluids, Volume 52, pp.449-527.*, 2006.

11. TAHARA, Y., 'A Reynolds-Averaged Navier-Stokes Equation Solver for Prediction of Ship Viscous Flow with Free Surface Effects', *NAPA User Meeting 2008, May 27-30, Helsinki, Finland*, 2008

12. KIJIMA, K., ASAI, S. and YAMAGAMI, Y., 'Estimation of ship manoeuvrability in deep condition', *Bulletin of The society of Naval Architects of Japan, No.668, pp.27-93.*, 1985

13. KOBAYASHI, E., ASAI, S., 'A Simulation Study on Ship Manoeuvrability at Low Speeds', *Paper NO.10 presented at the International Conference on SHIP MANOUEVRABILITY*, 1987

14. KIMURA, Y., KOBAYASHI, E., TAHARA, Y. and KOSHIMURA, S., 'A Basic Study on Estimation of Hydrodynamic Forces acting on a Ship Hull by CFD', *Conference Proceedings of the JASNAOE'10, Volume 10,G9-1*, 2010

8. AUTHORS BIOGRAPHY

Yasuhiro Kimura is a Lecturer at Shipping Department, Oshima College of Maritime Technology and is responsible for education and training of merchant cadets. He has studied toward a Ph.D. in coupling analysis of Tsunami and ship dynamics.

Eiichi Kobayashi is a Professor at Graduate School of Maritime Sciences, Kobe University. He has been responsible for a number of research projects and studies of Tsunami, maritime safety and ship manoeuvrability.

Yusuke Tahara is a Senior Researcher at Center for CFD Research, National Maritime Research Institute and is responsible for development of Aerodynamics and Hydrodynamics Designs, CFD and Simulation-Based Design.

Shunichi Koshimura is an Associate Professor at Disaster Control Research Center, Tohoku University and has worked in the fields of Tsunami engineering and civil engineering.

DECISION-SUPPORT SYSTEM FOR STS TRAINING ON SHIP HANDLING SIMULATOR BY MEANS OF VISUALIZATION OF DANGER ZONE DUE TO HYDRODYNAMIC INTERACTION

K Kirimoto, Norwegian University of Science and Technology, Norway
E Pedersen, Norwegian University of Science and Technology, Norway

SUMMARY

The authors have proposed a decision-support system for improved ship handling simulator-based training of the ship officers in charge of ship-to-ship (STS) operations. The basic concept is visualization of the danger of collision (steel-to-steel contact) from the viewpoint of manoeuvring ship based on hydrodynamic interaction in its final approach phase towards the constant heading ship. The danger of collision is evaluated based on hydrodynamic interaction acting on the ship hull and simulation of ship manoeuvring. The hydrodynamic interaction is estimated from the results of model test experiments with two tankers in various positions in the final approach phase of an STS lightering operation. The present approach succeeded to visualize the potential danger surrounding the ship-to-be-lightered, which can be considered to supply safety information of interaction effects to inexperienced mooring masters.

1. INTRODUCTION

A ship-to-ship (STS) operation is one of the most challenging maritime operations for the ship officers due to its complex hydrodynamic interaction phenomenon which can cause unpredicted and sudden change of ship motion when two ships are in close proximity. Owing to the accumulated knowledge of STS operation [1], it is nowadays regarded that STS is a middle risk operation.

To enhance the safety level of the operation from nautical aspects, one of the authors has proposed and developed a decision-support system (DSS) for the ship officer in charge of an STS operation [2]. This can provide intra-ship position and motion on 2-D view in real-time with adequate accuracy [3]. In this paper, it is focused on the hydrodynamic interactions of STS operation as an attempt to further improve the DSS. By visualizing special danger due to the interaction effects on the 2-D view as a function of the intra-ship position, an advanced DSS can be realized. The advanced system enables more effective STS simulator training for an inexperienced officer, making the officer be conscious of avoidance from the danger.

The present paper shows the concept of danger zone in STS operation due to the interaction effects. The effects are estimated based on available data from model test results, where two ships are in close proximity. An existing mathematical model for ship manoeuvre is applied to simulate the final approach phase. From static and dynamic analyses of the interaction effects, the danger zones which visualize the operational limitation and recommendation to escape from the danger are presented.

2. DANGER ZONE CONCEPT

A danger zone concept in STS operation is introduced. The difficulties of STS final approach phase from a nautical aspect are found in management of relative ship position and motion, and in avoidance from special danger due to hydrodynamic interaction. For the former difficulties, a decision-support system for the ship officer in charge of operation has been developed [2]. This system is based on the measurement of GPS and AIS data, which enables the application for real-ship operation.

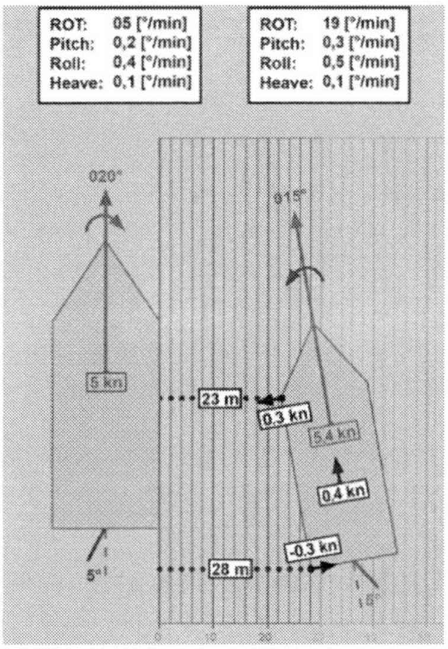

Figure 1: Graphical user interface of the decision-support system in STS approach phase [2].

Figure 1 shows an example of graphical user interface of the DSS in final approach phase. The DSS consists of the graphical and digital presentation, which copes with both visibility and accuracy. The ship-to-be-lightered (STBL) behaves as a course-keeping vessel and is positioned to the left of the interface. The service ship (SS) is approaching towards the STBL and is displayed on the right side of the figure. The shapes of both ships are modified with a certain linear expansion/reduction for graphical reasons. Moreover, directions of yaw motion and rudder are presented by colours corresponding to port/starboard signals. On the other hand, more specified values of intra-condition e.g. side-to-side distance or ship speed are shown by digital numbers inside highlight boxes. For further improvement of the DSS, the investigation of key information for the effective decision support is currently undertaken based on usability tests in ship handling simulator.

The authors have considered another possibility to improve the system, focusing on the effect of the hydrodynamic interaction which represents one of the difficulties in STS operation. Based on a real-time- or pre-analysis of the interaction effects, an advanced system which highlights a potential danger from the interaction on the 2-D diagram is suggested. In addition, the proposed system applied to simulator training for inexperienced ship officers can provide more effective training by being conscious of the avoidance from a danger due to interaction. From this viewpoint, estimation of the danger in STS operation is proposed.

The special dangers due to hydrodynamic interaction can be classified into static and dynamic effects of which the former can be evaluated by a balance of yaw moments affecting the hulls. Equation (1) formulates the static criterion.

$$\alpha \cdot |N_{R,\max}| < |N_{INT}| \quad . \tag{1}$$

Here, $N_{R,\max}$ and N_{INT} are the maximum rudder moment and external moments due to interaction effects, respectively. α ($\alpha \leq 1$) is a coefficient which represents a safety margin. Equation (1) represents the situation where the ship experiences loss of rudder control because of larger interaction moment. `No-Go Zone` can be displayed in 2-D view to highlight where the service ship must not enter since the interaction forces can be assumed as a function of intra-ship position.

The evaluation of dynamic effects includes several possibilities, e.g. case studies for some specified approaches. In this paper, the authors attempt to evaluate this effect as a function of intra-ship position to obtain comparability in both evaluations. A set of manoeuvre simulations with systematic change of the initial intra-ship position around STBL has been adopted.

Figure 2 shows a flow chart of the present proposal which consists of systematic manoeuvre simulations. Firstly, the initial intra-ship positions ξ_k and η_k are given by distances between two ship's reference positions, e.g. Centre of Gravity (COG). Secondly, manoeuvre simulation with selected scenario is executed and continued with the initial position changed until all pre-selected ship positions are examined. Manoeuvre actions undertaken by the ship officers are dependent on their own experiences, knowledge and decision, which results in unique response time or reaction in ship control. Therefore, the scenario should be decided based on a factor or aspect to be evaluated. After the implementation of a set of simulation, selected parameters are examined to define the `Potential Danger Zone` where it is not recommended to enter.

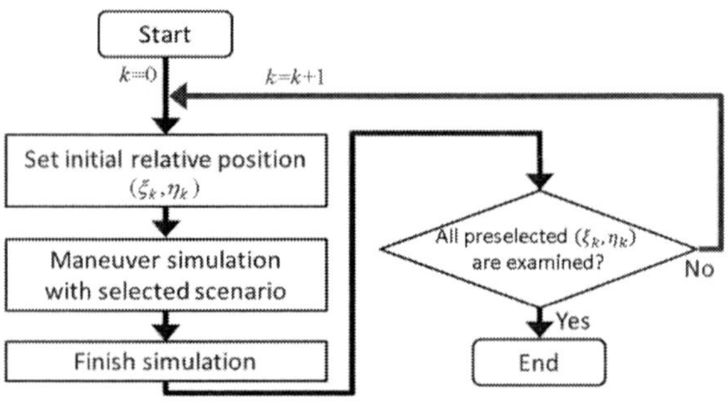

Figure 2: Flow chart to be used in evaluation of dynamic danger.

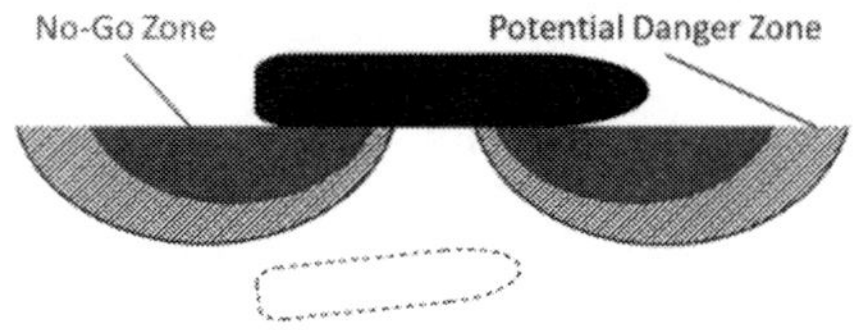

Figure 3: Schematic view of the concept of present analysis.

Figure 3 illustrates the concept with No-Go and Potential Danger Zones around the fore and aft of STBL. The integration of two Danger Zones with the current DSS would enhance, not only the operational safety, but also the effectiveness of STS simulator training.

3. MODELING OF STS INTERACTION EFFECTS BASED ON MODEL TESTS

In order to estimate the hydrodynamic interaction forces, analysis of data from available STS model test results has been performed. The data from the towing tank test programme with STS models positioned in close proximity were provided by Flanders Hydraulics Research and Ghent University, Belgium [4][5]. The available data consist of nearly 2 000 runs which include various measured data, parameters and situations. The present paper analyses a part of the available data.

Table 1: Main particulars of the ship models.

Ship type	*Aframax*		*VLCC*	
Scale	1	1/75	1	1/75
L (m)	231.4	3.085	320.0	4.267
B (m)	42.0	0.560	58.0	0.773
d (m)	15.0	0.200	20.8	0.277
C_b	0.833		0.809	

The main particulars of the ship models are shown in table 1 where L, B, d, and C_b are ship length between perpendiculars, breadth, draft and block coefficient, respectively. The ship types in question are an Aframax tanker and a VLCC, which is a common combination of STS operation. The size of the towing tank is (length × width × water depth = 77.5 × 7.0× 0.5 m^3). The VLCC (STBL) is located and fixed at a lateral position of the tank where distance between closest tank wall and ship side is about 2.1 *m*. Only the Aframax changes her lateral position to examine various intra-ship positions.

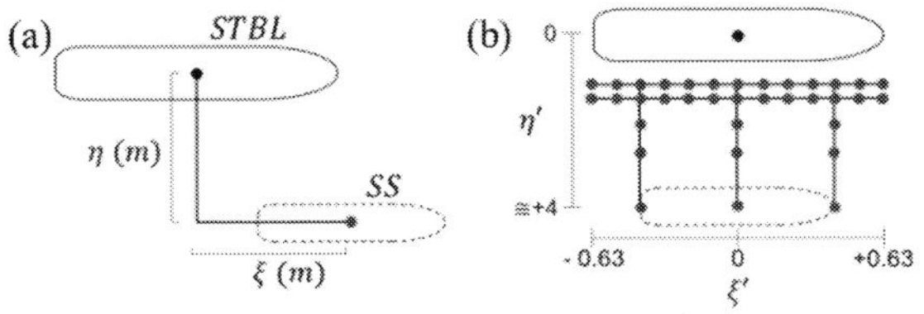

Figure 4: Configuration of the intra-ship positions in test programme; (a) definition of the relative position, (b) tested positions.

Figure 4 shows the test configuration where intra-ship position is expressed by longitudinal ξ and transverse η deviations measured by the distance between each ship's geometrical centres. Positive directions of ξ and η are given as shown in figure 4 (a). Figure 4 (b) shows tested intra-ship positions where $\xi' = \xi / ((L_{STBL} + L_{SS})/2)$ and $\eta' = \eta / ((B_{STBL} + B_{SS})/2)$. Here, the subscripts represent the ship identification. All test series are roughly classified into a few longitudinal and several transverse series of the intra-ship positions. The ships were towed at speeds corresponding to full-scale speeds of 2, 4, 5 and 6 knots in self-propulsion condition.

Table 2: Tested draft combinations of both ships.

Draft combination	*Aframax*	*VLCC*
d (m), (i=0, *1st load*)	7.5	20.8
H/d	5.0	1.8
d (m), (i=1, *2nd load–A*)	15.0	12.8
H/d	2.5	2.9
d (m), (i=2, *2nd load–B*)	7.5	12.8
H/d	5.0	2.9

Tested draft combinations and their relative water depth H/d are shown in table 2, where H and d represent the water depth and ship drafts, respectively. `*1$^{st}$ load*` and `*2$^{nd}$ load-B*` correspond to STS lightering operations where the STBL transfer cargo to SS while `*2$^{nd}$ load-A*` represents the opposite cargo transfer. In this paper, each draft combination is labelled by means of subscript *i*.

$$\begin{cases} X_{INT} = 0.5\rho C_x L d U^2, \\ Y_{INT} = 0.5\rho C_y L d U^2, \\ N_{INT} = 0.5\rho C_n L^2 d U^2 . \end{cases} \tag{2}$$

All measured forces are translated into hydrodynamic coefficients C_j (j = x, y, n) according to equation (2), where X_{INT}, Y_{INT} and N_{INT} are respectively the longitudinal- and transversal forces and yaw moment affecting the model ships. The subscript j is used to represent each force of equation (2) in following notes. Here, ρ and U are the density of water and towing velocity. The ship particular L and d are given by each ship's value.

The present analysis is to be applied to real-ship operation and it is therefore important to be conscious of the model scale effect. All tests were executed under the consistency of Froude number while Reynolds number was larger than real-ship scale. The thickness of the ships' wake in the model tests is therefore supposed to be relatively larger, thus decreasing the width of water channel. Hence, interaction forces measured in model tests will err on the high and safe side. The present paper concludes to use model test result without correction to the scale effect as basis for the discussion of safety of operation.

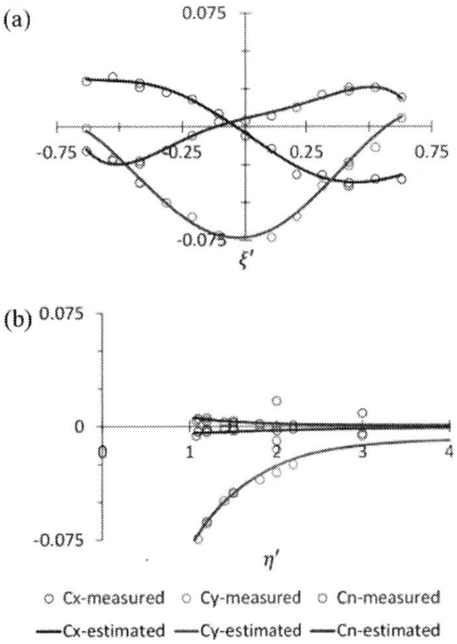

Figure 5: Approximation of the interaction forces affecting the service ship based on the model test results as a function of relative
(a) longitudinal position at $\eta' = 0.9$,
(b) transversal position at $\xi' = 0$.

Figure 5 shows measured hydrodynamic force coefficients and their approximations in the draft combination `*2$^{nd}$ load-A*: $i=1$`. Since only the condition $i=1$ includes a few longitudinal test series, modelling of the interaction effects are executed based on the case $i=1$ in the paper. The approximation of the coefficients is given by the following functions.

$$C_j(\xi', \eta') = f_j(\xi') \cdot g_j(\eta'). \tag{3}$$

$$\begin{cases} f_j(\xi') = a_5\xi'^5 + a_4\xi'^4 + a_3\xi'^3 + a_2\xi'^2 + a_1\xi' + a_0 , \\ g_j(\eta') = b_2 \cdot \exp(-b_1\eta') + b_0 . \end{cases} \tag{4}$$

Equation (3) represents that the coefficients are expressed by the product of two independent functions, f and g of intra-ship position ξ' and η'. Longitudinal changes of the coefficients $f_j(\xi')$ are approximated by fifth- or fourth order polynomial that refers to the characteristics on figure 5 (a). Transversal change $g_j(\eta')$ is also expressed by the exponential function similar to general confined water effects, e.g. shallow water/bank suction effects. The functions are consequently organized by parameters $a_5 \sim a_0$ and $b_2 \sim b_0$. All parameters are solved by the least-square method. In addition, $f_j(\xi')$ are solved based on figure 5 (a) while the combined solution $C_j(\xi',\eta')$ are solved by analysing transversal series of results, i.e. figure 5 (b) with $f_j(\xi')$ fixed at pre-solved parameters. It is confirmed by figure 5 that the present approximation has appropriate pursuance to test results.

Modelling of the other draft combinations assumes that the interaction coefficients on the other combination can be expressed by linear extrapolation of the characteristics in `$2^{nd}$ *load-A*: *i*=1` as shown in equation (5) where $C_j^{(i)}$ expresses the hydrodynamic coefficient C_j of the draft combination "*i*".

$$\begin{cases} C_j^{(i)} = p_i \cdot C_j \,, \\ p_i = 1 \quad ,(i = 1)\,, \\ p_i \neq 1 \quad ,(i \neq 1)\,. \end{cases} \tag{5}$$

The specified values of approximation coefficients given by the present analysis are summarized in table 3. Here, the longitudinal change of C_y are approximated by fourth-order polynomial because of its even characteristics as shown in figure 5 (a). Those expressions are integrated with a mathematical model to simulate ship behaviour in STS operation.

Table 3: Approximated parameters of the coefficients for interaction forces.

	SS-Cx	*SS-Cy*	*SS-Cn*	*STBL-Cx*	*STBL-Cy*	*STBL-Cy*
a5	-0.20415	0	-0.52103	0.45065	0	0.41637
a4	0.03526	-0.30168	0.07124	-0.05152	0.07064	-0.00585
a3	0.24593	-0.02676	0.16962	-0.38017	0.00218	-0.36204
a2	-0.00341	0.31353	-0.03745	0.03521	-0.15169	0.01581
a1	-0.11603	0.01810	0.04166	0.11959	0.00968	0.09453
a0	-0.00432	-0.07487	0.00559	0.01783	0.04512	-0.00094
b2	1.84568	3.98514	4.62808	1.11099	13.50030	3.82186
b1	0.57837	1.41018	1.57402	0.10424	2.64643	1.48689
b0	0.00077	0.10742	0.14168	0.00075	0.19600	0.21117
p0	1.68337	1.26932	1.06432	0.63266	0.35531	0.81036
p1	1	1	1	1	1	1
p2	1.06329	0.60534	0.66592	0.80458	0.53819	0.53212

4. MATHEMATICAL MODEL

The mathematical model adopted in this work is referred to as the "Kijima Model" [6] which consists of relatively brief expressions of each component of ship manoeuvrability.

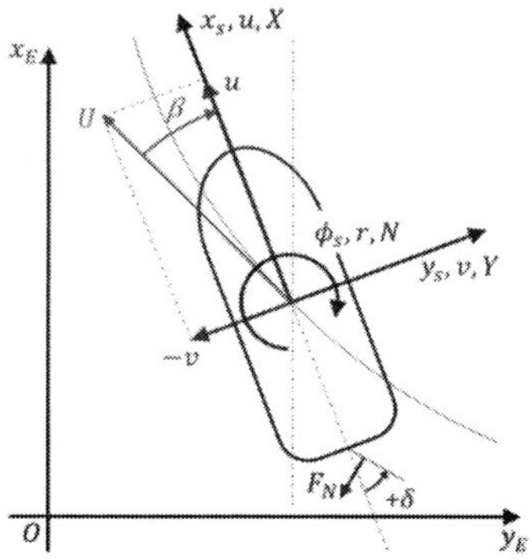

Figure 6: Coordinate system in manoeuvre equation.

Figure 6 shows the coordinate system of the mathematical model; earth-fixed x_E and y_E and ship-fixed x_s and y_s. When the ship moves with a certain heading ϕ_s, ship speed U, drift angle β and rate of turn r, the velocity components in the ship-fixed coordinate system are expressed by u and v. Moreover, external forces affecting the ship hull in the ship-fixed coordinate system are presented as X, Y and N. Rudder normal force is F_N with rudder angle δ.

$$\begin{cases} (m + m_x)\dot{u}_G - m v_G r = X + m_y v_G r \,, \\ (m + m_y)\dot{v}_G + m u_G r = Y - m_x u_G r \,, \\ (I_{zz} + J_{zz})\dot{r} \qquad = N. \end{cases} \tag{6}$$

Equation (6) represents manoeuvring equations of the present model where m and I_{zz} are the mass and moment of inertia of the ship with added mass and moment m_x, m_y and J_{zz} in each axis of x_s, y_s and ϕ_s. m_x, m_y and I_{zz} are estimated by the Motora Chart [7][8][9]. Here, $\dot{u}_G, \dot{v}_G$ and $\dot{r}_G$ are the accelerations defined on the COG of the ship [10] with subscript of "G". The right side of the equation (6) includes the external forces X, Y and N according to,

$$\begin{cases} X + m_y vr = X_H + X_P + X_R + X_{INT} \,, \\ Y - m_x ur = Y_H + Y_R + Y_{INT} \,, \\ N \qquad = N_H + N_R - x_G(Y_H + Y_R) + N_{INT} \,. \end{cases} \tag{7}$$

The right side of equation (7) shows that the external forces are given by linear summation of each module where the subscripts of each variable are `*H*-ship hull, *P*-propeller, *R*-rudder and *INT*-STS interaction`. In addition, the longitudinal position of COG is expressed by x_G. Each term on the right side is estimated according to the references [6][11][12]. In addition, the interaction effects are given by the above mentioned approximation as a function of intra-ship position.

5. ANALYSIS OF DANGER ZONE IN STS OPERATION

5.1 NO-GO ZONE

Based on the concept of danger zone and the present analysis mentioned above, the proposed DSS for STS simulator training is demonstrated. No-Go Zone defined by static evaluation to the interaction effects is shown at first. Concerning the static evaluation the interaction forces as a function of estimated intra-ship position are presented in figure 7.

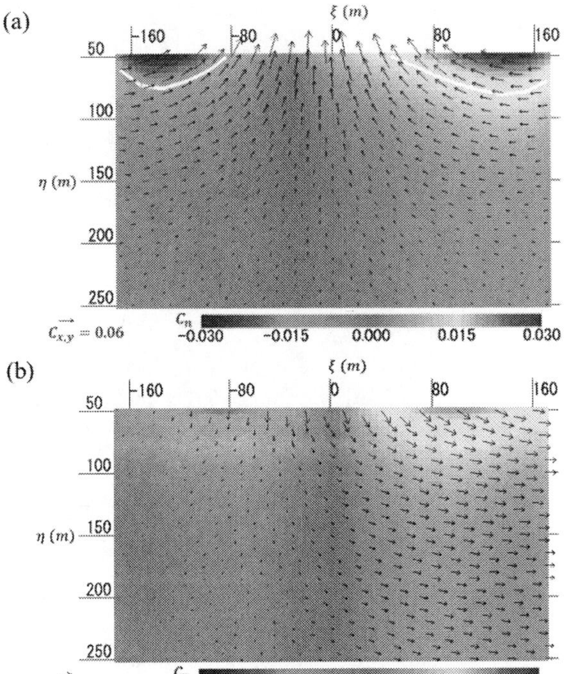

Figure 7: Approximated hydrodynamic interaction forces of the condition '*1st load*: *i*=0' effecting on (a) Service Ship, (b) Ship-to-be-Lightered.

Figure 7 shows the approximated interaction forces as a function of intra-ship positions in the condition '*1st load*: *i*=0'. Horizontal and vertical axes on the figure correspond respectively to longitudinal and transversal deviation between both ships, measured by each ship's geometrical centre in real ship scale. Then, the centre of STBL is positioned at the origin of the coordinate system. The figures give the hydrodynamic interaction force coefficients, as a vector, and the moment coefficients as colours that are affecting the SS (a) and the STBL (b). C_n is described by the colour scale and magnitude of C_x and C_y is given in the lower left of each figure. In addition, border lines of No-Go Zone, given by equation (1) with $\alpha = 1$, are illustrated on the figures by white lines.

It can be observed in figure 7 (a) that the SS tends to be sucked towards the region where it is located around mid-ships of STBL, side-by-side ($\xi = -80 \sim 80$ *m*, $\eta = 50$ *m*). The magnitude of the suction force is increased as the transverse distance between both ships decreases, figure 5 (b). On the other hand, the characteristics of interaction force vector on STBL shows its asymmetry in the longitudinal relative location of the SS. The STBL tends to be affected by large acceleration only when the SS is located at the fore side. However, the interaction yaw moment on STBL is almost symmetric, similar to that of the SS. In addition, magnitude of the moment coefficient is about a half of that of the SS, see colour scales on figure 7.

No-Go Zone given by the white lines in figure 7 is defined with reference to the geometrical centre of the SS. The maximum rudder moments are defined when the ship detects the rudder to 35 degrees when in a straight manoeuvre, i.e. zero drift angle. The interaction moment affecting the STBL is less than the maximum rudder moment in each examined intra-position. It is therefore confirmed that safe execution of an approach manoeuvre is depending on the control of the SS. It is considered advantageous to approach from the aft rather than to approach from the fore. This is because of the difference of the area of No-Go Zone. The OCIMF recommending manoeuvre [1] from parallel position of two ships is, of course, confirmed as the best way to approach.

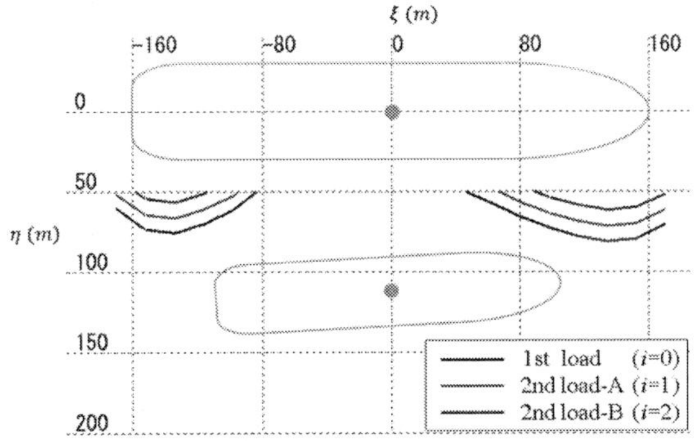

Figure 8: Examples of No-Go Zone given by the contour lines to maximum rudder moment in straight manoeuvre.

A comparison of visualized and quantified No-Go Zone between three different loading conditions is presented in figure 8, where corresponding ship shapes are also illustrated by green lines. Since the No-Go Zone doesn't appear in consideration to the balance of the moments on STBL, the zones are only dependent on the static evaluation to the moments on the SS. According to figure 8, the largest No-Go Zone appears in the condition '*1st load*: *i*=0'. This result suggests a possibility that the draft of STBL might be a considerable factor of the interaction effects on the SS. However, it should also be noted that the depth-to-draft ratio, *H/d*, of the STBL in condition

'*1st load*: *i*=0' is 1.8 (table 2), which is below the ratio when shallow water effects can be measured. Therefore, the characteristics in deep water condition and real ship operation have to be carefully considered.

5.2 POTENTIAL DANGER ZONE

The dynamic evaluations of STS operational safety are also discussed: The manoeuvre scenario is important and focus should be on the relationship between intra-ship position and time-delay of rudder reaction. In order to evaluate the effect, it is assumed that the STBL has advanced in a straight line, without any interaction effects, with the effects only on the SS being examined. The assumption is made because it is considered that the major part of difficulties of manoeuvre in STS final approach phase can be found in the control of the SS, figure 8. The interaction moment on the STBL is relatively weak, less than the rudder capacity as shown in figure 7 (b).

In the adopted scenario, the two ships start from a certain intra-ship position, and the SS takes a rudder from 0° to ±35° which direction is opposite to the interaction moment and keeps the rudder angle. A time from the start to the end when the SS starts to yaw toward rudder-corresponding direction is measured. Each simulated scenario is executed by systematic change in initial intra-ship positions. During a set of simulations, occurrence of steel-to-steel contact of both ship sides is also considered by transverse distance of two ships. When non-dimensional transverse distance η' is less than 1, in other words, $\eta <$ 50 m, the scenario in question is classified into the case of steel-to-steel contact.

Figure 9 presents the dynamic evaluation results by systematic manoeuvre simulation under the scenario in each draft combination. A time spent to start rudder-corresponding yaw motion is expressed by T_0 *s* as a function of intra-ship position and presented by a colour scale. In addition, white contour lines which correspond to specified values of T_0 are illustrated. The cases of steel-to-steel contact are distinguished by lack of data on the figure. It is observed that T_0 increases around No-Go Zone shown in figure 8, because of physical correlation between the response time and a ratio of the rudder to the interaction moment. In each figure, the SS gets a few response times less than 1 *s* when the two ships are parallel. Thus, it is confirmed that the recommended final approach from parallel position of two ships [1] allows the SS to manoeuvre without significant reduction of manoeuvrability. On the contrary, the other approaching way i.e. diagonal approach, similar to an ordinary ship to land berthing manoeuvre, requires a prediction of the interaction moment to control the SS without time delay. Compared with each draft combination, the condition '*2nd load-B*: *i*=1' shows that the largest response time in examined region, differ from the comparison of No-Go Zone in figure 8. It is considered that the largest moment of inertia in the condition '*2nd load-B*: *i*=1' results in the different characteristics.

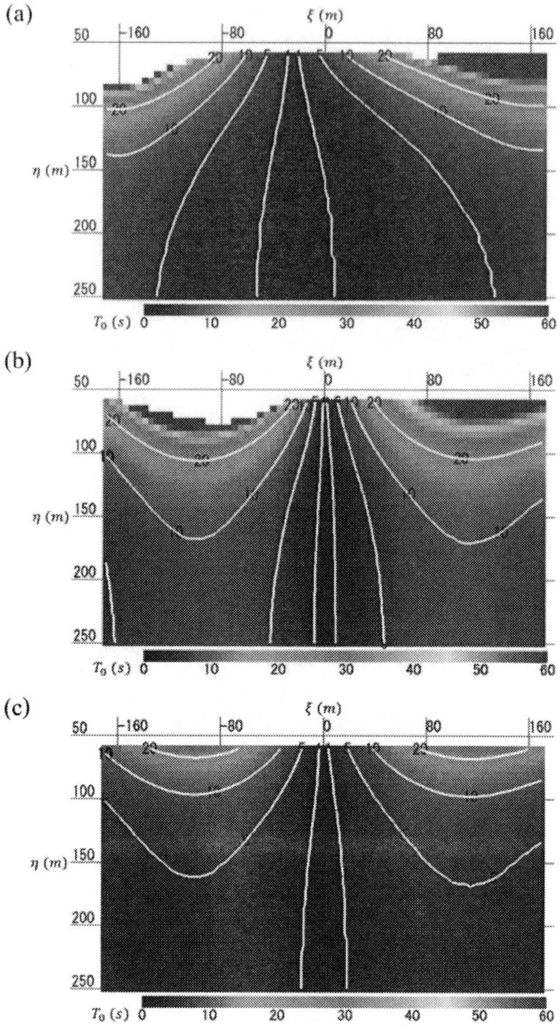

Figure 9: Reaction time characteristics as a function of intra-ship position, (a) '*1st load*: *i*=0', (b) '*2nd load-A*: *i*=1', (c) '*2nd load-B*: *i*=2'.

To define the Potential Danger Zone based on the figures, some criteria for the reaction time of the ship officer are required. The criteria should be classified into a several levels, reflecting the difference of skills/maturities of the ship officer and/or whether the DSS presented in figure 1 is available or not. This will be clarified by an analysis of a field observation of STS simulator training which will be implemented at the Ship Manoeuvring Simulator Centre in Trondheim, Norway, in March 2011.

5.3 ADVANCED EVALUATION FOR POTENTIAL DANGER BY MEANS OF SPECIFIED MODELING

In order to define the Potential Danger Zone presented in sub-section 5.2, it is concluded that a field observation is required to provide some criteria of the ship officer's reaction time and for modelling specified human factors in STS operation. An advanced evaluation of potential danger which enables to clarify how the interaction effects work to deteriorate the manoeuvrability can be proposed by modelling the human factor.

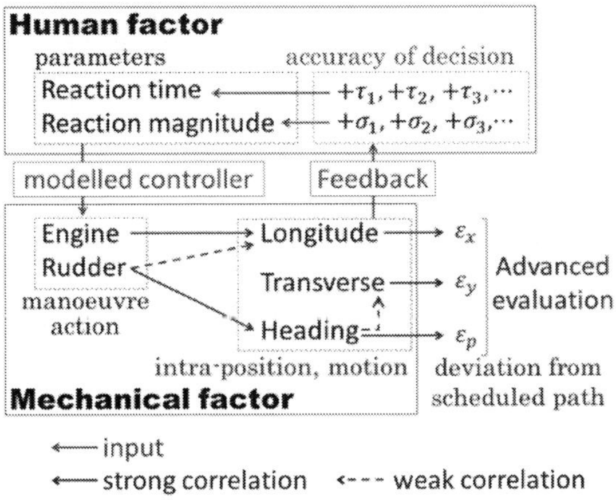

Figure 10: Schematic diagram of advanced potential danger evaluation for STS approaching manoeuvre based on specified modelling.

Figure 10 shows an overview of the advanced potential danger evaluation based on specified modelling. Human factor in manoeuvring is roughly modelled by two parameters of the reaction time τ and its error of the magnitude σ with unique accuracy corresponding to the ship officer's skills. At the same time, mechanical factor is classified into the engine and rudder action given by the ship officer. The manoeuvre actions result in the change of intra-ship position and motion with feedback to the ship officer's decision through an observation of the intra-condition. If it is assumed that the parameters depending on the ship officer's status i.e. skills or whether the DSS is available or not can be specified by field tests of STS simulator training, the specified value is integrated with feedback loop. In addition, by setting specific manoeuvring intra-ship path (manoeuvre scenario) with modelled ship motion controller, time series deviation $\varepsilon_x, \varepsilon_y, \varepsilon_p$ in intra-ship position/motion from the scheduled path is given. Analyses of the deviation enable the detailed evaluation of potential danger reflecting possible manoeuvre situation in STS operation.

Two alternatives of manoeuvring scenario are proposed in figure 11. Figure 11 (a) shows straight manoeuvre of the SS from various angles of approach. This analysis will show an acceptable deviation of the longitudinal intra-position to commence the final approach manoeuvre. Figure 11 (b) is approaches which consist of two steps: firstly, adjusting the longitudinal deviation of two ships and secondly, reducing the transverse deviation. This analysis is expected to give us appropriate distance between both ship's sides to execute the final approach. These above analyses can result in more specified decision-support how to commence the final approach manoeuvre in STS lightering operation reflecting the interaction effects.

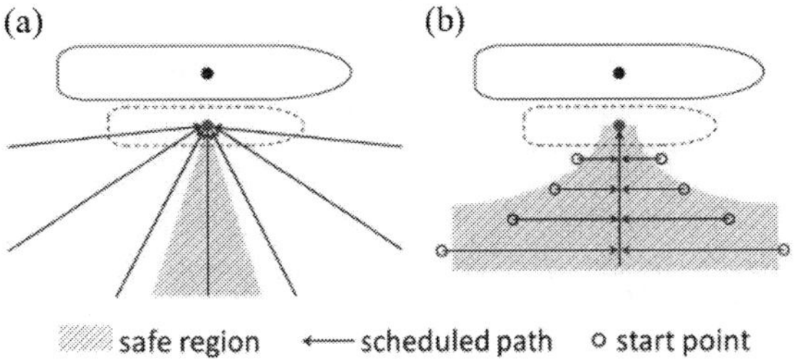

Figure 11: Schematic view of alternatives of scheduled ship paths in advanced analysis for the potential danger zone.

6. CONCLUSIONS

The concept of danger zone in STS operation due to hydrodynamic interaction is introduced. Available model test data have been analysed in order to estimate the hydrodynamic interaction forces. An existing mathematical model was adopted with integration of modelled interaction forces. 'No-Go Zone' and 'Potential Danger Zone' are presented based on the analyses of magnitude of the interaction effects and its influences to the response time. The danger in STS operation is visualised in 2-D view by the present approach as a proposal to further enhance the effectiveness of STS simulator training.

7. REFERENCES

1. OCIMF and ICS, 'Ship To Ship Transfer Guide, 4th Edition', 2005.

2. HUSJORD, D., PEDERSEN, E. and ØRISTRAND, T. A., 'Integration of Decision-Support System in STS Lightering Operation', *Proceedings of Techno Ocean 2010*, 2010, October 14-16, Kobe, Japan.

3. YOO, Y., PEDERSEN, E., TATSUMI, K., KOUGUCHI, N. and ARAI Y., 'Application of 3-D Velocity Measurement of Vessel by VI-GPS for STS Lightering', *Proceedings of 8th International Symposium on Marine Navigation (TransNav 2009)*, 2009, Gdynia, Poland.

4. LATAIRE, E. and VANTORRE, M., 'Investigating Hydrodynamic Aspects and Control Strategies for Ship-to-Ship Operation Report WP3 Real Time Models', 2008, Mod. 824, Flanders Hydraulics Research and Maritiem Technology, Ghent University, Antwerp, Belgium.

5. LATAIRE, E., VANTORRE, M. and DELEFORTRIE, G., 'Captive Model Testing for Ship-to-Ship Operations', *Proceedings of the 2009 International Conference on Marine Simulation and Ship*

Maneuverability (MARSIM '09), 2009, August 17-20, Panama City, Panama, pp. M-106-115.

6. KIJIMA, K., KATSUNO, T., NAKIRI, Y. and FURUKAWA, Y., 'On the maneuvering performance of a ship with the parameter of loading condition', *Journal of The Society of Naval Architects of Japan*, 1990, Vol.168, pp. 141-148.

7. MOTORA, S., 'On the Measurement of Added Mass and Added Moment of Inertia for Ship Motions', *Journal of the Society of Naval Architects of Japan*, 1959, No. 105, pp.83-92.

8. MOTORA, S., 'On the Measurement of Added Mass and Added Moment of Inertia for Ship Motions (Part 2. Added mass Abstract for the longitudinal motions)', *Journal of the Society of Naval Architects of Japan*, 1959, No. 106, pp. 59-62.

9. MOTORA, S., 'On the Measurement of Added Mass and Added Moment of Inertia for Ship Motions (Part 3. Added mass for the transverse motions)', *Journal of the Society of Naval Architects of Japan*, 1959, No. 106, pp. 63-68.

10. OGAWA, A., KOYAMA, T. and KIJIMA, K., 'On the Mathematical Model for Ship Maneuver (MMG Report 1)', *Journal of Society of Naval Architects of Japan*, 1977, No. 575, pp. 22-28.

11. SCHOENHERR, K. E., 'Resistance of Flat Surfaces Moving Through a Fluid', *Transaction of the Society of Naval Architects and Marine Engineers*, 1932, No. 40, pp. 279-313.

12. HASEGAWA, K., 'On a Performance Criterion of Autopilot Navigation', *Journal of the Kansai Society of Naval Architects*, 1980, No. 178, pp. 93-103.

8. AUTHORS BIOGRAPHY

Kensuke Kirimoto holds the current position of post-doctor researcher at Department of Marine Technology, Norwegian University of Science and Technology. His previous experience includes the doctor of engineering in the field of maritime sciences and an education for ship navigation officer.

Egil Pedersen is Professor of Marine Technology at the Norwegian University of Science and Technology. He is responsible for the education and research in Nautical Science at MSc and PhD levels, respectively. Dr Pedersen's research interests are at present on STS operations and multi-objective optimisation of vessel operation in the harsh environments of Arctic waters.

SHIP TO SHIP INTERACTION FORCES DURING LIGHTERING OPERATIONS

E Lataire, M Vantorre and **J Vandenbroucke** Ghent University – Maritime Technology Division, Belgium
K Eloot, Flanders Hydraulics Research, Flemish Government, Antwerp, Belgium

SUMMARY

For a better understanding of the hydrodynamic phenomena involved in ship to ship interaction during cargo transfer the research project entitled "*Investigating hydrodynamic aspects and control strategies for ship-to-ship operations*" has been initiated. In the frame of this project more than two thousand captive model tests were carried out in 2008 [1] in the Towing tank for manoeuvres in shallow water (co-operation Flanders Hydraulics Research – Ghent University) in Antwerp, Belgium. During these tests a scale model of a VLCC was attached to the main frame of the towing carriage while a model of an Aframax tanker was attached to the computer controlled planar motion carriage. Forces, moments and positions were measured on both models.

The current paper analyses the extensive model test data and the influence of different parameters on the manoeuvre. For training purposes with a ship manoeuvring simulator a mathematical model for lightering operations is proposed based upon [2].

NOMENCLATURE

B	[m]	*beam of the ship*
F_{NR}	[N]	*rudder normal force*
F_{TR}	[N]	*rudder tangential force*
g	[m/s²]	*earth's gravity acceleration*
h	[m]	*water depth*
L_{PP}	[m]	*length between perpendiculars*
K	[Nm]	*roll moment*
N	[Nm]	*yaw moment*
n	[rpm]	*propeller rate*
Q_P	[Nmm]	*torque on the propeller shaft*
Q_R	[Nmm]	*torque on the rudder shaft*
T_A	[m]	*draft aft (APP)*
T_F	[m]	*draft fore (FPP)*
T_P	[m]	*thrust of the propeller*
UKC	[-]	*under keel clearance*
V	[m/s]	*ship speed*
X	[N]	*longitudinal force*
x_{cc}	[m]	*longitudinal distance between midships sections (>0 if midship SS is ahead of STBL)*
Y	[N]	*sway force*
y_{bb}	[m]	*lateral distance between ship sides*
y_{cb}	[m]	*lateral distance between service ship centre line and side of ship to be lightered*
y_{cc}	[m]	*lateral distance between ship centre lines*
z_A	[m]	*sinkage aft*
z_F	[m]	*sinkage fore*
δ	[°]	*rudder angle*
Δ_{corr}	[-]	*correction term*
ξ	[-]	*longitudinal distance between the ships' midships section x_{cc} divided by a reference ship length*
ρ	[kg/m³]	*density*
∇	[m³]	*displacement volume*
ψ	[°]	*heading*

Subscripts:

SS	*service ship*
$STBL$	*ship to be lightered*
o	*own ship*
ref	*reference*
T	*target ship*
0	*earth bound coordinate system*

1. INTRODUCTION

Larger ship dimensions have economical advantages for most parties involved in maritime transport, but also cause restrictions when calling ports or when navigating in shallow water. Fully loaded ultra and very large crude oil carriers (ULCC, VLCC) can only reach harbours if the access channels have a sufficient width and depth. Alternatively, large tankers can be lightered underway, so that shore based handling equipment is not required. Typically, lightering takes place with the *ship to be lightered (STBL)*, mostly a VLCC (about 300 000 DWT), heading at a straight course maintaining a low, constant speed while the *service ship (SS)*, mostly an Aframax type tanker (about 100.000 DWT) performs a slow lateral approach. Lightering operations are not only carried out for transfer of crude oil; recently also liquefied natural gas (LNG) is regularly transferred from ship to ship.

The hydrodynamics of lightering manoeuvres are seldom discussed in literature. So far, most research has been focused on meetings (encounters), overtaking manoeuvres and passing of moored vessels. Various numerical [3], empirical [4] and semi-empirical [5] methods have been developed for the prediction of hydrodynamic forces and moments acting on both ships during such manoeuvres.

To some extent, these publications can contribute to a better understanding of the hydrodynamic phenomena occurring during lightering manoeuvres. Indeed, lightering can be considered as a special case of an overtaking manoeuvre during which both ships sail on parallel courses and with equal speed. For manoeuvring simulation applications, Flanders Hydraulics Research makes use of a mathematical model based on the results of a comprehensive captive model test program, [1]. However, it should be borne in mind that during the experimental program the speed difference between both ships was never less than four knots due to the length limitations of the towing tank. Therefore, it will be investigated whether this mathematical model can be used for lightering operations.

A validation of the mathematical model for the specific case of zero speed difference could be carried out after the execution of a series of captive model tests in 2008 focusing on lightering applications. During this program, carried out in the frame of a research project concentrating on ship-to-ship interaction effects during lightering operations, there was no forward speed difference between both ship models.

In the present paper it is shown that the existing mathematical model for overtaking manoeuvres is not sufficiently accurate for use in simulations of lightering operations. Therefore a new mathematical model is proposed specifically for the lightering manoeuvre, based upon the extensive model tests carried out in 2008.

2. MODEL TESTS

For a better understanding of the complex hydrodynamics involved in a lightering manoeuvre a research project entitled "*Investigating Hydrodynamic Aspects and Control Strategies for Ship-to-Ship Operations*", co-ordinated by MARINTEK (Trondheim, Norway) and supported financially by the Research Council of Norway was initiated. In the frame of this project, an extensive series of model tests was carried out in the Towing Tank for Manoeuvres in Shallow Water (cooperation Flanders Hydraulics Research – Ghent University) in Antwerp, Belgium. A description of these model tests as well as a very limited selection of test results is published in [1]. The test program is briefly summarised below.

2.1 TEST FACILITIES

At present the shallow water towing tank, with main dimensions 88 x 7 x 0.5 m³, is equipped with a planar motion carriage (consisting of a longitudinal or main carriage, a lateral carriage and a yaw table), a wave generator and an auxiliary carriage for ship-to-ship interaction tests with ship models towed at different forward speeds (overtaking, overtaken and meeting vessels). Thanks to full computer control, the facilities are operated in an unmanned way, so experimental programs are running day and night, seven days a week. An average of 35 tests a day can be achieved.

Figure 1 Planar motion mechanism with service ship (left) and new Rose Krieger construction with ship to be lightered (right) both attached to the main carriage of the towing tank.

For the lightering manoeuvre model test program, the tank facility was adapted to enable the attachment of two ship models to the main carriage. Indeed, it was decided not to make use of the auxiliary carriage that allows the execution of meeting and overtaking scenarios, in order to avoid any speed differences between the main carriage and the auxiliary carriage. The model of the service ship (Aframax type) was attached to the planar motion control mechanism (PMM), while the ship to be lightered (VLCC) was mounted to the main (longitudinal) carriage by means of a new rigid Rose-Krieger construction (Figure 1). Both models are free to heave and pitch but restricted with respect to the carriage mechanism in all other directions. This implies that the STBL can only perform a motion in the longitudinal direction of the tank, while the service ship can perform any motion in the horizontal plan applied by the PMM. The speed component of both ships in the direction of the axis of the tank is equal for both models. All forces, moments and motions measured on both ship models are summarised in Table 1.

Table 1 Measured forces, moments and motions

Hull:	SS	STBL	
Sinkage fore and aft	$z_{F,SS}$ $z_{A,SS}$		[m]
Longitudinal force	X_{SS}	X_{STBL}	[N]
Sway force	Y_{SS}	Y_{STBL}	[N]
Yaw moment	N_{SS}	N_{STBL}	[Nm]
Roll moment	K_{SS}		[Nm]

Propeller:			
Propeller thrust	$T_{P,SS}$	$T_{P,STBL}$	[N]
Propeller torque	$Q_{P,SS}$	$Q_{P,STBL}$	[Nm]
Propeller rate	n_{SS}	n_{STBL}	[rpm]
Rudder:			
Rudder normal force	$F_{NR,SS}$		[N]
Rudder tangential force	$F_{TR,SS}$		[N]
Rudder torque	$Q_{R,SS}$		[Nm]
Rudder angle	δ_{SS}		[deg]

2.2 SHIP MODELS

Two different ship models are selected for the model tests. The ship to be lightered is a model of a VLCC (scale 1/75). All geometric properties (hull, propeller, rudder) of this vessel are made available and published via [6] and [7]. This specific model is known as the *KVLCC2 Moeri tanker* and often used as a bench mark vessel by different towing tanks worldwide.

The smaller service ship is an Aframax type of vessel of about 100 000 DWT. Both models are tested with rudder and propeller attached. The main properties of both vessels are summarised in Table 2.

Table 2 Main properties of both ship models

	Service ship	STBL
Scale ( )	75	75
Hull		
L_{PP} (m)	231.4	320.0
B (m)	42.0	58.0
Design condition		
T_F (m)	15.0	20.8
T_A (m)	15.0	20.8
∇ (m³)	109139	312622
Ballast condition/2nd off load		
T_F (m)	7.5	12.8
T_A (m)	9.1	12.8
∇ (m³)	58456	182941

2.3 MODEL TESTS

Two different types of tests can be distinguished: steady state tests and dynamic tests. During the regime condition of a steady state test there is no relative speed difference between the ship to be lightered (STBL) and the service ship (SS) nor any parameter (rudder angle, propeller rate,...) of one of the models is changed in time. Three different types of dynamic tests are carried out: pure sway test, pure yaw test and a varying rudder angle test. During a pure sway test the service ship moves along a sinusoidal path towards and away from the ship to be lightered while the ship's forward speed and heading in the towing tank remain constant. A pure yaw test is analogue but the service ship performs a yawing motion along the same sinusoidal path with zero drift. During the third dynamic test the rudder angle of the service ship is varied in time (from -40 deg to +40 deg and back) while the forward speed is constant and the service ship does not yaw nor sway. At the end of the systematic program 1981 steady state tests and 162 dynamic tests were carried out.

Both ship models are tested at two different drafts. The service ship is tested in ballast condition and fully loaded while the STBL is tested fully loaded and with an even keel displacement similar to her displacement after the second off load. This makes four combinations but the combination SS fully loaded - STBL fully loaded is not tested because this does not occur in real life for obvious reasons.

Tests are carried out at two water depths and at speeds of 2.0, 4.0, 5.0 and 6.0 knots at full scale. The propeller rate of the STBL was always at the self propulsion point in open water i.e. without the influence of the service ship. The service ship was also tested at her open water self propulsion point and at the telegraph positions slow and half. The STBL never had a drift angle while the service ship had drift angles of 1.0, 2.0, 3.0 and 5.0 degrees, mostly with her bow towards the STBL. The smallest lateral distances y_{bb} between both vessels' sides are 4.0, 10.0, 25.0, 50.0 and 100.0 m full scale. 100 m is seen as the initial position of the approach stage of the lightering manoeuvre of the service ship towards the STBL, while 4.0 m is the clearance between both vessels when both are connected to each other, only separated by fenders.

The tests were carried out systematically at three relative longitudinal position of both vessels: the midship's section of the STBL at the same longitudinal location with the service ship's aft perpendicular (ξ = +0.5), midship section (ξ = 0) and fore perpendicular (ξ = -0.5). A limited amount of tests were carried out with a wider range of relative longitudinal positions.

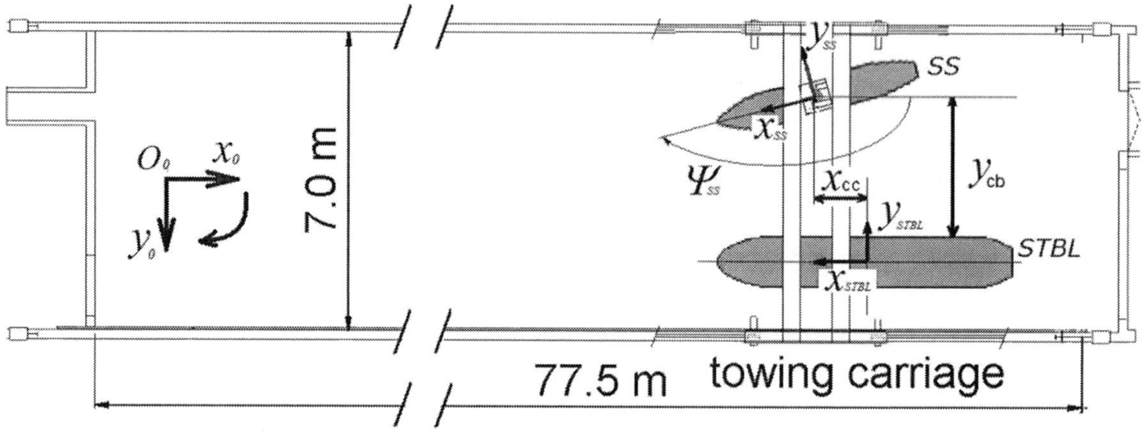

Figure 2 Test setup and coordinate systems of importance. In this figure x_{cc} is positive.

In Table 3 all parameters are summarized.

Table 3 Summary of all tested parameters

h	T_{STBL}	n_{STBL}	V	$T_{A\,SS}$	$T_{F\,SS}$	n_{SS}	y_{bb}	x_{cc}
[m]	[m]	[-]	[kn]	[m]	[m]	[-]	[m]	[m]
17.3	20.8	self prop	2.0	7.5	7.5	slow	4	0
20.3	12.8		4.0	15	15	half	10	$L_{SS}/2$
28.1			5.0			self	25	$-L_{SS}/2$
35.6			6.0				50	
							100	

3. MATHEMATICAL MODEL FOR OVERTAKING

In [8], the hydrodynamic longitudinal and lateral force components and the yawing moment measured during ship-to-ship interaction tests, including meeting and overtaking scenarios, are discussed. Based on these test results, a mathematical model for manoeuvring simulation was developed based on a regression analysis. Distinction was made between three cases according to the relative speed and direction of both vessels involved. The latter are, as usual, referred to as the *own ship* (i.e. the ship on which the acting forces are modelled) and the *target ship*. The three interaction cases are:

- Encountering manoeuvres;
- Overtaking manoeuvres, during which the own ship is overtaking the target ship;
- Overtaking manoeuvres, during which the own ship is overtaken by the target ship.

The general aspect of the longitudinal force (X), the lateral force (Y) and the yaw moment (N), and more specifically the number, magnitude and location of consecutive extreme values, depends on the interaction case and the ratio between the lengths of both ships involved. For target ships with a length not exceeding 1.5 times the length of the own ship, the number of extremes equals 2 for the longitudinal force, 3 for the lateral force and 4 for the yawing moment. The magnitude of the ith peak value is modelled in [8] as follows:

$$A^i = \frac{1}{2}\rho L_{pp} T \begin{bmatrix} A^i_{SS}(V_O+V_T)^2 + A^i_{SV}(V_O+V_T)(V_O-V_T) \\ +A^i_{VV}(V_O-V_T)^2 \end{bmatrix} \left(\frac{h-T_O}{T_O}\right)^{m^i_{h,A}} \left(\frac{h-T_T}{T_{Tref}}\right)^{m^i_{T,A}} \left(\frac{h-T_O}{T_{Oref}}\right)^{m^i_{O,A}} \left(\frac{y_{cc}}{B_O}\right)^{m^i_{y,A}} \tag{1}$$

with i = 1,2 for A=X; i = 1,2,3 for A=Y; i = 1,2,3,4 for A=N.

The relative position of the ith peak value is modelled as follows:

$$\xi^i = C^i_0 + C^i_1 \frac{V_O V_T}{g\frac{1}{2}(L_{PPO}+L_{PPT})} + C^i_2 \frac{V_O+V_T}{\sqrt{g\frac{1}{2}(L_{PPO}+L_{PPT})}} + C^i_3 \frac{V_O-V_T}{\sqrt{g\frac{1}{2}(L_{PPO}+L_{PPT})}} + C^i_4 \frac{h-T_O}{T_O} + C^i_5 \frac{T_T}{T_{T,ref}} + C^i_6 \frac{T_O}{T_{O,ref}} + C^i_7 \frac{y_{cc}}{B_O} \tag{2}$$

where ξ is the longitudinal distance between the own and target ships' midships section divided by a reference ship length L_{ref}, for overtaking the average of both ship lengths, for lightering the length between perpendiculars of the service ship:

$$\xi = \frac{x_{cc}}{L_{ref}} \tag{3}$$

The sign of ξ is defined in such a way that its value increases in time for all three interaction cases (overtaken, overtaking, meeting).

To investigate the validity of this model for a lightering manoeuvre, the latter can be considered as a special case of an overtaking manoeuvre during which the own ship and the target ship have the same speed ($V_O=V_T$). If equation (1) is applied to this special case ($V_O = V_T = V$), the magnitudes of the peak forces and moment only depend on V^2:

$$A^i = \frac{1}{2}\rho L_{pp} T A^i_{SS} V^2 \left(\frac{h - T_O}{T_O}\right)^{m^i_{h,A}} \left(\frac{y_{CC}}{B_O}\right)^{m^i_{y,A}} \quad (4)$$

while equation (2), resulting into the relative positions of the peak values is simplified to:

$$\xi^i_{V_O=V_T=V} = C^i_0 + C^i_1 \frac{V^2}{g\frac{1}{2}(L_{PPO}+L_{PPT})} + C^i_2 \frac{2V}{\sqrt{g\frac{1}{2}(L_{PPO}+L_{PPT})}} + C^i_4 \frac{h-T_O}{T_O} + C^i_5 \frac{T_T}{T_{T,ref}} + C^i_6 \frac{T_O}{T_{O,ref}} + C^i_7 \frac{y_{cc}}{B_O} \quad (5)$$

The longitudinal force on the service ship as measured during the model tests and based upon the mathematical model for overtaking is shown for one specific test condition in Figure 3. The decrease of the resistance is well predicted when the service ship is behind the STBL ($\xi<0$) but in the other positions tested ($\xi=0.0$; $\xi=0.5$) a clear difference in resistance force is seen. According to the mathematical model the maximal increase of the resistance occurs while the midship of the service ship is slightly in front of the midship of the STBL while according to the model tests the resistance increase peaks when the service ship is much more in front of the STBL.

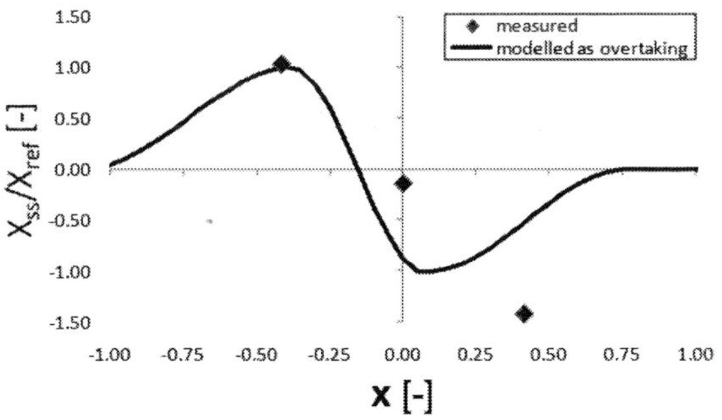

Figure 3 The longitudinal force on the service ship as measured during model tests and according to the mathematical model for overtaking.

Overall the sway force induced by the STBL on the service ship seems fairly predicted with the overtaking model (an example condition is shown in Figure 4) but the deviations between mathematical model and model test have an unacceptable order of magnitude (about 30% of the peak value).

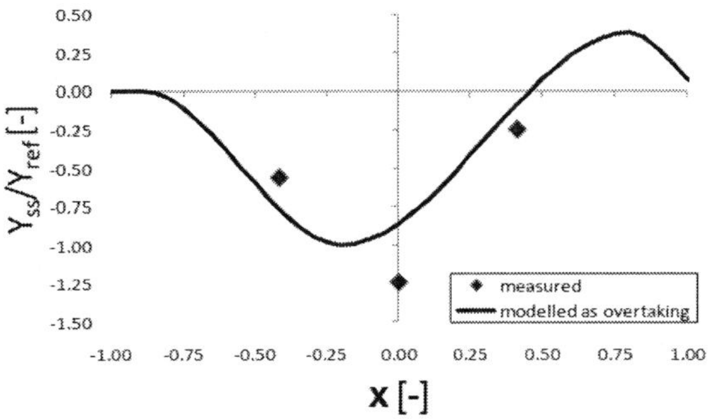

Figure 4 The sway force on the service ship as measured during model tests and according to the mathematical model for overtaking.

The comparison between yaw moment on the service ship as measured during the model test and according to the overtaking model is shown for three model tests in Figure 5. The prediction of the yaw moment is satisfying when the service ship is half a ship length $L_{PP,SS}$ in front of the STBL but totally unacceptable at all other longitudinal positions.

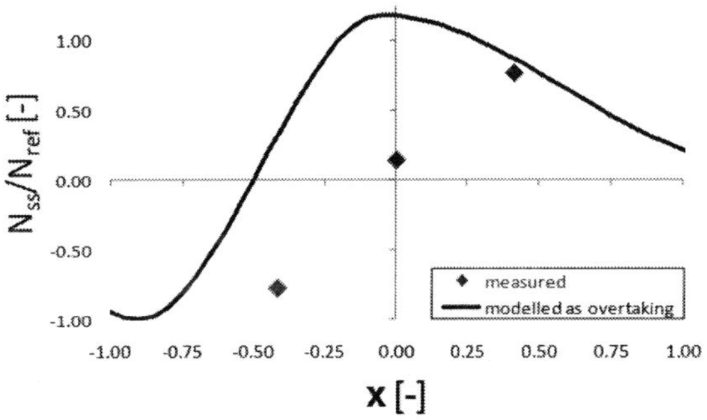

Figure 5 The yaw moment on the service ship as measured during the model tests and according to the mathematical model for overtaking.

Although Figure 3, Figure 4 and Figure 5 compare the results from the overtaking model with only three model tests the same conclusion can be made based upon all model tests. It is important to notice that the mathematical model for overtaking as proposed in [8] is accurate for a vessel overtaking another vessel with a significant speed difference but does not give satisfying results when the forward speed difference is very small or even zero.

Therefore a new mathematical model for hydrodynamic interaction forces of ships during lightering manoeuvres will be proposed, based on the model test results. The mathematical model for overtaking [8] will be used as a starting point, so that the newly developed lightering model will have common aspects with the original interaction model, but will also have its own peculiarities.

4. GENERAL ASPECTS

In the previous chapter it is shown that the assumption that a lightering manoeuvre (or any other manoeuvre characterised by a zero longitudinal speed difference between two vessels in close proximity of each other) can be modelled as an overtaking manoeuvre by extrapolating the mathematical models derived in [8] with both vessels at the same speed is not valid. Therefore a new mathematical model is proposed specifically for the lightering manoeuvre. This mathematical model is based on the extensive model tests as described in Chapter 2 and in [1].

4.1 RELATIVE LONGITUDINAL POSITION

Unlike an overtaking manoeuvre, lightering can be interpreted as a quasi-steady process because of the rather slow approach of the service ship towards the ship to be lightered.

A standard lightering manoeuvre takes place with the relative longitudinal position ξ around zero, i.e. with both midships at the same longitudinal position. This is because the manifolds for the oil transfer are located around the midship sections for both vessels. At a lateral distance of about 100 m, board to board, the service ship sails at the same heading and speed as the STBL. Slowly the service ship approaches the STBL with a small relative heading until both vessels are connected to each other with fenders of about 4 m diameter separating both hulls (Figure 6).

Figure 6 Picture taken from the service ship towards the STBL, the fenders in between both vessels clearly visible.

For training purposes it is important to cope with a wider range of relative positions between both vessels other than the common positions during a normal lightering manoeuvre. Therefore the relative longitudinal as well as lateral positions are changed during the model tests. As mentioned before, the reference ship length L_{ref} in formula (3) is for all ship to ship interaction tests $L_{PP\ SS}$.

Figure 7 shows the influence of the STBL on the longitudinal force acting on the service ship for different longitudinal positions. When the midship section of the service ship is aft of the midship section of the STBL ($\xi < 0$) the resistance of the service ship decreases. The service ship will be pushed forward by the proximity of the STBL. When the service ship's midship is in front of the STBL's midship ($\xi > 0$) the resistance of the service ship will increase or this vessel will decelerate. The influence on the longitudinal force becomes negligible when both midship sections are at the same (longitudinal) position. Although Figure 7 shows X_{SS} only for one speed and one lateral position y_{bb} (at a propeller rate which is self propulsion in open water) the same qualitative results are obtained at other speeds and lateral positions.

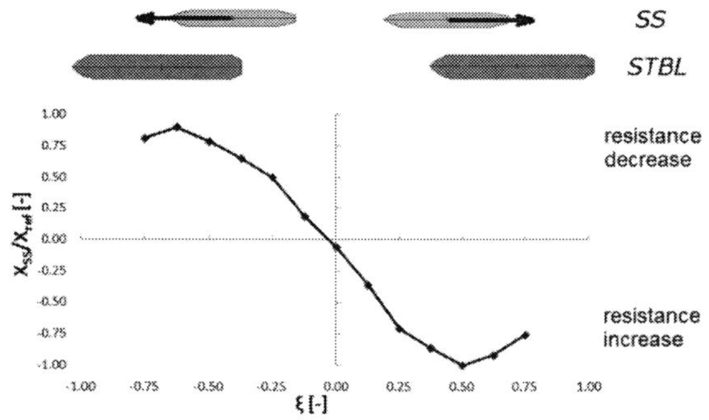

Figure 7 Longitudinal force on the service ship (made dimensionless via X_{ref}) induced by the proximity of the STBL with y_{bb} = 10.0 m at 4.0 knots full scale for a wide range of ξ with $L_{ref} = L_{PP,SS}$.

In Figure 8 the sway force on the service ship induced by the STBL is shown for a wide range of relative longitudinal positions ξ. When both midships are at the same longitudinal positions ($\xi \approx 0$), the attraction of the service ship towards the STBL peaks. Again, this is qualitatively valid for all speeds and lateral positions.

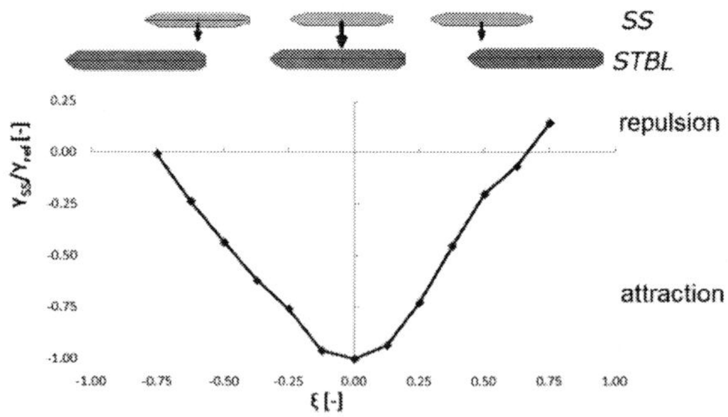

Figure 8 Lateral force on the service ship (made dimensionless via Y_{ref}) induced by the proximity of the STBL with y_{bb} = 10.0 m at 4.0 knots full scale for a wide range of ξ with $L_{ref} = L_{PP,SS}$.

The yaw moment on the service ship for a range of longitudinal positions ξ is shown in Figure 9. When the STBL is ahead the service ship (ξ < 0) the yaw moment on the service ship is negative, which implies that the bow of the service ship will be attracted towards the STBL as long as ξ values are smaller than -0.20. When both midships are more or less at the same longitudinal position, the yaw moment pushes the bow of the service ship away of the STBL while her stern is attracted towards the STBL.

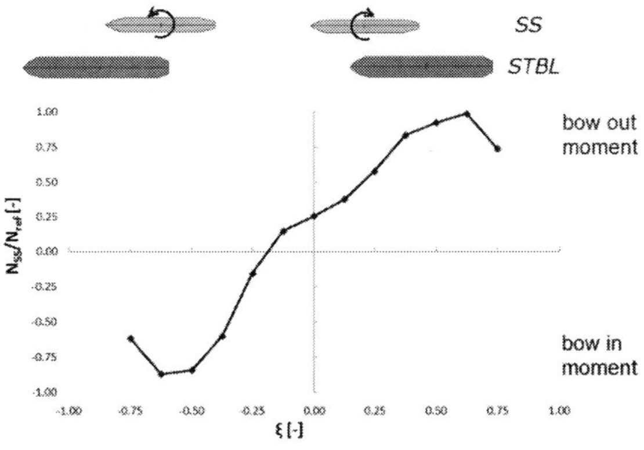

Figure 9 Yaw moment (made dimensionless via N_{ref}) on the service ship induced by the proximity of the STBL with y_{bb} = 10.0 m at 4.0 knots full scale for a wide range of ξ with $L_{ref} = L_{PP,SS}$

The longitudinal force X and moment N are not so unambiguous for $\xi = 0$. Around this position the sign changes and its magnitude is rather small compared to the extreme. Because of the importance of this force and moment on the manoeuvre both cannot be ignored. For the lateral force Y, its extreme occurs around $\xi = 0$ and decreases when both midships separate (in longitudinal direction). In the proposed new mathematical model the extreme will be modelled and a predefined function will solve the results for the values between these calculated extremes.

4.2 RELATIVE LATERAL POSITION

Figure 10 shows the magnitude of the longitudinal force on the service ship at a forward speed of 4.0 knots. The lateral distance y_{bb} between both ship sides is varied from 4.0 m (when both ships are attached to each other with fenders in between) and 100 m (when the approach of the lightering manoeuvre is expected to initiate). The absolute value of the longitudinal force obviously decreases with an increasing distance between both vessels. The influence on the resistance when both vessels' midship sections are at the same longitudinal position is almost unchanged and close to zero.

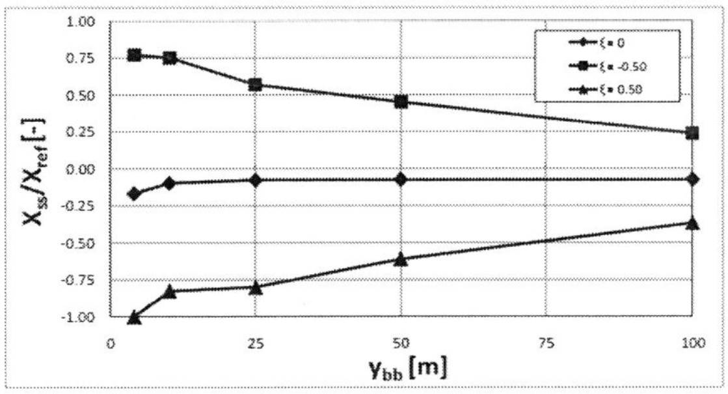

Figure 10 Longitudinal force on the service ship induced by the proximity of the VLCC at 4.0 knots full scale for a wide range of lateral distances y_{bb} and three longitudinal distances.

The lateral force is shown in Figure 11 at the same conditions as in Figure 10. Also this lateral force increases more than linear with an decreasing lateral distance between both vessels for all longitudinal positions.

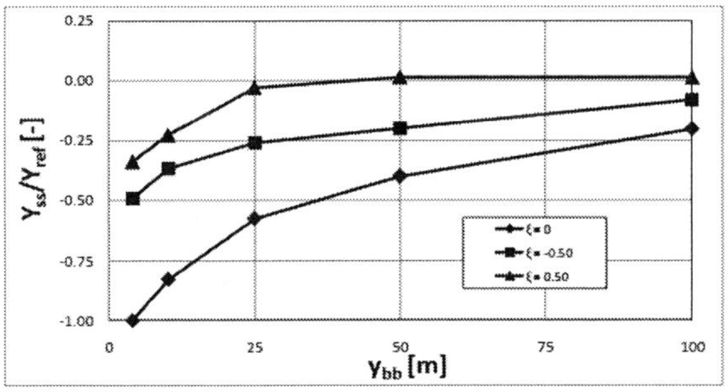

Figure 11 Lateral force on the service ship induced by the proximity of the VLCC at 4.0 knots full scale for a wide range of lateral distances y_{bb} and three longitudinal distances.

The yaw moment induced on the service ship by the STBL when both sail at 4.0 knots for three relative longitudinal positions and a wide range of lateral positions is shown in Figure 12. The drastic decrease in yaw moment when both midships are not at the same longitudinal location can be seen. When ξ is about zero the yaw moment decreases more gently and the magnitude of the yaw moment becomes negligible.

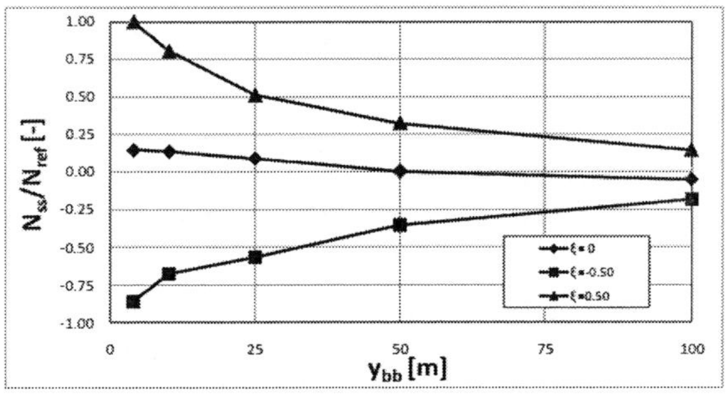

Figure 12 Yaw moment on the service ship induced by the proximity of the VLCC at 4.0 knots full scale for a wide range of lateral distances y_{bb} and three longitudinal distances.

4.3 DRAFT AND WATER DEPTH

As shown in Table 2 the service ship and the STBL are both tested at two different loading conditions. Furthermore the water depth was 35.6m full scale or corresponding to an UKC of 35% of the deepest draft. All tested combinations of drafts of both vessels and water depths are summarised in Table 4 together with the UKC for the service ship and STBL.

Table 4 Water depths and under keel clearances based upon the draft for both vessels.

Water depth	$T_{A,SS}$	$T_{A,STBL}$	UKC_{SS}	UKC_{STBL}
[m]	[m]	[m]	[-]	[-]
35.6	15	12.8	138%	178%
20.3	15	12.8	35%	58%
35.6	9.1	12.8	291%	178%
17.3	9.1	12.8	90%	35%
35.6	9.1	20.8	291%	71%
28.1	9.1	20.8	209%	35%

In Figure 13 the longitudinal force on the service ship is shown for three different longitudinal positions and for all tested under keel clearances with the STBL's draft at 12.8 m. In this case at six knots full scale, with 10 m between both ship sides (y_{bb}) and with a propeller rate according to self propulsion in open water. The absolute value of the longitudinal force decreases when the under keel clearance of the service ship increases or when more water can flow under the ship's hull.

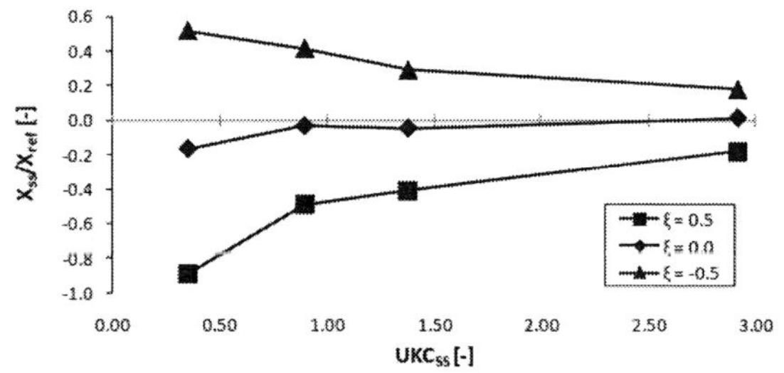

Figure 13 Longitudinal force on the service ship for different UKC_{SS} (with T_{STBL}=12.8 m; y_{bb} = 10 m; at 6.0 knots and propeller rate according to self propulsion)

In Figure 14 the sway force Y_{SS} measured when both midships are at the same longitudinal position ($\xi = 0$) is shown for different UKC_{SS} (all other conditions are the same as in Figure 13). As can be seen in Figure 8 at this relative longitudinal position the magnitude of the sway force peaks, so that the magnitude of the force decreases when the distance between keel and bottom increases.

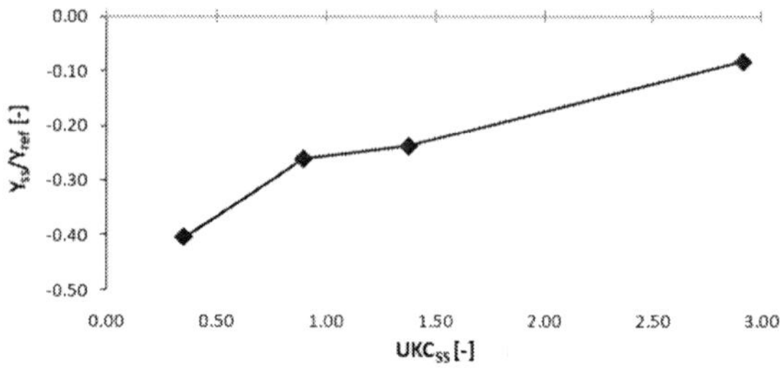

Figure 14 Sway force on the service ship for different UKC_{SS} (with T_{STBL}=12.8 m; y_{bb} = 10 m; at 6.0 knots and propeller rate according to self propulsion)

Finally for the same test conditions as in Figure 13 and Figure 14 the yaw moment N_{SS} induced by the vicinity of the STBL for different UKC_{SS} is plotted in Figure 15. The magnitude of the yaw moment decreases more than linear for an increasing UKC_{SS}.

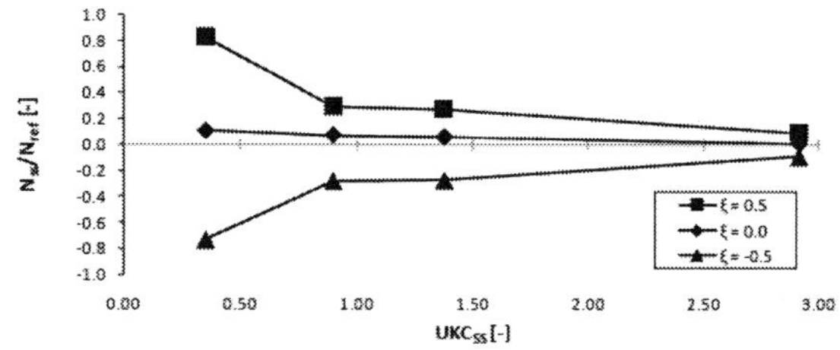

Figure 15 Yaw moment on the service ship for different UKC_{SS} (with T_{STBL}=12.8 m; y_{bb} = 10 m; at 6.0 knots and propeller rate according to self propulsion)

It can be concluded that an increasing water depth (or UKC) decreases all magnitudes of the forces and moments induced by the vicinity of a ship in close proximity.

4.4 FORWARD SPEED

For one specific test condition the relation between the longitudinal force, yaw moment and sway force and the square of the forward speed is shown in Figure 16, Figure 17 and Figure 18 for three relative longitudinal positions between the vessels. The tests are carried out with a water depth corresponding to a full scale water depth of 35.6 m, the service ship at design draft (T_{SS}=15.0 m) and the STBL with a draft 12.8 m. A constant distance of 10.0 m between both sides of the vessel was obtained and both sail with a propeller rate corresponding to the self propulsion point (for each speed) in open water.

As can be expected the magnitude of the longitudinal force induced by the STBL on the service ship is about linear to the square of the forward speed (Figure 16).

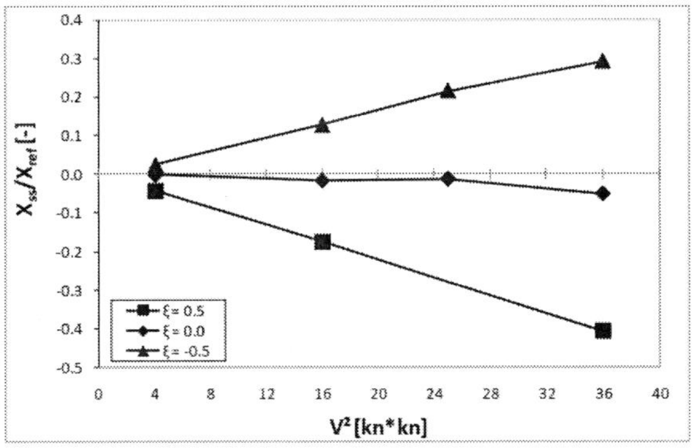

Figure 16 Relative longitudinal force on the service ship for different speeds (with T_{SS}=15.0 m; T_{STBL}=12.8 m; h=35.6 m; y_{bb} = 10 m and propeller rate according to self propulsion)

The same relation to the speed can be found for the lateral attraction force of the service ship towards the STBL. (Figure 17).

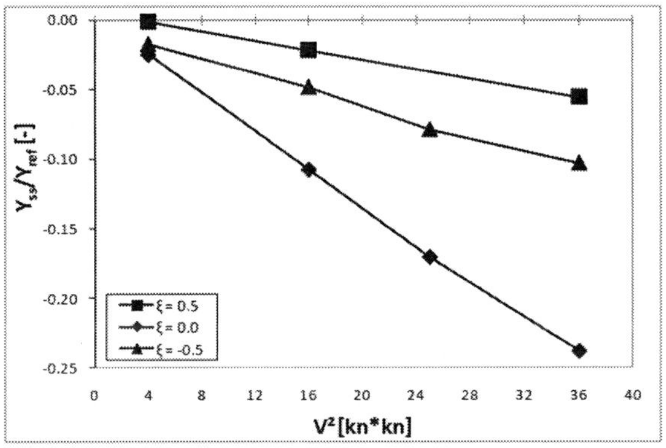

Figure 17 Relative sway force on the service ship for different speeds (with T_{SS}=15.0 m; T_{STBL}=12.8 m; h=35.6 m; y_{bb} = 10 m and propeller rate according to self propulsion)

The bow out and bow in moment on the service ship also increase quadratic with an increasing forward speed of both vessels, as illustrated in Figure 18.

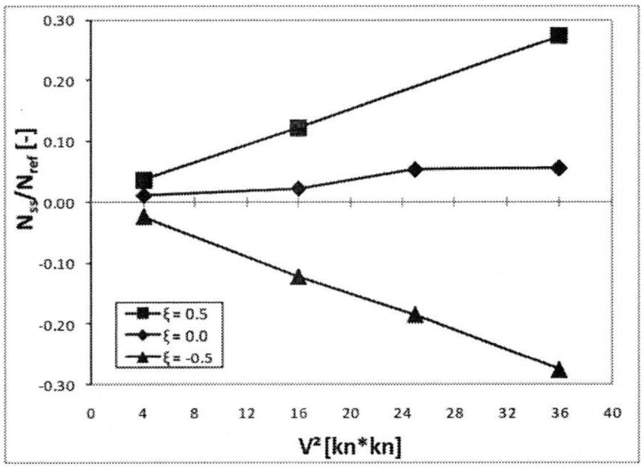

Figure 18 Relative yaw moment on the service ship for different speeds (with T_{SS}=15.0 m; T_{STBL}=12.8 m; h=35.6 m; y_{bb} = 10 m and propeller rate according to self propulsion)

5. LIGHTERING MODEL

A mathematical model is proposed for the longitudinal force, lateral force and yaw moment induced by the vicinity of a ship in close proximity and sailing at the same speed and heading as the ship under consideration. The model copes with a wide range of relative longitudinal and lateral positions between the vessels, forward speed and UKC's.

5.1 LATERAL FORCE Y

The lateral force on the service ship induced by the STBL during a lightering manoeuvre when both midships are at the same longitudinal position (ξ = 0) is modelled with:

$$Y_{\xi 0} = \frac{1}{2}\rho V^2 Lpp^2 A_Y \left(\frac{y_{cb}}{B_0}\right)^{m_y} \left(\frac{h-T}{T}\right)^{m_h} \Delta_{corr} \quad (6)$$

The lateral distance between both vessels is expressed as a function of the closest distance between both ship sides y_{bb} and the beam of the service ship.

$$y_{cb} = y_{bb} + \frac{B_{SS}}{2} \quad (7)$$

Δ_{corr} corrects the model for y_{bb} values lower than B_{SS} since the model tends to overestimate the lateral interaction forces in these circumstances.

$$0 < \frac{y_{bb}}{B_{SS}} < 0.1 \Rightarrow \Delta_{corr} = 0.85$$

$$0.1 < \frac{y_{bb}}{B_{SS}} < 1.0 \Rightarrow \Delta_{corr} = \frac{1}{6}\left(\frac{y_{bb}}{B_{SS}} + 5\right)$$

$$\frac{y_{bb}}{B_{SS}} > 1.0 \Rightarrow \Delta_{corr} = 1 \quad (8)$$

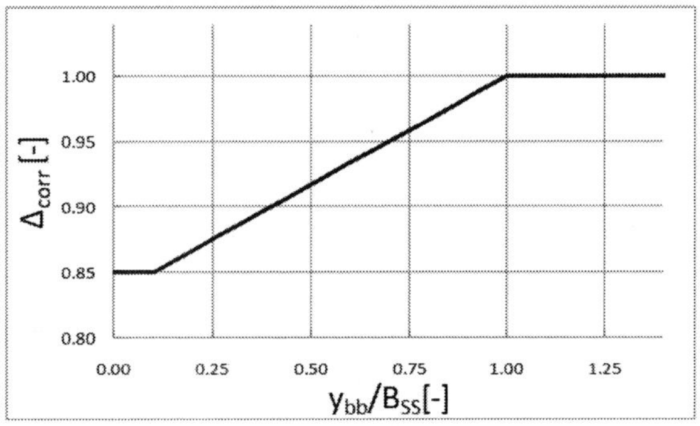

Figure 19 Correction term to counter the overestimation at very small y_{bb} values

Extra terms can be added to cope with a propeller rate shortly increasing (thrust based) or in order to correct hydrodynamic derivatives due to the presence of the second ship. As mentioned before Figure 8 shows extended measurements on the service ship during a lightering manoeuvre: a symmetric function with high attraction midships abeam is found for different lateral distances.

The proposed lightering model can be compared with the measurements and a good comparison is found.

The formulation for $Y_{\xi 0}$ can now be used to propose a formulation for a wide range of relative longitudinal positions based on functions mentioned in [8]. The lateral force can be fitted to:

$$Y = Y_{\xi 0} \cos(k_Y \pi \xi)\, e^{-\xi^2} \tag{9}$$

in which k_Y is defined by the extended lightering data and will be a fixed value. Figure 20 shows the approximation of the lateral force and measured values for a wide range of relative longitudinal positions.

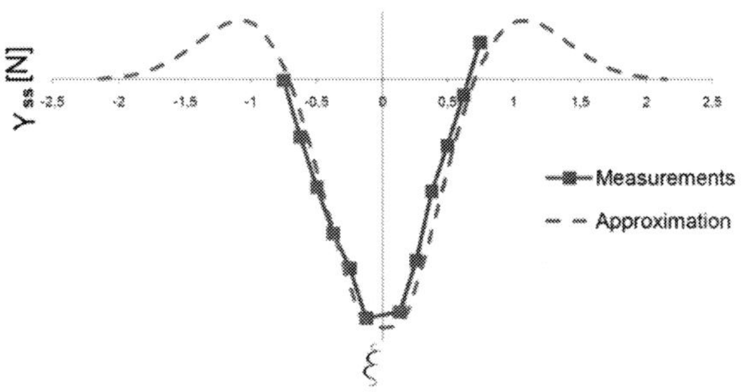

Figure 20 Comparison between the measured and modelled lateral force for a wide range of relative longitudinal positions between both vessels.

5.2 LONGITUDINAL FORCE X AND YAW MOMENT N

The peak values for the changing resistance and yaw moment are modelled as follows:

$$X_{\xi 50} = \frac{1}{2}\rho V^2 Lpp^2 A_X \left(\frac{y_{cb}}{B_0}\right)^{m_y} \left(\frac{h-T}{T}\right)^{m_h} \Delta_{corr} \tag{10}$$

and

$$N_{\xi 50} = \frac{1}{2}\rho V^2 Lpp^2 A_N \left(\frac{y_{cb}}{B_0}\right)^{m_y} \left(\frac{h-T}{T}\right)^{m_h} \Delta_{corr} \tag{11}$$

Both formulas are valid only for the relative longitudinal positions $\xi = 0.50$ and $\xi = -0.50$

As for the sway force these formula are extend to cope with all relative longitudinal positions by:

$$X = X_{\xi 50} \sin(k_X \pi \xi)\, e^{-\xi^2} \tag{12}$$

$$N = N_{\xi 50} \sin(k_N \pi \xi)\, e^{-\xi^2} \tag{13}$$

Again k_X and k_N are constant values defined based upon the results of the model tests.

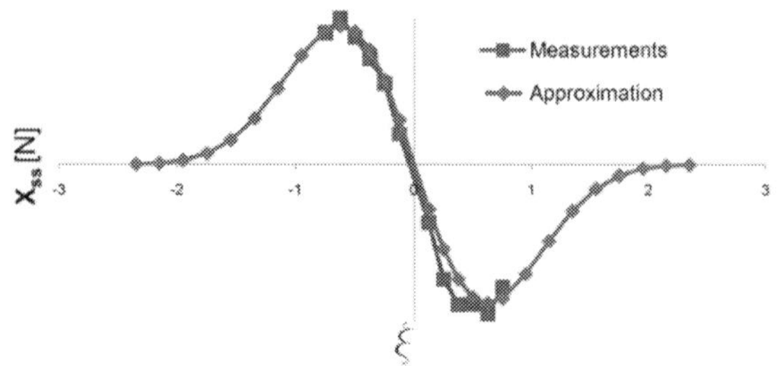

Figure 21 Comparison between the measured resistance and the result of the mathematical model for a wide range of relative longitudinal positions between both vessels.

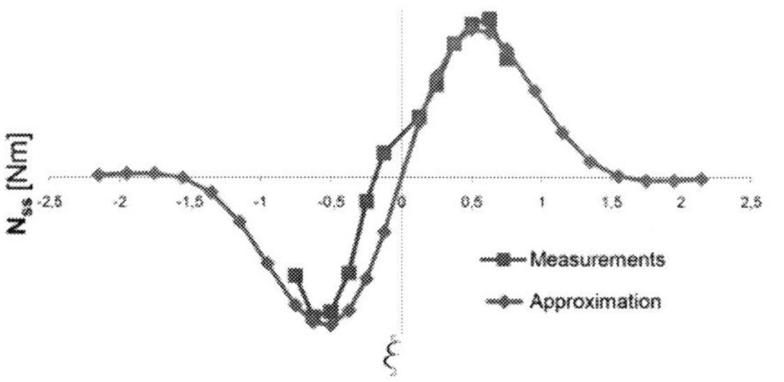

Figure 22 Comparison between the measured yaw moment and the mathematical model for a wide range of relative longitudinal positions between both vessels.

Figure 20 and Figure 21 show an acceptable correlation between the modelled and measured values for the longitudinal and lateral forces. The yaw moment shows a deviation between measured and modelled moment for ξ-values around zero. Unfortunately these are the most common and important positions during lightering. Different from overtaking or overtaken not only the magnitude and position of the peak values are important but also the exact progression of the moment between the peaks. Because this is not well predicted in the current mathematical model a new (tabular) model for the yaw moment will be proposed in the near future.

6. CONCLUSIONS

In recent years ship to ship interactions became more and more frequent as well as important. The ship to ship interactions during lightering take place with oil carriers but recently the transfer of gas from one ship to another is made. In 2008 an extensive set of model tests are carried out in the Towing tank for manoeuvres in shallow water (co-operation Flanders Hydraulics Research – Ghent University). These model tests are carried out without a forward speed difference between the model of the service ship (SS) and the ship to be lightered (STBL). These tests are compared with a mathematical model based upon older model tests for overtaken, overtaking and encountering manoeuvres [8]. It can be concluded that the mathematical model made for speed differences between ships cannot be extrapolated to ship to ship interaction without a forward speed difference between the vessels.

Therefore the influence of the relative longitudinal and lateral position between the ships as well as forward speed and UKC on the ship to ship interactions (here longitudinal force X, sway force Y and yaw moment N) are investigated. Based upon this research a new mathematical model is proposed specific for ship to ship interaction where both vessels have the same forward speed. This model is roughly based upon the overtaking model as published in [8]. The results of this new mathematical model comply for both longitudinal and transversal force. The yaw moment shows some discrepancy between model and test results when ξ is small or when the relative longitudinal position is small. Unfortunately this is the most important situation for lightering. Therefore in the near future model for the yaw moment will be more accurately modelled for small ξ values.

7. ACKNOWLEDGEMENTS

The project "KMB Investigating hydrodynamic aspect and control systems for ship-to-ship operations" is co-ordinated by MARINTEK (Trondheim, Norway) and financially supported by the Research Council of Norway.

8. REFERENCES

[1] LATAIRE, E., VANTORRE, M., DELEFORTRIE G., 'Captive Model Testing for Ship to Ship Operations', *International Conference on Marine Simulation and Ship Manoeuverability (MARSIM 2009),* Panamà city, Panamà, August 17-20 2009.

[2] VANDENBROUCKE, J., 'Krachtwerking op tankers tijdens lightering operaties', *Master thesis*, Ghent, Belgium, June 2010

[3] NEWMAN, J., TUCK E., 'Hydrodynamic Interaction between Ships.' *10th Symposium on Naval Hydrodynamics*, Pag 35-69. June 24-28 1974.

[4] VARYANI, K., VANTORRE, M., 'Development of New Generic Equation for Interaction Effects on a Moored Container Ship due to Passing Bulk Carrier.' *Transactions Vol. 147, International Journal of Maritime Engineering Part A2*, 2005

[5] BRIX, J.,' Manoeuvring Technical Manual' *Seehafen Verlag. Pag 50-52* 1993

[6] www.simman2008.dk/KVLCC/KVLCC2/tanker2.html

[7] SIMMAN 2008, *Workshop on Verification and Validation of Ship Manoeuvring Simulation Methods*, Copenhagen, Denmark, April 14th-16th 2008

[8] VANTORRE, M., VERZHBITSKAYA, E; LAFORCE, E., '*Model Test Based Formulations of Ship - Ship Interaction Forces'*. Ship Technology Research, 2002

9. AUTHORS BIOGRAPHY

Evert Lataire is currently assistant at the division of Maritime Technology at Ghent University and responsible for limited teaching tasks. He is making a PhD on the topic of bank effects mainly based upon model tests. His experience includes research on ship manoeuvring simulators and model tests with ship to ship interaction.

Marc Vantorre holds the current position of senior full professor of marine hydrodynamics at Ghent University, where he is head of the Maritime Technology Division. His research, in close co-operation with Flanders Hydraulics Research, Antwerp, mainly focuses on ship hydrodynamics in shallow and restricted waters, including captive model testing, simulation and access channel policy.

Joost Vandenbroucke holds a Master degree in Mechanical Engineering from KaHo Sint-Lieven in 2008, graduated in 2010 as an MSc. in naval architecture from Ghent University on the subject *"Interaction Forces on Tankers During Lightering Operations"* and is currently enrolled at the Vlerick Business School.

Katrien Eloot is co-ordinator of the nautical research group at Flanders Hydraulics Research and professor in ship hydrodynamics, specialism of ship manoeuvrability, at Ghent University, Belgium. This cooperation between FHR and UGent led in 2008 to the formation of the Knowledge Centre for Manoeuvring in Shallow and Confined water (www.shallowwater.be). Her experience includes modelling of ship behaviour based on model tests, fast- and real-time simulation and training through ship manoeuvring simulators.

AN EXPERIMENTAL RESEARCH ON THE HYDRODYNAMIC FORCES ACTING ON A TWIN-SCREW FERRY DURING UNBERTHING

S Lee, J Park, K P Rhee, Seoul National University, Korea
W Yoo, ADAFOS, Korea
S Kim, Daewoo Shipbuilding and Marine Engineering, Korea

SUMMARY

Unberthing operation of a twin screw ferry deserves special attentions in some cases. Crabbing ability of ships is important for Unberthing operation. Crabbing is induced by the use of a combination of main propellers, rudders and side thrusters. The speed of the ship is extremely low or even zero during unberthing. The operation of twin propellers and rudders is very peculiar, and the manoeuvring motion would be subjected to the effects of restricted water since it moves near the quay and/or in shallow water. The fluid dynamics during the unberthing is a very complicated phenomenon and requires a vast number of experiments with various environmental and operational parameters. In this study, a series of experiments are performed to find out how the hydrodynamic forces vary according to variation in water depth or the distance between the ship and the quay.

NOMENCLATURE

[Symbol]	[Definition] [(unit)]
ρ	Density of water (kg/m^3)
L	Ship length (m)
B	Ship breadth (m)
T	Ship draft (m)
U	Ship speed (m/s)
c	Distance between ship and quay (m)
h	Water depth (m)
g	Gravity acceleration (m/s^2)

1. INTRODUCTION

When ship is in unberthing operation, a strong suction force is induced between a quay and a ship. This suction force is induced by a complex turbulent flow around the ship that is generated by the propellers in the push-pull mode and the influence of the wall effect.

Recently, some research activities are being reported on the crabbing motion in berthing and unberthing operations. An experimental study on the hydrodynamic forces of a ferry is conducted [1]. And dynamics of ships and fenders during berthing are studied [2].

Although the suction force should be predicted accurately for safe unberthing operation, there are few available experimental results due to the difficulty of the model tests.

In this paper, the experimental approach was adopted to find out the interaction effects between various control units in the push-pull mode. A series of experiments was conducted with the twin-propeller, twin-rudder model of a ferry in 1:30 scale. The model ship was equipped with two bow thrusters and one stern thruster.

From the results of the experiments, the mathematical models for the crabbing motion are developed. A modular type mathematical model is developed to predict the crabbing motion in unberthing operation. The influences of the shallow water and the wall effect are considered in the mathematical model.

2. EXPERIMENTAL SETUP

2.1 TEST FACILITIES

For the present study, the model tests were performed in the Samsung Ship Model Basin (SSMB), which is 400 m long, 14 , wide, and 7 m deep. This towing tank is equipped with a main carriage, a wave generator, and an auxiliary carriage for ship-quay interaction tests. The facilities were operated with computerized control and data acquisition system.

2.2 MODEL SHIP

The model ship used in the present paper is a RoRo passenger ferry which is built in Samsung Heavy Industries. The scale ratio is 30.0 and Table 1 shows the principal dimensions of the ship.

Table 1. Principal dimensions of the ship

Particulars	*Value*
Lpp	198.0 [m]
Breadth	25.0 [m]
Draft at FP	6.6 [m]
Draft at AP	6.6 [m]
Wetted Surface Area	5,372 [m^2]
L/B	7.92
B/T	3.79

Figure 1. The model of RoRo passenger ferry

2.2 TEST CONDITION

The influences of the water depth and the suction force are of utmost importance when ship is in unberthing operation. So a series of captive tests were carried out for different water depth and the distance between ship and quay.
Table 2 shows the test condition of the water depth and the distance between the ship and quay.

Table 2. Test condition

Parameter	*Range*
c/B	0.1 ~ 4.0
h/T	2.0, 2.5, 3.0

In this paper, 3 Degree-Of-Freedom (DOF) motions such as surge, sway and yaw are considered.
Side thrust and lateral force due to the side thruster were measured with respect to various water depths and the distances between ship and quay. The lateral force and yaw moment were also measured in push-pull mode for unberthing.

3. MATHEMATICAL MODEL

3.1 OVERVIEW

Based on the results of the model tests, a mathematical model is developed for the crabbing motion in unberthing operation.
The used modular mathematical model is expressed in terms of hull force, propeller force, rudder force, side thruster force and the bank suction force.
In other words, the total force to which a ship is subjected is obtained by superposition:

$$F = F_H + F_P + F_R + F_T + F_Q \tag{1}$$

'F_Q' means the bank suction force which contains the effect of the interaction between quay and ship. Bank suction force is induced by the channel flow generated in push-pull mode. Channel flow is dependent on the hull shape, characteristics of propeller, water depth, and the distance between quay and ship.

In this paper, each component of the hydrodynamic force is modelled in consideration of unberthing operation.

3.2 COORDINATE SYSTEM

The coordinate system used in this study is shown in Figure 2. It consists of body-fixed coordinates, and inertial coordinates.
The origin of the body-fixed coordinate is located mid-ship, on the centreline. The z-axis is positive upward, and all angles are defined positive clockwise.

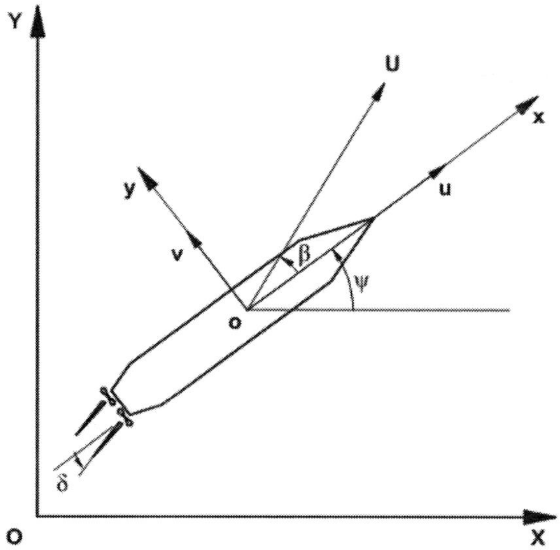

Figure 2. Coordinate system

3.3 NONDIMENSIONALIZATION

Generally, the hydrodynamic coefficients of the manoeuvring equations are nondimensionalized by use of the ship speed, U. However, another nondimensinalization method is needed since the ship speed is extremely low or even zero in unberthing operation.
Kose presented a new non-dimenization method which uses the acceleration of gravity for the low speed manoeuvring motion [3]. The non –dimension factor is proposed for force as $\rho L^2Tg/2$ instead of $\rho L^2U^2/2$. This is because to avoid the divergence on terms divided by U, when the ship speed is zero.
In this paper, the coefficients are nondimensionalized in the same way.
These nondimensionalized coefficients are as follows:

$$X' = \frac{X}{\frac{\rho}{2}L^2Tg}, \quad Y' = \frac{Y}{\frac{\rho}{2}L^2Tg}, \quad N' = \frac{N}{\frac{\rho}{2}L^3Tg}$$

$$x_G^{'} = \frac{x_G}{L},\ m^{'} = \frac{m}{\frac{\rho}{2}L^2T},\ I_{zz}^{'} = \frac{I_{zz}}{\frac{\rho}{2}L^4T}$$

$$\dot{u}^{'} = \frac{\dot{u}}{g},\ \dot{v}^{'} = \frac{\dot{v}}{g},\ \dot{r}^{'} = \frac{\dot{r}}{g/L}$$

$$u^{'} = \frac{u}{\sqrt{Lg}},\ v^{'} = \frac{v}{\sqrt{Lg}},\ r^{'} = \frac{r}{\sqrt{g/L}}$$

$$T_p^{'} = \frac{T_p}{\frac{\rho}{2}L^2Tg},\ T_T^{'} = \frac{T_T}{\frac{\rho}{2}L^2Tg},\ F_N^{'} = \frac{F_N}{\frac{\rho}{2}L^2Tg} \quad (2)$$

3.4 HULL FORCE

In the crabbing motion for unberthing, the forward speed is nealy zero. And sway velocity relatively increased due to the side thrusters and the propellers in push-pull mode. Then, because the forward velocity and the sway velocity become almost the same order, this makes it difficult to predict the lateral force with the conventional mathematical model.

The drift angle β becomes extremely larger and finally β reached $\pm 90^\circ$ while the turning rate r has small value. In such condition, the conventional expression can't be used.

So the mathematical model should be developed in consideration of extremely large drift angle for unberthing.

The static drift tests were conducted to measure the hull force with respect to the drift angle. However, a large range of drift angle was not covered: drift angles from 0 to 35 deg. So, the mathematical model for hull force is developed by referring the experimental results of Yasuo (2006) [4].

The equations of motion for static sway test can be written as the conventional expression as the following.

$$\begin{aligned} Y_H^* &= Y_v^* v^* + Y_{v|v|}^* v^* |v^*| \\ N_H^* &= N_v^* v^* + N_{v|v|}^* v^* |v^*| \end{aligned} \quad (3)$$

The non-dimensional expression of Eq. (3) becomes as the following.

$$Y_H^* = Y_H / \frac{\rho}{2} L d U^2$$

$$N_H^* = N_H / \frac{\rho}{2} L^2 d U^2, \quad \text{where, } v^* = v/U \quad (4)$$

From Figure 3, it could be known that the conventional expression is not appropriate to predict the hull force in unberthing operation.

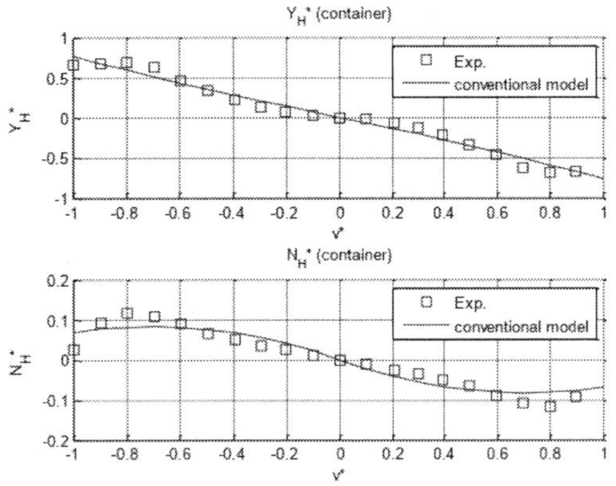

Figure 3. Regression curves of conventional mathematical model

In the above, transparent symbols mean the experimental results of Yasuo (2006) and the solid lines mean the calculated values by the conventional mathematical model. The discrepancy between the experimental results and the calculated value are non-negligible.

The mathematical model for hull force which is also applicable to crabbing motion is suggested for unberthing with two different approaches.

Firstly, the mathematical model is suggested to be applicable to drift angle from 0 to ±90deg at low speed. Suggested mathematical model for static sway test is formulated as follows.

$$\begin{aligned} Y_H^* &= Y_v^* v^* + Y_{v|v|}^* v^* |v^*| + Y_{vvv}^* v^{*3} + Y_{vvv|v|}^* v^{*3} |v^*| \\ N_H^* &= N_v^* v^* + N_{v|v|}^* v^* |v^*| + N_{vvv}^* v^{*3} + N_{vvv|v|}^* v^{*3} |v^*| \end{aligned} \quad (5)$$

Figure 4 shows the experimental results and calculated values by suggested mathematical models. From Figure 4, it could be known that suggested mathematical models are in better agreements with experimental results than conventional mathematical model for static sway test.

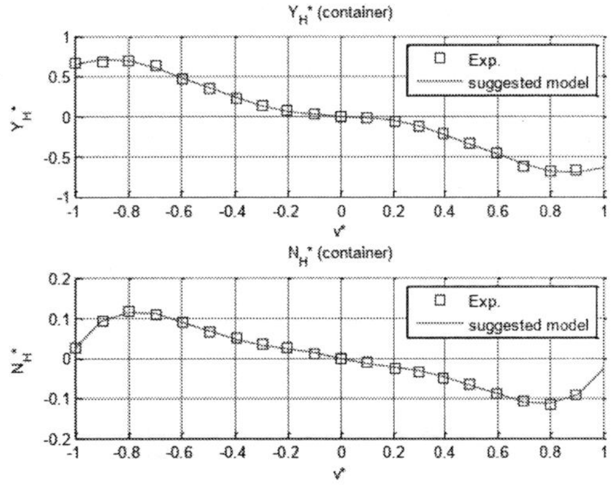

Figure 4. Regression curves of suggested mathematical model (first approach)

Secondly, the mathematical model which uses the conventional expression in the range of small drift angle is suggested. Suggested mathematical model for static sway test is formulated as follows.

$$\begin{aligned} Y_H^* &= Y_v^* v^* + Y_{v|v|}^* v^* \left|v^*\right|, \ where\ 0° \le \beta < 45° \\ &= Y_v^* v^* + Y_{v|v|}^* v^* \left|v^*\right| + Y_{vvv}^* v^{*3}, \ where\ 45° \le \beta \le 90° \\ N_H^* &= N_v^* v^* + N_{v|v|}^* v^* \left|v^*\right|, \ where\ 0° \le \beta < 45° \\ &= N_v^* v^* + N_{v|v|}^* v^* \left|v^*\right| + N_{vvv}^* v^{*3}, \ where\ 45° \le \beta \le 90° \end{aligned} \quad (6)$$

Figure 5 shows the experimental results and calculated values by conventional mathematical models and suggested mathematical models.

The mathematical model which can be used for unberthing is proposed with two different approaches. From the comparison between measured and calculated by the suggested mathematical model, it is found that the presented model can be well used for the prediction of very wide range of manoeuvring motion including unberthing operation.

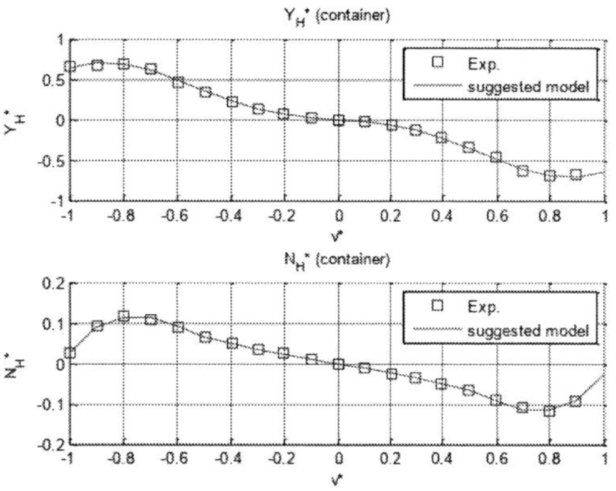

Figure 5. Regression curves of suggested mathematical model (second approach)

3.5 SIDE THRUSTER FORCE

The effects of the side thruster on the manoeuvring characteristics are considered as external force terms, such as longitudinal force X_T, lateral force Y_T, and yaw moment N_T. These external force terms are described as follows.

$$\begin{aligned} Y_T &= \left(1 - t_T\right) T_T \\ N_T &= x_T \left(1 - t_{N_T}\right) T_T, \ where\ T_T = K_{T_T} \rho n^2 D_T^{\ 4} \end{aligned} \quad (7)$$

Where T_T is thrust, t_T is thrust loss coefficient, and t_{N_T} is thrust loss coefficient relative to the yaw moment.

Figure 6. Side thruster test – front view

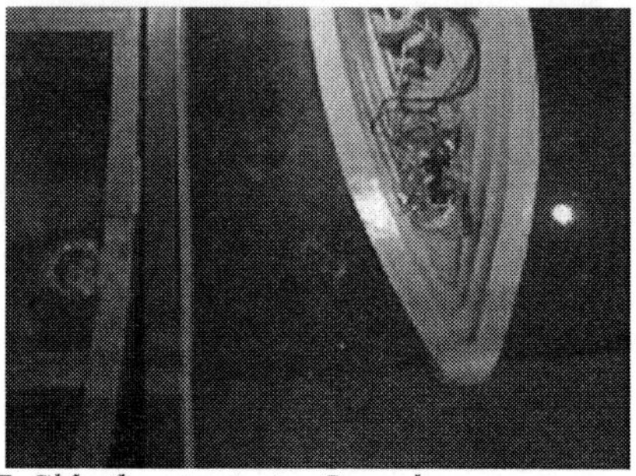

Figure 7. Side thruster test – Overview

Lateral force is interacted by water depth and the distance between ship and quay. So the model tests were performed to find out the effects of the water depth and quay on the side thrust. Figure 6 and Figure 7 show the flow pattern between ship and quay when side thruster was operated.

Figure 6 and Figure 7 show that turbulent flow and vortex were formed due to the interaction between side thruster-induced flow and reflected flow.

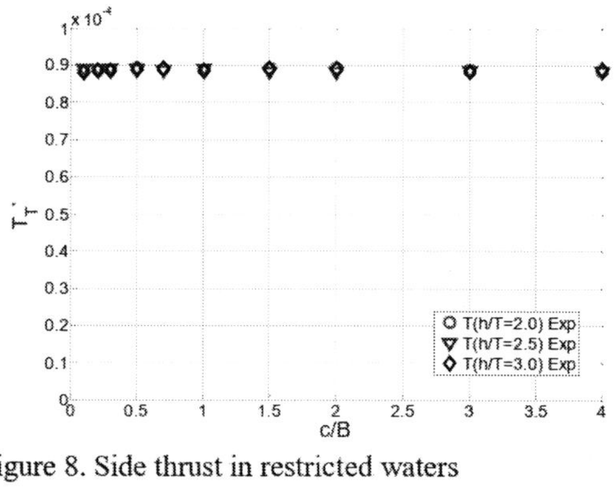

Figure 8. Side thrust in restricted waters

Side thrust changes according to various water depths and the distances between ship and quay are shown in Figure 8. All the tests were performed at constant RPS. The results show that the effect of the water depth and the distance between ship and quay on the side thrust is negligible.

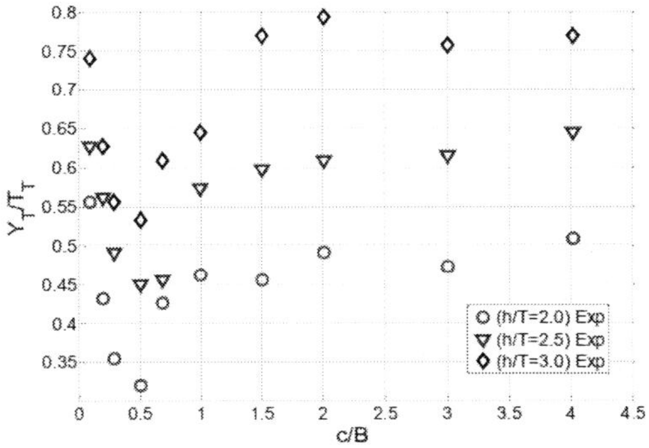

Figure 9. Lateral force due to the side thruster in restricted waters

Figure 9 shows the lateral force due to the side thruster according to various water depths and distances between ship and quay. In the Figure 9, ' Y_T / T_T ' means the ratio of the lateral force of ship and the side thrust. The thrust loss becomes larger when c/B is less than 0.5. However it decreases when c/B is 0.5 to 2. Then it becomes gradually constant when c/B becomes larger. And thrust loss becomes larger when h/T is getting smaller.

From Figure 8 and Figure 9, it could be known that thrust loss is significantly affected by the water depth and quay while the side thrust is nearly unaffected.

So lateral force due to side thruster is suggested as function of c/B, h/T by using experimental results. To build up the formulation, exponential function is multiplied by quartic of c/B so that thrust loss becomes constant when c/B becomes larger. A linear polynomial of h/T is multiplied by the formula to include water depth effects. Consequently the lateral force due to side thruster can be formulated as follows.

$$\frac{Y_T}{T_T}=\left(\begin{pmatrix} a_1(c/B)^4 \\ +a_2(c/B)^3 \\ +a_3(c/B)^2 \\ +a_4(c/B) \\ +a_5 \end{pmatrix}\cdot\exp\left[a_6\left(2-c/B\right)\right]\right)\cdot\left(a_7(h/T)+a_8\right) \tag{8}$$

Nonlinear least square method is used for obtaining 8 unknowns in the above formula. Nonlinear least square is the form of least squares analysis which is used to fit a set of data with a nonlinear model that contains a lot of unknowns.

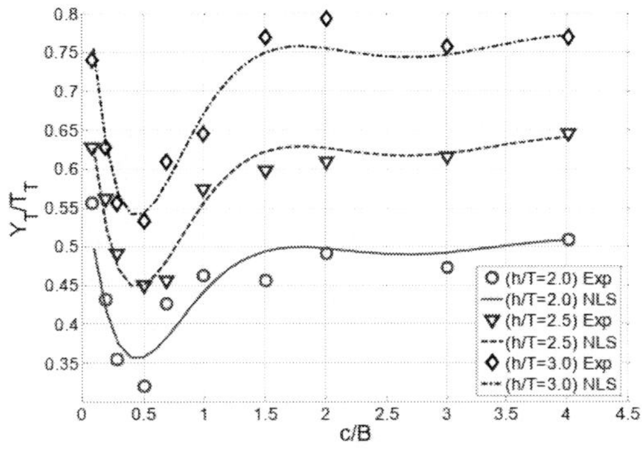

Figure 10. Regression curves of lateral force due to the side thruster in restricted waters

In the Figure 10, symbols mean the experimental results and lines are the calculated values with the suggested mathematical model. It is found that the regression curves approximately agree well with the experimental results. Table 3 shows the coefficients of side thrust model

Table 3. Coefficients of side thrust model

Coefficients	*Value*	*Coefficients*	*Value*
a_1	0.380	a_5	1.204
a_2	-1.854	a_6	0.954
a_3	4.503	a_7	0.041
a_4	-2.055	a_8	-0.003

3. PUSH-PULL MODE

In general, when ship speed is very low or nearly zero, a high lateral force is generated by means of the rudders, propellers and side thrusters. One main propeller astern is run at maximum power and the other ahead to balance the longitudinal force at the required value during unberthing. This is referred to as the push-pull mode, since one main propeller pushes the ship while the other propeller pulls the ship. This push-pull mode operation is illustrated in Figure 11.

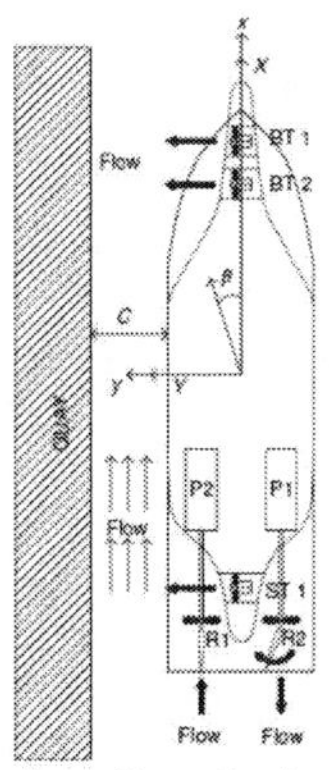

Figure 11. The push-pull mode for an unberthing[5]

The flow that streams from the reversing propeller produces channel flow between ship and quay, as shown in Figure 12.

Figure 12. Channel flow between ship and quay

This channel flow produces a suction force on the model ship, hence referred to as the bank suction effect. Bank suction effect, which produces a force to draw a ship to the quay, makes it more difficult to unberthing. In this study, a series of tests were performed to measure the forces and moments on the model ship in the push–pull mode. These experiments were performed at various water depths and intervals between hull and quay to investigate manoeuvring characteristics in the push–pull mode in restricted water.
In the push-pull mode, the thrusts generated from the two propellers should be balanced each other. In this study, balance propeller's RPM was adjusted to balance port propeller's thrust with starboard reverse propeller's thrust with constant RPM. The angle of the port rudder was 35°, while the angle of the starboard rudder was 0°.

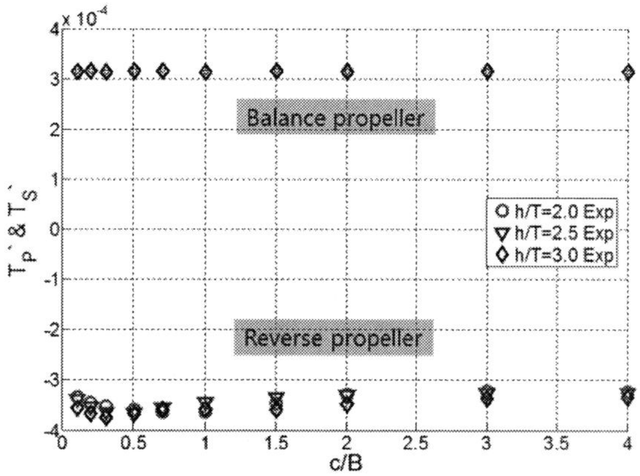

Figure 13. Thrust of propellers in restricted waters

Figure 13 shows the thrust of main propellers in restricted waters. Balance propeller's thrust had a positive value, while reverse propeller's thrust had a negative value. From the experimental results, it could be known that balance propeller's thrust is not affected by the water depth and the distance between ship and quay, on the other hand, reverse propeller's thrust is affected by the water depth and the distance between ship and quay. In the push-pull mode for unberthing, the change in thrust according to water depth and distance between ship and quay should be considered.

In the conventional manoeuvring mathematical model, the normal force on the rudder is considered, while the tangential force on the rudder is neglected. However, in the case of a ship that is berthing or unberthing, the tangential force acting on the rudder should be considered separately as following expression.

$$X_R = (1-t_R)(F_N \sin\delta + F_T \cos\delta)$$
$$Y_R = (1+a_H)F_N \cos\delta$$
$$N_R = (x_R + a_H x_H)F_N \cos\delta + (1-t_R)y_R F_N \sin\delta \qquad (9)$$

The results of the experiments about the normal force and the tangential force acting on the rudder are shown in Figure 14 and Figure 15.
When the c/B is less than 2.0, the normal force and the tangential force become larger as the distance to the quay becomes shorter. And the water depth affects the rudder force significantly when the c/B is less than 2.0.

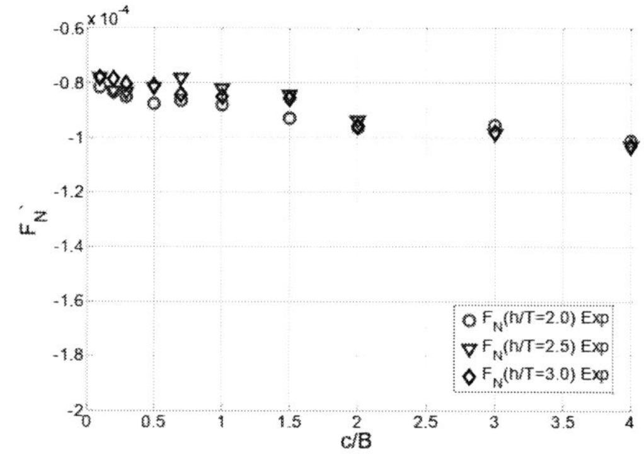

Figure 14. Normal force on the rudder behind balancing propeller

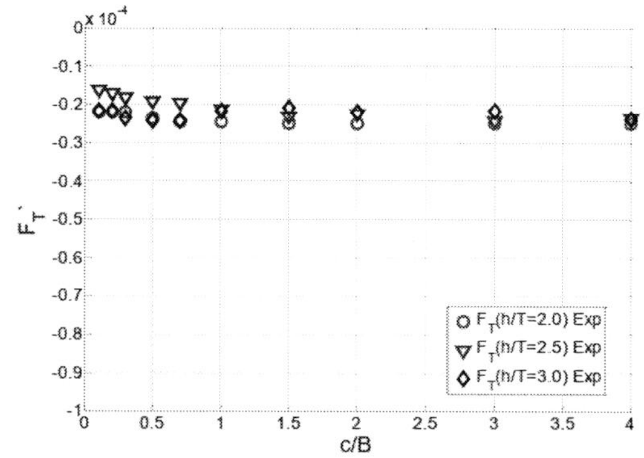

Figure 15. Tangential force on the rudder behind balancing propeller

Figure 16 shows the lateral force in the push–pull mode for an unberthing. The experiments were carried out to generate a large lateral force in the positive direction to unberth in the push–pull mode. When c becomes larger than 2.0B, the lateral force was generated in the positive direction using the push-pull mode operation. The direction of lateral force is reversed when c/B become smaller than 2. This is a typical example of the bank suction effect generated in narrow channel between a ship and the quay, which makes unberthing operation more difficult.

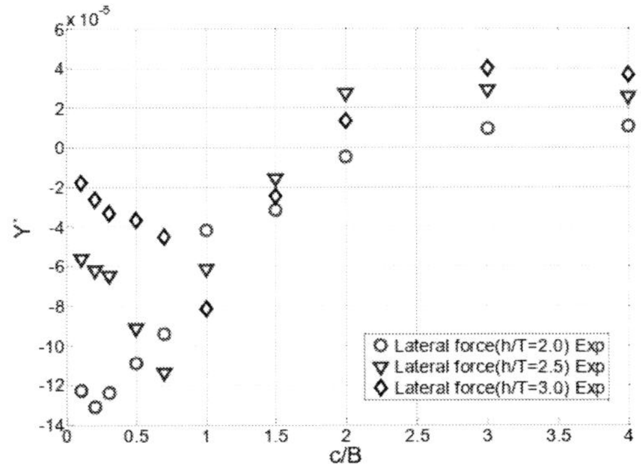

Figure 16. Lateral force in the push-pull mode for unberthing

From Figure16, it could be known that lateral force according to various water depths and distances between ship and quay has same tendency that of lateral force in Figure 9. However, the inflection point of lateral force is changed according to water depth in push-pull mode while that of lateral force due to the side thruster is not changed. So new variable k is adopted to consider the change of the inflection point. Consequently the lateral force in the push-pull mode for unberthing is formulated as follows.

$$Y' = \left(\begin{pmatrix} a_1 k^4 \\ +a_2 k^3 \\ +a_3 k^2 \\ +a_4 k \\ +a_5 \end{pmatrix} \cdot \exp\left[a_6(2-k)\right] \right) + \left(a_7 (h/T) + a_8\right)$$

$$, where\ k = \{(c/B) - (a_9 (h/T) + a_{10})\} \tag{10}$$

The unknowns are obtained by using the nonlinear least square method.

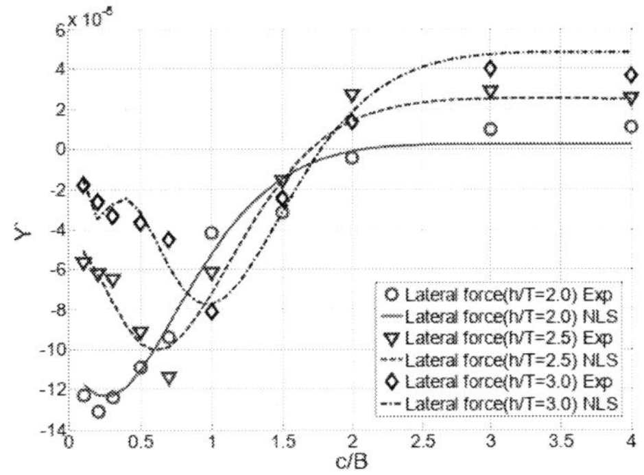

Figure 17. Regression curves of lateral force in the push-pull mode for unberthing

In the Figure 17, symbols mean the experimental results and lines are the calculated values with the suggested mathematical model. It is found that the regression curves are in good agreements with the experimental results. Table 4 shows the coefficients of lateral force model.

Table 4. Coefficients of lateral force model

Coefficients	*Value*	*Coefficients*	*Value*
a_1	3.625E-09	a_6	3.287
a_2	1.678E-08	a_7	4.604E-05
a_3	1.709E-08	a_8	-9.005E-09
a_4	-1.784E-08	a_9	0.7177
a_5	-2.794E-08	a_{10}	1.000E-03

Figure 18 shows a yaw moment in the push-pull mode in unberthing case. In the push–pull mode operation, yaw moment should be balanced, a negative moment must be produced to counterbalance the positive moment from the bow thruster. As the additional moment was generated, hence rotating the reverse propeller in a direction opposite to that of the balance propeller, a negative moment could be generated to counterbalance the positive moment of the bow thruster. Starboard turning moment was generated in when c/B is larger than 1.0 and the direction of turning moment is reversed when c/B is less than 1.0. Furthermore water depth affects the yaw moment significantly when c/B is less than 1.0.

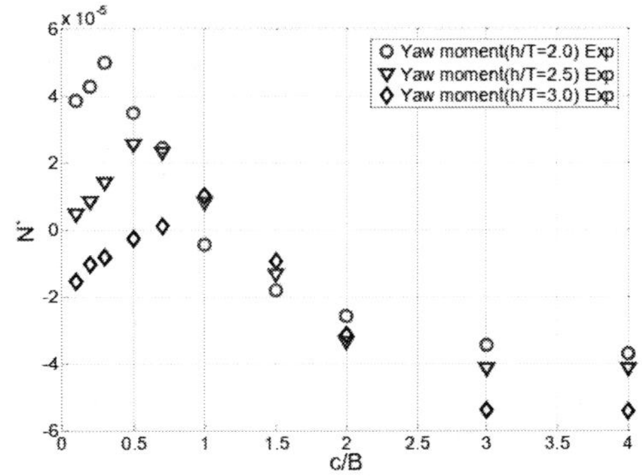

Figure 18. Yaw moment in the push-pull mode for unberthing

From Figure 18, it is found that the change of the yaw moment with respect to water depth has similar tendency with that of lateral force. So, yaw moment can be expressed similar with formula of lateral force.

$$N' = \left(\begin{pmatrix} a_1 k^4 \\ +a_2 k^3 \\ +a_3 k^2 \\ +a_4 k \\ +a_5 \end{pmatrix} \cdot \exp\left[a_6 (2-k) \right] \right) + \left(a_7 (h/T) + a_8 \right)$$

$$, where\ k = \{ (c/B) - (a_9 (h/T) + a_{10}) \} \qquad (11)$$

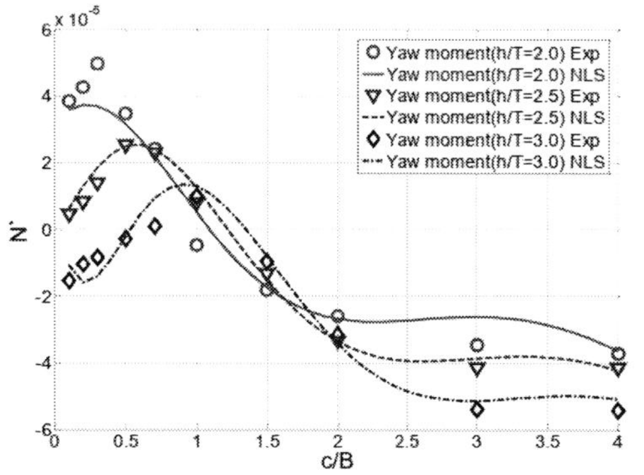

Figure 19. Regression curves of yaw moment in the push-pull mode for unberthing

Table 5. Coefficients of yaw moment model

Coefficients	*Value*	*Coefficients*	*Value*
a_1	1.242E-06	a_6	1.113
a_2	5.108E-06	a_7	-2.370E-05
a_3	8.906E-05	a_8	-1.229E-09
a_4	1.469E-05	a_9	0.685
a_5	1.695E-05	a_{10}	0.010

4. CONCLUSIONS

In this paper, a series of experiments were performed to investigate the hydrodynamic forces acting on a twin propeller/twin rudder ship during unberthing operation. The effects of the water depth and the quay are mainly studied. The different combinations of the test conditions allowed observing and quantifying the variation of the hydrodynamic forces due to the water depth and the distances to the quay.

In order to develop a modular type manoeuvring model, the forces by hull, propeller, rudder and side thruster are investigated.

The mathematical model for hull force which is applicable to drift angle from 0 to ±90deg at low speed is suggested for unberthing with two different approaches.

The empirical formula is proposed for the lateral force and yaw moment acting on the ship in push-pull mode during unberthing operation.

From the comparison between experimental results and calculated results by the suggested mathematical model, it is found that the presented model can be well used for the prediction of hydrodynamic forces in unberthing operation.

5. ACKNOWLEDGEMENTS

Some parts of this research were supported by the Marine Technology Education and Research Center of Brain Korea 21, the Research Institute of Marine Systems Engineering (RIMSE) at Seoul National University and Cruise Joint Research Project at Daewoo Shipbuilding and Marine Engineering.

6. REFERENCES

[1] Teresa Abramowicz-Gerigk, "Experimental study on the hydrodynamic forces induced by a twin-propeller ferry during berthing", Ocean Engineering, vol.35, pp. 323-332, 2008.

[2] Gokdeniz Neser and Deniz Unsalan, "Dynamics of ships and fenders during berthing in a time domain", Ocean Engineering, vol.33, pp. 1919-1934, 2006.

[3] K. Kose, "On a New Mathematical Model of Manoeuvring Motions of a Ship and its Applications", International Shipbuilding Progress (ISP), Vol. 29, 1982.

[4] Yasuo Yoshimura, Ikao Nakao, Atsushi Ishibashi, "Unified Mathematical Model for Ocean and Harbour Manoeuvring", Proceedings of the MARSIM'06, 2006.

[5] W-J. Yoo, B. Y. Yoo and K. P. Rhee, "An experimental study on the maneuvering characteristics of a twin propeller/twin rudder ship during berthing and unberthing", Ships and Offshore Structures, vol. 1, pp. 191-198, 2006.

[6] Harvald SA, "Wake and thrust deduction at extreme propeller loadings for a ship running in shallow water", Trans RINA, 1977.

[7] Kijima K, Yoshimura Y, Fukasawa T, "Manoeuvring characteristics in shallow water and at slow speed", Proceedings of the 4th Marine Dynamics Symposium, Japan, 1987.

[8] Lee SW, Hwang YS, Kim YS, "Crabbing simulation of ship with twin rudder and twin skeg", Proceedings of the Annual Spring Meeting, Society of Naval Architects of Korea, pp. 144-147, 2000.

7. AUTHORS BIOGRAPHY

Sungkyun Lee is PhD candidate of Dept. of Naval Architectures and Ocean Engineering, College of Engineering, Seoul National University. He is currently interested in ship manoeuvring simulation in shallow water.

Jongyong Park is master course student of Dept. of Naval Architectures and Ocean Engineering, College of Engineering, Seoul National University. He is making a master's thesis on the topic of crabbing motion simulation.

Key-Pyo Rhee is Professor of Dept. of Naval Architectures and Ocean Engineering, College of Engineering, Seoul National University. His research mainly focused on ship manoeuvrability and controllability including captive model testing and simulation, especially in waves.

Woojoon Yoo is graduated from College of Engineering, Seoul National University with a PhD degree in Dept. of Naval Architectures and Ocean Engineering and currently working for ADAFOS.

Sewon Kim is graduated from College of Engineering, Seoul National University with a master degree in Dept. of Naval Architectures and Ocean Engineering and currently working for Daewoo Shipbuilding and Marine Engineering.

INTERACTION-A PRACTICAL VIEWPOINT

G R Maxwell, Warsash Maritime Academy, UK

SUMMARY
This presentation focuses on the training of serving seafarers using manned models and how to increase awareness of the dangers of interaction as part of a general course of shiphandling which includes passing and overtaking in restricted waters with and without draught restrictions

1. INTRODUCTION

This paper intends to address the practical aspects of training in interaction and shallow water effects and how they affect the ship handler. It will be shown that whilst a theoretical perspective of the expected results can be obtained from books the seafarer can benefit from practical demonstrations which are not possible in real life.

Warsash Maritime Academy which is a faculty of Southampton Solent University has been delivering shiphandling and vessel manoeuvring courses for over thirty years using both computerised bridge simulators and scaled manned models. In this paper we will concentrate on the use of scaled manned models as a practical medium for providing training and for demonstrating the effect of shallow water and or interaction and the results observed.

Whilst modern simulators are good .with scaled manned models however, the results are purely dependent on hydrodynamic reaction and variable environmental forces hence are much less predictable and therefore closer to the "real world" experience.

The Ship Handling Centre currently use seven different hull types up to 11 metres in length which can be reconfigured to provide 12 different ship models on our own purpose designed lake. These resources can all be used to safely demonstrate the different effects that the mariner is likely to encounter either when manoeuvring in shallow water or when passing or overtaking within confined waterways such as a river or canal.

2. BACKGROUND

Whilst there is a great deal of research conducted regarding the cause and effect of both ship to ship and ship to shore interaction forces much of this escapes the serving mariner as there is no practical shiphandling training during certification. The aim here is to illustrate how training is conducted which may then lead to a better understanding of how practical seafarers look at and understand the problems and are thus better equipped to understand the results of research.

While mariners are believed to have known about interaction forces for a long time it is probable that in the past the cause and effect was not wholly understood. One of the first known and documented instances was the case of the Titanic departing from Southampton on the 10th April 1912.It is alleged that the ship passed close to the liner New York dragging the New York off the pier. A collision was only averted by the quick reactions of a tug. It is reported that the ships actually closed to within a distance of about two feet of each other. Although we will never know, could this be one of the first instances where training could have possibly averted the situation since Captain Smith was apparently only used to ships of half the size of the Titanic. It is well known that five days later the Titanic sank with the loss of over 1500 lives. [1]

While todays mariners have much more of a technical understanding of the effects of interaction, incidents continue to occur and there still exists the problem of how to get this across apart from textbooks or videos which do not have the dramatic impact of actually seeing it occur and thus observing the consequences . Whilst experiential learning is the most effective manner it is obviously impractical (from both a cost and a risk perspective) to use real ships for demonstrations and therefore probably the first thing that springs to mind is the use of computer simulators. But is this the most effective tool available?

Simulators

Whilst the latest generation of simulators are very good for many aspects of training perhaps one of their limitations is in the area of hydrodynamic reaction since the results can be predictable due to the nature of being programmed to make certain things happen under certain conditions. Although the the latest software can alleviate some of this there is still an element of predictability. Similarly experience with simulators indicates that when asking a simulator to simulate extreme shallow water effects i.e. 10% under keel clearance then the results are sometimes open to question. One of the reasons being that there is limited real and measured data on this situation which can then be incorporated into the computer programmes. Much of it has therefore to be extrapolated from deep water data.

Even with the latest generation of computer simulators which are very realistic, and in fact were used for the initial planning of the new manned model facility, it is our experience that some students are still somewhat sceptical and sometimes find it difficult to accept the results with a tendency to believe that it has been set up to produce the expected reaction.

Manned Models: The alternative

The other alternative which we at Warsash Maritime Academy support is the use of manned models since the effects are purely dependent on hydrodynamic reaction with no computer wizardry involved. All the lecturers on the course are either ex senior masters, ex pilots or serving Class 1 pilots.

The aim of training with scaled manned models is to enable mariners to be able to take the correct (though not necessarily the most obvious action) when encountering an interaction situation. An analogy may be that when a car enters a skid the natural action is to apply the brakes but training will inform that this is not necessarily the correct action.

3. SO WHAT IS A MANNED MODEL?

Figure 1: A typical Panamax size manned model

Many of you will be familiar with tank test models which are built to a high degree of precision to accurately reflect the properties of a real ship and the manned models are a further development of this but are built to 1:25 scale and capable of taking two students at any one time with one being the "captain" and the other being the "engineer/helmsman". A typical example is shown in figure 1. All are electrically powered with a Panamax size model being around 10 metres long with a beam of about 1.3 metres. The model shown above is also fitted with a bow thruster and a Schilling rudder and the speeds are scaled through a computer to accurately reflect what this would be in real life.

The concept originated in the 1970s when a major oil company was investigating the feasibility of sending larger ships through the Suez Canal and thus built a replica of the canal for experimental purposes. It was then realised that this concept could also be used for practical training without the risks involved in using full size ships. Currently there are only five such centres in the world of which three are in Europe, one in the USA and one in Australia.

Normally it takes about half a day for students to realise that everything that happens on a real ship will also happen on the model. The scaling effects mean however that it happens five times quicker: i.e. the square root of the scale. We do occasionally encounter some students who are initially unconvinced of the concept but normally find that this disappears after the first day as they realise that they are an excellent training tool for learning the principals of shiphandling which includes the cause and effect of interaction. It is so effective that some pilot organisations, particularly in the USA send their pilots back for revision every five years with some having been to us at least five times in their career.

Many facilities investigating interaction effects use test tanks with variable depths and different qualities of seabed. Our facilities are broadly similar though on a much larger scale and they include natural weather effects since it all takes place outdoors.

Whilst manned models are sometimes considered a rather crude training tool this is not the case as it is obvious from the start that there is no computer trickery involved. All the results obtained are purely the effect of naturally occurring environmental conditions and the shiphandlers response...

Thus when carrying out several runs involving interaction the delegates are able to observe how actions can vary with only a slight change in the conditions. Maybe they are slightly further apart or perhaps the relative speeds may be different.
In our experience this tends to be very beneficial to the learning experience and many are initially unaware of the differences this can make.

One of the benefits of using manned models as opposed to simulators is the ability to observe the manoeuvre/interaction from the bank. Being able to look down on the models as they pass is especially beneficial in demonstrating and highlighting the dangers of interaction and is something which can rarely be observed in real life

Due to the increasing need for training and the desire for more realistic and accurate scenarios the University is currently constructing a new 2.7 million pound facility for shiphandling training about 10 miles north of Southampton which has been carefully designed with a canal and banks that will give maximum but variable interaction dependent on the bank configuration whether it be vertical or sloping. [2] The canal has been designed with a mean average depth of a scale 20 metres whereas the models have a scale draught of 16-18 metres dependent on either being loaded or in ballast. The width has been designed at 150 scale metres of navigable water to intentionally force the models closer together to highlight the interaction effect although of course our instruction in real life is to avoid close passing whenever possible. This is generally in line with accepted recommendations [3]

Figure 2: Typical canal section

Although our students are generally either senior officers or pilots who come from all over the world from Melbourne right across to Hawaii, some with many years experience it is normally found that all will admit to having broadened their knowledge with many saying that initially interaction was not something they had really considered.

4. WHAT CAN BE PRACTISED?

Dramatic results can be obtained when ships are meeting end on with possibly the best known in the marine world being the Texas Chicken as carried out daily in the Houston Ship Channel. It is found that whilst our students are usually aware of the bow cushion effect they are often not aware of how strong this can be and how it will affect the ships but is very effectively demonstrated within the confines of the canal but in a safe environment. Other factors can be demonstrated such as whether having a bow thrust tunnel fitted will affect the results.

The classic meeting end on situation is how instruction normally starts to demonstrate the effect of the bow wave cushion and how the ships are pushed apart as further explained in The Shiphandlers Guide [4] which is often thought of as the bible of shiphandling.

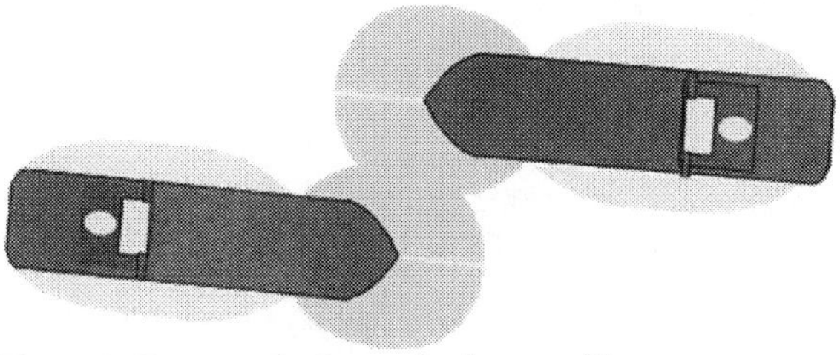

Figure 3: Bows pushed apart by bow cushion

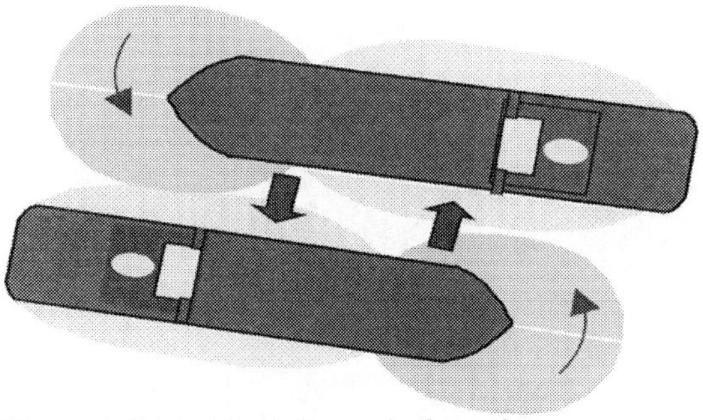

Figure 4: Both ships being sucked together

Although the effects are well known it is our experience that until someone has seen it actually happening it can be difficult to comprehend the extent of the forces involved particularly as to how the forces change due to the different pressure areas working together.

Other subjects covered include the effect of working a tug close to the bow when interaction has led to a number of serious or fatal incidents. The effects can be ably demonstrated using radio control tugs which are built to the same scale.

Overtaking is another area of interaction covered at length due to the increased time that the ships are exposed to one another. Most of our students are surprised at how realistic this can be and it is not uncommon to see one model being dragged along behind the other solely due to suction.

Another particularly effective demonstration is having one model pass another which has been deliberately moored with slack mooring lines. As most mariners are aware this is not an uncommon situation and when repeated several times is an especially good learning tool as the different effects of speed or distance off can be shown. [5]

In a similar vein ship to ship transfers are also covered since there is a large potential for interaction forces to cause an accident either during approach or on breakaway.

5. HOW DOES IT COMPARE WITH REAL LIFE?

As stated previously one of the most dramatic demonstrations of the power of interaction between ships is either in the overtaking manoeuvre or in one ship passing a moored vessel with slack lines

A typical example of an overtaking manoeuvre going wrong is shown in the following pictures with the first picture being a real life situation and the second being of a manned model situation. It can clearly be seen that both have very similar and dramatic results although with the manned models fortunately little damage is sustained apart from a few paint scratches

Figure 5: Cosco Hamburg/P&O Nedlloyd Finland

Figure 6: A similar incident but with models

Due to the shallow nature of the canal there are also squat effects to consider and as is often found in avoiding intership interaction they are led into the bank where they have to then contend with bank effect from being close. Results can be very dramatic and tend to form a lasting impression. In all cases the emphasis is very much on speed and with the need for pre-emptive and bold action [6].

6. PREVIOUS INCIDENTS

To emphasise the potential dangers we often refer to previous incidents and the tragic outcome than can happen using the following which are well known and documented but are worth highlighting due to the serious consequences

Probably one of the most serious cases which have been attributed to interaction is that of the Royston Grange and the Tien Chee on 11th May 1972 in the River Plate in which 69 people died. The official enquiry deduced that the Royston Grange was probably pushed towards a bank and then sheared off into the side of the tanker Tien Chee [7].

Similarly another collision took place between two vessels of the Royal Navy HMS Mermaid and HMS Fittleton on 20th Sep 1976 in which the ships were drawn close to one another by hydrodynamic forces eventually causing the smaller ship, a minesweeper, to capsize with the loss of twelve lives. [8]

More up to date is the incident involving the Cosco Hamburg and the P&O Nedlloyd Finland on the River Elbe in 2004 which occurred in an overtaking situation [9].

An example of a vessel being pulled from her moorings is the case in Savannah, Georgia in 2006 of the Golar Freeze alongside and being passed by the Charleston at excessive speed resulting in extensive damage to ship and wharf [10]. This can be ably repeated with the models.

Whilst the first two of these incidents took place a number of years ago they both resulted in serious loss of life and are the types of incident that we are trying to prevent happening in the future by making mariners aware of the dangers that can be posed in underestimating the forces involved by practical instruction and demonstrations.

7. RESEARCH

Although the new facility will not be completed until the spring of next year there are plans to undertake further research into the effects of interaction which with the canal being about 500 metres long or over 5 scale miles and with a significant bend it should be possible to conduct experiments which may not be possible in a towing tank.

8. CONCLUSIONS

By using manned models as an alternative to full mission simulators we continue to make the cause and effect of interaction available to serving seafarers through demonstrations and training. The serving mariner will then become more aware of the potential hazard involved and then will be more able to appreciate the results of continuing research into the subject.

9. REFERENCES

1. 101 things you thought you knew about the Titanic: Tim Maltin April 2010

2. Approach channels: A guide for design: PIANC June 1997

3. Guidelines for the safe design of commercial shipping channels: Fisheries and Oceans Canada

4. The Shiphandlers Guide: Capt RW Rowe FNI: The Nautical Institute

5 Report to Panama Pilots Technical committee 1998

6. DAND, I.W., 'Physical causes of interaction and its effects': *paper to NI conference on ship handling*, 1978:

7 Dept of Transport: 'Royston Grange Official Report: 26th June 1973', *Hansa Vol. 858: 28th June*, 1973.

8. www.mwassociation.co.uk

9. BSU: 'Collision between the Cosco Hamburg and the P&O Nedlloyd Finland: River Elbe: *report 45/2004*, 2004.

10 NTSB Docket 181 5th Nov 2008 and USCG appeal 2680

AUTHORS BIOGRAPHY

Gordon Maxwell holds the current position of Senior Lecturer at Warsash Maritime Academy. He is responsible for organising and operating a variety of shiphandling courses. His previous experience includes almost 20 years as harbour pilot and rigmove master with a major oil company in the Middle East.

NEW THINKING IN SHIP GENERATED HYDRODYNAMIC FIELDS: INTRODUCING CONCEPTS FOR PREDICTING BANK SUCTION AND REJECTION.

P J McARTHUR. Norwest Interaction Ltd, UK

SUMMARY

Whilst numerous papers have been written describing effects that vessels may experience when encountering one another or, indeed, when stationary but under the influence of a tidal stream or current, there is limited information on the causational dynamic that results in what we call 'interaction'.

This paper is, effectively, a work of four parts. It begins by outlining current teaching on the subject of marine hydrodynamics theory and highlights its particular shortcomings.

The second part introduces the Manchester Ship Canal as research venue and explains the benefits and peculiarities of the location.

The paper then introduces new theories on ship-generated hydrodynamic fields and provides detailed practical explanations that will be recognisable to experienced ship-handlers.

Finally, a new theory on the mechanics of bank suction and rejection, along with a means of predicting its occurrence, is proposed.

The ultimate aim is achievement of safer 'close-quarters' navigation and more predictable ship-handling.

1. INTRODUCTION

As an experienced marine pilot - accustomed to working at the boundaries of knowledge in relation to ship generated hydrodynamics - experience taught me that I could anticipate, with considerable certainty, what might happen between my ship and others operating in close proximity.

However, I still could not explain the 'science', the 'dynamic' or the 'why' of what I was doing to a concerned, often nervous, Captain. I felt that it was inappropriate to simply 'fob off' Masters with *'that's just the way it is'*.

If empirical knowledge allowed me to anticipate a hydrodynamic event then, logically, there had to be a rational explanation as to causation.

This research seeks to address the knowledge deficit that exists in respect of practical ship hydrodynamics and to provide the man on the bridge with a pragmatic understanding of the hydrodynamic forces that influence his vessel.

As an experienced mariner with an understanding of the watch-keeper's plight in coping with the realities of ship-handling in confined waters, in producing this work I have observed three simple rules:

- Firstly, one picture paints a thousand words. In a crisis, the man on the bridge can easily relate to diagrams. Years of experience has reinforced the concept that simplified explanations 'stick in the mind' and are both easily assimilated and quickly recalled.
- Secondly - in a crisis the watch officer does not concern himself with the numbers. He simply wants to know what mechanism is affecting his ship so that he can address the problem.
- Thirdly, keep it simple. Today's mariner is assaulted from every direction with situational information. The need to assimilate complex theories, complicated mathematical calculations or interpret complex diagrams, presents additional problems. What is really needed is an understanding of the dynamics that allows anticipation of interaction forces before they arise - thereby providing tools that permit proper compensation and management of the forces once they manifest.

2. CURRENT UNDERSTANDING AND TEACHING ON SHIP HYDRODYNAMICS

2.1 THE EARLIEST CONCEPT

My earliest understanding of ship generated hydrodynamic pressure systems derives from a basic concept of hydrodynamic pressure fields as three immutable 'blocks' of water which remained fixed in relation to the moving vessel (Figure 1).

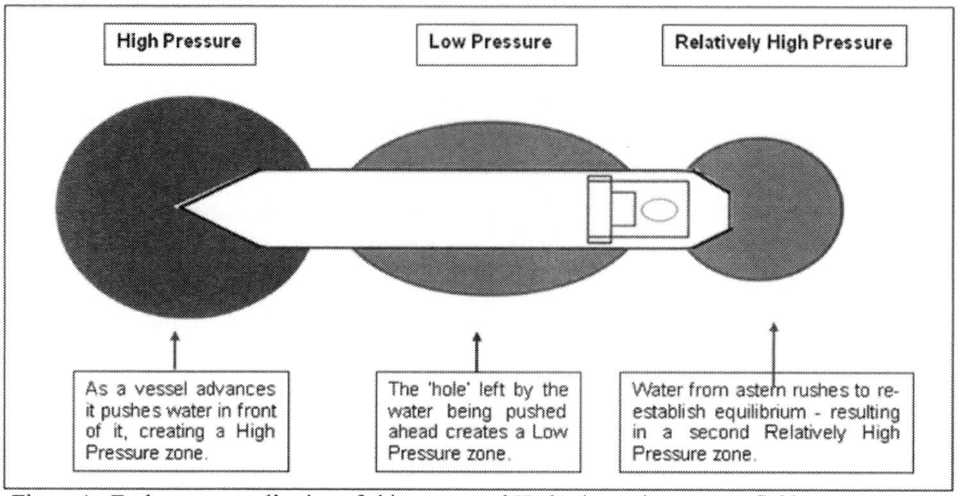

Figure 1: Early conceptualisation of ship generated Hydrodynamic pressure fields

This concept is still taught in some maritime schools - despite its obvious shortcomings, namely:

- The water moving ahead, and following astern, of the vessel is regarded as a 'solid' block with rigidly defined boundaries
- The high pressures zones are limited in horizontal extent to the immediate proximity of the vessel
- The low pressure fields amidships are limited in both longitudinal and lateral extent and, very poorly, reflect what is actually experienced.
- Application of the theory cannot predict that smaller vessels, especially those operating around the stern, might be subjected to: sudden and inexplicable sheer; loss of stability; loss of steering; poor position keeping capability; uncharacteristic manoeuvrability.
- Reliance on this theory meant the navigator had little appreciation for the distant influences his ship might be generating.

Whilst helpful in a limited way, this model could not be relied upon when operating in close proximity with other vessels. The principle benefit derived from continuing use of this concept is in explaining the historical development of theories leading to our current understanding of marine hydrodynamics.

2.2 DEVELOPMENT OF EARLY CONCEPTS

Later work by Dr Ian W Dand substantively modified this theory and refined our understanding of hydrodynamic pressure fields (Figure 2). This conceptualisation moved us away from the notion of pressure fields as a solidly definable mass impacting upon other vessels and structures. Instead, the different pressure zones are attributed with more amorphous qualities. The positive and negative fields are explained as a series of +'s and –'s to indicate their position relative to an advancing ship.

Significantly, this concept recognised areas of varying field intensity - a substantial improvement on previous ideas. Furthermore, the theory also introduces the prospect of *combining* pressure systems - in a manner similar to conventional wave theory - so that pressure fields might interact constructively or destructively.
It also introduced the concept of indefinable pressure zones that might represent a potential threat to the unwary mariner. Official nautical publications promulgated by national maritime authorities (for example, in the UK Merchant Shipping Notice M930, now superseded by Marine Guidance Notice MGN 199(M) '*Dangers of Interaction*') highlighted problems attributable to interaction, yet, they failed explain the underlying mechanism or causes – thereby leaving the seafarer none the wiser. Having recognised the potential hazards, lack of information as to causation still meant that the risk persisted and should be feared, therefore avoided so far as possible.

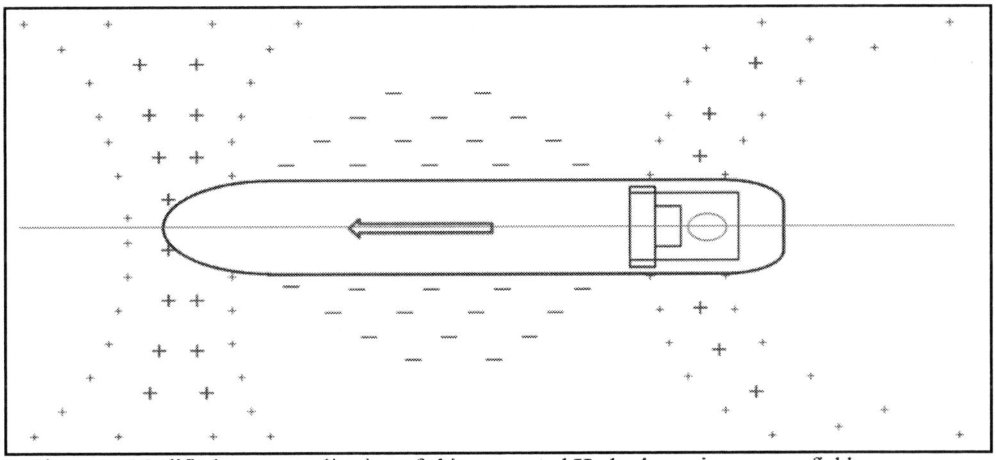

Figure 2: Modified conceptualisation of ship generated Hydrodynamic pressure fields

Despite the significant advancement introduced by this concept, there are a number of areas where this idea fails to serve the mariner:

- The fore and aft extent of the fields is limited and does not begin to explain distant influences
- The athwartships (side to side) extent of the pressure fields is limited
- The pressure fields relate to surface effects and do not address what is going on under the water.
- Effects are very localised, implying that vessels navigating only a short distance away will be 'safe'
- The diagram, and accompanying explanations, give no indication as to what might be happening in between the pressure fields or astern
- It deals only with a dynamic (moving) vessel and does not address effects experienced by a static vessel moored in a tideway or current.
- The diagram relates to open water pressure fields and does not take account of differences arising within enclosed waters.

In addition to developing the earlier concept, Dr Dand also described, in general terms, the resultant effects that interacting vessel's, or those operating in close proximity to land, submerged banks, or other obstructions might experience. Practical application under a variety of conditions verifies that those descriptions remain valid today. This concept forms the basis for most current teaching on marine hydrodynamics.

3. SCOPE AND FIELD OF RESEARCH

As a marine pilot who has to deal with the reality of hydrodynamic pressure systems on a daily basis, the earliest theory (Figure 1) provides little by way of assistance in carrying out my work. Dr Dand's theory (Figure 2) is significantly more useful. However, it too has its limitations in confined waters. Therefore, armed with years of practical ship-handling experience and a wealth of knowledge that informed me that something was *not quite right*, my research focussed on two objectives:

- Updating and remedying the deficiencies in current theory and teaching
- Trying to understand the initiating mechanisms that resulted in the effects described as 'interaction'.

The Manchester Ship Canal is unusual as a research venue for a number of reasons:

- The water is 'captive' – that is, confined in a single pond - and effects are amplified within the canal confines.
- There is no tide or current to complicate, or mask, findings
- Water levels are constantly maintained, therefore, serve as a good datum when measuring pressure waves.
- Meteorological influences can be limited or discounted in most instances.
- The controlled environment allows repetition of exercises and result re-verification.
- Accurate observations can be made several miles away from an advancing vessel thereby allowing water movement to be monitored over considerable distances.
- Blockage factors in the range of 10% to 80% are regularly experienced.
- Vessels pass at extreme close proximity (6 to 10 metres is not unusual).

- It is possible to discount influences from other craft operating nearby.
- Small details (eg slight changes in water level or flow) that may not be detected in scaled-down facilities become apparent when observed at full scale.
- Theories can be tested at full scale – there is no need for mathematical amplification of results.

The ship canal is, in effect, a full scale test tank with a closely controlled environment, stable water levels and limited (if any) water flow.

Based on a huge number of observations amassed over a 5 year period, this research produces two pressure field diagrams: the **first** (Figure 3) for vessels navigating in open water; the **second** (Figure 4) for vessels operating in narrow channels, estuaries and confined waterways.

As a guideline: *open water* is that where the blockage factor is approximately 10% or less.

4. HYDRODYNAMIC PRESSURE FIELD DIAGRAMS

4.1 THE BASIC (OPEN WATER) HYDRODYNAMIC FIELD DIAGRAM.

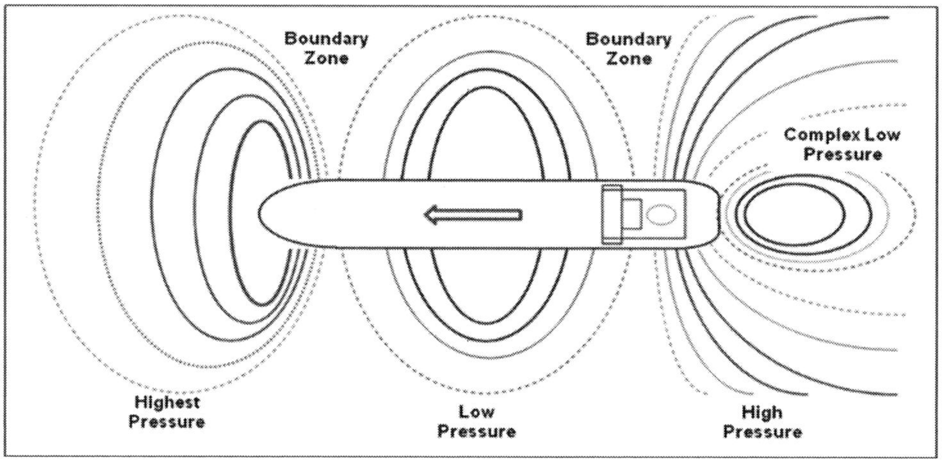

Figure 3: Ship-Generated Pressure fields for a vessel navigation in open water.

Explanations and factors to be considered when applying Figure 3.

- HIGH AND LOW PRESSURE fields relate to the hydrostatic 'head' or water-depth differential as measured against the normal (undisturbed) water datum. High pressure occurs where there is a discernable peak above normal datum, and low pressure where there exists a discernable trough, or dip below datum. For simplicity sake the convention used in these diagrams is: **high pressure fields** are portrayed in **red**, **low pressure fields** are coloured **blue**. No other significance should be attributed to the choice of colours.

- The relative intensity of the pressure field is indicated by the thickness of the lines.
  The measure of 'intensity' although somewhat simplistic in conception, is very practical in its application - being, roughly, the measure of effect that can be anticipated by an encountering vessel.

  The diagrams are meant to be easily visualised and quickly applied by mariners - who will understand that where there is no line, there is either 'no effect' or a 'neutral impact' in terms of forces that will influence ship-handling.

- FORWARD HIGH PRESSURE FIELD is not a purely localised phenomenon. Practically speaking, it is not necessarily confined to the immediate vicinity of the ship. For example, in a narrow channel it is not unusual for vessels to be affected by vessels manoeuvring several kilometres away. Surface water flow in high pressure fields, where observable (it is rarely obvious) is in the same direction of the moving craft.

Naval architect John Russel Scott, first introduced his 'wave of translation' or 'solitary' wave theory in 1834. He observed that the wave travelled at a constant velocity, did not lose power and was capable of proceeding independent of its originating mechanism.

To Russel's confined channel observations, several more can be added:

- There is a temporal element to development of the pressure fields. With continuing energy input, the field will build to a maximum over time.
- For any given channel once that maximum is reached, it requires considerable power input to overcome (for example, to hydroplane). Generally speaking, vessels with high block coefficients (i.e. tankers and bulk carriers) cannot generate this extra power - therefore are particularly susceptible and at risk (ie. from loss of manoeuvring capability – when altering speed or course) in confined waters.
- The maximum field intensity will, in part, be determined by waterway constriction, energy dissipation (through friction, drag, surface weather suppression and tidal/current flows, etc).
- A vessel moving at the same speed as its generated pressure field, risks being overcome by it, becoming directionally unstable and uncontrollable.
- Pressure fields generated by one vessel may cause others nearby, to exceed their *operational envelope* (see below for explanation) resulting in unexpected lateral displacement, sudden sheering or uncharacteristic handling.

- MIDSHIPS LOW PRESSURE FIELDS extend horizontally (in the fore and aft direction) over 60% of the vessel's length and are, essentially, attributable to Bernoulli's effect. The shape, extent, intensity and general configuration of this field component is typically determined by the vessel block co-efficient, although vessels of special construction and purpose may experience variations. The surface water-flow in the low pressure field is, very clearly, from forward to aft of the proceeding vessel. Careful observation shows that there is also a secondary, lesser, flow in towards the hull of the vessel. The two surface water flows combined may appear to produce a surface water flow the moves from some way off the bow, and in towards the hull at the after end.

  Experience has shown that the midships low pressure field can be felt some considerable distance from the side of the ship creating it. It is difficult to give precise measurements, but for a vessel proceeding at 4 knots (as a minimum) the influence can be detected over 3 ship lengths from the hull. Clearly, this has significant implications for large ships operating in confined waters, estuaries and channels.

- BOUNDARY ZONES are those areas between the high and low pressure field. On the surface, they can appear calm and undisturbed, but this is quite deceptive, therefore dangerous, especially to small craft. Within the boundary zone, pressure fields are changing from high to low, and vice-versa. The surface 'gradient', particularly close to a large or fast moving, vessel can be significant. The surface water flow will also change direction quite suddenly – flowing forward (in the direction of vessel progress) in the high pressure area, and towards the stern in the low pressure area. Furthermore, upwelling occurs in the high pressure area and may manifest as a 'p' force (the destabilising up-thrust experienced when a vessel *takes the blocks* during dry-docking) resulting in a small loss of statical stability for vessels working in the area. Clearly, this may present a particular hazard for vessels operating with reduced stability margins. Tugs, or small vessels, with their decks awash (therefore with reduced stability due to free surface effect), or small service craft, operating in the boundary zones, may suddenly find that they list unexpectedly, sheer suddenly, or are drawn in towards the hull and are unable to escape.

- AFTER LOW PRESSURE ZONE (astern of the ship) is extremely complex in nature. These diagrams do not address those complexities in detail. Briefly, the after low pressure zone comprises two separate systems. Firstly: a passive field (sometimes described as *a hole in the water*) resulting from the action of water flowing past the ships hull. Small vessels working near the stern of large vessels moored in a tideway often report 'falling into a hole' astern of the ship. They will sometimes deliberately seek out this area in rough weather as it offers an area of relative calm from the rougher surrounding waters. Secondly: when underway, an active element is generated by the action of the ships propeller. The active element, when combined with the passive system, can be one of the most dangerous areas around a ship – it is not unusual for vessels to be sucked into the *sternway* of a larger ship.

- BOW / STERN WAVES should not be confused with pressure fields. The normally visible ships wash is quite apparent under most circumstances. The positive and negative pressure fields described in this paper are more analogous to *tsunamis* in that, whilst they are rarely visible, they are both substantial and powerful.

The factors that most significantly influence the generation (and form) of these pressure fields are:

- The *speed of the vessel* relative to the water.
- The *blockage factor* (Relationship between the vessels *vertical cross-sec*tional area to that of the cross-sectional area of the waterway – expressed as a percentage (%))
- Block co-efficient will, generally, indicate where the highest blockage-factor occurs on the vessel. This provides some guidance on both the shape of the field and the location of the deepest midships low pressure trough.

NOTE: The 'operational envelope' for any vessel may be defined as: the physical extent within which the combination of resources, capabilities and assets available to the vessel allow it to function effectively and navigate safely in the prevailing conditions.

Factors which may influence the scope of a vessels operational envelope include, but are not limited to: maximum engine power capability; instantaneous power output compared with maximum output (engine setting); rudder type and efficiency; additional manoeuvring equipment available (i.e. thrusters, stabilisers etc); mechanical defects; weather restrictions; proximity of other vessels; OOW competence, alertness and ship-handling ability; command structure and leadership; general situational and hydrodynamic risk awareness. The integrated nature of a vessels operational envelope suggests that if any one factor changes significantly, then the effective operational envelope will change in response.

It is important that ship-handlers understand the concept of 'operational envelopes' - that they comprise not only the mechanical capabilities of the vessel, but also include the impact of personnel dynamics, physical capabilities, training, experience and prevailing environmental conditions.

If a vessel is operating at the limits of its operational envelope, the result of encountering additional unanticipated pressure fields generated by another vessel may mean that the ship is unable to cope (either mechanically or as a result of deficient human resources) with the additional factors. The result, particularly in confined or restricted waters, may be that the vessel is unable to determine what is happening when something 'strange' occurs, or manage / counter the effect. Experienced ship-handlers will often anticipate such possibilities, whether or not they understand the underlying dynamic, and plan accordingly.

4.2 ENCLOSED WATER HYDRODYNAMIC PRESSURE FIELDS

Part of the problem with previous theories is that they presented the pressure fields only for open water. Somewhat curiously, determination of open water hydrodynamic pressure systems was often verified using test facilities which, by their very nature, are confined. I use the word 'problem' because, invariably, the majority of interaction related difficulties arise when vessels are operating in confined waterways. Therefore, I considered it prudent to try and ascertain if there were particular differences between the form, or shape, of open and closed water pressure fields.

Another problem is that in open water (BF less than 10%) it becomes difficult assess boundaries and surface water flows. Monitoring mud plumes and using evidence gleaned from slight surface discolouration (usually after rainfall) made the task a little easier.

One further issue noted with earlier theories is that they tended to reflect only what was going on at the water surface. Repeated observation and testing indicates that effects are not purely confined to the surface, but extend throughout the entire depth of the water.

The equation for determining pressure at the bottom of a water column ($P = \rho gh$) suggests this to be the case. The argument being that, where there is clearly a hydraulic head of water, there must be a pressure differential at the bottom.

When constrained by boundaries, friction (or drag) plays a significant role in distorting the field shape -as in Figure 4 - a factor not explained or even suggested by previous theories. Whilst figure 3, and the short explanations provided, relate to the field propagation in open water, the general dynamic pertaining to the various field components is equally applicable to confined waterways.

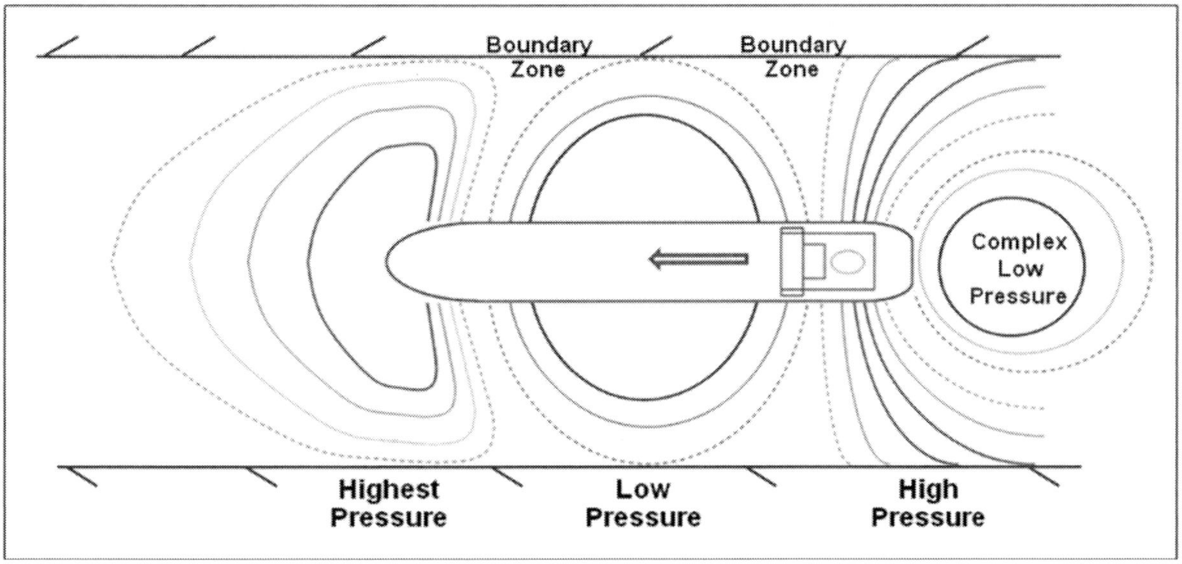

Figure 4: Ship-Generated Pressure fields for a vessel navigation in enclosed / confined water.

Broadly speaking, the main differences between open and enclosed water field shapes are:

- The preceding high pressure field is significantly more pronounced and, effectively 'squeezed' so that its effects are felt at much greater distances.
- Boundary zones are shorter (in the fore and aft line), more pronounced, and have a more dramatic effect on vessels working in their vicinity.
- The following high pressure field can be quite steep
- The low pressure zone astern is shorter, slightly broader, and can be significantly deeper. Passing vessels are quite definitely 'sucked' towards this area as they pass.
- The surface water flows in each zone is quite pronounced and easily visible.
- In steep sided water courses it is easy to appreciate the vertical form and intensity of pressure fields.

5. PRACTICAL IMPLICATIONS OF NEW FIELD THEORY

Applying the huge quantity of accumulated pressure field data affords the opportunity to begin explaining a range of hydrodynamic phenomena. For example; using a well known photo (depicting squat) that many mariners and pilots are familiar with (Figure 5).

Figure 5: Ship experiencing squat

And comparing it with a cross section of an open-water hydrodynamic field:

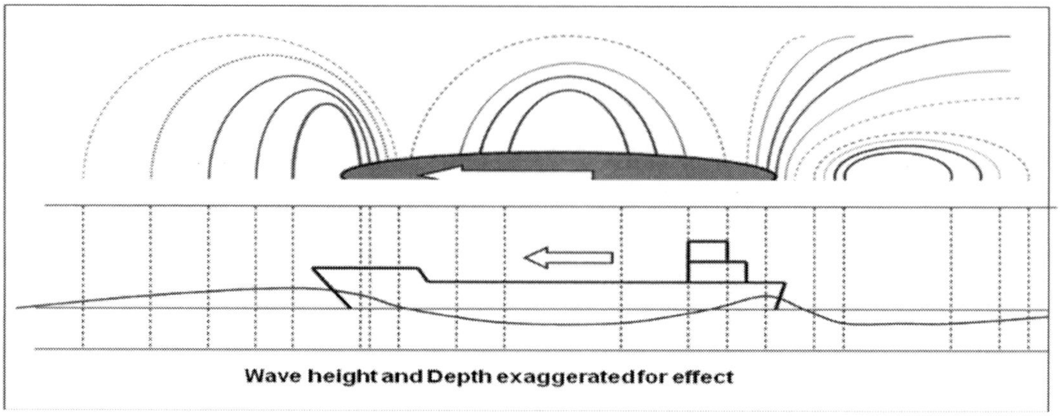

Figure 6: Cross-section of Ship Generated Open Water Hydrodynamic Pressure Field

Notwithstanding that the wave cross-section is slightly exaggerated for effect, clearly, there is more than a passing resemblance between the cross-sectional representation and the photo of the ship experiencing squat

6. ADVANCING TOWARDS A PRACTICAL APPLICATION

Having addressed the primary objective of '*updating and remedying the deficiencies in current theory and teaching*' and explaining basic mechanisms. The second objective of '*trying to understand the initiating mechanisms producing the effects described as interaction*' becomes feasible.

Few issues have vexed ship-handlers more than trying to understand the dynamic behind 'bank suction and rejection'

7. BANK SUCTION AND REJECTION RE-EXAMINED

Contemporary teaching on the cause of 'bank rejection' suggests that when a vessel is proceeding ahead and becomes orientated at an angle to an adjacent bank, that the primary operating factor is the wedge of water that inserts between the hull and the bank - thus acting as a cushion and forcing the vessel to bodily 'lift-off' from the bank.

Whilst this explanation seems plausible, many years of experience has shown that in extremely confined waterways, reliance upon this theory would prove both disastrous and costly. Current teaching ignores several realities, namely:

- When a vessel is moving ahead in confined waters, it will always generate a pressure field system. There is no exception to this rule. Ships displacing water, or water flowing past a static ships hull or a fixed object, will always result in some form of hydrodynamic mechanism being set up.
- The explanation does not accommodate asymmetries in hydrodynamic fields which can produce powerful effects - for example, imbalances in the pressure fields from one side of a vessel, to the other, will cause the vessel to move bodily towards the lower pressure side. The effect also manifests locally when, in a confined channel, one part of a vessel (eg, the bow) becomes offset causing an imbalance of pressure. In that instance, the bow will be cushioned by the higher pressure area and move towards the lower.
- Little allowance is made for the generation of localised Bernoulli effects (where a defined volume of water passes through a constriction, resulting in both an increased in flow speed and reduction in pressure).

The concept infers that any ship experiencing bank rejection will lift-off bodily, in a straight line, without deviation. However, the reality of bank rejection is rather different as, in practice, several factors contribute to the overall dynamic.

Figure 7 shows a vessel proceeding in a narrow channel - close to the 'mathematical middle'.

The mathematical middle is the 'centroid of the underwater cross-sectional area'. Notwithstanding collision regulation requirements to '*stay as near to the outer limits which lie on the Starboard side as is safe and practicable*', vessels proceeding with their fore and aft line over the mathematical middle will find that they are able to maintain position quite easily and require little helm to effect a course alteration. This is due to pressures on either side of the hull and rudder being equal and symmetrical.

In this example, and for explanation purposes, the vessel beam is 25% that of the channel width. The draft (D) is 50% of the maximum depth. The Blockage factor is therefore 25% (B) x 50% (D) = 12.5% (or 12%).

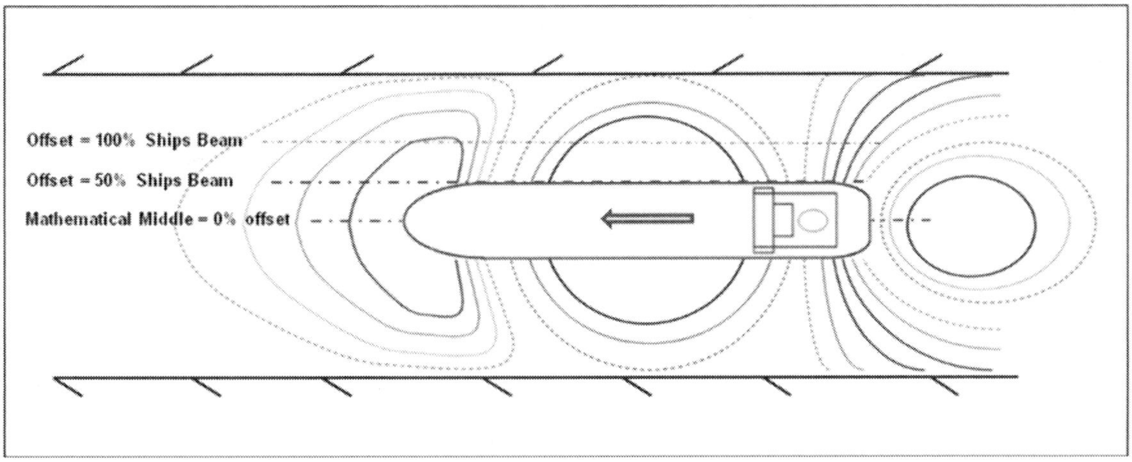

Figure 7: Vessel in a narrow channel, proceeding along the 'mathematical middle'. Showing 'offset' distances (relative to beam of ship) from middle position).

For demonstration purposes, two offset lines are shown, the first is offset from the middle by 50% of the ships beam and the second by 100%.

7.1 HYDRODYNAMIC EFFECTS ON A VESSEL OFFSET FROM MIDDLE

Once the vessel moves away from the mathematical middle and becomes offset, there is no longer hydrodynamic symmetry (Figure 8).

Conventional teaching suggests a 'wedge of water' will insert between the bank and the vessel and force the vessel to 'lift' bodily from the bank. However, this does not take account of what may happen as a result of the asymmetry in pressure fields.

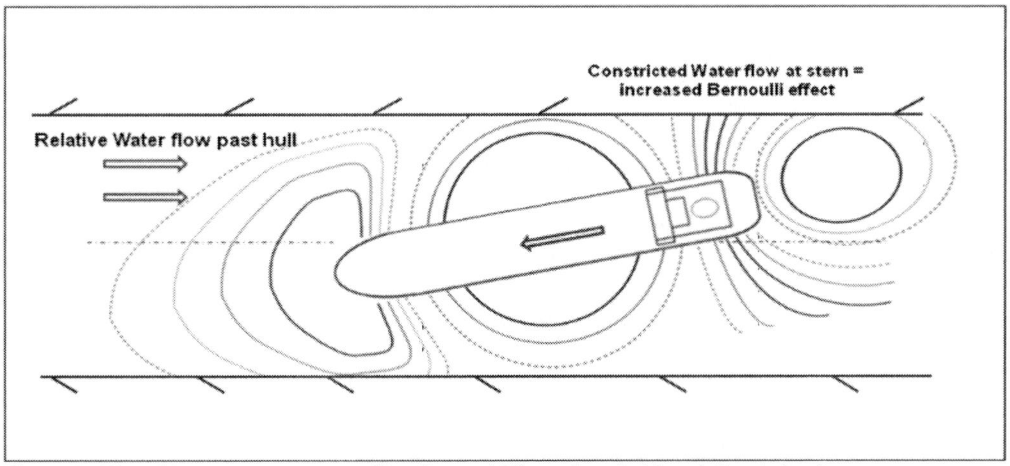

Figure 8: Vessel in a narrow - stern offset from middle and angled in relation to bank.

Applying only pressure field theory to figure 8, several effects may be anticipated:

- Increasing pressure at the bow should cause the bow to swing to the right (to Starboard)
- The high pressure imbalance at the stern should cause the stern to move to the left (to Port)
- The concentrated negative pressure field amidships should cause the vessel to draw slightly towards the right (starboard side) bank.

- The overall result is that, using only the macro pressure systems, the vessel should straighten up to re-align with the middle.

In an ideal world, this dynamic would mean that vessels following a narrow channel or fairway regulated themselves and always found the ideal position. Indeed, vessels navigating close to the middle and orientated no more than 15 to 20 degrees to the mathematical middle may experience such self-correction. However, this dynamic takes no account of localised Bernoulli systems that can develop and significantly affect the vessel.

7.2 LOCALISED MICRO SYSTEM EFFECTS

Experience has shown that the bow pressure systems will, eventually, cause the bow to swing, as indicated, however, the positive pressure system towards the stern is subjected to generation of a localised Bernoulli effect that is capable of 'overcoming' the macro pressure fields - see Figure 9.

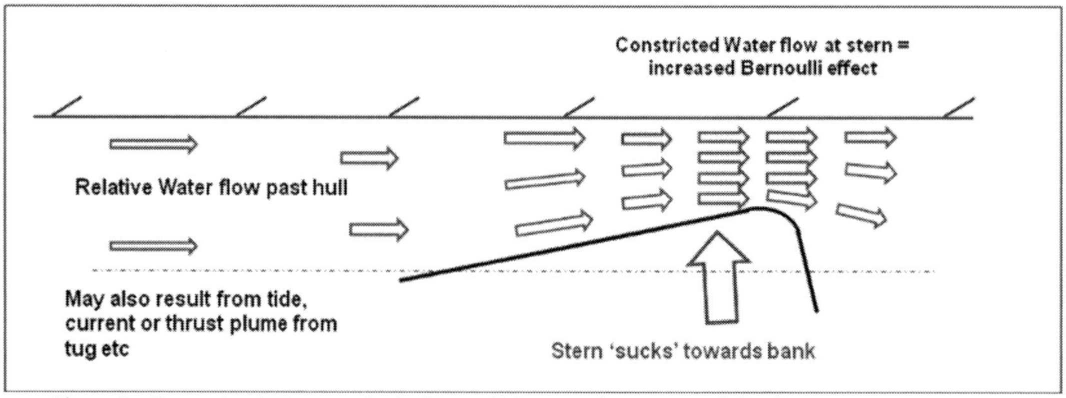

Figure 9: Dynamic of 'Overcoming' effect – Ship stern drawn towards bank

Where this dynamic is set up, rather than lifting away from the bank as might be expected if only the macro high pressure field is acting, the localised Bernoulli effect will cause the stern to 'suck' towards the bank.

7.3 RISK OF BERNOULLI EFFECT OVERCOMING POSITIVE PRESSURE FIELDS.

Significant testing on full sized ships, navigating in confined waters, indicates that it is possible to anticipate when, or if, the localised micro effect is likely to overcome the after high pressure system - resulting in bank suction. The results used in generating the "figure 10" graph derive from direct observations over a 5 year period when full-size vessels (both loaded and light) were deliberately offset from the waterway mathematical middle. The graph provides an indication of how much compensating helm (standard semi-balanced rudder) will be required to counter the effect. Vessel's having high efficiency rudders will overcome the effect more easily.

Testing took place over a range of channel widths ranging from 29 metres, 40 metres, 55 metres and 110 metres and depth ranges between 8 metres to 11.5 metres. These parameters compare with other major waterways, for example:

- Suez Canal width 200 metres depth 22.5 metres;
- Panama Canal width 150 to 300 metres but 70 metres (navigable) and depth 12 metres;
- Keil Canal width 102 – 204 metres depth 11 metres;
- Saint Lawrence Seaway width 61 metres depth 12.5 metres.

The effect has also been tested (to a lesser extent) in tidal channels up to 600 metres wide and 16 metres deep. Consequently, the graphs have general application and are not limited to the Manchester Ship Canal.

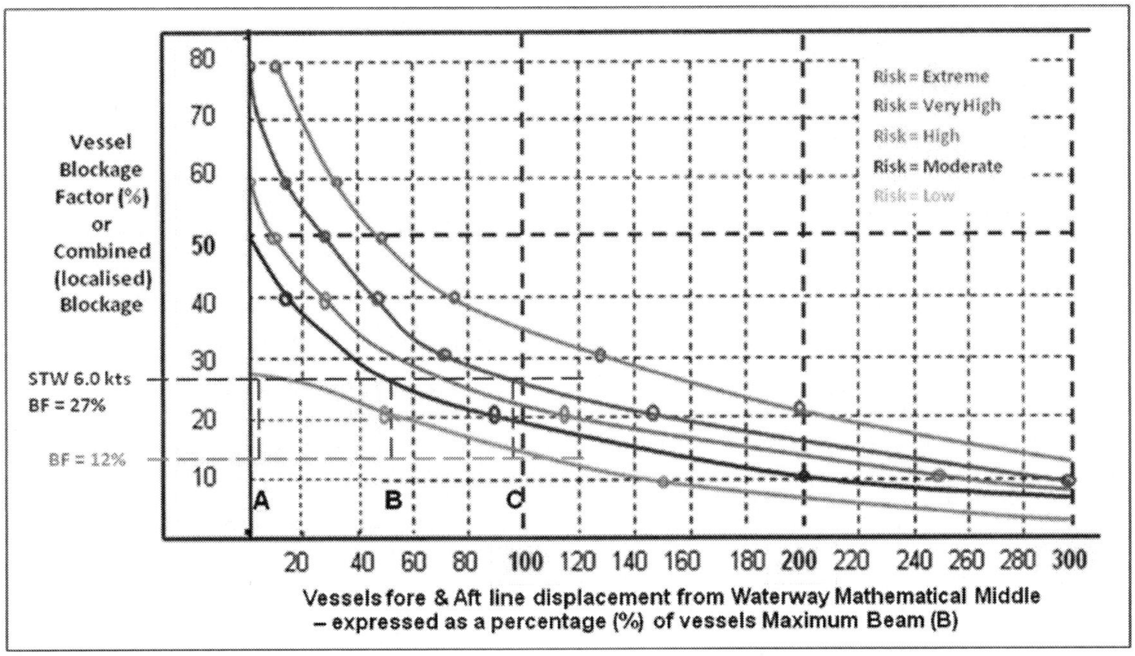

Figure 10: Graphs indicating the likelihood of 'overcoming effect' on moving vessel.

As a general guideline, the "figure 10" graph relates to a speed of 4 knots through the water (STW) although vessels with very high blockage factors are unlikely to proceed at this speed and prudent ship-handlers will regulate their speed according to conditions.

7.4 CHARACTERISTICS ATTRIBUTED TO EACH GRAPH

Extreme
The vessel is so close to the bank (or another vessel) that, in all likelihood, contact is imminent - or the vessel is already in proximate contact. Any positive pressure fields (singly or in combination) will be overcome by the localised Bernoulli effect. This situation is rarely encountered intentionally - although it may arise when docking in a strong tideway. If unintentional, tug assistance will probably be required to recover from the situation.

Very High
The vessel is extremely close to the bank (or another vessel). Under normal circumstances, compensatory measures should already have been taken to avoid closer approach. The 'overcoming' effect will be extremely apparent. The vessel may suddenly, and unexpectedly, sheer. Ship-handlers will experience difficulty in counteracting the movement. This situation may be encountered when entering locks or navigating within confined docks, canals or waterways. Considerable confidence and ship-handling knowledge, combined with extensive practical experience, will be required to deal with effects encountered at this level.

High
The vessel is close to the bank (or another vessel) and substantively offset from the waterway mathematical middle. Bank rejection will be extremely noticeable. Permanent helm will need to be applied simply to maintain a straight course. Any combination of the macro-positive pressure fields can be overcome by the locally generated Bernoulli effect. Within this range of risk, considerable uncertainty as to development of the 'overcoming' effect exists. Experienced ship-handlers operating close to banks in confined waters will, generally, have such risks in their contemplation.

Moderate
The vessel may be closing on an underwater bank (or another vessel) and substantively offset from the middle. Bank rejection may be very noticeable and permanent helm may already be applied simply to maintain a straight course. Such effects may be experienced when entering port, when navigating in an estuary or a reasonably wide (but constrained) waterway. Watch officers should recognise that additional caution must be exercised in such circumstances. Safe navigation (at reduced speed) and competent ship manoeuvring should be possible for experienced ship-handlers.

Low
Blockage factor will be in the region of 10% or less, and the vessel is a significant distance (relatively speaking) from potential hazards.

Normal navigational practices apply and the vessel should respond and handle within normal parameters. No exceptional measures are required. The knowledge and experience considered to be within the usual practices of seamen will usually be sufficient for vessels in this condition.

7.5 USING THE GRAPHS - A WORKED EXAMPLE

Using the vessel described in figure 7 (with a BF of 12%) as an example – indicated by the light blue dashed line. For speeds in excess of 4 knots, experience shows that for the graph to be applicable it is sufficient to multiply the BF by the square of the speed divided by 4 (i.e, for a B.F of 12% and a speed of 6.0 kts: Blockage factor = 12% x $(6.0 \div 4)^2$ = 12% x $(1.5)^2$ = 12 x 2.25) = **27%** (that is, the brown dashed line)

POSITION A: VESSEL AT, OR NEAR, MIDDLE
4 knots: The risk of bank suction is extremely small – below the **low** graph line.
6 knots: Effective BF is calculated to be 27%. The risk whilst navigating in the middle rises to **low.**

POSTION **B**: VESSELS IS OFFSET FROM MIDDLE BY 50% OF ITS BEAM.
4 knots: The risk of bank suction is still fairly low, but only a slight increase in speed or change in position will raise the level of risk.
6 knots: The risk immediately jumps from less than low, to **moderate**. Only a slight increase in speed, or change in position will raise the level of risk to **high**, or **very high**.

POSTION C: VESSELS IS OFFSET FROM MIDDLE BY 100% OF ITS BEAM.
4 knots: Risk is already at the category of **low.**
6 knots: The risk has suddenly, and dramatically, increased to **very high**. Any increase in speed, or further offset will quickly raise the risk to **extreme**.

Where the result equals, or exceeds, the **extreme** graph line, it must always be assumed that the localised Bernoulli effect will overcome the macro field effect.

7.6 SHIP FACTORS USED IN DEVELOPING THE GRAPHS

In constructing these graphs, the majority of observations relate to tankers and bulk carriers - with block co-efficients (Cb) in excess of 0.78. Vessels of lesser Cb were observed and the *overcoming effect* was found to be comparable. It is, however, felt that additional observations are required to verify the application in respect of lower Cb vessels. Therefore, vessels with block coefficients of less that 0.78 should exercise caution when employing the graphs.

7.7 SPEED THROUGH THE WATER

When calculating the likelihood of the macro-hydrodynamic fields being overcome by the localised Bernoulli effect, the mariner should use the water speed (STW) – not the GPS speed (SOG) as this will give a false result. In confined waterways, the difference between SOG and STW can generally be attributed to water's localised backflow velocity.

To be relevant, speed through the water should take account of:

- Vessels Speed over the ground (SOG or GPS speed)
- Tide or current flows
- Modifying effects of any inflows (i.e. from adjacent rivers etc)
- The potential of increased (river) current after rain, from snow melts etc.
- The effect of tug thrust plumes or wash from other craft operating in the immediate vicinity.

Note: Tug thrust plumes are included here because, although a very localised flow, experience shows that in confined conditions, their effect can significantly influence the intensity of overcoming effect.

7.8 LIMITATIONS OF GRAPH APPLICATION

The graphs apply to vessels navigating in confined waterways. Lower limits of 3% blockage factor and a maximum possible offset from middle of 300% is used. As a general guideline, vessels operating outwith these limits are deemed to be in open water.

8. CONCLUSIONS

This paper introduces a general overview of both open and confined water hydrodynamic fields and gives some guidance as to causation and the effects that might be anticipated by a moving vessel.

The purpose of this work is to add to the general corpus of knowledge so that mariners might: appreciate the unseen hydrodynamic forces that their vessels generate and which, either singly or in combination, affect ship-handling and, generally, impact upon the, normally anticipated, sea-keeping characteristics of vessels operating in confined waters.

Of all the effects encountered by vessels operating in confined waterways, bank suction and rejection are amongst those most commonly experienced, yet least understood or capable of anticipation. This paper goes some way towards addressing that lack of underpinning knowledge and contributes towards a practical comprehension of the effect.

9. ACKNOWLEDGEMENTS

The author gratefully acknowledges the support and encouragement of the Peel Ports Group Harbour Master's Dept (Manchester Ship Canal) who permitted unrestricted access to company properties for the purpose of undertaking research.

The author acknowledges the support of Captain Michael Robarts (Harwich Pilot) for his contributions into aspects of open water hydrodynamics.

10. REFERENCES

Dand I.W. *The Physical Causes of Interaction and its Effects*. Nautical Institute, Conference on Ship Handling, Plymouth, 1977.

Dand I.W. *Squat, Interaction, Manoeuvring*. Nautical Institute, Humber Branch Seminar, Hull, 1995.

http://www.britannica.com/EBchecked/topic/658890/Hydrodynamica Work of Daniel Bernoulli

J. Scott Russell. Report on waves, *Fourteenth meeting of the British Association for the Advancement of Science*, 1844.

(Figure 5: *from January 2008 UK Pilot Magazine, No 292*).

Source: http://www.pilotmag.co.uk/2008/02/01/squat-part-1/

11. AUTHORS BIOGRAPHY

Peter McArthur is the Director of Norwest Interaction Ltd, an organisation that offers consultancy and provides training for the maritime industry. He is also a practicing Manchester Ship Canal pilot. Peter's hands-on experience as a pilot has contributed significantly to his research into the practical aspects of interaction and the development of his theoretical understanding of ship generated hydrodynamic pressure fields in very confined waterways.

URANS STUDIES OF SHIP-SHIP INTERACTIONS IN CALM WATER AND WAVES FOR REPLENISHMENT AND OVERTAKING CONDITIONS

S M Mousaviraad, S H Sadat-Hosseini, P M Carrica, F Stern, IIHR-Hydroscience & Engineering, The University of Iowa, Iowa City, IA 52242-1585, USA

SUMMARY

Ship-ship hydrodynamic interactions are studied using URANS computations for Hope and Bobo in replenishment condition in calm water and waves and in overtaking maneuver in waves. The experiments were performed by MARIN and include a wide range of speed, heading, spacing, and configuration. The simulation matrix is selected to validate URANS capability and study the complex physics involved. Most of the simulations are completed and some are to be run as future work. The physics are explained using the extensive global and local flow information produced by current simulation. The calm water results are validated with EFD, average error being 21%*D*, and the interaction effects are investigated through comparison with single-ship computations. For replenishment in waves, the MARIN data was previously used to evaluate different simulation tools based on a range of methods from dynamic simulators to time domain panel method solvers. The average error values are 10%*DR* for current URANS, 15%*DR* for Rankine panel method, 32%*DR* for Green's function method, and 38%*DR* for dynamic simulators. The interaction effects on 0th and 1st harmonics are quantified and different effects compared. Sheltering is found to have most significant effect for oblique waves, 105% for the current conditions. The effects of spacing are found more significant for head waves than for 120 deg waves where sheltering effects are present, being 43% and 23%, respectively. An overtaking simulation in regular head waves is carried out where Hope (larger ship) overtakes Bobo with speed ratio of 1.6. The interaction effects are present from the beginning of the simulation when Hope is in the wake of Bobo, and are most significant during overtaking. The important effects include decreased motions and increased *Y*, *K*, and *N* as the ships pass by each other, being more significant for Bobo (the smaller vessel).

NOMENCLATURE

a	Wave amplitude
ak	Wave steepness
B	2nd Lt. John P. Bobo (T-AK 3008)
b-m	bow-to-amidships configuration
CFD	Computational Fluid Dynamics
D	Data value
DR	Dynamic Range (for data values)
E	Error value
EFD	Experimental Fluid Dynamics
f_e	Encounter frequency
f_n	Natural frequency
Fr	Froude number
Fr_c	Coincidence Fr
H	Bob Hope (TAKR 300)
K	Roll moment
L	Ship length
m-m	amidships-to-amidships configuration
N	Yaw moment
P	Nondimensional pressure
PF	Potential Flow
RANS	Reynolds Averaged Navier Stokes
RAO	Response Amplitude Operator
S	Spacing between ships
SM	Supplemental Material
X	Surge force
Y	Sway force
z	Heave motion
θ	Pitch motion
λ	Wavelength
σ	Sinkage
τ	Trim

1. INTRODUCTION

Ship-ship interactions are of interest for replenishment and maneuvering in close proximity. Comprehensive reviews of ship-ship interaction studies are included in 23rd (ITTC 2002) and the following ITTC Maneuvering Committee reports. Traditionally, the problems have been approached using experiments (e.g. [1]), simplified theoretical and computational methods (e.g. [2]), or empirical formulae (e.g. [3] and [4]). In calm water, interactions include increase or decrease of motions, forces and moments which depend mainly on the ship lengths and speed ratios. During overtaking or head-on maneuvers, interactions include consecutive increase and decrease of the forces and moments. In the presence of waves, incoming waves diffract and interact with the radiated waves from the ship motions and the steady wave patterns of each ship. The resulting complex wave systems significantly affect the motions, forces, and moments of each of the ships, especially in resonance conditions. Recent studies include potential flow (PF) computations and viscous computational fluid dynamics (CFD). 3D panel method for overtaking under hull drift was studied with a wake model implemented to take the effects of lift force into consideration [5]. 3D time domain panel code LAMP_Multi was used for Hope and Bobo in regular waves with comparison to data [6]. Chimera unsteady Reynolds averaged Navier Stokes (URANS) simulations for calm water ship-ship dynamic interactions in shallow water as well as in a channel are performed with good comparison against experiments [7]. The paper concludes that the free surface is important for the interaction between the two ships.

Present ship-ship study is based on extensive experiments performed at MARIN for replenishment and overtaking in calm water and waves for Bob Hope and Bobo free to pitch and heave [8]. The MARIN data were used to evaluate six computational tools for a selected range of conditions in the amidships-to-amidships configuration in waves [9]. The tools were based on different methods including dynamic simulators, time or frequency domain zero speed free surface Green's function, and time domain Rankine panel method. For the simulator types, the forces are obtained at each time step from tables of coefficients and impulse response functions. The tables are computed and provided by other programs at a series of speeds and headings. For all methods, only response amplitude operators (RAOs) and phases are obtained and 0th harmonics are not included in the studies. Overall, the differences of the potential codes from the model tests were from 10 to 30%*DR* for heave and pitch amplitudes, from 20 to 160%*DR* for surge and sway forces and roll and yaw moments, and from 10 to 25%*DR* for the phase correlations. The conditions were limited to replenishment in waves and no calm water or overtaking cases were performed. A limitation of the PF codes were reported to be the presence of resonant waves trapped in the gap between the ships, which occurs when the distance separating the ships is a multiple of half the wavelength of the incoming waves [10]. For such conditions, the standing transverse waves between the two ships and the forces associated with them can become unrealistically large, since the viscous damping is not included in the potential flow simulations (e.g. [11]). The MARIN tests did not include any resonant waves condition.

Previous URANS seakeeping studies for single-ship conditions are summarized in the recent Gothenburg 2010 workshop on CFD in ship hydrodynamics. Assessments are done separating capability for predicting 1st/linear vs. 2nd/nonlinear terms. For calm water, resistance is considered 1st order, whereas sinkage and trim 2nd order. This is justified since for slender ships X is 2nd order coupled to heave and pitch and confirmed by experiments, which generally indicate small sinkage and trim with relatively small effects on resistance. For head waves, resistance, heave and pitch are expanded in a Fourier series and consideration is given to 0th harmonic, streaming (difference between steady and 0th harmonic) and 1st harmonic amplitude and phase. Resistance 0th harmonic is 1st order and 0th harmonic heave and pitch are second order using same reasoning as for calm water. In Fourier series analysis differences between steady and 0th harmonics are called streaming and are in general 2nd/nonlinear; thus, added resistance, sinkage and trim are considered 2nd/nonlinear. For small amplitude ship motions/wave exciting force, heave and pitch are 1st order with 2nd order coupling with X. This is confirmed by G2010 data for KVLCC2 and KCS in head waves and also for the current Hope and Bobo in head and oblique waves. For the current study, the 0th and 1st harmonics of other force and moment components, i.e. Y, K, and N, are also found to be 2nd/nonlinear terms. Average error values for G2010 results are 13%*DR* for 1st/linear and 38%*DR* for 2nd/nonlinear terms, with 25%*DR* overall average.

Maximum response conditions are discussed for single-ship seakeeping based on resonance and maximum exciting forces. The coincidence Fr where both resonance and maximum exciting forces occur is defined and is expected to induce maximum response. The exciting forces and moments according to strip theory assumptions are estimated by the forward speed diffraction problem. Studies on exciting forces and moment is performed for model 5512 (Fr=0.19, 0.28, and 0.41) [12] and shows relatively small dependency on Fr and geometry through comparisons with data for Wigley hull at Fr=0.3 [13] and for a ship model at Fr=0.3 [14]. Irvine et al. [15] studied the monohull data from [13, 16, and 17] and their data for DTMB 5512, and stated that maximum pitch and heave response is fairly close to λ/L=1.33 for most cases. Assuming λ/L=1.33 as maximum exciting force condition, they introduced a formula for the coincidence Fr. For KCS containership, experiments ([18], also G2010 workshop) show that the maximum response occurs at the coincidence Fr for added resistance, but for both heave and pitch the transfer functions increase with speed. For heave, the maximum response for all three speeds occurs at resonance, while for pitch it occurs at different frequencies, all smaller than resonance. At fixed λ/L=1.33, local maximum response occurs for f_e~f_n for heave but not pitch. For high speed (Fr=0.45, 0.6 and 0.75) Delft catamaran, the results were similar to KCS [19]. The maximum response for added resistance occurred at the coincidence Fr while maximum heave response occurred for highest Fr. For pitch the maximum response occurred at medium Fr. The maximum heave response for all three speeds occurred at resonance, while for pitch it occurred at different frequencies, all smaller than resonance. At fixed λ/L=1.33, local maximum response occurred for f_e~f_n for pitch but not heave.

The objective of this paper is to perform detailed validation studies and assess capability for URANS computations of ship-ship interactions for wide range of conditions based on MARIN data, study the interaction effects, and investigate the complex physics using the high-fidelity results. Computations are performed using the CFDShip-Iowa code [20], which employs a single-phase level set method and a blended k-ε/k-ω turbulence model. URANS test matrix is selected based on studies of EFD and PF results and allows studying the effects of spacing, heading and sheltering/configurations. The results are extensive and are all included in a supplemental material (SM) document, available upon request. Discussions herein are based on the overall results, while only important results are included in tables and figures.

2. DATA AND SIMULATION CONDITIONS

MARIN tests were performed for the 45th scale models of the T-AKR 300 class (Bob Hope) and the T-AK 3008

(Bobo), with main particulars shown in Table 1. Both models were free to pitch and heave and constrained in all other motions, with z, θ, X, Y, K, and N measured. The ship-to-wave heading conditions included head to following waves with Hope on the weather side at the following headings: 180°, 150°, 135°, 120°, 90°, 30°, and 0° (where 180 ° is head and 0 ° is following). There were also a few runs made with Bobo on the weather side at headings of 330°, 270°, and 225°. Figure 1 shows the heading definitions. The tests were run for three separations and three speeds, i.e. Hope (larger vessel) at Fr=0.19, 0.12, 0.06, and 0.0, and Bobo at Fr=0.15, 0.09, 0.045, and 0.0. Configurations include amidships-to-amidships (m-m), as well as Bobo ahead of Hope (b-m) and overtaking conditions. Figure 2 shows the configurations.
The natural frequencies and coincidence conditions are summarized in SM-Table 12 based on formulae summarized in SM-Appendix A. The Fr values for the MARIN conditions are smaller and far from the coincidence Fr. Natural frequencies are similar for heave and pitch, being about 0.9 Hz for Hope and 1.0 Hz for Bobo.

Table 1: Main particulars of the ships (full-scale)

	H	B
L_{WL} (m)	279.16	181.689
B (m)	32.258	32.156
T_F (m)	7.83	5.19
T_A(m)	9.76	8.18
Δ (tons)	50396.711	30784.975
C_B	0.643	0.746

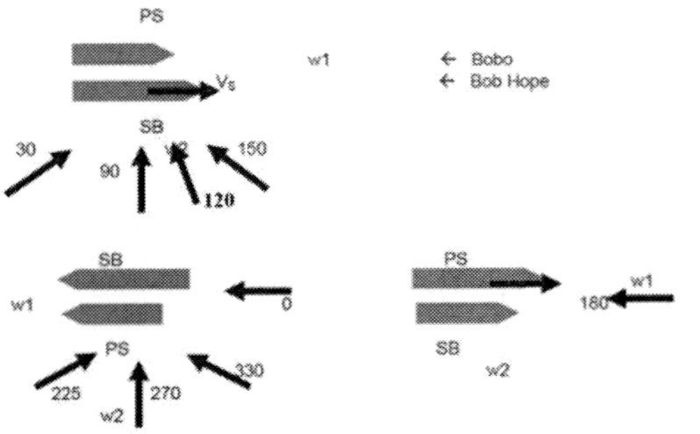

Figure 1: Wave heading definition

2.1 OVERVIEW OF EFD RESULTS

Selected experimental results are shown in SM-Figures 3 to 6. Most amplitudes peak in the near resonance region where encounter frequencies are close to the natural frequencies. Effects of sheltering (135 vs. 225 and 90 vs. 270) for zero speed, small spacing, and m-m configuration include slightly larger peaks for Hope sheltering Bobo for some components, as well as increased θ and X for Bobo when not sheltered by Hope. Effects of heading (0, 45, 90, 135, and 180) for Hope sheltering Bobo at zero speed, small spacing and m-m configuration are investigated. For 90 heading, z, Y, and K amplitudes are significantly higher, θ and X amplitudes are significantly lower, and N amplitudes are in the same order as other headings. Responses are similar for 0 and 180 heading, and for 45 and 135, with slight differences in the near resonance region. For 90 heading, unlike other headings, Y, K, and N amplitudes increase constantly with encounter frequency. Effects of heading are also considered for high speed, medium spacing and m-m configuration at 180, 150, and 120 heading. Resonance effects are generally maximum for 180 and minimum for 120, except for X amplitude of Hope, K amplitude of Bobo, and N amplitudes of both ships where resonance effects are maximum for 120 and minimum for 180. One significant effect is increased Y and K amplitudes of both ships at very high frequencies, which are largest for 120 heading. Since motion amplitudes are almost zero at high frequencies, this cannot be an effect of resonance. It is probably due to diffracted transverse waves in the gap between ships. Note that standing waves (resonant transverse waves) will occur for smaller wavelengths (higher frequencies) and therefore amplitudes probably will keep increasing with frequency up to the resonant wave conditions.

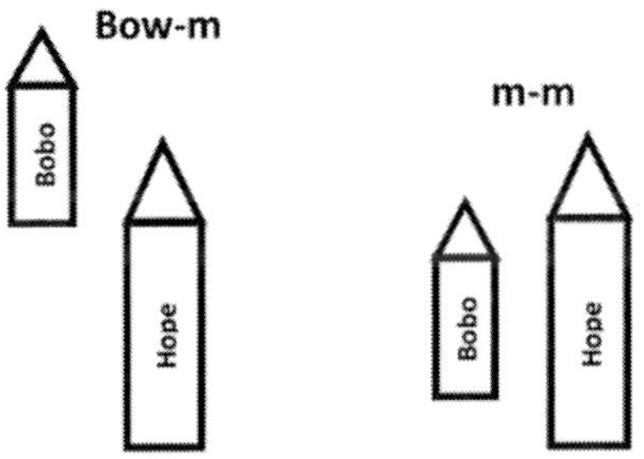

Figure 2: Definition of m-m and b-m configurations

2.2 CFD TEST MATRIX AND BASIS OF SELECTION

Table 2 shows the CFD test matrix selected to allow investigating different effects in calm water and waves, including spacing, heading, sheltering, configuration, and overtaking. Single-ship computations are performed to study the responses in the absence of second ship and obtain better physical understanding of the effects. Calm water cases include both high and low speed conditions. For wave cases, 180 and 120 headings at high speed are sufficient to include every effect, as shown in EFD results. For 120 heading, b-m configuration is included to study the effects of sheltering and to assess capability. One condition with low speed and small spacing is included to assess capability. Other headings are not included since EFD results show that heading effects are not significant. Overtaking conditions include calm water and wave simulations. Most of the simulations are complete, while a few are left to be done as future work.

Table 2: List of selected simulations, Red: not complete

Case No.	*Heading (deg)*	*Speed*		*Relative Position*	*Spacing (S)*			*Reason for Selection*
		Fr_H	Fr_B		S	S/L_H	S/L_B	
Calm Water; Replenishment								
1	Calm Water	0.15	0.19	H; B	∞			Single-ship
2	Calm Water	0.15	0.19	m-m	0.37m	0.059	0.091	Effects of spacing
3	Calm Water	0.047	0.059	m-m	0.067m	0.0107	0.0165	Low speed & small spacing
Regular Waves; Replenishment								
4	180	0.15	0.19	H; B	∞			Single-ship, 180 deg
5	120	0.15	0.19	H; B	∞			Single-ship, 120 deg
6	180	0.15	0.19	m-m	0.74m	0.12	0.18	Effects of spacing at 180 deg
7	180	0.15	0.19	m-m	0.37m	0.059	0.091	
8	120	0.15	0.19	m-m	0.74m	0.12	0.18	Effects of spacing at 120 deg
9	120	0.15	0.19	m-m	0.37m	0.059	0.091	Effects of heading
10	120	0.15	0.19	b-m	0.37m	0.059	0.091	Effects of sheltering/configuration
11	135	0.047	0.059	m-m	0.067m	0.0107	0.0165	Low speed & small spacing
Overtaking in Calm Water and Waves								
12	Calm Water	0.15	0.12	Overtaking	0.37m	0.059	0.091	Overtaking in calm water
13	180	0.15	0.12	Overtaking	0.37m	0.059	0.091	Overtaking in Waves
14	180	0.15	0.12	Overtaking	0.74m	0.12	0.18	Effects of spacing at 180 deg
15	120	0.15	0.12	Overtaking	0.37m	0.059	0.091	Effects of spacing at 120 deg
16	120	0.15	0.12	Overtaking	0.74m	0.12	0.18	

Table 3: Grid system

Hope BL	Hope Skeg	Hope Bilge Keel	Hope Rudder	Hope Total
783290	261744	260010	527512	1,832,556
Bobo BL	Bobo Skeg	Bobo Rudder	Bobo Total	
783290	261096	264880	1,309,266	
Hope Refinement	Bobo Refinement	Background		
3,059,700	2,290,156	3.399,708		
Total				
11,891,386				

2.3 URANS CODE AND SIMULATION CONDITIONS

CFDShip-Iowa Version 4.0 solves unsteady RANS equations for the liquid phase of a free surface flow using a blended k-ε/k-ω turbulence model. The domain is discretized using multiblock and overset structured grids. The overset capabilities are fully dynamic, allowing the code to simulate large-amplitude motions. Numerical methods incorporate finite difference discretization, with a second-order upwind scheme for convection terms and second-order centered for viscous terms. Temporal terms are discretized using a second-order backwards Euler scheme. Incompressibility is enforced by strong pressure/velocity coupling, using PISO or projection algorithms. The fluid flow is solved in an earth-fixed inertial reference system, while the rigid body equations of motion are solved in the ship coordinate system for each ship. Forces and moments are computed on the earth system, integrating separately on each ship, and projected into the corresponding ship coordinate system to perform the integration of the rigid body equations of motion, which are solved iteratively.

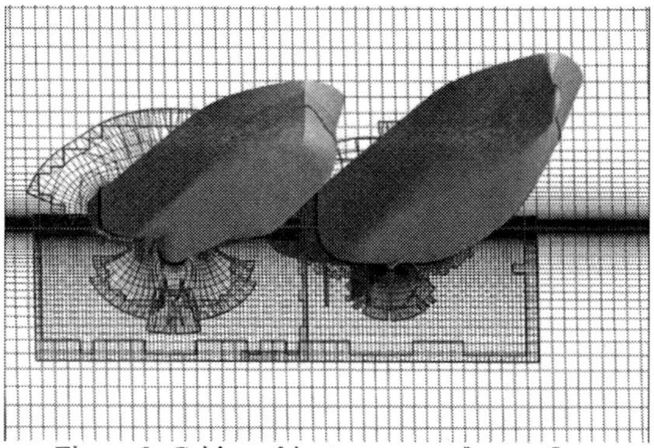

Figure 3: Grids and instantaneous free surface

The overset grid systems for each ship include double-O boundary layer grids, skeg, rudder, and bilge keel as summarized in Table 3. Refinement grids are included for both ships in the calm water calculations only. The Cartesian background grid extends enough to the sides and behind the ships to capture the free surface and wake flows. The boundary layer grid is designed to achieve $y^+<1$ for all Fr. Free surface grids are designed to include sufficient points per surface elevations and per wavelengths. Figure 3 shows the grids for ships, their refinements, and background. The domain size is different for different heading to capture the diffracted waves. The minimum domain size for 180 deg condition is $-0.75<x/L_B<3.65$, $-0.8<y/L_B<1.5$, and $-1.0<z/L_B<0.3$. Post-processing for replenishment conditions in waves are carried out similar to seakeeping post-processing procedures explained in G2010 workshop. The time traces of forces, moment, and motions are transferred into frequency domain using Fourier series to obtain the amplitudes and phases. 0th and 1st harmonics are presented and discussed herein. The error values are calculated as percent of dynamic range (*DR*) with *DR* values shown in SM-Table 15. *DR* values are obtained over the frequency range (6 wave frequencies for each condition) and are herein defined for 0th amplitudes as the maximum of the difference between the max and min amplitudes and the max amplitude, for 1st amplitudes as the max amplitude, and for phases as 360 deg.

Table 4: Validation of calm water CFD results: Fr_H=0.15, Fr_B=0.19, S=0.37m, m-m configuration

Motions	σ_H	σ_B	τ_H	τ_B	*Average*
E%D	41.55	27.15	31.61	30.30	32.65
Forces	X_H	X_B	Y_H	Y_B	
E%D	22.17	0.75	-21.96	-23.66	17.14
Moments	K_H	K_B	N_H	N_B	
E%D	12.37	20.42	18.97	-5.79	14.39
					21.39

Table 5: Effects of spacing on each of the ships for calm water simulations (same conditions as Table 4)

Percent difference	σ_H	σ_B	τ_H	τ_B	X_H	X_B
$\frac{S_{0.37}-S_{\infty}}{S_{\infty}}\times 100$	23.78	29.46	119.12	-50.55	-3.88	9.98
Average	55.73				6.93	
	31.33					

3. CALM WATER RESULTS

Calm water simulations are performed at the highest speed corresponding to Fr_H=0.15 and Fr_B=0.19. Single-ship results (SM-Table 9) show that sinkage, trim, and resistance are larger for Bobo than Hope due to higher *Fr*. For replenishment in m-m configuration with medium spacing, the validation results are shown in Table 4. Error values are generally large, with average 21.4%*D*, probably due to very small *Fr* and inadequate grid and/or domain size. Interaction effects (spacing effects) are quantified by calculating the differences between single-run (S=∞) and ship-ship (S=0.37m) simulation results as percent of single-ship results, shown in Table 5. Motions are affected significantly, 56%, while resistance is affected 7%. Sinkage is increased for both ships due to the presence of the second ship, and trim is increased for Hope and decreased for Bobo. For Hope, resistance is decreased by 4% due to presence of the second ship, while for Bobo it is increased by 10%. Figure 4 shows the percent difference in surface pressure for ship-ship vs. single-ship computations. The maximum differences are 10% for Hope and 7% for Bobo. The pressure is decreased for a large area at the bottoms of both ships which is the reason for the increase in sinkage. For Hope, the aft section has higher pressure than the fore, causing increased trim (bow-down) and decreased resistance. For Bobo, the increased bow pressure causes decreased trim and increased resistance. Note that unlike single-ship, the pressure distributions are not the same for the starboard and port sides, and therefore *Y*, *K*, and *N* are not zero, as in Table 4.

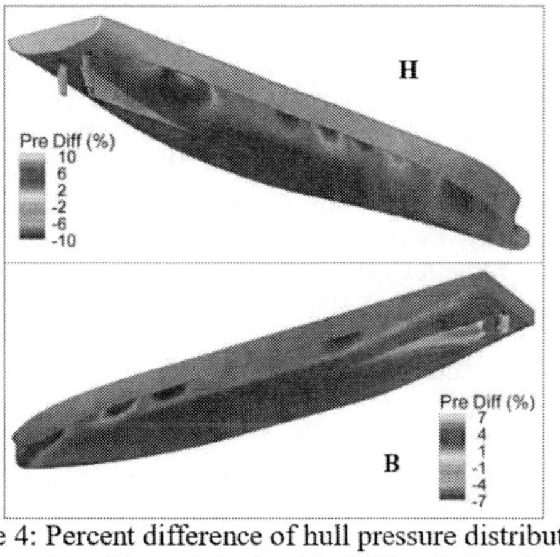

Figure 4: Percent difference of hull pressure distributions between ship-ship and single-ship CFD solutions in calm water normalized by max single-ship pressure magnitude

Table 6: Encounter frequency (model-scale), wavelength, and wave steepness for regular wave simulations

f_e *(Hz)* 180 deg	f_e *(Hz)* 120 deg	f_e *(Hz)* 135 deg	λ/L_H	λ/L_B	*ak*	$2a/\lambda$
0.57	0.50	0.46	1.38	2.12	0.012	1/257
0.75	0.65	0.58	0.88	1.36	0.019	1/165
0.96	0.80	0.71	0.61	0.94	0.029	1/110
1.19	0.97	0.84	0.45	0.69	0.038	1/83
1.43	1.14	0.98	0.34	0.53	0.049	1/64
1.82	1.42	1.19	0.24	0.38	0.069	1/45

4. WAVES RESULTS

Table 6 shows the encounter frequency, the wavelength, and the wave steepness for 180, 120, and 135 deg simulations in regular waves. Note that the natural frequencies for both ships fall into the frequency ranges for all headings. Figure 5 presents a snapshot of one of the 180 simulations showing the free surface and pressure distributions. Notice the complex dynamic wave system in the gap between ships that has significant effects on ship responses.

Single-ship (S_∞) RAOs are included in Figure 6. Since the current *Fr* for head waves simulations is much smaller than Fr_c, single ship RAOs do not show any peak and decrease with encounter frequency. For Hope, heave and pitch RAOs do not reach 1.0 in the simulation range since the maximum λ/L_H is 1.38. For Bobo, λ/L_B reaches up to 2.12 and RAOs approach 1.0 for small encounter frequencies. Both heave and pitch RAOs are larger for Bobo than Hope, probably due to *Fr* effects (Fr_H=0.15, Fr_B=0.19). For f_e larger than 1.2 Hz, RAOs are very small (close to zero) for both ships.

4.1 EFFECTS OF SPACING (S_∞, $S_{0.74}$, AND $S_{0.37}$) FOR 180 DEG HEADING

Head waves (180 deg) results for different spacing (S=0.74 and 0.37) are presented in Table 7 and the RAOs are shown in Figure 6. The most important physics for this condition is the resonance in ship motions in the resonance region, as seen in RAOs. Since the natural frequencies are close for two ships and they both have the same encounter frequency, the radiated waves due to motions of each ship further excite the other ship at its near resonant frequency. The average error values for 180 deg simulations are 15%*DR*. For 1st/linear terms the average error is 6%*DR* and for 2nd/nonlinear it is 25%*DR*. The effects of spacing are significant for the near resonance conditions (f_e=0.76, 0.96, and 1.19 Hz). Based on Figure 6 for motions and SM-Figures 11 and 12 for forces and moments, the effects of spacing on RAOs are as following. Compared to S_∞, heave RAOs increase significantly for both Hope and Bobo at f_e=0.96 *Hz*. This increase is larger for smaller spacing (S=0.37*m*) for Hope, and almost the same for Bobo. Note that CFD under-predicts heave RAOs for S=0.74 *m* for both Hope and Bobo. This may be due to inadequate resolution between hulls. At f_e=0.76 *Hz*, heave RAOs for both spacing conditions decrease for Bobo and remain the same for Hope. Effects of spacing are not significant for pitch motions. For X force, CFD predictions show insignificant spacing effects. There is no EFD for S_∞, but $S_{0.74}$ and $S_{0.37}$ results show peak for X amplitudes of Hope at f_e=1.19 *Hz* for $S_{0.74}$ and f_e=0.96 *Hz* for $S_{0.37}$, and Bobo at f_e=0.96 *Hz* for both spacing. Y amplitudes show peak at f_e=0.96 *Hz* for Hope at both spacing and Bobo at $S_{0.37}$ only. For K amplitudes, trough for Hope and peak for Bobo are observed at f_e=0.96 *Hz* for both spacing. N amplitudes show peaks for Hope at both spacing and a trough for $S_{0.74}$ and a peak for $S_{0.37}$ for Bobo at f_e=1.19 *Hz*.

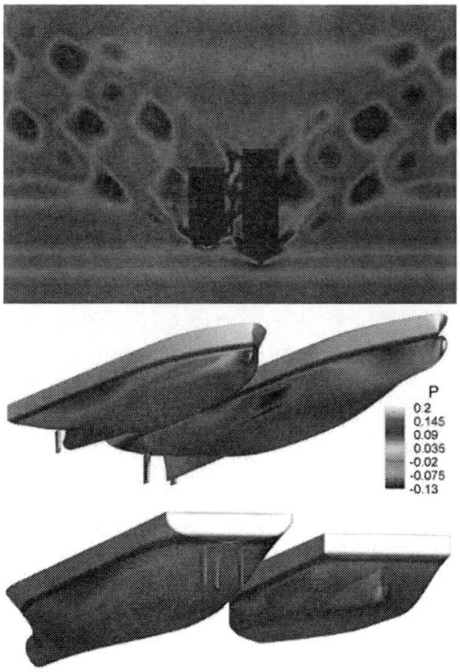

Figure 5: Free surface and hull pressure distributions at an instance during regular head wave simulation for f_e=0.96 *Hz*, Fr_H=0.15, Fr_B=0.19, S=0.37m, m-m

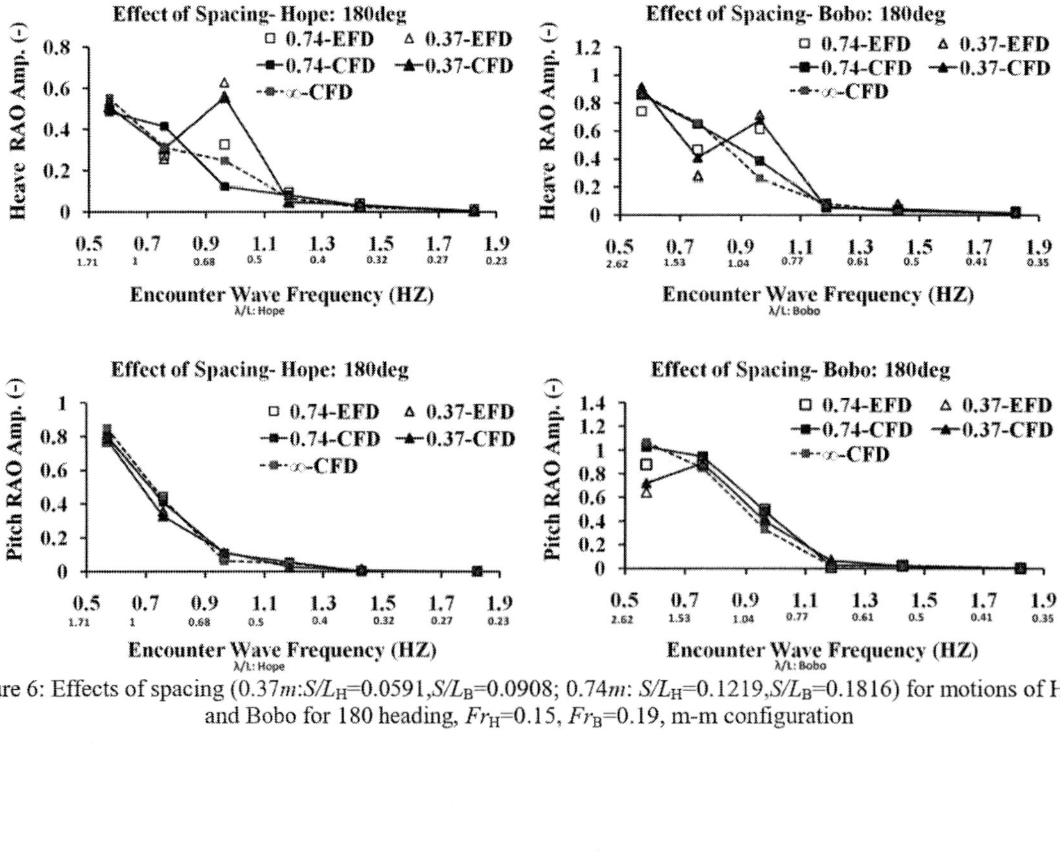

Figure 6: Effects of spacing (0.37*m*:S/L_H=0.0591,S/L_B=0.0908; 0.74*m*: S/L_H=0.1219,S/L_B=0.1816) for motions of Hope and Bobo for 180 heading, Fr_H=0.15, Fr_B=0.19, m-m configuration

Based on SM-Tables 24-26, the percent change in motions, forces, and moments due to spacing effects are as following. Overall effects compared to S_∞ are about 11% and 14% for $S_{0.74}$ and $S_{0.37}$, respectively. Motions are affected strongly by spacing, while X force changes less significantly. Largest effects are observed for 1st harmonic amplitudes. For motions, unlike X force, 0th harmonic amplitudes are also affected significantly. Effects of spacing are significant on both 1st/linear and 2nd/nonlinear terms. Generally, the percent effects are slightly larger for Hope than Bobo. If percent changes are considered for $S_{0.37}$ compared to $S_{0.74}$ condition, it is possible to study the effects on Y, K, and N. The results show that spacing effects are slightly larger on forces and moments compared to motions, being 42% and 38% respectively. 1st harmonic amplitudes show largest effects and for forces and moments the 0th harmonic amplitudes and 1st harmonic phases are also significantly affected. For Hope, the motions are affected more, while for Bobo the effects are more significant on forces and moments. The largest effects are for N (121%) and then Y (56%) for Bobo.

4.2 EFFECTS OF SPACING ($S_{0.74}$ AND $S_{0.37}$) FOR 120 DEG HEADING

At 120 deg heading, Bobo is sheltered since the incoming waves hit Hope (the larger ship) and diffract away. This will cause the spacing effects to be less significant compared to the head waves simulations. Another phenomenon observed here is elevated Y, K, and N amplitudes at highest frequency f_e=1.4 Hz. This is perhaps due to the wave system between the ships, similar to but not necessarily the same as the standing transverse waves explained in [10]. The overall average error for 120 deg simulations with m-m configuration is 12.5%DR, with 5%DR for 1st/linear terms and 20%DR for 2nd/nonlinear terms.

Figure 7 and SM-Figures 18 and 19 show the effects of spacing on RAOs for 120 deg heading. For motions, the effects are negligible and no resonance effect is observed. This is perhaps due to sheltering. For forces and moments, some effects are observed for the near resonance encounter frequencies. For X force, CFD results show a different trend than EFD with no significant difference between the two spacing. EFD results show peaks at f_e=0.97 Hz for both Hope and Bobo and both spacing. Y amplitudes for Hope show a peak at f_e=1.14 Hz for $S_{0.74}$ and a trough at f_e=0.97 Hz for $S_{0.37}$. For Bobo, peaks are observed for both spacing at f_e=0.8 Hz. Elevated Y amplitudes for $S_{0.37}$ at f_e=1.4 Hz are observed for both ships, perhaps due to the complex wave system in the gap between ships. Note that at this high frequency, the motions are almost zero and therefore the elevated force components are not related to increased ship motions. For K amplitudes of Hope, a peak is observed for $S_{0.74}$ at f_e=0.8 Hz and a trough for $S_{0.37}$ at f_e=0.97 Hz. Elevated amplitudes are observed at f_e=1.4 Hz for both spacing for Hope. For Bobo, peaks are observed at f_e=0.97 Hz for K amplitudes at both spacing. N amplitudes show peak for $S_{0.74}$ at f_e=0.8 Hz for Hope and f_e=0.97 Hz for Bobo, and for $S_{0.37}$ at f_e=0.97 Hz for Hope and f_e=1.14 Hz for Bobo.

Table 9 summarizes the percent change as effects of spacing for 120 deg simulations. The total average change is about 23%, which is significantly smaller than 180 deg heading being 43%. The effects are smaller for motions than forces and moments. Unlike 180 deg heading, spacing effects are much smaller on Hope than Bobo. The largest effect is for Y force of Bobo, being about 75%.

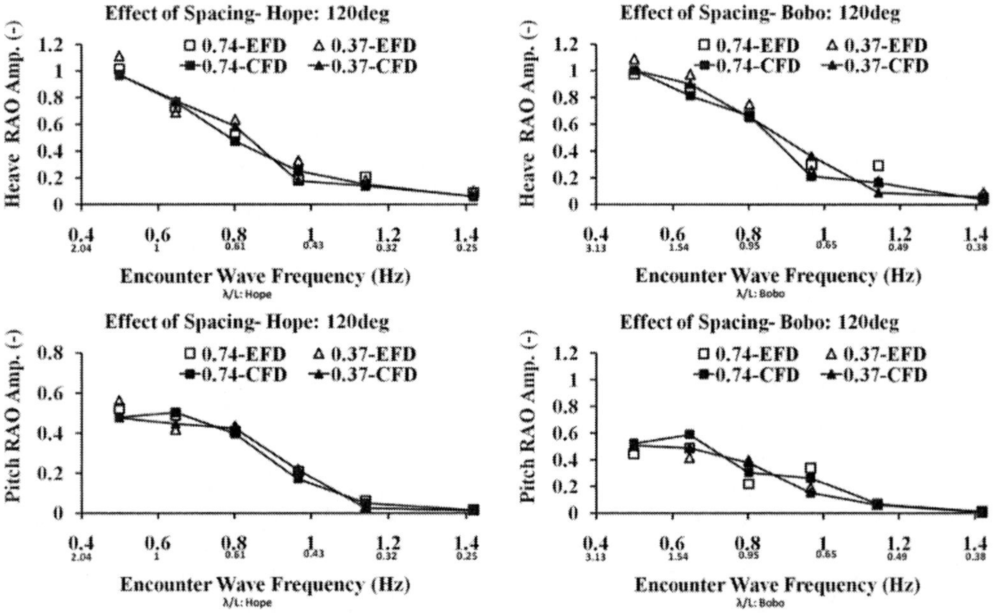

Figure 7: Effects of spacing (0.37m:S/L_H=0.0591,S/L_B=0.0908; 0.74m: S/L_H=0.1219,S/L_B=0.1816) for motions of Hope and Bobo for 120 heading, Fr_H=0.15, Fr_B=0.19, m-m configuration

Table 7: Validation (*E%DR*) summary for static relative position ship-ship simulations in waves

		Hope						Bobo						Average			
	Terms	X	Y	K	N	z	θ	X	Y	K	N	z	θ	forces	motions	1st/linear	2nd/nonlinear
180deg, $S_{0.74}$, m-m	0th	13.53	99.90	20.45	37.22	38.06	69.92	2.42	33.68	23.26	14.66	29.07	35.06	30.64	43.03		
	1st Amp	36.70	24.92	30.06	30.07	13.34	2.08	33.50	23.18	25.58	7.43	12.76	4.73	26.43	8.23		
	1st Phase	11.17	13.38	12.53	10.68	4.20	8.28	3.44	5.03	7.52	5.12	3.58	6.54	8.61	5.65	7.15	26.22
	Ave	20.47	46.07	21.01	25.99	18.54	26.76	13.12	20.63	18.79	9.07	15.14	15.44	21.89	18.97		
		28.39				22.65		15.40				15.29					
		25.52						15.35						16.68			
180deg, $S_{0.37}$, m-m	0th	13.72	90.07	14.47	41.77	31.02	14.72	1.98	50.42	112.00	4.30	21.64	9.05	41.09	19.10		
	1st Amp	33.14	20.35	22.12	14.21	5.05	1.07	35.59	14.64	27.41	9.33	5.45	2.27	22.10	3.46		
	1st Phase	5.09	4.59	8.41	3.22	3.64	7.23	4.16	5.29	10.00	5.20	1.72	7.79	5.74	5.09	4.99	23.55
	Ave	17.32	38.34	15.00	19.73	13.24	7.67	13.91	23.45	49.81	6.27	9.60	6.37	22.98	9.22		
		22.60				10.45		23.36				7.99					
		16.53						15.67						14.27			
120deg, $S_{0.74}$, m-m	0th	10.42	13.83	141.33	70.75	18.50	11.09	1.45	29.40	11.92	7.66	10.58	24.25	35.85	16.11		
	1st Amp	59.17	15.05	14.26	9.56	4.70	3.75	31.84	16.99	19.17	6.61	5.15	11.80	21.58	6.35		
	1st Phase	6.98	4.86	2.04	2.15	3.28	1.48	10.40	1.93	5.04	2.91	1.76	4.69	4.54	2.80	4.85	21.09
	Ave	25.53	11.25	52.54	27.49	8.82	5.44	14.56	16.11	12.04	5.73	5.83	13.58	20.66	8.42		
		29.20				7.13		12.11				9.71					
		18.17						10.91						12.97			
120deg, $S_{0.37}$, m-m	0th	11.04	31.84	54.11	46.59	18.80	38.93	2.27	35.12	19.04	8.18	8.91	14.46	26.03	20.27		
	1st Amp	46.63	18.89	15.64	12.52	7.74	3.92	28.14	15.50	23.67	6.01	7.52	4.69	20.87	5.97		
	1st Phase	8.63	5.76	6.82	2.02	3.29	4.30	6.54	7.13	6.90	4.38	2.45	3.08	6.02	3.28	5.03	18.89
	Ave	22.10	18.83	25.52	20.38	9.94	15.72	12.32	19.25	16.54	6.19	6.29	7.41	17.64	9.84		
		21.71				12.83		13.57				6.85					
		17.27						10.21						11.96			
120deg, $S_{0.37}$, b-m	0th	10.64	32.55	45.41	54.90	29.92	23.44	2.64	10.54	35.55	30.94	14.32	91.16	27.90	39.71		
	1st Amp	52.87	12.42	11.16	9.97	5.50	3.62	24.53	10.99	14.87	9.17	9.37	5.64	18.25	6.03		
	1st Phase	3.27	2.10	2.33	1.03	3.07	2.99	10.85	2.69	11.43	1.73	5.31	1.89	4.43	3.31	5.07	21.16
	Ave	22.26	15.69	19.63	21.97	12.83	10.02	12.67	8.07	20.62	13.95	9.66	32.90	16.86	16.35		
		19.89				11.42		13.83				21.28					
		15.66						17.56						13.11			
135deg, $S_{0.867}$, m-m	0th	16.33	16.58	27.94	19.75	43.09	145.59	15.90	5.85	19.00	14.17	39.33	74.03	16.94	75.51		
	1st Amp	18.30	18.43	19.92	13.55	6.62	5.40	22.52	27.91	16.79	17.69	19.23	15.41	19.39	11.67		
	1st Phase	12.82	8.05	3.48	6.34	7.33	7.43	14.88	5.25	8.59	5.17	6.31	7.09	8.07	7.04	10.71	24.04
	Ave	15.82	14.35	17.11	13.21	19.02	52.81	17.77	13.00	14.79	12.34	21.62	32.18	14.80	31.41		
		15.12				35.91		14.48				26.90					
		25.52						20.69						17.37			
Average	0th	12.62	47.46	50.62	45.16	29.90	50.61	4.44	27.50	36.80	13.32	20.64	41.33	29.74	35.62		
	1st Amp	41.14	18.34	18.86	14.98	7.16	3.31	29.35	18.20	21.25	9.37	9.91	7.43	21.44	6.95		
	1st Phase	7.99	6.46	5.94	4.24	4.14	5.28	8.38	4.55	8.25	4.09	3.52	5.18	6.24	4.53	6.30	22.49
	Ave	20.58	24.09	25.14	21.46	13.73	19.74	14.06	16.75	22.10	8.93	11.36	17.98	19.14	15.70		
		22.82				16.73		15.46				14.67					
		19.78						**15.06**						**14.39**			

4.3 EFFECTS OF SHELTERING FOR 120 DEG HEADING

Effects of sheltering due to different configurations (m-m vs. b-m, as shown in Figure 2) are studied for 120 deg heading and $S_{0.37}$ and the results are included in Table 9. For m-m configuration, Bobo is sheltered by Hope while for b-m configuration the sheltering effects are expected to be less significant. The sheltering reduces the resonance effects and the elevated amplitudes for lateral forces and moments at high frequencies. The error values for b-m configuration are shown in Table 7. The average error is about 13%, only slightly larger than m-m simulation. The difference seems to be due to higher error values for 0th amplitudes of motions for b-m configuration. Unlike m-m, the average error for Bobo is not smaller than Hope but in the same order, which makes sense due to reduced sheltering effects for b-m configuration.

Effects of sheltering on RAOs are shown in SM-Figures 23-25. Heave RAOs for m-m configuration show no peak and decrease with encounter frequency for both ships. For b-m configuration, heave RAOs peak at f_e=0.97 *Hz* for Hope, and at f_e=0.80 and 0.97 *Hz* for Bobo. For pitch motions, similar to other simulations, no peaks are observed and sheltering effects are not significant. For *X* force, sheltering effects seem insignificant in the resonance region and almost same amplitudes are obtained for both configurations. But at the highest frequency, elevated *X* amplitudes are observed only for b-m configuration. For *Y* force significant sheltering effects are observed since amplitudes are larger for b-m configuration in the resonance region and for m-m configuration at high frequency. For *K*, different effects are observed for Hope and Bobo. For Hope, *K* amplitudes are larger for b-m configuration in the resonance region and for m-m configuration at high frequency. For Bobo, *K* amplitude is larger for m-m configuration at f_e=0.97 and 1.42 *Hz* and for b-m at f_e=1.14 *Hz*. For *N*, sheltering effects are significant in the resonance region (f_e=0.97 *Hz*) for both ships where amplitudes are larger in m-m configuration for Hope and in b-m configuration for Bobo. At high frequency, sheltering effect is significant only for Bobo where *K* amplitude is larger for b-m configuration.

Sheltering effects are quantified by calculating the differences between m-m and b-m configuration results as percent of b-m configuration results (Table 9). The effects are significant, with overall effects being 105%. The maximum effects are observed for forces and moments, with largest components occurring for the 0th amplitudes. For 1st amplitudes, the effects for forces and moments are only slightly larger than motions, being 45% and 39% respectively. The effects are more significant for 1st/linear terms, being 183%, compared to 2nd/nonlinear terms being 27%. Overall, the sheltering effects are more significant for Hope (127%) compared to Bobo (83%). This is due to very large effects on 0th amplitudes of forces and moments. Considering 1st harmonics only, the sheltering effects are more significant on Bobo than Hope.

4.4 SUMMARY AND COMPARISON OF TOOLS

Comparing the interaction effects (Table 9), it seems that sheltering has much larger effects than spacing. Comparing the spacing effects for 180 and 120 heading, the effects are larger for 180 heading, especially for motions. This is perhaps due to sheltering caused by oblique waves, diminishing the effects of spacing. The effects of spacing on Hope and Bobo are similar for 180 deg heading, but smaller for Hope than Bobo for 120 deg heading since Hope is exposed to waves and is affected least by spacing size.

The CFD validations are summarized in Table 7 which include a small spacing ($s_{0.067}$) and small Fr (Fr_H=0.047, Fr_B=0.059) condition at 135 deg heading. The RAOs for this case are shown in SM-Figures 26 and 27. For this case of very small spacing, RAOs show peaks near resonance despite sheltering. For heave, both Hope and Bobo show peak at f_e=0.84 *Hz*. For Pitch, no effect is observed, similar to 180 deg heading. For X, CFD shows peaks for both ships at f_e=0.58 *Hz*, but no peak is observed for EFD. Y amplitudes peak at f_e=0.98 *Hz* for Hope and f_e=0.84 *Hz* for Bobo. For K amplitudes, at f_e=0.98 *Hz* peaks are observed for both ships, and at f_e=0.84 *Hz* a peak is observed for Hope and a trough for Bobo. N amplitudes are reduced significantly at f_e=0.84 *Hz* for both ships.

The overall prediction error (Table 7) is 14.4%*DR*, with 6.3%*DR* for 1st/linear and 22.5%*DR* for 2nd/nonlinear terms. The largest error components are for 0th amplitudes since they have small values. Considering only 1st harmonics, the errors are 5.7%*DR* for motions and 13.8%*DR* for forces and moments. Overall, the average error is slightly larger for Hope (the larger ship with smaller Fr) than Bobo.

Comparison of simulation tools ([9] and current) based on their computational methods is summarized in Table 8 for 1st amplitudes of motions, forces, and moments. The overall error values are reported for dynamic simulators using coefficients from tables (38%*DR*), time and frequency domain Green's function codes (32%*DR*), time domain Rankine panel method codes (15%*DR*), and current URANS study (10%*DR*). For all simulation types, the error values for forces/moments are significantly larger than motions.

Table 8: Summary of ship-ship validation results (*E%DR*) for different numerical methods

Simulation Type	Tools	Terms	forces	motions
Dynamic Simulator	MVS	1st Amp	68.21	35.63
		1st Phase	25.43	24.54
	MVTDS	Ave	46.82	30.09
			38.45	
Green's Function	AQWA	1st Amp	83.90	10.29
		1st Phase	21.90	10.47
	ShipMo3D	Ave	52.90	10.38
			31.64	
Rankine Panel Method	LAMP	1st Amp	22.68	7.02
		1st Phase	18.81	9.66
	AEGIR	Ave	20.75	8.34
			14.54	
URANS	CFDShip-iowa	1st Amp	21.44	6.95
		1st Phase	6.24	4.53
		Ave	13.84	5.74
			9.79	

Table 9: Ship-ship effects (spacing at 180 and 120 deg; sheltering at 120 deg and S=0.37m); Fr_H=0.15, Fr_B=0.19

				Hope						Bobo						Average			
			Terms	X	Y	K	N	z	θ	X	Y	K	N	z	θ	forces	motions	1st/linear	2nd/nonlinear
Spacing	**180deg, m-m**	$(S_{\infty}-S_{0.74})/S_{\infty}$	0th	2.91				17.76	68.82	5.10				17.93	17.33	1.00	30.46	13.31	13.73
			1st Amp	25.36				35.92	42.42	15.33				18.68	24.20	5.09	30.30		
			1st Phase	42.95				6.44	5.88	2.22				4.45	17.70	5.65	8.62		
			Ave	23.74				20.04	39.04	7.55				13.69	19.74	3.91	23.13		
				23.74				29.54		7.55				16.71					
				26.64						12.13						13.52			
		$(S_{\infty}-S_{0.37})/S_{\infty}$	0th	3.56				25.69	78.76	8.83				30.28	42.59	1.55	44.33	18.26	17.21
			1st Amp	24.39				43.34	40.76	17.08				48.59	47.01	5.18	44.92		
			1st Phase	13.22				8.33	10.33	3.74				3.49	11.10	2.12	8.31		
			Ave	13.72				25.78	43.29	9.88				27.45	33.57	2.95	32.52		
				13.72				34.54		9.88				30.51					
				24.13						20.20						17.74			
		$(S_{0.74}-S_{0.37})/S_{0.74}$	0th	1.63	38.70	20.58	33.54	6.80	14.48	5.19	69.04	75.87	98.56	10.64	35.67	42.89	16.90	48.94	37.47
			1st Amp	31.21	37.55	64.33	56.98	122.35	169.80	16.10	82.58	38.42	258.56	31.00	57.74	73.22	95.22		
			1st Phase	25.56	25.90	26.19	22.28	13.35	11.30	3.87	28.59	22.54	23.50	5.10	5.12	22.30	8.72		
			Ave	19.47	34.05	37.03	37.60	47.50	65.19	8.39	60.07	45.61	126.87	15.58	32.84	46.14	40.28		
				32.04				56.35		60.24				24.21					
				44.19						42.22						43.21			
	120deg, m-m	$(S_{0.74}-S_{0.37})/S_{0.74}$	0th	2.31	40.86	13.66	66.97	3.94	30.55	4.27	170.88	28.99	54.64	7.04	38.55	47.82	20.02	25.19	21.70
			1st Amp	28.75	41.34	34.70	27.11	11.07	19.53	26.62	51.52	32.91	32.51	31.20	23.26	34.43	21.26		
			1st Phase	11.03	6.74	19.55	1.67	10.24	3.25	3.49	3.60	15.26	23.78	1.75	10.66	10.64	6.48		
			Ave	14.03	29.65	22.63	31.92	8.42	17.78	11.46	75.33	25.72	36.98	13.33	24.16	30.97	15.92		
				24.56				13.10		37.37				18.74					
				18.83						28.06						23.44			
Sheltering	**120deg, $S_{0.37}$**	$(S_{b\text{-}\infty}-S_{m\text{-}m})/S_{b\text{-}\infty}$	0th	1.85	1741.03	766.33	77.56	31.63	6.80	1.16	189.79	88.88	1153.26	37.07	20.59	502.48	24.02	182.68	26.97
			1st Amp	67.90	47.07	27.56	35.90	48.86	20.90	19.02	77.70	52.85	27.79	53.37	28.42	44.47	37.89		
			1st Phase	8.71	14.18	6.43	7.66	8.00	5.59	3.63	19.62	12.13	26.87	4.67	12.43	12.40	7.67		
			Ave	26.15	600.76	266.77	40.37	29.50	11.10	7.94	95.70	51.29	402.64	31.70	20.48	186.45	23.19		
				233.51				20.30		139.39				26.09					
				126.91						82.74						104.82			

5. OVERTAKING RESULTS

During the overtaking simulations the relative location of the passing ship with the other ship is defined zero when amidships of one ship aligns with amidships of the other ship. SM-Table 30 describes the start and end location of the relative position of the overtaking ship in the overtaking window. Note that all overtaking results are in full-scale. Overtaking cases are studied in time domain to understand the instantaneous effects of one ship passing the other. 180 deg simulations are presented here which has no sheltering effects.

SM-Figure 28 shows the wave elevation at the gauge for CFD and EFD. Note that the gauge is far away and forward of the ships and ship waves do not affect the wave history. CFD waves match exactly the desired conditions in the experiment. The real waves in the experimental tank, however, do not have steady peaks and troughs. This can affect the results and be one of the reasons for difference between CFD and EFD. SM-Figure 29 shows EFD and CFD ship speed histories for both ships. The mean speed (full scale) is 10 knots for the Bobo and 16 knots for the Hope (which is thus overtaking the Bobo). Note that in EFD, at the starting and end condition the speed of the vessel is already the mean constant required speed, but the speeds of the two vessels can change as they pass by each other. In CFD however, ship speeds are not subject to change during the simulations. Sharp speed peaks (0.5 knots) are observed at around X=40 and X=160, which are most likely a measurement glitch in the carriage velocity measuring system ([8]) since it is present in most overtaking runs.

In Figure 8, heave and pitch motions for Hope and Bobo are compared with EFD during the overtaking event. Overall, CFD results agree very well with EFD. Hope motions follow the EFD closer than Bobo, probably because Bobo is smaller and more sensitive to ship-ship effects. For motions, Bobo is more influenced by Hope than vice versa. The important effects include decreased heave motions for Bobo as Hope approaches its amidships, caused by the bow and stern waves of the Hope, and decreased pitch motions after Hope passes Bobo's amidships. SM-Figures 31 and 32 show forces and moments compared to EFD. EFD results are very noisy and validation is not easy. However, it can be seen that the general trends agree well, especially for X and N for Bobo. The Y for Bobo increases significantly during the overtaking and decreases afterwards. A similar trend is observed for Hope for a short duration. K and N show similar trends as Y force: they increase during overtaking and decrease when the overtaking ship has passed, with a less dominating increase and decrease for Hope compared to Bobo. Snapshots of overtaking simulation are shown in SM-Figure 33.

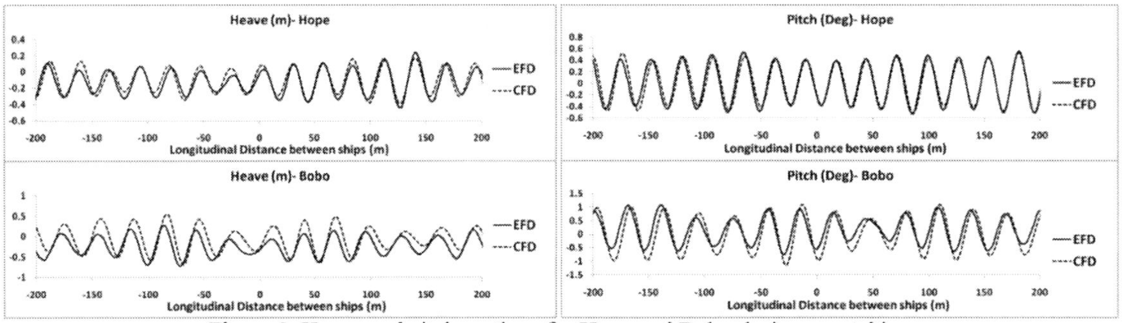

Figure 8: Heave and pitch motions for Hope and Bobo during overtaking

6. CONCLUSIONS AND FUTURE WORK

Generally good results are achieved for all simulations. For calm water, the average error is 21%D. The effects of second ship are explained through comparisons of hull pressure distributions in single-ship and ship-ship computations. Sinkage is increased for both ships due to reduced pressure at the bottoms. High pressure region at the aft of Hope causes increased trim (bow-down) and decreased resistance, while the high pressure region in the fore section of Bobo causes decreased trim and increased resistance. Average effects are 56% for motions and 7% for resistance. For waves simulations, the average error is 14%DR for 0th and 1st harmonics, 6%DR for 1st/linear terms and 22%DR for 2nd/nonlinear terms. For previous single-ship G2010 simulations, average error values were 13%DR for 1st/linear and 38%DR for 2nd/nonlinear terms. For cases with sheltering, the sheltered ship has smaller error value. Overall, the error values are in the same range for all heading, spacing, and configuration conditions. The error values for 1st harmonics are compared for different simulation tools. URANS achieves smallest error of 10%DR, followed by Rankine panel method with 15%DR, Green's function with 32%DR, and dynamic simulator with 38%DR. The interaction effects are studied for the RAOs. The most important physics include resonance where motions and consequently forces and moments increase near the ships' natural frequencies, and amplified forces and moments due to complex wave systems in the gap between ships at high frequencies where motions are nearly zero. The first phenomenon is significant in head waves and is diminished when sheltering is significant, while the second phenomenon is present only for oblique waves. The overall interaction effects are 105% for sheltering due to configuration, 43% for spacing in head waves (180 deg), and 23% for spacing in 120 deg waves.

Compared to single-ship computations (infinite spacing), the interaction effects in head waves are 14% for larger spacing and 18% for smaller spacing condition. For the overtaking simulation, the general trends agree well with data. The interaction effects are most significant during overtaking and include decreased motions and increased *Y*, *K*, and *N*. These effects are more significant for Bobo than Hope.
Future work will complete the designed simulations (red in Table 2) to further investigate the interaction effects. The calm water simulation will validate the URANS for very small *Fr* and small spacing. The single-ship simulations in 120 deg waves will allow studying the effects of heading and effects of sheltering. The overtaking simulation will include calm water and different spacing and heading in waves for a better understanding of spacing and sheltering effects in overtaking maneuvers.

7. ACKNOWLEDGMENTS

This work was sponsored by the Office of Naval Research under Grant N00014-01-1-0073, administered by Dr. Patrick Purtell. Computations were performed at the NAVY DoD Supercomputing Resource Center.

8. REFERENCES

1. DAND, I.W., 'Some Measurements In Interaction Between Ship Models Passing On Parallel Courses', *NMI R108*, 1981.

2. KAPLAN, P., SANKARANARAYANAN, K., 'Hydrodynamic Interaction Of Ships In Shallow Channels, Including Effects Of Asymmetry', *International Conference on Ship Manoeuvrability*, London, United Kingdom, 1987.

3. BRIX, J. (Editor), 'Manoeuvring Technical Manual', *Seehafen Verlag GmbH*, Hamburg, Germany, 1993.

4. VANTORRE, M., LAFORCE, E., VERZHBITSKAYA, E., 'Model Test Based Formulations Of Ship-Ship Interaction Forces For Simulation Purposes', *IMSF – 28th Annual General Meeting*, Genova, Italy, 2001.

5. DE KONING GANS, H.J., HUIJSMANS, R.H.M., PINKSTER, J.A., 'A Method to Predict Forces on Passing Ships under Drift', *9th International Conference on Numerical Ship Hydrodynamics*, Ann Arbor, Michigan, August 5-8, 2007.

6. ZHANG, S., WEEMS, K.M., LIN, W.M., 'Numerical Simulation and Validation of Ship-Ship Interactions in Waves', *9th International Conference in Numerical Ship Hydrodynamics*, Ann Arbor, Michigan, August 5-8, 2007.

7. CHEN, H.C., LIN, W.M., LIUT, D.A., HWANG, W.Y., 'An Advanced Flow Computation Method For Ship-Ship Dynamic Interactions In Shallow Water And Restricted Waterway', *MARSIM'03*, Kanazawa, Japan, 2003.

8. VAN'T VEER, A.P., VAN ENGELENBURG R., 'Multivessel Tests TAKR 300 (Bob Hope) & T-AK 3008 (2nd Lt. John P. Bobo)', *Computer Sciences Corporation/Advanced Marine Center*, Report No. 19037-3-SMB, 2006.

9. SILVER, A.L., HUGHES, M.J., CONARD, R.E., LEE, S.S., KLAMO, J.T., PARK, J.T., 'Evaluation of Multi-Vessel Ship Motion Prediction Codes', *Naval Surface Warfare Center Carderock Division*, Report NSWCCD-50-TR-2008/070, 2008.

10. MCIVER, P., NEWMAN, J.N., 'Trapping Structures In The Three-Dimensional Water-Wave Problem', *Journal of Fluid Mechanics*, No. 484, pp 283-301, 2003.

11. CHEN, G.R., FANG, M.C., 'Hydrodynamic Interactions Between Two Ships Advancing In Waves', *Journal of Ocean Engineering*, No. 28, pp 1053-1078, 2001.

12. GUI., L., LONGO, L., METCALF, B., SHAO, J., STERN, F., 'Forces, Moment, And Wave Pattern For Naval Combatant In Regular Head Waves—Part 2: Measurement Results And Discussions', *Experiments in Fluids*, 32, 27–36, 2002.

13. JOURNEE, J.M.J., 'Experiments And Calculations On Four Wigley Hull Forms', *Delft University of Technology*, Ship Hydromechanics Lab, Report No. 909, 1992.

14. GERRITSMA, J., BEUKELMAN, W., 'Analysis Of The Modified Strip Theory For The Calculation Of Ship Motions And Wave Bending Moments', *International Shipbuilding Progress*, 14, 1967.

15. IRVINE, M., LONGO, J., STERN, F., 'Pitch And Heave Tests And Uncertainty Assessment For A Surface Combatant In Regular Head Waves', *Journal Ship Research*, Vol. 52, No. 2, pp. 146-163, 2008.

16. LEWIS, E.V., 'Principles of Naval Architecture', *The Society of Naval Architects and Marine Engineers*, New York, 1989.

17. FONSECA, N., SOARES, C.G., 'Experimental Investigation Of The Nonlinear Effects On The Vertical Motions And Loads Of A Containership In Regular Waves, *Journal Of Ship Research*, 48, 118–147, 2004.

18. SIMONSEN, C.D., OTZEN, J.F., STERN, F., 'EFD and CFD for KCS Heaving And Pitching In Regular

Head Waves', *27th Symposium on Naval Hydrodynamics*, Seoul, Korea, 2008.

19. CASTIGLIONE, T., STERN, F., KANDASAMY, M., BOVA, S., 'Unsteady RANS Simulations For A Catamaran Advancing In Regular Waves', *10th International Conference on Fast Sea Transportation*, Athens, Greece, 2009.

20. CARRICA, P.M., WILSON, R.V., NOACK, R.W., STERN, F., 'Ship Motions Using Single-Phase Level Set With Dynamic Overset Grids', *Computers & Fluids*, Vol. 36, No. 9, pp. 1415-1433, 2007.

9. AUTHORS BIOGRAPHY

Maysam Mousaviraad holds the current position of postdoctoral research associate at IIHR-Hydroscience & Engineering at The University of Iowa where he is responsible for developing and using CFD methods for high-fidelity ship hydrodynamics computations. He received his PhD in Mechanical Engineering from The University of Iowa in May 2010 with his thesis entitled "CFD prediction of ship response to extreme winds and/or waves".

Hamid Sadat Hosseini is assistant research scientist at IIHR-Hydroscience & Engineering at the University of Iowa. His research includes 6DOF viscous ship hydrodynamics, computational fluid dynamics, and ship dynamic stability.

Pablo Carrica holds the position of associate professor of mechanical Engineering at The University of Iowa and research engineer at IIHR-Hydroscience & Engineering. His research includes multiphase flow, computational fluid dynamics, boiling and heat Transfers.

Frederick Stern holds the position of professor of mechanical engineering at The University of Iowa and research engineer at IIHR-Hydroscience & Engineering. He is the leader of ship hydrodynamics group at IIHR. His research includes computational and experimental fluid dynamics, uncertainty analysis, and CFD-based optimization.

DYNAMIC INTERACTION BETWEEN TOW AND TOWED SHIPS IN HEAD SEAS

Y. Nakayama and H. Yasukawa, Hiroshima University, Japan
J. Watanabe, Daihatsu Motor Co., Ltd. Osaka, Japan

SUMMARY

Concern for safe towing in waves, it is important to capture the interaction effects between tow and towed ship motions in waves. This paper discusses the vertical motion characteristics (including surge motion) of tow and towed ships and the dynamic interaction between those ships in regular head waves. A practical calculation method which can calculate the 3-DOF motions (surge, heave and pitch) of both tow and towed ships is presented. Using this method, the motion amplitude of tow and towed ships in various wave lengths are predicted. Moreover, the towing line length is selected variously. As a result, dynamic interaction components affecting each other are detected. Also, it is recognized that the magnitude of towing line length is deeply related to the ship's surge motion and the steady towing line tension affects the restoring force with respect to the ship's heave and pitch motions.

NOMENCLATURE

F_{Tx}	Surge force due to tension
F_{Tz}	Heave force due to tension
F_x	Surge force due to external force except tension
F_z	Heave force due to external force except tension
ℓ_t	Length between towing point and center of gravity of tow ship
ℓ_1	Length of towing line
ℓ_2	Length between towing point and center of gravity of towed ship
M_{Ty}	Pitch moment due to tension
M_y	Pitch moment due to external force except tension
R_w	Added resistance in waves
R_0	Resistance in still water
T	Tension acting in an axial direction of towing line
u_0	Averaged surge velocity in steady towing motion
$\|X_{tA}\|$	Surge motion amplitude of tow ship
$\|X_{2A}\|$	Surge motion amplitude of towed ship
$\|Z_{tA}\|$	Heave motion amplitude of tow ship
$\|Z_{2A}\|$	Heave motion amplitude of towed ship
α	Angle which denotes location of towing point
ζ_A	Wave amplitude
$\|\theta_{tA}\|$	Pitch motion amplitude of tow ship
θ_1	Inclination angle of towing line
$\|\theta_{2A}\|$	Pitch motion amplitude of towed ship
λ	Wave length
ω	Encounter frequency between waves and ships in steady towing motion
suffix t	Tow ship terms
suffix 2	Towed ship terms

1. INTRODUCTION

Towing with a towing line is widely employed for marine transportation and has been treated by many researchers as the problem to be considered. However, almost all studies discuss its horizontal motions on the water surface. And so, the vertical motions of tow and towed ships in waves are not very investigated before.

In this paper, the vertical motion characteristics of tow and towed ships and the dynamic interaction between those ships in regular head waves are discussed. Following assumptions are employed in formulation:

- Motions of tow and towed ships are expressed as 2D motions in vertical plane.
- Towing line is assumed to be a rigid and weightless single truss.
- Regular and small-amplitude waves are employed for the analysis.

Meanwhile, for easy understanding, the basic motion equations of tow and towed ships in waves are separated into steady and unsteady components, and also the unsteady components are linearized. Using this motion calculation method, the motions and interaction effects of a tug and a barge in various wave lengths are investigated. Moreover, the effect of the towing line length on those ship's motions is considered.

2. MOTION EQUATIONS

2.1 COORDINATE SYSTEMS

The coordinate systems used in formulation are shown in Fig.1. Tow and towed ships are assumed to be a rigid body and towing line to be a single truss, like the study of slewing motion in still water by Shigehiro[1].

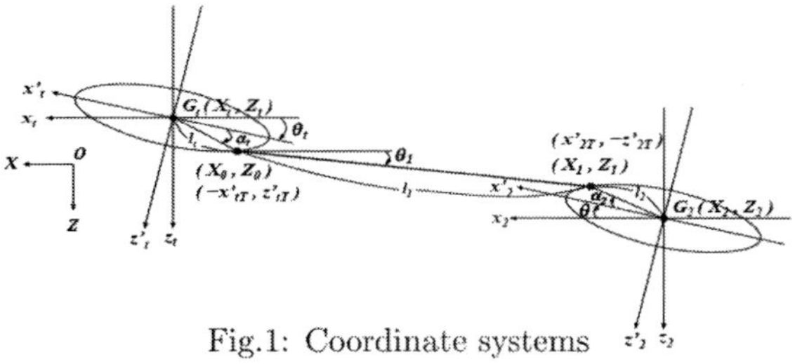

Fig.1: Coordinate systems

The coordinate systems fixed in the space $O-XZ$ and fixed to the ships at averaged position of wave-induced motion $G_t - x_t z_t$ and $G_2 - x_2 z_2$ which move with the towing motion in still water are employed. The towing point of tow ship side is $(-x'_{tT}, z'_{tT})$ of the coordinate system $G_t - x'_t z'_t$ which is rotated θ_t about G_t. Tow ship and towing line are connected by hinge connection. The length between the towing point and the center of gravity of tow ship is defined as $\ell_t = \sqrt{x'^{2}_{tT} + z'^{2}_{tT}}$ and the angle which denotes the location of the towing point is expressed as $\alpha_t = \tan^{-1}(z'_{tT}/-x'_{tT})$. Likewise, The towing point of towed ship side is $(x'_{2T}, -z'_{2T})$. Then, $\ell_2 = \sqrt{x'^{2}_{2T} + z'^{2}_{2T}}$ and $\alpha_2 = \tan^{-1}(-z'_{2T}/x'_{2T})$ are also defined. ℓ_1 means the length of towing line. In addition, considering the restraint condition of towing line, the center of gravity of towed ship (X_2, Z_2) is written as follows:

$$X_2 = X_t - \ell_t \cos(\theta_t + \alpha_t) - \sum_{j=1}^{2} \ell_j \cos\theta_j \tag{1}$$

$$Z_2 = Z_t + \ell_t \sin(\theta_t + \alpha_t) + \sum_{j=1}^{2} \ell_j \sin\theta_j \tag{2}$$

where $\theta_2 = \theta + \alpha_2$ is defined.

2.2 MOTION EQUATION OF TOW SHIP

In ship fixed coordinate system $G_t - x_t z_t$, the motion equations of tow ship in waves are expressed as:

$$(m^t + A^t_{11})\ddot{X}_t = F^t_x + F^t_{Tx} \tag{3}$$

$$(m^t + A^t_{33})\ddot{Z}_t + A^t_{35}\ddot{\theta}_t = F^t_z + F^t_{Tz} \tag{4}$$

$$(I^t_{yy} + A^t_{55})\ddot{\theta}_t + A^t_{53}\ddot{Z}_t = M^t_y + M^t_{Ty} \tag{5}$$

where m^t means the ship's mass, I^t_{yy} the pitch moment of inertia. A^t_{11}, A^t_{33} and A^t_{55} denote added masses for x-, z-direction and the added pitch moment of inertia, respectively. A^t_{35} and A^t_{53} denote the added mass coefficients in a broad sense in the coupled motion of heave and pitch. F^t_{Tx}, F^t_{Tz} and M^t_{Ty} mean the surge force, heave force and pitch moment acting on the ship due to tension caused by towing, respectively. Meanwhile, F^t_x, F^t_z and M^t_y mean the surge force, heave force and pitch moment acting on the ship due to other external forces respectively.

When the tension acting in an axial direction of towing line is denoted by T, the external forces due to tension F^t_{Tx}, F^t_{Tz} and M^t_{Ty} are written as:

$$\left.\begin{array}{lcl} F^t_{Tx} & = & -T\cos\theta_1 \\ F^t_{Tz} & = & T\sin\theta_1 \\ M^t_{Ty} & = & -T\ell_t \sin(\theta_t + \alpha_t - \theta_1) \end{array}\right\} \tag{6}$$

Also, the external forces due to others F^t_x, F^t_z and M^t_y are expressed as:

$$\left.\begin{array}{lcl} F^t_x & = & E^t_1 - R^t_0 - R^t_w + X_P \\ F^t_z & = & E^t_3 - B^t_{33}\dot{Z}_t - C^t_{33}Z_t - B^t_{35}\dot{\theta}_t - C^t_{35}\theta_t \\ M^t_y & = & E^t_5 - B^t_{53}\dot{Z}_t - C^t_{53}Z_t - B^t_{55}\dot{\theta}_t - C^t_{55}\theta_t \end{array}\right\} \tag{7}$$

where R^t_0, R^t_w and X_P denote the resistance in still water, the added resistance in waves and the propeller thrust, respectively. B^t, C^t and E^t mean the wave damping coefficient, the restoring coefficient and the wave exiting force respectively. As the present study is focused on the straight moving in regular head waves in vertical plane, rudder forces are not needed for the external forces acting on tow ship.

2.3 MOTION EQUATION OF TOWED SHIP

In ship fixed coordinate system $G_2 - x_2 z_2$, the motion equations of towed ship in waves are expressed as:

$$(m + A_{11})\ddot{X}_2 = F_x + F_{Tx} \tag{8}$$

$$(m + A_{33})\ddot{Z}_2 + A_{35}\ddot{\theta}_2 = F_z + F_{Tz} \tag{9}$$

$$(I_{yy} + A_{55})\ddot{\theta}_2 + A_{53}\ddot{Z}_2 = M_y + M_{Ty} \tag{10}$$

The definition of symbols such as m, A_{11}, F_x, etc. is the same as tow ship equation. The external forces due to tension F_{Tx}, F_{Tz} and M_{Ty} are expressed as:

$$\left.\begin{array}{lcl} F_{Tx} & = & T\cos\theta_1 \\ F_{Tz} & = & -T\sin\theta_1 \\ M_{Ty} & = & T\ell_2 \sin(\theta_1 - \theta_2) \end{array}\right\} \tag{11}$$

Also, the external forces due to others F_x, F_z and M_y are expressed as:

$$\left.\begin{array}{l} F_x = E_1 - R_0 - R_w \\ F_z = E_3 - B_{33}\dot{Z}_2 - C_{33}Z_2 - B_{35}\dot{\theta}_2 - C_{35}(\theta_2 - \alpha_2) \\ M_y = E_5 - B_{53}\dot{Z}_2 - C_{53}Z_2 - B_{55}\dot{\theta}_2 - C_{55}(\theta_2 - \alpha_2) \end{array}\right\} \tag{12}$$

where towed ship is assumed not to have the propeller and rudder and to be impossible to move by itself. The definition of symbols is the same as tow ship equation.

3. LINEARIZATION OF MOTION EQUATIONS

The motion equations mentioned above are too complicated to analyze the motions of tow and towed ships in waves. Therefore, to grasp those motion characteristics in waves, the simplified motion equations are derived by separating the motion equations into 2 groups where one is for steady towing motion and the other is for unsteady wave-induced motion, and also by linearizing the unsteady motion components.

3.1 ASSUMPTION OF LINEARIZATION

X_t, Z_t, θ_t, θ_1, θ_2 and T, which are unknown in the motion equations of tow and towed ships, are assumed to be expressed as sum of steady and unsteady components:

$$\left.\begin{array}{lcl} X_t & = & X_{0t} + \Delta X_t,\ Z_t = Z_{0t} + \Delta Z_t \\ \theta_t & = & \theta_{0t} + \Delta\theta_t,\quad \theta_1 = \theta_{01} + \Delta\theta_1 \\ \theta_2 & = & \theta_{02} + \Delta\theta_2,\ T = T_0 + \Delta T \end{array}\right\} \tag{13}$$

Here, the unsteady components ΔX_t, ΔZ_t, $\Delta\theta_t$, $\Delta\theta_1$, $\Delta\theta_2$ and ΔT are assumed to be sufficiently small, and so those square or more high-order terms are ignored. The approximation $\sin\Delta\theta \cong \Delta\theta$, $\cos\Delta\theta \cong 1.0$ are also employed. In addition, to simplify the equations, the averaged surge velocity of tow ship in steady towing motion is assumed to be given as u_0.

3.2 LINEARIZED MOTION EQUATIONS

Eqs.(3), (4) and (5), which are the motion equations of tow ship, are separated and linearized on the basis of the assumption mentioned above. Separating the equations into steady and unsteady motion terms, the following equations are obtained as,
Steady Components:

$$C_{33}^t Z_{0t} + C_{35}^t \theta_{0t} - T_0 \sin\theta_{01} = 0 \tag{14}$$

$$C_{53}^t Z_{0t} + C_{55}^t \theta_{0t} + T_0 \ell_t \sin(\theta_{0t} + \alpha_t - \theta_{01}) = 0 \tag{15}$$

Unsteady Components:

$$\begin{aligned}(m^t + A_{11}^t)\Delta\ddot{X}_t \\ = E_1^t + T_0 \sin\theta_{01}\Delta\theta_1 - \cos\theta_{01}\Delta T\end{aligned} \tag{16}$$

$$\begin{aligned}&(m^t + A_{33}^t)\Delta\ddot{Z}_t + A_{35}^t\Delta\ddot{\theta}_t \\ &+B_{33}^t\Delta\dot{Z}_t + B_{35}^t\Delta\dot{\theta}_t + C_{33}^t\Delta Z_t + C_{35}^t\Delta\theta_t \\ &\quad = E_3^t + T_0\cos\theta_{01}\Delta\theta_1 + \sin\theta_{01}\Delta T\end{aligned} \tag{17}$$

$$\begin{aligned}&A_{53}^t\Delta\ddot{Z}_t + (I_{yy}^t + A_{55}^t)\Delta\ddot{\theta}_t \\ &+B_{53}^t\Delta\dot{Z}_t + B_{55}^t\Delta\dot{\theta}_t + C_{53}^t\Delta Z_t + C_{55}^t\Delta\theta_t \\ &\quad = E_5^t - T_0\ell_t\cos(\theta_{0t} + \alpha_t - \theta_{01})(\Delta\theta_t - \Delta\theta_1) \\ &\quad -\ell_t\sin(\theta_{0t} + \alpha_t - \theta_{01})\Delta T\end{aligned} \tag{18}$$

The steady components coincide with the equations with respect to the attitude variation of tow ship caused by towing line tension. Here, the steady component of eq.(3) is omitted because the averaged surge velocity in steady towing motion is given as u_0. The unsteady components mean the wave-induced motions of tow ship. The equations show that not only the wave exciting forces but also the exciting force due to tension are acting on tow ship.

In the same way, eqs.(8), (9) and (10), which are the motion equations of towed ship, are separated and linearized. The following equations are obtained as,
Steady Components:

$$-R_0 - R_w + T_0\cos\theta_{01} = 0 \tag{19}$$

$$C_{33}Z_{02} + C_{35}(\theta_{02} - \alpha_2) + T_0\sin\theta_{01} = 0 \tag{20}$$

$$C_{53}Z_{02} + C_{55}(\theta_{02} - \alpha_2) - T_0\ell_2\sin(\theta_{01} - \theta_{02}) = 0 \tag{21}$$

Unsteady Components:

$$\begin{aligned}(m + A_{11})\Delta\ddot{X}_2 \\ = E_1 - T_0\sin\theta_{01}\Delta\theta_1 + \cos\theta_{01}\Delta T\end{aligned} \tag{22}$$

$$\begin{aligned}&(m + A_{33})\Delta\ddot{Z}_2 + A_{35}\Delta\ddot{\theta}_2 \\ &+B_{33}\Delta\dot{Z}_2 + B_{35}\Delta\dot{\theta}_2 + C_{33}\Delta Z_2 + C_{35}\Delta\theta_2 \\ &\quad = E_3 - T_0\cos\theta_{01}\Delta\theta_1 - \sin\theta_{01}\Delta T\end{aligned} \tag{23}$$

$$\begin{aligned}&A_{53}\Delta\ddot{Z}_2 + (I_{yy} + A_{55})\Delta\ddot{\theta}_2 \\ &+B_{53}\Delta\dot{Z}_2 + B_{55}\Delta\dot{\theta}_2 + C_{53}\Delta Z_2 + C_{55}\Delta\theta_2 \\ &\quad = E_5 + T_0\ell_2\cos(\theta_{01} - \theta_{02})(\Delta\theta_1 - \Delta\theta_2) \\ &\quad +\ell_2\sin(\theta_{01} - \theta_{02})\Delta T\end{aligned} \tag{24}$$

where

$$Z_{02} = Z_{0t} + \ell_t\sin(\theta_{0t} + \alpha_t) + \sum_{j=1}^{2}\ell_j\sin\theta_{0j} \tag{25}$$

$$\Delta X_2 = \Delta X_t + \ell_t\sin(\theta_{0t} + \alpha_t)\Delta\theta_t + \sum_{j=1}^{2}\ell_j\sin\theta_{0j}\Delta\theta_j \tag{26}$$

$$\Delta Z_2 = \Delta Z_t + \ell_t\cos(\theta_{0t} + \alpha_t)\Delta\theta_t + \sum_{j=1}^{2}\ell_j\cos\theta_{0j}\Delta\theta_j \tag{27}$$

The averaged surge velocity of towed ship in steady towing motion is the same as that of tow ship (u_0). Also, R_w means the averaged added resistance in waves when towed ship moves with steady velocity u_0 in regular head waves, and so is treated as the steady term.

By numerically solving eqs.(14), (15), (19), (20) and (21), steady attitude variation of ships and steady tension are obtained. These steady solutions are taken into account in solving eqs.(16), (17), (18), (22), (23) and (24), and then, unsteady motions of ships and unsteady tension are obtained. Additionally, a frequency response method is employed for getting the unsteady solutions. The unsteady solutions $\Delta X_t, \Delta Z_t, \Delta\theta_t, \Delta\theta_1, \Delta\theta_2$ and ΔT are assumed to have the same frequency as wave exciting forces $E^{(t)}$ and to be expressed as follows:

$$\left.\begin{aligned}\Delta X_t &= Re\left[X_{tA}e^{i\omega t}\right], \quad \Delta Z_t = Re\left[Z_{tA}e^{i\omega t}\right] \\ \Delta\theta_t &= Re\left[\theta_{tA}e^{i\omega t}\right], \quad \Delta\theta_{1,2} = Re\left[\theta_{1,2A}e^{i\omega t}\right] \\ \Delta T &= Re\left[T_A e^{i\omega t}\right], \quad E^{(t)} = Re\left[E_A^{(t)}e^{i\omega t}\right]\end{aligned}\right\} \tag{28}$$

where ω means the encounter frequency between waves and ships in steady towing motion. X_{tA}, Z_{tA}, θ_{tA}, $\theta_{1,2A}$, T_A and $E_A^{(t)}$ are the complex numbers which denote the amplitude and phase of motions and forces.

3.3 CONSIDERATION

Dynamic interaction between tow and towed ships is discussed by further simplification of the equations mentioned above.

Here, towing line between tow and towed ships is assumed to be parallel with still water. Then, steady attitude variation of ships are considered very small, and so $Z_{0t} = \theta_{0t} = \theta_{01} = 0$ and $\theta_{02} = \alpha_2$ are also employed for this consideration. In addition, α_t and α_2 which are the angles expressing the location of the towing point are assumed to be sufficiently small. From eqs.(19) and (22), the steady and unsteady component of tension (T_0, ΔT) are expressed as:

$$T_0 = R_0 + R_w \tag{29}$$

$$\Delta T = (m + A_{11})(\Delta\ddot{X}_t + \ell_t\alpha_t\Delta\ddot{\theta}_t + \ell_2\alpha_2\Delta\ddot{\theta}_2) - E_1 \tag{30}$$

The steady component of tension is given as sum of the resistance in still water and the added resistance in waves of towed ship and always takes positive value. The unsteady component of tension is expressed as difference of product of the apparent mass of towed ship and the surge motion acceleration of tow ship and the surge wave exciting force acting on towed ship, mainly. Meanwhile, from eq.(27), the unsteady component of inclination angle of towing line $\Delta\theta_1$ is expressed as:

$$\Delta\theta_1 = -(\Delta Z_t + \ell_t\Delta\theta_t - \Delta Z_2 + \ell_2\Delta\theta_2)/\ell_1 \quad (31)$$

The unsteady component of the inclination angle of towing line consists of the heave and pitch motions of respective tow and towed ships, and is inversely proportional to the towing line length. Using these equations and assumptions, the unsteady components of the motion equations of tow ship are expressed again as:

$$\begin{aligned}&(m^t + A_{11}^t + m + A_{11})\Delta\ddot{X}_t \\ &+(m + A_{11})(\ell_t\alpha_t\Delta\ddot{\theta}_t + \ell_2\alpha_2\Delta\ddot{\theta}_2) \\ &\quad = E_1^t + E_1 \end{aligned} \quad (32)$$

$$\begin{aligned}&(m^t + A_{33}^t)\Delta\ddot{Z}_t + A_{35}^t\Delta\ddot{\theta}_t + B_{33}^t\Delta\dot{Z}_t + B_{35}^t\Delta\dot{\theta}_t \\ &+(C_{33}^t + T_0/\ell_1)\Delta Z_t + (C_{35}^t + T_0\ell_t/\ell_1)\Delta\theta_t \\ &-(T_0/\ell_1)(\Delta Z_2 - \ell_2\Delta\theta_2) \\ &\quad = E_3^t \end{aligned} \quad (33)$$

$$\begin{aligned}&A_{53}^t\Delta\ddot{Z}_t + (I_{yy}^t + A_{55}^t)\Delta\ddot{\theta}_t + B_{53}^t\Delta\dot{Z}_t + B_{55}^t\Delta\dot{\theta}_t \\ &+(C_{53}^t + T_0\ell_t/\ell_1)\Delta Z_t \\ &+\{C_{55}^t + T_0\ell_t(1 + \ell_t/\ell_1)\}\Delta\theta_t \\ &+(m + A_{11})\ell_t\alpha_t\Delta\ddot{X}_t - (T_0\ell_t/\ell_1)(\Delta Z_2 - \ell_2\Delta\theta_2) \\ &\quad = E_5^t + E_1\ell_t\alpha_t \end{aligned} \quad (34)$$

Here, ΔX_t, ΔZ_t and $\Delta\theta_t$ mean the surge motion, heave motion and pitch motion of tow ship, respectively. ΔZ_2 and $\Delta\theta_2$ denote the heave and pitch motions of towed ship. The surge motion equation of tow ship includes the apparent mass of towed ship and the surge wave exciting force acting on towed ship. Also, the coupling terms with respect to the pitch motions of respective tow and towed ships exist in the equation. The restoring force terms in the heave motion equation of tow ship include the steady tension T_0. T_0 is always positive and contributes to the increase of the restoring coefficients of the heave and pitch motions. The coupling terms with respect to the heave motion and pitch motion of towed ships also exist in the heave equation. Similarly, the restoring force terms in the pitch motion equation of tow ship include T_0. T_0 always contributes to the increase of the restoring coefficients. In addition, there exist the coupling terms with respect to the surge motion of tow ship, the heave motion and pitch motion of towed ships and the surge wave exciting force acting on towed ship in the pitch equation.

In the same manner, the unsteady components of the motion equations of towed ship are expressed as:

$$\begin{aligned}&(m + A_{33})\Delta\ddot{Z}_2 + A_{35}\Delta\ddot{\theta}_2 + B_{33}\Delta\dot{Z}_2 + B_{35}\Delta\dot{\theta}_2 \\ &+(C_{33} + T_0/\ell_1)\Delta Z_2 + (C_{35} - T_0\ell_2/\ell_1)\Delta\theta_2 \\ &-(T_0/\ell_1)(\Delta Z_t + \ell_t\Delta\theta_t) \\ &\quad = E_3 \end{aligned} \quad (35)$$

$$\begin{aligned}&A_{53}\Delta\ddot{Z}_2 + (I_{yy} + A_{55})\Delta\ddot{\theta}_2 + B_{53}\Delta\dot{Z}_2 + B_{55}\Delta\dot{\theta}_2 \\ &+(C_{53} - T_0\ell_2/\ell_1)\Delta Z_2 \\ &+\{C_{55} + T_0\ell_2(1 + \ell_2/\ell_1)\}\Delta\theta_2 \\ &+(m + A_{11})\ell_2\alpha_2\Delta\ddot{X}_t + (T_0\ell_2/\ell_1)(\Delta Z_t + \ell_t\Delta\theta_t) \\ &\quad = E_5 + E_1\ell_2\alpha_2 \end{aligned} \quad (36)$$

Because ΔT was given from eq.(22), the motion equations except surge are shown above. The restoring force terms in the heave and pitch motion equations of towed ship include T_0. T_0 contributes to the increase or decrease of the restoring coefficients of the heave and pitch motions. Also, the coupling terms with respect to the motions of tow ship exist in those equations. Additionally, there exist the surge wave exciting force acting on towed ship in the pitch motion equation.

By the above considerations, it is dynamically-revealed that the motions of tow and towed ships in waves are coupled each other and interact apparently. Also the motions seem to have a close relation to the steady tension T_0. T_0 is not proportional to wave height because it consists of a sum of the resistance in still water and the added resistance in waves of towed ship. Therefore, the motions of tow and towed ships in waves are also not proportional to wave height, and it shows different trend from ordinary ship's motion in waves.

In addition, when towing line length ℓ_1 is assumed to be infinitely-long, all the coupling terms with respect to another ship motions and T_0 in the heave motion equations of tow and towed ships can be ignored. And then, the equations completely agree with the heave motion equations when tow or towed ship moves by itself in waves. Meanwhile, for the pitch motion equations of tow and towed ships, T_0 remains in only the restoring force term of the pitch motion of themselves.

4. SIMULATION RESULTS

Using the motion equations mentioned above, the motions of tow and towed ships in regular head waves are investigated. The averaged surge velocity in steady towing motion u_0 is assumed to be 7knot ($F_n = 0.147$, on a towed ship length basis).

4.1 SUBJECT SHIPS

A twin-propeller twin-rudder tug and a generic barge are employed as tow and towed ships. Table1 shows principal particulars of the tug and barge. Here, the propeller thrust X_P and the added resistance in waves R_w^t of the tug are not represented because the averaged surge velocity in steady towing motion is given in the motion simulation. The barge doesn't have the propeller and rudder, and so can't move by itself.

Table 1: Principal dimensions of a tug and barge

	tug	barge
Length $L(m)$	40.00	60.96
Breadth $B(m)$	9.00	10.67
draft $d(m)$	2.20	2.74
Volume $V(m^3)$	494.70	1646.21
block coefficient C_b	0.62	0.92
Position of LCB $x_G(m)$	2.23	-1.04
L/B	4.44	5.71
B/d	4.09	3.89
gyradius k_{yy}/L	0.25	0.33

The towing point of the tug side x_{tT} is located at AP of the tug and z_{tT} is 2.20m above still water. On the other hand, The towing point of the barge side x_T is located at FP of the barge and z_T is 2.74m above still water. The towing line length ℓ_1 is selected variously from $0.5L$ to $10.0L$, where L means barge length from now on.

The wave amplitude ζ_A is set to be 5% of L and stay constant. The ratio between wave length and barge length (λ/L) is employed from 0.3 to 2.0 for the analysis. The added masses, wave damping coefficients and wave exciting forces are obtained by a numerical calculation of New Strip Method[2], which is almost the same as Salvesen-Tuck-Faltinsen method[3], however the effects of wake and propeller slip stream generated by the tug on the barge are not considered for simplification. Besides, the added resistance in waves of the barge R_w is determined by the model test result shown in Fig.2.

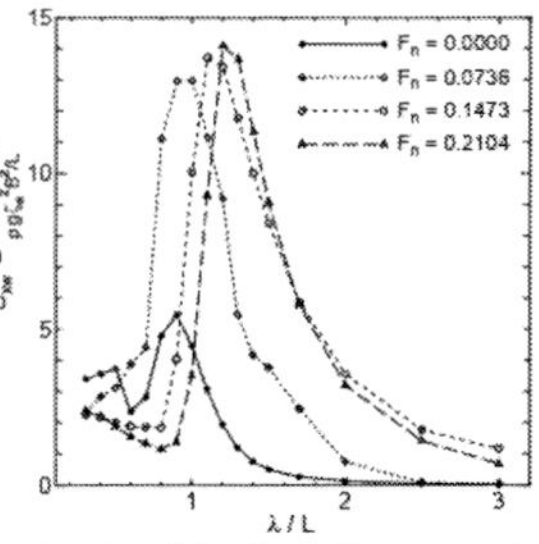

Fig.2: Coefficient of added resistance in waves for a barge

4.2 UNSTEADY SOLUTION (MOTION AMPLITUDE)

The tug and barge move periodically at the same frequency as the wave exciting forces. The center position of the periodic motion of the tug and barge coincides with the steady averaged position of wave-induced motion of those. Here, the calculated results of those motion amplitudes are shown below. Meanwhile, the results of the steady solution (steady averaged position of wave-induced motion) of the tug and barge are not indicated because the steady solutions have a little effect on the motion amplitude. Furthermore, as the towing line is assumed to be a single truss, the steady solutions don't change so much in various calculation conditions.

4.2 (a) Effect of Towing Line Length

In case of $\ell_1 = 0.5L, 2.0L$ and $10.0L$, the calculated results of the motion (surge, heave and pitch) amplitude are shown. In addition, the motion amplitude when the tug or barge moves by itself in the same wave condition (ζ_A=5%L) are also shown for comparison. The surge and heave motion amplitudes are non-dimensionalized by the wave amplitude ζ_A respectively, and the pitch motion amplitude is non-dimensionalized by the wave slope $k\zeta_A$. k means wave number and is obtained from ω^2/g, where g denotes acceleration of gravity.

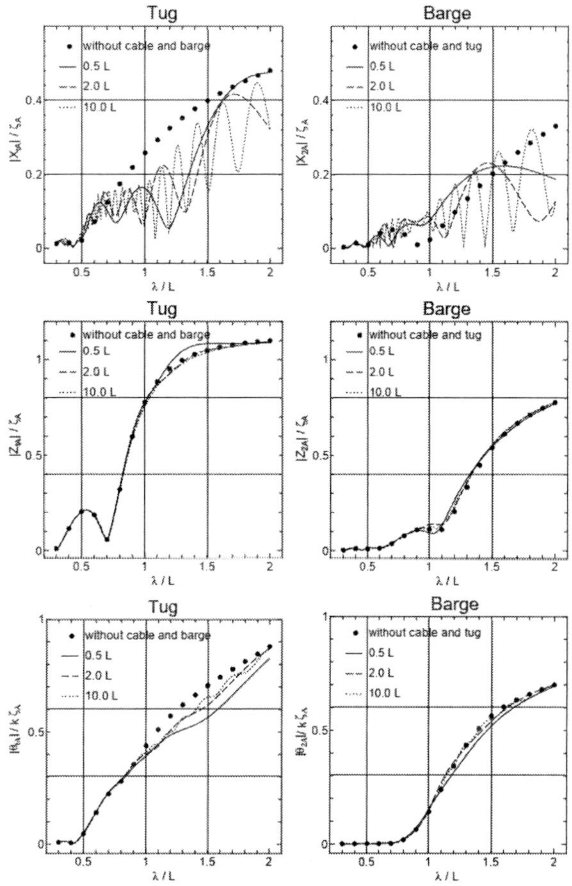

Fig.3: Effect of towing line length on motion amplitude of tug and barge (line modelled as a single truss)

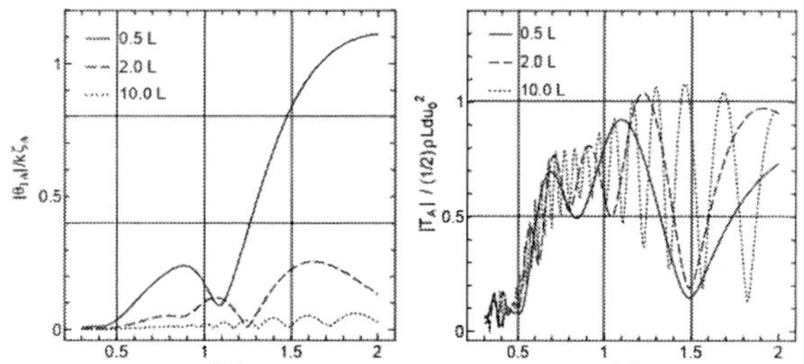

Fig.4: Effect of towing line length on amplitude of inclination angle of towing line and tension (line modelled as a single truss)

Fig.3 shows the calculated results of the motion amplitudes of the tug and barge versus λ/L. $|X_{tA}|$, $|X_{2A}|$, $|Z_{tA}|$, $|Z_{2A}|$, $|\theta_{tA}|$ and $|\theta_{2A}|$ are the absolute values of the complex number and mean the motion amplitudes of the tug and barge. Also, Fig.4 shows the calculated amplitudes of unsteady component of the inclination angle of towing line and the tension. $|T_A|$ is non-dimensionalized by $(1/2)\rho L d u_0^2$, where ρ means the water density and L, d and u_0 denote the barge length, draft and averaged surge velocity, respectively.

The surge motion amplitude of the tug in towing motion in waves has a different trend from that when the tug moves independently in waves, and increase with increase of λ/L showing some local maximal and minimal values. The reason for this tendency is that the surge motion of the tug depends mainly on both the surge wave exciting forces acting on the tug and barge as shown in eq.(32). When both the phases of the surge wave exciting forces agree, the surge motion of the tug is amplified, on the other hand, when there is a half wave length difference between one phase of that and another of that, the surge motion of the tug is damped. Moreover, when the length between the center of gravity of the tug and that of the barge is denoted by ℓ_{GG}, the conditions of appearance of local maximal and minimal values are written as follows:

$$\ell_{GG} = n\lambda \ \text{(max)} \ , \ \ell_{GG} = (n - 1/2)\lambda \ \text{(min)}$$
$$(n = 1, 2, \cdots, \quad \lambda = 0.3L, \cdots, 2.0L) \qquad (37)$$

where n is positive integer. If ℓ_{GG} is $1.0L$, the combinations of n and λ which meet the condition of appearance of maximal values are ($n = 1$, $\lambda = 1.0L$), ($n = 2$, $\lambda = 0.5L$) and ($n = 3$, $\lambda = 0.33L$). The longer ℓ_{GG} is, the more the combination of those increase. Also, ℓ_{GG} almost depends on the towing line length. Therefore, the number of times that local maximal and minimal values appear increase with increasing the towing line length.

The heave and pitch motion amplitudes of both the tug and barge roughly agree with those when the tug or barge moves by itself in waves, however, in case of $\ell_1 = 0.5L$, have a different trend from the result of other towing line length in the range over λ/L=0.9. As shown in eqs.(33)~(36), the reason for this result is that the effects of the restoring force terms and coupling terms with respect to T_0 become relatively-large when the towing line length is short. Also, magnitude of T_0 depends largely on the added resistance in waves of the barge shown in Fig.2 and so increase rapidly over λ/L=0.9. In addition, as discussed in previous consideration, the heave motion amplitudes of the tug and barge show more good agreement with those when the tug or barge moves by itself in waves with increasing towing line length. Also, in case of $\ell_1 = 10.0L$, some minute local maximal and minimal values appeare in the calculated results of the pitch motion amplitudes of both the tug and barge. This is because the coupling terms with respect to the surge motion acceleration of the tug and the surge wave exciting force acting on towed ship affects the pitch motion of the tug and barge as shown in eqs.(34) and (36).

For the amplitudes of the inclination angle of towing line and the tension, as shown in eqs.(30) and (31), it is confirmed that $|\theta_{1A}|$ depends largely on the towing line length ℓ_1, and $|T_A|$ has a close relation to the surge motion amplitude of the tug.

4.2 (b) Effect of Towing Line Model

Towing line was treated as a single truss so far, but, actual towing line has its own weight and slack in a catenary line. In addition, when the line is immersed under water plane, the fluid resistance acting on the line should also be considered. So, by applying 2D lumped mass method proposed by Yasukawa et al.[4] to the towing motion problem in waves, the calculated results of the motions of the tug and barge with high accuracy are shown. In the lumped mass method, towing line is modelled as what a finite number of lumped masses are connected by trusses. Therefore, the method can consider the effects of gravity, motion of towing line, and resistance acting on the line.

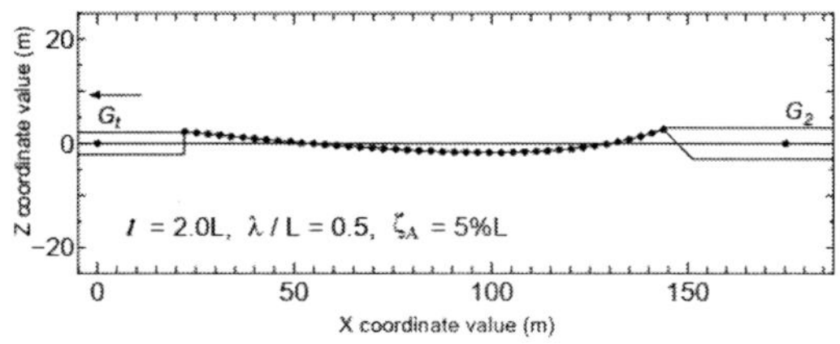

Fig.5: Steady towing line catenary obtained by lumped mass method

Fig.5 shows the calculated towing line catenary and positional relation between the tug and barge when the center of gravity of the tug G_t is assumed to be zero on X coordinate. Unlike a single truss method, lumped mass method can express the steady catenary, namely the averaged position of wave-induced motion of towing line. Next, The motion amplitude of the tug and barge calculated by a single truss method and lumped mass method respectively are compared below. For towing line length, $\ell = 2.0L$ is employed. And the other parameters are the same as the previous calculations.

Fig.6 shows the calculated results of the motion amplitude of the tug and barge. Also, Fig.7 shows the calculated amplitudes of unsteady component of the inclination angle of towing line and the tension. Although there is a little quantitative difference in the results of the surge motion amplitudes of the tug and barge and the pitch motion amplitude of the barge respectively, the calculated results of both methods mostly agree. It seems that $|T_A|$ is the main factor in the difference of the surge motion amplitudes of the tug and barge and also $|\theta_{1A}|$ is the main factor in the difference of the pitch motion amplitude of the barge. $|\theta_{1A}|$ and $|T_A|$ calculated by lumped mass method include the coupling terms with respect to the towing line motion and the resistance caused by its motion, and so the method can consider those towing line motion effects. However, for the understanding of the motion characteristics and interaction effects of tow and towed ships in waves, the present method (a single truss method) is useful enough.

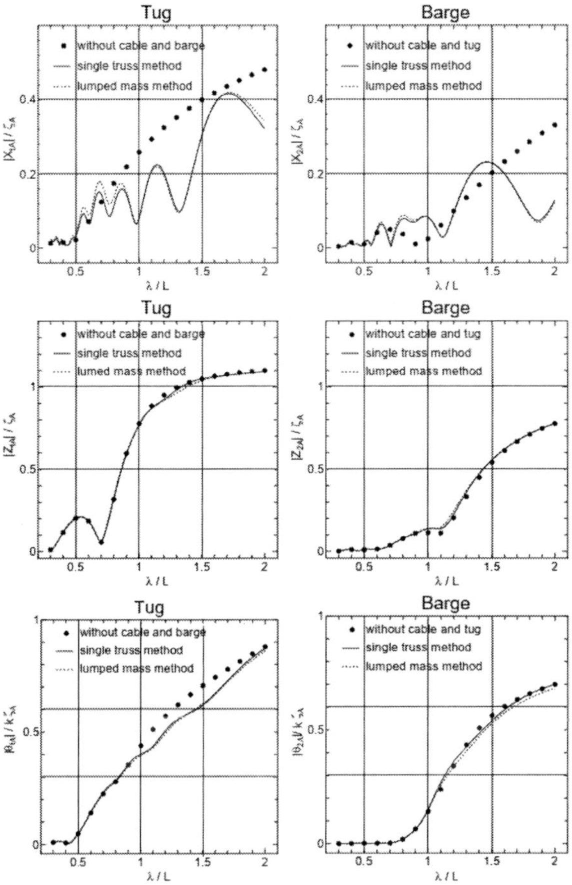

Fig.6: Effect of towing line model on motion amplitude of tug and barge

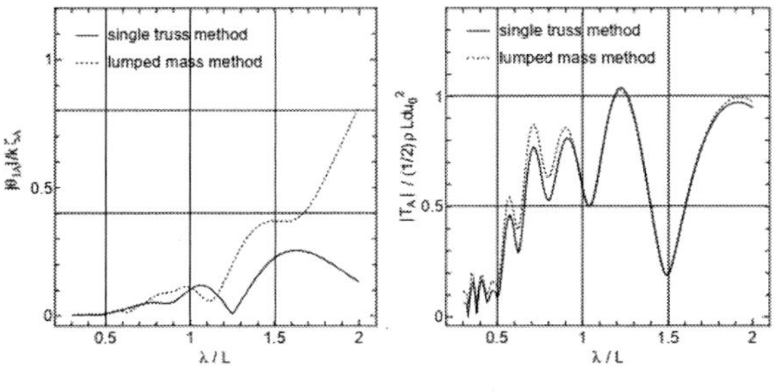

Fig.7: Effect of towing line model on amplitude of inclination angle of towing line and tension

5. CONCLUDING REMARKS

In this paper, for understanding of the motion characteristics of tow and towed ships in waves, a motion calculation method that basic motion equations of tow and towed ships in waves are separated into steady and unsteady components was proposed. Using the equations obtained by linearization of the unsteady components, the motion amplitudes of tow and towed ships in waves and interaction effects of those were investigated in various conditions. As the results, some knowledge were obtained as follows:

- The surge motion of tow ship is significantly affected by the apparent mass of towed ship and the surge wave exciting force acting on towed ship.
- Steady tension T_0 is given as sum of the resistance in still water and the added resistance in waves of towed ship and affects the restoring force terms of the heave and pitch motions of tow and towed ships respectively. Especially, when towing line length is short, the effects of T_0 become large.
- Effect of towing line length appears remarkably in the surge motions of tow and towed ships. This is because the phase of the surge wave exciting force acting on towed ship depends on the length between tow and towed ships.
- The heave motions of tow and towed ships show more good agreement with those when tow or towed ship moves by itself in waves with increasing towing line length.
- The pitch motions of tow and towed ships are affected by the coupling terms with respect to the surge motion acceleration of tow ship and the surge wave exciting force acting on towed ship, when towing line length is long.

As our future work, following additional investigations are needed:

- Accuracy validation of the present method by comparison with model test
- Consideration of wake and propeller slip stream of tow ship

6. REFERENCES

[1] Shigehiro, H., *A Mathematical Model for the Maneuvering Motions of Tow and Towed Vessels*, Journal of the Kansai Society of Naval Architects, Japan, Vol.230 (1998), pp.153-164 (in Japanese).

[2] Takaishi, T. and Kuroi, M., *A Practical Calculation Method of Ship Motions in Waves*, The 2nd Seakeeping Symposium, The Society of Naval Architects of Japan (1977), pp.109-133 (in Japanese)

[3] Salvesen, N., E. O. Tuck and O. Faltinsen, *Ship motions and sea load,* TSNAME, Vol.78 (1970), pp.250-287

[4] Yasukawa, H., Nakamura, N., Hirata, N. and Matsumoto, Y., *Simulations of Slewing Motion of a Towed Ship*, Journal of the Japan Society of Naval Architects and Ocean Engineers, Vol.4 (2006), pp.137-146 (in Japanese)

7. AUTHORS BIOGRAPHY

Yoshiyuki NAKAYAMA is doctor course student in Hiroshima University. He is responsible for a development of towing motion simulation method in waves.
Hironori YASUKAWA holds the current position of professor in department of vehicle and environmental systems, Hiroshima University. He is responsible for research on the ship dynamics and hydrodynamics. Now, he is maneuvering committee member of ITTC.
Junichi WATANABE is car engineer at Daihatsu Motor Co., Ltd. He was studying on the numerical calculations of towing motion in waves as the master course student in Hiroshima University.

DECISION SUPPORT SYSTEM IN SHIP-TO-SHIP OPERATIONS USING HIGH ACCURACY DGPS

E. Okuyama and H. Oda, Akishima Laboratories (Mitsui Zosen) Inc. JAPAN
E. Shimizu, Tokyo University of Marine Science and Technology, JAPAN
S. Sakakibara and S. Yamada, The Yokohama Rubber Co., Ltd. JAPAN

SUMMARY

Berthing and ship-to-ship operations are the most difficult and delicate phase for tanker and LNG carrier navigation. From the view point of the safety operation, to watch the vessel motions of such large carriers at the good accuracy real-time is very important, because it is increasingly required to avoid dangerous conditions for the large tanker and LNG carrier while berthing, To answer such demand, authors have developed Decision Support System (DSS) of ship-to-ship operation using high accuracy differential GPS, which is possible to measure the vessel position at the accuracy of sub meter and real-time. DSS works efficiently at going safety operation under the risky environment and the operators can control the ship by watching own and other vessels' motions watching mobile and handy monitor PDA. This paper presents the outstanding characteristics of DSS and shows the full-scale experiments and also the evaluation of DSS.

1. INTRODUCTION

It is important to watch vessel's motion with dangerous load e.g. large tanker or LNG carrier under berthing and ship-to-ship manoeuvre at the good accuracy real-time from a viewpoint of safety operation. The berthing and ship-to-ship manoeuvring are the most delicate phase of vessel's navigation. To answer such demand, Decision Support System (DSS) of ship-to-ship operation using high accuracy differential GPS is possible to measure the vessel position at the accuracy of sub meter and real-time. DSS consists of a few modes, e.g. approaching mode, berthing mode and ship-to-ship operation mode on monitoring. In ship-to-ship operations, instrumental measurements and human operation senses lead to safety ship-to-ship operations. This paper introduces the outstanding characteristics of DSS, full scale experiments and also the evaluation of DSS with ship manoeuvring simulation of ship-to-ship operations.

2. DGPS AUGMENTATION SYSTEM

2.1 StarFire-WADGPS [1]

The StarFire Wide Area DGPS (WADGPS) has been in progress from a set of regional DGPS networks which provided high accuracy service over independent continental areas to a robust unified global network offering unprecedented accuracy. This global network provides uniform sub decimetre real time service over almost the entire Earth. It is based on technology developed by the Jet Propulsion Laboratory (JPL) for the National Aeronautics and Space Administration (NASA).

2.2 MSAS-DGPS

MTSAT Satellite-based Augmentation System (MSAS) provides GPS augmentation information to aircraft through MTSAT (Multi-functional Transport Satellite) located at 36000km above the ground over the equator.

3. SHIP-TO-SHIP OPERATIONS SYSTEM

3.1 SYSTEM OVERVIEW

DSS has efficient functions of good performance to watch positions and motions of the ship utilizing the GPS real-time. Depending on the function of StarFire-WADGPS, it is possible to measure the position at the accuracy of sub meter. DSS consists of GPS receiver and fiber optical gyroscope and portable monitor unit [2]. The block diagram of DSS is shown in **Figure 1**.
DSS is installed in Shuttle while approaching to VLCC. Information of VLCC that navigates by an almost constant speed and course are gained from AIS (Automatic Identification System).

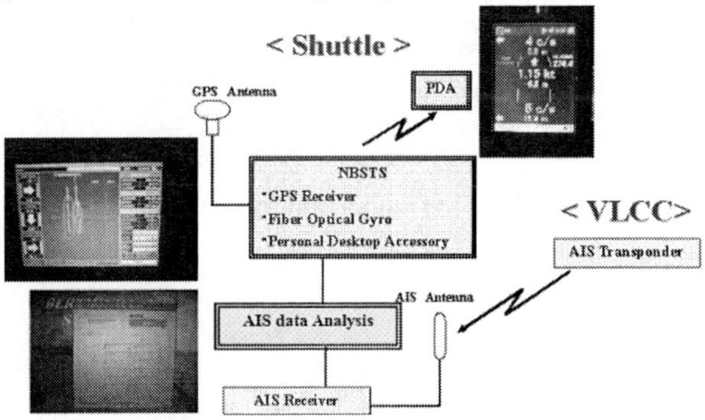

Figure 1 Block diagram of DSS

3.2 BASIC FUNCTION AND OPERATION

DSS with applied StarFire-WADGPS [3] is shown in **Figure 2**. The main functions of DSS as GPS reception, communication between ship-to-ship are described below.

DSS provides the following information;
a. Relative distance and speed between ship-to-ship,
b. Final docking position and recommendation route,
c. Dangerous manoeuvring zone and alarm,
d. Absolute speed and heading of each ships.

DSS performs;

a. High accuracy in sub meter order,
b. Ship position, speed can be monitored at real time,
c. The object vessels data can be recorded,
d. Data can be monitored with PDA at the wing.

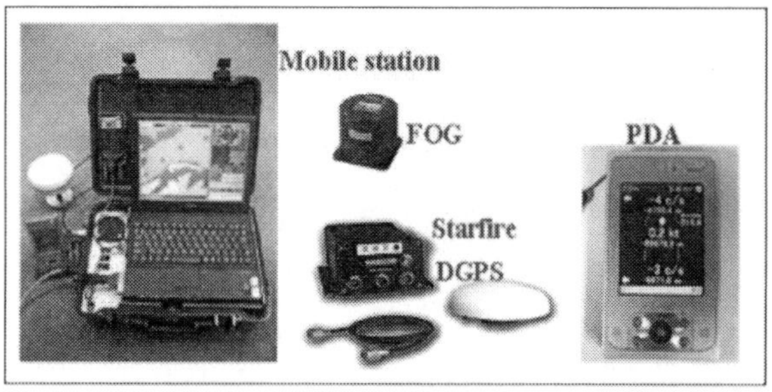

Figure 2 Components **of Decision Support System**

3.3 APPLICATIONS OF DSS

Applications of ship-to-ship operations for cargo transfer are supposed to increase in the future. Future development of Northwest Russian onshore and offshore fields increasingly need ship-to-ship transfer from a fleet of smaller ice strengthened vessels to standard vessels for transport of oil and gas products to customers in Europe and US. The images of the situation of ship-to-ship transfer are shown in **Figure 3**.

Figure 3 Applications of DSS

4. FULL-SCALE EXPERIMENTS

The full scale experiments were conducted with VLCC 'APOLLONIA' and shuttle tanker 'CHALLENGER'. The principal particulars of the actual ships are shown in **Table 1**. **Figure 4** shows the situation of ship-to-ship operations experiments and operation area at Mexico Gulf. [4]

Table 1 Principal particular of actual ships

Ship	APOLLONIA	CHALLENGER
Length	333 m	240 m
Breadth	58m	42m
DW	308,200 t	105,000 t

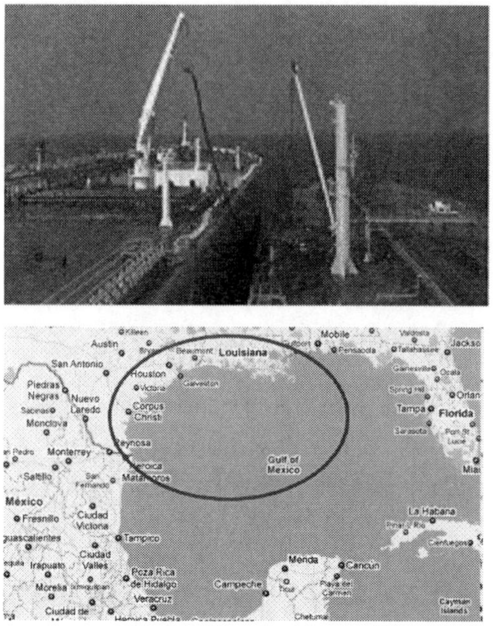

Figure 4 Ship-to-ship Operation Experiments in the Mexican Gulf ca. 60 miles from Freeport, USA.

The monitoring of DSS while the experiment (Docking) is shown in **Figure 5**. Upper figure shows positions of target and tanker which is original monitoring system of berthing support system. Lower figure shows image of docking mode with relative speed and distance between VLCC and shuttle tanker [5].

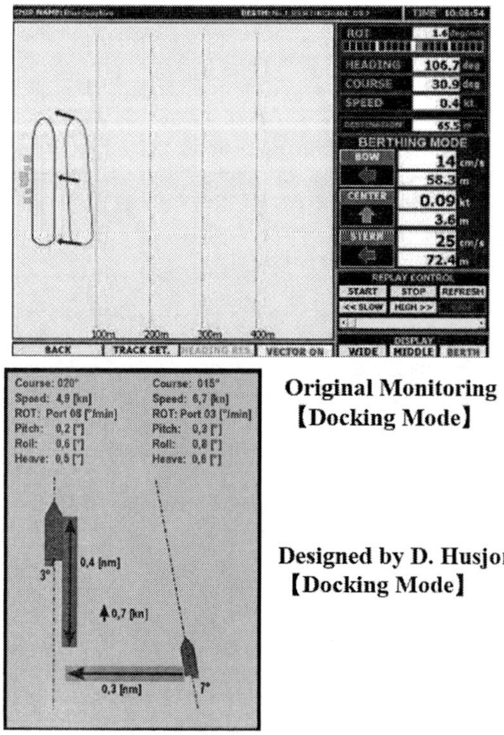

Original Monitoring
【Docking Mode】

Designed by D. Husjord
【Docking Mode】

Figure 5 Monitoring of DSS

The time histories of speed and distance are shown in **Figure 6**. Approach speed shows absolute speed of the Shuttle tanker in upper side of **Figure 6**. This result means that the steady approach speed is about 5.4 (kts). Docking speed and distance in centre and lower of **Figure 6** shows relative speed and distance of bow side. The result of docking steady relative speed is about 10(cm/sec.). The result of steady relative distance depends on the position where GPS antenna is set up and accuracy of the input value of fender size for the cause of a negative distance. It does not indicate a collision condition, meaning the safe distance condition.

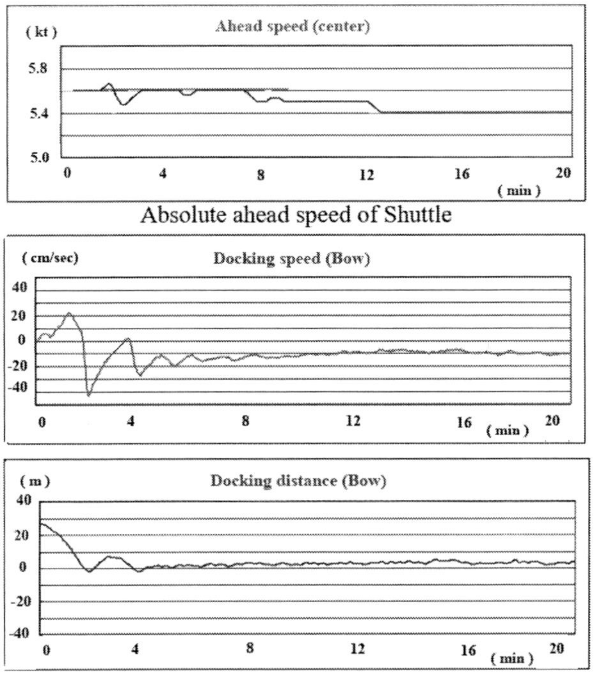

- Relative docking speed and distance -

Figure 6 Results of Measurement

5. EVALUATION OF DSS

DSS worked efficiently with the checking calculation ability of relative velocity and performance of displaying two vessels to carry out actual sea test. This chapter introduce evaluations of this system.

The processes of evaluations are conducted from the following three characteristics;

a. functions of DSS on ship-to-ship operation,
b. necessary information for the decision of DSS,
c. functions to recognize the mutual relative movements of the ships on PDA's graphics.

To evaluate DSS, this system connected to the simulator of SMSC (Ship Manoeuvring Simulation Centre) located in Trondheim in Norway. We implemented the tests to ask the operators to utilize DSS for the training of STS operations tests. Actually we obtained many precious comments and advices regarding the DSS operations.

Based on these actual operators' advices, we made some improvements on the previous DSS, especially adjusting the graphics and connected the system to the simulator of SMSC. First, we requested SMSC to output each vessel's absolute position and heading simulated on the simulator through the local area network of simulator.
Second, we added a function to DSS to receive information sent from the simulator and to substitute them for the system.
By means of these two improvements, when DSS is connected to the simulator using LAN, this system changed to receive ships' information sent from the simulator, and calculate and display relative movement of two vessels in real time.

In respect of a new version of DSS, this system does not need GPS, FOG and AIS receiver or AIS connection. Therefore the facilities are only consisted of one PC calculating and displaying relative motion and PDA shown in **Figure 7**.
To achieve this subject, we already solved the problems and the connection tests between both at SMSC have succeeded in September 2010 and we confirmed that DSS could provide relative speed and distances using absolute information of each vessels simulated on the simulator in real time. Therefore we could test and achieve the efficiency of this system in cooperation with the operators.

We explain how to evaluate graphics of PDA. DSS is usually set and used by operators on the bridge of a ship at the first approach. When two vessels are closing in, at the berthing approach, the operators go outside of bridge and order manoeuvring of the ships on the wing. Using a PDA, operators can know the relative movement of the ships for long time. PDA is useful and important equipment for the operators onboard when two vessels are closed.

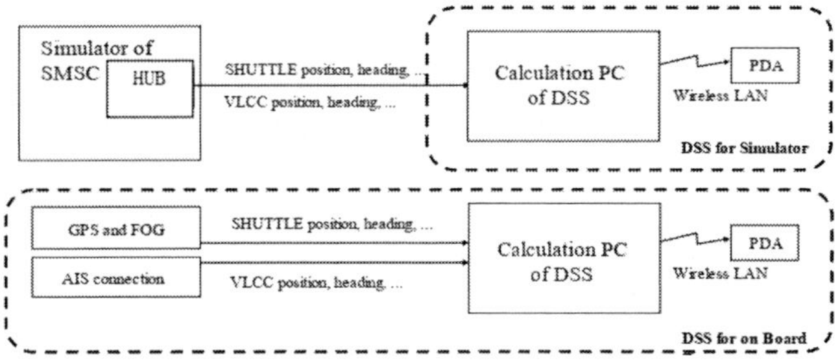

DSS (upper side) & Former type (lower side)

Figure 7 Block diagram of DSS for the simulator

Figure 8 shows some ideas of graphical displays of PDA. The display on the left hand side is a former type. The operator checks relative information indicated by PDA at a glance while observing delicate motion of two vessels at real time. Graphical display of PDA informs several important data which are relative speed and distance of bow-side and aft-side.

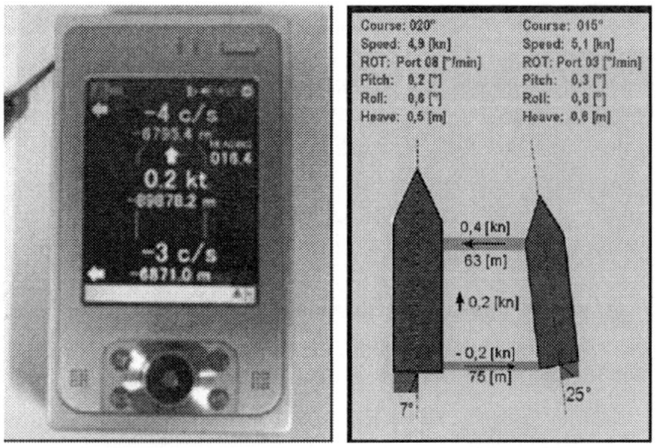

Former type New type [5]

Figure 8 Graphical displays of PDA

The graphical display on the right hand side was designed by Dr. D.Husjord of NTNU [5], which shows the relative positions of two ships on PDA. He has many experiences as a shipmaster, mooring master and realized visual information was important to notice relation of two ships and added some value of speed and distance.

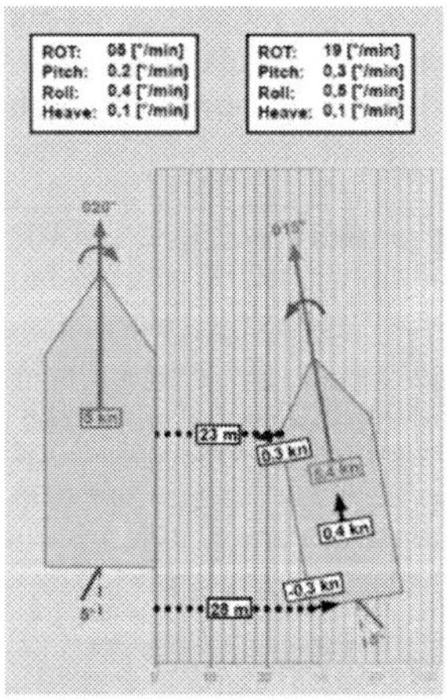

Figure 9 shows another bird eye view that a grid gives the distance between the vessels and long arrows representing the course, give a better indication of relative angle between the vessels. In this display, rudder is presented with red (port) or green (starboard) line, which indicates if the rudder status is to port or starboard.

Figure 9 Graphical displays during evaluate [5]

We have a plan of ship-to-ship experiment using new type PDA for the purpose of the further evaluate of graphical displays. It could be more important repeatedly to use this system and feedback them to the operators.

6. CONCLUSIONS

It is important to watch the vessel motions such as the large tanker and LNG carrier to keep safety under ship-to-ship operations.

DSS with applied global high accuracy GPS is possible to offer the sufficient and necessary information such as relative distance and approach speed of two vessels under lightering to the officers and the crew to control their ship. DSS can provide a precious database with users for not only supporting the vessel's safety operation but also the vessels managing system.

It is convinced that DSS contributes to more helpful and safety operations. The first feature is situated in approach manoeuvring and cargo handling operation, the second is for vessel management concerning ship-to-ship cargo transfer and crisis-management.

7. ACKNOWLEDGMENT

A part of this research has been funded by the Norwegian Research Council through the international joint-industry research programme "Investigating Hydrodynamic and Control Aspects of Ship-to-Ship Operations". The authors appreciate Subagent Petro Trans Inc. in Houston, USA, for the practical measurement programs onboard the tanker in their fleet in November 2009. The authors are grateful to staffs of Skuagen Petro Trans Inc. Also authors thank to crew for their help in implementing the DSS and their support in the field testing.

8. REFERENCES

1. R. Hatch, 'StarFire: A Global High Accuracy Differential GPS System', *Proc. of GPS Symposium 2002, Japan Institute of Navigation*, 2002

2. H.Oda, 'New Berthing Support System with StarFire DGPS', *International Symposium on GNSS/GPS 2004*, 2004

3. H.Oda, 'New Berthing and Ship To Ship operation support system (NBSTS)', *Asia Navigation Conference*, 2009

4. E. Shimizu, 'Development of Decision Support System for Ship-to-Ship Operation', "Navigation" of *Japan Institute of Navigation, No.173,* 2010

5. D. Husjord, 'A guidance and decision-support display for STS operations', *NTNU Note,* 2010

9. AUTHORS BIOGRAPHY

Etsuro Okuyama

Mr. Etsuro Okuyama received master's degree of Naval Architecture and Ocean Engineering, Graduate School of Engineering, Osaka University at 2003. He is the chief researcher of Akishima Laboratories (Mitsui Zosen) Inc. He specializes in ship performance in wave field and ship manoeuvring. He is a member of the Japan Society of Naval Architects and Ocean Engineers.

Hiroyuki Oda

Dr. Hiroyuki Oda received his bachelor's degree in Physics from the Sophia University and his Doctor's degree of engineering from Osaka University in 1992. He developed the rudder roll stabilization system of ships. He is currently the general manager of Akishima Laboratories (Mitsui Zosen) Inc. He is a member of the Japan Society of Naval Architects and Ocean Engineers, the Japan Institute of Navigation, the Society of Instrument and Control Engineering of Japan.

Etsuro Shimizu
Associate Professor Etsuro Shimizu received the Ph.D. degree of Control Engineering from Tokyo Institute Technology in 1999. He is an associate professor of Tokyo University of Marine Science and Technology from 2003. From March 2005 to March 2006, He was a visiting professor of Norwegian University of Science and Technology. His Fields of research are nonlinear control theory, modelling and control of vessels. He is a member of IEEE and Ocean Machinery Research Committee of the Japan Institute of Marine Engineering.

Shigeki Sakakibara
Dr. Shigeki Sakakibara received master's degree of Mercantile Marine, Graduate School of Kobe University of Mercantile Marin in 2003 and Doctor's degree of engineering from Osaka University in 1995. He is the senior engineer of the Yokohama Rubber Co., Ltd. He specializes in effect of mooring system on ship motions due to waves at ship-to-jetty and ship-to-ship. He is a member of the Japan Institute of Navigation and Japan Society of Civil Engineering.

Shuu Yamada
Mr. Shuu Yamada received master's degree of Naval Architecture and Ocean Engineering, Graduate School of Engineering, Yokohama National University in 1999. He is the chief engineer of the Yokohama Rubber Co., Ltd.
He specializes in marine rubber products, fendering system and mechanical engineering.

PREDICTION OF HYDRODYNAMIC INTERACTION EFFECTS OF VESSELS IN PORTS

J.A. Pinkster, PMH bv, Netherlands

SUMMARY

A review is given of two computational procedures to compute the effects of a passing vessel on moored vessels. The procedures are based on 3-dimensional potential flow and the flow equations are solved using panel methods. The first method is based on double-body flow assumptions i.e. the free-surface is rigid and flat. The second method solves the low-frequency waves generated by the passing vessel as it sails through the port. In this paper some results of model tests are shown and compared results of computations. Some applications of a multi-domain approach which can be applied to cases with water-depth changes are shown. Finally some results, as yet not validated by model tests, are shown which indicate some of the effects which need to be accounted for in the passing vessel problem

1. INTRODUCTION

The effects of hydrodynamic forces which a passing vessels exerts on a moored vessel have been the focus of considerable amount of attention in recent years. While the dominant characteristics of the hydrodynamic forces on moored ships has been known for a long time, i.e. draw-down generation and suction forces on the moored ship as the passing vessel sails by, up to now only limited attention has been given to researching the many factors, besides passing distance , passing speed and water depth, which influence the forces and the response of the moored vessel in terms of motions and forces in the mooring system.

At the same time, in spite of the incidents which have occurred as a results of passing vessel effects, there is a lack of guidelines how such effects should be evaluated and incorporated in mooring system design and in operational practices. The California State Lands Commission's development of the MOTEMS document in which besides treating a great number of important factors regarding safety and integrity of Oil terminals also has taken the unique step to include passing vessel effects in the loads acting on marine oil terminals. Most likely other authorities will have some guidelines but these are often given in the form of limiting passing speeds and have little basis in hydrodynamic analyses.

With respect to passing vessel effects, beside the possibility of personal injury , cargo spillage, disruption in loading/unloading operations and damage to mooring equipment , the mooring terminal and the vessels are focal points of interest.

In recent years the spectacular increase in the size of container vessels and the advent of LNG terminals on inland waterways have been a major driving factor in the acceleration in the development of numerical techniques for the prediction of passing vessel forces and moored vessel response. Large vessels such as tankers have been making use of ports and inland water ways for many years. Passing vessel effects due to tankers and experienced by moored tankers and other vessels have received some attention in the past both as part of the design process of a jetty or , in some cases, when incidents have occurred. No comprehensive, up to date internationally recognized guidelines on how such effects should be evaluated and treated have , however , been issued by any of the regulatory bodies active in the marine field.

While in the past important steps were made in the analysis of passing vessel effects and the experimental validation of such analyses, see, for instance [1] through [6] , most were aimed at analysing relatively uncomplicated cases i.e. forces due to a vessel sailing past a moored vessel which is lying parallel to the passing vessel in horizontally unrestricted waters. Much insight was gained from these developments but applicability to real cases of ships moored in ports was limited due, among others, to the presence in real life situations of quays, water depth differences between the fairway and the terminal, currents etc. For an adequate understanding of the passing vessel effects under such conditions the availability of validated methods of analysis are an important asset. In the past analytical or semi-analytical methods were developed for the simple cases, (Seelig [6]) , which are used for more complicated case by the application of correction factors based on ,among others , model test data. However, if we are to account for real life situations in a systematic way, full numerical methods need to be applied which can take into account such aspects as irregular port geometry, variations of water depth, multiple ships (moored and /or passing and current. See for instance [7] through [11].

In this paper some of these aspects will be highlighted using a 3-d potential flow based numerical tool which has been developed over the past number of years. The method has been partially validated based on existing and new experimental results both from full scale measurements and from model tests. A brief summary of the computational procedures will be given. More details are given in previous papers which are referred to in the text and listed in the reference section.

2. COMPUTATION OF PASSING VESSEL FORCES

The method used is based on 3-dimensional potential theory. The fluid is assumed incompressible ,irrotational and inviscid. The fluid equations are solved using a zero-order panel method i.e. the mean wetted surface of all vessels and the port geometry are described by quadrilateral or triangular panels. Two computational methods based on these assumptions have been developed. The first, a so called 'double-body flow' method assumes that, due to the relatively low passing speed of vessels in ports, free surface effects can be neglected. The free surface is assumed to be flat and rigid while the flow between the bottom of the fairway and the flat surface remains 3-dimensional. The second method takes into account the long waves created by the passing vessel when its primary flow field interacts with the surrounding port geometry. These long waves create additional loads on moored vessels besides the well-known draw-down effect.

In the following the capabilities of the computational models will be discussed briefly and subsequently some comparisons will be made between results obtained these models and with results of model tests and full scale measurements. For both models, restricted water depth is taken into account. We refer to Pinkster[8] for a more detailed description of the numerical aspects of both methods.

2.1 THE DOUBLE-BODY MODEL

The model is similar to that described by Korsmeyer et al [7] in that it is based on 3-dimensional potential flow. Differences between the methods can be found in the basic formulations for the hydrodynamic forces due to the passing vessel. For the double-body flow model, the potentials describing the flow are based on the Rankine source formulation taking into account restricted water depth and a rigid still water level. The double-body flow model is suitable for computing interaction forces in 6 D.O.F. on multiple vessels, taking into account the harbour or fairway geometry. Vessels which are sailing are assumed to sail on a straight course at constant speed. The computations make use of 3-d panel models of the vessels and the harbour geometry.

2.2 THE FREE-SURFACE MODEL

A comprehensive description of the computational procedure is given in [8] and [14]. These papers describe the theoretical and numerical background to the computations, validation with results of full scale measurements and model tests, and a number of illustrative examples. In this section we give a short overview with the main aspects of the method.

The dominant part of the hydrodynamic forces on the moored vessel which results in mooring forces and relatively low frequency motions originates from the so-called "primary flow" around the passing vessel. The characteristics of the primary flow are a high pressure area ahead of the bow of the vessel which increases near the bow, a pressure drop at the side of the vessel and a high pressure area at the stern. The flow velocities induced by the passing vessel forward of the bow are predominantly in the direction of motion of the vessel , directed aft between the bow and stern and directed forward aft of the stern.

The primary flow around a vessel can be computed based on 3-dimensional potential theory using a panel method. In this method each panel describing the hull form of the passing vessel represents a source with strength determined by the no-leak condition on the hull. Such methods are based on the double-body flow which assumes that the free-surface remains flat and horizontal. Using this assumption it is possible to take into account the influence not only of the passing vessel itself on the flow but also the additional influence due to the presence of the boundaries formed by the harbour geometry and moored vessels which are also described by panel models.

The present computations are carried out in two phases: i.e. phase 1 in which the source strengths of all sources (represented by panels) on the passing vessel are determined using the 3-d double-body model. These source strengths are a function of time since the passing vessel sails through changing geometry (channel, terminal with moored LNG carrier). This phase also yields the classic double-body results for the interaction forces. For the second phase involving the free-surface, the source strengths on the passing vessel are used to determine the three velocity components and the pressure at the grid points of the panel models for the terminal and the moored vessel. These velocities and pressures represent the "incoming disturbance" for the terminal and the moored vessel.

Since the incoming disturbance for the terminal and the moored vessel are time-dependent due to the passing vessel and that the boundaries of the terminal and moored vessel also have to comply with the no-leak condition, the response to the disturbance will also be time-dependent leading to the generation of long-period waves , or seiches.

The response of the terminal and moored vessel are computed using the results of a standard multi-body, frequency domain, 3-d diffraction program. The input into this program are the frequency-domain components of the time-dependent disturbances which are obtained by applying the Fast Fourier Transform (FFT) method to the time records of the disturbances. The frequency-domain results in terms of pressures, velocities, forces and wave elevations (the sum of the double-body results and the diffraction results) are transformed back to the time-domain using the Inverse Fast Fourier Transform (IFFT) technique finally yielding the required time-domain forces on the moored vessel. The diffraction computations do not include the model of the passing vessel It is assumed that the passing vessel does not modify significantly the seiches it sets up due to their long periods. The passing vessel will ride with these long waves without significant diffraction effects.

The computational method described here computes only the low-frequency disturbance due the passing vessel. For the speeds considered, these are by far the dominant disturbance. Other disturbances are due to the secondary or wash-wave system of the passing vessel. However, wash waves of large, slow moving vessels are very low and relatively short and as a consequence, do not result in significant loads on large moored vessels or lead to large motions and mooring loads. Wash wave effects are mostly of importance when the passing vessel speed is high and the moored vessel is relatively small.

3. VALIDATION OF THE DOUBLE-BODY MODEL

In this section we shall apply the double-body model to the surge and sway forces and yaw moment on a moored tanker in shallow water in order to show that, for the case of a vessel moored in open water, this model gives a good approximation for the passing ship effects. See also Huang and Chen [8].

The results of computations will be compared with results of model tests carried out by Remery [4].

The results apply to a moored loaded 100 kdwt tanker being passed by a loaded 30 kdwt tanker at a distance, measured board-to-board, of 30 m. The set-up is shown in Figure 1. The main particulars of the vessels are given in Table 1.

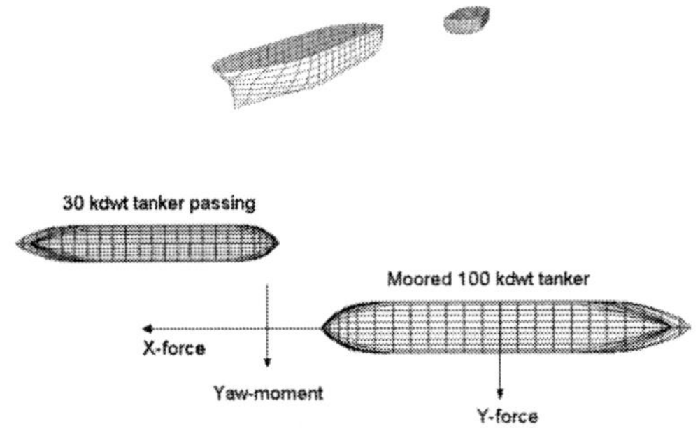

Figure 1: Passing vessels

The water depth amounted to 18.05 m being 1.15 times the draft of the moored tanker. The model tests were carried out at a scale of 1:60 in the former Wave and Current Basin of MARIN. This facility measured 60 m x 40 m so the passing manoeuvre can be considered as having taken place in horizontally unrestricted waters. The passing speed in the model test corresponded with 7 kn. full scale.

The computations were carried out using panel models for the tankers which had the same main dimensions as the models used in the tests. Actual body plans were not available so the panel models do not conform exactly with the model hull shapes. The total number of panels on each of the tankers amounted to 388. By modern standards this is not a particularly high number but is adequate for the purpose at hand.

The results of the model tests and computations are given in Figure 2 through Figure 4 in the form of the relative position dependent surge and sway forces and yaw moment on the moored vessel. The relative position has been made non-dimensional on the basis of the mean length of both vessels and the forces and moment have been divided by the square of the velocity of the passing vessel. A negative distance indicates that the passing vessel is approaching the moored vessel. At zero the mid-ships are abreast.

The results shown these figures show that the double-body model gives a good approximation of the forces on the moored vessel. The measured amplitudes of the surge force and the yaw moment are slightly larger than the computed values. Remery [4] carried out a comprehensive set of tests in which the passing ships, passing distances and ship's speeds were varied. We have shown comparisons for just one case. Korsmeyer et al.[7] have also shown comparisons of the double-body model with measured results by Remery which are in keeping with our findings. The conclusion is then that for the open water case, the double-body model gives a good prediction of the passing ship forces.

		Passing tanker	Moored tanker
Length between pp.	m	183.0	257.0
Breadth	m	26.1	36.8
Draft	m	10.5	15.7
Displacement	m^3	40,200	118,800

Table 1 : Vessel dimensions

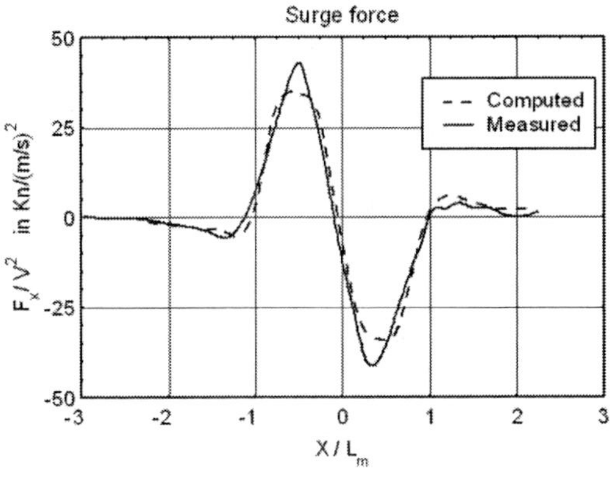

Figure 2 : Surge force on moored tanker

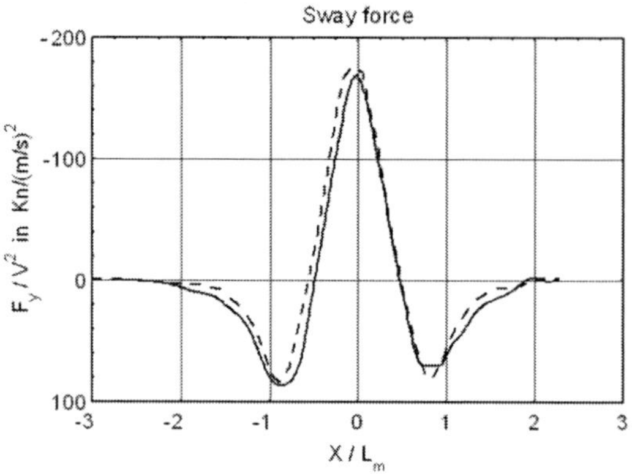

Figure 3 : Sway force on moored tanker

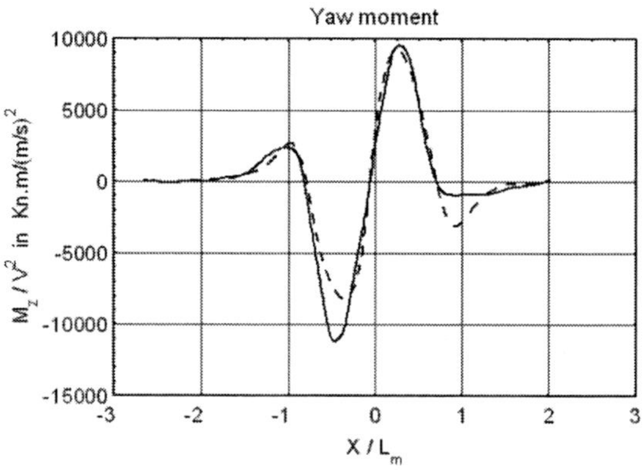

Figure 4 : Yaw moment on moored tanker

4. VALIDATION OF THE PREDICTION OF PASSING FORCES FOR A VESSEL MOORED TO A VERTICAL QUAY

Model tests were carried out by MARIN on behalf of the Port of Rotterdam on the passing forces on a moored Panamax container vessel due to a passing 12,500 TEU container vessel, see van Wijhe and Pinkster [12]. The investigation formed part of a study on the required width of the entrance channel to the Maasvlakte II extension in the Port of Rotterdam. The model tests were carried out at a scale of 1:38 in the Shallow Water Basin for a full scale water depth of 17 m. The basin width corresponded to the entrance channel width of 600 m. The vessel particulars are shown in Figure 5 and Figure 6 and Table 2 and Table 3.

In Figure 7 a close-up view is given of both vessels and a part of the side of the channel which is in the form of a vertical quay wall. Near to the moored vessel the panel density on the quay wall is increased to account for sharper flow gradients.

In Figure 8 the sign conventions of forces on the moored vessel are given as well as the location of the points at which the current velocities u and v and the wave elevations 1,2 and 3 were determined both by measurements and computations.

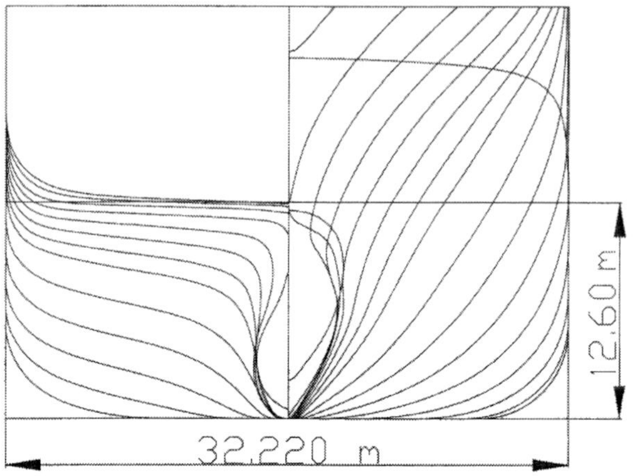

Figure 5 : Body plan of Panamax Container Vessel

Length between perpendiculars	278.0 m
Length on waterline	282.3 m
Breadth	32.2 m
Draught (even keel)	12.6 m
Displacement	66245 m^3

Table 2 : Main particulars of the Panamax Container Vessel

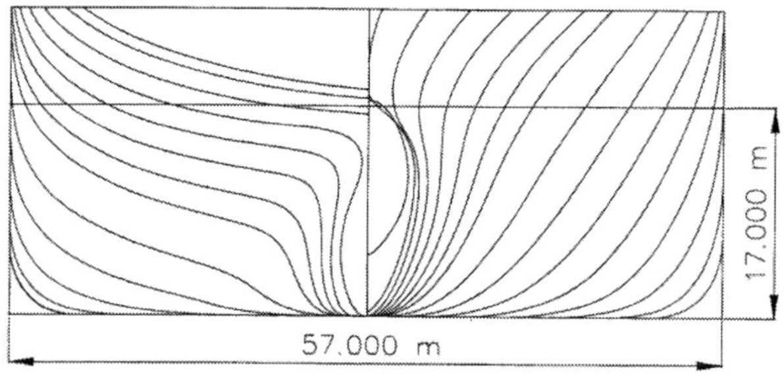

Figure 6 : Body plan of 12500 TEU Container Vessel

Length between perpendiculars	366.0 m
Length on waterline	363.6 m
Breadth	57.0 m
Draught (even keel)	17.0 m
Displacement	215121 m^3

Table 3 : Main particulars of the 12500 TEU Container Vessel

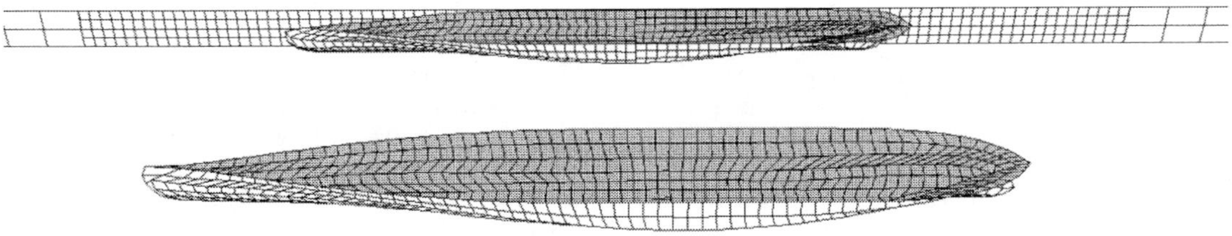

Figure 7 : Panel models of the moored and passing vessels

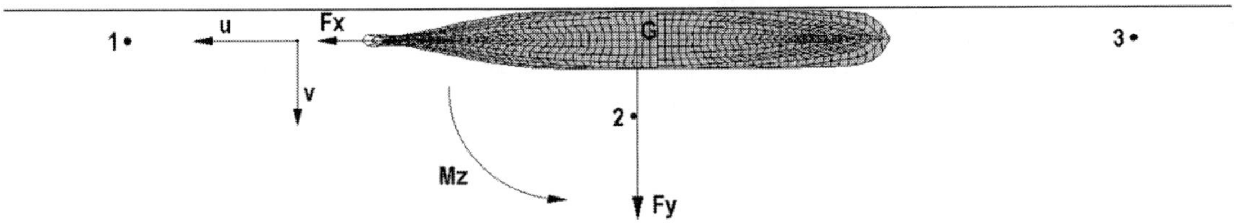

Figure 8 : Locations of measurements of wave elevation and current

In this section some results of the model tests are compared with results of the computational method including free-surface effects described in the previous section. We have selected one case i.e. the passing speed is equal to 5.5 kn and the passing distance between the moored and the passing vessel amounts to 75 m measured side-to-side. The results are shown in Figure 9. The solid lines refer to the computed data.

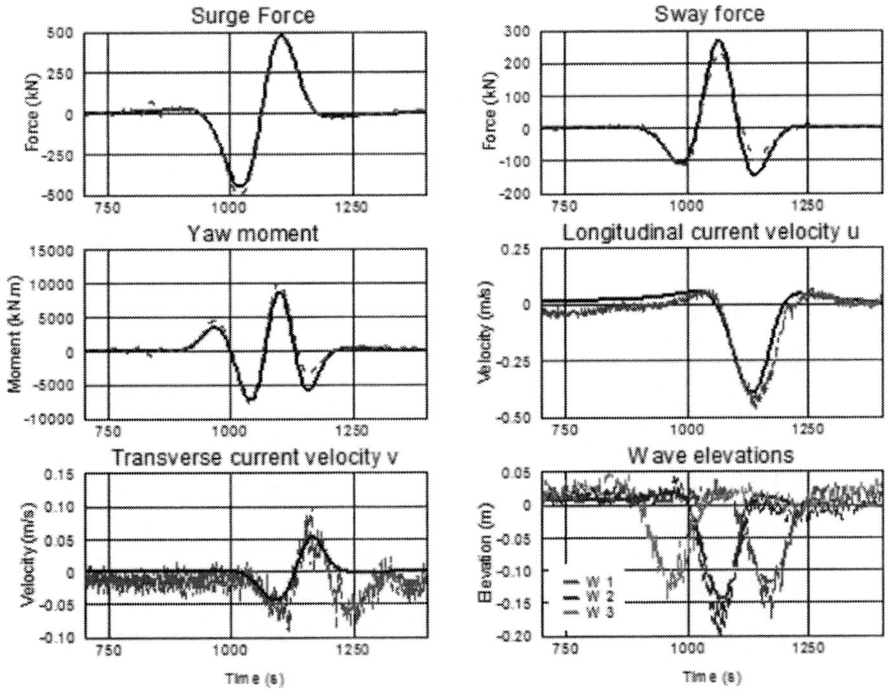

Figure 9 : Forces, Current velocities and Wave elevations; Passing speed 5.5 kn. 75 m distance

A feature of interest in the passing forces is the relatively small value of the peak sway force compared with the peak surge force. In open water, i.e. with no quay present beside the moored vessel, the sway forces tend to be significantly larger and the surge forces less. This is related to the fact that the current generated by the passing vessel is restricted by the presence of the vertical quay wall so that near the quay wall longitudinal current is increased and the transverse current is reduced and even zero at the quay when compared with the case without a quay wall. See the previous case and also Pinkster [8]. The results shown on the longitudinal and transverse currents confirm the fact that the former is significantly larger than the latter.

The wave elevations measured and computed before, mid-ship and aft of the moored vessel show that the negative peak of the wave elevation, also denoted as 'draw-down' , is largest for the mid-ship location. This reflects the restricted cross-sectional area of the channel at this point due to the presence of the moored vessel combined with the effect due to the measuring point being closed to the passing vessel.

Comparison of the results of the computed and measured forces and yaw moment show that these are well predicted by the computational method. The predicted values of the current velocities are also close to the measured data.

Computed wave elevations correspond well with the measured data when the passing vessel is closest i.e. when the negative peak elevation values occur. Before and after the vessel has passed, differences are somewhat larger. This may be due to different starting-up procedures of the passing vessel model compared with the computations. The acceleration phase of the passing vessel tends to generate long waves or seiches in the basin and also in the computations. In both cases efforts are made to reduce such effects to a minimum. They are, however, not completely avoidable.

5. SOME COMPLICATING ISSUES

In the afore going it has been shown that computations of passing vessel forces based on 3-d potential flow methods give results which are in good agreement with measured data. In all cases the results were situations without current and with constant water depth. Although such results are encouraging, there still remain a number of issues which need to be resolved before we can apply potential flow methods to arbitrary situations of moored vessels in ports with confidence.

Some of these unresolved issues are :

- Influence of water depth differences in the fairway
- Influence of current
- Influence of hydrodynamic reaction forces (added mass and damping of the moored ship)

5.1 DIFFERENCES IN WATER DEPTH

In many locations the water depth at the location of the moored vessel is significantly different to the water depth in the main channel adjacent to the terminal. An example is shown in the Figure 10 which is typical of , for instance , the Port of Rotterdam. The water depth in the main channel is about 21 m while the water depth in the terminal is 12 m and 17 m respectively. Results of computations of the sway forces due to a passing tanker acting on a vessel moored in the 17 m deep part (Case_1) or in the 12 m deep part (Case_2) are shown in Figure 12. The results show a significant difference in the force which will certainly impact on the mooring line force differences. While the computational procedures can cope with this geometry, the question remains with respect to the correlation between results of computations and model tests. The method employed to treat water depth differences is given in Pinkster [14] and involves subdividing the total fluid domain in interconnected domains with each their own water depth. The flow equations are solved taking into account equality of pressure and continuity of normal velocity at the interfaces between the domains. These interfaces are shown in Figure 10 and Figure 11 as dotted red lines.

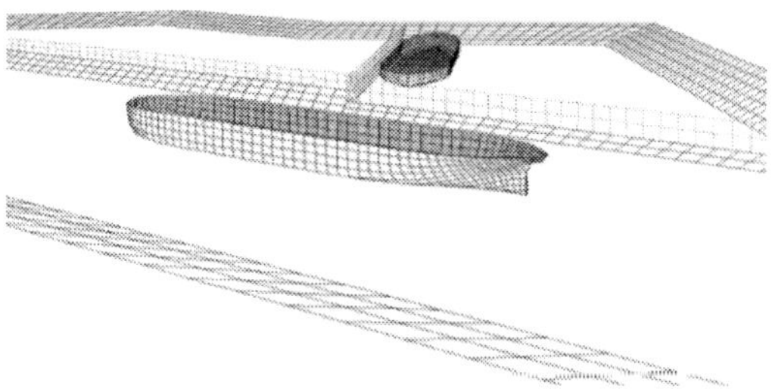

Figure 10 : LNG carrier moored in deeper part of terminal . Case_ 1

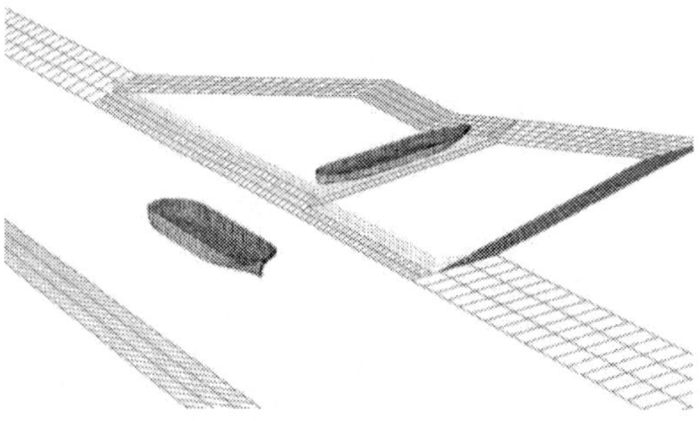

Figure 11 : LNG carrier moored in shallow part of terminal. Case_2

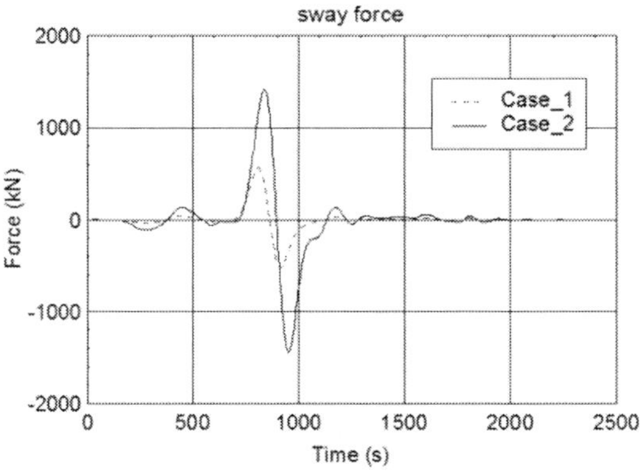

Figure 12 : Sway forces on the LNG carrier moored at two locations

It should be mentioned that water depth differences can also be accounted for by using horizontal panels on the area with reduced water depth. This can , however , result in large number of panels being required. In some cases such solutions are not applicable, for instance, in cases involving a pit in the fairway floor. In such cases not the pit but the whole fairway would need to be paneled possibly resulting in unacceptable number so panels. In such cases a two-domain solution can be applied. As an illustration we consider the, fictitious, case of a tunnel section lowered into a trench in the fairway floor. The question is ; what is the effect of a vessel sailing overhead, i.e. what forces are exerted on the tunnel section ? The modeling for such a case is shown in Figure 13

In this figure, the passing vessel, the tunnel section in the trench and the trench itself are shown. The domain interface is also shown in this figure. It consists of a double set of panels just above the fairway floor. The top set of panels 'belongs' to the domain of the fairway and the lower to the trench. The water depth in the fairway is 15 m and 50 m in the trench. The passing vessel in a cruise vessel of 49,270 m^3 displacement and a draft of 7.8 m passing at 9 kn. The results of the passing vessel forces on the tunnel section are shown in Figure 14. From such results the possibility of the tunnel section being moved before final installation can be assessed.

5.2 INFLUENCE OF CURRENT

In many locations where vessels are moored along the banks of rivers, the effect of current on the passing vessel loads are also an issue to be considered. The influence of current on the forces due to a passing vessel is not to be confused with the current force acting on the moored vessel. What we consider here are the changes that occur in the passing vessel forces due to the presence of current. An example is shown in Figure 15, which applies to Case_2 of Figure 11. The passing vessel, in this

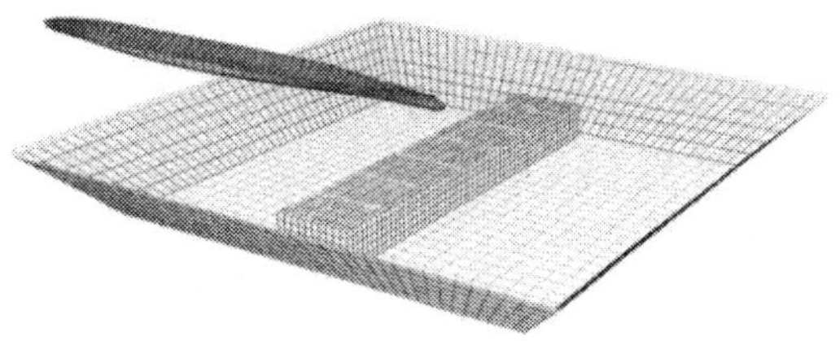

Figure 13 : Cruise vessel passing a trench with a tunnel section. Domain interface shown as red dotted grid.

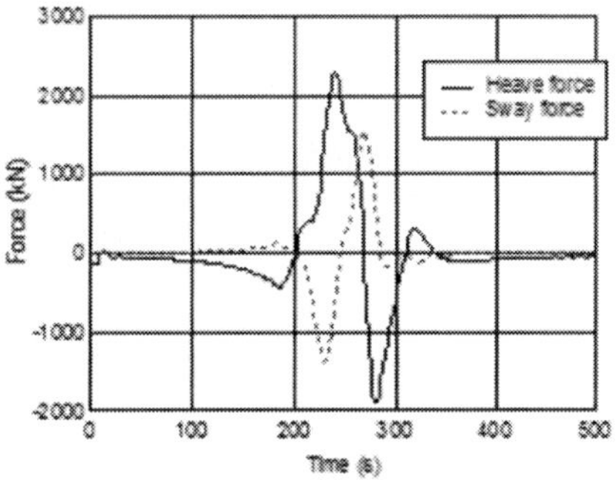

Figure 14 : Forces on the tunnel section in the trench

case a large tanker, is proceeding past the terminal at a constant speed through the water of 6 kn. Including a head or following current changes the passing speed of the vessel relative to the moored vessel which is reflected in the frequency changes is the force. Due to the relative speed change, the fluid accelerations generated by the passing vessel at the location of the also change , leading to an increase in the force level at the higher relative passing speed. The predicted effects of current on the passing vessel forces are significant and need to be taken into account. At this time no experimental data is available to validate the computed results.

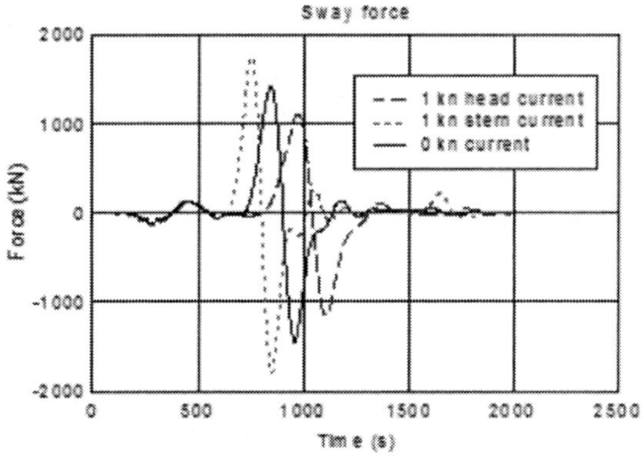

Figure 15 : Influence of current on the forces on the moored LNG carrier of Case_2

5.3 HYDRODYNAMIC REACTION FORCES

The standard procedure to obtain the mooring loads and motions of the moored vessel is to apply the computed passing vessel loads along with wind and current forces on the model of the moored vessel and to solve the equations of motions. This approach allows mooring systems to be systematically varied without having to take recourse to re-computing the passing excitation forces for every mooring system investigated. Time domain simulations of motions and mooring loads take far less computation time than computation of the excitation forces. The assumption behind this approach is that hydrodynamic excitation forces and hydrodynamic reaction forces can be superposed i.e. linear hydrodynamics. Motion simulations can , however, take into account any non-linearities in the mooring line and fender configuration and also any additional hydrodynamic effects such as viscous damping.

Up to now we have focussed attention on the passing vessel forces. In a complete analysis of the mooring loads it is necessary to consider the motion response of the moored vessel as dictated by the excitation forces on the one hand and the mooring system stiffness characteristics and the hydrodynamic reaction forces (added mass and damping) on the other hand. This data is used in dynamic analyses of the moored vessel response. As has been shown by Headland and Smith [13], static analyses of mooring system loads are inadequate for the prediction of such effects as snatch loads in mooring lines and fenders.

Mooring system stiffness characteristics are related to mooring line materials, size, number of lines , pretension etc. as well as the fender characteristics. This will not be treated further here although in the final analysis these aspects are of great importance also.

We are interested here in the frequency characteristics of the added mass and damping of a moored vessel since these are also important for the motion responses and subsequent loads in the mooring system. We compare in Figure 16 the added mass and damping of the surge and sway motions of a 9000 ton barge in shallow water for two cases i.e. barge in open water and barge in a dock to the side of the main channel. As can be seen from the mentioned Figure, the effects of the dock geometry on the hydrodynamic added mass and damping are considerable and of importance for the results of dynamic analyses. These data are based on 3-dimensional diffraction computations.

The Figures show that with the barge moored in a dock, hydrodynamic reaction forces are significantly affected by reflection from the dock sides. Typically reflection effects show saw-tooth characteristics such as most clearly seen in the added mass for surge a11.

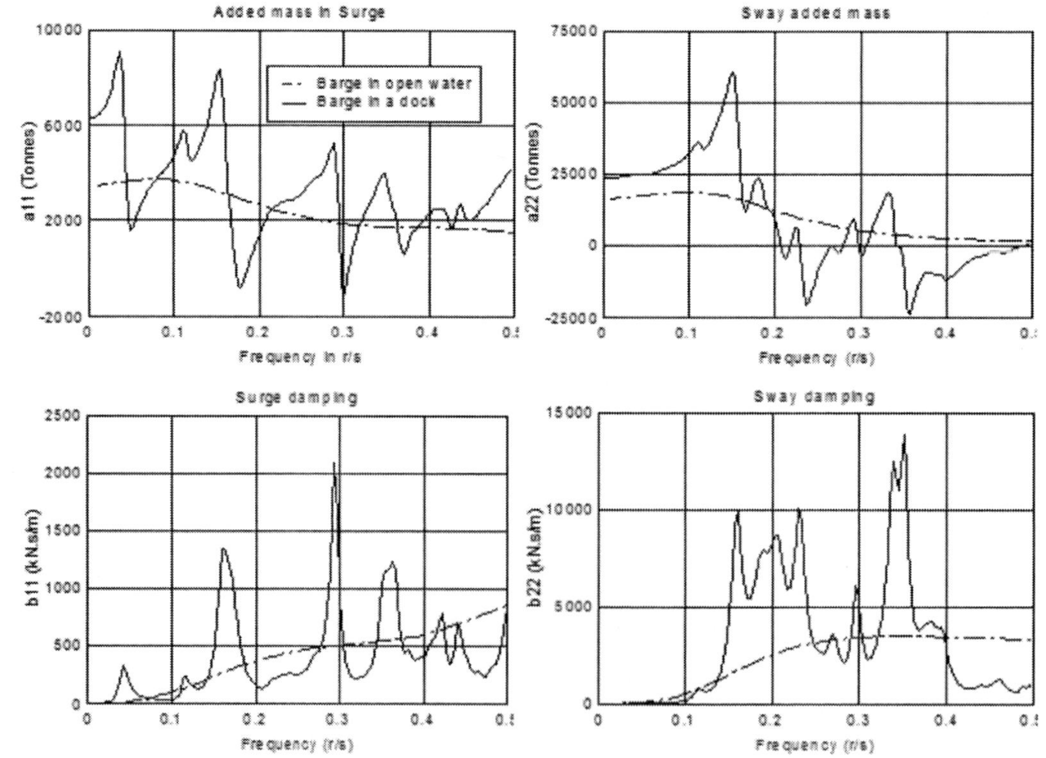

Figure 16 : Effect of port geometry on the added mass and damping of a barge

6. CONCLUSIONS

In this paper we have treated the effects of passing vessels by showing a number of examples from scientific research projects also for real life cases. Where measured data was available we have shown the correlation between two numerical methods based on potential flow, i.e. a double-body flow method and a method including free-surface effects. In general , the correlation has been shown to be adequate for the prediction of passing vessel forces for practical cases. This does not necessarily mean that such methods will be sufficiently accurate for more complicated cases , for instance , involving complex port geometry, current , passing vessel sailing under a drift angle , etc.

In order to gain sufficient insight in those areas where knowledge is lacking, more research will be carried out. Such research should preferably include experimental validation for some of the mentioned cases and most probably also for other cases not mentioned here.

7. REFERENCES

1. WANG, SHEN. "Dynamic Effects of Ship Passage on Moored Vessels." ASCE, Journal of the Waterways,Harbors and Coastal Engineering Division, WW3, pp. 247-258, Aug. 1975

2. KING, G.W."Unsteady Hydrodynamic Interactions Between Ships." Journal of Ship Research, Vol.21, No. 3, Sept. 1977.

3. VARYANI, K.S." Estimation of Loads on Mooring Ropes of a Moored Ship due to a Passing Ship." Proceedings International Conference on Safety and Operations in Canals and Waterways, SOCW 2008, Glasgow, 2008

4. REMERY, G.F.M. "Mooring Forces Induced by Passing Ships", OTC 2066, 1974.

5. KRIEBEL,D. "Mooring Loads due to Parallel Passing Ships" Technical Report TR-6056-OCN, Naval Facilities Engineering Service Center, Port Hueneme, 2005

6. SEELIG, W. "Passing Ship Effects on Moored Ships" Technical Report TR-6027-OCN, Naval Facilities Engineering Service Center, Port Hueneme, 2001.

7. KORSMEYER, F.T., LEE, C.-H. and NEWMAN,J.N. "Computation of Ship Interaction Forces in Restricted Waters." Journal of Ship Research, Vol. 4

8. PINKSTER, J.A. "The Influence of a Free Surface on Passing Ship Effects," International Shipbuilding Progress, 51, No. 4, 2004.

9. HUANG, E.T. , CHEN, H-C." Passing Ship Effects on Moored Vessels at Piers." Proceedings Prevention First 2006 Symposium, Long Beach, California, 2006

10. FENICAL, S., ET AL. "Numerical Modeling of Passing Vessel Impacts on Berthed Vessels and Shoreline." Paper No. 1234, International Coastal Engineering Conference 2006, San Diego, 2006

11. PINKSTER, J.A. AND RUIJTER, M.N. DE "The Influence of Passing Ships on Ships Moored in Restricted Waters", Paper OTC 16719, Offshore Technology Conference, Houston , 2004

12. WIJHE, H.J.VAN, AND PINKSTER J.A. "The Effects of Ships passing Moored Container Vessels in the Yangtzehaven, Port of Rotterdam" Proceedings SOCW 2008 Conference, Glasgow, 2008

13. HEADLAND, J.R., SMITH, E.D." Dynamic Analysis of Moored Ships Exposed to Passing Vessels", PIANC Annual Meeting , U.S. Section, Portland, Oregon, 2003

14. PINKSTER,J.A.:"Suction, Seiche and Wash Effects of Passing Ships in Ports", Proceedings of the SNAME Annual meeting, Providence, Rhode Island, 2009

8. AUTHORS BIOGRAPHY

Johannes A. Pinkster is a former research engineer and head of the department of Offshore engineering projects, MARIN, professor emeritus of Offshore and Ship Hydromechanics of the Delft University of Technology, the Netherlands and at present active as consultant based in Rotterdam.

USING DYNAMIC POSITIONING FOR SIDE-BY-SIDE CARGO TRANSFER OPERATIONS

F.H.H.A. Quadvlieg and **R. Hallmann,** MARIN, Netherlands

G. Hughes, Oceaneering International, Inc., USA

D.H. Harris, Maritime Advanced Physics Corporation, USA

SUMMARY

A rather unconventional, but nevertheless practical method is presented to keep two large ships in close proximity to allow side-by-side operations. This method is developed in the ONR - STLVAST project, in support of the sea basing concept adopted by the US Navy. This paper describes the particular application and the relevant (hydro)dynamic phenomena the feasibility of which was demonstrated using model tests and theoretical methods. Model tests were carried out with both vessels under either DP or dynamic tracking control. Full scale measurements were carried out in 2010. The results of full scale measurements demonstrated that the engineered concept was feasible.

1. INTRODUCTION

Often, transfer of cargo at sea takes place when ships are moored to each other. The connection of lines and mounting of fenders may however take up too much time and decreases flexibility. The mooring lines may suffer from considerable snap loads. Thus, there are a couple of reasons why mooring lines and fenders are not desired for sea basing. This is explained in [1] and [2].

The focus area of application is in this case consisting of (at least) two large ships. The cargo to be transferred is consisting of heavy rolling material, including for example trucks, humvees but also people. In order to do this, cranes or ramps are considered. Special ramps and cranes are developed for this purpose (see the LVI Lo/Lo project on www.youtube.com/watch?v=JvId272itb0).

Cargo transfer at sea is not a new area. Cargo transfer for navies takes place using replenishment at sea. Cargo transfer in the offshore takes place in skin-to-skin or tandem mooring.

Replenishment at sea is a side-by-side cargo transfer operation that has been applied for a long time. During these operations, hawsers and ropes are connected between the large ships, and cargo is transferred in this way (see Figure 1). Steering and manoeuvring of the vessels during this operation is a demanding job. Usually, one ship sails on autopilot (the largest ship or the tanker). The other vessel is manually steered and controlled using very strict procedures for approach and breakaway. The challenging hydrodynamics involved in this are already noted before the second world war. Typical speeds are in the range of 12 to 16 knots, a typical distance between the ships is 50 meters.

In the offshore industry, skin-to-skin or tandem offloading operations are carried out enabling hawser connections and hence cargo transfer. Also these types of operation are already generally acknowledged and often applied. These ships are very large, the speeds are very low (usually zero, apart from the current). In a skin-to-skin arrangement, the ships are separated with fenders resulting in a distance of about 5 meter. In a tandem offloading, the ships are at typically 100 meters from each other. Tandem offloading is left out of consideration for the moment. During the skin-to-skin operations, the ships remain next to each other, moored by ropes and fenders.

Figure 1: Replenishment at Sea (from http://en.wikipedia.org/wiki/Underway_replenishment)

The project described in this paper is called STLVAST, an acronym for Small To Large Vehicle At Sea Transfer. This project is funded by the Office of Naval Research (ONR) to support the sea basing development. The objective is to come to a procedure that is a mix of the above sketched technologies, but also with a couple of exceptions:

- Ships should be in very close proximity to allow for example crane operations.
- Ropes are not allowed, alternatively dynamic positioning will be used.

- The relative motions should remain within a square box of 3 meter.
- The ship speeds are from 0 to 8 knots.
- The target ships are very large (ranging from 150 to 300 meter).
- Seaways are up to sea state 4.

The ships under consideration are on one hand, an LMSR, a Large Medium Speed Roll-on Roll off ship, with a length of 270 meter and a breadth of 32.25 meter. On the other hand, a heavy lift ship, such as the Mighty Servant 3 or the Transshelf from Dockwise, with a length of 180 meter and a breadth of 40 meter.

DP (Dynamic Positioning) or DT (Dynamic Tracking) is a good method that may be able to perform that job. DP is the zero speed application, whereas DT described the ability to follow a point at a ship speed.

DP and DT have been around for a long time, but traditional DP or DT may not be enough to achieve the objectives. Alternative DP or DT methods had to be investigated, demonstrated on model scale and on full scale.

2. (HYDRO)DYNAMIC PHENOMENA

The following aspects are known to have a significant impact on ships sailing side-by-side at a forward speed.

- Wave forces on the ships, where the leeward ship may be shadowed by the weather side vessel. These wave forces are thought to consist of first order (wave frequent) and second order (mean and low frequent) wave forces. Wave interaction, caused by wave diffraction between the two vessels is an important aspect of this.
- Ship-to-ship suction in calm water.
- The traditional manoeuvring forces (typically for low frequent oscillations): the ships are generating lift and drag.
- Motion reaction forces at wave frequent oscillations (typically the wave frequent added mass and damping). These are traditionally calculated using potential flow theory.
- Steering particulars: rudders, propellers and thrusters are generating forces.
- The implementation of relative DP in the DP control system.

2.1 WAVE FORCES, INCLUDING INTERACTION

Typically, diffraction theories are used to get an impression of the wave excitation forces on vessels. This diffraction theory is a linear potential flow panel method and these theories are in use for already a long time (see Pinkster [5]). These theories are applicable to ships in close proximity, but there are a couple of issues. When the ships are in close proximity, the waves that are working between the two vessels are of considerable magnitude. This can be observed in Figure 2. These wave height amplifications are causing significant forces that will push the ships away from each other. These wave forces are a direct function of the relative wave height between the vessels.

In previous validation work for a side-by-side offloading system (see references [6] and [7]), it was found that a straightforward diffraction calculation of the two-body system leads to unrealistically high local velocities of the water in the gap between the two ships. This type of unrealistically resonant wave oscillations can occur due to the absence of viscous flow effects in the potential flow diffraction analysis. In reality the viscous effects of the high velocity flow around bilges and bilge keels will result in a significant amount of damping. Unfortunately it is not possible to simply add viscous damping to the equation of motion (as can be done for viscous roll damping), because the problem occurs inside the solution method of the potential flow calculation itself. It was found that this unrealistic behaviour could be suppressed by placing a numerical "lid" on the free surface in the gap. This technique is summarised in reference [7], [8] and [9]. A further explanation of the technique is published in reference [6]. The use of a free surface lid in the multiple-body diffraction analysis for the side-by-side offloading system resulted in an important improvement of drift force prediction and resulting relative sway and yaw motions.

Figure 2: Wave interaction between 2 vessels (photograph by Mark Scholma)

This numerical lid method is developed for small clearances between vessels: 5 meter mutual distance for vessels with a length of 270 meter. This is therefore also applied in the present calculations.

Until recent, it was always recommended to perform a model test to quantify the permeability of the lid. The recommended permeability has only recently be established (see [7]).

For the application under consideration, these wave forces are considered important. The ships will often work in bow or bow quartering waves. At these wave conditions, the interaction forces are considerable.

Furthermore, the wave forces are causing wave frequent and low frequent oscillations, and hence the DP system has to cope with these forces.

Figure 3 shows the calculated wave drift forces (mean second order). This shows that this repulsion force is higher at forward speed and small distances. Figure 3 shows only the mean second order wave drift force. The slowly varying wave drift forces may even be more important for the performance of the DP system. These forces increase with the same order of magnitude.

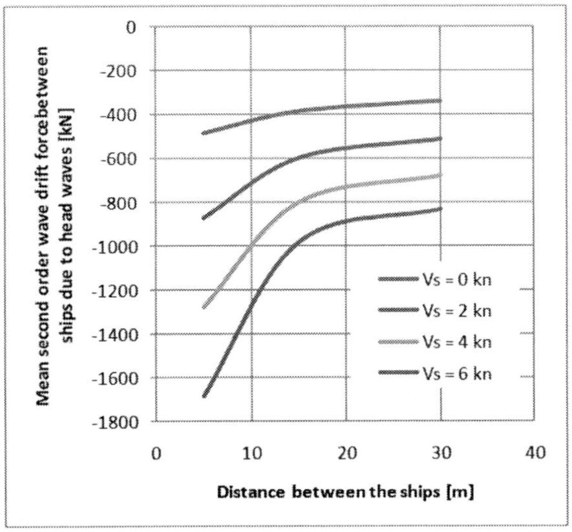

Figure 3: Mean second order wave drift force in head waves (180°) for different forward speeds and different mutual distances, for sea state 4 (Hs=2.5m and Tp=8.8s)

2.2 SHIP-TO-SHIP SUCTION

In early years, publications on these forces were scarce, but these suction forces are treated in recent years (as can be observed in [10] and [11]). The calculation possibilities have increased considerably.

For ships in the side-by-side arrangement, the ships are sailing for a long time at the same mutual position (at least: that is the objective). At this time, the suction forces between the ships are of a constant magnitude. The ship velocities are ranging between 0 and 8 knots. At 0 knots, the expected suction forces are negligible, and obviously at 8 knots, the forces are larger. The fact that these forces are relatively constant makes is easier for a DP system. As long as the magnitude of the suction forces is smaller than the steering forces, the DP system should be able to deal with this. These ship-to-ship suction forces are an order of magnitude smaller than the 2nd order wave forces.

This is derived in a couple of calculations. Calculations were performed using double body calculations based on [13] and [14]. Also when we compare to other measurements and calculation methods (see [10], [11] and [12]), although most methods are focussing on overtaking or encountering methods.

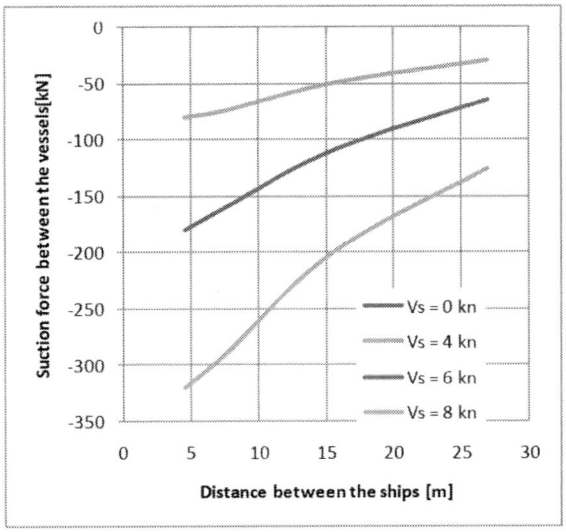

Figure 4: Estimated suction force as function of speed and distance between the vessels

The suction forces are related to the velocity squared, and much depending on the distance between the vessels, both in transverse direction and in longitudinal position. The water depth is an extremely important factor for the magnitude of these suction forces. The present foreseen activities are in deep water. Figure 4 gives the estimation of the suction forces. These forces are an order of magnitude smaller than the 2nd order wave forces.

2.3 MANOEUVRING FORCES

When the ships are moving obliquely through the water they experience lift and drag. Knowledge about the magnitude of these forces is traditionally obtained through captive model tests. In recent years, with very promising results, CFD has achieved increased attention as a method to predict these forces. As the ships are oscillating in their motions, these forces are slowly oscillating as well. These forces are quadratic related to the speed of the vessel: like the suction forces these will be more important at 8 knots than at zero knots.

The control system may use the possibility to compensate the suction and mean wave forces with the manoeuvring forces. These forces are an order of magnitude smaller than the 2nd order wave forces.

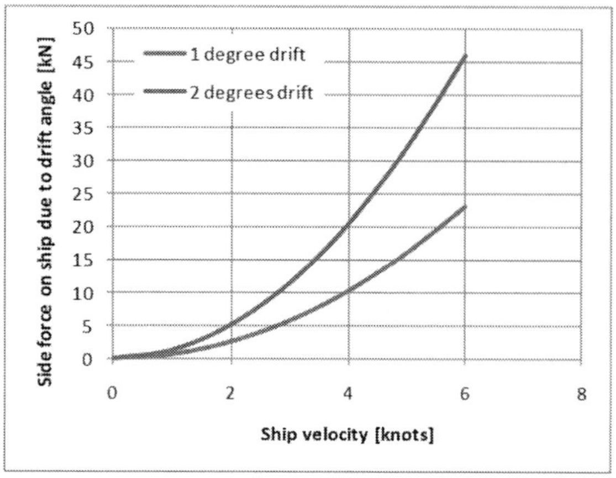

Figure 5: Manoeuvring force as function of drift angle and speed

2.4 FORCES DUE TO PROPULSORS

Different propulsor types have been considered. The propellers and rudders, which are present on both vessels, may be used at larger speeds, or at least larger propeller loadings.

Azimuthing thrusters are commonly used for DP, but the disadvantage is that the time to react - to change the direction of the force - is considerably large. Time domain simulations have been used to demonstrate that faster thrusters would significantly improve the DP/DT performance. These were found in the Voith Schneider propellers. In the kinematic mode, these can change from full thrust starboard to full thrust portside in approximately 3 seconds. Because of this quality, VSP's are used to obtain improved DP performance.

Another important factor is the efficiency of the thrusters at speed. The target speed may range from 0 to 8 knots, and hence the decrease of thrust of traditional azimuthing thrusters with increasing speed will be a limiting factor for the DP performance.

2.5 DP CONTROLLERS

The controller (DP or DT) is very important dynamic contributor to the performance of the side by side arrangement. This is therefore treated separately in the next chapter.

2.6 METHOD FOR ASSESSMENT OF PERFORMANCE

All previously mentioned force and moment components are in reality much more complex to describe than the given indicative numbers. A pure theoretical description of all force components is not available, although the physical insight in these force components is growing significantly, certainly in the past years. A mathematical model encompassing all these components in great detail, great robustness and to the highest possible accuracy would not fit within the time scale of the project.

The typical time frame to come up with an appraisal of the recommended configurations, the risks and any study of possible failure modes needs typically to be investigated in something like 4 months. Free tests, with all aspects "singing and dancing" at the same time would contain all physical aspects and consequently be a very convincing demonstration of the concept. It is very well possible to carry out these tests from a technical point, because all components are operational. Free running model tests are therefore used.

3. DP CONTROLLER FOR RELATIVE DP

3.1 DP IN GENERAL

DP is achieved by a traditional feed-back control system. The measured deviation from a set-point is used as input to determine the forces that the actuators have to generate. Feed forward is used in a limited way to account for wind. This is achieved by measuring the instantaneous wind velocity, calculate the wind force and directly use this information in the DP control loop.

At the end of the eighties, MARIN performed the first DP model tests on drillships and semi-submersibles. The amount of DP model tests has increased ever since and the number of applications has grown. For steering the models during model tests, MARIN uses a home-grown software package called Runsim. This software computes on-line the required control settings for free sailing models. The models are moving as a consequence of the propulsors and the new position of the model forms the input for the DP controller. And hence, it becomes a closed-loop system.

The DP controller are traditionally made for zero speed. For dynamic tracking, the DP controller may use the manoeuvring forces, by pushing the ship in a drifting motion, the hull force can be used within the control system.

A particular item of a DP controller is the filter. This is in general an Extended Kalman Filter. Amongst others, this filter separated low frequent and high (wave) frequent disturbances. The Kalman Filter has the advantage that the amount of phase delay is lower than the phase delay by conventional filters. This is discussed in many papers. Software that describes the possibilities is described in [15]

3.2 RELATIVE DP

For relative DP, there is another issue. For ships under DP in a side-by-side arrangement, a master vessel and a slave vessel are appointed. The master vessel acts under normal DP, in a global system of coordinates. The slave vessel has to react on the motions of the master vessel, it reacts on the mutual distance between the master and the slave. The slave ship is steered in such a way that a point of the slave ship accurately follows the position of a point on the master ship. The challenge for the Kalman filter is that where a normal Kalman filter contains a mathematical model of the ship, an ideal Kalman filter for side-by-side DP would have to contain a mathematical model of the slave ship and the master ship. This is not always practical. The issue has become on how to implement relative motions as input for the filter and how to handle the high frequent motions of the master vessel.

3.3 MEASUREMENT SET-UP

MARIN performs these tests in the Seakeeping and Manoeuvring Basin, the Shallow Water Basin and in the Offshore Basin using actual ship models equipped with thrusters and propellers and of course, with a fully functional, automatic DP system, just as in real life. The DP system in model tests contains all the components that a real-life DP system has: position measurement systems, a Kalman filter, a controller and an allocation algorithm. Figure 6 illustrates how the DP system works for such a side-by-side arrangement. The purpose of the Kalman filter is to let the DP system and thrusters respond only to low frequency (drift) motions of the vessel and not to the wave frequency (oscillating) motions..

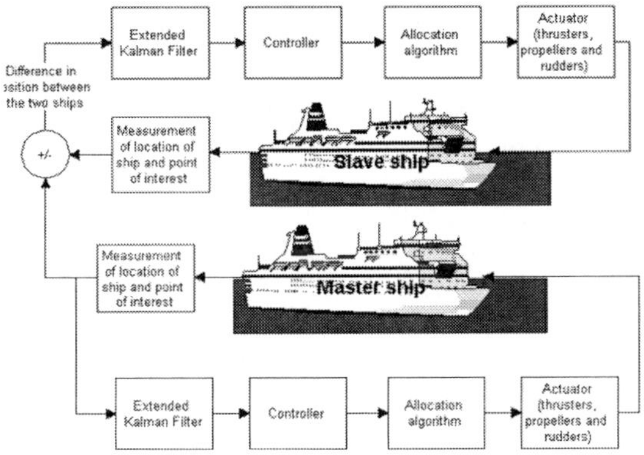

Figure 6: DP controller for absolute and relative DP as used in the model basin.

Model tests with DP simultaneously working on two models has been applied successfully since 2007 (see [1], [2] and [3]). Figure 2 shows two ships under DP in close proximity, in sea state 4. For the tests in Figure 7, the master (on the left) is equipped with traditional twin propeller twin rudder arrangement with two bow thrusters. The second ship (the slave, on the right) uses two main propellers and 4 Voith Schneider propellers. These operate in kinematic mode. All actuators of all ships are steered by their DP controller.

Figure 7: Example of ships operating side-by-side at forward speed at relative DP (STLVAST project). The model size is between 6 and 10 meter.

3.4 FURTHER IMPROVEMENTS TO DP

When more knowledge would be present, the DP controller would be able to perform better. Therefore, Wave Feed Forward was explored as a technique to be able to reduce the excursions by anticipating to the forces before they could realise an excursion.

4. MODEL TESTS PROGRAMME

Free sailing model tests were carried out with a variety of ships, ranging in size from 150 to 300 meters. Both models were steered by DP controllers or DT controllers. The rapid prototyping approach was used during the tests. Rather than having a fixed model test schedule, the target of the model tests program was to improve DP step by step by constant modifications and improvements. These constant modifications and improvements were based on the insights obtained during the testing days. During these model tests the DP controllers were improved and refined. The observations from the model tests showed that the forces due to the waves which are entrapped between the two ships are indeed dominant for the performance.

As part of the project several DP configurations were tested:

- DP using 4 Voith Schneider propellers on the slave vessel
- DP using 5 azimuthing thrusters on the slave vessel
- Improved DP controllers
    - Different feed back systems and alternative Kalman filters
    - Wave feed forward
- Different control techniques for the master vessel
- Different relative DP technique
    - Normal relative DP
    - DP pushing
- External DP controller (Kongsberg)

Several of these techniques have been discussed in different papers by the authors. Wave feed forward is extensively discussed in [4]. Alternative methods for DP were discussed in [3]. The improvements achieved by DP-ing using Voith Schneider propellers are described in [1].

Figure 8 shows a footprint of a selection of the results, for one particular case. During the model tests, always in similar wave trains, improved DP controllers could be tested, and their performance could be quantified objectively. In Figure 8, the position of the slave vessel is plotted versus the motions of the master vessel: only the relative motions are plotted.

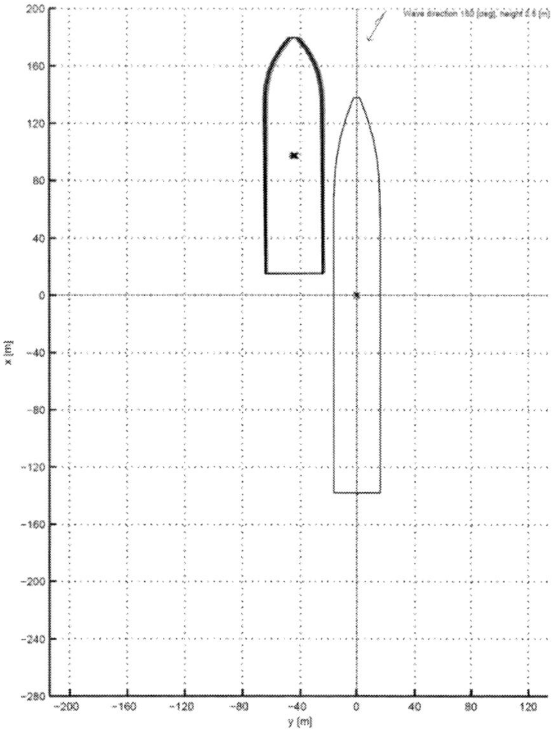

Figure 8: Bird-eye view of relative position of slave with respect to master, the relative position of the small ship is plotted every 180 seconds during a 3 hours test

Particularly unique is the coupling of the Kongsberg DP controller to the MARIN model basin equipment. So, instead of the universal MARIN controller, the Kongsberg DP controller that was going to be used at full scale, could already be used and tested.

5. FULL SCALE TRIALS

Full scale trials (the so-called VTAST trials) were carried out with 2 ships side-by-side. These full scale trials served as a demonstration of the feasibility of the concept.

The goal of the STLVAST part of the VTAST sea trial was to bring the Mighty Servant 3 (outfitted with 5 portable 1.5MW Azimuthing Thrusters) close alongside the LMSR, using DP, and maintain the 'operational distance' safely for two hours at both zero knots and at 3 to 4 knots ahead speed. Figure 9 shows the 2 vessels. All this to occur in sea state 3. Figure 10 illustrates the 'operational distance' which is touching the rear fender and yawed out 2 degrees at the bow. This arrangement is selected instead of a parallel position because the MS3's starboard thruster porch is forward of the front fender, and therefore not well protected. A comprehensive DP Operators Manual was developed that included Project Specific Operating Guidelines (PSOG), a step by step procedure for the experiment and an escape logic, in the event of a DP failure.

Figure 9: Full scale tests under DP showing the LMSR and the Mighty Servant 3 at 'bridge distance'

Bringing the MS3 and LMSR to the 'operational distance' (about 5 meter apart, just touching the fenders) required:

- Extensive testing of multiple new vessel configurations.
- Relative position sensors.
- An improvement in Kongsberg 'follow target' code, that followed from the model tests.
- Upgrading and instrumentation of two 4.5 m fenders on MS3, so that the fender loads during the operations were measured continuously.

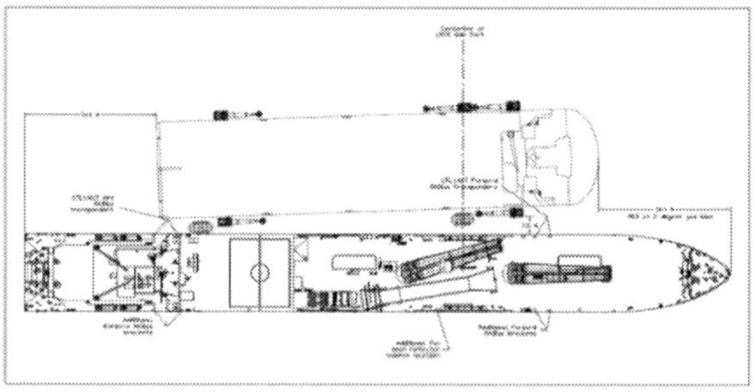

Figure 10: Plan view of the 'operational distance' used during the VTAST sea trial

All of the required equipment and software upgrades were successfully completed in time for the sea trial which began in late January 2010. The trial area was west of Tampa in the Gulf of Mexico. From late afternoon on Saturday 13th Feb until the evening of Sunday 14th Feb the MS3 and LMSR conducted Close-in Proximity DP tests. Saturday testing in sea state 4 allowed checking and calibration of the various relative position sensors. During this phase, the vessels did not approach closer than 30m (see Figure 9). On Sunday, in sea state 3, the MS3 was safely manoeuvred to the 'operational distance' and safely maintained it's close proximity to the LMSR for 2 hours, at zero knots, and then the test was repeated at 3 knots ahead speed (see Figure 11 and Figure 12). In both tests the MS3's DP system kept the relative watch circle and relative bearing between the vessels, within the previously defined PSOG limits of 2.5m and 1° yaw from the initial 2° yaw-out. Figure 13 and Figure 14 record the relative watch circle for the two vessels during the two hours of close proximity. The results were excellent, with the tight watch circles validating Kongsberg's improvement in the "follow target" and the skill of the C-MAR DPO's.

The improvement in the watch circle seen as we moved ahead is consistent with the MARIN tank results and occurs as a result of the effective biasing of the thrusters. In the test at 3 knot forward speed, the five thrusters produce the ahead thrust to keep the MS3 next to the LMSR, in addition to producing the athwartship thrust to maintain the separation to the LMSR. This has the effect of cycling the thrusters thru about 90 degrees of azimuth, a relatively tight range that makes them more responsive to any position error between the vessels. In the zero knots test, with the only 'load' coming from the position error the thrusters were repeatedly required to azimuth thru 180 degrees, which slowed their response, and increased the watch circle.

Figure 11: View forward during sea trial, vessels are in position for a crane transfer, and their relative watch circle is 1.5m

Figure 12: View aft during sea trial, vessels in position, aft fender touching, underway at 3.7 knots

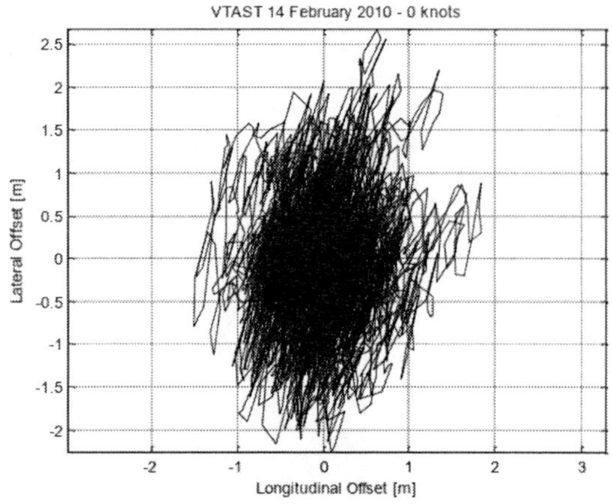

Figure 13: Relative Watch Circle between MS3 and LMSR, 0 Knots

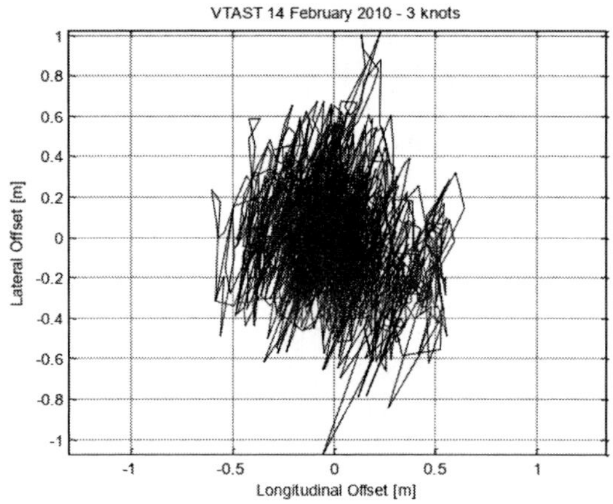

Figure 14: Relative Watch Circle between MS3 and LMSR, 3 knots

6. CONCLUSIONS

As part of the STLVAST program several techniques have been investigated that enable to position two very large ships very close to each other.

It was demonstrated that many physical aspects are playing a role. The wave forces are a considerably important force component. The wave interaction forces are the most dominant forces for the presently sketched configuration.

To investigate the feasibility, free running model tests were used, with simultaneously two models fully equipped and sailing at zero speed and at forward speed next to each other. During these tests, not only the hydrodynamics are modelled as realistic as possible, but also the control systems. Two vessels were acting under DP simultaneously. Both the MARIN in-house and the Kongsberg DP controllers were used as DP controller. Rapid prototyping enabled us to perform various improvements and explore various techniques to improve the DP performance (were DP performance was rated based on the dimensions of the footprint of the relative position of the vessels).

After establishment of a working arrangement in the model basin, full scale experiments were carried out in the Gulf of Mexico. These experiments showed that the concept was feasible and that it is possible to keep the ships in very close proximity using Dynamic Positioning techniques.

7. ACKNOWLEDGEMENTS

The STLVAST project was initiated and carried out for ONR with Oceaneering as main contractor. Their stimulation for publication of the results is gratefully acknowledged.

8. REFERENCES

[1] Hughes, G, R. Harris and F.H.H.A. Quadvlieg; A Tighter Watch Circle at Higher Speeds; Dynamic Positioning Conference, Houston, 2008

[2] Hughes, G., R. Harris, F.H.H.A. Quadvlieg, and R. Hallmann, "Close-In Precision DP using Wave Feed Forward: STLVAST Phase 2 & 3", Marine Technology Society Dynamic Positioning Conference, Houston, 2009

[3] Hughes, G, R. Harris, F.H.H.A. Quadvlieg and R. Hallmann; DP Advances in the USN Seabase: STLVAST Sea Trial and Completion; Dynamic Positioning Conference, Houston, 2010

[4] Quadvlieg, F.H.H.A., Hallmann, R., Hughes, G. and Harris, D.H.; Improved dynanic positioning using wave feed forward. Proceedings of the ASME 2011 30th International Conference on Ocean, Offshore and Arctic Engineering. OMAE2011. June 19-24, 2011, Rotterdam, The Netherlands

[5] Pinkster, J.A.; Low Frequency second order wave exciting forces on floating structures; Ph.D. Thesis, 1980

[6] Huijsmans, R.H.M., Pinkster, J.A. and Wilde, J.J. de, Diffraction and Radiation of Waves around Side-by-side Moored Vessels. ISOPE Conference, Stavanger, 2001

[7] Buchner, B., Wilde, J.J. de and Dijk, A. van, Numerical Multiple-body Simulations of Side-by-side Mooring to an FPSO. ISOPE Conference, Stavanger, 2001

[8] Mamoun Naciri (SBM); Olaf Waals and Jaap de Wilde (MARIN), Time Domain Simulations of Side-by-side Moored Vessels - Lessons Learnt from a Benchmark Test. OMAE Conference, San Diego, 2007

[9] Bunnik, T.H.J., Pauw, W. and Voogt, A.J., Hydrodynamic analysis for side-by-side offloading. Nineteenth International Offshore and Polar

Engineering Conference (ISOPE), Osaka, Japan, 2009

[10] Vantorre M., E. Verzhbitskaya, E. Laforce "Model test based formulations of ship-ship interaction forces", Schiffstechnik Bd.49–2002/Ship Technology Research Vol.49, pp124-141, 2002

[11] Lataire, Vantorre, M. and Delefortrie, G.; Captive model testing for ship to ship operations. MARSIM 2009, Panama

[12] Varyani, K, Thavalingam, A and Krishnankutty, P.; New generic mathematical model to predict hydrodynaic interaction effects for overtaking maneuvres in simulators. J.Mar Sci Technology (2004) 9:24-31

[13] Tuck, E.O. and Newman, J.N.; Hydrodynamic Interactions Between Ships , Proc. of the 10th Symp. on Naval Hyd., Cambridge, Mass., USA, pp. 35–70.

[14] Abkowitz, M.A., G.M. Ashe, and R.M. Fortson; Interaction Effects of Ships Operating in Proximity in Deep and Shallow Water. Proceedings of the 11th Symposium on Naval Hydrodynamics, London, UK, March 28–April 02, pp. VII.37–VII.57

[15] J.J. Serraris, "Time Domain Analysis for DP Simulations" 28th International Conference on Ocean, Offshore and Arctic Engineering, Honolulu, Hawaii, May- June 2009

9. AUTHORS BIOGRAPHY

Frans Quadvlieg holds the position of Senior Project manager and Knowledge Coordinator at MARIN. His field of expertise is the manoeuvrability and controllability of surface ships and submarines.

Rink Hallmann holds the position of project manager at MARIN. He is responsible for various projects in the area of manoeuvring and seakeeping of special ships. He has carried out multiple projects using dynamic positioning, using time domain simulations and model tests.

Greg Hughes holds the position of program manager at Oceaneering. He is responsible for advanced projects in the maritime field. He is the program manager of the STLVAST project.

Rick Harris holds the position of Principal Engineer at MAPC. His field of expertise is the ride control systems, model testing, hydrodynamics and manoeuvrability and controllability of ships.

All 4 authors worked extensively on the SLVAST project, which lasted from 2007 to 2011.

URANS STUDIES OF SHIP-SHIP INTERACTIONS IN SHALLOW-WATER

Hamid Sadat-Hosseini, IIHR - Hydroscience & Engineering, The University of Iowa, Iowa City, IA, 52242, USA.
Ping-Chen Wu and **Yasuyuki Toda,** Department of Naval Architecture and Ocean Engineering, Osaka University, Osaka, Japan.
Pablo Carrica and **Frederick Stern,** IIHR - Hydroscience & Engineering, The University of Iowa, Iowa City, IA, 52242, USA.

SUMMARY

The purpose of the present study is to investigate the interaction between two different tankers Aframax and KVLCC2 free to heave and pitch advancing in shallow-water with same speed and with a fixed separation distance using URANS simulation. The results are validated against EFD conducted in Flanders Hydraulics Research (FHR) towing tank. The present work will focus on the test cases that have the smallest under-keel clearance and water depth among all EFD cases (case A and B). Also, effects of longitudinal alignment of the ships are investigated for case B. For both cases A and B, the wave pattern, forces and moments, and motions are predicted and overall show fairly good agreement with experimental data. The simulations also reveal several influences of the ship-ship interactions such as the suction force and asymmetric ship wake. Also, it is found that the longitudinal alignment has significant impact on the yaw moment.

NOMENCLATURE

β	Ship drift angle (°)
δ	Rudder angle (°)
δx	Longitudinal distance between mid-ships (m): >0 if Aframax is ahead
δy	Separation (m): distance between ship sides
k	Turbulent kinetic energy (m^2 s^{-2})
ν_t	Turbulent kinematic viscosity (Ns m^{-2})
σ	Ship sinkage (mm)
τ	Ship trim (°)
Ψ	Ship heading (°)
ω	Specific dissipation rate (s^{-1})
∇	Ship displacement volume (m^3)
B	Beam length of the ship (m)
C_p	Pressure coefficient
D_P	Propeller diameter (m)
D	Experimental data
$E\%D$	Comparison Error=100(D-S)/D
F_h	Froude number based on water depth
F_n	Y-direction force on rudder (N)
Fr	Froude number based on ship length
F_t	X-direction force on rudder (N)
h	Water depth (m)
K	Roll moment (N m)
KG	Ship's center of gravity above keel (m)
K_{XX}, K_{YY}, K_{ZZ}	Radius of gyration (m)
L_{PP}	Ship length between perpendiculars(m)
N	Yaw moment (N m)
n	Propeller rate (rpm)
P	Propeller pitch
p	Pressure in flow field (N m^{-2})
Q	Propeller torque (N mm)
Q_r	Rudder shaft torque (N mm)
R	Resistance (N)
Re	Reynolds number
S	Simulation result
T	Ship draft (m)
T_F, T_A	Ship draft fore and aft (m)
T_P	Propeller thrust (N)
U	Ship longitudinal velocity (m s^{-1})
u,v,w	Velocity in flow field (m s^{-1})
UKC	Under-keel clearance (m)
V	Ship lateral velocity (m s^{-1})
X	X-direction total force (N)
Y	Y-direction total force (N)
y_0	Ship lateral position (m): the position of ship center line respect to the center line (y=0) of the towing tank
z_F	Ship sinkage fore (mm)
z_A	Ship sinkage aft (mm)

1. INTRODUCTION

The hydrodynamic interaction between two ships with reduced distance has large impact on the maneuvering of both ships. The magnitude of the interaction forces depends on the speed, size, geometry of the ships, lateral distance between the ships, and the water depth. In restricted water, the interaction is more critical as the waves generated by each ship have more influence on the other one even though the Froude number is small [1]. Also, the magnitude of the interaction force is several times bigger such that the ships could easily lose steerage efficiency and alter course without change in rudder helm. In the present work, the interaction between two different tankers Aframax and KVLCC2 advancing in shallow-water with same speed and with a fixed separation distance is investigated using URANS (Unsteady Reynolds Averaged Navier-Stokes) simulation and the results are validated against experimental data.

Traditionally, the ship-ship interactions in shallow-water have been investigated by conducting model test in the towing tank. In the Towing Tank for Manoeuvres in Shallow Water (cooperation Flanders Hydraulics Research and Ghent University), Vantorre *et al.* [2] studied several different conditions of ship-ship interactions: ships encounter and overtake each other.

Four 1/75 scale ship models were used. The ship speed range was 0~16 knots. The different length, draft and under-keel clearance of ships, the lateral distance between ships and water depth were considered. The target ship was a bulk carrier free to heave and pitch. The longitudinal/lateral force and vertical displacement on ships, rudder and propeller performance were measured on the second ship which was one of a container, a tanker or a small tanker. They showed that the resistance would decrease and then increase during the ships encounter. The repulsion happens in the beginning and end and the attraction forms in the middle. In the case of target ship overtaking second ship, the resistance increases first and then decreases. It has the same changing on the repulsion and attraction with the ships encounter. Lastly, when the second ship overtakes the target ship, the resistance decreases slightly and then increases. They showed that the forces and moments increase with the increasing water depth or decreasing lateral distance. Pinkster and Ruijter [1] introduced the full scale tests carried out in Noordzee canal at Amsterdam. The surge force, horizontal water velocity and wave elevation were measured on a moored barge passed by a bulk carrier at 5.8 knots or a cruise liner at 8.9 knots. The result indicated that as the bigger vessel passes, the relatively smaller barge would be sucked toward it and then pulled backward. As the bigger vessel moves more close, the lower water level would form around the barge. The ship-generated solitary waves were also studied. When the barge moves in the canal, the single wave crest and trough with the almost constant shape was detected in the far-upstream. Kriebel [3] investigated the mooring loads resulting from parallel and perpendicular passing ships. The experiments were conducted in the wider basin instead of the towing tank in order to eliminate side wall effect. The surge and sway forces and yaw moments were measured on the mooring ship. Ship models used in the test were two Series 60 models with the same length (5ft) but different beams, drafts and block coefficient. The draft-to-depth ranges from 0.24 to 0.9. The separation ratio between two ships is from 0.3 to 1.0 moored ship length. When the ship is passing, the stern-directed surge force for moored ship shifts to the bow direction. The suction force increases and then becomes repulsion after passing. The yaw moment would first push the bow toward the pier and then pull the bow away from the pier. Lataire *et al.* (2009) performed captive model tests in shallow-water with finite width in the towing tank of FHR. The 1/75 scaled model of a VLCC and an Aframax tanker moving in parallel with self-propulsion were tested. They showed that the yaw moment changes with the different longitudinal distance between two ships. When the shorter Aframax model is ahead, a bow-out moment would exert on it. When the Aframax model is behind, a bow-in moment would generate. The detailed test conditions and results will be discussed later as these data are used to validate the computations.

Previous URANS studies in shallow-water mainly focus on single-ship computations and only few included ship-ship interactions. Chen *et al.* [5] investigated the multi-ship interactions in restricted water by the chimera RANS method developed based on finite-analytic method. The two-layer turbulence model was employed. The effects of ship speed, wind, wharf distance, sheltering, superposition and bottom clearance were considered. In the small bottom clearance, the high-frequency force oscillations on the hull were induced by passing ships. Simonsen *et al.* [6] performed CFD simulations for KVLCC2 in deep and shallow-water. Low speed simulations (Fr=0.0643) are carried out without free-surface for straight ahead (β=0°) and static drift with β=4° for h/T ranging from 1.2 to 10. For the β=4° case with h/T=1.2, simulations are also carried out with and without free-surface for higher speed (Fr=0.1424). Validation and verification (V&V) studies are carried out for β=4° shallow-water (h/T=1.2) with low speed (Fr=0.0643). Validation is achieved for Y force, but for X force (E=18.4%D) and N moment (E=5.4%D) validation is not achieved. At low speed in shallow-water, blockage seemed to play an important role, while the influence from free-surface and squat were less significant. Sakamoto *et al.* [7] performed simulations for Wigley hull with fixed motions in shallow-water (h/T=1.2 and 3) for Fr=0.392. A relative-frame no-slip boundary condition was proposed for the seafloor. V&V study was performed for the h/T=3 case for total resistance, with 1.8M grid points for the fine grid, which showed monotonic convergence but validation was not achieved (E=17.0%D). Also, the result showed that the frictional resistance is less significant than pressure resistance in shallow-water. Jachowski [8] used commercial code FLUENT to evaluate the ship squat in shallow-water. The computations were performed for KRISO Container Ship (KCS) in scale 1:26 with 700,000 structured grids. Four different values of depth to draft (h/T=1.2, 1.5, 2.5 and 22.4) and ship speed (4~21.5 knots) were tested. The result showed good agreement with the experimental data. Zlatev et al. [9] studied the Delft catamaran free to heave and pitch with h/T=8.21 (deep) and 1.5 using 4.67M grid points. Results were compared with EFD only for deep water conditions.

The purpose of the present study is to investigate the interaction between two different tankers Aframax and KVLCC2 free to heave and pitch advancing in shallow-water with same speed and with a fixed separation distance using URANS simulation. The results are validated against EFD conducted in Flanders Hydraulics Research [4]. The computation is performed by using the CFDSHIP-IOWA V4.5. A single-phase level set method and a blended k-ε/k-ω turbulence model are employed. Relative inertial coordinates are used and the domain is discretized using multi-block and dynamic overset structured grids. Body force model will be used for propellers. Rudders are also included in the grid system capable of moving dynamically for the future work.

Table 1: Main dimensions of the models.

	KVLCC2 (STBL)	Aframax (SS)
Hull		
L_{PP} (m)	$4.267 = L_{KV}$	$3.085 = L_{AF}$
B (m)	$0.773 = B_{KV}$	$0.560 = B_{AF}$
Design condition		
T_F / T_A (m)	0.277 / 0.277	0.2 / 0.2
∇ (m^3)	0.741	0.259
KG (m)	$0.248 = KG_{KV}$	0.199
K_{XX} (m)	$0.4B_{KV}$	$0.28B_{AF}$
K_{YY} (m)	$0.25L_{KV}$	$0.25L_{AF}$ ※1
K_{ZZ} (m)	$0.25L_{KV}$	$0.25L_{AF}$ ※1
Ballast/loading condition		
T_F / T_A (m)	0.171 / 0.171	0.1 / 0.121
∇ (m^3)	0.434	0.140
KG (m)	$0.75KG_{KV}$ ※1	0.082
K_{XX} (m)	$0.4B_{KV}$ ※1	$0.28B_{AF}$ ※1
K_{YY} (m)	$0.25L_{KV}$ ※1	$0.25L_{AF}$ ※1
K_{ZZ} (m)	$0.25L_{KV}$ ※1	$0.25L_{AF}$ ※1
Propeller		
Type	Fixed	Fixed
Model	Wageningen B-series	NMRI model
Rotating direction	Right	Right
Number of blades	5	4
Diameter, D_P (m)	0.091	0.131
Hub ratio	0.169 ※1	0.155
Pitch ratio, P/D_P	0.744	0.7212
Area ratio	0.610	0.431
Thickness (m) ※2	6.539×10^{-3}	1.768×10^{-2}
Rudder		
Type	Horn	Ocean Mariner Schilling
Wetted surface (m^2)	0.048	0.028
Lateral area (m^2)	0.024	0.010
※1 Evaluated by the authors for CFD usage. ※2 For the propeller model in CFD.		

2. EFD METHOD

The experiments are conducted for captive models with scale of 1/75 in FHR towing tank. The tank is 88 m long, 7 m wide and 0.52 m deep. The ship to be lightered (STBL) is a KVLCC2 and the service ship (SS) is an Aframax tanker. All tests are carried out with the Aframax model sailing at the starboard side of the KVLCC2. Table 1 shows the main dimensions of both models. The Aframax model is attached to the computer controlled planar motion carriage and the KVLCC2 model is attached to the main carriage of this mechanism while both are free to heave and pitch. Both models are appended with rudders and dynamic propeller. The self-propulsion test has been executed to obtain the propeller rate. Two types of tests are considered: steady state tests, during which the main tests parameters (ships' speed, relative longitudinal and lateral position, propeller rates, drift angle of the Aframax tanker, rudder angle) are kept constant, and dynamic tests, characterized by a varying rudder angle, lateral distance and/or heading. Horizontal forces and moments, propeller thrust and torque, and vertical motions are measured on both ship models, while the vertical motions of the free-surface are monitored at three fixed points in the towing tank. One wave gauge is installed between the ship models, one at the port side of the KVLCC2 and one at the starboard side of the Aframax. For five steady state tests and one dynamic test the measured data are made available as shown in Table 2. In case C, two ships move very close to each other and to the towing tank wall. In case D, two ships move forward with an angle. In case E, the Aframax is ahead. Case F has dynamic rudder angle. The present work will focus on the shallow-water effect. Thus, the study on case A and B is proposed here. Among all cases, both of them have the smallest water depth and under-keel clearance.

Table 2: Test conditions.

	Aframax / KVLCC2 (SS / STBL)					
	A	**B**	**C**	**D**	**E**	**F**
U	0.356	0.297	0.238	0.237	0.356	0.356
V	0	0	0	0.008	0	0
Fr	0.06 / 0.055	0.054/ 0.046	0.043/ 0.037	0.043 / 0.037	0.06 / 0.055	0.06 / 0.055
F_h	0.24	0.18	0.11	0.12	0.078	0.24
h	0.23	0.27	0.475	0.374	0.475	0.23
T ※	0.1 / 0.171	0.2 / 0.171	0.2 / 0.171	0.1 / 0.277	0.1 / 0.277	0.1 / 0.171
h/T	2.3 / 1.345	1.35 / 1.579	2.375 / 2.77	3.74 / 1.35	4.75 / 1.71	2.3 / 1.345
UKC	0.13 / 0.059	0.07 / 0.099	0.275/ 0.304	0.274 / 0.097	0.375/ 0.198	0.13 / 0.059
δy	0.667	0.667	0.053	0.333	0.333	0.133
$\delta y/B$	1.19 / 0.863	1.19 / 0.863	0.095/ 0.069	0.595 / 0.431	0.595/ 0.431	0.237 / 0.172
ψ	180 / 180	180 / 180	180 / 180	178 / 180	180 / 180	180 / 180
δx	0	0	0	0	1.543	0
y_0	-0.327/ 1.0065	-0.327/ 1.0065	0.287/ 1.0065	0.007/ 1.0065	0.007/ 1.0065	0.207/ 1.0065
n	588 / 344	384 / 287	432 / 249	273 / 249	410 / 345	410 / 345
δ	0 / 0	0 / 0	0 / 0	0 / 0	0 / 0	-40 ~ +40
※ T=0.1 for Aframax implies its ballast condition i.e. T_F=0.1 and T_A=0.121.						

In the cases A and B, the Aframax model has two different drafts: design and ballast condition and the KVLCC2 model only has one draft: loading condition. The water depth is different for case A (0.23 m) and B (0.27 m). The separation distance δy between two ships is fixed, but the KVLCC2 is more close to the one side of the channel. Also, no longitudinal distance existed between the mid-ships. The Froude number Fr for Aframax / KVLCC2 is Fr=0.06 / 0.055 for case A and Fr=0.054 / 0.046 for case B.

3. CFD METHOD

The code CFDShip-Iowa v4.5 [10] is used for the CFD computations. CFDShip-Iowa is an overset, block structured CFD solver designed for ship applications using either absolute or relative inertial non-orthogonal curvilinear coordinate system for arbitrary moving but non-deforming control volumes. Turbulence models include blended k-ε/k-ω based isotropic and anisotropic RANS, and DES approaches with near-wall or wall functions. A single-phase level set method is used for free-surface capturing. Captive, semi-captive, and full 6DOF capabilities for multi-objects with parent/child hierarchy are available. Numerical methods include advanced iterative solvers, higher order finite differences with conservative formulation, PISO or projection methods for pressure-velocity coupling, and parallelization with MPI-based domain decomposition. Dynamic overset grids use SUGGAR to compute the domain connectivity.

3.1 COMPUTATIONAL DOMAIN, GRID SYSTEM, AND BOUNDARY CONDITIONS

The length of the upstream and downstream of the computational domain is $1.5L_{KV}$ for the self-propulsion and is $3.0L_{KV}$ for both cases A and B, as shown in Figure 1(a). The domain extends from z=$0.2L_{KV}$ in the air to z=-0.23 m for case A and z=-0.27 m for case B, same as the water depth. The real width of the shallow-water towing tank is 7 m. However, the present work would use wider domain (17 m) to speed up the numerical convergence. Note that the ratio of both ship beams to towing tank width is about 10% and both are located about the middle of tank so that the effects of walls should be negligible in the experiment.

The grid systems are generated by Gridgen V15.15. There are three different grid blocks for each ship: hull, tail and rudder. The grids are generated with a hyperbolic grid generator using a double-O topology, one each for the starboard and the portside. The tail part is used to improve the grid resolution on the complicated geometry around the stern and shaft. An independent grid block is used on the rudder part to describe its geometry extending out of the hull. Thus, it is capable to simulate the operating rudder. Since no wall function is used in this study, the minimum size of the grid cell is refined to 10^{-6} on the boundary layer area to capture the detailed fluid property due to the turbulence. The dimensions ($imax \times jmax \times kmax$) of grids for half (e.g. starboard) of the hull, tail, and rudder for both the KVLCC2 and the Aframax are the same and are 154×50×144, 55×50×40 and 44×43×70, respectively. The background is built using H-type grid topology with dimension of 201×101×151. The fine grids are distributed near the free-surface and surrounding the ships to capture the ship waves. Figure 1(b) shows the grid systems for the stern, tail and rudder of two ships and background. The total number of grid points is 8,470,411.

For the numerical computation, the boundary conditions are specified as described in Table 3. The name of the location follows the setting of grid systems. No-slip conditions are applied on the solid surface of two ships, including hull, tail and rudder. A relative-frame no-slip condition is required on the bottom of the background because the ships and the background move relatively. On the two sides of background i.e. the side walls of the

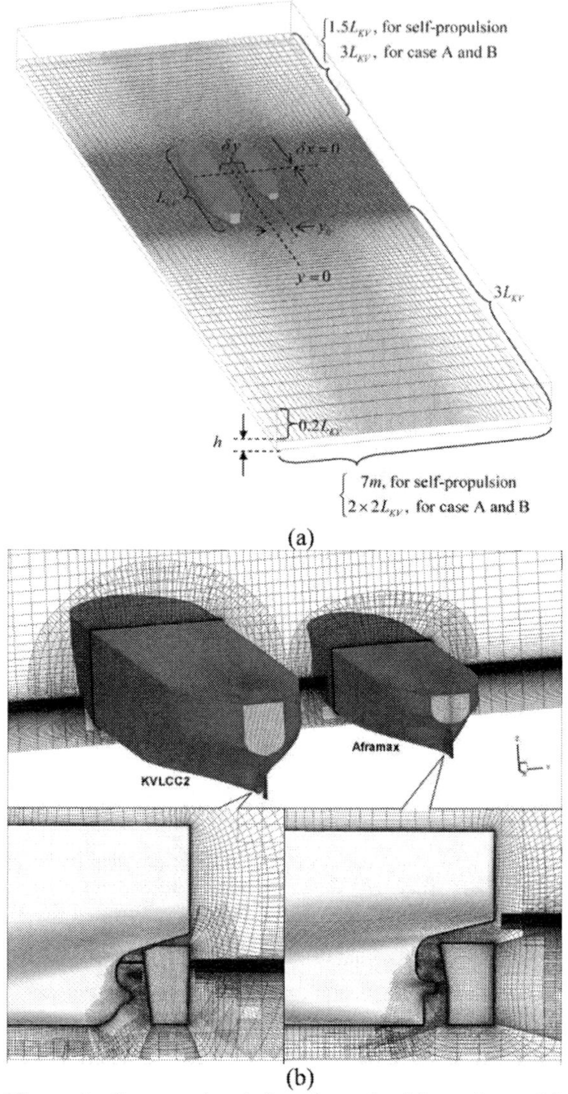

Figure 1: Computational domain and grid topology. (a) The setting of the domain; (b) Overset grid systems.

Table 3: Boundary conditions.

Type/Location						
u	v	w	p	k	ω	ν_t
Inlet/Upstream of background						
u_∞	v_∞	w_∞	$\nabla p=0$	$k_{fst}=10^{-7}$	$\omega_{fst}=9.0$	$\nu_{t,fst}$
Exit/Downstream of background						
$\nabla^2 u=0$	$\nabla^2 v=0$	$\nabla^2 w=0$	$\nabla p=0$	$\nabla k=0$	$\nabla \omega=0$	$\nabla \nu_t=0$
Far field #2/Top (for all cases) Two sides of background (for test case A and B)						
u_∞	v_∞	w_∞	$\nabla p=0$	$\nabla k=0$	$\nabla \omega=0$	$\nabla \nu_t=0$
Absolute-frame no-slip/Hulls, tails and rudders						
0	0	0	$\nabla p=0$	0	$\frac{60}{Re \cdot \beta_1 \Delta y^2}$	0
Relative-frame no-slip/Bottom of background						
$\dot{x}$	$\dot{y}$	$\dot{z}$	$\nabla p=0$	0	$\frac{60}{Re \cdot \beta_1 \Delta y^2}$	0
Zero gradient/Two sides of background (for self-propulsion case)						
$\nabla u=0$	$\nabla v=0$	$\nabla w=0$	$\nabla p=0$	$\nabla k=0$	$\nabla \omega=0$	$\nabla \nu_t=0$

towing tank, zero gradient boundary condition is used for the self-propulsion cases. For the case A and B, wider domain is used to speed up the numerical convergence. Thus, the far-field boundary conditions are specified.

4. RESULTS AND DISCUSSIONS

4.1 CFD SELF-PROPULSION

Self-propulsion is computed free to sink and trim in calm water, using a speed controller to reach the target speed for each model. The predicted propeller rpm is then prescribed for the captive ship-ship simulations at the same nominal Fr. The propeller is modeled by a body force applied on a disk plane or thin cylinder. The open water curves and thickness of the propeller are required by the body-force propeller. According to the available open water test data [11,12], the regression equations of thrust coefficient K_T and torque coefficient K_Q are

$$\begin{cases} K_T = -0.1544J^2 - 0.2837J + 0.3371 \\ 10K_Q = -0.2489J^2 - 0.1928J + 0.3611 \end{cases}, \text{ for Aframax}$$

$$\begin{cases} K_T = -0.1385J^2 - 0.2753J + 0.2931 \\ 10K_Q = -0.2045J^2 - 0.1856J + 0.3071 \end{cases}, \text{ for KVLCC2}$$

where J is the advance ratio defined by U / nD_P. The detailed information of the propellers is listed in Table 1.

One self-propulsion case is performed as a check to evaluate the body force model. The simulation is performed for the Aframax with the draft T=0.1 m in the water depth h=0.23 m. Figure 2(a) indicates the trend of rpm and ship speed during self-propulsion test. The ship speed converges to the target speed Fr=0.055 and the propeller rate also reaches 555 rpm indicating very good agreement (E=5.6%D) with EFD. Thus, the rpm used in EFD can be applied in the following computations for ship-ship interactions.

4.2 RESULTS FOR CASES A AND B

The both test cases A and B are simulated but the simulation of case A suffered the severe difficulty because of more restricted condition. According to Table 2, the KVLCC2 of the case A has the shallowest water condition and its under-keel clearance is extremely small. Thus, for case A, the ship motion of the KVLCC2 would not be solved.

The iteration history is described in Figure 2(b) indicating that the forces and moments exerting on the two ships converge to the steady state value. Because of ignoring the ship motion of KVLCC2, the convergence of case A is faster than of case B.

Table 4 shows the comparison of the variables between CFD results and experimental data. The results are presented in EFD ship-fixed coordinate system in which x positive upstream, y positive to starboard with z pointing downward and the free-surface at rest lies at

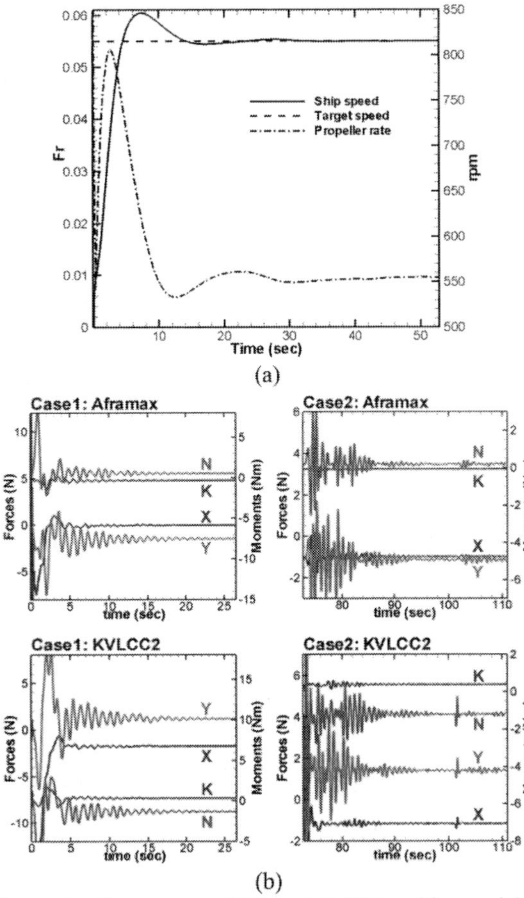

Figure 2: Convergence history (a) CFD self-propulsion with Aframax(SS) for Fr=0.06, h/T=2.3; (b) Left: case A with Fr for Aframax / KVLCC2 =0.06 / 0.055, F_h=0.24; Right: case B with Fr for Aframax / KVLCC2 =0.054 / 0.046, F_h=0.18.

z=0. In case A, σ of the Aframax is predicted by -10%D and the resistance has error of -36%D. For the KVLCC2, the error of Y is -41%D and the disagreement of resistance might result from neglecting of ship motion. In case B, Y and N for the Aframax are predicated by 27%D and -11%D but the resistance R is predicted with larger error -60%D. For the KVLCC2, a very good agreement was obtained for Y with -0.7%D. Its resistance R is predicated by -27%D. Overall, the CFD over-predicts resistances which might be related to the pressure resistances, especially for the Aframax in case B. The flow field analyzed later will reveal the sources of the differences. In both cases, the positive side force on KVLCC2 and negative one on Aframax show an attraction force between the ships. The propeller thrust and torque show large difference between CFD and experiment. The body-force model might not be suitable in this kind of restricted water condition as the propeller performance based on the open water test would be different in shallow-water. The predicted forces on rudders are very small compared with the total forces and moments such that rudder effects should be negligible.

Table 4: Steady integral variables.

	Case A						Case B					
	Aframax (SS)			KVLCC2 (STBL)			Aframax (SS)			KVLCC2 (STBL)		
	D	*S*	*E* (%*D*)	*D*	*S*	*E* (%*D*)	*D*	*S*	*E* (%*D*)	*D*	*S*	*E* (%*D*)
R	-0.7	-1.32	-36	-0.96	-2.97	-209	-0.87	-1.39	-60	-1.47	-2.0	36
Y	-0.30	-1.51	-403	0.85	1.2	-41	-1.55	-1.13	27	1.41	1.42	-0.7
N	1.38	0.57	59	3.44	-1.33	139	0.19	0.21	-11	-0.68	-1.19	-75
K	0.00	-0.29		-	0.35		0.01	-0.06	700	-	0.40	
z_F/z_A	1.1/4.3			-			3.0/-0.3			-		
σ	2.72	2.99	-10	-	0.0		1.37	2.92	-113	-	2.39	
τ	0.059	0.005	92	-	0.0		-0.061	-0.03	51	-	-0.003	
T_P	0.87	1.28	-47	2.44	1.22	50	0.62	0.41	34	1.68	0.85	49
Q	25	14.62	42	37	20.94	43	15	5.04	66	26	14.53	44
F_t	-0.05	-0.12	-140	-	-0.05		-0.06	-0.03	50	-	-0.03	
F_n	-0.09	-0.19	-111	-	0.0085		-0.11	-0.07	36	-	0.01	
Q_r	0.18	0.31	-72	-	-0.005		-0.17	-0.12	29	-	0.08	

-: no data from experiment

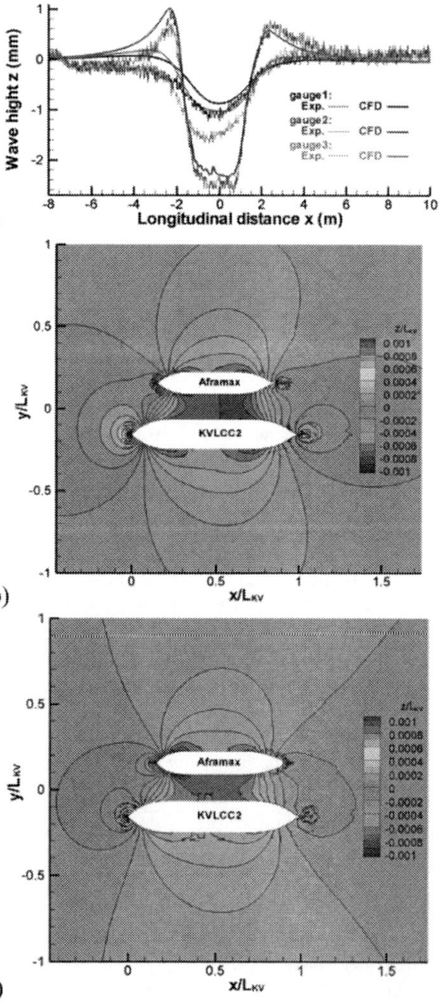

(a) (b) (c)

Figure 3: (a) Wave height detected by three gauges for case B with Fr for Aframax / KVLCC2 =0.054 / 0.046, F_h=0.18; (b) Wave pattern for case A with Fr for Aframax / KVLCC2 =0.06 / 0.055, F_h=0.24; (c) Wave pattern for case B with Fr for Aframax / KVLCC2 =0.054 / 0.046, F_h=0.18.

Figures 3(a), (b) and (c) present the free surface elevation. Figure 3(a) shows the wave height distribution traced in the three locations where the three gauges are placed in the experiment. Gauge1: y_0=-1.834 m, gauge2: y_0=0.595 m and gauge3: y_0=2.166 m in the EFD earth-fixed coordinate (fixed to towing tank). The minus x-direction distance points toward the upstream. The result of gauge2 matches the experiment data, especially the wave trough in the mid-ship and the crest in the stern. The gauge2 locates between the two ships. Thus it detects lowest wave trough due to the lower pressure caused by narrower channel between two ships. The gauge3 obtains the lower wave trough than the gauge1 does because it is closer to the ship. The CFD result also presents this kind of trend but has some error. Since the far-field boundary condition was applied on both sides of the background, the error might be from the wall effect such as the wave reflection or the channel between ship and wall. The wave pattern is shown in Figure 3(b) and (C) for cases A and B, respectively. For both cases, the Froude number based on water depth (F_h=0.24 for case A and F_h=0.18 for case B) is far less than 1.0 which means they are at subcritical speed i.e. ship speed is slower than the phase speed of waves. Thus, the diverging wave is dominant and there are no transverse waves and Kelvin envelope. This is the feature of the solitary wave in shallow-water. The URANS simulation successfully models this phenomenon. The diverging waves transmit to the two sides with the angle around 180 degrees. The lowest wave trough is at the area between two ships and the depth of wave trough increases as being close to the ship. This conclusion is consistent with the result in Figure 3(a). Also, higher crest for bow and stern wave, and lower trough in the mid-ship and between two ships are observed for case A in comparison with those for case B as the Froude number based on ship length Fr is high for case A.

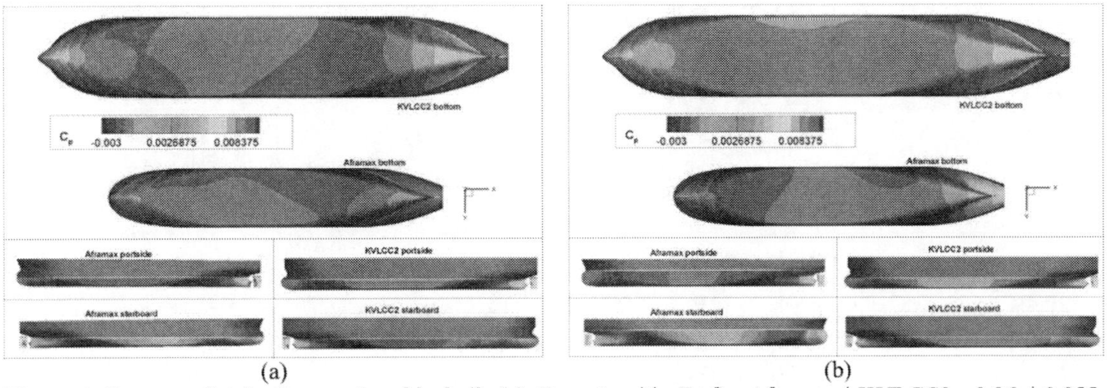

Figure 4: Pressure distribution on the ship hull. (a) Case A with *Fr* for Aframax / KVLCC2 =0.06 / 0.055, F_h=0.24; (b) Case B with *Fr* for Aframax / KVLCC2 =0.054 / 0.046, F_h=0.18.

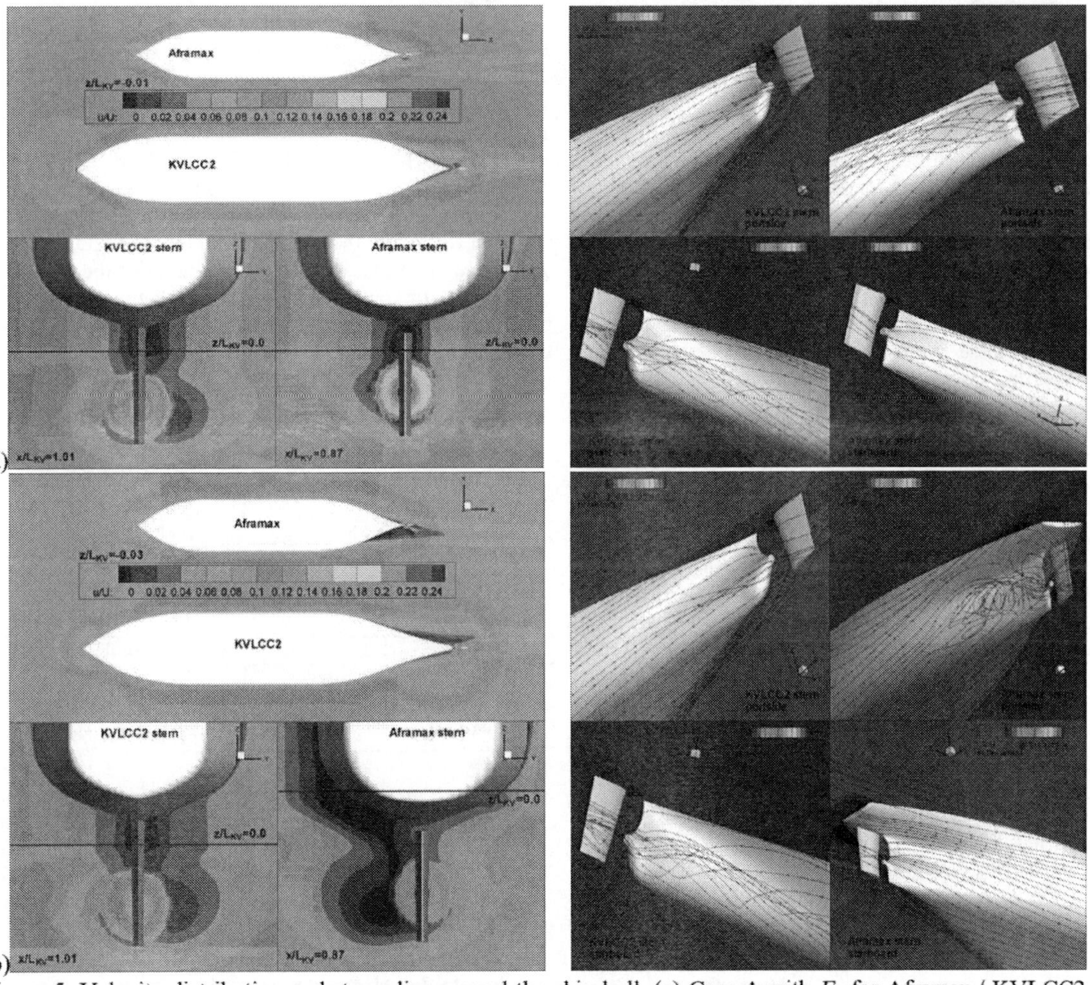

Figure 5: Velocity distribution and streamline around the ship hull. (a) Case A with *Fr* for Aframax / KVLCC2 =0.06 / 0.055, F_h=0.24; (b) Case B with *Fr* for Aframax / KVLCC2 =0.054 / 0.046, F_h=0.18.

Figure 4 shows the distribution of pressure coefficient on the ship hull explaining the negative *Y* on the Aframax and positive *Y* on the KVLCC2 in Table 4. As shown, the low pressure which forms between two ships causes the suction forces on them.

Figure 5 shows the velocity distribution and streamline in the flow field around the ship hull. The thick boundary layer, huge low-speed areas and asymmetric ship wake are presented obviously on the portside of stern of the Aframax and starboard of stern of the KVLCC2. They

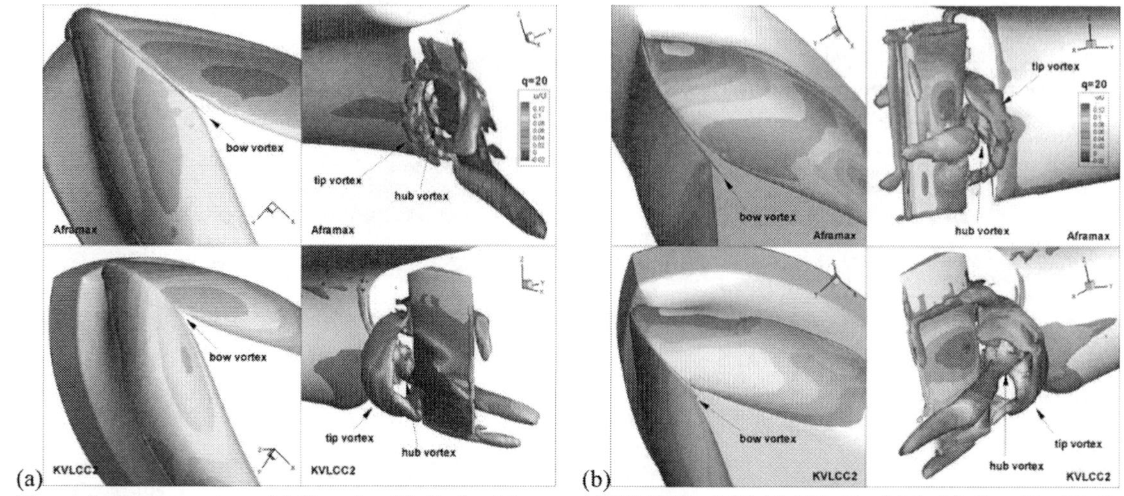

Figure 6: Vortex system. (a) Case A with Fr for Aframax / KVLCC2 =0.06 / 0.055, F_h=0.24; (b) Case B with Fr for Aframax / KVLCC2 =0.054 / 0.046, F_h=0.18.

are even larger than the radius of propellers. That might be the reason of over-predicting the resistances as shown in Table 4. Especially, the Aframax has a very large low-speed area near the free-surface and under its stern. The streamline points out that the strong recirculation zone attaches at this area which causes strong low-pressure area. The possible reason might be the deep draft condition of the Aframax. Fluid close to the Aframax's stern flows from the starboard to the portside which is not observed under the shallower draft condition such as both ships in case A and the KVLCC2 in case B. Their sterns are above the free-surface and rudders are surface-piercing. Therefore, the lower pressure would be "released" by changing the wave profile because the free-surface is a free interface. However, the lower part of the stern and the whole rudder of the Aframax in case B are totally submerged. Since the lower part of the stern of the Aframax is a rigid interface the lower-pressure area cannot be "released" and would attach to it.

Figure 6 shows the velocity contour on the iso-surface of q=20. The vortex systems induced by propellers are observed including tip and hub vortices and the vortex shedding to the downstream. The downwash flows of propellers also cause the high-velocity area on the rudder on the part shown in yellow-red color. Because of the low Froude number in two cases, the bow vortices on the two ships are not significant.

4.3 EFFECTS OF LONGITUDINAL ALIGNMENT

In this section, the influence of different longitudinal distance δx is investigated. The CFD simulation named case B2 is performed for δx=0.657 m, about 21%L_{AF}, and compared with the results for case B which has δx=0.0. In fact, no longitudinal distance existed between two ship bows in case B2 while the mid-ships are aligned in case B. The flow condition, domain setting and geometry of case B2 are the same with these of case B (Fr for Aframax / KVLCC2 =0.054 / 0.046, F_h=0.18).

Table 5 shows the comparison of the variables between case B2 and B, presented in EFD ship-fixed coordinate. The strongest influence by changing δx is on the yaw moment. When the Aframax is in front of the KVLCC2, the positive N of the Aframax increases significantly. This trend is consistent with the changing of the bow-out moment in the experiment [4]. Also the resistance R of the Aframax increases. However, N on the KVLCC2 varies from negative to positive and R decreases and τ increases dramatically.

Table 5: Comparison of steady integral variables between case B2 (δx=0.657m) and B (δx=0.0m).

	Aframax (SS)			KVLCC2 (STBL)		
	Case B2	Case B	※(%)	Case B2	Case B	※(%)
R	-2.06	-1.39	48	-1.41	-2.0	-30
Y	-0.89	-1.13	-21	1.67	1.42	18
N	1.25	0.21	495	0.61	-1.19	-151
K	-0.11	-0.06	83	0.42	0.40	5
σ	2.71	2.92	-7	2.52	2.39	5
τ	-0.02	-0.03	-33	-0.01	-0.003	233
※Difference=[(Case B2)-(Case B)]/(Case B)						

Figure 7(a) presents the comparison of wave heights measured on three gauges between case B2 and B. The condition and setting are the same with Figure 3(a). The amplitude of the waves are similar because of the same ship length-based Froude number (Fr=0.054 / 0.046 for Aframax / KVLCC2). However, on gauge 1 and 2 the wave troughs have phase shifts since those two gauges are close to the Aframax and are affected by longitudinal alignment of the ships. Figure 7(b) shows the same solitary-wave pattern at subcritical speed with Figure 3. The only difference is the more symmetric bow waves radiating to the upstream due to the same x-position of two ship bows.

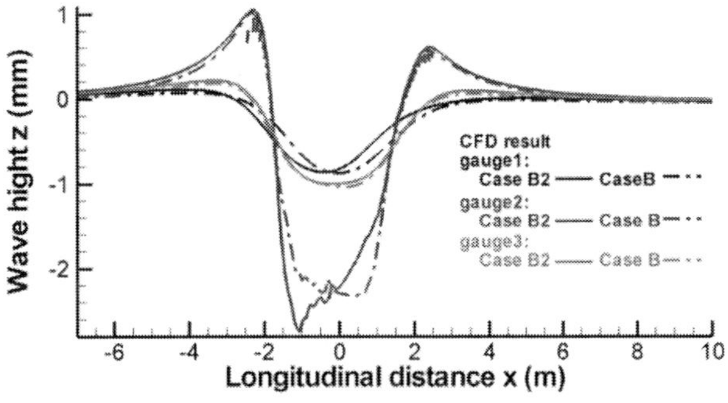

(a)

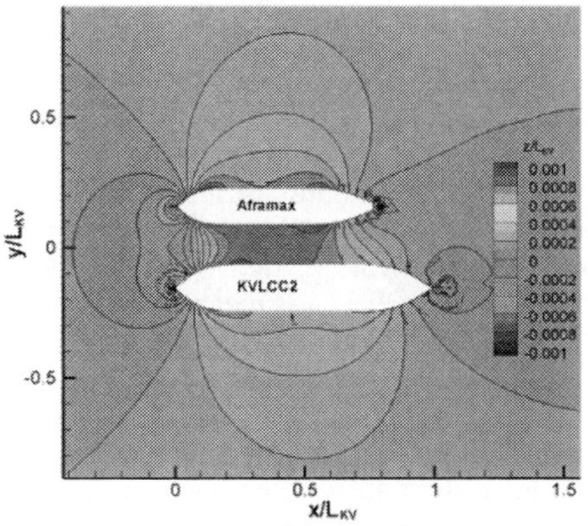

(b)

Figure 7: (a) Comparison of wave height detected on three gauges between case B2 (δx=0.657m) and B (δx=0.0m). (b) Wave pattern of case B2.

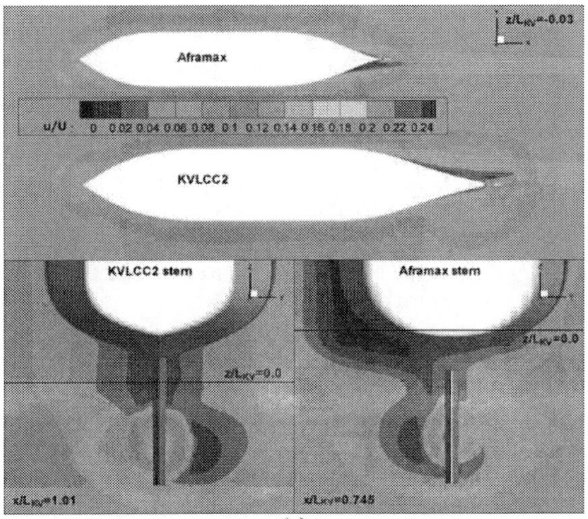

(a)

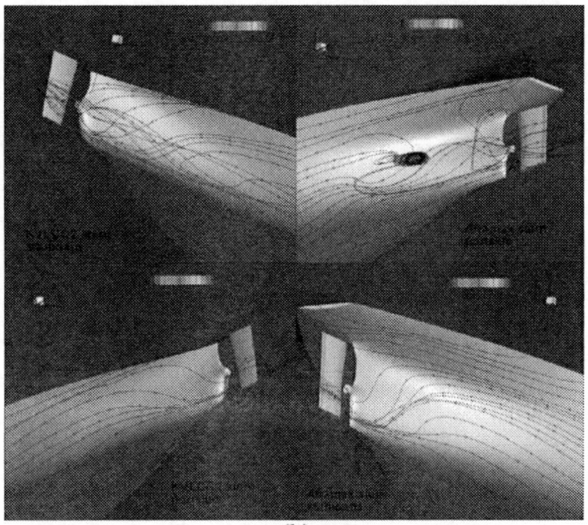

(b)

Figure 8: Velocity distribution and streamline around the ship hull for case B2.

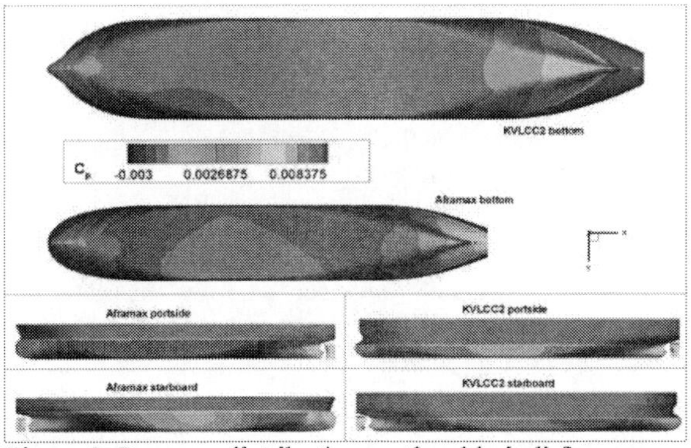

Figure 9: Pressure distribution on the ship hull for case B2.

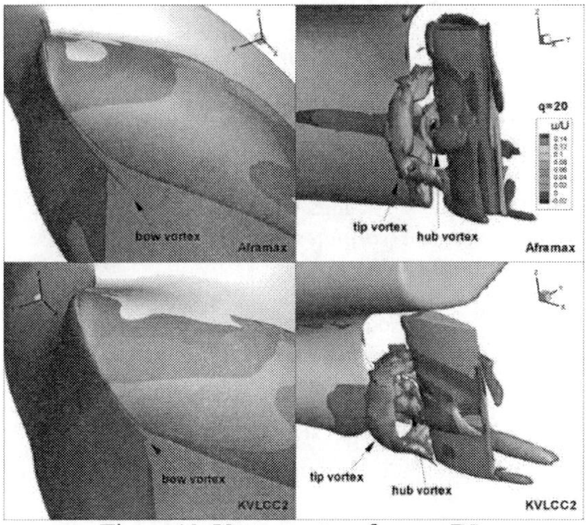

Figure 10. Vortex system for case B2.

Figure 8 shows the velocity distribution and streamline around the ship hull indicating thick and asymmetric boundary layer and ship wake similar to those observed in Figure 5. The strong recirculation zone is also attached to the portside of the stern of the Aframax similar to Case B. Figure 9 indicates the pressure distribution on the ship hull. It also reveals the attraction force between two ships. Figure 10 presents the tip/hub vortex and vortex shedding formed by propellers and indicates insignificant bow vortex similar to the vortex systems of Figure 6.

5. CONCLUSION AND FUTURE WORK

The ship-ship interaction in shallow water is investigated for the test cases A and B as both of them have the smallest under-keel clearance and water depth among all EFD cases. CFDSHIP-IOWA successfully simulated the wave pattern and the results show fairly good agreement with experimental data. The simulation for case A encountered some difficulty to reach numerical convergence as the KVLCC2 in this case has the smallest under-keel clearance among these cases. Therefore the ship motion is ignored for the KVLCC2 in case A. The simulations reveal several influences of the ship-ship interactions such as the suction forces and asymmetric ship wakes. It seems that the over-predicted resistance is induced by the large low-speed area on the stern of each ship. Also, it is found that the longitudinal alignment has significant influence on the yaw moment. In the future work, the grid will be improved for cases with small under-keel clearance and more cases will be simulated to study the effects of water depth and ship separation distance on the flow field surrounding both ships. Also, the propeller performance in shallow-water needs to be investigated.

6. ACKNOWLEDGEMENTS

The research was sponsored by the US Office of Naval Research, grants N00014-01-1-0073 under administration Dr. Patrick Purtell. The CFD simulations were conduced utilizing DoD HPC.

7. REFERENCES

1. Pinkster, J.A., and Ruijter, M.N., 'The Influence of Passing Ships on Ships moored in Restricted Waters', *Svasek Hydraulics Source Offshore Tech. Conf.*, Houston, 2004.
2. Vantorre, M., Verzhbitskaya, E., and Laforce, E., 'Model Test Based Formulations of Ship-Ship Interaction Forces', *Ship Tech. Research Vo.49*, 2002.
3. Kriebel, D., 'Mooring Loads Due to Parallel Passing Ships', *TR-6056-OCN, Naval Facilities Engng Service Center, Port Hueneme, California*, 2005.
4. Lataire, E., Vantorre M., and Delefortrie, G., 'Captive Model Testing For Ship-to-ship Operations', *Intl Conf. on Marine Simulation & Ship Maneuverability, Panama*, 2009.
5. Chen, H.-C., Lin, W.-M., and Hwang, W.-C., 'Application of Chimera RANS Method for Multiple-Ship Interactions In a Navigation Channel', *Proc. 25th Intl Offshore & Polar Engng. Conf., Kitakyushu, Japan*, 2002.
6. Simonsen, C.D., Stern, F., and Agdrup, K., 'CFD with PMM test validation for maneuvering VLCC2 tanker in deep and shallow water', *MARSIM2006, Netherlands*.
7. Sakamoto, N., Wilson, R.V., and Stern, F., 'Reynolds-Averaged Navier-Stokes Simulations for High-Speed Wigley Hull in Deep and Shallow Water', *J. Ship Research, 51, 3, pp. 187–203*, 2007.
8. Jachowski J., 'Assessment of Ship Squat in Shallow Water Using CFD', *Archive of Civil & Mech. Engng., Vol. 8, No. 1*, 2008.
9. Zlatev, Z., Milanov, E., Chotukova, V., Sakamoto, N., Stern, F., 'Combined Model-Scale EFD-CFD Investigation of the Maneuvering Characteristics of a High Speed Catamaran', *FAST2009*, Greece.
10. Carrica, P.M., Wilson, R.V., Noack, R.W., and Stern, F., 'Ship motions using single-phase level set with dynamic overset grids', *Computers & Fluids, Vol. 36, No. 9, pp. 1415-1433*, 2007.
11. Bernitsas, M.M., Ray D., and Kinley P., 'K_T, K_Q and Efficiency Curves for the Wageningen B-Series Propellers', *Report No. 237, Naval Arch. & Marine Engng., University of Michigan, Ann Arbor*, 1981.
12. KLVCC2 Propeller Openwater Data: http://www.simman2008.dk/PDF/KVLCC_POT_N MRI.zip, *SIMMAN2008*.

8. AUTHORS BIOGRAPHY

Hamid Sadat-Hosseini is assistant research scientist at Iowa Institute of Hydraulic Research (IIHR) at the University of Iowa, Iowa City, IA, USA. His research includes 6DOF viscous ship hydrodynamics and CFD.

Ping-Chen Wu is a PhD student at the Department of Naval Architecture and Ocean Engineering, Osaka University, Japan. His main study topic is about CFD.

Yasuyuki Toda is professor at the Department of Naval Architecture and Ocean Engineering, Osaka University, Japan. He is the leader of the laboratory for hull form design. His research includes ship hydrodynamics, CFD, under water vehicle and PIV measurement.

Pablo Carrica is associate professor of Department of Mechanical and Industrial Engineering at the University of Iowa and research engineer at Iowa Institute of Hydraulic Research (IIHR), Iowa City, IA, USA. His research includes multiphase flow and CFD.

Frederick Stern is professor of Department of Mechanical and Industrial Engineering at the University of Iowa and research engineer at Iowa Institute of Hydraulic Research (IIHR), Iowa City, IA, USA. He is the leader of ship hydrodynamics group at IIHR. His research includes ship hydrodynamics, CFD, EFD, UA and multi-criteria CFD-based optimization.

ADVANCED SHIP MANOEUVRING AND MOORING SUPPORT SYSTEM AT SHIP-TO-SHIP TRANSFER OPERATIONS BY FENDER MONITORING SYSTEM OF PNEUMATIC FENDERS

S. Sakakibara and S. Yamada, The Yokohama Rubber Co., Ltd., Japan
E. Shimizu, Tokyo University of Marine Science and Technology, Japan
E. Okuyama and H. Oda, Akishima Laboratories (Mitsui Zosen) Inc., Japan

SUMMARY

The floating type pneumatic fender is an energy absorber for safe ship berthing and mooring used in ship-to-ship oil and gases transfer operations offshore. We proposed a quick fender selection method by using the ordinary ship berthing energy calculation method based on the OCIMF guidelines and simulation software which consists of frequency domain analysis for hydrodynamic forces and time domain analysis for motions of ships. However, it is required for safety to monitor the performance and workability of the fenders, mooring lines and ship behaviours on site. In this paper, a basic study on a new fender monitoring system for the floating type pneumatic fender, which uses radio waves, and an advanced ship manoeuvring and mooring support system by combining with the fender monitoring, are introduced. Several validation tests are demonstrated to evaluate the effectiveness and accuracy of the system.

1. INTRODUCTION

The ship-to-ship (STS) operation transferring crude oil and LNG/LPG has become an important operation in the offshore marine industry [1,2]. Furthermore the operation is also taking place at offshore LNG-terminals and LNG-regasification plants. During the operations, the floating pneumatic rubber fender plays an important role to keep the operations for ship berthing and mooring safe. The ISO standard: ISO17357 was issued to identify and maintain the minimum required quality in performing the operations [3].

However, in the actual operations, it is required for safety to monitor the performance and workability of the fenders, mooring lines and ship behaviours on site. We have developed a fender monitoring system for the floating pneumatic fender, which uses radio waves. The system consists of a wireless air pressure transmitter installed inside the fender body and a receiver of the pressure signal at a required place [4].

In this paper, as an integrated stage for performing safe STS operations, an advanced ship manoeuvring, berthing and mooring operations support system combining two ship positioning system by AIS, measuring system of ship behaviours by digital compass gyro and DGPS, and the fender monitoring system of pneumatic fenders as well as mooring line tension monitoring system, is demonstrated by several laboratory validation tests and field observations.

2. FENDER MONITORING SYSTEM FOR PNEUMATIC FENDER

2.1 CONSTRUCTION OF FLOATING TYPE PNEUMATIC FENDERS

The floating type pneumatic fenders are composed of outer rubber, reinforcement synthetic tire cords and inner rubber as shown in Fig.1.

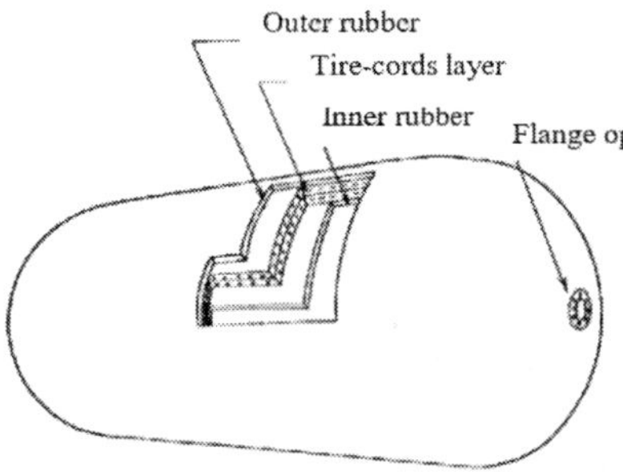

Figure 1: Basic construction of the floating type pneumatic fender

The pressurized air is filled inside the fender body to achieve the rated performance. The material, performance (reaction force and energy absorption/ minimum endurable pressures), dimensions and inspection methods are specified in the international standard: ISO17357 [3,5].

2.2 PRINCIPLE OF FENDER MONITORING SYSTEM, -UNIQUENESS IN PERFORMANCE OF PNEUMATIC FENDER-

In the fender monitoring system, the unique property of the pneumatic fenders is applied. The monitored pressure is dependant not only on the fender deflection but also on the reaction force and the energy absorption of the fender. The pneumatic fender has the unique property as shown in Fig.2.

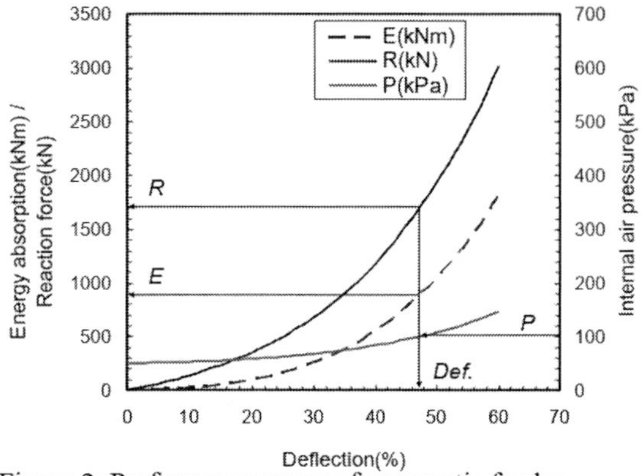

Figure 2: Performance curve of pneumatic fender

2.3 BASIC SYSTEM OF FENDER MONITORING SYSTEM

The floating type pneumatic fenders are used in operation as shown in Fig.3. Thus, fastening a fixed tube for direct measurement of pressure in the fender body under operation is hazardous. In order to measure the internal air pressure, an air pressure transmitter and receiver for radio waves are applied as shown in Fig.4 [4]. The receiver is a handheld device which can display the internal air pressure of the pneumatic fender and its position. This FenderWatch system is considered as an application of the TPMS (Tire Pressure Monitoring System) [6] and it is a basic part of the system of the fender monitoring system.

Figure 3: Floating type pneumatic fender usage at STS

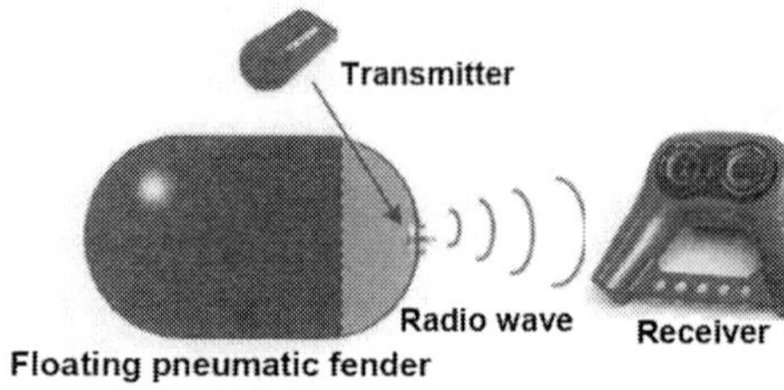

Figure 4: Basic system for measuring the internal air pressure of the floating type pneumatic fender

2.4 CONTROL PROGRAM OF FENDER MONITORING SYSTEM

Fig.5 shows the flow chart of the control program for the fender monitoring system installed in a computer. As shown in Fig.5, there are several functions: initial internal air pressure checking before usage, fender monitoring during usage, alarm indication and recording of the monitored data.

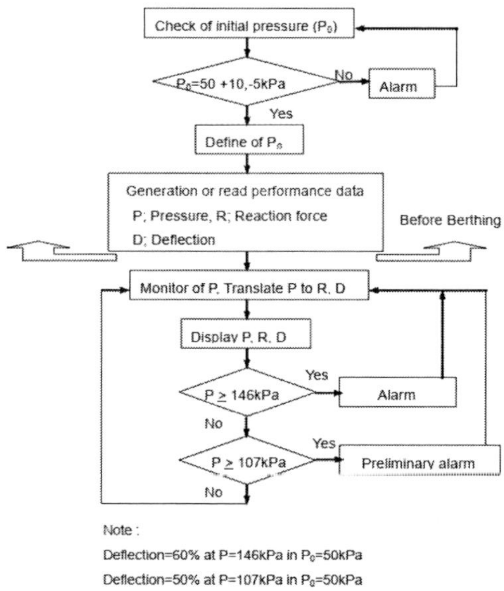

Figure 5 Control program of fender monitoring system

2.5 CONFIGRATION OF THE TOTAL SYSTEM

Fig.6 shows the configuration, process stages and display of an advanced ship manoeuvring, berthing and mooring operations support system combining with the fender monitoring system. As shown in Fig.6(a), the system consists of AIS, pressure sensor and transmitter, and receiver. A digital compass gyro and load cells can be added to the system to measure the ship motions and the mooring rope tensions simultaneously as option. Fig.6(c)-(e) shows the displays of the each process stage in Fig.6(b).

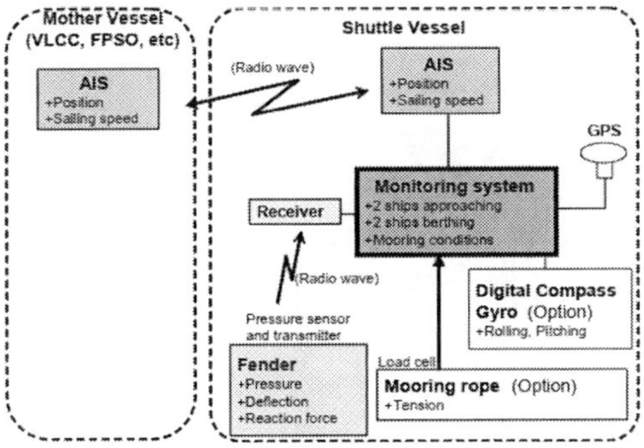

(a) Configuration

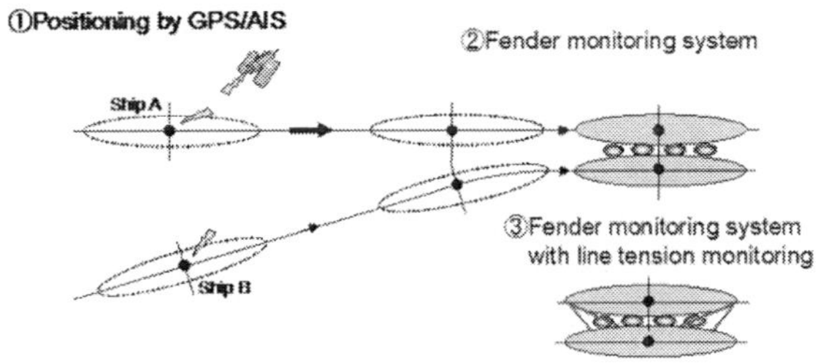

(b) Process stages

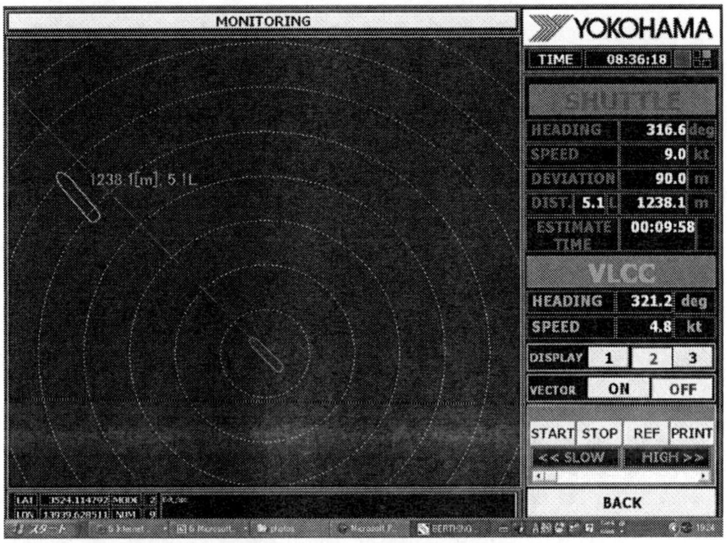

(c) Display at manoeuvring and approaching (1st stage)

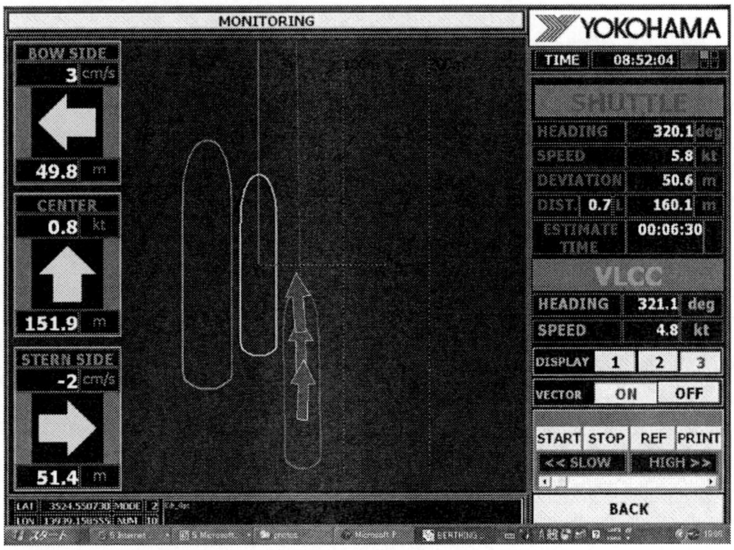

(d) Display at berthing (2nd stage)

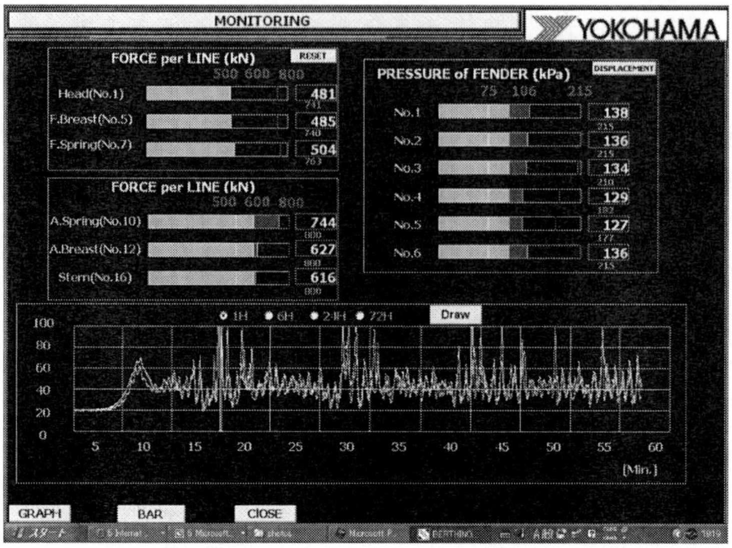

(e) Display at mooring (3rd stage)

Figure 6: Configuration, process stages, and display of the system

3. VALIDATION TEST FOR ACCURACY

3.1 COMPRESSION TEST

In order to confirm the accuracy of pressure data obtained by the pressure transmitter and receiver with radio waves, a compression test was conducted by a 5000kN compression test apparatus for the floating type pneumatic fender: 500mm diameter x 1000mm length. The measured items are compression displacement, reaction force and internal air pressure. The internal air pressure was measured not only by tubing from the fender body directly but also with the pressure transmitter and receiver as shown in Fig.7. The compressions were conducted until 250mm displacement (50% deflection) at the initial internal air pressure: 80kPa. Fig.8(a) shows the test results as the time series of the internal air pressures by direct tube measurement and by radio wave transmission during the compressions. In Fig.8(b), comparisons of the design performance and the test results for the reaction force and deflection derived from the internal air pressure by radio waves are shown. As seen from Fig.8, fairly good agreement between both internal air pressures by tube directly and radio waves is confirmed, and the performance can be monitored by the internal air pressure by radio waves properly.

(a) Test apparatus and fender

(b) Fender monitoring system

Figure 7: Compression test of the floating type pneumatic fender: 500mmD x 1000mmL

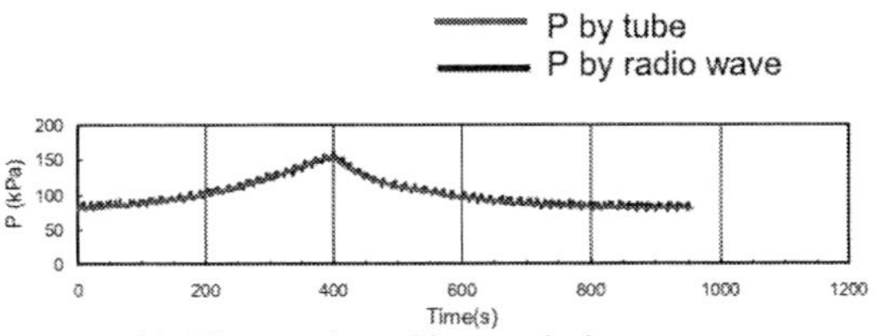

(a) Time series of internal air pressure

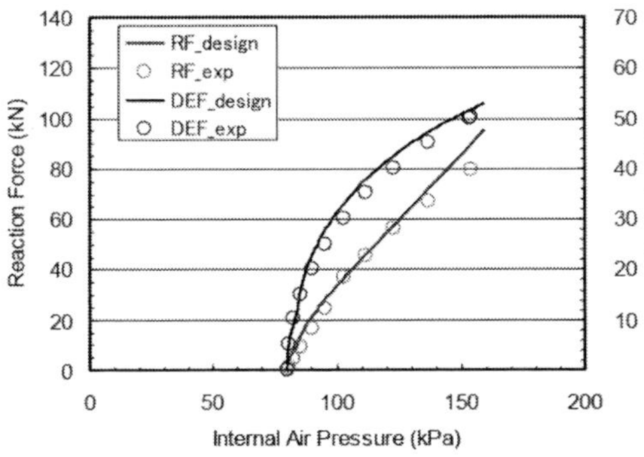

(b) Fender performance based on internal air pressure

Figure 8: Confirmation on accuracy of the fender monitoring system by test results of the floating type pneumatic fender: 500mmD x 1000mmL

3.2 COMMUNICATION TEST

In order to measure the receiving distance between the transmitter and the receiver, a communication test was conducted in our factory using the floating type pneumatic fender: 3300mm diameter x 6500mm length as shown in Fig.9. From this test result, it is confirmed that the communication distance is measured as 400m.

Figure 9: Communication test of fender monitoring system using the floating type pneumatic fender: 3300mmD x 6500mmL

3.3 INVESTIGATION ON WORKABILITY

In order to confirm the workability of the fender monitoring system during STS operations, the covered area by the communication distance: 400m for VLCC and Aframax tanker is shown in Fig.10. Two kinds of the covered areas are shown in Fig.10 where the four fenders are installed along the VLCC or the Aframax tanker. As shown in Fig.10, the fender monitoring system can be applied at ship berthing and approaching properly.

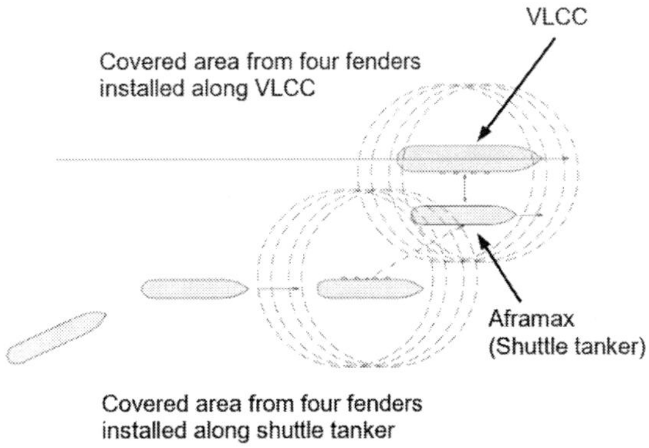

Figure 10: Workability of the fender monitoring system during STS operations

4. CONCLUSIONS

We have developed a fender monitoring system for floating-type pneumatic fenders at STS operations. In this paper, as the basic confirmation of the functions, the effectiveness and accuracy has been investigated by several validation tests for the performance and the workability based on the laboratory test. We have proceeded to the final acceptance test by using the actual tankers for the advanced ship manoeuvring and mooring support system based on the fender monitoring system. We wish that this system will contribute to keep the STS operations safe.

5. REFERENCES

1. OCIMF, 'Ship to Ship Transfer Guide (Petroleum), 4th Edition', 2005.

2. OCIMF, SIGTTO, 'Ship to Ship Transfer Guide (Liquefied Gases), 2nd Edition', 1995.

3. ISO17357, 'Ships and Marine-Technology, High-Pressure Floating Pneumatic Rubber Fenders', International Standard, 2002.

4. Sakakibara, S. and Kubo, M., 'Ship Berthing and Mooring Monitoring System by Pneumatic-type Fenders', Int. Journal of Ocean Engineering, Vol. 34, pp.1174-1181, Elsevier, 2007.

5. Sakakibara, S. and Wan, M., 'Prototype Test of Floating Pneumatic Rubber Fenders based on ISO17357', Proc. of ISOPE, 13th, pp.427-434, 2003.
6. Tanoshita, K., Nakatani, K. and Yamada, Y., 'Electric Field Simulations around a Car of the Tire Pressure Monitoring System', Int. Symposium on Antennas and Propagation, IEICE Trans. Communication, 2006.

6. AUTHORS BIOGRAPHY

Shigeki Sakakibara
Dr. Shigeki Sakakibara received master's degree of Mercantile Marine, Graduate School of Kobe University of Mercantile Marine in 1989 and his Doctor's degree of engineering from Osaka University in 1995. He is the senior engineer of the Yokohama Rubber Co., Ltd. He specializes in effect of mooring system on ship motions due to waves at ship-to-jetty and ship-to-ship. He is a member of the Japan Institute of Navigation and Japan Society of Civil Engineering.

Shuu Yamada
Mr. Shuu Yamada received master's degree of Naval Architecture and Ocean Engineering, Graduate School of Engineering, Yokohama National University in 1999. He is the chief engineer of the Yokohama Rubber Co., Ltd. He specializes in marine rubber products, fendering system and mechanical engineering.

Etsuro Shimizu
Associate Professor Etsuro Shimizu received the Ph.D. degree of Control Engineering from Tokyo Institute Technology in 1999. He is an associate professor of Tokyo University of Marine Science and Technology from 2003. From March 2005 to March 2006, he was a visiting professor of Norwegian University of Science and Technology. His fields of research are nonlinear control theory, modelling and control of vessels. He is a member of IEEE and Ocean Machinery Research Committee of the Japan Institute of Marine Engineering.

Etsuro Okuyama
Mr. Etsuro Okuyama received master's degree of Naval Architecture and Ocean Engineering, Graduate School of Engineering, Osaka University in 2003. He is the chief researcher of Akishima Laboratories (Mitsui Zosen) Inc. He specializes in ship performance in wave field and ship manoeuvring. He is a member of the Japan Society of Naval Architects and Ocean Engineers.

Hiroyuki Oda
Dr. Hiroyuki Oda received his bachelor's degree of Physics from the Sophia University and his Doctor's degree of engineering from Osaka University in 1992. He developed the rudder roll stabilization system of ships. He is currently the general manager of Akishima Laboratories (Mitsui Zosen) Inc. He is a member of the Japan Society of Naval Architects and Ocean Engineers, the Japan Institute of Navigation, and the Society of Instrument and Control Engineering of Japan.

RELATIVE DISTANCES AND SPEEDS MEASUREMENT SYSTEM FOR STS LIGHTERING OPERATIONS USING STEREO-CAMERAS

E Shimizu, Tokyo University of Marine Science and Technology, JAPAN
E Pedersen, Norwegian University of Science and Technology, NORWAY

SUMMARY

A Ship-to-Ship (STS) lightering operation is carried out and controlled by human operators in order to transfer crude oil or LNG from one vessel to another. The final approach phase of two STS vessels with speed underway is particularly critical in order to avoid steel-to-steel contact. Operational safety and efficiency require sufficient instrumental measurements and adequate human operations. A major objective in safe navigation and operation with the ongoing research activity on STS operations is to develop and implement a guidance and decision-support system for the ship navigation officers. In order to develop and implement such a system, the measurement of the relative distances, heading and speeds are important. In this paper, a measurement system using stereo vision is proposed.

1. INTRODUCTION

Applications of Ship-to-Ship (STS) operations for cargo transfer are expected to be increasing. Currently, about 1/4 of all oil imported to the USA comes through lightering operations from the US Gulf, West and East coasts. It is anticipated that further development of the Northwest Russian onshore oil and gas fields will increase the need for STS transfer from a fleet of smaller ice-strengthened vessels to standard vessels for transport of oil and gas products to customers in Europe and the USA. An ongoing international joint-industry research programme on STS operations with focus on STS lightering has a major objective to develop a guidance and decision-support system for the key operative personnel.

STS operations are individually different because of variations in the environmental conditions and the manoeuvring characteristics of ships. The final approach phase is particularly critical in order to avoid steel-to-steel contacts. The officer in charge of an STS lightering, the mooring master, has currently no equipment at his disposal for determining the relative distances and speeds with sufficient accuracies and the decision of adequate manoeuvring orders is thus mainly based on visual observations.

In order to analyze actual lightering operations, two field observation programmes had been carried out by Shimizu et al. [1]. The data recorded by the voyage data recorder (VDR) equipped on the vessel was used to analyze the operation during the approach phase. Relative distances between the ships were measured by the radar equipped on the approaching vessel while relative speeds could not be measured due to lack of time for equipment set-up.

Relative distances and, in particular, speeds are important information for operational safety. If relative speeds can be measured, the ship officer is able to estimate the vessel motion easily and manoeuvre the vessel to avoid steel-to-steel contact. Velocity Information GPS and DGPS have been proposed and introduced to measure relative motion in STS operations, Yoo et al [2], Oda et al [3]. This equipment requires a communication system such as wireless LAN, radio wave, etc. for real-time measurements. However, it is an operative challenge to set-up measurement equipment prior to commence the approach phase.

In this paper, a new measurement system of relative distances, heading and speeds is proposed by application of stereo vision. This is a technology that is widely applied in mobile robots to measure the distance between own robot and objects. The benefit with stereo vision is that relative distances and speeds can be measured by the system equipped on own vessel only. In other words, the setup time of measurement system on the other vessel is not required. This paper discusses how to apply stereo vision to measure relative distances, heading and speeds for STS lightering operations.

2. STS LIGHTERING OPERATION

STS lightering where two vessels, normally referred to as a service ship (SS) and ship-to-be-lightered (STBL), are underway takes normally place in the speed range of 4 - 6 (kts). (Another type of STS lightering is where the SS approaches an anchored STBL.) In the initial approach phase, the SS approaches the STBL with a speed surplus of 1 - 2 (kts) and frequent changes in engine and rudder orders in order to come in position for the final approach, i.e. side-by-side with about 150 (m) separation and identical speeds. Although the STBL is ordered to maintain a steady course and speed, practice is occasionally that this proves difficult to achieve due to various reasons, e.g. miscommunication between the Mooring Master and the STBL's Captain, lack of skills by the helmsman of the STBL's senior officers. Manual steering is the recommended steering mode in order to enable a quick response to any new orders.

It is critical that the STBL maintains a steady course in the final approach phase. The SS closes in by rudder orders with an overall objective to land on all fenders simultaneously to minimize risk of steel to steel contact and with a longitudinal position that is in accordance with the position of the manifolds. The approach requires that the STS has a relative approach angle (up to 5 deg.). However, landing on the bow fender only is not uncommon. Practice has shown that if the bows are not together after the initial landing, it becomes difficult to bring the ships together for mooring and the approach must be aborted.

3. REVIEW OF STEREO VISION

In this section, the fundamentals of stereo vision are briefly explained. Image figure of stereo vision is shown in figure 1. Two (Left and Right) cameras, displaced horizontally from one another, are used to obtain differing views on a scene. By comparing these two images, the relative depth information can be obtained, in the form of a disparity map, which is inversely proportional to the distance to the object. The calculation algorithm to obtain the 3D position $P_0 = (x_0, y_0, z_0)$ of the object is as follows.

$$x_0 = \frac{x_l \cdot d}{x_l - x_r},\ y_0 = \frac{y_l \cdot d}{x_l - x_r},\ z_0 = \frac{f \cdot d}{x_l - x_r}. \qquad (1)$$

$P_l = (x_l, y_l)$	Position on the image obtained by left camera.
$P_r = (x_r, y_r)$	Position on the image obtained by right camera.
d	Horizontally displacement distance between cameras.
f	Focal length.

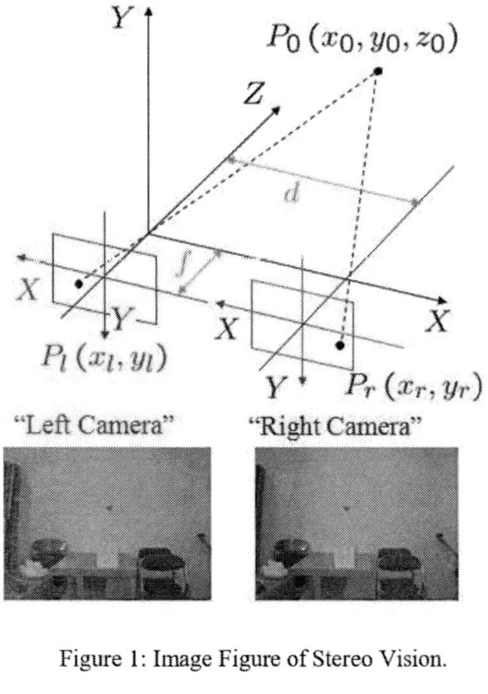

Figure 1: Image Figure of Stereo Vision.

Stereo vision is able to extract information about the relative position of 3D objects without any sensors on objects.

4. SYSTEM SETUP

The application to lightering operation is divided into two phases due to the limitation of the visual field, precision etc. of the camera. Examples of the visual field of the camera are shown in figures 2 and 3.

In figure 2, it is difficult to recognize the precise position of the STBL from SS. However, both sides of the bridge wing of STBL are easy to recognize from the camera. Night time may be easier than day time because navigation lights are turned on. The phase that the positions of both sides of the bridge wing can be observed is defined as the approaching phase in this paper. In this phase, one stereo camera unit is enough to measure the relative distance, heading and speed between ships.

On the other hand, it is difficult to recognize both sides of the bridge wing at the situation shown in figure 3 when the vessels are closing in.

Figure 2: Approaching Phase [4].

Figure 3: Final Phase [4].

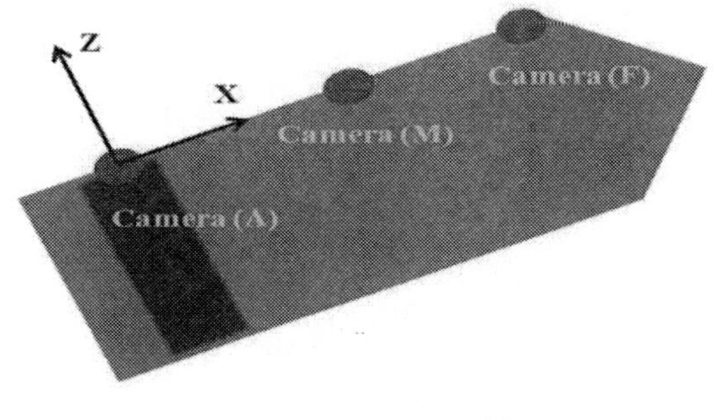

Figure 4: Camera Positions.

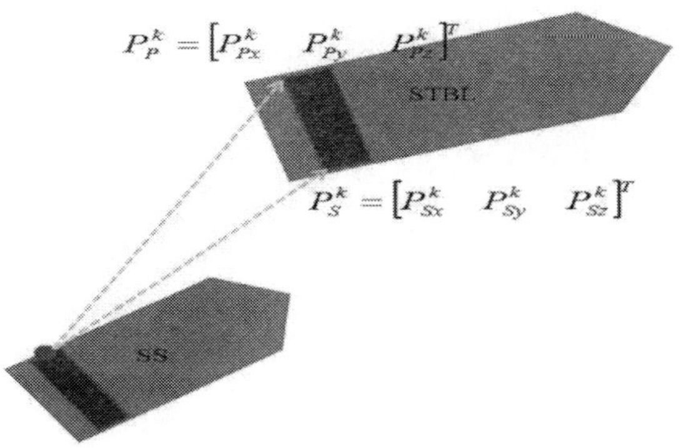

Figure 5: Definition of Notations for Approaching Phase.

This phase is defined as the final phase in this paper. From the limitation of the visual field, one camera unit is not sufficient to measure the relative heading between two ships because two or more points are required to calculate the relative heading. Therefore, two or more stereo camera units and communication between cameras are required. Some numbers of places of hull, manifold etc have to be observed by each camera.

It is assumed that the position of these units on SS are shown as red marks in figure 4 and altitudes of camera (F) and camera (M) are the same. Camera (A) is placed on the bridge wing. The coordinate system of the camera unit is the same as figure 1, i.e., x axis for longitudinal, y axis for altitudinal and z axis for lateral direction. Algorithms to calculate the relative position and speed between ships are shown in the next section.

5. CALCULATION ALGORITHM

5.1 APPROACH PHASE

It is assumed that the position of the port and starboard side of the bridge wing can be obtained by the stereo camera unit using equation (1). Symbols are defined as follows:

P_P^k	3D position of the port side of the bridge wing at the step k.
P_S^k	3D position of the starboard side of the bridge wing at the step k.
$P_{*?}^k$	Distance from SS to STBL in ? axis.

If there are no roll motions, the altitudes of both wing have to be the same distance. In order to simplify the problem, this condition is also assumed.

5.1 (a) Relative Distance and Heading

The relative distance from the camera on SS to the centre of the bridge can be easily obtained by the following equation.

$$d_A^k = \sqrt{\left(\frac{P_{Sx}^k - P_{Px}^k}{2}\right)^2 + \left(\frac{P_{Sz}^k - P_{Pz}^k}{2}\right)^2} \tag{2}$$

d_A^k	Relative distance between ships at the step k.

The gradient between P_P^k and P_S^k can be calculated by

$$a^k = \frac{P_{Sz}^k - P_{Pz}^k}{P_{Sx}^k - P_{Px}^k} \tag{4}$$

a^k	Gradient between P_P^k and P_S^k.

Therefore, the relative heading between ships can be calculated by the following equation.

$$\phi^k = \tan^{-1}\left(-\frac{1}{a^k}\right) = \tan^{-1}\left(-\frac{P_{Sx}^k - P_{Px}^k}{P_{Sz}^k - P_{Pz}^k}\right) \tag{5}$$

ϕ^k	Relative heading from SS to STBL at the step k.

5.1 (b) Relative Linear and Angular Speed

In order to calculate the relative linear and angler speed from position data, many methods are considered. One method to obtain the relative linear and angular speed is the following equation:

$$v_A^k = \frac{d_A^k - d_A^{k-1}}{\Delta t} \tag{4}$$

$$r^k = \frac{\phi^k - \phi^{k-1}}{\Delta t} \tag{5}$$

v_A^k	Relative linear speed between ships at the step k.
r^k	Relative angler speed between ships at the step k.
Δt	Sampling step time.

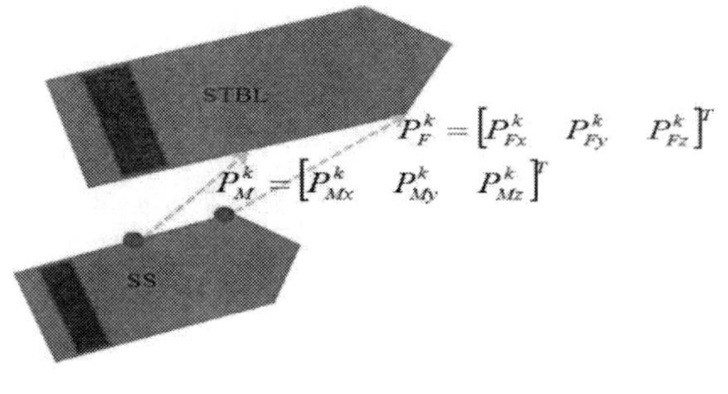

Figure 6: Definition of Notations for Final Phase.

5.2 FINAL PHASE

It is assumed that the position of the front and middle marked points can be obtained by the stereo camera unit using equation (1). Symbols are defined as follows.

P_F^k 3D position of the front marked point at the step k.

P_M^k 3D position of the middle marked point at the step k.

$P_{*?}^k$ Distance from SS to STBL in ? axis.

It is also assumed that altitude of P_F^k and P_M^k are identical.

5.2 (a) Relative Distance and Heading

The relative distance from the camera on SS to the front and middle marked position on STBL can be obtained by following equation.

$$d_*^k = \sqrt{(P_{*x}^k)^2 + (P_{*z}^k)^2} \quad (6)$$

d_*^k Relative distance between SS and the * (front or middle) marked point of STBL at the step k.

The relative heading between ships can be calculated by the following equation:

$$\phi^k = \tan^{-1}\left(\frac{P_{Fz}^k - P_{Mz}^k + \Delta P_z}{P_{Fx}^k - P_{Mx}^k + \Delta P_x}\right)$$

(7)

ϕ^k Relative Heading from SS to STBL at the step k.

ΔP_* Position deviation between Camera (F) and Camera (M).

5.2 (b) Relative Linear and Angular Speed.

The relative linear and angular speed can be calculated by using the following difference equation:

$$v_*^k = \frac{d_*^k - d_*^{k-1}}{\Delta t} \quad (8)$$

$$r^k = \frac{\phi^k - \phi^{k-1}}{\Delta t} \quad (9)$$

v_*^k Relative linear speed between SS and the * (front or middle) marked point of STBL at the step k.

r^k Relative angler speed between ships at the step k.

Δt Sampling step time.

6. CONCLUDING REMARKS

In this paper, a new measurement system of relative speeds and distances using stereo vision has been proposed. The most attractive point of the stereo vision is that relative distances and speeds can be measured by the system equipped on own vessel only. No external equipment on STBL is required for the measurement. The algorithm to obtain relative distances and speeds were proposed. The true positions, heading and speeds from obtained relative distances, heading and speeds can be obtained by addition-subtraction of own position and motion.

7. ACKNOWLEDGEMENTS

A part of this research has been funded by the Norwegian Research Council through the international joint-industry research programme 'Investigating Hydrodynamic and Control Aspects of Ship-to-Ship Operations'.

8. REFERENCES

1. SHIMIZU, E., PEDERSEN, E. and TAKANO, Y., 'Field Observation on Actual Lightering Operations', *Proceedings of OCEANS '10 IEEE Sydney Conference and Exhibition*, 2010.

2. YOO, Y., HUSJORD, D., SHIMIZU, E., KOUGUCHI, N., ARAI, Y. and PEDERSEN, E., 'Field Observations and STS Lightering with Comparative Analysis Results of Relative Speed', *International Conference on Marine Simulation and Ship Maneuverability & IMSF General Meeting (MARSIM '09)*, 2009.

3. ODA, H., OKUYAMA, E. and SHIMIZU, E., 'Ship to Ship Operations Monitoring System using High Accuracy DGPS', *8th IFAC Conference on Control Applications in Marine Systems (CAMS 2010)*, 2010.

4. SPT Inc. Home Page, http://www.sptmts.com/

SEABASE MODEL TESTS AND SURFACE EFFECT SHIP CODE EVALUATION FOR T-CRAFT

A L Silver, M J Hughes, R C Bishop, S S Lee, D W Wundrow, Naval Surface Warfare Center, Carderock Division, USA
J F O'Dea Computer Sciences Corporation, USA

SUMMARY

The US Navy's Office of Naval Research (ONR) has been funding development of a new type of ship that transforms from a surface effect ship (SES) to an air cushion vehicle, referred to as the T-Craft. One of the projected uses of the T-Craft is to be one of the ship-to-shore connectors of an offshore seabase. The T-Craft program has sponsored two model tests of potential seabase configurations and a comparison study of multiple vessel SES motion prediction simulation codes to the model test results. The model tests were conducted to determine the relative motion between the two ships composing the seabase and to compare the wave induced forces on the connections of the ramps in a variety of seaways and headings. The SES code evaluation used the results from the model tests in regular and irregular waves to correlate with the code output by performing a qualitative graphic assessment and a quantitative percent difference analysis. This paper shows some of the results of both the model tests and evaluation of three of the multiple vessel SES seakeeping prediction codes using regular wave model test results.

NOMENCLATURE

ACV – air cushion vehicle
LMSR – Large Medium Speed Roll-on Roll-off Ship
ONR – Office of Naval Research
RAOs – Response Amplitude Operators
SES – Surface Effect Ship
SSA – single significant amplitude
T-Craft – Transformable Craft
DLL – Dynamic Link Library

1. INTRODUCTION

Under the Innovative Naval Prototype (INP) program, the Office of Naval Research (ONR) has been funding development of a transformable craft (T-Craft) that can operate as a surface effect ship (SES) and an air cushion vehicle (ACV) as a new type of ship-to-shore connector. One of the critical missions of the T-Craft is to be able to support the transfer of military vehicles and equipment from another ship in SES mode at sea. To support this mission, the T-Craft program funded two seabase model tests. One of these model tests measured the relative motion between the T-Craft and a large medium speed roll-on roll-off (LMSR) ship in a variety of seaways, headings and speeds. The other model test measured the loads at the ramp connections between the two ships in the same seaways and headings.

The other major effort funded by the T-Craft program was an evaluation of ship motion prediction codes that predict the motions of a SES near another ship in a seaway. Of the many ship motion prediction codes that exist, six were chosen for this study. Of the six codes, the results of three codes are presented in this paper. They are the Advanced Quantitative Wave Analysis (AQWA) by ANSYS Inc, Wave Analysis MIT (WAMIT) by WAMIT Inc., and Multi-Operational Structural Engineering Simulator (MOSES) by Ultramarine Inc. The evaluation was performed by correlating the output from each of the codes with the results of the model tests. The codes were compared both qualitatively and quantitatively. The qualitative assessment is accomplished by establishing trends from plots of the data. The quantitative assessment is done by calculating the percent difference between the model test results and the simulation output.

This paper describes the two model tests and the process and outcome of evaluating each of the SES motion prediction codes.

2. DESCRIPTION OF THE MODEL TESTS

2.1 SEABASE SEAKEEPING TEST

A model test of two ships in close proximity was performed in the Maneuvering and Seakeeping Basin at the Naval Surface Warfare Center, Carderock Division [1]. The models selected for the test were the USNS GORDON (TAKR-296), a Large Medium Speed Roll-on Roll-off (LMSR) ship, and a US Navy designed T-Craft in SES mode. The models were built to a scale ratio of 30.209. The LMSR had stern and sideport ramps that were built to the same scale ratio. The purpose of this test was to obtain an understanding of the dynamics between the two ships in different seaways and ship operating conditions while tethered together in four different configurations; tandem with the stern ramp, med-moor with the sideport platform ramp, side-by-side with no ramp connection, and hinged to the transom of the LMSR with no ramp. The models were tested in regular waves to obtain response amplitude operators (RAOs), and irregular waves to measure single significant amplitude (SSA) motions.

The basic test setup had the LMSR model mounted to a heave staff that was attached to the basin carriage, as shown in the photos in Figure 1. The T-Craft model was connected to the LMSR in four different ways referred to as Side-by-Side, Tandem with ramp, Med-Moored, and Tandem Hinged. This setup allowed the LMSR model to pitch, heave and roll but constrained in all other modes. In the Tandem and Med-Moor configurations, the connection of the ramp to the T-Craft model allowed the T-Craft model to heave, pitch, roll, and a limited amount of surge, but constrained it in sway and yaw. In the Side-by-Side configuration, the T-Craft was held at a specified distance from the LMSR by soft spring attachments. The soft springs allowed the T-Craft to heave, pitch and roll and a very limited amount of surge, sway and yaw. The Tandem Hinged connection of the T-Craft to the transom of the LMSR allowed the T-Craft to pitch, heave and roll with the LMSR but constrained it for all other motions. The hinge connection of the T-Craft to the LMSR restricted relative vertical motion. Sensors installed on each of the models measured all degrees of freedom for each model to account for the motions that may occur in each configuration.

The test matrix included 307 unique environmental and operational conditions to be run. It included runs in regular waves, Sea State 3 waves (with three different modal periods), and Sea State 4 waves (with three different modal periods). For the two tandem cases, the test matrix included runs at two speeds (0 and 4 knots) and two ship-to-wave headings (head and bow seas 20 degrees off head). The side-by-side configuration was tested only at 4 knots, and the med-moor configuration was tested only at 0 knots with a T-Craft ship-to-wave heading of 110° (20° ahead of beam of LMSR and 20° off of head sea of the T-Craft).

One example of results from this test is shown in Figure 2. Here the T-Craft was tested in a Side-by-Side and Tandem with ramp connection in various Sea State 4 combinations in bow seas. As shown, the pitch motion of the LMSR was very small, but increased with increasing modal period. The pitch motion for the T-Craft varied over the three modal periods with the 8.2 modal period waves producing the largest pitch amplitude. Also, the T-Craft pitch amplitude at 8.2 second modal period differed significantly between the two configurations. The pitch single significant amplitude (SSA) motion was 1.04° for the Side-by-Side configuration, and 1.4° for the Tandem with ramp configuration in the same wave conditions.

The results of this model test provided some insight into the complex hydrodynamics of the seabase, and were used for correlating with SES motion prediction codes.

2.2 RAMP LOADS SEABASE SEAKEEPING TEST

A second model test - using the same models as the seabase seakeeping test - was conducted to determine the loads on the pin connections of the ramps that connected the two ships [2]. The test matrix specified 685 conditions including the same operating conditions and similar wave conditions to the previously described test, with the addition of a Sea State 3 bi-modal spectrum where the two modal periods were 7.5 and 15 seconds. The wave periods of the bi-modal spectrum were chosen based on the frequency of the largest response of the pitch and roll RAOs for each of the two ships. Another addition to the test matrix was the addition of four different T-Craft and ramp loading conditions. The T-Craft and LMSR seabase was tested (1) unloaded, (2) with an Abrams M1A1 Tank on the ramp and the T-Craft unloaded, (3) with four Abrams M1A1 Tanks on the T-Craft and the ramp empty, and (4) the four Abrams M1A1 Tanks on the T-Craft and one tank on the ramp.

As in the previous test, three seabase configurations were tested with the two models – Tandem with ramp, Side-by-Side, and Tandem Hinged. However, in this test, the Side-by-Side configuration had the T-Craft connected to the LMSR with a sideport ramp and the Tandem Hinged configuration employed a ramp between the models as shown in Figure 3. The LMSR was connected to the basin carriage in the same way as in the previous test by a heave post. The T-Craft was connected to the LMSR with ramps that had connections that restricted surge, sway, and yaw motions. The new ramps built for this test had strain gauges mounted at each of the connection pins to measure the longitudinal, transverse and vertical loads at the ramp ends. The load cell pins allowed the T-Craft to heave and pitch relative to the LMSR, while roll was governed by the torsional flexibility that was built into the ramps.

One example outcome from the test is shown in Figure 4, depicting how the axial loads on the pins of the ramp differed between each of the three seabase configurations as a function of sea state. In this figure, the measured longitudinal loads on the port ramp pin at the top of the ramp is plotted to show the dependence on loading condition and sea state. For the Tandem with ramp configuration, the two Sea State 3 conditions were virtually the same and in Sea State 4 the longitudinal loads were significantly higher. The Hinged Tandem configuration longitudinal loads were very low for all three wave conditions, and the Side-by-Side configuration had significantly higher longitudinal forces with the bi-modal Sea State 3 than the Sea State 3 7.5 second spectrum.

Results of this test have been made available to the user community that is developing tools for the T-Craft program.

3. SES MOTION CODE EVALUATION

3.1 DESCRIPTION OF EVALUATED CODES

As mentioned in the introduction, the T-Craft program funded an evaluation of several ship motion prediction codes that are of interest to the US Navy. Three of the programs in this evaluation were AQWA [3], MOSES [4] and WAMIT [5]. Each of these codes computes the hydrodynamic forces using a three-dimensional boundary element method, and assumes potential flow theory with any important viscous forces included using empirical corrections. The panel method in each of the codes uses the zero forward speed Green's function in which singularities are placed on each hull surface, automatically satisfying a linearized boundary condition on the water free surface. Each of these codes solves a set of frequency domain problems to obtain a set of hydrodynamic force coefficients for the added mass, wave damping and wave exciting forces. These coefficients can be used to solve the equations of motion in the frequency domain to obtain RAOs that are valid where the ship response is linear with respect to the amplitude of the incoming waves. AQWA and MOSES use the frequency domain coefficients to compute the hydrodynamic forces at each time step during a time domain simulation, in which case some non-linear forces can be included [6]. WAMIT, exclusively a frequency domain code, includes a utility to compute the impulse response functions, which could be used as input to a time domain simulator.

The original software for each of the codes did not support SES type ships directly. They each have unique approaches to solve this problem. Modeling the full details of the system would require models for the lift fans that pump air into the cushion, deformation of the flexible seals at the bow and stern, leakage of air past the bow and stern seals and under the side hulls. Since the T-Craft model used in the tests had a transverse seal, there were two cushions that needed to be modeled in the simulation codes.

WAMIT solved the problem of modeling an air cushion under the hull by placing a set of panels on the water surface inside the air cushion, which are allowed to move and deform relative to the vessel with specified generalized modes [7]. The cushion pressure in WAMIT can be modeled as either spatially constant or varying within each air cushion. In the current WAMIT calculations, the cushion pressure was assumed to be spatially constant in each cushion with no air flow in or out of each cushion. The effect of cushion compressibility was modeled by a linearized adiabatic gas equation to represent the "air spring" of the two T-Craft cushion compartments.

AQWA does not account for the air cushion in the frequency domain calculations, and the wave diffraction problem is solved using only the wetted side hulls. The air cushion is accounted for in the time domain portion of these two codes. In AQWA, the SES is modeled using an additional dynamic link library (DLL) developed by Alion Science and Technology Corporation [8]. The DLL computes the steady free surface depression from the air cushion, the deformation of the bow and stern seals, the air leakage out of the cushion and air flow into the cushion through the fans. The pressure is computed at each time step and assumed to be spatially constant within the air chamber and a term is included which accounts for the wave damping from the air cushion. The T-Craft was modeled within the AQWA DLL as a single air chamber.

MOSES is a general purpose simulation environment which, unlike WAMIT or AQWA, uses its own scripting language and command-line interpreter. To model the air cushions for a SES, the T-Craft was modeled as an air cushion vehicle elevated by pressurized compartments. These compartments were modeled as rigid–walled rectangular boxes placed under the wet deck, between the side hulls and opened to the water surface. No attempt was made to model the forward or stern skirts of the T-Craft. Instead, a rigid inclined bow and rigid vertical transom were placed between the side hulls. Initially two compartments under the hull were modeled as the T-Craft model had a transverse seal. It was found to be advantageous to add a second divider along the hull centerline because it eliminated errors encountered by MOSES's built-in auto-ballasting macro. The lift fan dynamics and air leakage under the seals were not computed in MOSES.

The way each of these codes models the multiple ship, or seabase, hydrodynamics is different as well. Although AQWA and MOSES both use three-dimensional radiation/diffraction potential theory, each code has a different methodology to achieve the required results. AQWA computes the hydrodynamic interaction of two ships, one of which is a SES ship by first calculating the forces on the wet part of the side-hull using the frequency domain module AQWA-LINE. Those results are used to run the time domain modules, AQWA-DRIFT and AQWA-NAUT, along with the Alion-developed DLL to compute the motions of air cushion vehicles. The AQWA software suite uses a linear hydrodynamics approach and fully accounts for hydrodynamic interactions between hulls, including not only the radiation coupling but also the shielding effects as well. In the time domain, AQWA can also handle simple mechanical articulations between structures such as the hinges between vehicles.

WAMIT also fully accounts for all hydrodynamic interactions between bodies, when it solves the coupled set of 6*NBODY equations (NBODY=2 for the seabase). Non-hydrodynamic interactions (such as mooring or ramp connections) are included by specifying external spring and damping matrices to be used when the equations of motion are solved.

MOSES provides rudimentary support for modeling interacting multi-body systems. This is done by defining "obstacles" in the water with which a floating body will have hydrodynamic interactions. An obstacle must "belong" to one of the bodies in the system and will mimic that body's mean motion. The obstacle does not, however, mimic the body's unsteady motion and so it does not generate a radiated wave field. As such, the modeled hydrodynamic interaction only captures those effects communicated through incident and diffracted waves around the obstacle. For this study, the two-body configuration had to be run twice; first with the LMSR as the floating body and the T-Craft as the obstacle, and second with the T-Craft as the floating body and the LMSR as the obstacle.

All three codes require descriptions of the ship geometries. The underwater geometries of the T-Craft and LMSR were represented for AQWA by 592 panels and 2762 panels respectively. AQWA uses both triangular and quadrilateral panels for pressure plates. In an AQWA data input file, the elements which are below the still water line must be denoted as diffracting elements. Although AQWA can handle mixed models of both diffracting and non-diffracting elements, the top row of diffracting elements must have their top edges aligned with the still water line.

The panel descriptions of the hull geometries used by MOSES were developed from station-off set data. For the two ship configurations, a total of 3151 panels were used. Also included as part of the hull descriptions were the vertical location of the center of gravity and the roll, pitch and yaw radii of gyration.

The input to WAMIT used the hull surface modeling program MultiSurf which produces files that describe the underwater surface of the two hulls. The hull surfaces are represented by higher-order surface patches. Each of these patches can have curvature, and the potential calculated by WAMIT on each is represented by B-splines [ref #24 from WAMIT Manual]. In the current study, the higher order method was used. Other input files specify the wave conditions (frequency, direction) to be computed, inertial properties of the vehicles that are needed in the equations of motion, and external files that represent spring connections and empirical damping.

3.2 EVALUATION OF THE CODES

From the model test referenced in [1], five regular wave and seven random wave runs were chosen to be compared with the output from each of the motion codes. This paper shows the results of comparing two regular wave runs as shown in Table 1. These runs were for the tandem and med-moor configurations. For each configuration, the codes were run with the LMSR restricted in surge, sway and yaw, and the ramp connection was modeled by the codes to basically reflect the conditions in the model test.

The methodology used to evaluate the codes is outlined in Silver and Hughes [9]. Briefly, each of the three codes was run for the same conditions that were selected from the tests. The output from the codes was then plotted against the model test data for a qualitative comparison. To obtain a quantitative difference between the model test and simulation results, a percent difference calculation was used. This percent difference calculation between the model test and the simulation output was calculated by determining the difference between the model test and simulation results and dividing that difference by the largest value attained in the model test for the condition tested.

The comparison between the model test and the output from each of the three codes include plots of the heave and pitch RAOs of both the LMSR and the T-Craft in SES mode. Heave motion is non-dimensionalized by wave amplitude and pitch is non-dimensionalized by wave slope. The model test matrix had 10 runs of the same condition executed so that an uncertainty analysis could be performed on the expected variability between runs. The plots of the model test results are shown here with error bars representing the 95% confidence level. Figure 5 shows the regular wave qualitative results for the tandem case with the ramp at 4 knots and 200° heading. Here only MOSES and AQWA are compared because there was no model test data taken at zero speed.

Looking at these plots, there are some notable observations that can be pointed out regarding the trend of the data. First, all of the data from the model test and the two simulations had the same general trend of having higher values at the lower frequencies and lower values at the higher frequencies. Also, for T-Craft pitch and heave, MOSES tends to consistently produce higher results than the model test. For the Med-moor case, the T-Craft was at a 200° heading to the waves and the LMSR was at a 110° heading. The regular wave results for the Med-moor configuration are shown in Figure 6. Here the trend of the data is, again, generally high for the lower frequencies and lower for the higher frequencies. For heave, WAMIT follows the model test data well, with MOSES slightly higher and AQWA slightly lower. For pitch, the amplitudes of each of the codes generally fall within a reasonable range of the model data.

The codes were also compared quantitatively using a percent difference calculation. For this calculation, the difference between the magnitude of the motion from the prediction code and the model test is taken and then divided by the largest model test value over all headings and frequencies for a specific speed. This division normalizes the result which reduces the occurrence of large percent differences when the response is very small. The results of the percent difference calculation are shown in Figures 7 and 8. Looking at the Tandem configuration, it is clear that both codes predict the heave and pitch motions of the LMSR much better than the T-Craft. This is to be expected as the LMSR is a conventional displacement monohull whereas the T-Craft is a SES. It is interesting to note that MOSES generally over predicts heave motion and AQWA under predicts heave motion. For pitch, MOSES over predicts by the same amount as heave, but AQWA's predictions are closer to the model test results.

For the Med-Moor configuration, the same trend is true between MOSES and AQWA for heave motion of the T-Craft. For T-Craft pitch motion, however, both MOSES and AQWA under predict with respect to the model test data except at the lower frequencies in which AQWA over predicts the motion. WAMIT predicts higher values for both heave and pitch though not as large a difference as the other two codes. For the LMSR, again, the output of both heave and pitch motions for all three codes are closer to the model test data.

4. CONCLUSIONS

In summary, two model tests and a multiple vessel seakeeping prediction code comparison were performed to support the cargo transfer between a SES ship, the T-Craft, and a LMSR. The model test data aided an understanding of the complex hydrodynamics between the two ships and the resulting loads on the ramp connectors. The seakeeping code prediction comparison provided some insight in interpreting the output from each of the codes.

Figure 1. Photos of the model test setup of the four configurations (clockwise from upper left) side-by-side with 4.6 meter separation, tandem with ramp, med-moor and tandem with hinged connection.

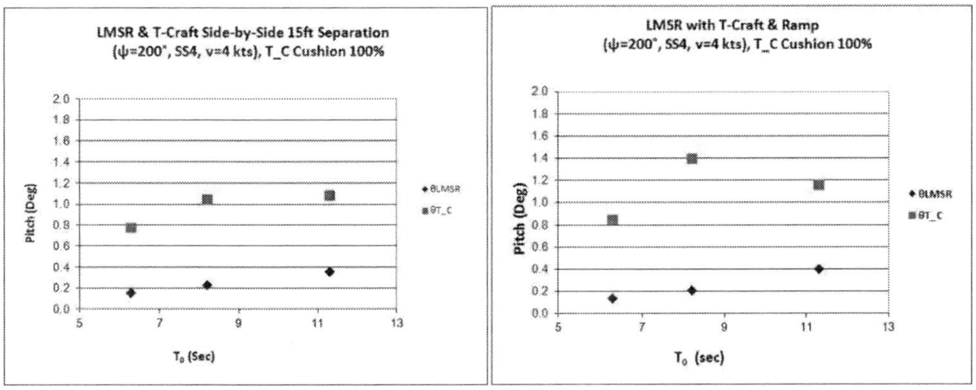

Figure 2. Comparison of pitch motion (SSA) between the Side-by-Side 4.6 meter separation and tandem with ramp configurations in three different Sea State 4 conditions (modal periods of 6.3, 8.2 and 11.3 seconds) at 4 knots and heading of 200°. (◆ =LMSR, ■ = T-Craft).

a. b. c.

Figure 3. Ramp loads model test configurations. a. Tandem with ramp, b. Tandem hinged, and c. Side-by-Side.

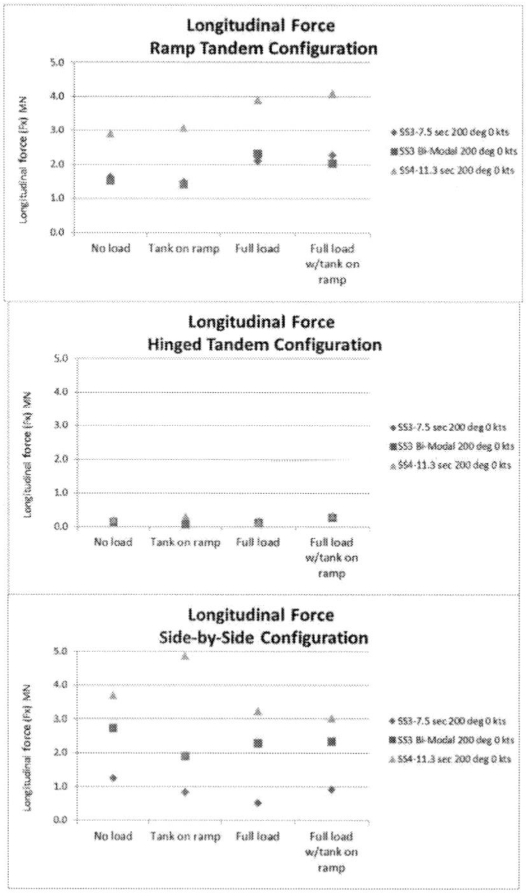

Figure 4. Comparison of Longitudinal Forces on Ramp Pin Connection at the LMSR side over T-Craft Load Conditions, Seabase Configurations, and Seaway.
(◆ SS3 7.5 sec, ■SS3 Bi-Modal, ▲SS4 11.3 sec)

Table 1. Regular Wave Conditions Selected from T-Craft Seabase Seakeeping Model Test

Model Configuration	Heading (deg)	Speed (kts)	Frequency range
Tandem LMSR Stern to T-Craft Bow	200	4	0.4-0.9 rad/sec
LMSR Med-Moored with T-Craft	200 (T-Craft)	0	0.4-0.9 rad/sec

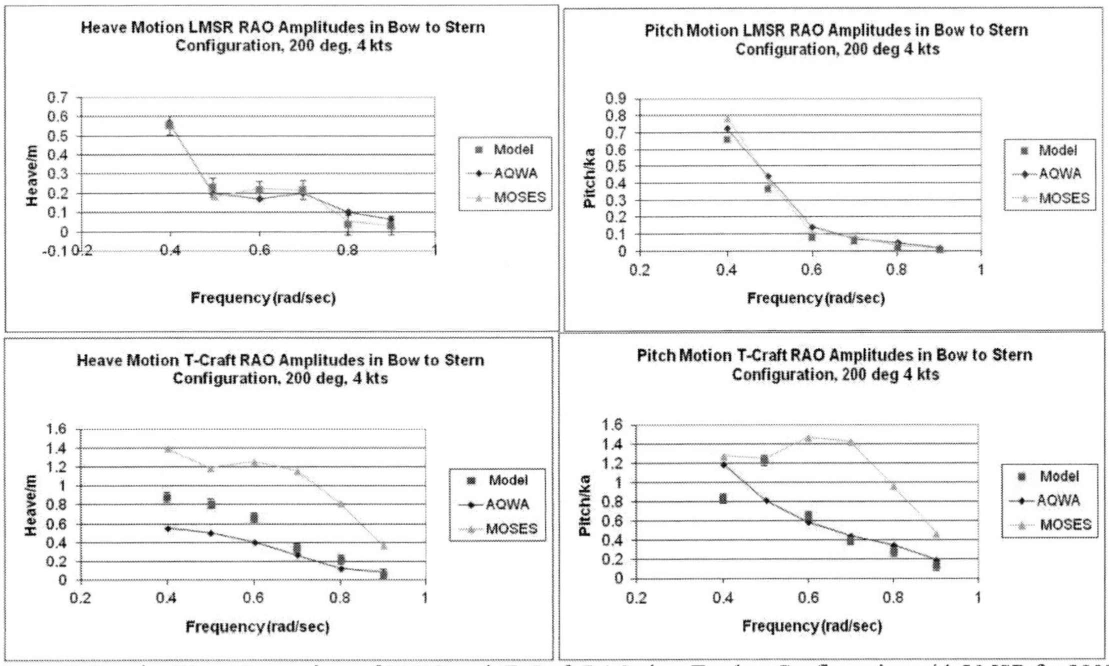

Figure 5. Regular Wave Comparison of LMSR and T-Craft RAOs in a Tandem Configuration with LMSR for 200° Heading and 4 knots Conditions.

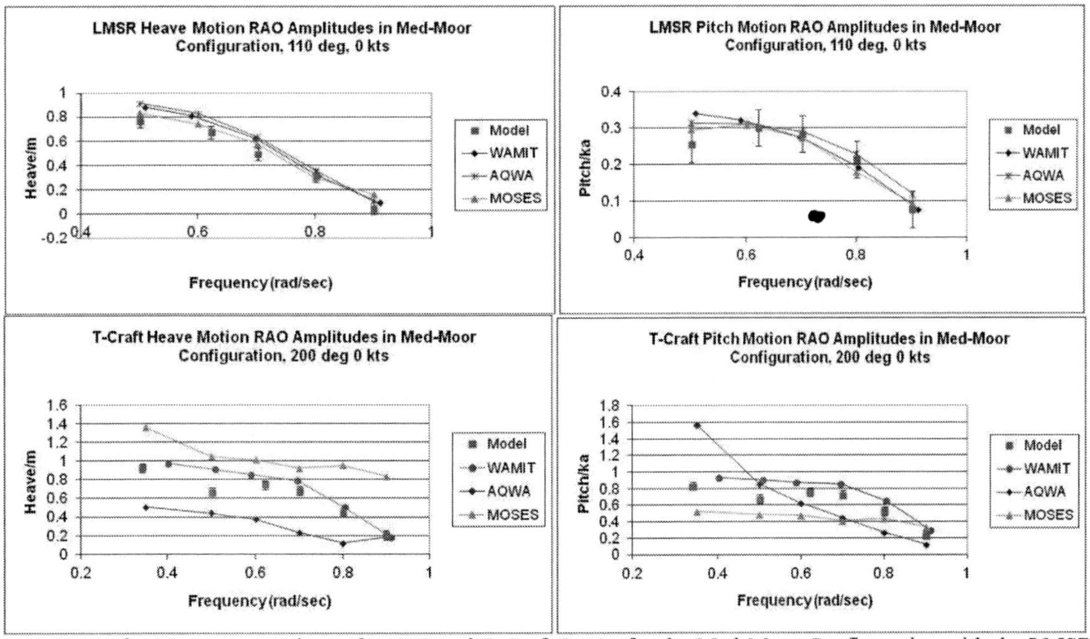

Figure 6. Regular Wave Comparison of LMSR and T-Craft RAOs for the Med-Moor Configuration with the LMSR at 110° and T-Craft at 200° heading to the waves and 0 knots

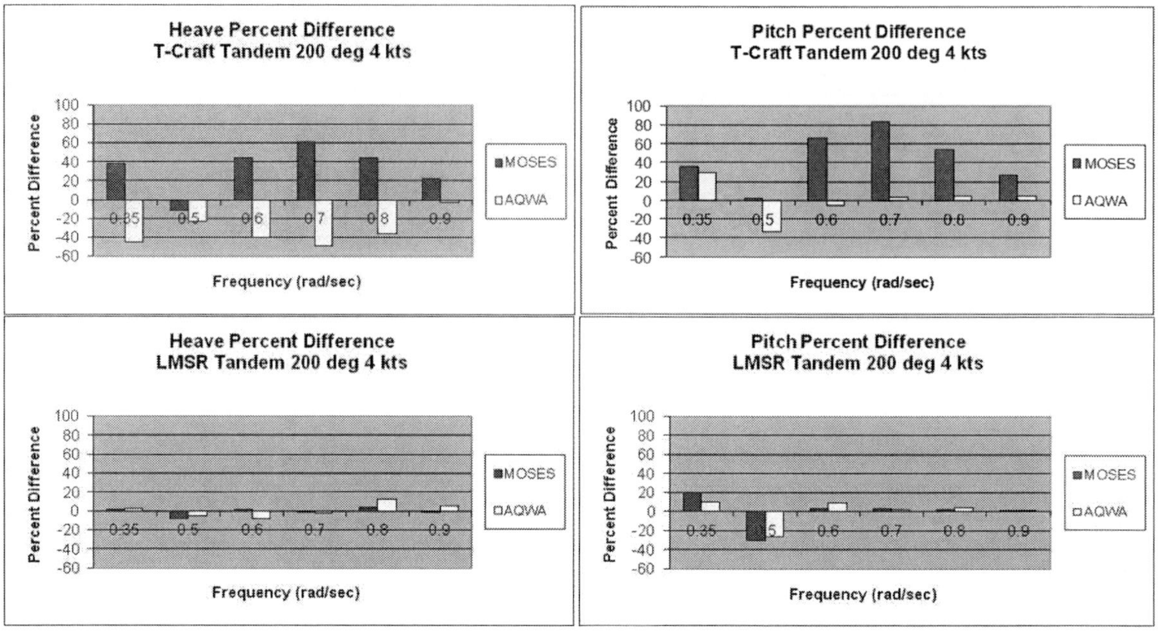

Figure 7. Heave and Pitch Percent Differences for the LMSR and T-Craft Heave and Pitch Motions in a Tandem Configuration

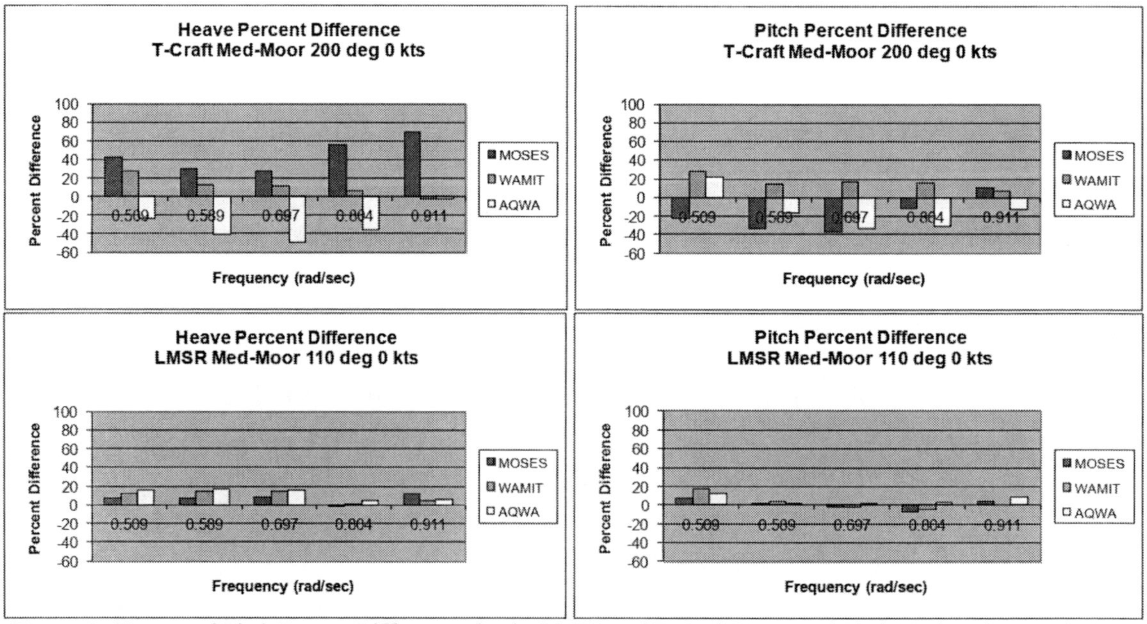

Figure 8. Heave and Pitch Percent Differences for the LMSR and T-Craft Heave and Pitch Motions in a Med-Moor Configuration

5. REFERENCES

1. Bishop, R.C.; A.L. Silver; D. Tahmasian; S.S. Lee; J.T. Park; L.A. Snyder, 'T-Craft Seabase Seakeeping Model Test Data Report Update', NSWCCD-50-TR-2010/062, 2010.

2. Bishop, R.C.; A.L. Silver; D. Tahmasian; S.S. Lee; L.A. Snyder and J.T. Park, 'T-Craft Seabase Ramp Loads Model Test Data Report', NSWCCD-50-TR-2010/073, 2010.

3. ANSYS, Inc., "AQWA Reference Manual Version 5.7D" , 2007.

4. Ultramarine, Inc., "Reference Manual for MOSES," June 1989 and July 2008.

5. WAMIT Inc., "WAMIT User Manual 6.4," WAMIT, Inc 2008.

6. Cummins, W.E., 'The Impulse Response Function and Ship Motions', Schiffstechnik, Bd. 9, Heft 47, 1962.

7. Lee, C.H. and J.N. Newman, 'Wave Effects on Large Floating Structures with Air Cushions', VLF'99 Conference, 1999.

8. Jiang, C.B.; C.B. McKesson; M. Musatow, and R. Sher, 'Time-domain Nonlinear Surface Effect Ship Motion Simulation', Proceedings of the 27th Symposium on Naval Hydromechanics, 2008.

9. Silver, A.L. and M.J. Hughes, 'A Method of Evaluating Multi-Vessel Surface Effect Ship Motion Prediction Codes', Grand Challenges in Modeling and Simulation Conference, 2010.

6. AUTHOR BIOGRAPHIES

Andrew L. Silver is a naval engineer in the Modeling and Simulation Branch of the Seakeeping Division at the Naval Surface Warfare Center, Carderock Division. Through his career, beginning in 1980, Mr. Silver has developed operator guidance software tools for predicting ship motions in shallow water. These tools have been employed by the Navy to predict underkeel clearance of deep draft ships in shallow channels, and by the Corps of Engineers to determine the required dredge depths for the entrance channels to ports of the United States. Mr. Silver has also led three different teams to evaluate multiple ship response simulations with model test data aid in the transfer military equipment and personnel within a seabase. This work has led to a better understanding of the capabilities and use of these simulations.

Dr. Michael J. Hughes is a naval architect in the Modeling and Simulation Branch of the Seakeeping Department at Naval Surface Warfare Center, Carderock Division, where he has worked since February 2005. His work at NSWCCD has focused on the computational methods for predicting the dynamic stability and seakeeping performance of ships. Prior to coming to NSWCCD, he had worked at Analytical Methods, Inc. (AMI) from December 1995 through February 2005 as a Senior Research Scientist. His primary responsibilities at AMI involved the improvement and development of computational prediction methods for hydrodynamics problems. He also used these methods to perform studies of ship resistance and powering, seakeeping performance, high-speed ferry wake wash, propeller performance and two body interaction problems. From 1994 to 1995, Dr. Hughes worked at the Institut für Schiffbau in Hamburg, Germany with the support of a research fellowship from the Alexander von Humboldt Foundation. During this fellowship he developed a higher order panel method for predicting ship wave resistance and analyzing steady free surface flow problems. While attending MIT, Dr. Hughes was employed from 1988 to 1993 at MIT as a research assistant, teaching assistant and post-doctoral associate, where his research work was concentrated in propulsor hydrodynamics.

Richard C. Bishop is an engineer in the Model Scale Experimentation Branch of the Seakeeping Division at the Naval Surface Warfare Center, Carderock Division. His work experience has included design and execution of scale model tests in the Harold E. Saunders Maneuvering and Seakeeping (MASK) facility as well as participation in full scale seakeeping and maneuvering trials. He is a graduate of the George Washington University where he earned his Bachelor's degree in Mechanical Engineering in 1977.

Dr. Sang Soo Lee is an engineer in the Modeling and Simulation Branch of the Seakeeping Division at the Naval Surface Warfare Center, Carderock Division. He has been working on various fluid mechanics areas including numerical simulation of ship motion, nonlinear wave interaction, laminar-to-turbulent flow transition and nonlinear instability of shear flow. He also worked in the aero-acoustics area where he designed the phased-array system for jet noise detection.

Dr. David Wundrow joined the Numerical Simulations Branch of the Seakeeping Division at the Naval Surface Warfare Center, Carderock Division in 2007. Since his arrival, Dr. Wundrow has been involved in numerous ship-motion simulation studies ranging from dynamic stability risk assessments to the analysis of at-sea transfer technologies. His work in the area of seabasing has addressed hydrodynamic interactions between both mono-hulled and multi-hulled vessels. Dr. Wundrow has been a participant in several ship-motion code evaluations created to determine the ability of current software to address the Navy's future seabasing needs.

Dr. John O'Dea spent most of his career at NSWC Carderock Division, working on various Navy projects in hydrodynamics and ship dynamics. He specialized in analyzing the seakeeping performance of advanced Navy ship designs, including multi-hull ships, and the hydrodynamic interaction of multiple ships in close proximity, in support of Seabasing concepts. After retiring from Carderock, he joined CSC Advanced Marine Center as a Senior Engineer, and continues to support Carderock Division on various Navy design projects. Dr. O'Dea graduated from Webb Institute with a bachelor of science degree in naval architecture and marine engineering, and received a Ph.D. in hydrodynamics from M.I.T.

CFD BASED PREDICTION OF SHIP-SHIP INTERACTION FORCES ON A TUG BESIDE A TANKER

C. D. Simonsen, FORCE Technology, Denmark
C. K. Nielsen, FORCE Technology, Denmark
J. F. Otzen, FORCE Technology, Denmark
K. Agdrup, FORCE Technology, Denmark

SUMMARY

The work covers a study of a quasi steady ship-ship interaction problem taken from a model test conducted in FORCE Technology's towing tank, where an ASD type tug is located next to a tanker, while advancing with constant speed. The tug is located in a number of longitudinal and transverse positions along the tanker with different drift angles. The RANS code Star-CCM+ is used to simulate a subset of cases taken from the experimental test program in order to study the local flow field and the surge and sway forces and the yaw moment acting on the tug due to the presence of the assisted ship. The computations are compared with experimental data and the CFD results generally look promising. Though, there are some cases, which seem to be challenging for the CFD tool and more worked is planned to study this further.

1. INTRODUCTION

In connection with manoeuvring simulation, effects from ship-ship interaction play an important role. The interaction effects have traditionally been determined in the towing tank, but this approach typically requires that physical scale models are built and that a special test setup is made. This is expensive and time consuming and is consequently only done on rare occasions.

Computational Fluid Dynamics (CFD) is currently being used for many marine applications and the question is whether the numerical tools can be used to quantify ship-ship interaction with sufficient accuracy to be used for manoeuvring simulations involving ship-ship interaction. The long-term perspective is whether a numerical tool could be both accurate and fast enough to be running on-line as an integrated part of the simulation loop. This part, however, is not within the scope of the present work.

In the present work a quasi steady ship-ship interaction problem is studied. The setup is taken from a previous model test conducted in FORCE Technology's towing tank. In the test an ASD type tug is located next to a conventional tanker, while advancing with constant speed. The tanker is locked in all degrees of freedom, while the tug is free to heave, pitch and roll but otherwise fixed. The tug is located in a number of longitudinal and transverse positions along the tanker with different drift angles. For each case the tug and tanker positions are locked relative to each other.

The present CFD simulations, which are conducted with the RANS code Star-CCM+, cover a subset of cases taken from the experimental test program and is chosen to study the influence of drift angle and transverse and longitudinal positions on the surge and sway forces and the yaw moment acting on the tug due to the presence of the assisted ship. The results are compared with experimental data to investigate how well the CFD code can capture trends and absolute values of the forces and moments.

2. TEST CASES AND CONDITIONS

The study is conducted for the bare hulls of a tug and a tanker with the dimensions listed in Table 1. The computations are made in model scale using a scale of 1:25, to match the experiments. Depending on the orientation of the tug, two speeds are considered namely 4 and 6 knots in full scale.

		Tug	Tanker
Lpp	[m]	25.6	186.2
B	[m]	11.0	31.6
T	[m]	4.6	10.3

Table 1: Main dimensions

The test program included variations of longitudinal and transverse relative positions and drift angles between tug and tanker for different speeds. Relevant parts of the test program were repeated without the tanker in order to evaluate the force contribution due to the tanker, i.e. the pure ship-ship interaction effect. The positions of the tug and the tanker during the experiments are illustrated in Figure 1.

It should be noted that the CFD cases were a subset of the experimental program, the purpose being to validate the CFD method rather than produce data for the full range of parameters. Cases were selected to study three different parameter variations:

A) Transverse position – from gap 0.25 m to 10 m
B) Longitudinal position – from "Aft Shoulder" to "Forward Shoulder"
C) Drift angle – from 0 deg (parallel) to 40 deg

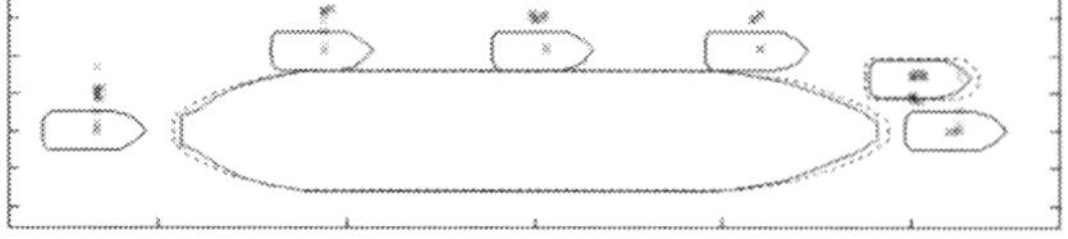

Figure 1: Illustration of relative positions of tug and tanker during experiments. Green crosses: tug drift angle equal zero; red crosses: tug drift angle different from zero (not all were included in CFD).

2. EXPERIMENTAL SETUP

Experiments were performed in FORCE Technology's deep water towing tank with dimensions 240 m x 12 m x 5.5 m. The full test program was also carried out at lower water depths in the shallow water basin, but these cases have not been calculated by CFD yet.

Figure 2: Photograph taken during experiments, tug at midships, drift angle 0 deg.

Figure 3: Photograph taken during experiments, tug at forward shoulder and non-zero drift angle.

The model of the tanker was attached underneath the carriage, fixed in all degrees of freedom. The tug model was placed in various angles and positions around the tanker, both inside and outside the pressure field created by the tanker. In order to measure the surge and sway forces and yaw moment acting on the tug, the model was attached by two strain gauges positioned fore and aft in the model. The model of the tug was free to heave, roll and pitch, and thus ballasted to the corresponding full scale GM_T.

Photographs from the tests are shown in Figure 2 and Figure 3.

3. COMPUTATIONAL METHOD AND GRID

The computations were performed with the Reynolds Averaged Navier-Stokes (RANS) solver Star-CCM+. The code solves the RANS and continuity equations on integral form on an unstructured mesh by means of the finite volume technique. For the considered steady state calculations the solver is advanced in time using a temporal first order Euler difference and only one inner iteration. Spatial discretization is performed with second order schemes for both convective and viscous terms. The pressure and the velocities are coupled by means of the SIMPLE method. Closure of the Reynolds stress problem is achieved by means of the isotropic blended k-ε/k-ω SST turbulence model with an all Y^+ wall treatment. The free surface is modeled with the two phase volume of fluid technique (VOF). Further details about the code can be found in [1].

With respect to the orientation of the tug during the simulation, it is positioned and locked according to the measured dynamic sinkage, trim and heel. On the boundaries of the computational domain the following conditions are applied: Upstream of the ships a velocity inlet condition is applied, i.e. the volume fractions of water and air plus the velocities are prescribed. On the bottom, top and side of the domain slip walls are applied, i.e. there is no flow through the walls $v=0$, while zero gradients are used for u,w,p and the volume fractions. Finally, on the outlet boundary zero gradients are applied to the velocity and volume fractions while pressure is set to hydrostatic pressure.

Concerning the mesh, the computations are performed with unstructured grids generated in Star-CCM+. Both ships are modeled with a trimmed mesh, which is a hexa-dominant polyhedral mesh. The mesh contains approximately 8.9 million cells, see Figure 4. The near wall spacing of the grids on no-slip surfaces are in the range from $y^+\approx1$ to $y^+\approx30$.

With respect to the extension of the computational domain, the width and the depth of the domain match the dimensions of the towing tank in which the model test was conducted. The inlet boundary is located 3 tanker lengths upstream of the models, while the outlet is 4 tanker lengths downstream of the ships. Local refinements are used in the free surface region to resolve the wave field around the ships, plus in the gap region between the ships, see Figure 4.

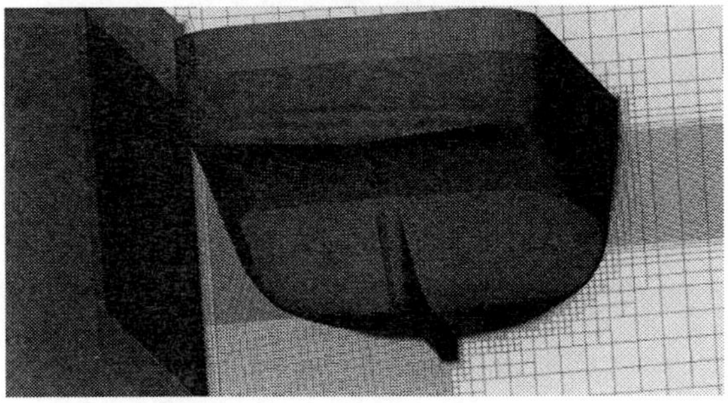

Figure 4: Computational mesh (fine)

4. VERIFICATION OF THE COMPUTATIONAL METHOD

To check the grid sensitivity of the solution, a grid refinement study was made with a grid refinement factor of $r_G = \sqrt{2}$. Error estimates with unstructured grids are not possible, since systematic refinement cannot be performed. However, by changing the reference length used to scale the cell sizes in the mesh generation tool, a semi-systematic refinement was done in order to check the trend of the solution changes with the fineness of the mesh. Figure 5 shows the coarse mesh containing around 1.8 million cells.

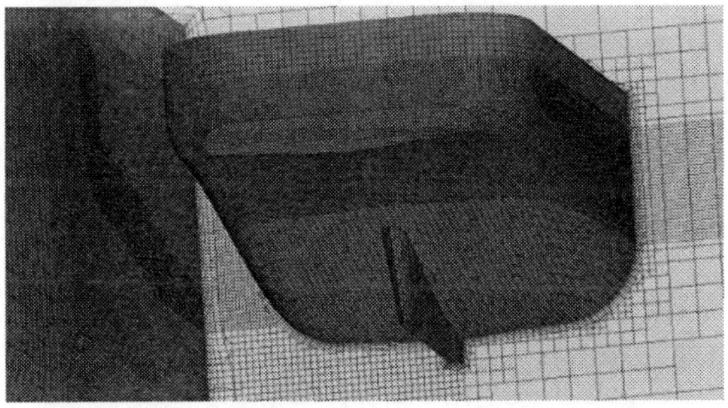

Figure 5: Computational mesh (coarse)

No error estimates are made, but the convergence ratio $R_G = \varepsilon_{21}/\varepsilon_{32}$ is calculated in order to judge the grid convergence. ε is the change in solution between meshes, with subscripts 1, 2 and 3 indicating fine, medium and coarse meshes, respectively.

Coefficient	Convergence ratio
C_X	-1.18
C_Y	0.07
C_N	-0.53

Table 2: Convergence ratios for force coefficients.

Depending on the value of R_G three conditions can occur: i) $0 < R_G < 1$, grid convergence, ii) $R_G < 0$, oscillatory convergence and iii) $R_G > 1$, grid divergence. From Table 2 it is seen that convergence is obtained for the sway force, while the surge force and the yaw moment show oscillatory convergence, i.e. none of the quantities diverges.

5. RESULTS FOR VARIATION IN TRANSVERSE DISTANCE BETWEEN TUG AND TANKER

The first results to be discussed are obtained for the tug's longitudinal position fixed at midships, while sailing straight ahead at different transverse distances between tug and tanker. The presentation of the results will initially cover the properties of the flow field around the tug and in the gap between the tug and the tanker. No experimental data is available for the flow field quantities, so this study is purely based on CFD results. Afterwards, the integral quantities are considered, i.e. the surge and sway forces plus the yaw moment acting on the tug. For these quantities the trends will be discussed based on the experimental values and a comparison between EFD and CFD will be made in order to learn how well the CFD code predicts the forces and moments. To get an idea about how potential theory based numerical methods performs compared to measurement and RANS based CFD, results obtained with the code AEGIR taken from [2] will be included in the final comparison.

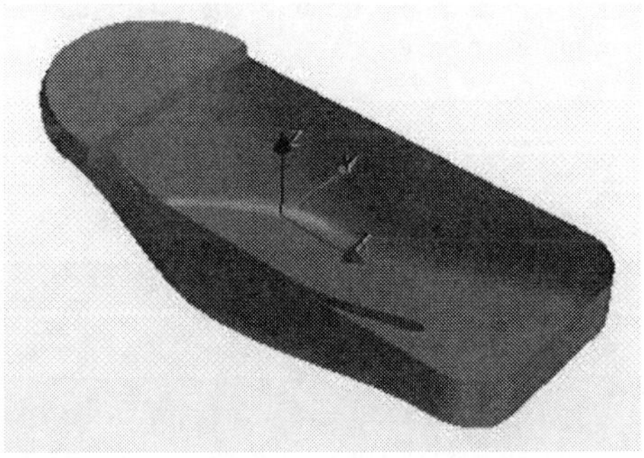

Figure 6: Applied coordinate system in tug.

The forces and moments will be presented in non-dimensional form using the following coefficients:

$$C_X = \frac{X}{½\rho U^2 \nabla_{\text{Tanker}}^{1/3} \nabla_{\text{Tug}}^{1/3}}$$

$$C_Y = \frac{Y}{½\rho U^2 \nabla_{\text{Tanker}}^{1/3} \nabla_{\text{Tug}}^{1/3}}$$

$$C_N = \frac{N}{\frac{1}{2}\rho U^2 \nabla_{Tanker}^{1/3} \nabla_{Tug}^{1/3} L_{PP,Tug}}$$

where X, Y and N are the surge force, the sway force and the yaw moment of the tug, respectively. ρ is the density of water, U is the ship speed, V_{Tank} is the displacement of the tanker, V_{Tug} is the displacement of the tug and $L_{pp,Tug}$ is the perpendicular length of the tug. Signs of forces and moments are given according to the coordinate system shown in Figure 6. Finally, all forces and moments are total forces acting on the tug and not increments due to interaction.

Figure 7 shows the free surface around the tug when it is located 2.0 meter away from the tanker. At starboard shoulder downstream of the bow wave, it is seen that a relatively large wave trough is formed, which is typical for this type of ship. Further, towards the aft part of the ship, the stern wave is seen. On port side of the tug, i.e. in the gap, the trough and the wave along the tug are larger than on the open port side. Based on the velocity field in the gap, it is found that 1) the flow is accelerated when it passes between the two ships, as a consequence of the constraint formed by the two geometries and 2) the boundary layers on the ships do not meet. This again is reflected in the pressure field on the tanker, where a large low pressure region is observed where the flow is accelerated, see Figure 8. Correlating this to the free surface it is revealed that the significant shoulder wave trough of the tug is located just above the low pressure region. Further, the pressure field on the tanker also shows the signatures of the bow and stern stagnation zones coming from the tug.

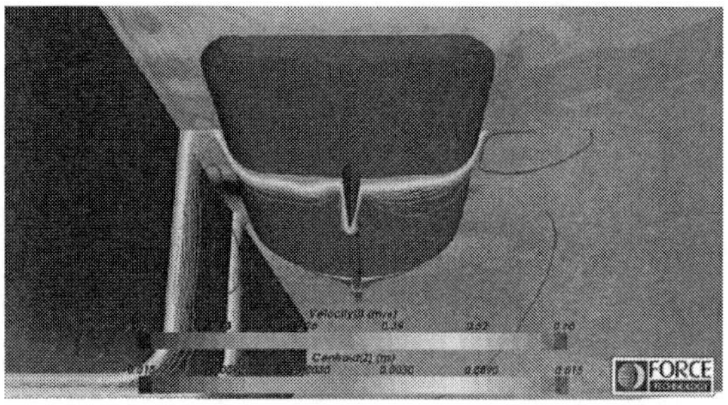

Figure 7 : Free surface and axial velocity around tug, midships, y=2.0 m and β=0 degrees.

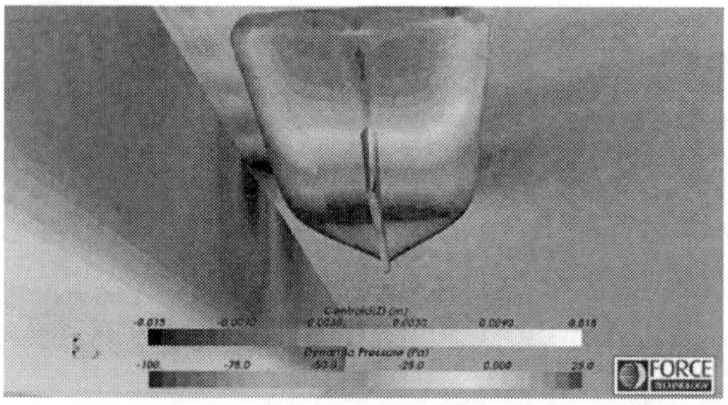

Figure 8 : Hull surface pressure, midships, y=2.0 m and β=0 degrees.

When the tug is moved closer to the tanker, now y=0.25 meter, large changes occur. With the short distance between the two ships, the flow becomes dominated by viscous effects. Consequently, the flow is slowed down in the gap and in some regions massive separation occurs, as seen in Figure 9. Since the flow cannot pass between the two ships due to the narrow gap, it seems that the flow over starboard side of the bottom is accelerated. The disturbed flow field on port side can also be seen in the hull surface pressure on the ships; for instance in the finger shaped pressure field on the tanker showing the signatures of the vertices in the flow, see Figure 10.

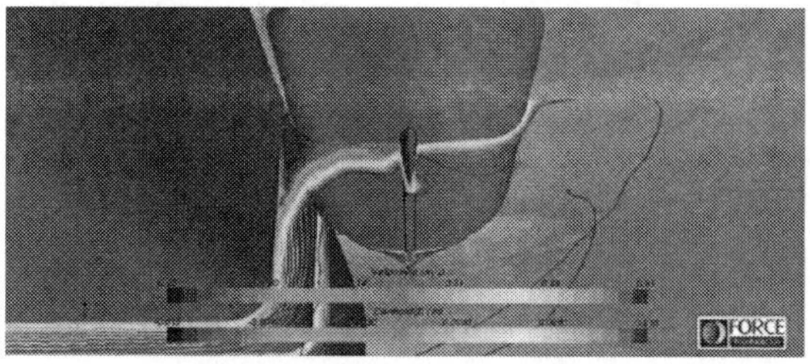

Figure 9 : Free surface and axial velocity around tug, midships, y=0.25 m and β=0 degrees.

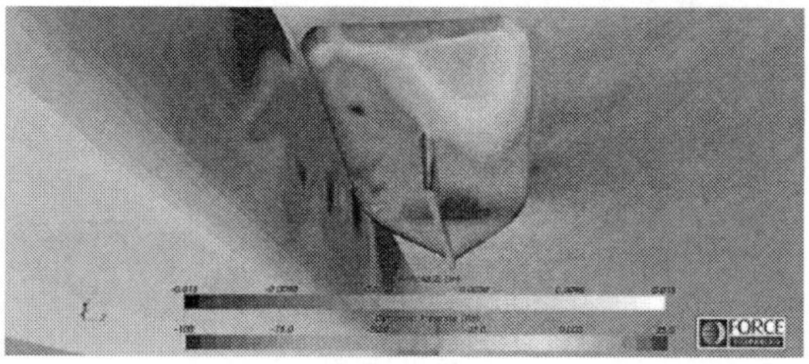

Figure 10 : Hull surface pressure, midships, y=0.25m and β=0 degrees.

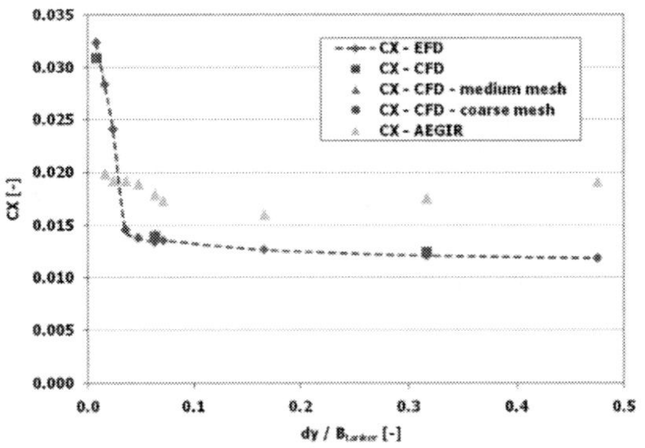

Figure 11 : Surge force on tug at midships and β=0 degrees.

The large changes in the flow field due to the width of the gap are naturally reflected in the hydrodynamic forces and moments acting on the tug. Figure 11 shows the surge force on the tug as function of gap width. The force appears to be almost constant for the large y values. When the width of the gap between tanker and tug decreases, the force starts to increase very slowly, corresponding to the situation where the flow is just accelerated in the gap. Finally, with the tug being very close to the hull the gap flow breaks down and the surge force increases significantly.

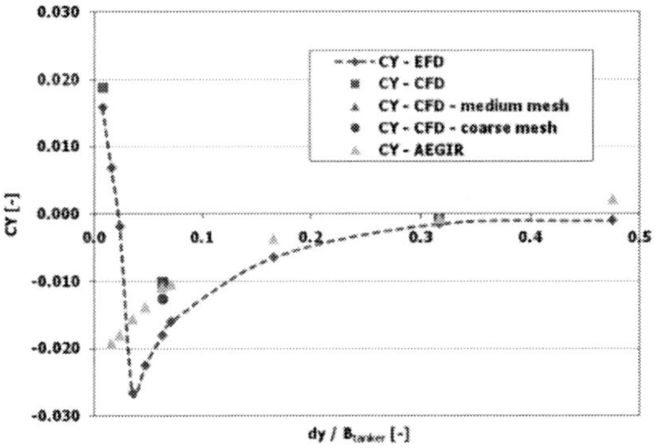

Figure 12 : Sway force on tug at midships and β=0 degrees.

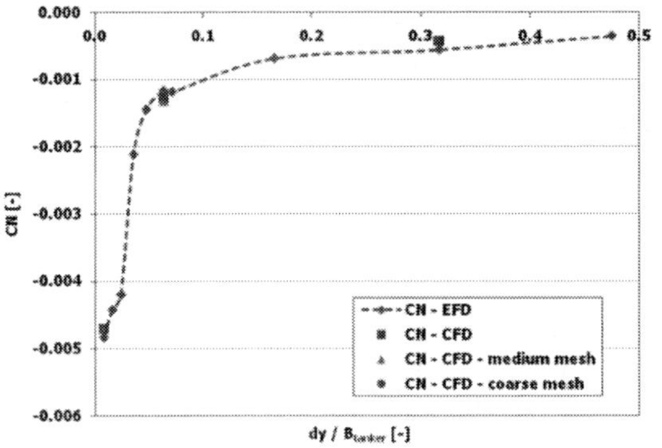

Figure 13 : Yaw moment on tug at midships and β=0 degrees.

For the sway force, drastic changes are also observed. Figure 12 shows that the force is small and almost constant zero for large Y values. As the tug moves closer to the tanker, the magnitude of the negative sway force increases. Due to the flow acceleration in the gap and the resulting pressure decrease, the tug is pulled towards the tanker. However, this only happens down to a certain distance after which the sway force increases significantly and becomes positive, which means that the tug is pushed away from the tanker.

Figure 13 shows the yaw moment acting on the ship. Here it is seen that the moment is negative throughout, meaning that the bow is pushed away from the tanker. The absolute value of the yaw moment increases initially slowly as the tug moves closer to the tanker, but as for the forces this changes at a certain y-position after which a strong increase is observed.

In addition to the trends in the forces and yaw moment the three figures described above also show the comparison between measured (EFD) and computed (CFD) results. On a qualitative level it appears that the CFD results follow the same trends as found in the measurements. When it comes to quantitative agreement between EFD and CFD fairly good agreement is obtained for the surge force and the yaw moment, while the sway force is more difficult to calculate, particularly in the peak region where the force is somewhat under-predicted. The reason for this is not known, but it could be that the wall flow in the CFD model is missing some details to model the flow properties correctly. Another issue could also be the y setting in the measurement. As previously noted the sway force has large gradients at the considered y-position, which means that even small deviations in the distance will give large changes in the force. Misalignment of the tug may also influence the measurement.

Finally, comparison between results from the RANS and the potential theory based methods has also been made. For the surge force it is found that potential theory has problems capturing the level and the trend compared to the RANS and EFD results, but this may be explained by the missing viscosity in potential theory. For the sway force the pictures is different, since the RANS and potential theory results follow the same trend and level for some of the y-distances. However, for very small gap sizes, potential theory fails to capture the sign change in the sway force, which probably again can be explained by the missing viscous effects in potential theory. Results from potential theory are only available for the surge and sway forces.

6. RESULTS FOR VARIATION IN LONGITUDINAL POSITION ALONG TANKER

In the next setup the tug's longitudinal position is varied with the transverse position fixed, while sailing straight ahead.

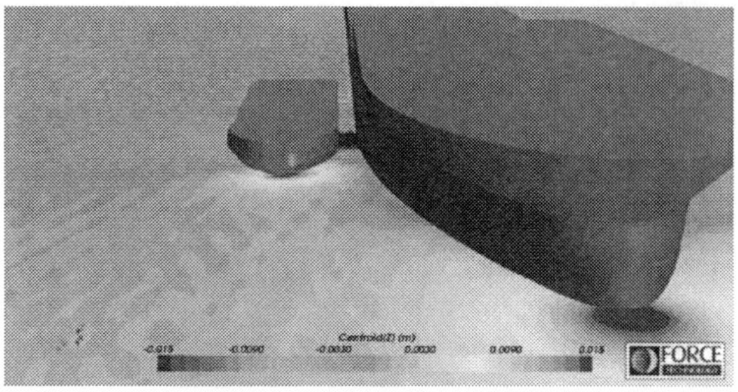

Figure 14 : Free surface around tug, forward shoulder, y=2.0 m and β=0 degrees.

Figure 14 and Figure 15 show how the tug is positioned at the forward and aft shoulders of the tanker, respectively. Typically these regions are interesting from a free surface point of view due to the wave troughs being located in these regions, but at the low Froude number of the tanker the wave making is not large. Focusing on the free surface and the axial velocity around the tug in the two positions, see Figure 16 and Figure 17, the overall flow pattern looks much the same as at the mid-ship position (Figure 7). Comparing the forward and aft positions, the velocity is a little higher at the forward shoulder of the tanker and consequently the wave trough at the port side shoulder of the tug is deeper. The higher velocity makes sense in the forward end of the tanker, since the flow is accelerated in order to pass around the bow of the tanker.

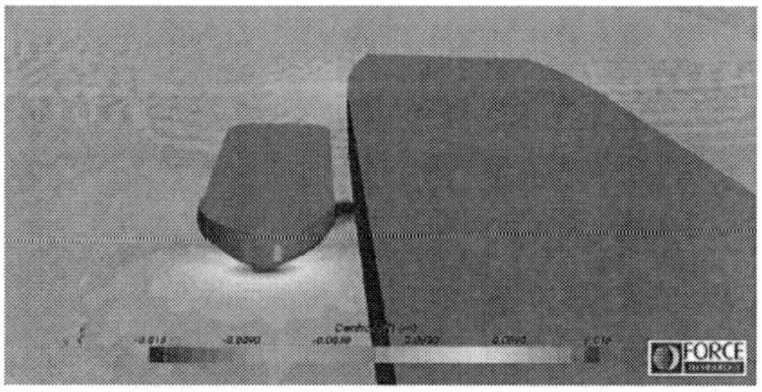

Figure 15 : Free surface around tug, aft shoulder, y=2.0 m and β=0 degrees.

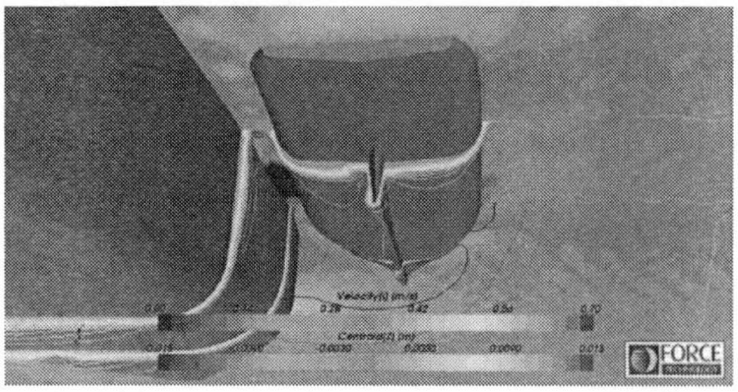

Figure 16 : Axial velocity, forward shoulder, y=2.0 m and β=0 degrees.

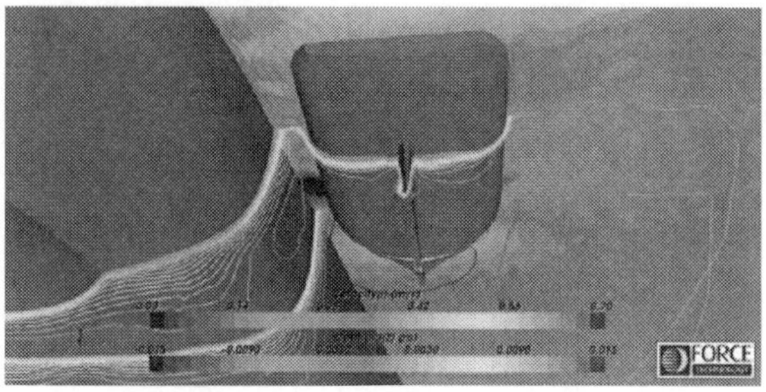

Figure 17 : Axial velocity, aft shoulder, y=2.0 m and β=0 degrees.

The hull surface pressure fields are shown in Figure 18 and Figure 19. Compared to the pressure in the mid-ship position (Figure 8) the forward shoulder position is dominated by a low pressure zone in the gap and a high pressure zone at the stern of the tug. This again is different from aft shoulder position where high pressure is located at the bow of the tug instead of the stern, since the narrowing in of the tanker hull reduces the pressure signature from the tug's stern on the tanker. Finally, the low pressure region is smaller at the aft shoulder.

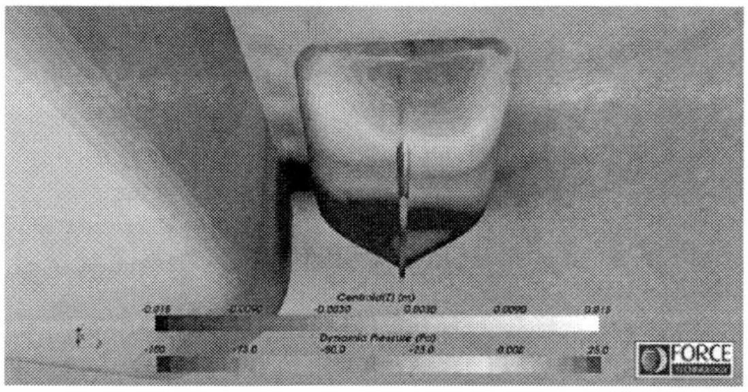

Figure 18 : Hull surface pressure, forward shoulder, y=2.0 m and β=0 degrees.

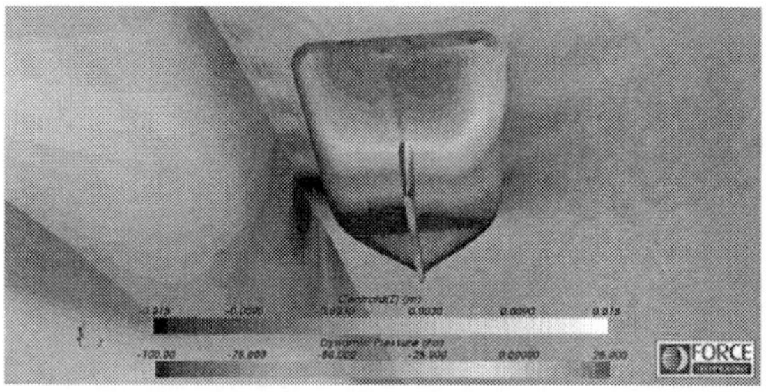

Figure 19 : Hull surface pressure, aft shoulder, y=2.0 m and β=0 degrees.

When it comes to the forces acting on the tug, the surge force in Figure 20 shows that the maximum force occurs at midships for this y-position. In the figure, *dx* is the longitudinal distance from midship position on the tanker to the midship location of the tug. Negative *dx*-values correspond to the bow.

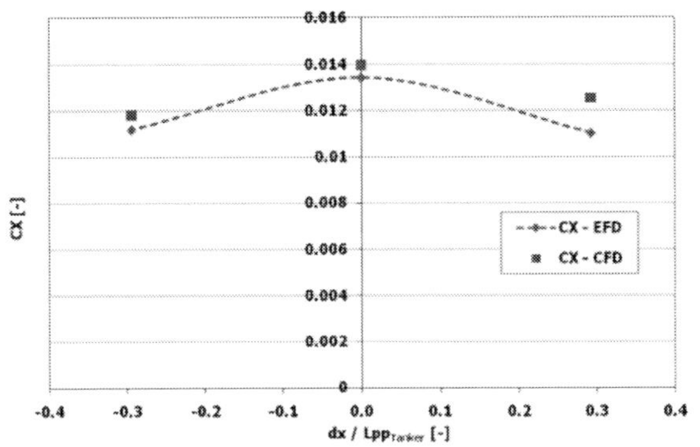

Figure 20 : Longitudinal variation of surge force on tug at y=2.0m and β=0 degrees.

It appears that the CFD results follow the same trend as the measured data, though the computation seems to over predict the surge forces for all the x-positions.

The computed and measured sway forces are shown as function of the longitudinal position in Figure 21. Negative x-values correspond to the bow. The largest sway force seems to be present at the bow of the tanker, which makes sense, since the shape of the tanker's bow accelerates the flow around the shoulders and forces it outwards towards the tug. Minimum force occurs at mid-ship and for all positions the tug is pulled towards the tanker. Again it is seen that CFD captures the trend of the force with change in x-position. Generally the CFD appears to somewhat under-predict the size of the force. Why this relatively large under-prediction occurs is difficult to say. When the influence of the y-position, with the tug at mid-ship tanker position, was studied earlier in the paper, it was found that y=2.0 meter was the most difficult position to match the measured sway forces. The same seems to be the case with the tug at aft and fore shoulder of the tanker. The reason could be that the tug is positioned based on measured data including dynamic sinkage and trim of the tug and motion due to the wave elevations from the tanker' overall wave pattern. More work should be put into studying the accuracy of the measured position used for positioning of the tug.

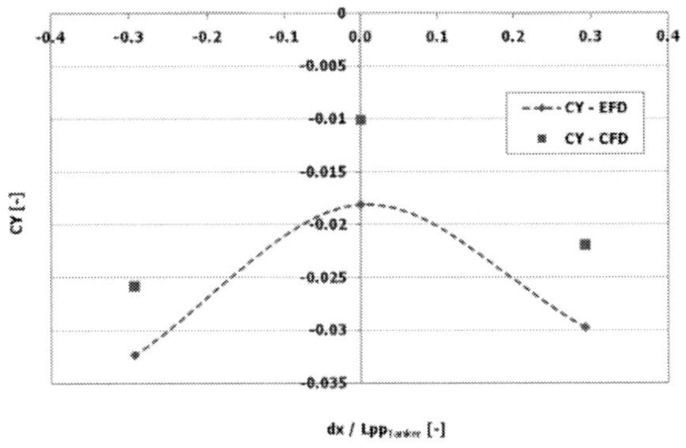

Figure 21 : Longitudinal variation of sway force on tug at y=2.0m and β=0 degrees.

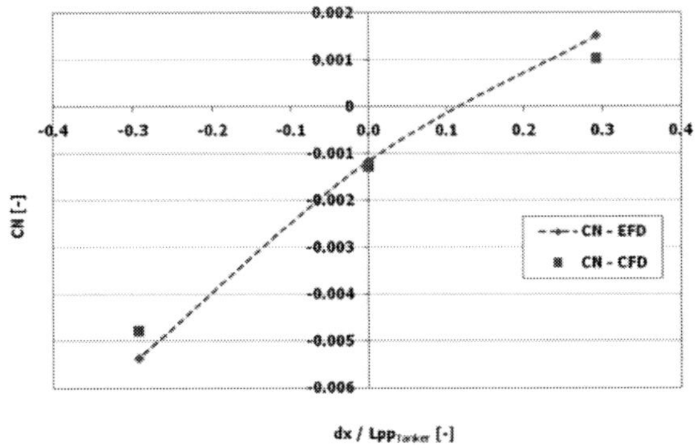

Figure 22 : Longitudinal variation of yaw moment on tug at y=2.0m and β=0 degrees.

Regarding the yaw moment, illustrated in Figure 22, it is seen to be largest at the forward shoulder of the tanker where the tanker bow shape forces the flow outwards and into the bow of the tug. This increases the yaw moment and pushes the tug's bow outwards. As the tug is moved downstream along the tanker, it is seen that the yaw moment increases from negative values and it actually changes sign at the aft shoulder, where the tug's bow is now pulled towards the tanker. The exact same behaviour is seen in the experimental data, meaning that the CFD picks up the trend in the measured data. Quantitatively the computation agrees well with the measurement at mid-ship position, but at the aft and forward shoulders the yaw moments are slightly under-predicted. More work is required to investigate this further.

7. RESULTS FOR VARIATION IN TUG DRIFT ANGLE

In the final setup the longitudinal and transverse positions of the tug were fixed. The tug was positioned at mid-ship with a gap between the tanker and the tug of 10.0 meters. For this investigation the vessel speed is reduced from 6 to 4 knots. The drift angle of the tug was then varied in order to investigate the capabilities of CFD to capture the effect of different drift angles. In Figure 23 the axial velocities and the free surface can be seen with a drift angle of 20 degrees. Focussing on the axial velocities a long vortex can be seen extending from the bow region. The vortex could be expected to have some dynamic motion but since the calculation is not run transient the extent of the motion cannot be quantified in the present study. Looking at the free surface a diffraction pattern can be seen between the tanker and the tug resulting in a larger wave elevation on the portside of the tug than on the starboard side.

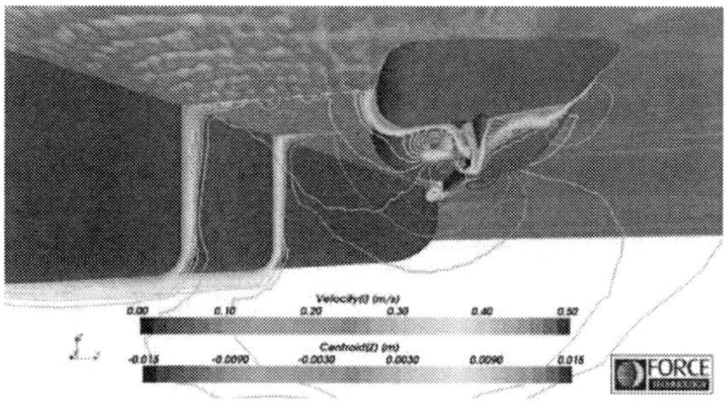

Figure 23 : Axial velocity and free surface, midships, y=10.0 m and β=20 degrees.

In Figure 24 the pressure distribution on both the tanker and the tug can be seen. Due to the drift angle there are evidently large differences between the pressures on port and starboard side of tug. At the side of the tanker a slight pressure decrease can be seen where the water is accelerated in the gap between the tanker and the tug. The pressure decrease is however much smaller than seen above for smaller gaps. On the tug a large high pressure zone can be seen on the aft port side corner which is due to the low velocities present at this leeward corner. At the forward part of the skeg where a low pressure region can be seen, water is sucked around the skeg. Further, small low pressure zones can be seen on the portside bottom of the tug. These are probably a consequence of the earlier described vortex originating from the bow region.

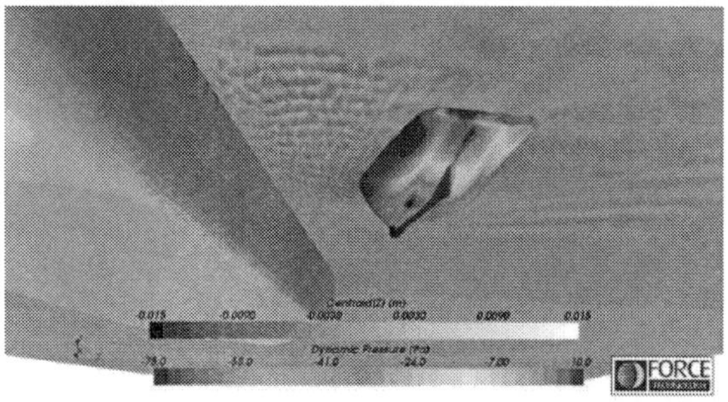

Figure 24 : Hull surface pressure, midships, y=10.0 m and β=20 degrees.

Looking at Figure 25 it is seen that the surge force generally tends to increase with the drift angle within the range of investigated drift angles. However the measurements indicate a decrease in surge force for 40 degrees drift angle and then a large increase towards 60 degrees drift angle. Comparing the calculated and measured surge force, Figure 25 shows that CFD gives good agreement for both 0 and 20 degrees drift angle with a gap of 10.0 m. However, at 40 degrees drift angle the CFD calculation over-predicts the surge force. At large drift angles the presence of separation in the flow and the separation is difficult to capture with CFD since the calculations in the present study is not run transient.

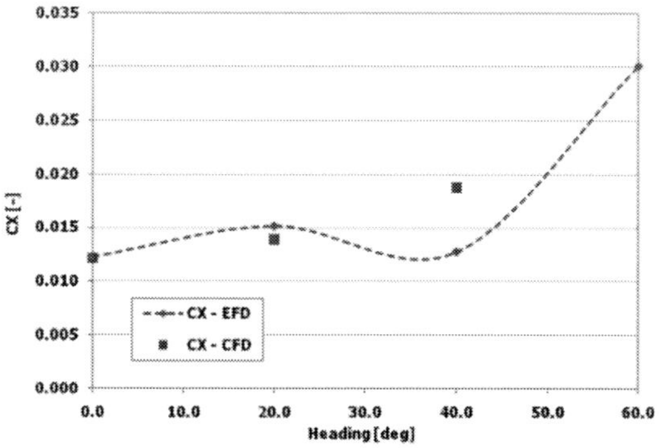

Figure 25 : Variation of surge force on tug with drift angle at midships and y=10.0m.

The magnitude of the sway force, which is illustrated in Figure 26, increases with the drift angle within the tested range. Moreover, it is seen that the tug is pulled towards the tanker for all investigated drift angles. The CFD is seen to under-predict the sway force for all drift angles compared to the measurements.

The yaw moment in Figure 27 is increasing up to around 40 degrees and after which it decreases. The yaw moment is within the investigated range only positive, meaning that the bow is pushed towards the tanker. Comparing the CFD with the measurements shows that the CFD tend to under-predict the yaw moment for larger drift angles. CFD seems to predict the yaw moment quite well up to 20 degrees, which is within the normal manoeuvring range for tug operations at speed. This is encouraging since the yaw moment is the key element in manoeuvring simulation.

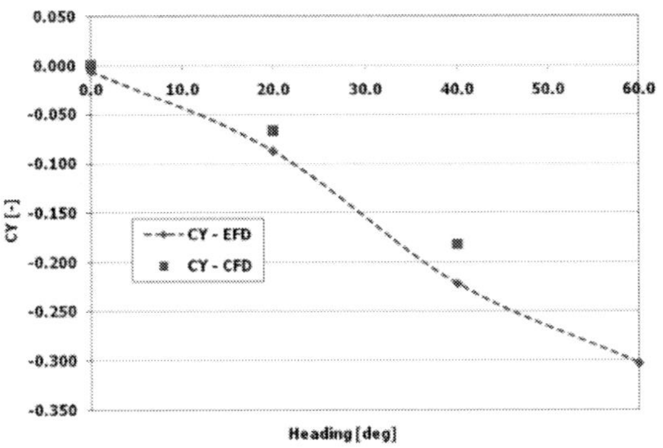

Figure 26 : Variation of sway force on tug with drift angle at midships and y=10.0m.

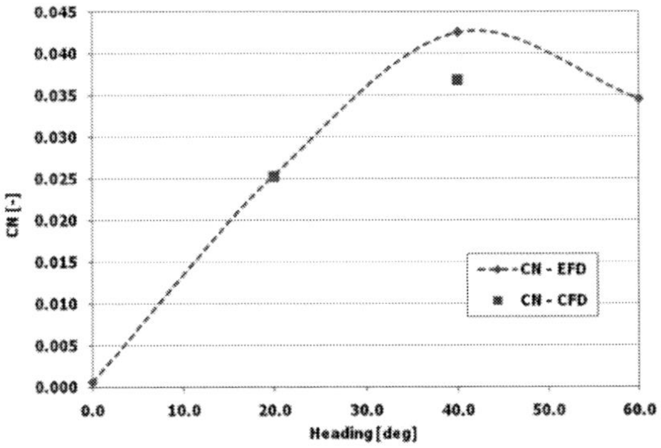

Figure 27 : Variation of yaw moment on tug with drift angle at midships and y=10.0m.

8. CONCLUDING REMARKS

In the present work a quasi steady ship-ship interaction problem is investigated by means of CFD. The test case is an ASD type tug located next to a conventional tanker, while advancing with constant speed. The tug is located in a number of longitudinal and transverse positions along the tanker with different drift angles. For each condition the tug and tanker positions are locked relative to each other. The setup is taken from a previous model test conducted in FORCE Technology's towing tank. In the model test the tanker was locked in all degrees of freedom, while the tug was free to heave, pitch and roll and otherwise fixed. In the CFD simulation, which is conducted with the RANS code Star-CCM+, the tug is positioned and locked according to the measurement. The considered conditions are chosen from the experimental test program in order to study the influence of drift angle, transverse and longitudinal positions on the flow field and on the surge and sway forces and the yaw moment acting on the tug. The computed results are compared with experimental data to investigate how well the CFD code can capture trends and absolute vales of the forces and moments. Comparison of data obtained with a potential theory based tool and the RANS code is also done to check how these two methods perform compared to each other.

On the flow field level, the work shows that the CFD results give a detailed picture of the flow features around the tug and in the gap between the tanker and the tug. Qualitatively it is possible to learn about how the two ships mutually influence the velocity field and the vortex structures as the position of the tug is changed and how this correlates with the behaviour of the hull surface pressure. Further, it seems that the flow field behaviour to a large degree can be correlated to the behaviour of the forces and moments. However, no measured field quantities are available, so on the quantitative level it is not possible to say how well the CFD tool predicts the flow features.

For the forces and moments it is found that the RANS code is capable of picking up the trends in the quantities as the position of the tug changes, but quantitatively, deviations between measurement and computation are observed for some of the cases.

For fixed x-position at midships and zero drift angle the variation with y-distance is predicted quite well for the surge force and the yaw moment, but for the sway force under-prediction appears close to the peak region of the force. Concerning the performance of the RANS code compared to the potential theory based method it seems that the RANS code is in better agreement with the measurement than the potential code. This indicates that inclusion of viscous effects is important for the present type of application, particularly with narrow gaps where viscous effects are dominating the flow field.

When y-position and drift angle are fixed and the x-position is varied, the surge force and yaw moment are again predicted fairly well, but there seems to be a general and relatively significant under-prediction of the sway force by the CFD code. According to the y-distance study, the y-position used for longitudinal variations is in the critical range, since it is close to the rather narrow peak of the sway force, which could explain why this position is generally difficult to handle. In order to conclude on this, a y-distance study should be performed with the tug at aft and forward shoulder of the tanker. However, this is outside the scope of the present work.

Finally, for the drift angle study it appears that the CFD code predicts the forces and moments in fair agreement with the measurement for the moderate drift angle of 20 degrees. For the large drift angle of 40 degrees larger deviations are observed for all quantities, which is not surprising, since extensive separation and unsteady effects are seen to occur at such drift angles.

Based on the above work RANS-based CFD seems to be a promising tool in connection with studies of the ship-ship interaction problems. However, there are still un-resolved issues related to under-prediction of the forces in some conditions, which need to be investigated further in order to apply the CFD tool for generation of ship-ship interaction data for the manoeuvring simulator.

9. FUTURE ACTIVITIES

In the presented work the CFD simulations were made for the tug sailing in a locked position taken from the experiment, but in future work, the tug should be free to heave, pitch and trim. This means that future simulations should be done with the 6DOF solver enabled in the CFD code. Further, it is planned to continue this study by investigating similar cases at 1) higher speeds where the tankers wave elevations are higher influencing the tug and 2) in shallow water, to study shallow water effects and compare with experimental data, which is already available.

10. ACKNOWLEDGEMENTS

The present research is sponsored by FORCE Technology, Denmark. The experimental campaign was sponsored by Den Danske Maritime Fond.

11. REFERENCES

[1] *StarCCM+ Users manual*

[2] Shin, Keun Woo, Bingham, Harry B. and Andersen, Poul. "Numerical analysis of ship-ship interaction by a high-order potential flow code with ship motions", proceedings, Numerical towing Tank Symposium. Brest, France, 2008

12. AUTHORS BIOGRAPHY

Claus D. Simonsen: Employed at FORCE Technology, Denmark, as CFD Senior Specialist in the Department for Hydro- and Aerodynamics since 1997. He has a M.Sc. degree in Naval Architecture and a PhD degree within application of RANS for manoeuvring related flow simulations from the Technical University of Denmark (DTU). He is responsible for the activities related to application of viscous CFD to both R&D and practical assignments within hull form design and optimization, resistance- and self-propulsion, manoeuvring and sea-keeping.

Christian Klimt Nielsen: Employed at FORCE Technology, Denmark, as Project Manager in the Department for Hydro- and Aerodynamics since 2008. Has a M.Sc. degree in Naval Architecture from the Technical University of Denmark (DTU) specializing in ship hydrodynamics. Primarily responsible for hull form design and optimization based on viscous CFD resistance, self-propulsion and manoeuvring calculations. Also conducting resistance, self-propulsion and sea-keeping tests in the FORCE Technology towing tank.

Janne Flensborg Otzen: Employed at FORCE Technology, Denmark, as Project Engineer in the Department for Hydro- and Aerodynamics since 2007. Has a degree in M.Sc. in Naval Architecture from the Technical University of Denmark (DTU). Her primary responsibilities at FORCE are manoeuvring tank tests (PMM) and development of mathematical models of ships for the simulator department.

Kristian Agdrup: Employed at FORCE Technology, Denmark, as Project Manager in the Department for Hydro- and Aerodynamics since 1997. Has a degree in M.Sc. in Naval Architecture from the Technical University of Denmark (DTU). His primary responsibilities at FORCE are manoeuvring tank tests (PMM) and development of mathematical models of ships for the simulator department. Is member of the Manoeuvring Committee of the 26th International Towing Tank Conference (ITTC).

INVESTIGATING SHIP MANEUVERS AROUND A FLOATING STRUCTURE UNDER THE INFLUENCE OF A UNIFORM CURRENT IN DEEP AND CALM WATER

R Skejic, MARINTEK, Norway
M Breivik, Centre for Ships and Ocean Structures (CeSOS), Norwegian University of Science and Technology, Norway
T E Berg, MARINTEK, Norway

SUMMARY

This paper considers the topics of hydrodynamic interaction load effects and ship motion control system design associated with maneuvering a ship around a floating object such as an offshore cargo storage facility in deep and calm water under the influence of a uniform current. In particular, the hydrodynamic interaction loads in the time domain are estimated by using generalized Tuck and Newman theory [1]. These loads are estimated within the combined seakeeping and maneuvering model of Skejic and Faltinsen [5, 6] and Skejic [7]. The applicability of the model is discussed from a hydrodynamic point of view and also evaluated through numerical simulations. In addition, a novel ship motion control system is developed to make it possible for a ship to safely maneuver around moored or drifting objects despite being exposed to significant hydrodynamic interaction effects. The motion control system consists of a guidance system which instructs a course control system which subsequently commands the ship rudder to enable circular path following around any specified object. Two close-proximity maneuvering scenarios are simulated using a scaled 'MARINER' ship and a hexagonal prism-shaped cargo storage facility which is stationary or drifting under the influence of a uniform current. The hydrodynamic interaction effects and effectiveness of the maneuvers are then investigated for these scenarios. Finally, aspects concerning the marine inspection operations are discussed from the perspective of navigational safety.

1. INTRODUCTION

The main objective of this paper is to investigate the hydrodynamic interaction effects which occur when a ship operates in close proximity of a floating object which is stationary or drifting under the influence of a uniform current field. It can be expected that the ship's maneuvering behavior will be significantly affected, which will require a good understanding of the hydrodynamic loads in order to accomplish safe and efficient operations.

From a practical point of view, this scenario may arise in marine towing operations when it is required to inspect the status of a floating object. Such a requirement could occur during towing preparations or as a result of an unexpected breakage of towing lines.

In the past decades, much work has been done to predict the hydrodynamic interaction loads acting on a ship moving in proximity of another stationary or movable object in confined or deep water using slender-body theory, see for example Abkowitz et al. [2], Kijima and Yasukawa [3], Yeung [4], Wang [12], Yeung and Tan [15]. All these theories are based on potential flow, which means that the viscous effects are neglected. Furthermore, calm water conditions are assumed.

The present paper uses generalized Tuck and Newman far-field theory [1] to estimate the hydrodynamic interaction loads between two floating objects at moderate speeds in deep water. Owing to the safety distance requirement imposed on a maneuvering inspection ship in the vicinity of a floating object, the far-field theory provides satisfactory prediction of the interaction loads affecting both objects.

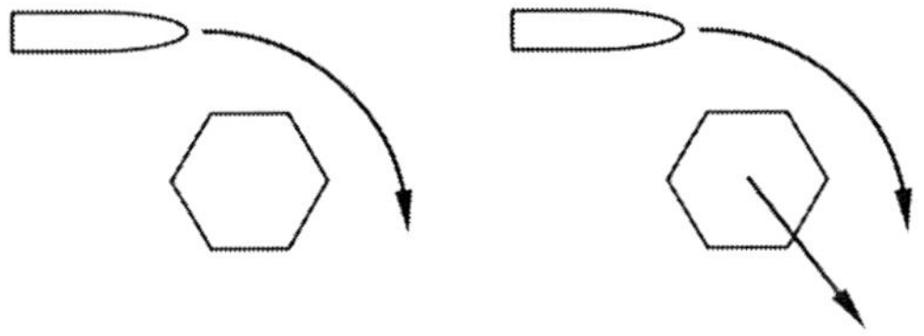

Figure 1: Examples of maneuvering scenarios which involve an inspection vessel and a stationary or drifting floating object.

In order to evaluate the navigational safety aspects of the considered close-proximity operation, two distinctive maneuvering scenarios will be studied as shown in Figure 1. This kind of analysis will provide closer insight into the behavior of the hydrodynamic interaction loads and corresponding maneuver safety and effectiveness.

As far as we are aware, no similar investigations have previously been carried out. The main contributions of this paper include a generalization of the Tuck and Newman theory [1] as well as the development of a motion control system which enables precise and safe circling maneuvers.

2. MANEUVERING MODEL

In this section, a unified model generally formulated on two time scales by Skejic and Faltinsen [5, 6] and Skejic [7] has been extended to include the influence of uniform current as described by Artyszuk [11]. The model is applied to analyze the marine inspection operation involving an inspection ship and a cargo storage facility. Although the general capability of the unified model is to provide a combined seakeeping and maneuvering analysis of two advancing ships, the present study uses the model to investigate the maneuvering behavior of the inspection ship in calm water when the intended maneuver is performed in close proximity to both a stationary and drifting storage facility, where the drift is induced by a uniform current.

The Cartesian right-handed maneuvering body-fixed coordinate systems $O_{Ci}x_{Ci}y_{Ci}$ shown in Figure 2 with positive z_{Ci} – axis pointing upwards are used for the maneuvering scenarios, where index $i \in [A, B]$ represents the inspection ship A and the cargo storage facility B of a hexagonal prism shape, respectively.

The centre of gravity (COG) of each object is located in the origin O_{Ci} of the coordinate system $O_{Ci}x_{Ci}y_{Ci}$ which is located in the lateral symmetry plane of each body. $OX_EY_EZ_E$ is an Earth-fixed coordinate system with positive Z_E – axis pointing upwards. X_i and Y_i are hydrodynamic forces defined positively in the positive x_{Ci} – and y_{Ci} – directions, respectively. N_i denotes the yaw moment about the z_{Ci} – axis, defined positively as shown in Figure 2. Also, u_i and v_i are the x_{Ci} – and y_{Ci} – components of the total speed $U_i = \sqrt{u_i^2 + v_i^2}$ of the involved objects. The angular rate is denoted by r_i, while ψ_i, β_i = atan($-v_i/u_i$) and δ_i are respectively the yaw (heading), drift (sideslip) and rudder angles. The cargo storage facility does not posses any kind of propulsive or steering devices.

The longitudinal and transverse distances between the centres of gravity of the ship and the floating structure s_A and e_A are defined with respect to the body-fixed coordinate system of the inspection ship A. The lengths of the maneuvering objects are denoted L_i for $i \in [A, B]$.

The maneuvering module within the unified-theory model is given by

$$\begin{bmatrix} M & 0 & 0 \\ 0 & M & 0 \\ 0 & 0 & I_{66} \end{bmatrix}_i \begin{bmatrix} \dot{u} \\ \dot{v} \\ \dot{r} \end{bmatrix}_i + \begin{bmatrix} 0 & -Mr & 0 \\ 0 & 0 & Mu \\ 0 & 0 & 0 \end{bmatrix}_i \begin{bmatrix} u \\ v \\ r \end{bmatrix}_i =$$
$$= \begin{bmatrix} X_{\dot{u}} & 0 & 0 \\ 0 & Y_{\dot{v}} & Y_{\dot{r}} \\ 0 & N_{\dot{v}} & N_{\dot{r}} \end{bmatrix}_i \begin{bmatrix} \dot{u} \\ \dot{v} \\ \dot{r} \end{bmatrix}_i + \begin{bmatrix} 0 & 0 & 0 \\ 0 & Y_v & Y_r \\ 0 & N_v & N_r \end{bmatrix}_i \begin{bmatrix} \tilde{u} \\ \tilde{v} \\ r \end{bmatrix}_i +$$
$$+ \begin{bmatrix} 0 & -Y_{\dot{v}}r & -Y_{\dot{r}}r \\ 0 & 0 & X_{\dot{u}}\tilde{u} \\ 0 & -X_{\dot{u}}\tilde{u} & 0 \end{bmatrix}_i \begin{bmatrix} \tilde{u} \\ \tilde{v} \\ r \end{bmatrix}_i + \begin{bmatrix} -R(\tilde{u}) + (1-t)T(\tilde{u}) \\ 0 \\ 0 \end{bmatrix}_A +$$
$$\begin{bmatrix} -R(\tilde{u}) \\ 0 \\ 0 \end{bmatrix}_B + \begin{bmatrix} X_\delta \\ Y_\delta \\ N_\delta \end{bmatrix}_A + \begin{bmatrix} 0 \\ Y_{CF} \\ N_{CF} \end{bmatrix}_i + [INT]_i \,, \qquad (1)$$

which is consistent with the ITTC [8] nomenclature and represents a version of the generalized surge-sway-yaw slender-body theory by Söding [9]. The unified model has a validity within the range of moderate Froude numbers, i.e., $Fn <\approx 0.2$, which means that steady wave generation is implicitly assumed to be small. The formulation of the maneuvering model is carried out by using a modular concept, where the forces and moments in equation (1) due to resistance R, propulsion T, rudder δ, and nonlinear viscous cross-flow CF form separate modules. Details about each module are further explained by Skejic and Faltinsen [5, 6] and Skejic [7].

In (1), the ship and cargo storage object mass and products of inertia are denoted as M and I_{66}, respectively. The maneuvering coefficients associated with the speeds and accelerations are obtained from the strip theory of Salvesen-Tuck-Faltinsen [10] for zero encounter

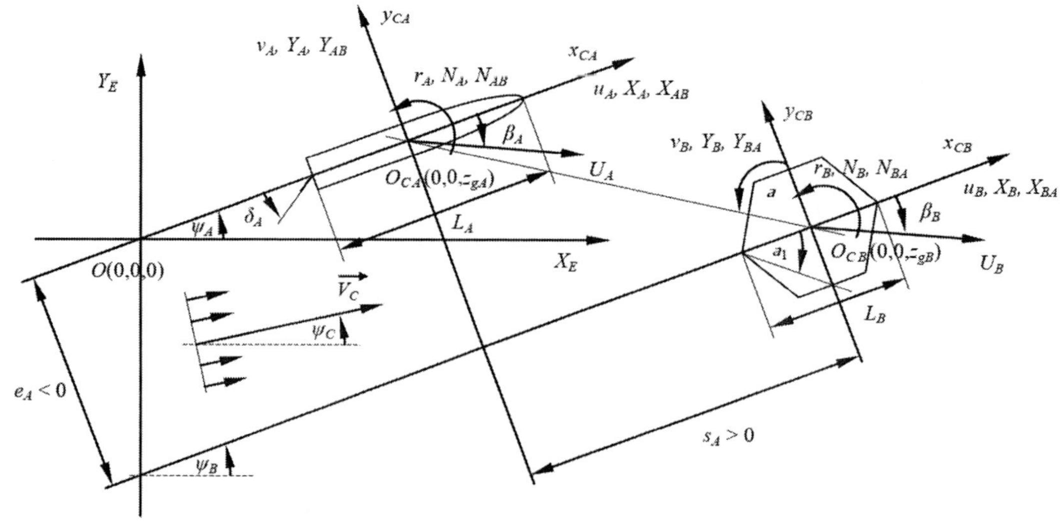

Figure 2: Coordinate systems for inspection ship A and cargo storage facility B in a uniform current field.

frequency ω_e and referenced to the maneuvering body-fixed coordinate systems of Figure 2. More details about the applied procedure can be found in Skejic and Faltinsen [5, 6] and Skejic [7]. In addition, the hydrodynamic maneuvering loads for the cargo storage object are estimated by methods presented by Wendel [13] and Korotkin [14].

The influence of a uniform current is accounted for in (1) by using the relationships

$$\begin{bmatrix} \tilde{u} \\ \tilde{v} \end{bmatrix}_i = \begin{bmatrix} u \\ v \end{bmatrix}_i - \begin{bmatrix} \cos\psi & \sin\psi \\ -\sin\psi & \cos\psi \end{bmatrix} \begin{bmatrix} |\vec{V}_C|\cos\psi_C \\ |\vec{V}_C|\sin\psi_C \end{bmatrix} \quad (2)$$

$$\begin{bmatrix} \dot{\tilde{u}} \\ \dot{\tilde{v}} \end{bmatrix}_i = \begin{bmatrix} \dot{u} \\ \dot{v} \end{bmatrix}_i, \quad (3)$$

where the symbol '~' designates values relative to the water, and where $\vec{V}_C$ represents the uniform current velocity positively defined by its orientation angle γ_C with respect to the Earth-fixed coordinate system $OX_EY_EZ_E$.

The matrix $[INT]|_i$ in (1) represents hydrodynamic interaction loads, namely surge and sway forces as well as yaw moments between the two involved objects. They are predicted by modified Tuck and Newman theory [1], given as follows:

for i = A; $[INT]|_{i=A} = [X_{AB}\ Y_{AB}\ N_{AB}]^T$
for i = B; $[INT]|_{i=B} = [X_{BA}\ Y_{BA}\ N_{BA}]^T$.

Here, X_{AB} and Y_{AB} represent the surge and sway interaction forces acting on the inspection ship A due to the proximity of the cargo storage object B, while N_{AB} denotes the interaction yaw moment on ship A due to the proximity of the storage object B, and vice versa. The positive direction of the interaction loads is defined with arrows as indicated in Figure 2.

It should also be noted that the interaction surge and sway forces on the cargo storage object B are of equal magnitude when it is stationary. On the other hand, the interaction yaw moment N_{BA} can be equal to or different from zero depending on the symmetry of the moving ship A which encircles the object B, as shown in Figure 3.

In particular, the hydrodynamic interaction loads on both objects involved in the maneuvering analysis are estimated as follows:

Inspection ship A

$$X_{AB} = \rho\frac{U_B^2}{8\pi}\int_{L_A} S_A'(x_A)\left\{\int_{L_B}\frac{S_B'(x_B)(x_B - x_A + s_A)}{\left[(x_B - x_A + s_A)^2 + e_A^2\right]^{3/2}}dx_B\right\}dx_A + \rho\frac{U_B^2}{8\pi}S_A(x_A^{AFT})\left\{\int_{L_B}\frac{S_B'(x_B)(x_B - x_A^{AFT} + s)}{\left[(x_B - x_A^{AFT} + s_A)^2 + e_A^2\right]^{3/2}}dx_B\right\} \quad (4)$$

$$Y_{AB} = -\rho\frac{U_B e_A}{4\pi}\int_{L_A}\left[U_B S_A'(x_A) + (U_B - U_A)A_A'(x_A)\right]\left\{\int_{L_B}\frac{S_B'(x_B)}{\left[(x_B - x_A + s_A)^2 + e_A^2\right]^{3/2}}dx_B\right\}dx_A - \rho\frac{U_B^2 e_A}{4\pi}\left[S_A(x_A^{AFT}) + A_A(x_A^{AFT})\right]\left\{\int_{L_B}\frac{S_B'(x_B)}{\left[(x_B - x_A^{AFT} + s_A)^2 + e_A^2\right]^{3/2}}dx_B\right\} \quad (5)$$

$$N_{AB} = -\rho\frac{U_B^2 e_A}{4\pi}\left\{-\left[S_A(x_A^{AFT}) + A_A(x_A^{AFT})\right]x_A^{AFT}\left\{\int_{L_B}\frac{S_B'(x_B)}{\left[(x_B - x_A^{AFT} + s_A)^2 + e_A^2\right]^{3/2}}dx_B\right\}\right\} + \rho\frac{U_B e_A}{4\pi}\int_{L_A}\left\{U_B S_A'(x_A)x_A + (U_B - U_A)A_A'(x_A)x_A + U_B\left[S_A(x_A) + A_A(x_A)\right]\right\}\left\{\int_{L_B}\frac{S_B'(x_B)}{\left[(x_B - x_A + s_A)^2 + e_A^2\right]^{3/2}}dx_B\right\}dx_A \quad (6)$$

Cargo storage facility B

$$F_Y = -\rho\frac{U_A e_A}{4\pi}\int_{L_B}\left[U_A S_B'(x_B) + (U_A - U_B)A_B'(x_B)\right]\left\{\int_{L_A}\frac{S_A'(x_A)}{\left[(x_A - x_B - s_A)^2 + e_A^2\right]^{3/2}}dx_A\right\}dx_B - \rho\frac{U_A^2 e_A}{4\pi}\left[S_B(x_B^{AFT}) + A_B(x_B^{AFT})\right]\left\{\int_{L_A}\frac{S_A'(x_A)}{\left[(x_A - x_B^{AFT} - s_A)^2 + e_A^2\right]^{3/2}}dx_A\right\}$$

$$F_X = F_Y\tan(\alpha)$$

$$X_{BA} = F_X\cos(\psi_A - \psi_B) - F_Y\sin(\psi_A - \psi_B) \quad (7)$$

$$Y_{BA} = F_X\sin(\psi_A - \psi_B) + F_Y\cos(\psi_A - \psi_B) \quad (8)$$

$$F_{IN} = \sqrt{(X_{BA})^2 + (Y_{BA})^2} = \sqrt{(F_X)^2 + (F_Y)^2} \quad (9)$$

$$\begin{aligned} N_{BA} = -\Bigg\{ & \rho\frac{U_A^2 e_A}{4\pi}\left\{\left[S_B(x_B^{AFT}) + A_B(x_B^{AFT})\right]x_B^{AFT}\left\{\int_{L_A}\frac{S_A'(x_A)}{\left[(x_A - x_B^{AFT} - s_A)^2 + e_A^2\right]^{3/2}}dx_A\right\}\right\} \\ & + \rho\frac{U_A e_A}{4\pi}\int_{L_B}\left\{U_A S_B'(x_B)x_B + (U_A - U_B)A_B'(x_B)x_B + U_A\left[S_B(x_B) + A_B(x_B)\right]\right\} \left\{\int_{L_A}\frac{S_A'(x_A)}{\left[(x_A - x_B - s_A)^2 + e_A^2\right]^{3/2}}dx_A\right\}dx_B\Bigg\} \\ & + \tan^2(\alpha_1)\tan(\alpha)\Bigg\{\rho\frac{U_A^2 e_A}{4\pi}\left\{\left[S_B(x_B^{AFT}) + A_B(x_B^{AFT})\right]x_B^{AFT}\left\{\int_{L_A}\frac{S_A'(x_A)}{\left[(x_A - x_B^{AFT} - s_A)^2 + e_A^2\right]^{3/2}}dx_A\right\}\right\} \\ & + \rho\frac{U_A e_A}{4\pi}\int_{L_B}\left\{U_A S_B'(x_B)x_B + (U_A - U_B)A_B'(x_B)x_B + U_A\left[S_B(x_B) + A_B(x_B)\right]\right\} \left\{\int_{L_A}\frac{S_A'(x_A)}{\left[(x_A - x_B - s_A)^2 + e_A^2\right]^{3/2}}dx_A\right\}dx_B\Bigg\} \end{aligned} \quad (10)$$

where α represents the angle between the distances s_A and e_A, while α_1 denotes the angle between the x and y coordinates of the floating object, see Figure 2. Furthermore, F_{IN} represents the instantaneous force on the cargo storage object B, see Figure 3.

In addition, $S_i(x_i)$ is the cross-sectional area of the double bodies, $A_i(x_i)$ is the cross-sectional sway added mass of the double bodies, while $S_i'(x_i)$ and $A_i'(x_i)$ are their ordinary derivatives. Furthermore, ρ is the fluid density while U_i and L_i are the speeds and lengths of the involved bodies. The superscript 'AFT' means aftermost transom stern cross section (concerning flow separation, see Skejic [7]), which implies for example that $S_B(x_B^{AFT})$ represents the cross-sectional area of the aftermost cross section of the double body B.

A complete derivation of the above expressions is not included in this paper because it takes too much space, but will be included in a separate publication.

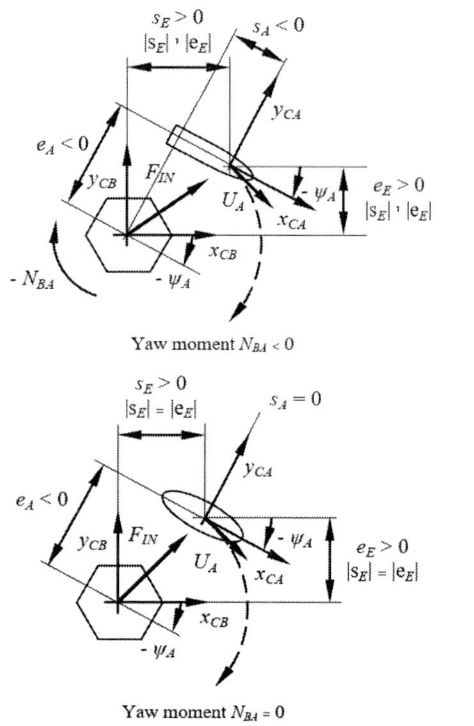

Figure 3: Behavior of the hydrodynamic interaction yaw moment N_{BA}. The instantaneous force on the cargo storage object B is defined with F_{IN} as stated in (8), while the values s_E and e_E are defined relative to the Earth-fixed coordinate system.

The application of the Tuck and Newman far-field theory [1] depends on the body speeds, geometries and separation distances. It should be noted that this theory has certain limitations since it is derived on a basis which only accounts for lower-order terms in the slenderness parameter. Further details about the limitations are discussed by Skejic and Faltinsen [5, 6] and Skejic [7].

The maneuvering equation system (1) is solved with a time integration algorithm based on the explicit Runge-Kutta fourth-order method (RK-IV) with constant time steps. Also, the ship and cargo storage facility positions in the Earth-fixed coordinate system $OX_EY_EZ_E$ can be obtained by transforming the body-fixed speeds $(u, v, r)_i$ into the Earth-fixed speeds $(\dot{x}, \dot{y}, \dot{\psi})_i$ by using

$$[\dot{x} \;\; \dot{y} \;\; \dot{\psi}]_i^{\mathrm{T}} = J_{3\times3_i} [u \;\; v \;\; r]_i^{\mathrm{T}}, \quad (11)$$

where $J_{3\times3_i}$ is the Euler angle matrix

$$J_{3\times3_i} = \begin{bmatrix} \cos\psi & -\sin\psi & 0 \\ \sin\psi & \cos\psi & 0 \\ 0 & 0 & 1 \end{bmatrix}_i \quad (12)$$

for $i \in$[A, B].

3. MOTION CONTROL SYSTEM

This section develops a motion control system which enables an inspection ship to encircle a floating target object at sea. In particular, the encirclement scenario must be solved by considering that:

- The ship moves forward with a constant rpm and therefore cannot control its speed.
- Only the position and velocity of the ship and target object are available for control. Acceleration and jerk are not measured.
- The interaction forces between the ship and the floating object are not known.

The basic architecture of the suggested motion control system is shown in Figure 4. As can be seen, it is divided into two main parts, consisting of a guidance system and a course control system.

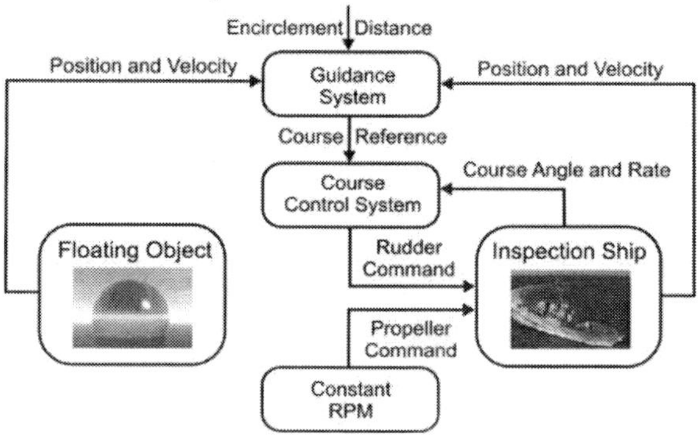

Figure 4: A reference course is computed for the inspection ship by comparing the position and velocity of the floating object with those of the ship, also taking into account a desired encirclement distance. The course control system then computes a rudder command to achieve the course reference and thereby keep the desired distance to the floating object while moving forward at a constant rpm.

In what follows, the inspection ship A is represented by its planar position $\mathbf{p}(t) \triangleq [x(t), y(t)]^{\mathrm{T}} \in \mathbb{R}^2$ and velocity $\mathbf{v}(t) \triangleq \mathrm{d}\mathbf{p}(t)/\mathrm{d}t = \dot{\mathbf{p}}(t) \in \mathbb{R}^2$, omitting the subscripts 'A' and 'B' for notational ease. These motion variables are stated relative to the Earth-fixed coordinate frame, where the x-axis points east, the y-axis points north, and the z-axis points up from the plane, see Figure 2. The ship velocity can also be represented by its polar coordinates speed

$$U(t) \triangleq |\mathbf{v}(t)| = \sqrt{\dot{x}(t)^2 + \dot{y}(t)^2} > 0 \quad (13)$$

and course

$$\chi(t) \triangleq \mathrm{atan2}(\dot{y}(t), \dot{x}(t)) \in \mathbb{S} \triangleq [-\pi, \pi], \quad (14)$$

where $\mathrm{atan2}(y, x)$ is the four-quadrant version of $\arctan(y/x) \in \langle -\pi/2, \pi/2 \rangle$. Similarly, the floating target object B is represented by its position $\mathbf{p}_t(t)$, while the target velocity $\mathbf{v}_t$ is constant such that the target speed U_t and course χ_t are both constants. A fundamental constraint of the circling scenario is that the ship speed must be larger than the target speed, satisfying $U(t) > U_t$ at all times. Details about the guidance and course control systems will now be given.

3.1 GUIDANCE SYSTEM

Our objective is to design a desired velocity $\mathbf{v}_d(t)$ for the ship which enables encirclement of the target object at a fixed distance $r_c > 0$. In particular, the desired velocity can be chosen to be composed of two distinct parts as

$$\mathbf{v}_d(t) = \mathbf{v}_t + \mathbf{v}_a(t), \tag{15}$$

where $\mathbf{v}_t$ is a feedforward of the target velocity and $\mathbf{v}_a(t)$ is the relative approach velocity between the ship and the target, see Figure 5. Since the ship speed is uncontrollable due to a constant rpm, the velocity control problem reduces to a pure steering problem because (15) can also be written as

$$\mathbf{v}_d(t) = U(t)[\cos\chi_d(t), \sin\chi_d(t)]^T, \tag{16}$$

where $U(t)$ is the (uncontrollable) ship speed and $\chi_d(t)$ is the desired course angle required for encirclement. The challenge then becomes how to derive $\chi_d(t)$, which can be calculated as

$$\chi_d(t) \triangleq \operatorname{atan2}\left(\mathbf{v}_{d2}(t), \mathbf{v}_{d1}(t)\right), \tag{17}$$

where from (15) we have that

$$\mathbf{v}_{d1}(t) = U_t\cos\chi_t + U_a(t)\cos\chi_a(t) \tag{18}$$

and

$$\mathbf{v}_{d2}(t) = U_t\sin\chi_t + U_a(t)\sin\chi_a(t). \tag{19}$$

Since we already know U_t and χ_t, the challenge of deriving $\chi_d(t)$ corresponds to the challenge of deriving $U_a(t)$ and $\chi_a(t)$ associated with the approach velocity

$$\mathbf{v}_a(t) = U_a(t)[\cos\chi_a(t), \sin\chi_a(t)]^T. \tag{20}$$

Considering Figure 5, we can define a circular path with radius $r_c > 0$ around the target object. Thereafter, we can define a path-tangential reference frame with origin located at the direct projection of the ship position $\mathbf{p}(t)$ onto the circular path. Specifically, the x-axis of this reference frame is rotated a positive angle

$$\chi_c(t) = \chi_e(t) + \lambda\frac{\pi}{2} \tag{21}$$

relative to the x-axis of the Earth-fixed reference frame, where

$$\chi_e(t) \triangleq \operatorname{atan2}\left(y(t) - y_t(t), x(t) - x_t(t)\right), \tag{22}$$

is the encirclement angle, and where $\lambda \in \{-1, 1\}$. For a coordinate system where the x-axis points east, the y-axis north and the z-axis up from the plane, $\lambda = -1$ corresponds to clockwise motion and $\lambda = 1$ to counterclockwise motion.

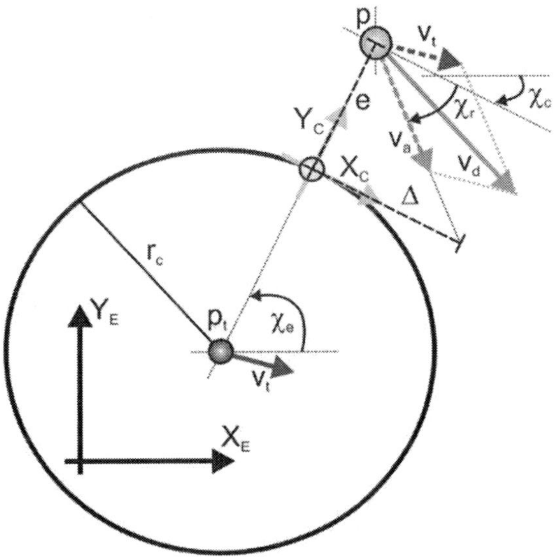

Figure 5: Geometry of the kinematic velocity assignment $\mathbf{v}_d$ which enables a vehicle located at $\mathbf{p}$ to encircle a target object located at $\mathbf{p}_t$ which moves forward with constant velocity $\mathbf{v}_t$.

The distance from the ship to its projection onto the path is called the cross-track error $e(t)$, which in this case can be calculated as

$$\begin{aligned} e(t) &= \lambda\left(r_c - |\mathbf{p}_t(t) - \mathbf{p}(t)|\right) \\ &= \lambda\left(r_c - \sqrt{(x_t(t) - x(t))^2 - (y_t(t) - y(t))^2}\right). \end{aligned} \tag{23}$$

For a ship moving forward at some positive speed, the encirclement control objective corresponds to reducing the cross-track error to zero, which can be achieved by steering control through an assignment which is separated into two parts

$$\chi_a(t) = \chi_c(t) + \chi_r(e), \tag{24}$$

where $\chi_c(t)$ is the path-tangential angle (21), while

$$\chi_r(e) = \arctan\left(-\frac{e(t)}{\Delta}\right) \tag{25}$$

is a velocity-path relative angle which ensures that the ship velocity is directed toward a point that is located a lookahead distance $\Delta > 0$ ahead of the direct projection of its position $\mathbf{p}(t)$ onto the path.

Hence, we have now obtained the course angle $\chi_a(t)$ related to the approach velocity (20). It therefore remains to calculate $U_a(t)$. For this purpose, we can consider the relationship (15), which also can be stated as

$$|\mathbf{v}_d(t)| = |\mathbf{v}_t + \mathbf{v}_a(t)| \quad (26)$$

where $|\mathbf{v}_d(t)|$ is equal to the (uncontrollable) ship speed $U(t)$. Hence, we have that

$$U(t) = \sqrt{(\mathbf{v}_t + \mathbf{v}_a(t))^T(\mathbf{v}_t + \mathbf{v}_a(t))} \quad (27)$$

which is equal to

$$U(t)^2 = \mathbf{v}_t^T\mathbf{v}_t + \mathbf{v}_t^T\mathbf{v}_a(t) + \mathbf{v}_a(t)^T\mathbf{v}_t + \mathbf{v}_a(t)^T\mathbf{v}_a(t), \quad (28)$$

where

$$\mathbf{v}_t^T\mathbf{v}_t = U_t^2, \quad (29)$$

which is known. Likewise,

$$\mathbf{v}_a(t)^T\mathbf{v}_a(t) = U_a(t)^2, \quad (30)$$

which is unknown. Also,

$$\begin{aligned}\mathbf{v}_t^T\mathbf{v}_a(t) &= \mathbf{v}_a(t)^T\mathbf{v}_t \\ &= U_a(t)\left(U_t \cos\chi_t \cos\chi_a(t) + U_t \sin\chi_t \sin\chi_a(t)\right),\end{aligned} \quad (31)$$

which is partly known. Hence, (26) becomes

$$\begin{aligned}U(t)^2 &= U_t^2 \\ &+U_a(t)2\left(U_t \cos\chi_t \cos\chi_a(t) + U_t \sin\chi_t \sin\chi_a(t)\right) \\ &+U_a(t)^2\end{aligned} \quad (32)$$

which can be rewritten as the second-order equation

$$ax(t)^2 + b(t)x(t) + c(t) = 0 \quad (33)$$

with $x(t) = U_a(t)$ and

$$a = 1 \quad (34)$$

$$b(t) = 2\left(U_t \cos\chi_t \cos\chi_a(t) + U_t \sin\chi_t \sin\chi_a(t)\right) \quad (35)$$

$$c(t) = U_t^2 - U(t)^2. \quad (36)$$

The standard analytic solution to (33) then gives

$$x(t) = \frac{-b(t) \pm \sqrt{b(t)^2 - 4ac(t)}}{2a} \quad (37)$$

which can be compacted to

$$U_a(t) = \frac{-b(t) + \sqrt{b(t)^2 + 4\left(U(t)^2 - U_t^2\right)}}{2} \quad (38)$$

since we know that $U_a(t) > 0$, i.e., only the positive solution makes sense. This equation also explicitly shows that the ship must have a greater speed than the target for a valid solution to exist.

Hence, $\chi_d(t)$ can now be calculated from (17) with $\chi_a(t)$ as in (24) and $U_a(t)$ as in (38). The task of the course control system is then to command the ship rudder such that the actual ship course $\chi(t)$ converges to the desired course $\chi_d(t)$ and consequently that the cross-track error $e(t)$ is reduced to zero. The design of this control system will be detailed in the following.

3.2 COURSE CONTROL SYSTEM

We now develop a course control system which translates the steering command from the guidance system into a rudder command for the ship such that it is able to traverse the circle surrounding the moving target at the desired distance.

Since a positive rudder command $\delta_c > 0$ gives a negative yaw rate $r < 0$, we introduce the transformed variable $\delta_t = -\delta_c$ such that $\delta_t > 0$ corresponds to $r > 0$. The maximum rudder $\delta_{t,max}$ is considered at 35°, which is equal to 0.6109 radians.

Exploring the yaw-rate maneuverability of the scaled 'MARINER' type of ship detailed in Section 4 results in the maneuverability vector

$$\begin{aligned}r &= [0, 0.430, 0.788, 1.358, 1.809, 2.146, 2.409]^T \\ &\cdot 10^{-2}[rad/s]\end{aligned} \quad (39)$$

for the scaled rudder input

$$\delta_{t,scaled} = [0, 0.1, 0.2, 0.4, 0.6, 0.8, 1.0]^T \quad (40)$$

which is related to the actual rudder input through $\delta_t = \delta_{t,max}\delta_{t,scaled}$. Hence, a maximum rudder $\delta_{t,max}$ of 35° gives a maximum turning rate r_{max} of 1.38 °/s, which means that it takes 260.82 s to turn around 360°.

Then assuming a symmetric rudder behavior, curve fitting of the maneuverability data to obtain an $r-\delta$ map is done using the third-order polynomial

$$\sigma(r) = \sigma_3 r^3 + sgn(r)\sigma_2 r^2 + \sigma_1 r + sgn(r)\sigma_0, \quad (41)$$

where $sgn(\cdot)$ is the *sign* function, such that $sgn(r) > 0$ if $r > 0$ and vice versa. Specifically, the variable $\sigma(r)$ represents a normalized rudder assignment $\sigma(r) \in [0,1]$ which takes yaw rate r as an input. The parameters σ_i with $i \in [0,3]$ are calculated based on least-squares curve fitting using (39) and (40), resulting in $\sigma_3 = 35585$, $\sigma_2 = -164.77$, $\sigma_1 = 24.76$ and $\sigma_0 = -0.002$.

Hence, we can now design a feedforward (FF) steering control law as

$$\delta_{\mathrm{ff}}(r_{\mathrm{d}}(t)) = \delta_{\mathrm{t,max}}\sigma(r_{\mathrm{d}}(t)) \tag{42}$$

which will be able to achieve a constant desired yaw rate $|r_{\mathrm{d}}(t)| \leq r_{\max}$ to within small errors in steady state. Feedback (FB) control must also be used to completely remove such errors, which can be caused by modeling uncertainties, changing operating conditions, etc. Consequently, the total rudder deflection can be commanded as

$$\delta_{\mathrm{t}}(r_{\mathrm{d}}(t), \tilde{r}(t)) = \delta_{\mathrm{ff}}(r_{\mathrm{d}}(t)) + \delta_{\mathrm{fb}}(\tilde{r}(t)) \tag{43}$$

where

$$\tilde{r}(t) \triangleq r_{\mathrm{d}}(t) - r(t) \tag{44}$$

is the yaw-rate control error, and where

$$\delta_{\mathrm{fb}}(\tilde{r}(t)) = k_{\mathrm{p},\tilde{r}}\tilde{r}(t) + k_{\mathrm{i},\tilde{r}}\int_0^t \tilde{r}(\tau)d\tau \tag{45}$$

is the proportional-integral feedback term which ensures that the steady-state yaw rate will be kept, using the control gains $k_{\mathrm{p},\tilde{r}} > k_{\mathrm{i},\tilde{r}} > 0$. Hence, the total FF + FB control law (43) will be able to achieve a desired constant yaw rate in the presence of modeling uncertainties and varying environmental conditions.

We now proceed to define a desired yaw rate $r_{\mathrm{d}}(t)$ which takes the desired steering angle $\chi_{\mathrm{d}}(t)$ as input. Specifically, we can use

$$r_{\mathrm{d}}(t) = r_{\max}\tanh\left(k_{\mathrm{p},\tilde{\chi}}\tilde{\chi}(t)\right) \tag{46}$$

where $\tilde{\chi}(t)$ is the steering error, $r_{\max}$ is the maximum yaw rate achievable by the ship, and $k_{\mathrm{p},\tilde{\chi}} > 0$ is a control gain which shapes the transient convergence behavior of $\tilde{\chi}(t)$. Also, instead of calculating $\tilde{\chi}(t)$ as

$$\tilde{\chi}(t) = \chi_{\mathrm{d}}(t) - \chi(t), \tag{47}$$

the error can rather be calculated from

$$\tilde{\chi}(t) = \mathrm{atan2}\left(\sin\tilde{\chi}(t), \cos\tilde{\chi}(t)\right), \tag{48}$$

where

$$\sin\tilde{\chi}(t) = \sin\chi_{\mathrm{d}}(t)\cos\chi(t) - \cos\chi_{\mathrm{d}}(t)\sin\chi(t) \tag{49}$$

and

$$\cos\tilde{\chi}(t) = \cos\chi_{\mathrm{d}}(t)\cos\chi(t) + \sin\chi_{\mathrm{d}}(t)\sin\chi(t) \tag{50}$$

to avoid any possible wraparound problems associated with (47), see also [17] and [19].

In addition, since we do not have access to acceleration and jerk measurements, higher-order derivatives of the desired course angle $\chi_{\mathrm{d}}(t)$ cannot be calculated, which again means that perfect circle following cannot be achieved. However, introducing integral action in the steering law by using

$$\chi_{\mathrm{r}}(e(t)) = \arctan\left(-k_{\mathrm{p,e}}e(t) - k_{\mathrm{i,e}}\int_0^t e(\tau)d\tau\right) \tag{51}$$

instead of (25), the inspection ship will be able to track the virtual circle surrounding the target object with only small errors. Note that $k_{\mathrm{p,e}} = 1/\Delta$, see also [18], while $k_{\mathrm{i,e}} < k_{\mathrm{p,e}}$. More detailed discussions regarding the course control system will be made in a separate publication.

Finally, in order to achieve the correct yaw rate, the commanded rudder must be calculated according to $\delta_{\mathrm{c}} = -\delta_{\mathrm{t}}$, where δ_{t} is given by (43). The combined guidance and course control system will now be used to control the scaled 'MARINER' ship in what follows.

4. CASE STUDIES

The two case studies investigated in this paper concern marine inspection operations performed in calm and deep water. The main particulars of the considered scaled 'MARINER' type of ship (scale 1:4) and cargo storage facility are given in Figure 6.

	'Storage facility'	*'MARINER-(1:4)'*
Hull particulars		
Length between perpendiculars, L_{PP}	50 m	40.23 m
Breadth, B	43.3 m	5.78 m
Draft (even keel), T	5 m	1.867 m
Block coefficient, C_B	0.75	0.61
Displacement	8118.99 t	268.298 t
Centre of gravity, CG (x_g, y_g, z_g)	(0, 0, 0) m	(0, 0, -0.696) m
Centre of buoyancy, CB (x_b, y_b, z_b)	(0, 0, -2.3) m	(0, 0, -0.87) m
Radius of gyration, pitch/yaw k_{yy}/k_{zz}	16.14 m	10.06 m
Nominal (design) speed, U	0 knots	10 knots
Propeller(s) particulars		
No. of propellers	N/A	1
Type	N/A	Wagen. B-ser.
Diameter, D	N/A	1.198 m
Pitch/Diameter ratio, P/D	N/A	1.13
Blade area ratio, A_E/A_0	N/A	0.45
No. of blades, Z	N/A	5
Direction of rotation	N/A	right-handed
Rudder(s) particulars		
No. of rudders	N/A	1
Rudder type	N/A	semi-balanced
Rudder area, A_R	N/A	2.84 m^2
Rudder aspect ratio, A	N/A	1.23
Max. design rudder angle, δ	N/A	37° - 40°
Max. design rudder rate, $\dot{\delta}$	N/A	2.5 °/s – 3.7 °/s

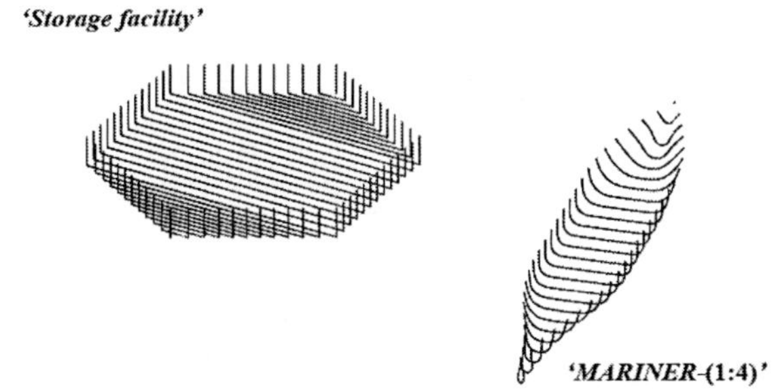

Figure 6: Principal characteristics of the scaled 'MARINER' ship and the cargo storage facility, as well

as an example of the cross-sectional distribution of the submerged parts at even keel on the selected drafts.

The motion control system parameters are chosen as $\Delta = 2.5L_A$ which corresponds to $k_{p,e} = 0.0099$, while $k_{i,e} = 0.00007$. Furthermore, $k_{p,\tilde{\chi}} = 2$, $r_{max} = 0.024$, $k_{p,\tilde{r}} = 12$ and $k_{i,\tilde{r}} = 3$. In addition, $r_c = 60$ and $\lambda = -1$, which means that the ship should circle the center of the floating object at a distance of 60 m in a clockwise manner.

4.1 CIRCLE MANEUVER AROUND A STATIONARY STORAGE FACILITY

This subsection provides analysis of the main maneuvering parameters and hydrodynamic interaction loads on the selected 'MARINER' type of ship and cargo facility in case of a starboard circle maneuver around the floating but stationary object as shown in Figure 7. The mentioned maneuver is carried out by considering calm and deep water conditions. The starting positions for the maneuvering ship and floating object are (– 100, 100) m and (10.07, – 3.17) m, respectively. Since in this case the floating storage object is assumed to be tightly anchored, its velocity is set to zero during the simulation. Furthermore, it has been assumed that the loading conditions of the involved objects do not change during the simulation of the maneuver.

As can be seen from Figure 7, the ship starts the maneuver with an approach speed of 6 knots, which gradually decreases until it reaches a steady value at around 1100 s. In parallel, the oscillating behavior of the transverse and longitudinal distances decreases towards steady values, indicating that the maneuvering ship reaches her stabilized circular maneuver with a turning radius of $-e_A = -e_{A,d} = 60$ m. Until 2600 s, the ship remains within the steady turn maneuver. The time interval is chosen to give sufficient insight into the behavior of the hydrodynamic interaction effects.

Due to the presence of the maneuvering ship, the floating stationary object experiences interaction loads which are changing in time, depending on e_A and s_A. The magnitude of the hydrodynamic interaction surge and sway forces is

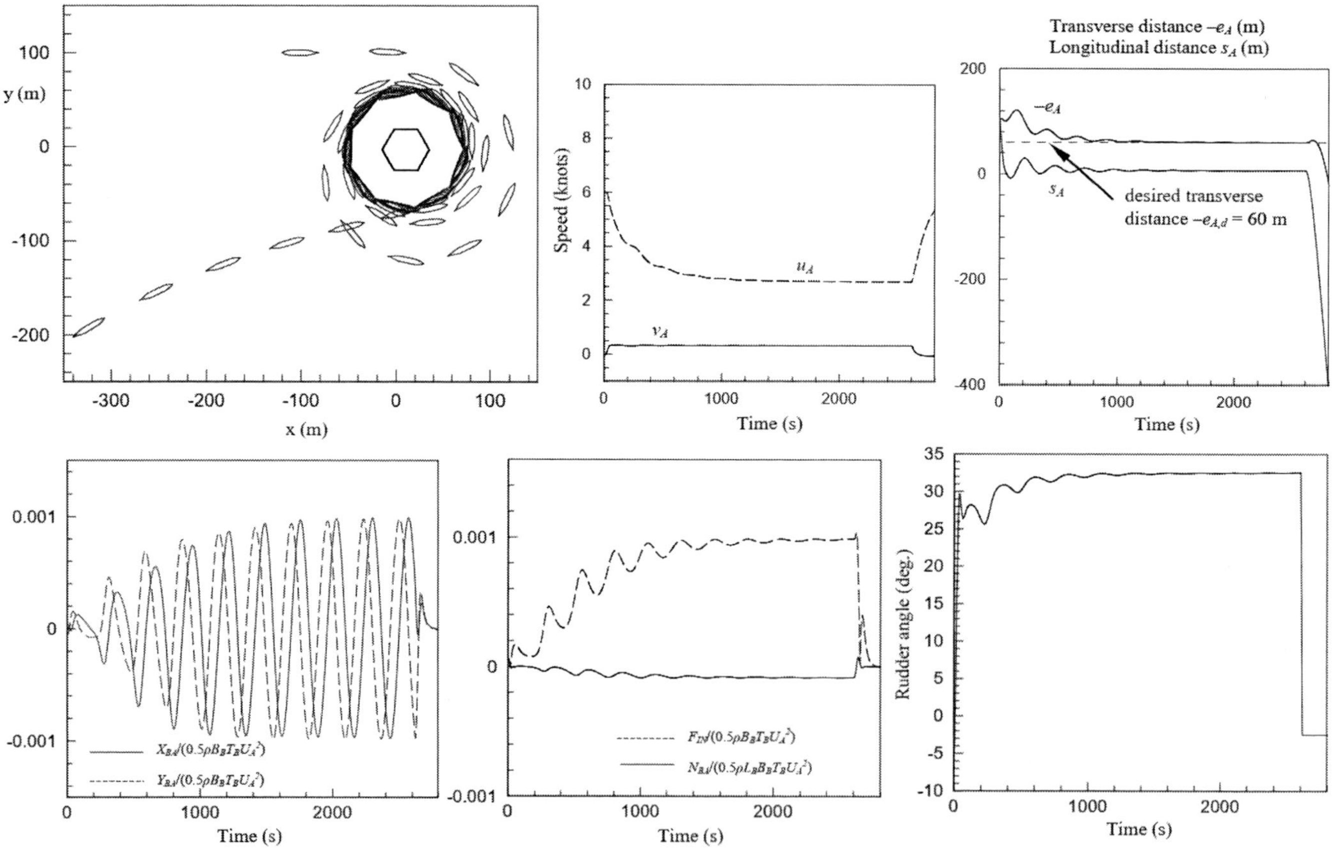

Figure 7: The scaled 'MARINER' performing a starboard turn maneuver at an initial forward speed of 6 knots, with steady turning radius $-e_{A,d} = 60$ m around the stationary storage facility. The simulated ship trajectory is plotted at every 30 s.

gradually increasing until they reach their maximum values at approximately 1800 s. This behavior is to be expected since the maneuvering ship simultaneously establishes a steady starboard-turn maneuver. After the initial time phase, the surge and sway interaction forces are stabilized to maximum equal values due to a stabilized ship turning trajectory and symmetry of the floating structure. The distribution of these forces is shifted in time because the maneuvering ship is encircling the cargo facility.

As Figure 7 shows, the floating facility experiences a negative yaw moment after the initial time phase, which is in accordance with the general behavior of the yaw moment in case of two moving or one moving and one stationary object operating in close proximity, see for example Tuck and Newman [1], Yeung [4], Wang [12], Yeung and Tan [15].

At time 2600 s, the ship executes a disengagement maneuver lasting for 200 s where the ship rudder is deflected with a turning rate of 2.5 °/s until it reaches a preselected angle of 2.5°. During this maneuver, the surge speed of the ship starts increasing again while the floating object experiences typical repulsion behavior from the hydrodynamic interaction surge and sway forces accompanied with positive ('bow in') yaw moment.

A direct consequence of slenderness and other assumptions used in the formulation of the Tuck and Newman theory [1] is that the maneuvering ship experiences zero hydrodynamic interaction loads. This behavior is unrealistic from a physical point of view, but can be alleviated by using a three-dimensional (3D) formulation of the loads, see for instance Xiang and Faltinsen [16].

4.2 CIRCLE MANEUVER AROUND A DRIFTING STORAGE FACILITY

The considered maneuver is now carried out in deep water by taking into account the influence of a uniform current acting from a specific direction defined with respect to the Earth-fixed coordinate system. In particular, the uniform current speed is chosen to be $|\vec{V}_C|$ = 0.2916 knots from the direction γ_C = 315°. As before, the approach speed of the ship is set to 6 knots. Hence, the ship moves about 20 times as fast as the floating object. The task of the motion control system is to ensure a stable starboard-turn maneuver without collision with the floating storage facility which now drifts in the direction of the uniform current.

After approximately 1100 s, the surge and sway speeds reach steady oscillatory values due to the effect of the uniform current, see Figure 8. Similar behavior is associated with the transverse e_A and longitudinal s_A distances since both the maneuvering ship and floating object are operating in the same uniform field. A certain steady 'symmetrical' pattern can be observed in the response of the hydrodynamic interaction loads on both the ship and the floating object. Due to the effect of uniform current, the instantaneous force F_{IN} on the object shows periodical interchange of steady maximum and minimum magnitude corresponding to the transverse and longitudinal distances as well as the speeds. A similar response is observed for the hydrodynamic interaction moment, as can be seen in Figure 8. The hydrodynamic interaction loads on the maneuvering ship also resemble those of the drifting object.

The disengagement phase of the maneuver starts at 2600 s and lasts until 2800 s when the ship rudder is deflected identically to the previous scenario. During this time, the repulsion effects associated with the hydrodynamic interaction forces are experienced by both objects, accompanied by the 'bow in' yaw moment on the floating object and the 'bow out' yaw moment on the maneuvering ship as seen in Figure 8. The figure also shows the relative position of the ship and floating structure, where it can be seen that the latter rotates due to the interaction forces during the maneuver.

In comparison with the previous maneuvering scenario, the maneuvering ship now experiences a more realistic behavior, owing to the fact that the floating object has a drifting speed being exposed to a uniform current field. However, it should be noted that the Tuck and Newman theory [1] applied here is essentially a slender-body theory and its application is strictly speaking 'overstretched' in the present work. This will normally have consequences on the quantitative prediction of the hydrodynamic interaction loads, while it is believed that the qualitative prediction will not be significantly changed in comparison with the use of 3D theory.

4.3 NAVIGATIONAL SAFETY ASPECTS

As already shown, differences in the current conditions affect the behavior of the floating storage facility as well as the maneuvering performance of the inspection ship while executing the encircling maneuver.

From a navigational safety point of view, the behavior of the hydrodynamic interaction loads (surge and sway forces and yaw moments) is of primary concern if similar kind of maneuvers are to be performed under minimum collision risk requirements.

By taking a closer look at figures 7 and 8 the differences concerning the distribution of the loads are obvious. In Figure 7 they are reaching a steady constant magnitude, while in Figure 8 they have steady oscillatory patterns with more and less pronounced peaks which repeat themselves after a certain time period. It can be concluded that the situation presented in Figure 8 is far more dangerous than the one shown in Figure 7. Besides being dependent on the transverse and longitudinal distances, the presence of peaks (low and high extremes) also strongly depend on the speeds of the objects

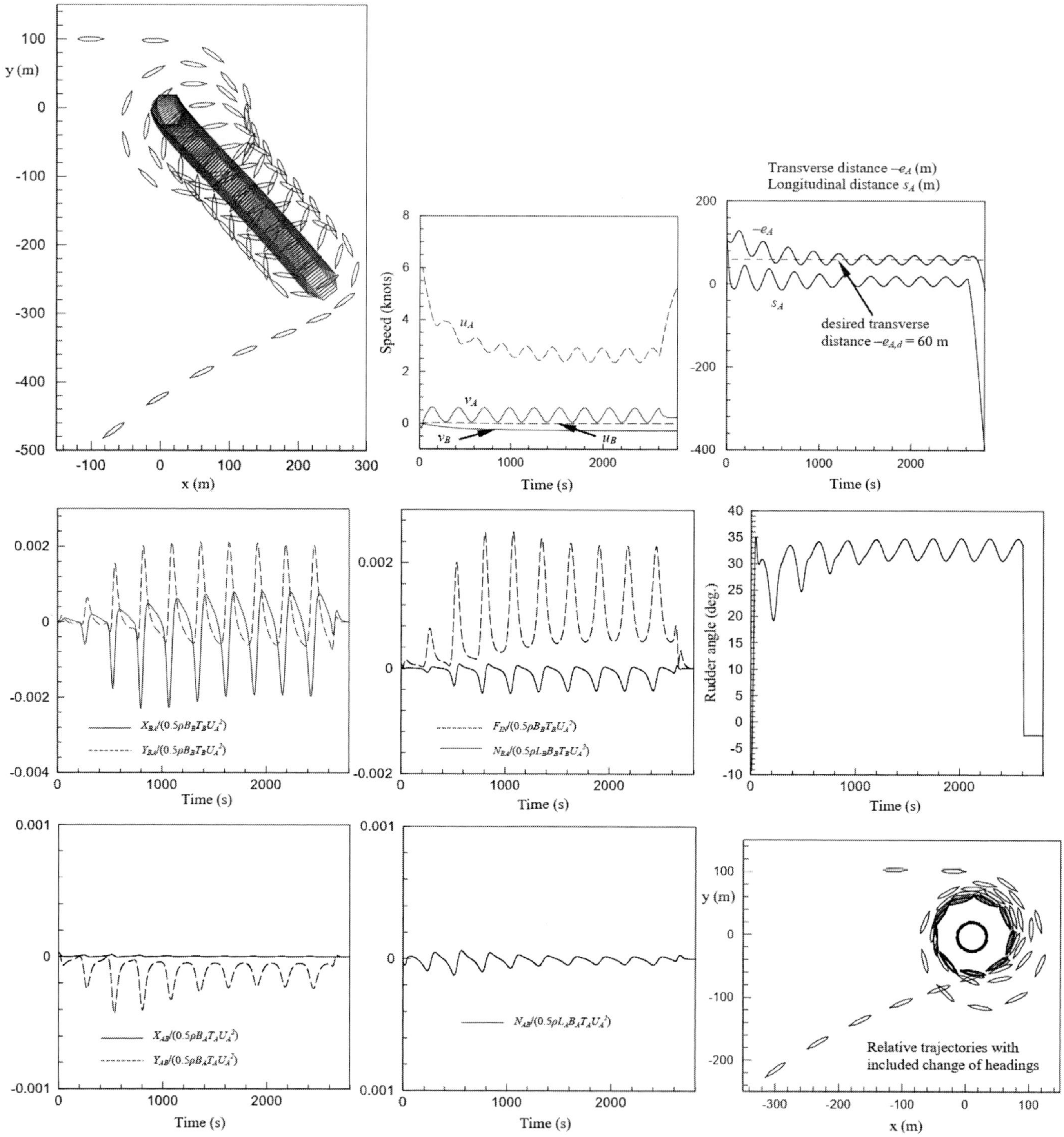

Figure 8: The scaled 'MARINER' performing a starboard turn maneuver around a drifting storage facility under the influence of a uniform current with speed $\left|\vec{V}_C\right|$ = 0.2916 knots coming from the direction γ_C = 315°. The simulated trajectories of the ship and drifting object are plotted at every 30 s.

involved. These factors require an adequate type of education and simulator training of the captain and navigational officers responsible for such operations in order to execute minimum collision risk maneuvers. The responsible personnel should be familiar with the risk levels and the possibly sudden and unexpected change of maneuvering performances of objects with forward speeds operating in close proximity of each other.

5. CONCLUSIONS

A numerical study investigating the effects of ship maneuvers around a floating structure in deep water has been carried out by using a two time scale unified seakeeping and maneuvering model of two interacting bodies. The unified model is based on a modular concept and accounts for the propulsion and resistance, rudder loads and nonlinear viscous loads commonly used within the maneuvering research field. In addition, the interaction effects between the maneuvering ship and the floating cargo object are estimated by using generalized Tuck and Newman far-field theory [1].

The unified model also includes a newly developed motion control module applied on a maneuvering ship. The module plays the role of an 'experienced' helmsman whose skills ensure that the maneuvering ship encircles the stationary or moving floating object at a preselected distance and also that the maneuver is performed efficiently with a minimum collision risk.

By considering two different scenarios of a maneuvering ship which encircles a symmetrical floating object, the behavior of the main maneuvering parameters and hydrodynamic interaction loads are analyzed. In particular, two cases of the same type of turn maneuver are carried out in calm water with and without the presence of a uniform current field. It has been shown that the importance of the loads and in particular their transient distribution upon the floating object, which is of a design commonly expected in the offshore industry, cannot be neglected.

It must be emphasized that the quantitative estimations of the hydrodynamic interaction loads are undervalued due to the fact that the basic assumptions behind Tuck and Newman theory [1] are strictly not satisfied by the floating storage facility. Furthermore, it should be noted that these underestimation effects can be removed through the application of 3D theory. Nevertheless, the present work gives new insight into the behavior of hydrodynamic interaction loads. The results can therefore be of significance for projects within the offshore industry concerning mooring and anchor-line dynamics, as well as for offshore simulators interested in similar maneuvers as those considered here.

6. ACKNOWLEDGEMENTS

MARINTEK expresses gratitude to the Research Council of Norway for their financial support to the KMB project "Investigating Hydrodynamic Aspects and Control Strategies for Ship-to-Ship Operations". Part of this work was also supported by the Research Council of Norway through the Centre for Ships and Ocean Structures at the Norwegian University of Science and Technology.

7. REFERENCES

1. Tuck, E.O. and Newman, J.N., 'Hydrodynamic Interaction Between Ships' In Proceedings of the 10th Symposium on Naval Hydrodynamics, Cambridge, MA, USA, pp. 35–70, 1974.

2. Abkowitz, M.A., Ashe, G.M. and Fortson, R.M., 'Interaction Effects of Ships Operating in Proximity in Deep and Shallow Water' In Proceedings of the 8th Symposium on Naval Hydrodynamics, Rome, Italy, pp. 671–691, 1970.

3. Kijima, K. and Yasukava, H., 'Manoeuvrability of Ships in Narrow Waterway' J. Soc. Naval Architects of Japan, 23:25–37, 1985.

4. Yeung, R.W., 'On the Interaction of Slender Ships in Shallow Water' J. Fluid Mech., 85:143–159, 1978.

5. Skejic, R. and Faltinsen, O.M., 'A Unified Seakeeping and Maneuvering Analysis of Two Interacting Ships' In Proceedings of the 2nd International Conference on Marine Research and Transportation, Ischia, Italy, pp. 209–218, 2007.

6. Skejic, R. and Faltinsen, O.M., 'A Unified Seakeeping and Maneuvering Analysis of Ships in Regular Waves' Journal of Marine Science and Technology, 13:371–394, 2008.

7. Skejic, R., 'Maneuvering and Seakeeping of a Single Ship and of Two Ships in Interaction' PhD thesis, Norwegian University of Science and Technology, Trondheim, Norway, 2008.

8. ITTC, 'The Specialist Committee on Esso Osaka. Final Report and Recommendations to the 23rd ITTC' In Proceedings of the 23rd ITTC, Italian Ship Model Basin (INSEAN), Venice, Italy, pp. 581–617, 2002.

9. Söding, H., 'Prediction of Ship Steering Capabilities' Schiffstechnik, 29:3–29, 1982.

10. Salvesen, N., Tuck, E.O. and Faltinsen, O.M., 'Ship Motions and Sea Loads' Trans. SNAME, 78:250–287, 1970.

11. Artyszuk, J., 'A Uniform Current in Ship Manoeuvring Mathematical Model' Annual of Navigation, 8:67–75, 2004.

12. Wang, S., 'Forces and Moment on a Moored Vessel Due to a Passing Ship' J. Waterways and Harbours, Coastal Eng. Div. Proc. A.S.C.E. (WW3), 101:247–258, 1975.

13. Wendel, K., 'Hydrodynamische Massen und Hydrodynamische Massenträgheitsmomente' Jahrb. d. STG, 44:207–255, 1950.

14. Korotkin, A.I., 'Added Masses of Ship Structures' FLUID MECHANICS AND ITS APPLICATIONS, Vol. 88, Springer Science + Business Media B.V., 2009.

15. Yeung, R.W. and Tan, W.T., 'Hydrodynamic Interactions of Ships with Fixed Obstacles' J. of Ship Res., 24(1): 50-59, 1980.

16. Xiang, X. and Faltinsen, O.M., 'Maneuvering of Two Interacting Ships in Calm Water' In Proceedings of the 11th International Symposium on Practical Design of Ships and other Floating Structures, Rio de Janeiro, Brazil, 2010.

17. Breivik, M., Hovstein, V. E. and Fossen, T. I., 'Straight-Line Target Tracking for Unmanned Surface Vehicles' Modeling, Identification and Control, 29(4):131-149, 2008.

18. Breivik, M. and Fossen, T. I., 'Guidance Laws for Autonomous Underwater Vehicles' In A. V. Inzartsev (Ed.), Underwater Vehicles, pp. 51-76. IN-TECH Education and Publishing, 2009.

19. Haugen, J., 'Guidance Algorithms for Planar Path-Based Motion Control Scenarios' MSc thesis, Norwegian University of Science and Technology, Trondheim, Norway, 2010.

8. AUTHORS BIOGRAPHY

Renato Skejic obtained his PhD degree in marine hydrodynamics at the Department of Marine Technology at NTNU. He is currently employed as a research scientist at the Norwegian Marine Technology Research Institute – MARINTEK. His current research interests include combined seakeeping and maneuvering of single and multiple bodies operating in various environmental conditions at forward speed.

Morten Breivik received his PhD degree in control engineering at the Department of Engineering Cybernetics at NTNU. He is currently employed as a researcher at CeSOS. His current research interests include guidance and motion control system design for unmanned aerial and surface vehicles.

Tor E. Berg obtained his Dr.ing. degree in marine hydrodynamics at the Department of Naval Architecture at the Norwegian Institute of Technology (now NTNU). He is currently employed as a principal research engineer at MARINTEK and Professor II at NTNU. His current research and project activities include topics related to ship maneuvering, ship-to-ship interactions, marine emergency operations and safety at sea in Arctic waters.

THE INFLUENCE OF DIFFERENT VESSEL SPEEDS OF NAVIGATION ON MEETING IN NAVIGABLE CHANNELS OF HYDRO SYSTEM DANUBE-TISA-DANUBE

V Škiljaica and **I Škiljaica** Faculty of Technical Sciences at the University of Novi Sad, Serbia
M Živković Public Water Management Company "Vode Vojvodine" in Novi Sad, Serbia

SUMMARY

It is known that navigation of vessels on the waterways of limited dimensions is accompanied with a number of side effects in comparison to waterways with sufficient widths and depth.
At the present in the world literature there are many different methods for determining the critical speed of navigation on the waterways of limited dimensions. In this paper the authors present several methods for calculating critical speed as well as several methods for calculating the height of the generated waves, which were applied on one channel direction within the Danube-Tisa-Danube system. The procedure for calculation is realized on the examples for three vessels which differ in their constructive and exploitation characteristics and are built for navigation on the channels of the Danube-Tisa-Danube system
The goal of this paper is to present the way of determining the critical and the real speed of navigation as well as height of the vessel generated waves.

NOMENCLATURE

B_{cwl}	Width at CWL (m)
B_K	Width of the channel on the water line (m)
b_K	Width of the channel at the bottom (m)
CWL	Constructive water line
$F_{\otimes}$	Midship cross section area (m^2)
F_K	Area of the channel's cross section (m^2)
$F_{K(s)}$	Channel's cross section in the depression zone (m^2)
g	Earth's gravitational acceleration (m/s^2)
H	Height of the side (m)
ΔH	Average lowering of water level in the zone of a vessel (m)
ΔH_{max}	Maximal value of the lowering of water level in the zone of a vessel
H_K	Depth of water in the channel (m)
h	Height of the vessel generated waves (m)
h_b	Height of vessel generated waves at the bow (m)
h_s	Height of waves in the zone of the stern of the vessel (m)
L_{cwl}	Length at CWL (m)
L_{oa}	Length overall (m)
n_k	Channel's characteristic number
Q	Vessel capacity (t)
R_h	Hydraulic radius of canal's cross section with the vessel
R_K	Hydraulic radius of canal's cross section
T_0	Minimum draught (m)
T_{cwl}	Maximum draught (m)
U_K	Length of wet perimeter of channels cross section (m)
$U_{\otimes}$	Length of wet perimeter of cross section of vessels main frame (m)
V	Displacement of the vessel (m^3)
v	Speed of navigation in the channel (m/s);
v_{avh}	Speed of water flow around vessel's hull (m/s)
v_{cr}	Critical speed of navigation (m/s)
v_{lim}	Limiting speed of navigation of vessels in the canal (m/s)
v_{opt}	Optimal speed of navigation (m/s)
v_{re}	Real speed of navigation (m/s)
$v_{re(max)}$	Real speed of navigation according to A. B. Michailov S. N. Levacov method (m/s)
θ	Angle between the slopes of the channel and horizontal plane (degrees).
Δv	Average change of speed of water flow around vessel's hull (m/s)
Δv_{max}	Maximal value of the speed of water flow in the zone of the vessel (m/s)
δ	Block coefficient
α	Coefficient of the CWL area
β	Midsihp cross section coefficient

1. BASIC CHARACTERISTICS OF THE HYDRO SYSTEM DANUBE-TISA-DANUBE

Inland waterways of Serbia belong to the Danube river basin and are basically made of rivers that have been made more or less navigable and the network of navigable channels of Hydro system Danube-Tisa-Danube.

The Hydro system Danube-Tisa-Danube is a unique system for irrigation, drainage, navigation, flood control, fishery, providing industry with water, forestry, tourism and recreation. Existing basic channel network consists of the main channels, hydro engineering and supporting objects. The total length of the channels connected into the basic channel network is 960 km of which 600 km are navigable, while 360 is used for navigation and drainage. On the channel network of the Hydro system Danube-Tisa-Danube there are 17 navigation locks.

Navigational depth on the channel network is defined based on the lowest navigation level in such a way that for the maximum draught of the authoritative vessel an additional under keel clearance of 0,2 m has been added.

For determining the size of navigation locks the dimensions of the vessel which carries 1000 tons have been used.

The speed of navigation on the channels must be adjusted in such a manner so that the navigation of the vessels does not endanger the stability and security and that it doesn't damage the banks and hydro engineering objects on the channel, as well as other vessels that navigate or stand on the channel. The maximum allowed speed of navigation on channels is 8,0 km/h [1], [2] and [3].

2. BASIC CHARACTERISTICS OF MOTOR CARGO BOATS AND CHOSEN CHANNEL FOR THE CALCULATION OF THE CRITICAL SPEED OF NAVIGATION

For the calculation of the critical speed of navigation and wave heights tree types of vessels - motor cargo boats have been chosen which were built for navigation on the channels of Hydro system Danube-Tisa-Danube marked with MT 700, MT 850 and MT 1000. Their characteristics are shown in the table 1, while their appearance is shown in the figures 2, 3 and 4.

For the calculation of the critical speed of navigation of vessels and wave heights a part of the Novi Sad - Savino Selo channel has been chosen, from the mouth into the river Danube, km 0,0 to km 22,0, the part from the navigation lock in Novi Sad to the joint with the channel Bački Petrovac - Karavukovo (km 4,3 to km 22,0).

Dimensions of the chosen sector of the adopted part of the route are:

- Width of the channel on the water line B_K=55,0 m;
- Depth of water in the channel H_K=4,0 m;
- Area of the channel's cross section F_K=148,0 m^2
- Width of the channel at the bottom b_K=19,00 m;

The appearance of the cross section of the adopted part of the route for which the calculation of the critical speed and height of the vessel generated waves is shown in the figure 5 [1], [2] and [3].

Information on the dimensions of the channel has been obtained from project documentation of the channel from Public Water Management Company "Vode Vojvodine"

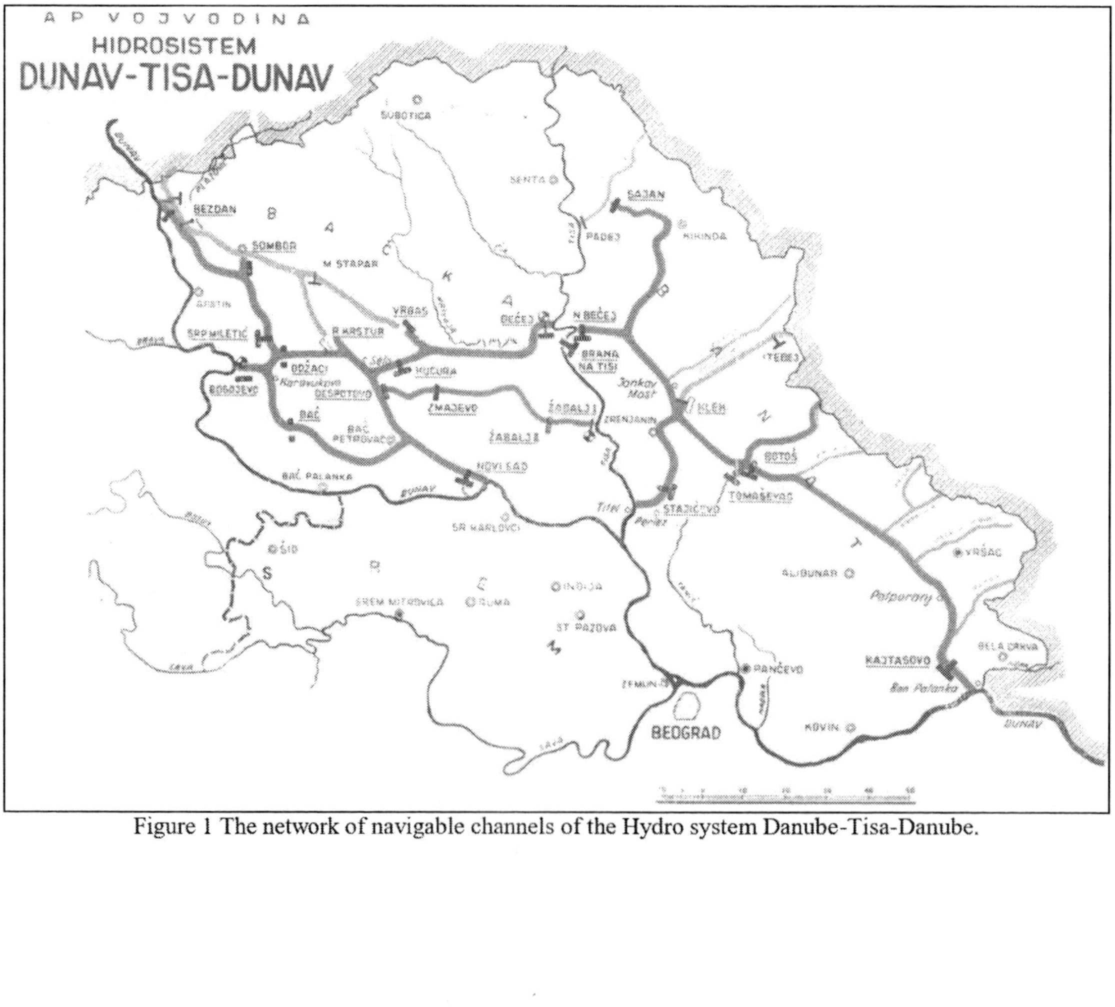

Figure 1 The network of navigable channels of the Hydro system Danube-Tisa-Danube.

Table 1. Basic characteristics of motor cargo boats

Characteristics of motor cargo boats			MT700	MT850	MT1000
Length overall	L_{oa}	(m)	63,05	69,70	77,00
Length at CWL	L_{cwl}	(m)	60,05	69,63	75,60
Width at CWL	B_{cwl}	(m)	8,61	8,48	10,80
Height of the side	H	(m)	2,22	2,82	2,60
Minimum draught	T_0	(m)	0,44	0,59	0,80
Maximum draught	T_{cwl}	(m)	2,05	2,15	2,20
Displacement of the vessel	V	(m^3)	903,0	1054	1450
Vessel capacity	Q	(t)	720	890	1000
Midship cross section area	$F_{\otimes}$	(m^2)	17,65	18,23	23,76
Block coefficient	δ	–	0,830	0,846	0,810
Midsihp cross section coefficient	β	–	0,980	0,995	0,990
Coefficient of the CWL area	α	–	0,925	0,945	0,950

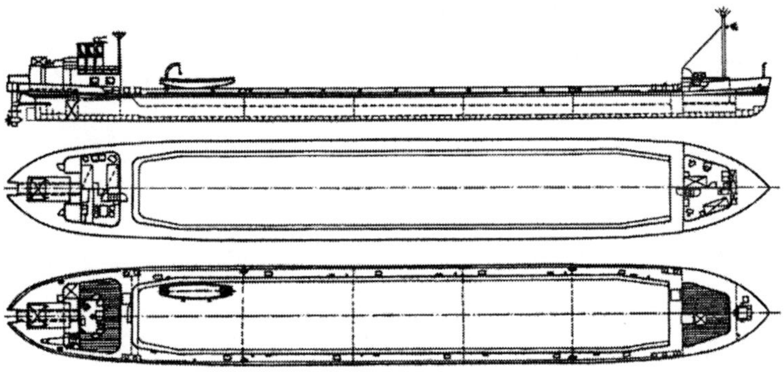

Figure 2. General plan of the vessel MT 700

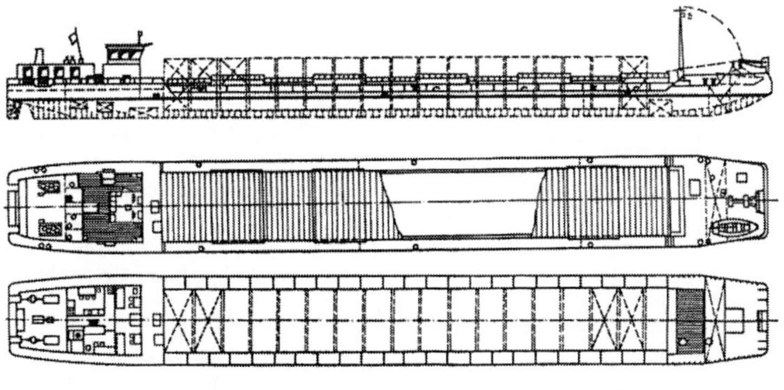

Figure 3. General plan of the vessel MT 850

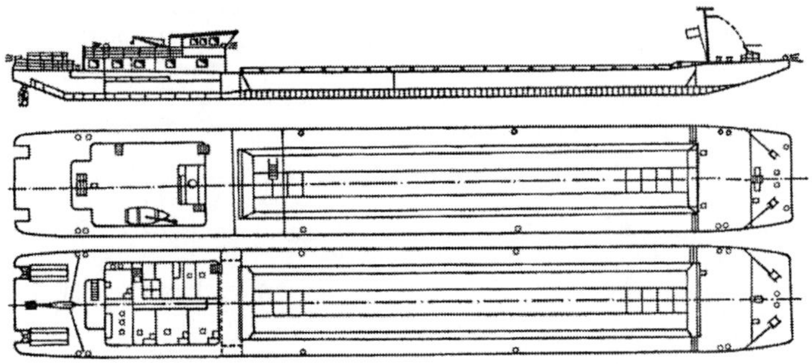

Figure 4. General plan of the vessel MT 1000

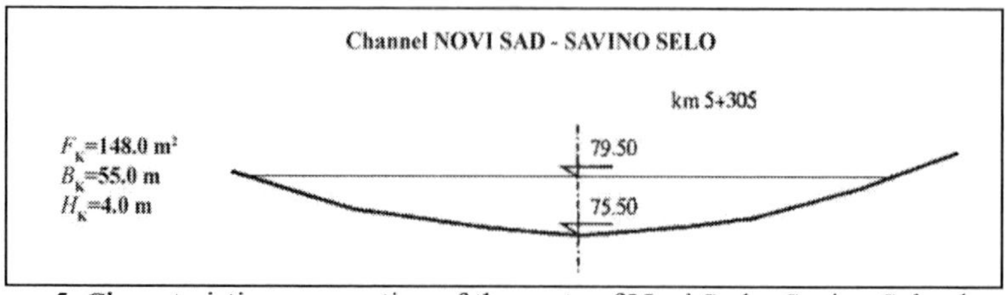

Figure 5. Characteristic cross section of the route of Novi Sad – Savino Selo channel

3. METHODS FOR CALCULATING THE CRITICAL OR LIMITING SPEED OF NAVIGATION

As mentioned, there are numerous different methods for determining the critical speed of navigation of vessels on waterways of limited area of their cross section and on canals. The authors have decided to present several methods for calculating the critical speed of navigation.

3.1 SCHIJF'S METHOD

$$\frac{v_{\lim}}{\sqrt{g \cdot R_K}} = \left(\frac{2}{3}\right)^{\frac{3}{2}} \cdot \left(1 - \frac{F_{\otimes}}{F_K} + \frac{v_{\lim}^2}{2 \cdot g \cdot R_K}\right)^{\frac{3}{2}} \quad (3.1)$$

For this method to be used it is necessary for the condition of the method to be fulfilled which is $2<(B_K/B_{cwl})<12$, where B_{cwl} is the width of the vessels hull on the constructive water line (m), and B_K is the width of the channel at the water line. Hydraulic radius of canal's cross section, R_K is obtained from project documentation of the channel from Public Water Management Company "Vode Vojvodine" [5].

3.2 PROFESSOR MLADEN JOVANOVIĆ'S METHOD

According to this method critical speed of navigation in a navigable channel is defined according to the following equitation [4]:

$$v_{cr} = \sqrt{g \cdot R_h} = \sqrt{g \cdot \frac{F_K - F_{\otimes}}{U_K + U_{\otimes}}} \quad (3.2)$$

$U_{\otimes}$ is the length of wet perimeter of cross section of vessels main frame (m) which is calculated as:

$$U_{\otimes} = B_{cwl} + 2 \cdot T_{cwl}$$

3.3 A. B. MICHAILOV – S. N. LEVACOV METHOD

$$v_{cr} = F \cdot \sqrt{g \cdot \frac{F_K}{B_K}} \quad (4.3)$$

where

$F<1,0$ is a coefficient that changes as a function of channel's characteristic number, $F=f(n_K)$, where channel's characteristic number is calculated as [6]:

$$n_K = \frac{F_K}{F_{\otimes}}.$$

4. REAL (OPTIMAL) SPEED OF NAVIGATION

In order to avoid possible damage by vessel generated waves on the banks of the channels speed of navigation of vessels in the channels is limited according to the expressions (5.1), (5.2) and (5.3).

4.1. SCHIJF'S METHOD

Real speed of navigation according to Schijf's method is calculated as [5]:

$$v_{re} = 0.90 \cdot v_{\lim} \quad (4.1)$$

4.2 MLADEN JOVANOVIĆ'S METHOD

Optimal speed of navigation according to M. Jovanović is calculated as [4]:

$$v_{opt} = c \cdot \frac{R_h}{R_K} \cdot v_{cr} \quad (4.2)$$

where

c is a coefficient which is between 0,85 and 0,95.

4.3 A. B. MICHAILOV – S. N. LEVACOV METHOD

Real speed of navigation according to A. B. Michailov – S. N. Levacov method is calculated as [6]:

$$v_{re(\max)} = 0.90 \cdot v_{cr} \quad (4.3)$$

5. LOWERING OF WATER LEVEL IN THE ZONE OF A VESSEL

Average lowering of water level in the zone of a vessel is calculated by using the Schijf's method according to the following expression [5]:

$$\Delta H = \frac{v_{re}^2}{2 \cdot g} \cdot \left[a_1 \cdot \left(\frac{F_K}{F_{K(s)}}\right)^2 - 1\right] \quad (5.1)$$

where

a_1 is a coefficient calculated as:

$$a_1 = 1{,}4 - 0{,}4 \cdot \frac{v_{re}}{v_{\lim}} \quad (5.2)$$

$F_{K(s)}$ is channel's cross section in the depression zone (m^2) which is calculated as:

$$F_{K(s)} = [b_K \cdot (H_K - \Delta H) + (H_K - \Delta H)^2 \cdot m - F_{\otimes}] \quad (5.3)$$

where

m is a coefficient calculated as: $m = cot(\theta)$,

θ is the angle between the slopes of the channel and horizontal plane.

Maximal value of the lowering of water level in the zone of a vessel (ΔH_{max}) depends on average value (ΔH) and for motor cargo vessels is calculated according to following expressions:

for $\frac{B_K}{L_{cwl}} > 1{,}5$

$$\Delta H_{max} = \Delta H \cdot \left[3 - 4 \cdot \left(\frac{F_K - F_{\otimes}}{F_K} \right) \right] \quad (5.4)$$

for $\frac{B_K}{L_{cwl}} < 1{,}5$

$$\Delta H_{max} = \Delta H \cdot \left[2 - 2 \cdot \left(\frac{F_K - F_{\otimes}}{F_K} \right) \right] \quad (5.5)$$

6. SPEED OF WATER FLOW AROUND VESSLE'S HULL

6.1 A. B. MICHAILOV – S. N. LEVACOV METHOD

The speed of water flow around vessel's hull (m/s) is calculated according to the following expression [6]:

$$v_{avh} \approx \frac{v_{re(max)}}{n_k - 1} \quad (6.1)$$

6.2 SCHIJF'S METHOD

Schijf's method calculates average change of speed of water flow around vessel's hull Δv (m/s) according to following expression [5]:

$$F_k \cdot v_{re} = F_{K(s)} \cdot (v_{re} + \Delta v) \quad (6.2)$$

Same author gives an expression for calculating maximal value of the speed of water flow in the zone of the vessel Δv_{max} (m/s) for motor cargo vessels is calculated according to the following expressions:

for $\frac{B_K}{L_{cwl}} > 1{,}5$

$$\Delta v_{max} = \Delta v \cdot \left[2{,}50 - 3 \cdot \left(\frac{F_K - F_{\otimes}}{F_K} \right) \right] \quad (6.3)$$

for $\frac{B_K}{L_{cwl}} < 1{,}5$

$$\Delta v_{max} = \Delta v \cdot \left[1{,}50 - \left(\frac{F_K - F_{\otimes}}{F_K} \right) \right] \quad (6.4)$$

7. METHODS FOR CALCULATING THE HEIGHT OF WAWES

7.1 BALANIN AND BYKOV METHOD

According to this method the height of vessel generated waves h (m) is calculated with the following expression [4]:

$$h = \frac{1{,}25 \cdot v^2}{g} \left\{ 1 - \left[1 - (4{,}2 + n_K)^{-\frac{1}{2}} \right] \cdot \left(\frac{n_K - 1}{n_K} \right)^2 \right\} \cdot \frac{2 + \sqrt{\frac{B_K}{L_{cwl}}}}{1 + \sqrt{\frac{B_K}{L_{cwl}}}} \quad (7.1)$$

where

h is the height of the vessel generated waves, (m);

v is the speed of navigation in the channel, (m/s);

n_k is the characteristic channel number which is calculated as $n_K = \frac{F_K}{F_{\otimes}}$;

B_K is the channel width at the water line, (m) ;

L_{cwl} is the length of the vessel on its constructive waterline, (m);

7.2 P.A. SANKIN METHOD

Height of the waves (m) at the bow of the vessel is [5]:

$$h_b = \frac{v^2}{2 \cdot g} \cdot \left(0{,}65 + 3{,}20 \cdot \frac{T_{cwl} \cdot B_{cwl}}{B_K \cdot H_K} \right)^2 \quad (7.2)$$

Height of the waves (m) in the zone of the stern of the vessel is calculated according to the expression (7.3) [5].

$$h_s = h_b \cdot \left[1 + \sqrt{0{,}73 \cdot \frac{L_{cwl}}{B_K}} - 0{,}027 \cdot \left(1 + \sqrt{0{,}73 \cdot \frac{L_{cwl}}{B_K}} \cdot \sqrt[4]{1{,}37 \cdot B_K - L_{cwl}} \right) \right] \quad (7.3)$$

7.3 SCHIJF'S METHOD

Height of the transversal waves (m) at the stern of the vessel is [5]:

$$h_s = \Delta H_{max} \qquad (7.4)$$

Height of the waves (m) at the bow of the vessel is [5]:

$$h_b = 0{,}1 \cdot \Delta H + \Delta H_{max} \qquad (7.5)$$

8. THE RESULTS OF THE CALCULATIONS

As stated before in the research the vessels, motor cargo boats built for navigation on the channels types MT 700, MT 850 and MT 1000 have been used. The calculation of the critical speed of navigation (and by it the real, recommended, maximal, optimal speed of navigation of a vessel in a channel), height of the waves as well as speed of water around vessel's hull has been done for the Novi Sad – Savino Selo channel, on the sector upstream from the navigation lock. The calculated values of the critical and the real speed of navigation, height of the waves created by vessels while navigating in channels and the speed of water around vessel according to adopted methods are shown as tables in following tables.

Table 2 presents the results of calculations of limiting speed of navigation or critical speed of navigation for different methods.

Results are given for navigation of vessels while navigating towards each other and in cases when vessels pass by each other parallel.
Table 3 presents real (recommended, maximal, optimal) speed of navigation, again, according to different methods for its calculation for cases when vessels are navigating towards each other and in cases when vessels pass by each other parallel.

Table 4 presents heights of vessel generated waves. According to different methods these heights are calculated for cases when vessels are navigating towards each other and in cases when vessels pass by each other parallel. It has to mentioned that P.A. Sankin's method can't be used for calculation the wave heights for two vessels, hence there are no values for that method in the part of the table with the results when two vessels pass by each other.

Table 5 presents speeds of water around vessels hull according to different methods for both cases of vessels' navigation in channels, for the case when they navigate individually and when they pass by each other.

Table 2 Critical or limiting speed of navigation

	Method/Equitation	Critical – limiting speed, $v_{cr\,(lim)}$ (m/s)		
		MT 700	MT 850	MT 1000
Vessels navigating towards each other	Schijf's (3.1)	2,985	2,956	2,681
	Mladen Jovanović (3.2)	4,314	4,302	4,137
	A. B. Michailov – S. N. Levacov (3.3)	3,005	2,954	2,646
Vessels passing by each other parallel	Schijf's (3.1)	1,734	1,706	1,452
	Mladen Jovanović (3.2)	3,685	3,662	3,377
	A. B. Michailov – S. N. Levacov (3.3)	2,260	2,183	1,926

Table 3 Real (recommended, maximal, optimal) speed of navigation

	Method/Equitation	Critical – limiting speed, $v_{cr\,(lim)}$ (m/s)		
		MT 700	MT 850	MT 1000
Vessels navigating towards each other	Schijf's (4.1)	2,686	2,660	2,413
	Mladen Jovanović (4.2), c=0,9	2,710	2,695	2,492
	A. B. Michailov – S. N. Levacov (4.3)	2,704	2,658	2,381
Vessels passing by each other parallel	Schijf's (4.1)	1,560	1,535	1,306
	Mladen Jovanović (4.2), c=0,9	1.977	1,953	1,660
	A. B. Michailov – S. N. Levacov (4.3)	2.260	2,183	1,926

Table 4 Height of vessel generated waves

	Method/Equitation	Height of vessel generated waves, h (m)		
		MT 700	MT 850	MT 1000
Vessels navigating towards each other	Balanin and Bykov (7.1)	0,189	0,195	0,323
	P.A. Sankin (7.2)	0,093	0.095	0,112
	P.A. Sankin (7.3)	0,169	0,176	0,218
	Schijf's (7.4)	0,112	0,121	0,169
	Schijf's (7.5)	0,111	0,112	0,157
Vessels passing by each other parallel	Balanin and Bykov (7.1)	0,265	0,272	0,309
	P.A. Sankin (7.2)	–	–	–
	P.A. Sankin (7.3)	–	–	–
	Schijf's (7.4)	0,115	0,118	0,151
	Schijf's (7.5)	0,123	0,125	0,160

Table 5 Speed of water around vessels hull

	Method/Equitation	Speed of water flow around vessel's hull, v_{awh} and Δv (m/s)		
		MT 700	MT 850	MT 1000
Vessels navigating towards each other	A. B. Michailov – S. N. Levacov (6.1)	0,366	0,373	0,455
	Schijf's (6.2)	0.408	0,431	0,644
Vessels passing by each other parallel	A. B. Michailov – S. N. Levacov (6.1)	0,637	0,642	0.820
	Schijf's (6.2)	0,679	0,706	0,964

9. FINAL CONSIDERATIONS

- Depending on the method that is used a diversity of names for the speed of navigation on channels can be noticed. The speed of navigation that should not be in any case exceeded is called: *the first critical speed;, critical speed; limiting speed* (by achieving it destruction of channel banks and bed occurs as well as sudden increase of resistance). The speed of navigation at which there is no destruction of the banks and the bed of the channel and is suitable from the resistance view-point is called: *real speed; recommended speed; maximal speed; optimal speed.*
- By analyzing the results it can be noticed that the limiting (critical) speed of navigation (and with it the real-recommended-maximal-optimal speed) decreases with the increase of vessels draught (T), midship cross section area $F_{\otimes}$ or the displacement of the vessel, and they increase in the same manner with the decrease of draught (reduction of the displacement of the vessel). The calculations show that the values of the real speed of navigation for all three types of vessels – motor cargo boats exceed, even significantly, the value of speed of navigation (8,0 km/h) which is allowed according to The regulations of the ways and conditions of use of the basic channel network of Hydro system Danube-Tisa-Danube for navigation, except in cases when vessels pass by each other.
- Wave heights increase with the size of the vessels as well as the speed of navigation. In cases when two vessels pass by each other the height of the waves also increases. In cases when the real speed of navigation must be significantly reduced for the passing by maneuver then the height of the waves does not increase.
- Based on these reasons the speed of navigation of vessel in channel should be determined for each vessel individually, based on its' real exploitation characteristics (draught and displacement) as a function of the size of channel, type of bank protection and its' real condition. At the same time the authors consider that protecting speed of navigation on channels of Danube-Tisa-Danube system (8,0 km/h) should not be applied-limited at the same time to all types of vessels neither on all channels of the Danube-Tisa-Danube system.

10. REFERENCES

1. ČOLIĆ, V., ŠKILJAICA, V., RADMILOVIĆ, Z., "Navigational possibilities on the channel network of Hydro system Danube-Tisa-Danube", monograph, *Faculty of Transport and Traffic Engineering at the University of Belgrade, Belgrade*, 2000.

2. ŠKILJAICA, V. – project leader "Elaboration feasibility of international waterway opening trough the Hydro system Danube-Tisa-Danube and measures for its reinstatement", *University of Novi Sad, Faculty of Technical Sciences, Department for Traffic Engineering, Novi Sad*, 2002.

3. ŠKILJAICA, V – project manager, "The research of the navigating conditions on the network of channels of the Hydro system Danube-Tisa-Danube with determining the most suitable functioning parameters of the channel transportation within the transportation system of Serbia", *University of Novi Sad, Faculty of Technical Sciences, Department of traffic*, 2006.

4. JOVANOVIĆ, M., 'Designing vessels',(in Serbian language) *Faculty of Transport and Traffic Engineering at the University of Belgrade, Belgrade*, 2002.

5. MUŠKATIROVIĆ, D., 'Inland waterways and berths',(in Serbian language) *Faculty of Transport and Traffic Engineering at the University of Belgrade, Belgrade*, 1993.

6. MICHAILOV, A. B., LEVACOV, S. N., 'Waterways and ports', (in Russian language) *Transport, Moscow 1982.*

11. AUTHORS BIOGRAPHIES

Vladimir Škiljaica PhD is a full professor at the Faculty of Technical Sciences at the University of Novi Sad as well as at Faculty of Transport and Traffic Engineering at the University of Belgrade. He is in charge of the department for technologies of transportation systems at the Faculty of Technical Sciences at the University of Novi Sad. He teaches "Designing and building of vessels" and "Technology of water transport". He mentored numerous Master and PhD papers and is author of large number of expert and scientific papers.

Ivan Škiljaica M.Sc. transport engineer in water sector is employed as a teaching assistant the Faculty of Technical Sciences at the University of Novi Sad at the subject "Technology of water transport". He is a PhD student at the second year of studies. Coauthored over 20 papers published in Serbia and internationally.

Mirjana Živković M.Sc. transport engineer in water sector is employed at Public Water Management Company "Vode Vojvodine" in Novi Sad as a leading engineer for navigation at the sector Hydro System Danube-Tisa-Danube. She coauthored several expert and scientific papers.

THE USE OF A BOUSSINESQ-TYPE WAVE MODEL TO DETERMINE SHIP-INDUCED WAVES OVER NON-UNIFORM BATHYMETRIES

A. J. van der Hout, **O. M. Weiler**, and **M. Borsboom**, Deltares, The Netherlands

SUMMARY

Waves generated by passing vessels are a significant issue for mooring in modern port waterways and for the protection of river banks. To simulate passing-vessel effects, it is in many situations necessary to take the effect of a complex geometry and non-uniform bottom into account. This study aims to evaluate the use of a time-domain Boussinesq-type wave model to predict ship-induced waves over a non-uniform bathymetry. The Boussinesq-type wave model TRITON has been used for the simulations. The numerical model has been validated using scale-model test data obtained at Flanders Hydraulics Research. After validation, we evaluate two methods to generate a wave signal at the model boundary representing ship-induced waves in the Boussinesq-type model. These methods use a simple approach based on the main dimensions of the passing vessel.

NOMENCLATURE

α	Bank slope (-)
Δ	Displacement (m^3)
ε	wave height to water depth ratio (-)
θ	Angle in polar coordinates (rad)
ψ	Stream function (m^2/s)
Λ	Source strength (m^2/s)
μ	water depth to wavelength ratio (-)
μ_s	wave length to length scale of changes in bottom level ratio (-)
ζ	Surface elevation (m)
B	Beam (m)
C_b	Block coefficient (m)
g	Gravitational acceleration (m/s^2)
h	Water depth (m)
L_{eff}	Equivalent ship length (m)
L_{oa}	Length over all (m)
L_{pp}	Length between perpendiculars (m)
r	Distance in polar coordinates (m)
u	Flow velocity in x-direction (m/s)
v	Flow velocity in y-direction (m/s)
V	Ship velocity (m/s)
V_∞	Undisturbed flow velocity (m/s)
W_{eff}	Effective channel width (m)
x	x-coordinate (m)
y	y-coordinate (m)

1. INTRODUCTION

Hydrodynamic effects due to passing vessels can be of importance for moored ships and for the protection of river banks. In the past incidents have taken place, which in some cases resulted in severe damage to mooring facilities. Breaking-line incidents and emergency disconnections in case of offloading of hazardous cargo (i.e. LNG) can lead to potentially dangerous and unacceptable situations.

The issues caused by these passing ships can originate from suction effects or from the shorter secondary wash waves, or from a combination of the two. To predict these hydrodynamic effects numerous approaches have been applied. In general, the effects of a passing ship on a moored ship can be split in several components. Table 1 shows this process schematically for a four-step approach to predict moored vessel response. Some methods have been developed to solve all steps integrally; other methods simulate only a part of the process. It is quite common and often convenient to make a division between the effect of a passing vessel on water, and the effect of water on the moored ship.

Step 1: Ship wave generation
⇓
Step 2: Wave propagation
⇓
Step 3: Computation of hydrodynamic forces
⇓
Step 4: Computation of vessel response

Table 1: Schematic overview of the steps involved in predicting the effect of passing ships on moored ships.

The method described by Seelig [1] is based on empirical relationships found in model tests and can be used to directly predict hydrodynamic forces on moored vessels due to passing vessels. An example of a method based on basic physical relationships is described by Schijf et al. [2]. With this method the water-level elevation around a passing vessel in restricted water is determined; computation of hydrodynamic forces on the moored vessel is not included using this approach and should be obtained by a coupling to another model.

More sophisticated numerical methods are often based on potential theory. Some methods evaluate the hydrodynamic forces on a moored vessel resulting from the suction effects of a passing vessel, like DELPASS (Pinkster et al. [3]) which is based on the assumption of double-body flow. Other methods take also nonlinear free surface effects into account and are able to simulate wash waves. An example of such a method is RAPID (Raven [4]), which has been primarily developed for hull design optimization, but can be used for step 1 of the approach shown in Table 1 since it does simulate ship-generated waves.

Most numerical models currently existing are limited in their application in the sense that they are based on linear theory. It is possible to resolve suction effects in a non-uniform geometry, but without nonlinear free-surface effects. Nonlinear models, such as RAPID, are able to simulate free-surface effects, but only for a limited domain and in a uniform geometry.

Recently, problems due to passing-ship effects have been reported on a beach near Vlissingen, the Netherlands. The shipping lane, which is used by vessels travelling between Rotterdam and Antwerp, lies very close to this beach. It has been reported that when a ship passes this beach, water retracts from the beach and long after the ship has passed a breaking bore comes rushing in from the side. Recreants on the beach have been surprised by this effect, caught by this wave and flushed against obstacles on the beach. An incident which resulted in one person injured was reported in the newspapers. Figure 1 shows the result of a simulation for this situation computed with the Boussineq-type wave model TRITON.

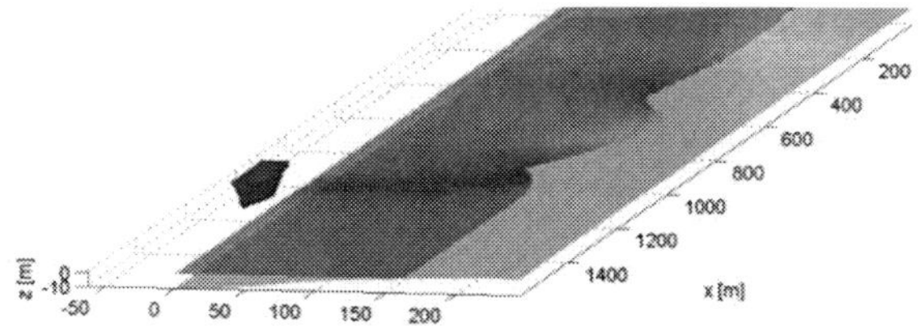

Figure 1: Example of the use of Boussinesq-type model TRITON to simulate wave run-up/run-off on a beach.

This example shows that ship waves can interact with complex bathymetries/geometries in a way not described by linear theory. It is therefore necessary to have a model that can simulate wave propagation including not only linear wave behaviour, like shoaling, refraction, and diffraction but also (to some extent) nonlinear shallow-water effects, like wave-wave interaction, wave breaking, and wave run-up. Input for such a wave propagation model is the wave disturbance created by a passing vessel that is given as boundary condition.

In the example shown above, measurements of ship-induced waves have been used as boundary condition for the wave propagation model. In many situations measurements will not be available. Applying a model like RAPID (Raven [4]) that simulates ship-induced waves to generate a boundary condition for the wave propagation model can then be a valid approach. However, these computations can be quite computational intensive. In many applications it is sufficient to have a simple, realistic estimate of the effect of primary waves (suction effects) for the range of ships that are expected to pass. The simulation of a wave pattern for one specific hull form in high detail will not be necessary in these cases. In many situations it is more practical to have a method to predict passing-ship effects based on only the main dimensions of the passing ship and the geometry of the channel.

In this paper it is proposed to use a Boussinesq-type model for ship-induced wave propagation. A Boussinesq-type model is able to simulate wave propagation in shallow water and can include nonlinear effect. The objective of this paper is to validate this approach with measurements conducted by the Flanders institute [5]. Furthermore, two simple methods have been evaluated that can be used to generate a wave signal to be used as a boundary condition for the Boussinesq-type model, based on the main dimensions of the passing vessel.

In the next section the proposed approach will be explained in more detail. The features and properties of the applied Boussinesq-type model are discussed in Section 3. A short description of the scale model tests is given in Section 4. In Section 5 the Boussinesq-type model is validated with the results of the scale model tests. Two alternative methods for modelling the generation of ship-induced waves are evaluated in Section 6. In Section 7 the results are discussed and the conclusions are summarized.

2. PROPOSED METHOD

As described in the introduction the use of a Boussinesq–type model can have advantages over the more conventional panel models, especially in complex geometries. The Boussinesq-type model that will be used in this study is the time-domain model TRITON developed at Deltares (Borsboom et al., [6]). The advantage of a time-domain Boussinesq-type model is that the primary waves (suction effect), the secondary waves (wash effect) and (non)linear interactions are solved simultaneously.

TRITON has been developed for wave propagation in nearshore applications, but has been used in the past for propagation of ship-induced waves. Since the main purpose of the model is to correctly simulate wave propagation, the hydrodynamic forces on a moored ship are not solved within the model. For this a coupling to panel model DELMULTI (Pinkster [7]) is available, which uses the surface elevation and water velocities to determine the resulting forces. This coupling has been described and validated by Wenneker et al. [8]. The computed hydrodynamic forces can be used as input to a vessel response model, which will provide resulting vessel motions and the corresponding line loads. Bingham [9] also argues that a coupling between a Boussinesq-type model and a vessel response model based on potential theory provides an efficient means of predicting the non-linear response of a moored structure in restricted water.

The coupling described above takes care of the output side of the methodology schematized in Table 1. On the input side, a method which describes the waves generated by the passing vessel is required. Several models that compute the wave pattern of a sailing ship are available. A coupling between RAPID and TRITON has proven to give good results and has already been validated by Doorn et al. [10]. In this approach RAPID is used to compute the stationary wave pattern of a vessel sailing at steady velocity. The wave profile at a certain distance from the vessel is determined and this wave profile is prescribed at the boundary of the TRITON domain, whilst being translated along the boundary with the vessel speed.

By prescribing the waves generated by a sailing ship as boundary condition along the model domain, it is possible to model one side of a channel/fairway. The coupling between ship and fairway side is one-sided, i.e., waves reflected off the fairway banks are not coupled back to the ship and to the main part of the fairway. Since this effect is usually small, the omission is valid for most applications.

The objective of this study is to validate the prediction of wave propagation (Step 2 of Table 1) over a non-uniform bathymetry using a Boussinesq-type model. To this end, a comparison with scale model measurements is made. Instead of input from a wave generation model (step 1), measured water levels are used as input for TRITON, eliminating any uncertainty due to the coupling of two models.

Once validated, two methods are evaluated that can be used as an alternative for a complex model in step 1, like RAPID, using a simplified approach. The aim is to develop a robust method that computes the primary wave of a passing ship fairly accurately, using only the main characteristics of the ship. This will allow for a quick assessment of potential ship-induced wave issues at a certain location for a wide variety of ship types.

3. APPLIED BOUSSINESQ-TYPE MODEL

TRITON is a Boussinesq-type wave model developed by Deltares. Boussinesq-type wave models are based on the assumption of incompressible, constant-density flow. By assuming the flow to be irrotational in the two horizontal directions (no recirculation in vertical planes), the flow structure can be expanded in a series expansion in vertical direction. This allows the construction of computationally efficient formulations by the elimination of the vertical dimension.

The development of TRITON aimed at a Boussinesq-type model with reasonable accuracy for near-shore applications with the following properties:

- conservation of mass and momentum,
- a formulation independent of the vertical reference level for bottom topography and free-surface elevation,
- dispersion and shoaling modelled in a compact way,
- reasonable modelling of nonlinear wave-wave interactions to allow the handling of fairly large wave heights.

In the derivation of the model, an expansion was made in terms of

- μ, a typical water depth to wavelength ratio: a measure of the relative importance of linear dispersion,
- ε, a typical wave height to water depth ratio: a measure of the relative importance of nonlinear effects, and
- μ_s, a typical wave length to length scale of changes in bottom level ratio: a measure of the relative importance of shoaling.

In the truncation of the expansion all terms of $O(\mu^4)$ (and higher) were neglected and some terms of $O(\varepsilon\mu^2)$ (and all higher) were neglected. To improve the model stability and accuracy for short waves the model has been extended with a [2/2] Padé approximation of linear dispersion and first-order wave shoaling, including effects of $O(\mu^4)$ and $O(\mu^4\mu_s)$. Considering the model's formulations and truncations, Borsboom et al. [6] argues that the model should perform reasonably well for $\varepsilon<0.25$, $\mu<4$ and $\varepsilon\mu^2<0.25$.

The model has been complemented with a run-up procedure and a breaker model (Borsboom, [11], Borsboom et al., [12]).

4. DESCRIPTION OF THE MODEL TESTS

The model test results used in this study have been obtained by captive model testing in the Towing tank for manoeuvres in shallow water (cooperation Flanders Hydraulics Research - Ghent University). We have made use of a selection of the measured model test data that has been made available for validation of numerical calculations. The measurements are described in detail by Lataire et al. [5].

The towing tank in which the tests have been conducted has main dimensions of 88 x 7 x 0.5 m^3 and is fully automated. Captive towing tests have been performed for several bank geometries. For this study results for two different bank geometries have been used. Figure 2 shows the layout of the selected tests: Test C and Test I. Only the side of the towing tank on which the bank has been built is shown, the other side of the towing tank has a flat bottom with a vertical wall at the end.

The vessel particulars are shown in Table 2 and the selected test conditions are shown in Table 3. The water level elevation has been measured by three wave gauges installed at different lateral but at the same longitudinal position. The data obtained at these wave gauges were used for comparison with the computed water level elevation.

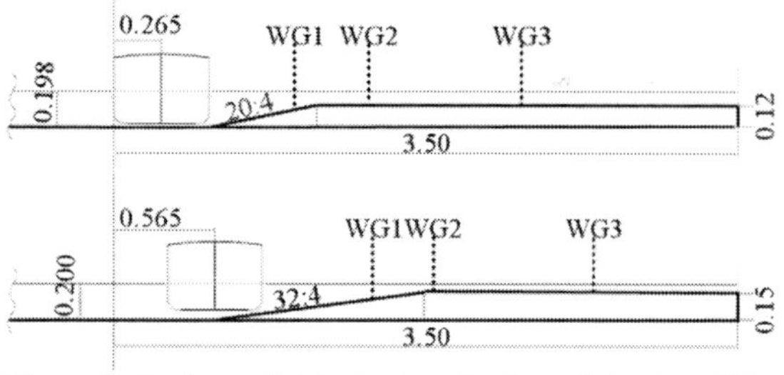

Figure 2: Scale model test setup for Test C (top) and Test I (bottom).

	8000 TEU container carrier	
	Test C	Test I
L_{oa} [m]	4.332	
B [m]	0.530	
T [m]	0.180	0.149
C_b [m]	0.66	0.65

Table 2: Vessel particulars.

		Test C	Test I
V	[m/s]	0.458	0.572
h	[m]	0.198	0.200
Lateral ship pos.	[m]	0.265	0.565
Wave gauge 1	[m]	1.080	1.440
Wave gauge 2	[m]	1.480	1.780
Wave gauge 3	[m]	2.315	2.680

Table 3: Test conditions.

5. VALIDATION OF THE BOUSSINESQ-TYPE MODEL

5.1 MODEL SETUP

In order to show that the Boussinesq-type model is capable of propagating ship-generated waves over a non-uniform bathymetry, the water-level elevation measured at wave gauge 1 (located at approximately 1.5 x Beam from the centreline of the ship) is used as input signal for the model.

The model has four boundaries. One boundary is the incoming-wave boundary along which the waves generated by the ship are prescribed. All other boundaries are closed boundaries. The wave signal along the incoming wave boundary is obtained from the measured water-level elevation. This signal is applied to each boundary grid-point, applying a time-shift derived from the vessel speed. The surface elevation prescribed at the boundary is propagated through the model domain by TRITON, taking the interaction with the (complex) bathymetry into account.

All computations have been carried out on model scale. The domain considered in the computations was a part of the towing tank, with a length of 31 m. A rectangular equidistant grid was used with a grid-cell size that, depending on the run, was between 5 and 6 cm in both transversal and longitudinal direction. A time step of 0.05 s was used, which is sufficiently small to meet accuracy and stability criteria

5.2 VALIDATION RESULTS

The comparison between measured and computed water-level elevations is shown in the Figures 3a and 3b. For both tests the match between the measurements and computations is very good for the primary wave.

The main difference between Test C and Test I is that for Test C the ship's velocity is relatively low and the water depth on the shallow flat is relatively large, compared to the faster and shallower Test I. This results in quite different wave patterns.

In Test C no secondary waves are observed: both the measurements and the computations show that the passing ship only causes a water-level depression. The shape and magnitude of the computed water-level depression at wave gauge 2 and wave gauge 3 compares well to the measured water levels. The results show that TRITON is capable of simulating the propagation of this primary wave.

In test I the water-level depression at the incoming-wave boundary is larger than for Test I and the shape of the primary wave is much more influenced by the shallow water. At wave gauge 3, it can be seen that at the end the water-level depression has become steeper. This shallow-water effect is also captured by TRITON, although there is a slight underestimation of the depth of the wave trough. Secondary waves are present as well at the position of wave gauge 1 in Test I. At wave gauge 2, the secondary waves are underestimated by TRITON compared to the measurements. At wave gauge 3, the amplitude of the secondary wave is still slightly underestimated, but the correspondence with the measured wave height is better. The computed phases of the secondary waves however are shifted, compared to the measurements. It is also observed that in the measurement the secondary waves disappear quicker than in the computations.

A closer inspection revealed that the very short secondary waves have a value of μ of about 5 to 6 and are hence slightly outside the range of applicability of TRITON (cf. Section 3). This would explain the secondary-wave phase shift in the computations. Spurious reflections at the ship-wave boundary of the model domain may have caused the waves in the computations to be 'trapped'. This would explain why in the computations the decay of the amplitude of the secondary waves is too slow. Note that the ship-wave boundary is of course artificial; it does not exist in the measurements, where reflected secondary waves can move freely out of, and quickly disappear from, the model domain. Further research is required to reduce the effect of the boundary on the secondary waves. The prediction of the primary wave is apparently not affected by the boundary.

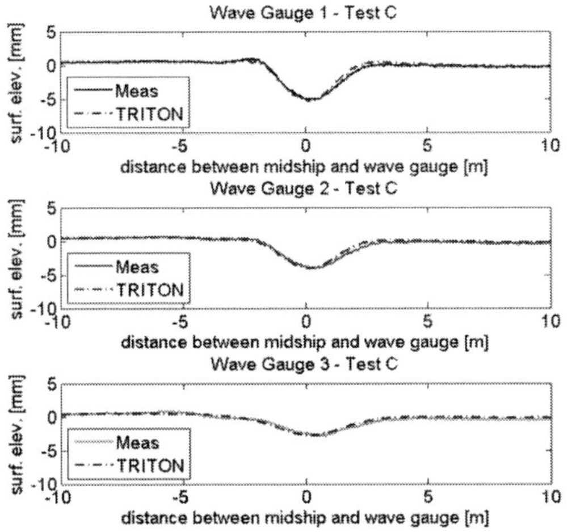

Figure 3a: Comparison between measurements and TRITON results for Test C.

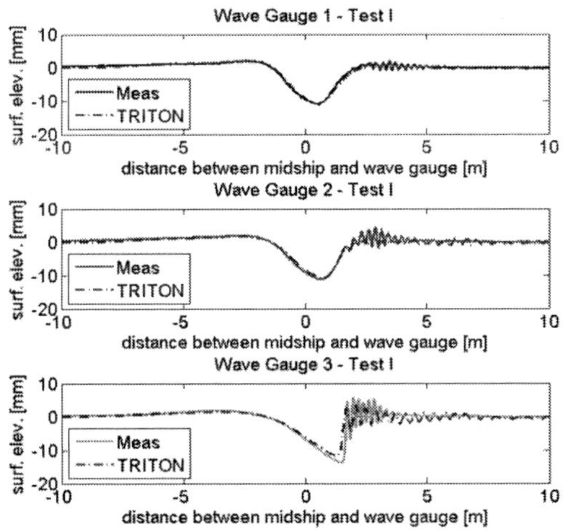

Figure 3b: Comparison between measurements and TRITON results for Test I.

Although some differences between the computations and the measurements have been found, the general agreement is quite good. The primary wave created by a passing ship will in general cause the largest forces on a moored vessel. It has been shown that the propagation of this primary wave over a non-uniform bathymetry is simulated adequately by TRITON. These results also show that the one-sided coupling approach is valid for the situation considered here.

6. TWO METHODS FOR MODELLING SHIP-INDUCED WAVES

6.1 METHOD 1 – BASED ON DISPLACEMENT

The wave propagation model has to be forced with a signal describing the disturbance caused by the passing vessel. Since in TRITON only a water-level elevation can be prescribed at the wave boundary, and not the actual boat itself, a relation between water level and the disturbance caused by a vessel has to be found. A simple and straightforward approach is chosen. The method presented here is based on the work described by Wenneker [8].

A travelling boat generates a primary wave by pushing water away over the first section of the hull, and drawing water back towards the stern. The amount of water that is displaced depends on the variation of the volume of the submerged part of the hull along the length of the ship. For this method the distribution of the submerged cross-sectional area over the length of a vessel is used as starting point.

Every computational time step the position of the vessel is determined along the incoming wave boundary. This position is compared to the ship's position in the previous time step. The cross-sectional area curves are then subtracted for the old and the new position, resulting in a net change in submerged volume along the boundary.

The change in submerged volume still has to be translated into a water-level elevation. For this the change of volume is assumed constant over the water depth. Applying long-wave theory a relation for the water level can now be found by:

$$vh = \zeta\sqrt{gh}$$
$$\zeta = \sqrt{\frac{h}{g}}\frac{ds}{dt} = \sqrt{\frac{h}{g}}\frac{dV/(dx\cdot h)}{dt} \qquad (6.1)$$

In this equation ζ is the water-level elevation, dt one time step, v the flow velocity perpendicular to the incoming wave boundary, and the virtual average displacement of water ds is the virtual displaced volume of water dV divided by the water depth h and the width of a grid cell dx.

In Figure 4, a computed water-level signal is shown. First a wave crest is observed where water is pushed into the domain at the bow (source), followed by a wave trough where water flows out of the domain at the stern (sink). This signal is assumed to be representative at the centreline of the ship.

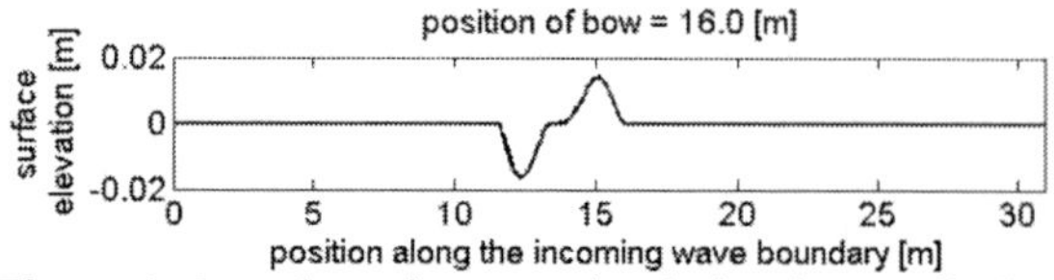

Figure 4: Snapshot of a water-level signal prescribed at the incoming wave boundary.

The assumption of a constant water displacement averaged over the water depth is of course a crude approximation. The flow around and close to the hull will therefore not be simulated correctly. However, it is expected that the wave disturbance at a larger distance from the vessel is represented sufficiently accurate, since the first-order parameters like displacement and vessel velocity are modeled correctly. An advantage of this approach is that it is possible to model waves generated by boats that sail with fluctuating velocities. This would not be possible using most other methods.

The assumption underlying the method is that the ship displaces water primarily normal to its sailing direction. This assumption is reasonable for very slender hulls, but less adequate for ships with a higher block coefficient, such as the ship considered here. This will be confirmed by the results.

6.2 RESULTS FOR METHOD 1

Figure 5 shows the comparison between the measurements and the computations using method 1. Only the results for Test I are presented here, since this is the most challenging test to simulate accurately.

The correspondence between measurements and simulations is not good. Using method 1 a large accumulation of water in front of the boat is created. This results in increase of water level instead of depression of the water level as measured.

Wenneker et al. [8] have shown that this method provides good results for ship-type waves generated by wave paddles. These results were obtained in open water, simulating ship-induced waves at a larger distance from a passing vessel. However, in the measurements in the wave basin, the passing vessel was simulated using the same wave signal to steer the wave paddles as used in the computations. In other words, the suitability of approximating a passing ship with a surface elevation obtained with method 1 was not investigated. On the basis of the present results it is concluded that method 1 is not suited to estimate ship-induced waves, with the exception perhaps for slender ships. A comparison between Figure 4 and the first Figures 3a and 3b shows a systematic error. By construction, the average surface displacement of method 1 is exactly equal to zero, cf. Figure 4. In reality, the average surface displacement caused by a passing ship is negative, cf. the first Figures 3a and 3b.

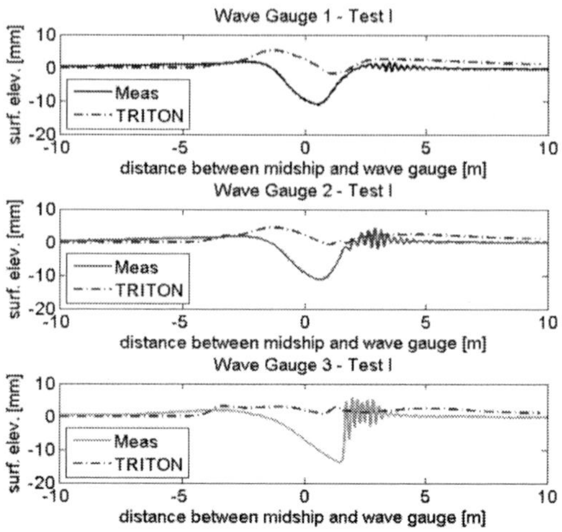

Figure 5: Comparison between measurements and TRITON results for Test I using method 1.

6.3 METHOD 2 – BASED ON POTENTIAL SOURCE-SINK FLOW

Due to the disappointing results of method 1, a second approach is proposed in which two-dimensional potential flow is used as starting point to describe the flow around the hull of a vessel. By placing a source and a sink in a uniform flow field, it is possible to create a limiting streamline which represents the contour of a ship. Using potential flow theory, the velocity field around a ship can be solved for a uniform flow around a still ship or, equivalently, for a ship sailing at a certain velocity in still water. Using Bernoulli's principle the pressure distribution around the ship's contour can be determined. From this pressure distribution the surface elevation next to the ship can be computed, assuming hydrostatic equilibrium.

We consider an incompressible stream-function ψ:

$$\frac{d\psi}{dy} = u \qquad \frac{d\psi}{dx} = -v \tag{6.2}$$

in which u and v are the flow velocities in x- and y-direction, respectively. Note that a special property of the stream function ψ is that the spatial derivative in x-direction gives the flow velocity is y-direction, and vice versa. The equation for the stream function generated by a source and a sink in a uniform flow of speed V_∞ aligned with the source-sink line and formulated in polar coordinates r and θ, is given by [13]:

$$\psi(r,\theta) = V_\infty r \sin\theta + \frac{\Lambda}{2\pi}(\theta_1 - \theta_2) = const \tag{6.3}$$

In this equation are θ_1 and θ_2 the angles at the source and the sink of the triangle formed by the point (r,θ), the position of the sink, and the position of the source. The line ψ=0 defines a dividing streamline that has an oval shape (Rankine oval) whose size is determined by Λ. See [13] for details. The region inside this oval can be replaced by a solid body and the region outside the body can be interpreted as the inviscid, potential, incompressible, two-dimensional flow around the body. By selecting the appropriate source strengths and source locations, the dividing streamline can be adjusted to an oval with a length and width that are representative of the vessel under consideration. Hereby it is important to note that not the actual length of the ship (L_{pp}) is used but an effective length (L_{eff}), based on:

$$L_{eff} = \Delta / BhC_b \quad (6.4)$$

in which Δ is the volume displacement of the ship, B the beam of the ship, h the water depth, and C_b the block coefficient. This expression assures that the correct volume of water is displaced by the oval approximation of the passing ship, correcting for the two-dimensional approach. Figure 6 shows an example of the streamlines around the hull of the ship that is obtained in this way. Note that, instead of adjusting the length of the ship, the correct water displacement can also be ensured by adjusting the beam. A combination of both is possible as well and will be a subject of further study. By representing the hull form with only one source-sink pair, it is noted that the shape of the diving streamline does not yet reflect the distribution of displaced volume along the length of the ship.

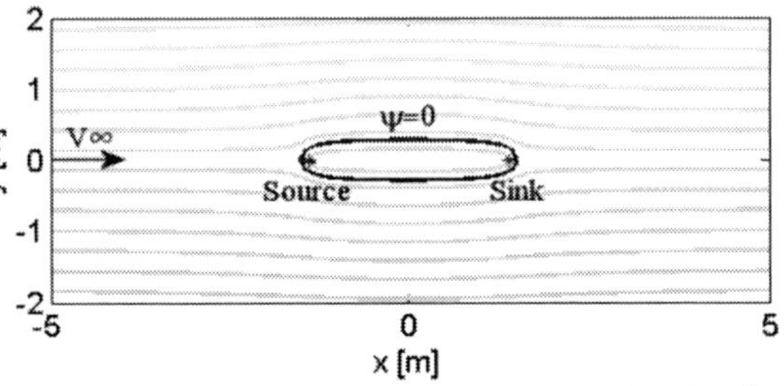

Figure 6: Streamline function ψ. In black the limiting streamline ψ=0 is plotted, representing the hull. The asterisks indicate the source and sink locations.

The approach also allows taking into account the restricted width of the channel, by assuming a symmetry plane at a distance from the centreline of the vessel. Such a symmetry plane is created by placing a second source-sink pair at the same distance at the other side of this symmetry plane. Adding the contributions of both source-sink pairs represents the flow in a restricted channel.

In the computations presented here a symmetry plane is placed at the effective channel width from the centreline of the vessel. The effective channel width is determined by assuming that only the water above the slope contributes to the effective channel width; the flat shallow part is assumed not to be of influence on the water-level depression at the side of the vessel. The effective channel width is determined by:

$$W_{eff} = \frac{\frac{1}{2} h(\alpha h)}{h} \quad (6.5)$$

In this equation h is the water depth at the side of the vessel and α is the slope of the bank. For Test I, α has a value of 8.

As mentioned before, the surface elevation can be computed from the pressure distribution around the hull. The water-level elevation at the side of the ship is prescribed at the grid points along the incoming-wave boundary of TRITON, with a time-shift derived from the uniform velocity of the ship. Figure 7 shows the surface elevation at the side of the vessel. The shape of the water-level depression is related to the shape of the limiting streamline, which has a blunt front and end.

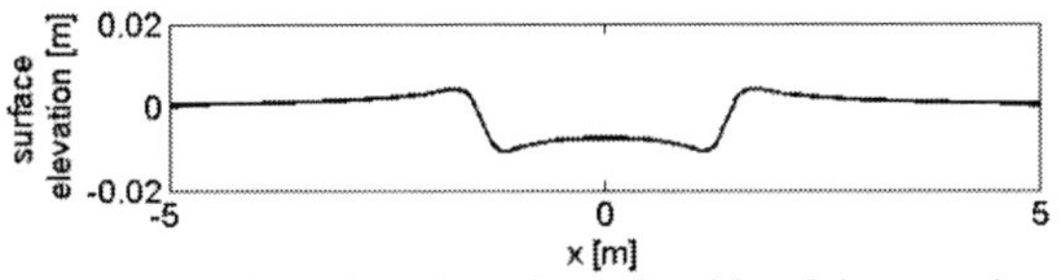

Figure 7: Surface elevation along the side of the vessel.

6.4 RESULTS FOR METHOD 2

The results obtained with method 2 are much better than using method 1, as is shown in Figure 8. At the first wave gauge, the timing and the length of the water-level depression are simulated quite accurately compared to the measurements. The magnitude of water-level depression however is somewhat underestimated. As a result the correlation for wave height at wave gauge 2 and 3 is poor. At wave gauge 3 the wave trough is not as steep as in the measurements.

Although the results compared to the measurements are not perfect yet, as a first step the results are promising. Representing a hullform with only one source and one sink is crude, but already provides reasonable results considering length and shape of the primary wave disturbance. If the prediction of the magnitude of the water level depression can be further improved by refining the proposed method, the considered Boussinesq-type model is capable of accurately propagating the primary wave over non-uniform bathymetries, as shown in Section 5.

A comparison between Figure 4 and Figure 7 shows that method 2 does not have the same drawback as method 1. Unlike method 1, method 2 provides an approximation of the surface displacement whose average is negative, in agreement with the measured ship-induced wave signal (first Figures 3a and 3b). This better behaviour of method 2 is due to the nonlinear relation between the pressure distribution around the oval approximating the hull and the water-level elevation, which clearly results in a better model for ships with higher block coefficients.

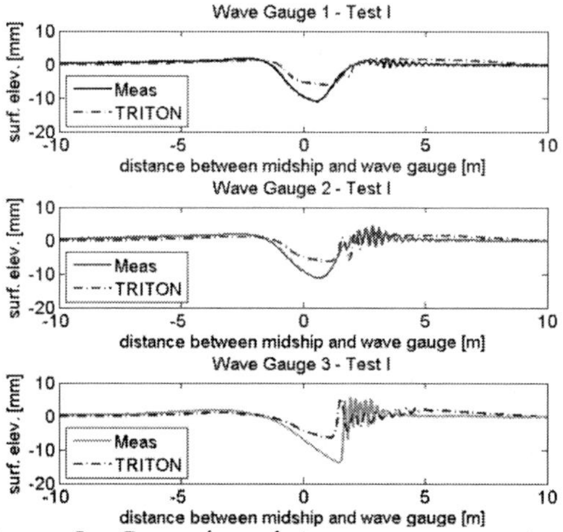

Figure 8: Comparison between measurements and TRITON results for Test I using method 2.

7. DISCUSSION AND CONCLUSIONS

In this paper the use of a Boussinesq-type model is proposed to simulate ship-induced wave propagation over non-uniform bathymetries.

In Section 5 it has been shown that if the incoming wave signal is accurate, the applied Boussinesq-type is capable of adequately simulating the propagation of ship-induced waves over a non-uniform bathymetry.

Of the two proposed simple methods to determine the wave pattern created by a passing ship in the absence of any data, the method based on a potential source proves to be the most promising. Although this method does not yet provide results of sufficient quantitative accuracy, qualitatively the results agree well. It is expected that the results can be improved by extensions of the approach, which will be the subject of further study. Another approach that may be worth investigating, is to approximate the ship directly into the Boussinesq-type model. By adding time-dependent source terms to the momentum equations, the sailing speed and corresponding water displacement of a moving ship can be enforced. Although a rather crude approximation, the promising results of method 2 suggest this to be a viable approach for the prediction of ship-induced waves at some distance away from the ship. Such a method would allow for arbitrary sailing speed and direction. It would also avoid the problem of reflections at the boundary that we now need to impose the ship-induced wave signal (cf. Section 5.2), since such a boundary would not be present.

On the basis of the presented results, it can be concluded that the use of a Boussinesq-type model to simulate wave propagation of ship-induced waves can be a valuable step in determining hydrodynamic forces on moored ships. The approach described in this paper will be favourable to more conventional methods for situations in which nonlinear shallow water effects are important.

9. ACKNOWLEDGEMENTS

We thank Flanders Hydraulics Research for making the model test data available for validation of numerical calculations.

10. REFERENCES

1. SEELIG, W., 'Passing Ship Effects on Moored Ships', Technical Memorandum TM-6027-OCN, Facilities Engineering Service Center, 2001.

2. SCHIJF, J.B., JANSEN, P.P., 18th International Navigation Congress (Rome), Section I, Communication 1, p.p. 175-197, 1953.

3. PINKSTER, J.A., NAAIJEN, P. 'Predicting the Effect of Passing Ships', 18th International Workshop on Water Waves and Floating Bodies, Le Croisic, France, 2003.

4. RAVEN, H.C., 'A solution method for the nonlinear ship wave resistance problem', PhD-thesis, Delft University of Technology, 1996.

5. LATAIRE, E., VANTORRE, M., ELOOT, K., 'Systematic model tests on ship - bank interaction effects', International Conference on Ship Manoeuvring in Shallow and Confined Water: Bank Effects, Antwerp, Belgium, 2009.

6. BORSBOOM, M.J.A., DOORN, N., GROENEWEG, J., VAN GENT, M.R.A., 'A Boussinesq-type wave model that conserves both mass and momentum', Proc. 27th ICCE Conference, Sydney, 148 – 161, 2000.

7. PINKSTER, J.A., 'The influence of a free surface on passing ship effects', *International shipbuilding progress, 51(4),* 313-338, 2004

8. WENNEKER, I., BORSBOOM, M., PINKSTER J., WEILER, O., 'A Boussinesq-type wave model coupled to a diffraction model to simulate wave-induced ship motion', 31th PIANC Congress, Estoril, Portugal, May 14-18, 2006.

9. BINGHAM, H.B., 'A hybrid Boussinesq-panel method for predicting the motion of a moored ship', Coastal Engineering 40 1–38, 2000.

10. DOORN, N., GROENEWEG, J., WEILER, O., BORSBOOM, M.J.A., 'Numerical modeling of ship-

induced wave propagation', Proc. 28th Conference on Coastal Eng., Cardiff, 689 – 701, 2002.

11. BORSBOOM, M., 'Wave run-up in TRITON', WL|Delft Hydraulics, Delft, H4387.31, 2005.

12. BORSBOOM, M., DOORN, N., GROENEWEG J., VAN GENT, M.R.A., 'Nearshore wave simulations with a Boussinesq-type model including breaking', Proc. of Coastal Dynamics 2001 Conference, Lund, 2001.

13. ANDERSON, J. D., 'Fundamentals of Aerodynamics, Third edition', McGraw-Hill, 2001.

11. AUTHORS BIOGRAPHY

Arne van der Hout holds the current position of researcher/advisor at Deltares. He is involved in consultancy work on harbour related projects and research in the field of numerical wave modelling. His previous experience includes validation of a Boussinesq-type model and various research projects on moored ships and (low-frequency) waves.

Otto Weiler holds the current position of Expert Advisor Hydraulic Engineering at Deltares. His experience includes a wide range of port related projects, including manoeuvring, moored ship and passing ship issues. Based on this project involvement he has been stimulating developments in numerical models for flow and waves and for the interaction of flow and waves with ships in confined waters. .

Mart Borsboom holds the current position of specialist at the Deltares Software Centre of Deltares. He is involved in activities related to the development and maintenance of the Deltares simulation software, in particular the numerical and mathematical-physical aspects. His personal interest is in the development and application of accurate and efficient methods for detailed wave simulations, on which subjects he is frequently consulted. His research topics include improved Boussinesq-type formulations, accurate non-reflecting boundary conditions, higher-order discretization methods, and fast iterative solvers.

CALCULATION OF FORCES ON MOORED SHIPS DUE TO PASSING SHIPS

W van der Molen and **J Moes**, CSIR, South Africa
P B Swiegers, WSP Africa, South Africa
M Vantorre, Ghent University, Belgium

SUMMARY

Ships moored along channels in ports experience forces due to ships passing in the channel, which can lead to breaking of mooring lines or excessive ship motions. A numerical model to predict these passing ship effects is described in this paper. The model, which is based on boundary-integral equations, is validated against model tests for a container ship passing a bulk carrier. The results focus on the effect of quay and channel walls on the forces on the moored ship, and on the non-linear effects of the forces at high (but subcritical) passing speeds.

1. INTRODUCTION

Ships sailing in a fairway or a confined channel generate a disturbance that interacts with ships moored alongside the waterway. This can lead to breaking of mooring lines or delays in the loading process due to ship motions. A good understanding of the phenomenon is essential to provide sufficient passing distance in the design of new terminals and waterways, and to provide operational regulations with respect to speed limits and minimum passing distances.

Physical model tests were carried out by Remery [1] and Muga & Fang [2] on VLCC tankers in open water. These results were used by Seelig [3] and Flory [4] to provide empirical relationships for the maximum forces due to a passing ship. Vantorre et al. [5] used data from an extensive model test program on a bulk carrier, a container ship and two tankers to derive similar empirical relationships. More model test data on two tankers are available from Lataire et al. [6]. These empirical relations can be used to obtain quick estimates in early design phases. However, the formulae are very generic with respect to hull shapes and the effect of other structures and a bottom topography are not included.

Besides the growing number of physical model data, advances were also made in the numerical modelling of passing ship effects. These models can be used in further design stages. Pinkster [7] has developed a panel model to calculate the disturbance due to a passing ship including free surface effects, such as the generation of oscillations in partly enclosed basins alongside the channel. Huang & Chen [8] have developed a RANS model to calculate the flow in arbitrarily shaped channels and basins, which provides accurate results but is difficult to set up and requires long computation times.

The panel model presented in this paper is relatively straightforward to set up and run, and can be applied for open jetties, quay walls and confined channels. The model is validated against physical model tests for these three situations and for different passing speeds.

2. MODEL DESCRIPTION

The model to determine the flow around a passing ship and the forces on a moored ship is a 3D potential flow model. The model is based on the double-body flow assumption with restricted water depth, similar to Korsmeyer et al. [9] and the double-body model of Pinkster [7]. However, a few terms relative to the depth-related Froude number have been added to make the results more accurate at higher passing speeds. The double-body flow assumption implies that water level fluctuations can be computed, but the flow pattern is not affected by these fluctuations. Hence, inertia effects of the changing flow pattern are neglected and the flow is calculated for a horizontal free surface. This assumption also implies that the computation focuses on the disturbance due to the primary return flow due to the passing ship. Shorter so-called wash waves are neglected. Wash waves become important at near-critical or supercritical speeds, such as for high-speed ferries (Van der Molen et al. [10]). For simplicity, the case, as shown in Figure 1, is considered here where the passing ship travels parallel to the moored ship at a constant speed and without the effect of an ambient current.

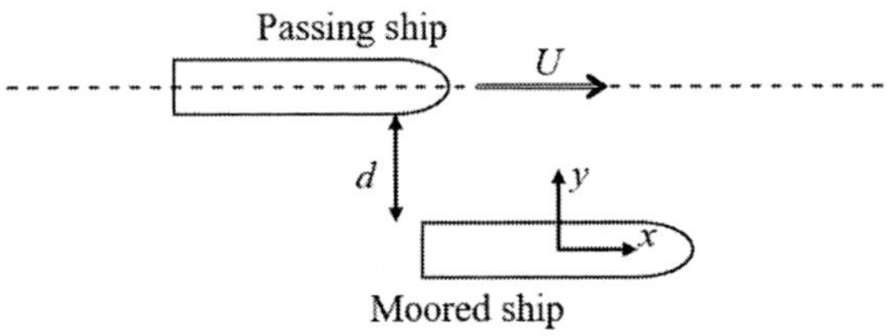

Figure 1: Definition sketch for a ship passing the moored ship in the same direction with speed U at distance d.

The fluid is considered incompressible, irrotational and homogeneous with a density ρ. The model domain $\mathcal{D}$ is bounded by the body surfaces $\mathcal{H}_P$ and $\mathcal{H}_M$ of the passing ship and the moored ship respectively, the surface of any nearby structures $\mathcal{W}$, the fluid surface $\mathcal{F}$ and the sea bed $\mathcal{B}$. The sea bed may have an arbitrary bathymetry above an ambient horizontal sea bed with a water depth h. The wetted surface of all bodies ($\mathcal{H}_P$, $\mathcal{H}_M$, $\mathcal{W}$ and $\mathcal{B}$) is denoted as $\mathcal{S}$. The coordinate system $Oxyz$ is a right-handed system with the origin at the still water level, the x-axis in the passing direction and the z-axis positive upwards.

The fluid flow around the bodies due to the steady forward speed U of the passing ship is described by means of a velocity potential ϕ, which is defined such that the gradient of ϕ is equal to the flow velocity. The velocity potential has to meet the following conditions:

$$\nabla^2\phi = F_h^2\phi_{xx}\,, \quad (1)$$

$$\phi_n = Un_x \text{ on } \mathcal{H}_P, \quad (2)$$

$$\phi_n = 0 \text{ on } \mathcal{H}_M, \mathcal{B} \text{ and } \mathcal{W}, \quad (3)$$

$$\phi_z = 0 \text{ on } z = 0 \text{ and } z = -h, \quad (4)$$

$$\phi = 0 \text{ for } R \to \infty, \quad (5)$$

where $F_h = U/\sqrt{(gh)}$ is the depth-related Froude number (g is the acceleration of gravity), $\vec{n}$ is the unit normal vector directed out of the fluid domain and R is a horizontal distance. The subscripts denote partial derivatives. The Laplace equation (Eq. 1) has been modified to account for deformations of the free surface due to the flow around the passing ship in a similar manner as in slender body, shallow water theory (see e.g. Tuck [11] and Gourlay [12]).

The velocity potentials are solved using a boundary-integral equation method based on the source potential formulation:

$$\phi(\vec{x}) = \iint_S G(\vec{x},\vec{\xi})\sigma(\vec{\xi})dS_\xi\,. \quad (6)$$

The source strengths σ follow from the integral equation:

$$2\pi A\sigma(\vec{x}) + \int_S G_{n_x}(\vec{x},\vec{\xi})\sigma(\vec{\xi})dS_\xi = \begin{cases} Un_x \text{ on } H_P \\ 0 \text{ elsewhere.} \end{cases} \quad (7)$$

The right-hand side of this equation denotes the boundary condition on the body surface. A is a parameter (close to 1) which accounts for the modification of the Laplace equation (Eq. 1) and the subsequent modification of the Green function $G(\vec{x},\vec{\xi})$, which is the potential in $\vec{x}$ due to a unit source in $\vec{\xi}$ with a constant source strength:

$$G = \frac{1}{r'} + \frac{1}{r_2'} + 2\int_0^\infty \frac{e^{-kh}}{\sinh kh}\cosh k(z+h)\cosh k(\zeta+h)J_0(kR')dk \quad (8)$$

where $r' = \sqrt{[(x-\xi)^2/(1-F_h^2)+(y-\eta)^2+(z-\zeta)^2]}$ is the modified radial distance between (x,y,z) and (ξ,η,ζ), r_2' is the modified radial distance to the image source below the bottom and R' is the modified horizontal distance. This Green function is the zero-frequency limit of the free surface Green function. The Green function and its derivatives are computed using algorithms described in Newman [13]. For $R/h < ½$ the integral is expressed by the contribution of an array of Rankine image sources above the free surface and below the sea floor. For larger R/h the Green function is approximated by a Fourier series in terms of modified Bessel functions. The determination of A is given in the Appendix.

The source strengths are solved numerically by discretising the mean wetted body surfaces with a large number of quadrilateral or triangular panels with the collocation points of the source strengths at the panel centroids. The velocity potentials are computed for subsequent positions of the passing ship as it passes the moored ship. In this process, it is assumed that the changes of the flow pattern are slow, so that for each time step the flow can be considered as a steady-state solution.

The fluid pressure p due to the passing ship can be determined based on Bernoulli's equation:

$$p = -\rho U\frac{\partial\phi}{\partial X_P} - \frac{1}{2}\rho|\nabla\phi|^2\,, \quad (9)$$

which is given here in terms of the potentials ϕ and the position of the passing ship X_P. The first term refers to the (almost hydrostatic) pressures in the primary wave generated by the passing ship. The second term refers to the pressure reduction due the flow velocity squared. The derivatives with respect to X_P in the first term are evaluated after the potentials have been determined for all passing ship positions using a polynomial approximation of the potentials with respect to the passing ship position. The fluid velocities $\nabla\phi$ are easily evaluated using the gradients of the Green function in Eq. 6. The force $\vec{F}$ and moment $\vec{M}$ on the moored ship are obtained by integration of the pressures over the submerged hull:

$$\vec{F} = \iint_{H_M} p\vec{n}dS\,, \quad (10)$$

$$\vec{M} = \iint_{H_M} p(\vec{x}-\vec{x}_G)\times\vec{n}dS\,, \quad (11)$$

where x_G is the position of the centre of gravity. These forces can be input in a moored ship response model to determine the motions of the moored ship and the forces in mooring lines.

3. PHYSICAL MODEL TESTS

3.1 TEST SET-UP

Model tests were conducted in a large basin of 55 m by 36 m at the Hydraulics Laboratory of the CSIR in Stellenbosch, South Africa. The passing ship was towed along a straight track parallel to the moored ship. The passing ship was connected to a trolley rolling over a long rail. The wheels on one side of the rail are transversely fixed. Two hollow vertical pipes are mounted on the trolley. These pipes guide two vertical rods that are able to slide inside the pipes and are connected to the deck of the passing ship. With this set-up the ship is firmly fixed to the trolley in the horizontal directions, while it is able to move freely in heave and pitch. A rope is connected to the trolley. The speed is regulated by an electromotor and set at a specified constant speed for each test.

The moored ship is firmly connected to the wharf with a fixed force frame. The forces in the force frame are measured with three force transducers aligned such to measure the horizontal forces and moment on the ship in surge, sway and yaw. The force transducers are connected to the ship with thin rods. This allows the force transducers to measure both compression and tension forces, while the ship is free to move in heave, roll and pitch.

3.2 SHIPS

The tests were conducted for a Post-Panamax container ship as the passing ship and a Panamax bulk carrier as the moored ship at a scale of 1:100. Both ships were fully laden to the same draught. The main ship dimensions are given in Table 1.

Table 1: Ship dimensions

Ship type			Container	Bulk
Position			Passing	Moored
Length betw. perps.	L	m	260.0	243.4
Breadth	B	m	40.0	32.0
Draught	T	m	14.0	14.0
Displacement	Δ	m^3	85 989	90 412
CoB forw. midship	LCB	m	–6.7	5.5

Figure 2: Model test set-up with the passing container ship in the rail and the moored bulk carrier in the force frame along a straight quay wall.

3.3 TEST CONDITIONS

The tests were carried out for the container ship passing the bulk carrier in parallel but opposite directions along port side of both ships, as shown in Figure 2. The tests were done (among others) for different passing speeds between 4 kn and 14 kn in open water, for a ship moored at a quay wall and for a ship passing in a confined channel. The quay wall was inserted as a long straight wall leaving a gap of 4 m between the wall and the moored ship. Another wall was added to create a 192 m wide channel. The tests were conducted on a horizontal floor with a water depth of 16.8 m, for a depth-to-draught ratio of 1.2. The passing distance is 64 m, defined as the board-to-board distance between the two ships.

4. VALIDATION RESULTS

The numerical model was validated against the results of the physical model tests. The hulls of the ships were discretised in the model by 1464 panels for the container ship and 1430 panels for the bulk carrier, as shown in Figure 3. The basin was sufficiently large to neglect the effects of the basin boundaries. The quay wall and the channel walls were modelled as long arrays of panels (two rows of panels over the water depth away from the moored ship and three rows in the gap between the moored ship and the quay). The flat bottom is represented in the formulation of the Green function (Eq. 8).

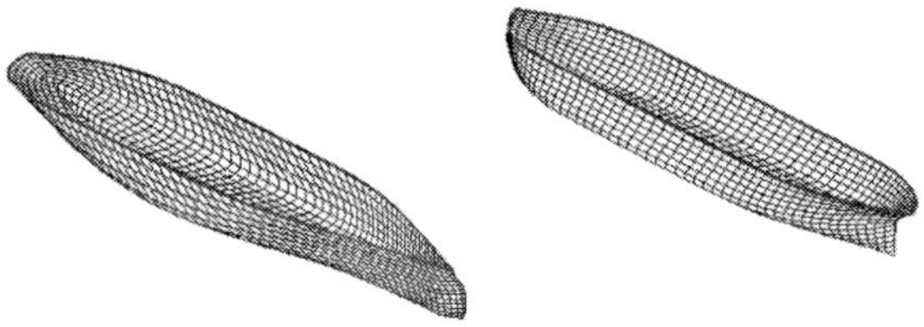

Figure 3: Panel descriptions of the container ship (left) and the bulk carrier (right).

The time series of the forces on the moored ship, for the ship passing with a speed of 6 kn in open water, along a quay wall and in a channel, are given in Figure 4. The position of the passing ship X_P is given with respect to the position where the ships are side by side. The forces and moment are made non-dimensional with respect to the displacement of the moored ship Δ_M and the depth-related Froude number F_h.

The open water case yields the smallest surge forces, but the largest sway forces and yaw moments. The surge forces are significantly larger in the channel, but they are comparable to the case with only a quay wall for sway and yaw. The correspondence between the measured and calculated results is very good for the open water case. For the case with a channel, the surge forces are slightly underestimated and there is a phase shift in the yaw moment.

The amplitudes of the forces on the moored ship for passing speeds of 4 kn to 14 kn (F_h = 0.16 to 0.56) are given in Figure 5. The surge force and yaw moment

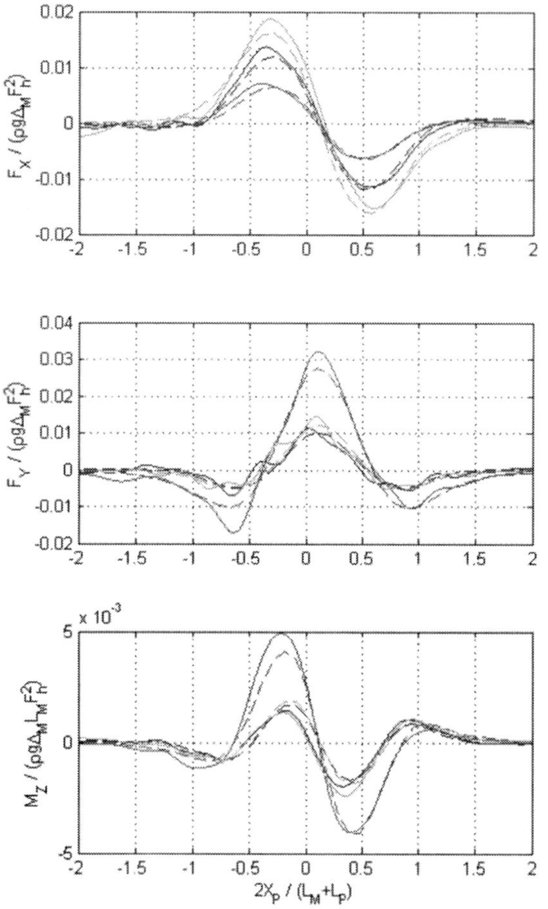

Figure 4: Measured (solid) and calculated (dashed) forces on the moored ship for surge (top), sway (middle) and yaw (bottom) in open water (red), along a quay wall (blue) and in a channel (green) for a ship passing at 6 kn.

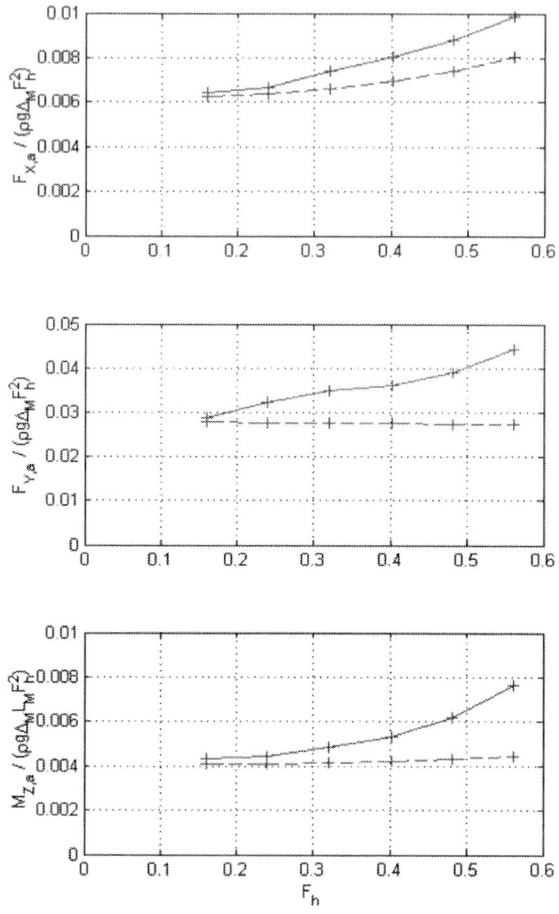

Figure 5: Amplitudes of measured (solid) and calculated (dashed) forces on the moored ship for surge (top), sway (middle) and yaw (bottom) as function of the Froude number.

amplitudes are defined as the average of the positive and negative extremes, and the sway force amplitude is the maximum positive sway force. Because the plotted forces are divided by the Froude number squared, it is easy to observe that the forces are more than proportional to the passing speed squared. The further increase of the forces is about 50% for a Froude number $F_h = 0.5$. This effect is stronger for larger passing distances (not shown here). However, for low speeds the assumption that the forces are proportional to the passing speed squared is valid.

The nonlinear effect of the forces for high speeds is also observed in the calculated results, albeit not as strongly as for the measured forces and only significantly for surge. Some nonlinear effects were not considered, such as more confined flows underneath the passing ship due to squat. Another possible explanation for the differences at higher speeds is the neglection of turbulence and viscous effects in the numerical model, which is based on potential flow theory. E.g. the flow underneath the passing ship will be affected by shear stresses, so that more water is pushed to the sides leading to stronger flows adjacent to the passing ship and around the moored ship. These effects are relatively stronger at higher speeds. This could also partly clarify the differences observed for the ship moored at a quay wall, where the flow in between the ship and the quay experiences friction effects. A different approach of the numerical model, e.g. using a direct diffraction potential formulation instead of the indirect source potential formulation, may lead to improved nonlinear behaviour of the sway forces and yaw moments. This is a subject for further research.

5. CONCLUSIONS

A numerical model has been developed to calculate the forces on a moored ship induced by a passing ship. The model is based on boundary-integral equations which makes it relatively easy to set up for different types of ships and to include nearby structures and bottom topographies, such as quay walls and channel slopes. The model has been validated against physical model tests. The model provides good results in open water as well as including a quay wall and channel walls. The calculated results show the same nonlinear pattern of the forces on the moored ship at higher speeds as observed in the model test measurements, albeit less strongly and mainly for the surge forces.

The assumption that the forces on a moored ship due to a passing ship are proportional to the passing speed squared is valid for low speeds. However, the forces can be significantly larger for larger Froude numbers. The forces significantly divert from this assumption roughly for a depth related Froude number $F_h > 0.25$.

REFERENCES

1. REMERY, G.F.M., 'Mooring Forces Induced by Passing Ships', OTC Paper 2066, *Offshore Technology Conf.*, Richardson, TX, 1974.

2. MUGA, B.J. and FANG, S.T., 'Passing Ship Effects from Theory and Experiment', OTC Paper 2368, *Offshore Technology Conf.*, Richardson, TX, 1975.

3. SEELIG, W.N., 'Passing Ship Effects on Moored Ships', Technical Report TR-6027-OCN, *Naval Facilities Engineering Command, East Coast Detachment*, Washington, DC, 2001.

4. FLORY, J.F., 'The Effect of Passing Ships on Moored Ships', *Prevention First Symp.*, Long Beach, CA, 2002.

5. VANTORRE, M., VERZHBITSKAYA, E. and LAFORCE, E., 'Model Test Based Formulations of Ship-Ship Interaction Forces', *Ship Techn. Res.*, 49, 124–141, 2002.

6. LATAIRE, E., VANTORRE, M. and DELEFORTRIE, G., 'Captive Model Testing for Ship-to-Ship Operations', *Proc. MARSIM '09 Conf.*, Panama City, Panama, 2009.

7. PINKSTER, J.A., 'The Influence of a Free Surface on Passing Ship Effects', *Int. Shipbuilding Progr.*, 51(4), 313–338, 2004.

8. HUANG, E.T. and CHEN, H.C., 'Influences of Site Specifics on Passing Ship Effects', *Proc. 17th ISOPE Conf.* (Vol. IV, pp. 2356–2363), Lisbon, Portugal, 2007.

9. KORSMEYER, F.T., LEE, C.H. and NEWMAN, J.N., 'Computation of Ship Interaction Forces in Restricted Waters, *J. Ship Res.*, 37(4), 298–306, 1993.

10. VAN DER MOLEN, W., WENNEKER, I. and BORSBOOM, M., 'Moored Ship Motions due to Passing Ships', *Proc. Port-Maritime Innovation and R&D Conf.*, Rotterdam, the Netherlands, 2005.

11. TUCK, E.O., 'Shallow Water Flows past Slender Bodies', *J. Fluid Mech.*, 26, 81–95.

12. GOURLAY, T., 'Sinkage and Trim of Two Ships Passing Each Other on Parallel Courses', *Ocean Eng.*, 36(14), 1119–1127, 2009.

13. NEWMAN, J.N., 'The Approximation of Free Surface Green Functions', *Wave Asymptotics* (pp. 107–135), 1992.

APPENDIX. DETERMINATION OF THE SOURCE STRENGTH MODIFICATION PARAMETER

The parameter A is introduced in Eq. 7 to account for the effect of reduced source strengths for a non-zero Froude number. The value of A can be determined from the integral of the radial derivative of the principal singularity of the Green function, $1/r'$ (not considering the mirror sources above the free surface and below the bottom), over the surface of a sphere:

$$A = \frac{1}{4\pi}\int_0^{2\pi}\int_{-\frac{\pi}{2}}^{\frac{\pi}{2}} \frac{\partial}{\partial r}\frac{r^2}{r'}\cos\varphi d\varphi d\theta \qquad (12)$$

using a spherical coordinate system $Or\varphi\theta$ with the origin at the source point. A nonlinearity parameter, $\varepsilon = 1/(1-F_h^2) - 1$, is defined such that the modified radial distance can be written as $r' = \sqrt{(r^2+\varepsilon x^2)}$. Substituting this in Eq. 12, leads to

$$A = \frac{1}{4\pi}\int_0^{2\pi}\int_{-\frac{\pi}{2}}^{\frac{\pi}{2}} \frac{1}{\sqrt{1+\tan^2\varphi+\varepsilon\cos^2\theta}} d\varphi d\theta . \qquad (13)$$

The integral is equal to 4π for $\varepsilon = 0$, so that Eqs. 6 and 7 reduce to a set of equations for the flow between two horizontal plates (one at $z = 0$ and one at $z = -h$). The integral can be easily evaluated numerically for non-zero ε.

AUTHORS BIOGRAPHIES

Wim van der Molen is a research engineer at the Hydraulics Laboratory of the CSIR in Stellenbosch, South Africa. Recent research activities at the CSIR include the development of numerical models for wave-ship interaction, passing ship effects and moored ship response, and the development of video monitoring techniques for waves and moored ship motions in a laboratory. He holds a PhD at the University of Delft, the Netherlands, with the subject to couple wave propagation models and ship motion models for the calculation of moored ship response in harbours.

Pierre Swiegers is a coastal engineer at WSP in Stellenbosch, South Africa. He has a BSc degree in Civil Engineering at the University of Stellenbosch. He has conducted the physical model tests and numerical model calculations for passing ship effects as part of his coastal engineering MSc graduation project at the same university.

Hans Moes is a (now retired) port and coastal engineer at the CSIR. Research activities over the past 35 years among others include physical and numerical modelling of waves and moored ship response as well as ship navigation in port entrance channels. He is an active member of PIANC with participation in several working groups.

Marc Vantorre holds the position of senior full professor of marine hydrodynamics at Ghent University, where he is head of the Maritime Technology Division. His research, in close co-operation with Flanders Hydraulics Research, Antwerp, mainly focuses on ship hydrodynamics in shallow and restricted waters, including captive model testing, simulation and access channel policy.

SQUAT COMPUTATIONS FOR CONTAINER SHIPS IN RESTRICTED WATERWAYS

A. von Graefe, University Duisburg-Essen, Duisburg, Germany
O. El Moctar, University Duisburg-Essen, Duisburg, Germany
V. Shigunov, Germanischer Lloyd AG, Hamburg, Germany
H. Söding, Technical University Hamburg-Harburg, Hamburg, Germany
T. Zorn, Germanischer Lloyd AG, Hamburg, Germany

SUMMARY

The paper presents results of dynamic squat computations in restricted waterways with two potential flow solution methods and a solver of Reynolds-averaged Navier-Stokes equations (RANSE). Computational results are compared with model tests and full-scale measurements for two large post-panamax class container carriers. Ways of computational time reduction are discussed.

1. INTRODUCTION

Optimal exploitation of the available fairway depths is crucial for some internal sea ports, e.g. Hamburg. This issue becomes even more important because of larger ships being built presently. Minimisation of dredging costs (for river Elbe, depth increase costs about $1.0 mio per 1.0 cm of additional depth without maintenance costs [5]) is possible if reliable squat predictions could be offered to the shipping companies and fairway authorities. The estimation of dynamic sinkage and trim in restricted waterways is a difficult task. Despite the large number of available empirical formulae, experience shows that none of them can predict squat with sufficient accuracy for all ship sizes and forms and channel configurations [1]. A more reliable approach is to use model tests; a cheaper alternative are computational fluid dynamics methods.

As a classification society, Germanischer Lloyd is committed to providing reliable assessment methods concerning all aspects of ship safety. In this paper, several numerical methods are used for the prediction of dynamic trim and sinkage in restricted waterway: a potential method *GL Rankine*, employing Rankine sources and patch method for steady and periodic flow problems, another potential method based on slender-body theory and matching asymptotic expansions, and a finite-volume RANSE solver *Comet*. Computational results are compared with model tests and full-scale measurements for two post-panamax class container carriers. Efficient methods for computational time reduction are proposed.

2. SHIP AND MEASUREMENT DATA

Two post-panamax class container ships are used in the study, referred to as PPM1 and PPM2; their main dimensions are shown in Table 1.

PPM1 is a modern container ship with capacity of about 7500 TEU. Full-scale measurements were carried out for this ship along the 120 km-long part of the river Elbe from Hamburg to the North Sea [5]. In this paper, measurements at two stations referred to as km650 and km660 were used. At each of these stations, measurements were carried out along a several kilometre

Table 1: Main dimensions of the used container ships

Dimensions	PPM1	PPM2
length overall, m	320.0	350.0
length between perpendiculars, m	300.0	330.0
draught at forward perpendicular, m	12.65	13.8
draught at aft perpendicular, m	12.50	13.8
breadth at water line, m	42.80	46.0

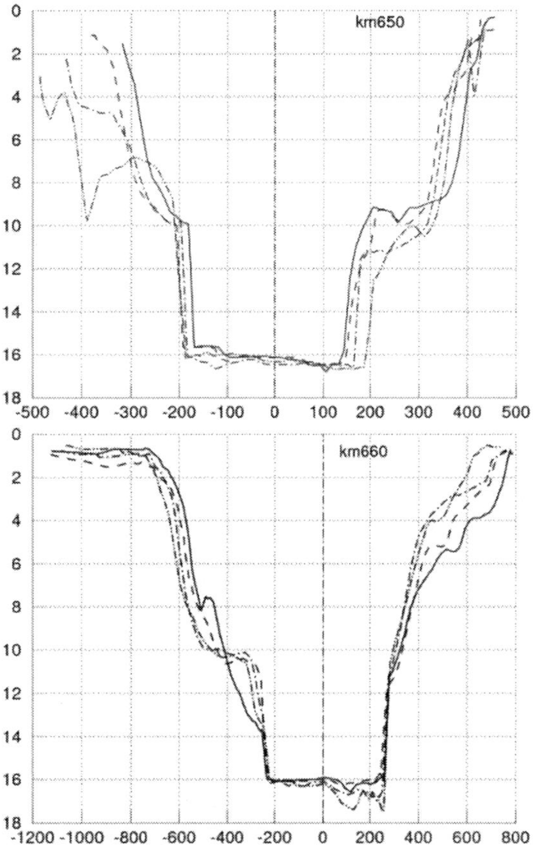

Figure 1: Section form for the stations km650 and km660 (ship position corresponds to transversal coordinate 0.0)

long river portion; Figure 1 shows the (slightly varying) form of the cross-sections along these portions for stations km650 (top) and km660 (bottom); the ship location corresponds to zero along the horizontal axis. For the selected stations, the effect of the finite depth is significant while the influence of the limited breadth is marginal.

Measurement data were provided by BAW (Federal Waterways Engineering and Research Institute) Hamburg, including time histories of the ship position and attitude measured with a Precise Differential Global Positioning System (PDGPS), bottom topography as a digital model, and current speed and water density measured from a support boat at 6 locations along the river.

The measurement procedure described in [5] provides the overall accuracy of squat data within ±5.0 cm; note however that comparison with computations may be significantly influenced by such not controlled factors as e.g. local variations of the water level and current speed, short-term changes of bottom topography and, more important, dynamic effects due to longitudinal acceleration or deceleration of the ship. The latter might explain the observed differences between the measurements for km650 and km660, see Figure 3, since the influence of side restrictions for both these stations is marginal and the measurements are expected to be closer to each other.

For the second ship PPM2 considered here, model tests were carried out by BAW Hamburg at scale 1:40 for 9 speeds. Figure 2 shows different sections along the model path and the model position within the section.

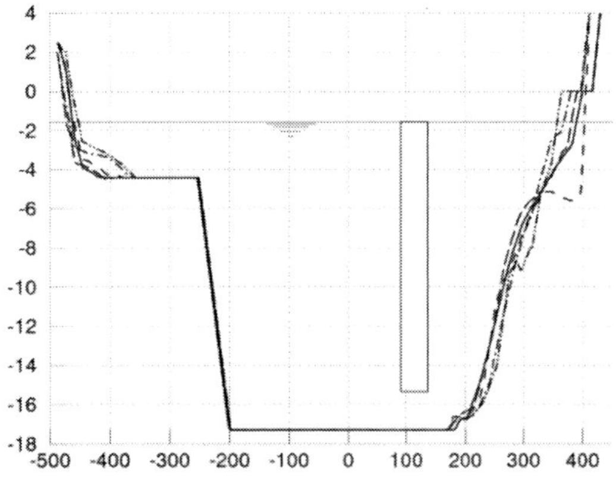

Figure 2: Section form and ship position in the model tests for PPM2 ship; numbers refer to full scale

3. COMPUTATIONAL METHODS

3.1 FINITE-VOLUME RANSE SOLVER

The solution method is based on conservation equations in integral form with appropriate initial and boundary conditions. The solution domain is subdivided into a finite number of control volumes which can be of an arbitrary polyhedral shape and are locally refined in regions of rapid variation of flow variables. The time interval of interest is subdivided into time steps of appropriate size. The governing equations contain surface and volume integrals, as well as time and space derivatives. These are approximated for each control volume and time level using finite-difference approximations, leading to an algebraic equation system.

The flow is assumed to be governed by the Reynolds-averaged Navier-Stokes equations, in which turbulence effects are included via an eddy-viscosity model (k - ε model was used here). The mass conservation equation

$$\frac{\mathrm{d}}{\mathrm{d}t}\int_V \rho \,\mathrm{d}V + \oint_S \rho\vec{v}\cdot\vec{n}\,\mathrm{d}S = 0\,, \tag{1}$$

momentum conservation equation

$$\frac{\mathrm{d}}{\mathrm{d}t}\int_V \rho\vec{v}\,\mathrm{d}V + \oint_S \rho\vec{v}(\vec{v}\cdot\vec{n})\,\mathrm{d}S = \oint_S \left(\hat{T} - p\hat{I}\right)\cdot\vec{n}\,\mathrm{d}S + \int_V \rho\vec{b}\,\mathrm{d}V \tag{2}$$

and two equations for turbulence properties are solved. The latter have the form of a generic transport equation for scalar quantities:

$$\frac{\mathrm{d}}{\mathrm{d}t}\int_V \rho\phi\,\mathrm{d}V + \oint_S \rho\phi\vec{v}\cdot\vec{n}\,\mathrm{d}S = \oint_S \Gamma\nabla\phi\cdot\vec{n}\,\mathrm{d}S + \int_V \rho b_\phi\,\mathrm{d}V\,. \tag{3}$$

In the above equations, ρ stands for fluid density and $\vec{v}$ is the fluid velocity vector; $\vec{n}$ is the unit vector normal to control volume surface with area S and volume V. $\hat{T}$ stands for the stress tensor (expressed in terms of velocity gradients and eddy viscosity), p is the pressure, $\hat{I}$ is the unit tensor, ϕ stands for some scalar variable (volume fraction, k or ε), Γ is the diffusivity coefficient, $\vec{b}$ is the vector of body forces per unit mass and b_ϕ represents sources or sinks of ϕ.

All integrals are approximated by midpoint rule, i.e. the value of the function to be integrated is evaluated at the centre of the integration domain (control volume face centres for surface integrals, control volume centre for volume integrals, time level for time integrals) and then multiplied by the integration range (face area, cell volume or time step, respectively). These approximations are of second-order accuracy for smooth functions, irrespective of the shape of the integration region.

Since variables are computed at control volume centres, interpolation is needed to compute values at face centres. Convective fluxes are computed by blending first-order upwind with second-order central difference scheme.

Free-surface motion is modelled using multiphase description of the water-air flow by solving transport equation (3) without diffusion and source terms for the volume fraction of air. In order to resolve the sharp interface between water and air without numerical instability or excessive numerical diffusion, a high-resolution interface capturing (HRIC) scheme [6] is used for the interpolation of the volume fraction of water.

The solution of the Navier-Stokes equations is accomplished in a segregated iterative manner, in whereby the momentum component equations are solved first using prevailing pressure and mass fluxes through cell faces (inner iterations), followed by solving the pressure-correction equation derived from the continuity equation (SIMPLE-algorithm, more detailed in [2]). Thereafter equations for air volume fraction and turbulence quantities are solved; the sequence is repeated (outer iterations) until all equations are satisfied within a prescribed tolerance, after which the process advances to the next time level.

The equations of ship motions are solved together with the flow equations, allowing for accurate simulation of transient ship motions in the time domain. In order to capture ship motions with respect to the bottom, deforming grids are used. The local grid around the ship keeps unchanged and dynamically follows ship motions over time, while the outer grid portion remains fixed to the bottom and to the non-moveable outer boundary of the computational domain. The connecting grid part between the outer and local grids is deformed over time.

3.2 GL RANKINE

The program *GL Rankine* predicts, besides others, steady and slowly varying flows around floating bodies. The fluid is assumed inviscid, incompressible and irrotational; thus a velocity potential φ exists. It satisfies the Laplace equation (conservation of mass) in the fluid domain and a kinematic boundary condition (no flow through the surface) at the body and channel surfaces as well as at the free surface. At the latter also a dynamic boundary condition (constant pressure) is satisfied:

$\Delta\varphi = 0$ within the fluid domain (4)

$(\nabla\varphi - \vec{U})\cdot\vec{n} = 0$ on the body boundary (5)

$\nabla\varphi\cdot\vec{n} = 0$ on the channel wall (6)

$(\nabla\varphi - \vec{U})\cdot\vec{n} = 0$ on the free surface (7)

$g\zeta = \vec{U}\nabla\varphi - |\nabla\varphi|^2/2$ on the free surface (8)

$\vec{U} = [u,0,0]^T$ is the ship velocity, $\vec{n}$ the surface normal and $\zeta = \zeta(x,y)$ the free surface elevation. In addition, a radiation condition (no incoming waves) has to be satisfied on the free surface.

The boundary of the solution domain is discretised following the boundary element method (BEM). Unstructured triangular grids are used on the submerged ship surface, and block-structured quadrilateral grids are employed on the free surface. The channel can be modelled either directly using triangular panels or employing image sources (for rectangular cross sections).

Rankine point sources are located at points $\vec{\xi}_j$ outside of the fluid at suitable distances from the panel centres. An additional point source is located high above the water surface; it is used to satisfy the condition that the sum of all source strengths is zero. The use of point sources works only in the context of the patch method [10] which satisfies the boundary condition not at collocation points, but on the average over panels.

The conditions at the free surface are nonlinear. Further nonlinearities arise from changes of the waterline along the body and from the dynamic squat. Thus iterations are performed to solve for the source strengths. Within these iterations the body and the free-surface panel mesh are adapted to the current position of the boundaries. Designating the flux through a panel i as

$$r_i = \int_{\Gamma_i}(\nabla\varphi - \vec{U})\cdot\vec{n}\,dS = \sum_{j=1}^{n} q_j \int_{\Gamma_i}\nabla G(\vec{x},\vec{\xi}_j)\vec{n}\,dS - \int_{\Gamma_i}\vec{U}\vec{n}\,dS, \qquad (9)$$

the changes of source strenghts Δq_j are determined from the equations $r_i = 0$ for all panels by a Newton iteration

$$\sum_{j=1}^{n}\frac{dr_i}{dq_j}\Delta q_j + r_i = 0. \qquad (10)$$

After each iteration step, the source strengths are updated as $q_j^* = q_j + \omega\Delta q_j$ with a relaxation parameter $0 < \omega < 1$. On the free surface, the expression dr_i/dq_j is evaluated as $\frac{dr_i}{dq_j} = \frac{\partial r_i}{\partial q_j} + \frac{\partial r_i}{\partial\zeta}\frac{\partial\zeta}{\partial q_j}$ where $\partial r_i/\partial q_j$ designates the derivative for fixed panel heights ζ whereas the second term takes account of the changes of ζ depending on the change of source strength q_j.

In order to fulfil the radiation condition, the sources above the free surface panels are shifted downstream by one panel length, [4]. In regions where a steep forward wave slope is developing the free-surface conditions are modified to account for wave breaking.

After determining the potential, the forces and moments acting on the ship are computed by pressure integration and used to determine dynamic trim and sinkage. Further, the free-surface shape is updated, and the body and free-surface grids are correspondingly adjusted.

3.3 PROGRAM SQUAT

The program computes inviscid flow around a ship in shallow water; the present version considers steady motion and constant water depth without side restrictions. Linear boundary conditions are used on the free surface. The program is based on the slender-body method described in [12] assuming that ship breadth, ship draft and water depth are much smaller than ship length. The method matches an outer with an inner flow field at their common boundary, i.e. the exterior limit of the inner flow field and the interior limit of the outer flow field. In the outer flow field the flow is, asymptotically, confined to the horizontal plane. The free-surface condition is taken into account by satisfying the modified two-dimensional Laplace equation

$$(1 - F_d^2)\varphi_{xx} + \varphi_{yy} = 0 \qquad (11)$$

where $F_d = u/\sqrt{gd}$ is the Froude number based on the water depth d. By scaling the x coordinates this equation can be reduced to the normal Laplace equation. The flow is generated by a distribution of sources along the centre line of the ship. The source density follows from the longitudinal derivative of the transverse section area curve of the ship. The pressure in the inner flow field, which is identical to the pressure at the interior limit of the outer flow field, determines the vertical section force and the trim moment. It determines also the squat and the change of height of the free surface at the interior limit of the outer flow field, which in turn change the immersed section area. Thus iterations are performed to determine the final solution.

4. SENSITIVITY STUDIES

4.1 CHANNEL SECTION FORM INFLUENCE

Figure 3 shows squat at the forward (top) and aft (bottom) perpendiculars depending on the relative speed through water for PPM1. Note that for this ship, the relative speed is the sum of the ship velocity over ground and the tidal current in the river. RANSE results for the asymmetrical flow with a real cross section (O) and for the symmetrical flow with a rectangular cross section of the same area (△) show that the influence of the cross section form on squat is only about 1% for this case because of low blockage effect (the ratio of the cross sectional area of the channel to the mid-ship cross section area is $n \approx 30$).

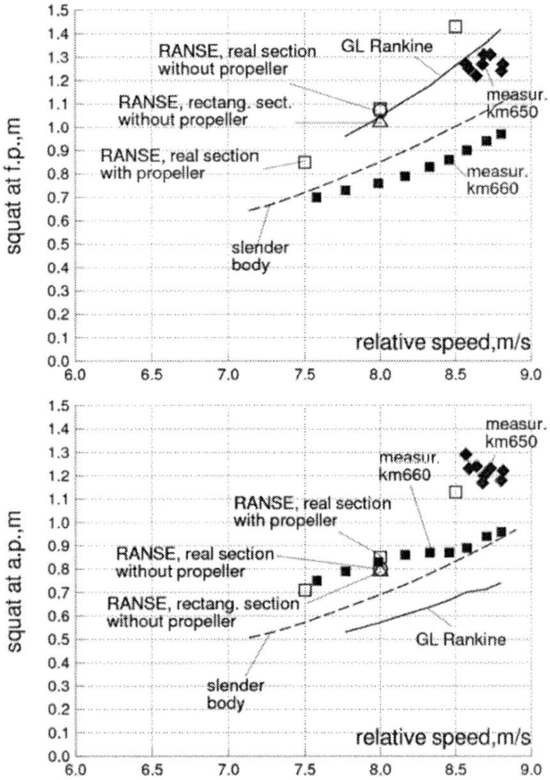

Figure 3: Squat at the forward (top) and aft (bottom) perpendiculars for PPM1 according to full-scale measurements for km650 (◆) and km660 (■), *GL Rankine* (—) and slender-body theory (- - -) computations, and RANSE simulations for real section with (□) and without (O) propeller as well as for a rectangular cross-section without propeller (△)

On the other hand, for the second test case PPM2 with significantly stronger blockage effect ($n \approx 14$), the difference between RANSE results for the asymmetrical flow with the actual cross-section (▲) and a symmetrical flow with rectangular cross-section with the same area (△) is significantly larger as shown in Figure 4: the simplification of the cross section reduces the squat at the forward perpendicular by about 17% and the squat at the aft perpendicular by about 30%.

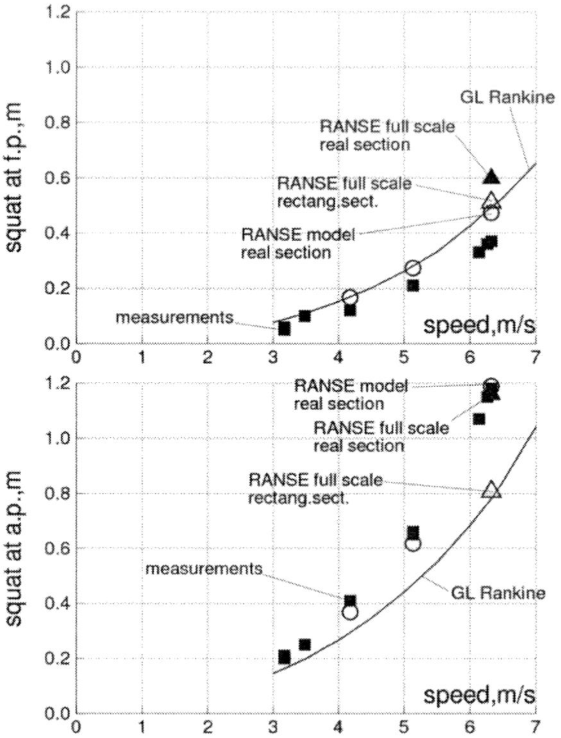

Figure 4: Dynamic squat at the forward (top) and aft (bottom) perpendiculars for PPM2, computed with RANSE for actual section in model scale (O), actual section in full scale (▲) and symmetrical rectangular cross section in full scale (△) in comparison with model test (■) and *GL Rankine* (—) results

4.2 PROPELLER INFLUENCE

Figure 3 compares RANSE simulations for PPM1 with (□) and without (O) propeller. The propeller has no visible effect on the squat at the forward perpendicular and increases the squat at the aft perpendicular by about 6%; the reason for such a small influence may be the relatively large under-keel clearance (UKC) of 4 to 5 m for this case; studies of propeller influence for smaller values of UKC are under way.

4.3 MODEL VS. FULL SCALE

Comparison of RANSE simulations for PPM2 in Figure 4 between full scale (▲) and model (O) shows that squat in full scale is about 27% larger than for the model at the forward perpendicular while almost the same as in the model scale at the aft perpendicular. This means that the ship has a larger average (between the forward and aft perpendiculars) squat and a deeper submerged bow in the full scale compared to the model.

5. VALIDATION

5.1 RANSE SOLVER

The computed squat with RANSE for PPM1 (results for real section with propeller in Figure 3, □) at the forward perpendicular is larger than in the measurements. Squat at the aft perpendicular agrees well with the measurements for km660 for the two lower relative speeds used in the RANSE simulations, 7.5 and 8.0 m/s, and with the measurements for km650 for the largest relative speed 8.5 m/s used. The not understandable difference between the measurements for the stations km650 and km660 does not allow for more certain conclusions about the accuracy of RANSE results.
The computed squat with RANSE for PPM2 (results for real section in model scale in Figure 4, O) agrees very well with the measurements.

5.2 GL RANKINE

Figure 3 shows *GL Rankine* results (solid lines) for PPM1. Symmetrical flow in a channel with rectangular cross-section was used because in this case the influence of the cross-section form is weak. The computed squat at the forward perpendicular is overestimated when compared to the km660 measurements but fits well both RANSE predictions and km650 data. Squat at the aft perpendicular is underestimated by *GL Rankine*; the reason is under study.
Figure 5 compares computed wave ordinates with *GL Rankine* and RANSE for PPM1 at the relative speed of 8.0 m/s for symmetrical flow with rectangular cross-section. The bow wave is slightly under-predicted by *GL Rankine*; still the prediction is in general accurate.

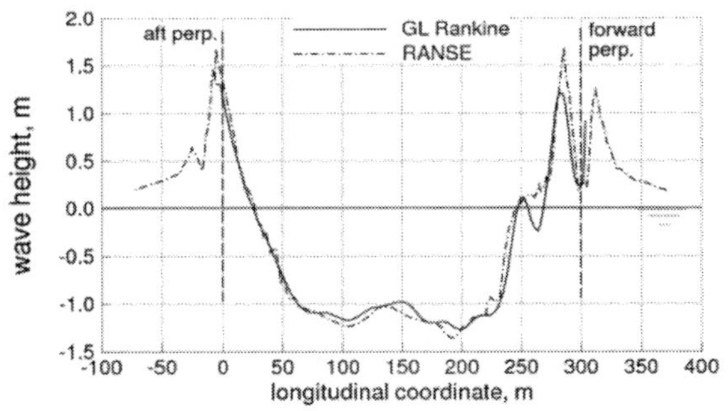

Figure 5: Wave ordinates along the waterline of PPM1 ship computed with *GL Rankine* in comparison with RANSE results

Comparison of *GL Rankine* results with full-scale RANSE simulations for PPM2 for a symmetric flow in a rectangular cross-section (solid lines and empty triangle, respectively, in Figure 4) shows that squat both at the forward and aft perpendiculars agrees very well between these two methods.
Calculations with *GL Rankine* for a real cross-section require significantly larger computational time than for a rectangular cross-section: for the latter problem, mirror images of Rankine sources with respect to the bottom and side walls are used, while for the former, the channel bottom has to be discretised into panels with corresponding Rankine sources in the same way as the ship hull surface. Numerical algorithms for the reduction of computational time discussed in the next section are currently being implemented in order to tackle complex cross-section forms.

5.3 SLENDER-BODY METHOD

Because side restrictions play marginal role for PPM1 measurements at the stations km650 and km660, slender-body theory can be used. Comparison of its results (dashed lines in Figure 3) with RANSE predictions shows the under-prediction of squat by only 20 and 13% respectively at the forward and aft perpendiculars.

6. NUMERICAL EFFICIENCY

Finite-volume solution of RANS equations is the most expensive approach among those used here with respect to both preparation (generation of numerical grids) and carrying out numerical simulations (computational time). Despite significant progress in pre-processing software and numerical solution algorithms in RANSE solvers over the last two decades (iterative solution procedures, multigrid and parallel computing), preparation of the computational setup requires one day to one week and numerical simulation requires two days to one week per speed (on eight processors) for such cases as considered here, depending on the difficulty of the problem. The difficulty depends on the Froude number, symmetry, the size of the computational domain and the under-keel clearance.
Using the slender-body theory is by far the least expensive method both in terms of input preparation (the method requires the distributions of the submerged sectional area and waterline breadth over the ship length) and carrying out computations (computing time is less than 1 s per speed).
Preparation of the input data for *GL Rankine* is rather quick. The most expensive part of the preparation is the hull surface description, which can be read either in IGES- or STL-format. Computing time is short for calculations without channel (about 15 min. per speed) but grows significantly for computations of restricted flows, because

- using mirror images to take into account channel bottom for shallow-water problems doubles the number of Green function evaluations, while taking into account bottom and side walls for rectangular cross-sections increases this number by an order of magnitude

- discretisation of channel with panels for complex cross-sections increases the number of unknowns and corresponding equations up to ten times
- small Froude numbers require very fine discretisation of the free surface: the required panel size decreases with the Froude number squared
- it is necessary to expand the free surface grid about $2L_{pp}$ forward and about L_{pp} behind the ship, which results in a much larger free surface grid compared to computations in unrestricted water

With the usual direct solution methods for algebraic equation systems arising from the discretisation of the boundary conditions, the required computational time grows with second (calculation of the coefficients of the system) to third (solution of the system of equations) power of the number of unknowns. For a small number of unknowns, the calculation of the matrix coefficients dominates the total computational time, while for large number of unknowns the solution of the system requires the most of computational time. Even on modern computers the number of unknowns is limited by about ten or fifteen thousand for dense equation systems, because the required computational time and memory become too large.

In *GL Rankine*, the Fast Multipole Method (FMM) is being implemented to accelerate the computations and to reduce the memory consumption. In this method [3,8], the product of the BEM matrix with an arbitrary vector is approximated to speed up the computation. All sources are clustered in a cluster tree. Boxes on the same level at a certain (large) distance to each other are called well separated. Boxes on the same level which are not well separated are called near neighbours. All evaluation points in boxes which are well separated from the box which includes a certain source point are called the *far field* of this source. All other evaluation points are called the *near field*. The matrix-vector product of the BEM matrix is split up into two parts: one for the near field and one for the far field; the same is done for the evaluation of the potential.

For every box on the finest level, a multipole expansion is computed which describes the potential due to all sources within this box. The expansion is only valid in the far field. Recursively, the multipole expansions of the child nodes in the cluster tree are translated to their parent nodes (upward pass). On the coarsest level, the multipole expansions are converted to local expansions which describe the influence from all sources in the far field on all points within the box. These local expansions are then translated recursively to the finest level (downward pass). This structure is illustrated in comparison with the standard BEM in Figure 6.

At every evaluation point within a box, the potential of the far field can be evaluated using the local expansion. For the evaluation, the local expansion on the coarsest possible level is used. The patch method integrals are calculated using numerical integration formulae to make use of the FMM; the integrals within the near field are solved using analytical formulae. The corresponding coefficients are stored in the so called near field matrix.

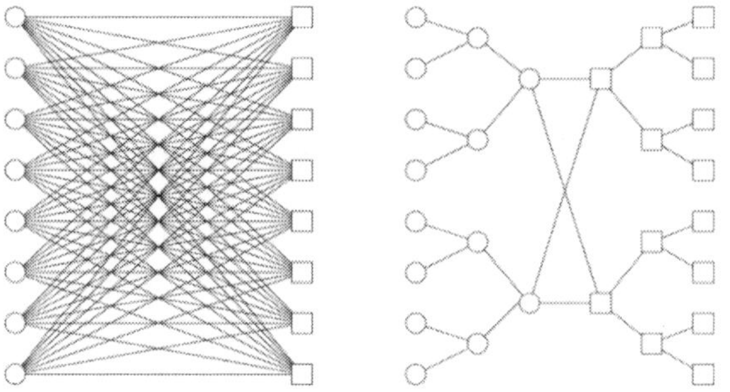

Figure 6: Interaction between sources (○) and panels (□): standard (left) and FMM-accelerated (right) BEM

Since the multipole expansion is an infinite series, it has to be approximated. In the present examples, the first six to eight summands were used.

Due to the hierarchical structure of the cluster tree and the properties of the multipole expansion, the setup of the cluster tree as well as every matrix-vector multiplication have an almost linear time complexity, which might be much faster than standard BEM. Only for small problem dimensions the standard BEM might be faster due to the overhead of the FMM for clustering etc. The memory consumption is almost linear as well.

Since the full BEM matrix is not assembled, only iterative solvers employing fast matrix-vector product can be used. Here, the Krylov subspace method GMRES was used. An incomplete LU factorisation of the near field matrix was used as preconditioner, similar to [9]. This solves the problem of the ill conditioned matrix resulting from the source shifting on the free surface. For the assembly and evaluation of the far field, the library described in [7] is used.

Figure 7 shows computational time per Newton iteration with *GL Rankine* when using the usual direct solver of the algebraic system and the iterative solver with FMM-acceleration vs. the total number of panels on the ship hull and the free surface.

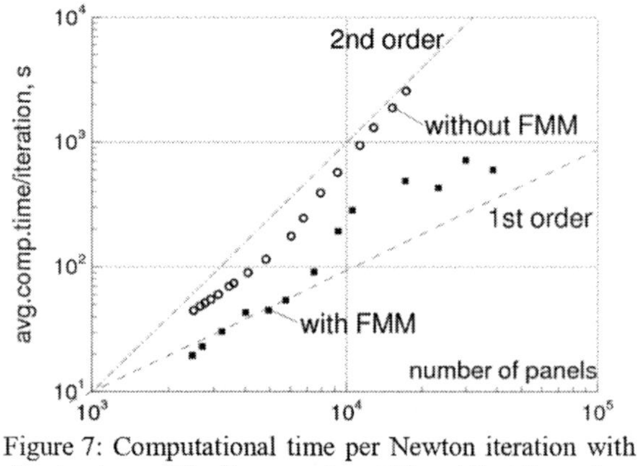

Figure 7: Computational time per Newton iteration with direct solver of the linear system (○) and iterative solver with FMM-acceleration (■) vs. the total number of panels on the ship hull and free surface; lines show the first-(– – –) and second-(– · – · –) order dependencies

For a typical squat computation, about ten thousand panels and about twenty Newton iteration steps are needed to achieve sufficient accuracy.

Figure 8 shows the portions of computational time per Newton iteration required for setting up the matrices of the algebraic equation system, its pre-conditioning and solution together with the total computing time for the FMM-accelerated solver. The total computing time increases with changing rate due to the varying number of GMRES iterations and the varying time consumption of the preconditioner.

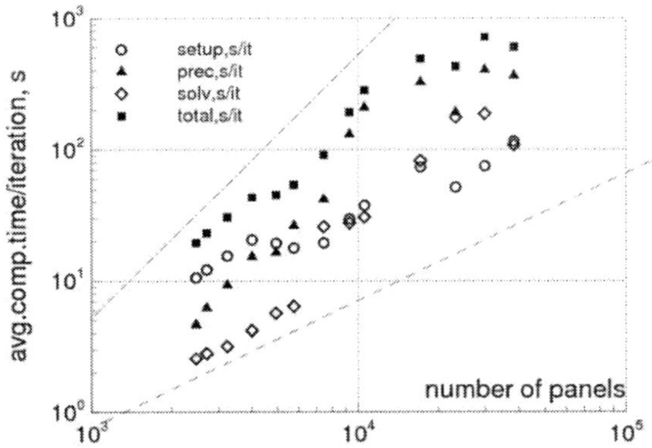

Figure 8: Total computing time per Newton iteration (■) and its components for setting-up equation system (○), its preconditioning (▲) and solution (◇) vs. the total number of panels when using the FMM-accelerated *GL Rankine* method; lines show the first-(– – –) and second-(– · – · –) order dependencies

Figure 9 shows required computational time for the iterative GMRES solution per internal iteration vs. the total number of panels. The nearly linear dependency can be clearly observed.

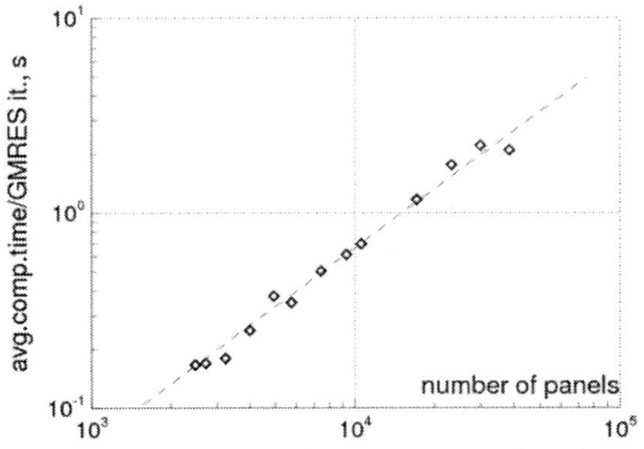

Figure 9: Computational time per GMRES iteration vs. the total number of panels on the ship hull and free surface; line shows first-order dependency

7. CONCLUSIONS

Results of potential flow solution methods *GL Rankine* and *Squat* moderately differ from each other, from the measurements and from RANSE results; their accuracy requires improvement. An encouraging result is the good agreement between *GL Rankine* and RANSE predictions of the wave elevation along the hull. RANSE simulation is the most general approach, showing here the most feasible squat estimations. However, the application of RANSE is also most expensive both in preparation and computations, especially for the computation of ship motions in restricted waterways: here, grid generation and dynamic grid deformation to account for free trim and sinkage are challenging tasks.

The study shows no significant influence of the propeller on squat for relatively large under-keel clearance (4-5 m); propeller influence for smaller values of under-keel clearance is under study.

An alarming result is the difference between squat in model and full scales: RANSE computations show that for the considered case, the average (between the forward and aft perpendiculars) squat in the full scale is larger than for the model, and that the full-scale ship trims more to the bow than the model.

8. ACKNOWLEDGEMENTS

Our sincere gratitude to BAW (Federal Waterways Engineering and Research Institute) Hamburg and personally to Dr. Klemens Uliczka for the fruitful collaboration, insightful discussions and measurement data. Our thanks to Prof. Olaf Steinbach and Dr. Günther Of, Technical University Graz for exciting discussions and their help with the development of the FMM-acceleration methodology.

9. REFERENCES

1. BRIGGS, M., DEBAILLON, P., ULICZKA, K. & DIETZE, W., 'Comparison of PIANC and numerical ship squat predictions for rivers Elbe and Weser', *Proc. 3rd Squat-Workshop: Nautical Aspects of Ship Dynamics, Elsfleth, Germany*, pp. 133-154, 2009

2. FERZIGER, J. H. & PERIC, M., 'Computational Methods for Fluid Dynamics'. – Springer-Verlag, Berlin, 1997

3. GREENGARD, L. & ROKHLIN, V. 'A fast algorithm for particle simulations', *J. Comput. Phys.* **73,** 325-348, 1987

4. JENSEN, G., SÖDING, H. & MI, Z.-X., 'Rankine source methods for numerical solutions of the steady wave resistance problem', *Symp. Naval Hydrodynamics*, Berkeley, 1986

5. MAUSHAKE, C. & JOSWIG, S., 'Messung von Squat, Trimm und Krängung sehr großer Containerschiffe im Rahmen von Grundsatzuntersuchungen auf der Elbe', *Hydrographische Nachrichten* **62**(2) 2004

6. MUZAFERIJA, S. & PERIC, M., 'Computation of free-surface flows using interface tracking and interface-

capturing methods', in *Nonlinear Water Wave Interaction*, pp. 59-100. Comp. Mech. Publ., 1998

7. OF, G., 'BETI-Gebietszerlegungsmethoden mit schnellen Randelementverfahren und Anwendungen', *PhD Thesis*, University Stuttgart, 2006

8. ROKHLIN, V., 'Rapid solution of integral equations of classical potential theory', *J. Comput. Phys.* **60**(2) 187-207, 1985

9. SERTEL, K. & VOLAKIS, J. L., 'Incomplete LU preconditioner for FMM implementation', *Microwave and Optical Technology Letters* **26**(4) 265-267, 2000

10. SÖDING, H., 'A method for accurate force calculation in potential flow', *Ship Technology Research/ Schiffstechnik* **40,** 176-186, 1993

11. SÖDING, H. & BERTRAM, V. 'A 3-d Rankine source seakeeping method', *Ship Technology Research/ Schiffstechnik* **56,** 50-66, 2009

12. TUCK, E., 'Shallow-water flows past slender bodies', *J. Fluid Mech.*, 81-95, 1966

13. ULICZKA, K. & KONDZIELLA, B., 'Dynamic response of very large container ships in extremely shallow water', *Proc., 31st PIANIC Congr.*, Estoril, Portugal, 2006

14. VON GRAEFE, A., SHIGUNOV, V. & ZORN, T., 'Squat computation for a containership with potential and viscous methods', *Proc. 13th Numerical Towing Tank Symposium*, Duisburg, Germany, 2010

9. AUTHORS BIOGRAPHY

Alexander von Graefe is a PhD student at the Institute of Ship Technology, Ocean Engineering and Transport Systems of the University Duisburg-Essen. He works in a joint project together with Germanischer Lloyd, developing and implementing numerical algorithms and software.

Bettar el Moctar is professor at the University of Duisburg-Essen and head of department Fluid Dynamics at Germanischer Lloyd. His publications cover various aspects of ship hydrodynamics

Vladimir Shigunov is a project manager at the Fluid Dynamics department of Germanischer Lloyd. His experience includes development and application of potential theory methods to ship theory problems.

Heinrich Söding is professor emeritus of the University Hamburg-Harburg. He has developed many numerical methods for the prediction of ship resistance and seakeeping, rudder forces etc.

Tobias Zorn is a project manager at the Fluid Dynamics department of Germanischer Lloyd. His experience includes application of viscous flow methods to ship hydrodynamics problems.

HYDRODYNAMIC INTERACTION LOADS BETWEEN TWO SHIPS DURING LIGHTERING OPERATION IN CALM WATER

X Xiang, Centre for Ships and Ocean Structures (CeSOS), Norwegian University of Science and Technology, Norway
R Skejic, MARINTEK, Norway
O M Faltinsen, Centre for Ships and Ocean Structures (CeSOS), Department of Marine Technology, Norwegian University of Science and Technology, Norway
T E Berg, MARINTEK, Norway

SUMMARY

The present paper presents the study of hydrodynamic interaction loads between two tanker types of ships involved in typical underway lightering operation. It was assumed that the execution of complete lightering maneuver takes place in open and deep waters by considering calm water conditions.
The hydrodynamic interaction effects in time domain are estimated by using three dimensional (3D) potential flow theory based method recently developed by Xiang and Faltinsen [1]. The method is further compared against near-field theoretical results given by Newman [2] and Wang [3, 17]. In addition, based on the comparison of interaction forces/moments estimated for a close proximity scenario of two different bodies of identical length/beam ratio the applicability of the near-field theories is discussed. In particular, it is shown that the near-field theories tend to over-predict the interacting loads between slender bodies with blunt ends. Thus, 3D method should be a preferable choice when studying hydrodynamic interaction effects between ships of modern hull forms.
The 3D hydrodynamic interaction loads solver is implemented with the combined seakeeping and maneuvering model by Skejic and Faltinsen [4, 5 and 6] in calm water conditions. A lightering operation is simulated using Aframax and KVLCC (type 2 – Moeri tanker) tanker type of ships on selected drafts. Typical relative positions between the two ships and development of the lightering maneuver were chosen in accordance to Skejic and Berg [7, 8]. Finally, the future requirements and recommendations concerning the qualitative and accurate prediction of the hydrodynamic interaction loads in the real time are discussed from the perspective of typical lightering operations.

1. INTRODUCTION

Lightering operations which are usually consider as ship-to-ship (STS) transfer operations of crude oil and its products involve typically two tanker type of ships of different size [10]. Larger ship is commonly named as ship to be lightered while; the smaller one is called the lightering ship.

Most of the lightering operations are carried out in open waters areas. Depending on the nature of operation, lightering can be performed when two ships are stationary at anchor, moored together while, drifting or moving underway in calm water or under influence of moderate environmental condition (current, waves, wind).

Focus of the present study is on an investigation of underway lightering operation. Since the involved ships operate at close vicinity of each other it is normally to be expected that their maneuvering performances will be affected on a way significantly different from the case of a maneuvering single ship. The reason for this is presence of the hydrodynamic interaction effects which may initiate an unexpected and sudden deviation in the desired course of one or both of the ships. This on the other hand may compromise the navigational safety and effectiveness of operation due to increased collision hazard.

The abeam sequence in typical underway lightering operation of involved ships is shown in Figure 1. The same figure shows guidance [10] how to execute the

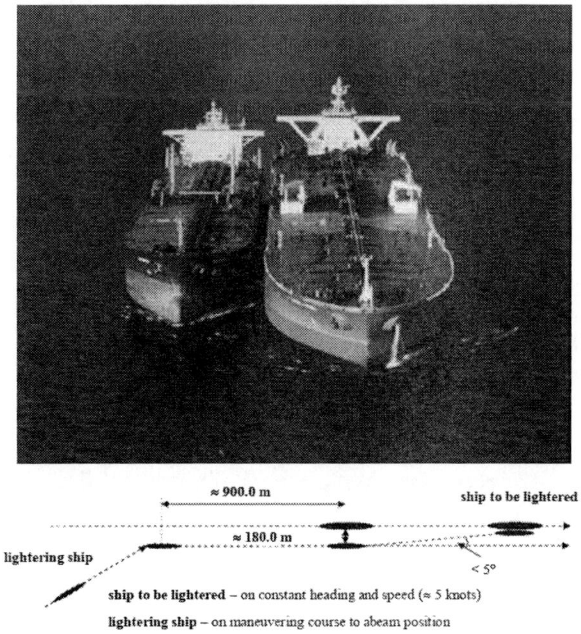

Figure 1: Typical abeam configuration during underway lightering operation [10] of the lightering ship and ship to be lightered. Courtesy of Skaugen PetroTrans Inc.

complete approach phase of the maneuver satisfying at the same time minimum collision risk requirement. As can be seen, ship to be lightered (usually larger) approaches to lightering by keeping the steady speed (< 5.0 knots) and heading. The lightering ship (usually smaller) starts to approach a ship to be lightered from starboard stern side on distance around 900.0 m. As time goes on both ships reach their abeam position by matching their speeds and transversal and longitudinal distance. Transfer of liquefied cargo can begin after both advancing ships are moored together, by using fenders (usually 6 rubber bumpers) which ensure desired lateral (transverse) separation up to 5.0 m in diameter.

Current work numerically investigates the underway lightering maneuver of two ships in calm and deep water. Both ships are advancing on specified preselected course as shown in Figure 2.

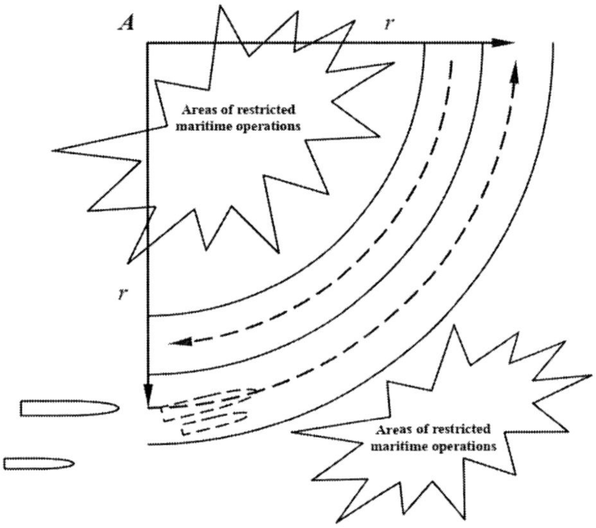

Figure 2: Investigated case scenario of underway lightering operation in calm and deep water.

It has been assumed for this study that the underway lightering operation should take place between the areas of restricted maritime operations following predefined part of circular paths by the maritime authorities (see Figure 2.). Furthermore, the radius r of shown part of the circular path with the origin at point A is defined to be of significantly larger magnitude than the length of a larger ship involved in the lightering operation. On this way, the effects of heading change on maneuvering capabilities on both ships is negligible and, what is of main concern, sustainable by the ship to be lightered which, by the way, leads the underway lightering formation.

Our modeling of the underway lightering operation described on Figure 2 is carried out in a way that mimics as much as possible, realistic situations which may happen under imposed restrictions and requirements. For that purpose, a unified seakeeping and maneuvering model based on modular concept by Skejic and Faltinsen [4, 5], and Skejic [6] has been used. The hydrodynamic interaction loads are obtained by the method of Xiang and Faltinsen [1]. The method is capable of predicting the hydrodynamic interaction loads between two arbitrary bodies in infinite fluid with Froude numbers Fn lower than ~0.2, in order for the steady wave generation to be secondary and a rigid free-surface condition can be assumed in a potential-flow formulation. The mentioned theory is based on potential flow; i.e., the viscous effects are not accounted for. The method is further discussed in section 3 of this study. In addition, the model includes a motion control system by Skejic et al. [11] employed on both advancing ships. This way, the subjective decisions of human operators which may lead to collision situation are completely eliminated. Further this means that the investigated underway lightering operations is carried out under imposed requirement of zero collision risk tolerance.

In complex STS operations, such as this one considered here, the present practice from the perspective of navigational safety requires adequate theoretical and practical knowledge from the personnel involved in operation. This is commonly achieved through theoretical knowledge and training activities performed in marine simulators such is for instance the Ship Maneuvering Simulator Centre (SMSC) in Trondheim, Norway. From this point of view, it is believed that our theoretical study contributes to better understanding of physical nature of the hydrodynamic interaction effects which can further improve marine simulators related to STS operations.

The time domain simulation of underway lightering maneuver in calm and deep water is shown in section 4 of this paper. The presented set of maneuvering parameters on both ships is given on a way which will allow easy recognition of critical and potentially dangerous phases for both ships from a navigational safety point of view.

2. MATHEMATICAL MODEL

A unified seakeeping and maneuvering model by Skejic and Faltinsen [4, 5] and Skejic [6] is applied to analyze the underway lightering maneuver of two ships advancing on a general course in calm and deep water. Since the present study is focused on an investigation of maneuvering behavior of two advancing ships in calm water the estimation of the wave field effects is correspondingly set to zero, i.e. neglected.

The Cartesian right-handed maneuvering body-fixed coordinate systems $O_{Ci}x_{Ci}y_{Ci}$ shown in Figure 3 with positive z_{Ci} – axis pointing upwards are used to analyze the underway lightening maneuver between the lightering ship A and the ship B which needs to be lightered. The index $i \in [A, B]$ is used to represent both ships respectively. The centre of gravity (COG) on each

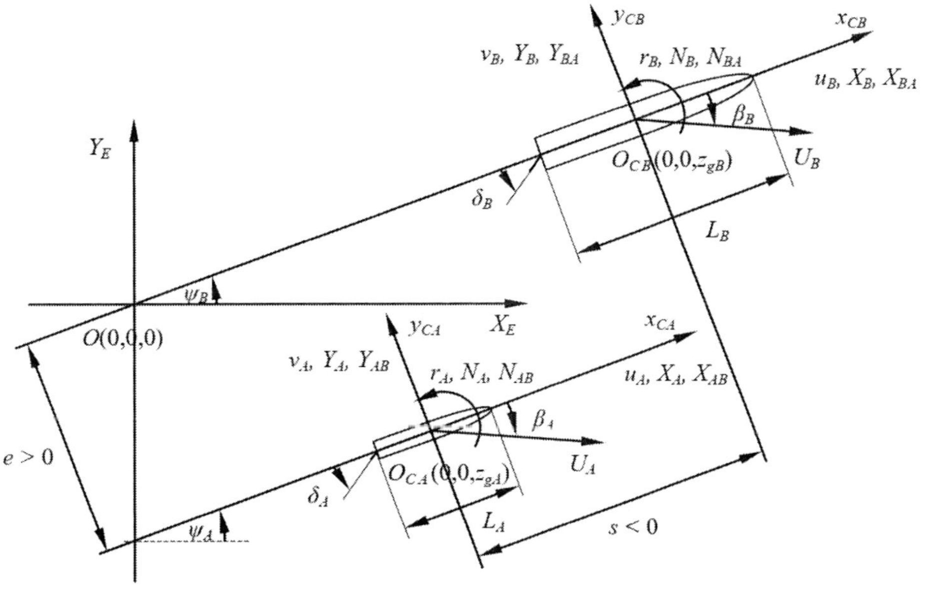

Figure 3: Coordinate systems on lightering ship (ship A) and ship to be lightered (ship B).

ship is in the origin O_{Ci} of the coordinate system $O_{Ci}x_{Ci}y_{Ci}$ in the lateral symmetry plane of each body. $OX_EY_EZ_E$ is an Earth-fixed coordinate system with positive Z_E – axis pointing upwards. X_i and Y_i are hydrodynamic forces defined positively in the positive x_{Ci} and y_{Ci} directions, respectively. N_i denotes the yaw moment about the z_{Ci} – axis, defined positively as shown in Figure 3. u_i and v_i are x_{Ci} and y_{Ci} – components of the total ships' speed $U_i = (u_i^2 + v_i^2)^{0.5}$. The yaw angular speed is defined by r_i. ψ_i, δ_i and $\beta_i = \text{atan}(-v_i/u_i)$ are respectively the yaw (heading), rudder and drift angles from the maneuvering analysis. The longitudinal and transverse distance between the centres of gravity (COG) of the ships (O_{CA} and O_{CB}) are s and e, respectively. s is negatively defined for the initial time step t = 0 s, while e is defined in respect to coordinate system of ship A. The position of ships and ship lengths between perpendiculars, defined with L_i for $i \in$[A, B], on Figure 3 are arbitrary.

The maneuvering module within the unified theory mathematical model is given by the equation system (1) which is consistent with the ITTC [12] nomenclature and conforms surge-sway-yaw version of 6 DOF nonlinear maneuvering equation system given by Imlay [13].

The formulation of the maneuvering model is carried out by using a modular concept. The forces/moments in (1) due to rudder (subscript R), resistance R and propulsion T, and nonlinear viscous cross-flow (subscript CF) form the separate modules where details about each module are further explained by Skejic and Faltinsen [4, 5] and Skejic [6].

The equation system (1) is presented in the matrix form where the subscript A and B denote the ship to be lightered and lightering ship, respectively (see Figure 3).

$$\begin{bmatrix} \mathbf{M}^A_{3x3} & \mathbf{0}_{3x3} \\ \mathbf{0}_{3x3} & \mathbf{M}^B_{3x3} \end{bmatrix} \begin{bmatrix} \dot u_A \\ \dot v_A \\ \dot r_A \\ \dot u_B \\ \dot v_B \\ \dot r_B \end{bmatrix} + \begin{bmatrix} \mathbf{K}^A_{3x3} & \mathbf{0}_{3x3} \\ \mathbf{0}_{3x3} & \mathbf{K}^B_{3x3} \end{bmatrix} \begin{bmatrix} u_A \\ v_A \\ r_A \\ u_B \\ v_B \\ r_B \end{bmatrix} = \begin{bmatrix} \mathbf{A}^A_{3x3} & \mathbf{0}_{3x3} \\ \mathbf{0}_{3x3} & \mathbf{A}^B_{3x3} \end{bmatrix} \begin{bmatrix} \dot u_A \\ \dot v_A \\ \dot r_A \\ \dot u_B \\ \dot v_B \\ \dot r_B \end{bmatrix}$$

$$+ \begin{bmatrix} \mathbf{B}^A_{3x3} & \mathbf{0}_{3x3} \\ \mathbf{0}_{3x3} & \mathbf{B}^B_{3x3} \end{bmatrix} \begin{bmatrix} u_A \\ v_A \\ r_A \\ u_B \\ v_B \\ r_B \end{bmatrix} + \begin{bmatrix} \mathbf{S}^A_{3x3} & \mathbf{0}_{3x3} \\ \mathbf{0}_{3x3} & \mathbf{S}^B_{3x3} \end{bmatrix} \begin{bmatrix} u_A \\ v_A \\ r_A \\ u_B \\ v_B \\ r_B \end{bmatrix} + \begin{bmatrix} X_{R_A} \\ Y_{R_A} \\ N_{R_A} \\ X_{R_B} \\ Y_{R_B} \\ N_{R_B} \end{bmatrix} + \begin{bmatrix} -R(u_A)+(1-t_A)T(u_A) \\ 0 \\ 0 \\ -R(u_B)+(1-t_B)T(u_B) \\ 0 \\ 0 \end{bmatrix}$$

$$+ \begin{bmatrix} 0 \\ Y_{CF_A} \\ N_{CF_A} \\ 0 \\ Y_{CF_B} \\ N_{CF_B} \end{bmatrix} + [INT]_{6x1} + \begin{bmatrix} -X_{\dot u}|_A \dot u_A + C_{TN_A} Y_{\dot v}|_A r_A v_A + Y_{\dot r}|_A r_A r_A \\ -Y_{\dot v}|_A \dot v_A - Y_{\dot r}|_A \dot r_A - X_{\dot u}|_A u_A r_A \\ -N_{\dot v}|_A \dot v_A - N_{\dot r}|_A \dot r_A + X_{\dot u}|_A u_A r_A \\ -X_{\dot u}|_B \dot u_B + C_{TN_B} Y_{\dot v}|_B r_B v_B + Y_{\dot r}|_B r_B r_B \\ -Y_{\dot v}|_B \dot v_B - Y_{\dot r}|_B \dot r_B - X_{\dot u}|_B u_B r_B \\ -N_{\dot v}|_B \dot v_B - N_{\dot r}|_B \dot r_B + X_{\dot u}|_B u_B r_B \end{bmatrix} \quad (1).$$

The ships mass and moment of inertia are defined as M and I_{66} and present in the matrices **M** and **K** as follows

$$\mathbf{M}^A_{3x3} = \begin{bmatrix} M & 0 & 0 \\ 0 & M & 0 \\ 0 & 0 & I_{66} \end{bmatrix}_A \qquad \mathbf{M}^B_{3x3} = \begin{bmatrix} M & 0 & 0 \\ 0 & M & 0 \\ 0 & 0 & I_{66} \end{bmatrix}_B$$

$$\mathbf{K}^A_{3x3} = \begin{bmatrix} 0 & -Mr_A & 0 \\ 0 & 0 & Mu_A \\ 0 & 0 & 0 \end{bmatrix}_A \qquad \mathbf{K}^B_{3x3} = \begin{bmatrix} 0 & -Mr_B & 0 \\ 0 & 0 & Mu_B \\ 0 & 0 & 0 \end{bmatrix}_B .$$

The maneuvering derivatives, i.e. the added mass and damping coefficients are expressed in the following matrices

$$\mathbf{A}_{3x3}^{A}=\begin{bmatrix} X_{\dot u} & 0 & 0 \\ 0 & Y_{\dot v} & Y_{\dot r} \\ 0 & N_{\dot v} & N_{\dot r} \end{bmatrix}_A \qquad \mathbf{A}_{3x3}^{B}=\begin{bmatrix} X_{\dot u} & 0 & 0 \\ 0 & Y_{\dot v} & Y_{\dot r} \\ 0 & N_{\dot v} & N_{\dot r} \end{bmatrix}_B$$

$$\mathbf{B}_{3x3}^{A}=\begin{bmatrix} 0 & 0 & 0 \\ 0 & Y_v & Y_r \\ 0 & N_v & N_r \end{bmatrix}_A \qquad \mathbf{B}_{3x3}^{B}=\begin{bmatrix} 0 & 0 & 0 \\ 0 & Y_v & Y_r \\ 0 & N_v & N_r \end{bmatrix}_B$$

$$\mathbf{S}_{3x3}^{A}=\begin{bmatrix} 0 & -C_{TN_A}Y_{\dot v}r_A & -Y_{\dot r}r_A \\ 0 & 0 & X_{\dot u}u_A \\ 0 & -X_{\dot u}u_A & 0 \end{bmatrix}_A \qquad \mathbf{S}_{3x3}^{B}=\begin{bmatrix} 0 & -C_{TN_B}Y_{\dot v}r_B & -Y_{\dot r}r_B \\ 0 & 0 & X_{\dot u}u_B \\ 0 & -X_{\dot u}u_B & 0 \end{bmatrix}_B .$$

Furthermore, the zero matrix **0** and the hydrodynamic interaction loads matrix **INT** are defined as

$$\mathbf{0}_{3x3}=\begin{bmatrix} 0 & 0 & 0 \\ 0 & 0 & 0 \\ 0 & 0 & 0 \end{bmatrix} \qquad [INT]_{6x1}=\begin{bmatrix} X_{AB} \\ Y_{AB} \\ N_{AB} \\ X_{BA} \\ Y_{BA} \\ N_{BA} \end{bmatrix}=\begin{bmatrix} F_1^{(1)} \\ F_2^{(1)} \\ L_3^{(1)} \\ F_1^{(2)} \\ F_2^{(2)} \\ L_3^{(2)} \end{bmatrix}=\begin{bmatrix} X_{NLH}^{(1)} \\ Y_{NLH}^{(1)} \\ N_{NLH}^{(1)} \\ X_{NLH}^{(2)} \\ Y_{NLH}^{(2)} \\ N_{NLH}^{(2)} \end{bmatrix} .$$

The additional details related to the hydrodynamic interaction loads, i.e. X_{AB}, Y_{AB}, N_{AB}, X_{BA}, Y_{BA} and N_{BA} terms are discussed in subsection 3 with further explanations given in Xiang and Faltinsen [1]. The positive direction of the interaction loads is defined with the arrows as indicated in Figure 3.

In addition, it should be noted that the nondimensional coefficients C_{TN_A} and C_{TN_B} which modify the term $Y_{\dot v}r$ what potential flow theory predicts at moderate speeds take value of 1.0 in the present simulation of underway lightering maneuver since both ships are moving mainly in a joint parallel configuration.

The maneuvering model given with equation (1) requires the evaluation of the maneuvering coefficients (referenced to the maneuvering body-fixed coordinate systems shown on Figure 3) associated with the velocities and accelerations. In particular, the evaluation of maneuvering added mass coefficients is based on the limiting values of coefficients for encounter frequency ω_e = 0 s^{-1} and obtained from Xiang and Faltinsen [1] method. In case of ships with transom stern, the maneuvering damping derivatives (two dimensional (2D) lifting terms, see Table 1) for the same limiting value of encounter frequency ω_e are evaluated by using the strip theory of Salvesen-Tuck-Faltinsen [14].

$X_{\dot u}=-A_{11}$	$Y_v=-ua_{22T}$
$Y_{\dot v}=-A_{22}$	$Y_r=-ux^{(T)}a_{22T}$
$Y_{\dot r}=-A_{26}$	$N_v=-uA_{22}-ux^{(T)}a_{22T}$
$N_{\dot v}=-A_{62}$	
$N_{\dot r}=-A_{66}$	$N_r=-uA_{62}-u\left[x^{(T)}\right]^2 a_{22T}$

Table 1: The maneuvering added mass and damping derivatives.

The maneuvering equation system (1) is solved with a time integration algorithm based on the explicit Runge-Kutta fourth-order method (RK-IV) with constant time steps.

The vessels positions in the Earth-fixed coordinate system $OX_EY_EZ_E$ (see Figure 3) can be obtained by transformation of the velocities $(u, v, r)_i$, which are defined with respect to the maneuvering ship-fixed coordinate systems $O_{Ci}x_{Ci}y_{Ci}$, into the velocities $\left(\dot X_E, \dot Y_E, \dot\Psi_E\right)_i$, which are defined with respect to the Earth-fixed coordinate system $OX_EY_EZ_E$. The transformation is carried out by using the finite Euler angles matrices and can be expressed as

$$\left[\dot X_E \quad \dot Y_E \quad \dot\Psi_E\right]_i^T = J_{3\times3_i}\left[u \quad v \quad r\right]_i^T \tag{2}$$

where dot means the time derivative d/dt, while $J_{3\times3_i}$ are the Euler angle matrices given as

$$J_{3\times3_i}=\begin{bmatrix} \cos\psi & -\sin\psi & 0 \\ \sin\psi & \cos\psi & 0 \\ 0 & 0 & 1 \end{bmatrix}_i \tag{3}$$

for $i\in$[A, B]. From (2) it follows that the vessels positions in the Earth-fixed coordinate system $OX_EY_EZ_E$ can be found from

$$\left[X_E \quad Y_E \quad \Psi_E\right]_i^T=\left[X_E^0 \quad Y_E^0 \quad \Psi_E^0\right]_i^T+\left[\int_0^t \dot X_E dt \quad \int_0^t \dot Y_E dt \quad \int_0^t \dot\Psi_E dt\right]_i^T \tag{4}$$

where superscript '0' means initial value for time t = 0 s.

2.1 MOTION CONTROL MODULE

The underway lightering maneuver of two advancing ships engaged on general course in calm water requires modeling of the speed and rudder controller on both ships involved in the maneuver. This is done through application of the motion control module by Skejic et al. [11]. In short, the module ensures fulfillment of operator preselected parameters:

- desired transverse distance e,
- desired relative longitudinal distance s,
- desired heading angle ψ_d,
- zero relative speed.

The complete control system is acting as a unit and it can be shortly described as follows. Requested transverse (lateral) distance e is given as an input to the steering module, which ensures transverse alignment. In the same time requested longitudinal distance s is given as an input to the synchronization module, which computes the speed control reference signals that ensure longitudinal alignment. Further particular details about the implemented system can be found in Skejic et al. [11].

3. HYDRODYNAMIC INTERACTION LOADS IN CALM AND DEEP WATER

The calm water non-lifting hydrodynamics interaction loads are calculated based on the 3D boundary element method (BEM) method given in Xiang and Faltinsen [1], which studied an overtaking maneuver between one 'MARINER' and a scaled 'MARINER' type of a ships. The hydrodynamic loads given by **INT** matrix in the previous section (section 2) are derived in maneuvering body-fixed coordinate systems of two bodies, thus they account for the hydrodynamic loads on two interacting bodies in arbitrary motions in the horizontal plane (surge, sway and yaw):

$$X_{NLH}^{(1)} = -\frac{d}{dt}\left(\sum_{j=1}^{6} A_{1j}u_j\right) + u_3\sum_{j=1}^{6} A_{2j}u_j - \rho\iint_{S_H^{(1)}}(0.5\boldsymbol{V}\cdot\boldsymbol{V}n_1 - V_1V_n)dS$$

$$Y_{NLH}^{(1)} = -\frac{d}{dt}\left(\sum_{j=1}^{6} A_{2j}u_j\right) - u_3\sum_{j=1}^{6} A_{1j}u_j - \rho\iint_{S_H^{(1)}}(0.5\boldsymbol{V}\cdot\boldsymbol{V}n_2 - V_2V_n)dS$$

$$N_{NLH}^{(1)} = u_2\sum_{j=1}^{6} A_{1j}u_j - u_1\sum_{j=1}^{6} A_{2j}u_j - \frac{d}{dt}\left(\sum_{j=1}^{6} A_{3j}u_j\right) - \rho\iint_{S_H^{(1)}}[0.5\boldsymbol{V}\cdot\boldsymbol{V}n_3 - (x_1V_2 - y_1V_1)V_n]dS \quad (5)$$

$$X_{NLH}^{(2)} = -\frac{d}{dt}\left(\sum_{j=1}^{6} A_{4j}u_j\right) + u_6\sum_{j=1}^{6} A_{5j}u_j - \rho\iint_{S_H^{(2)}}(0.5\boldsymbol{V}\cdot\boldsymbol{V}n_4 - V_1V_n)dS$$

$$Y_{NLH}^{(2)} = -\frac{d}{dt}\left(\sum_{j=1}^{6} A_{5j}u_j\right) - u_6\sum_{j=1}^{6} A_{4j}u_j - \rho\iint_{S_H^{(2)}}(0.5\boldsymbol{V}\cdot\boldsymbol{V}n_5 - V_2V_n)dS$$

$$N_{NLH}^{(2)} = u_5\sum_{j=1}^{6} A_{4j}u_j - u_4\sum_{j=1}^{6} A_{5j}u_j - \frac{d}{dt}\left(\sum_{j=1}^{6} A_{6j}u_j\right) - \rho\iint_{S_H^{(2)}}[0.5\boldsymbol{V}\cdot\boldsymbol{V}n_6 - (x_2V_2 - y_2V_1)V_n]dS \quad (6)$$

In expressions (5) and (6) for body 1 (A) and body 2 (B) (see Figure 3) the hydrodynamic interaction is considered via the fluid potential. The ship velocities used in the expressions are in correspondence with the maneuvering model (1) as follows: $u_1 = u_A$, $u_2 = v_A$, $u_3 = r_A$ and $u_4 = u_B$, $u_5 = v_B$, $u_6 = r_B$. Please refer to the **INT** matrix for the correspondence of notations for forces/moments. $\boldsymbol{V}$ is the total fluid velocity vector, V_i (i = 1, 2, 3) the components and V_n is the normal fluid velocity on the ship hull. The n_i (i = 1,…, 6) are the unit normal vector components on the bodies in correspondence with six body motion modes of two ships, i.e., i=1,2,3 for surge,sway and yaw motion modes of ship (body) 1 and i.e., i=4,5,6 for surge,sway and yaw motion modes of ship (body) 2, Xiang and Faltinsen [1]. The additional details related to given expressions can be found in Xiang and Faltinsen [1]. One should notice that in the original expressions given in [1] there was a missing V_n term in the integral part of moments.

In our early studies (ibid.) we have shown that our 3D method agrees well with the near-field (Wang, [3]) and far-field (Tuck and Newman, [9]) slender body theories when two non-lifting slender bodies are efficiently near and far from each other. The studied body surfaces are defined by the x_I – dependent radius $r_I(x_I) = R_I[1 - (x_I/L_I)^2]$ for $|x_I| \le L_I$ where $R_I(I = 1, 2)$ are the maximum radius of the bodies and $2L_I$ are the body lengths. In the following analysis we identify this body as slender body A – type. Now we introduce a slender body B – type whose body surface is defined as $r_I(x_I) = R_I[1 - (x_I/L_I)^2]^{0.5}$ and meaning of notation is defined in the same way as for body A – type. The slenderness parameter of the considered bodies is given as $\varepsilon = R_I/(2L_I)$. Both body types are with pointed ends (no transom stern) and the longitudinal and transverse distances between their centers of gravity is denoted with x and $2p$, corresponding to e and s in the maneuvering model, respectively. In the following examples the slenderness parameter for both body B – type bodies is chosen the same as body A-type studied in Xiang and Faltinsen [1], i.e., $\varepsilon = 0.05$. The illustration of the two types of slender bodies is given in Figure 4 and Figure 5.

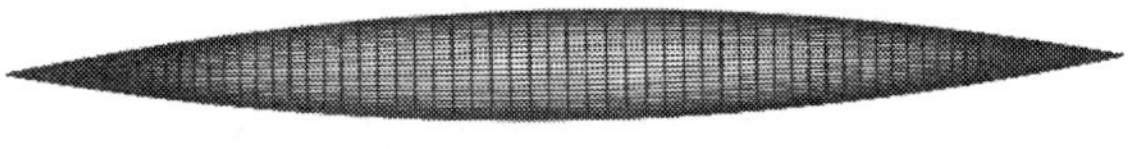

Figure 4: Slender body A – type studied earlier in Xiang and Faltinsen[1], slender body with sharp point ends.

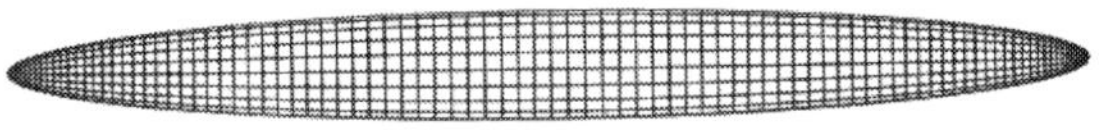

Figure 5: Slender body B – type studied in the present paper, slender body with relatively blunt, but point ends.

The first case (see Figure 6) examines two identical B – type axis-symmetric bodies with no stagger (x = 0) that advance with the same constant speed $u_1 = u_4$. Because the bodies have pointed ends, there is no lifting effect. We consider the transverse force $Y_{NLH}^{(1)}$. The force comes from the term with a surface integral over the wetted hull surface $S_H^{(1)}$ which expresses a suction force between the two bodies. Our predictions are compared with the analytical results from Newman [2] or Wang [17] based on near-field slender body theory (SBT). The non-dimensional transverse force $Y_{NLH}^{(1)}/[\rho u_1^2(2L_1)^2\varepsilon^3]$ on body 1 is presented as function of non-dimensional transverse distance p/R_1 between the two body axes. The similar case was presented in Xiang and Faltinsen [1] for two identical A – type slender bodies with good agreement to the near-field slender body theory when two bodies are close. However, the agreement for B – type slender bodies are not satisfactory.

This is explained as follows. The small slenderness parameter ε cannot assure that the basic assumption of slender body theory is always strictly fulfilled on the present B – type slender bodies. Classic slender body

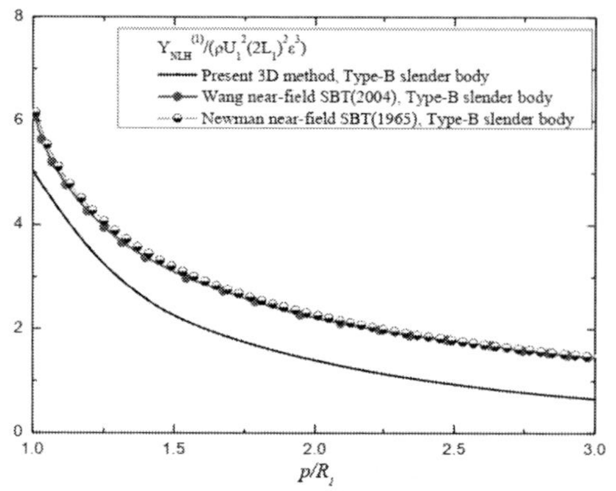

Figure 6: Lateral force $Y_{NLH}^{(1)}$ acting on one of two identical B-type slender bodies with no stagger advancing with constant forward speed $u_1 = u_4 = U_1$ versus non-dimensional lateral distance between the bodies. The slenderness parameter $\varepsilon = 0.05$. Calculated values by present 3D method, the near-field results by Newman [2] and Wang [17].

theory (Salvesen et al. [14], Faltinsen, [15]) requires that $n_x << n_y$, n_z, along the whole body, but this is not strictly satisfied by slender bodies with blunt ends, resulting in the so called 'end effect' in slender body theory (Newman, [2]). A comparison can be made for the sectional load distribution of an A – type, B – type slender bodies applying formula given in Newman [2]. His formula for the force shows that a major contribution comes from the body ends for B-type slender bodies. Because the blunt ends of the B-type slender bodies do not satisfy $n_x << n_y$, n_z, , we can understand why the slender-body theory fails. Further, due to the fact that the slender-body theory overpredicts the force contribution from ends we can understand why we obtain lower values with a 3D theory.

In the second case we study the same two identical type-B axis – symmetric slender bodies. However, body 1 is stationary (moored) while body 2 advances with constant speed u_4. The non-dimensional transverse distance between the two bodies is $p/R_1 = 1.5$, i.e., they are very close to each other. The non-dimensional transverse force $Y_{NLH}^{(1)}/[\rho u_4^2(2L_1)^2\varepsilon^3]$ and yaw moment $N_{NLH}^{(1)}/[\rho u_4^2(2L_1)^3\varepsilon^3]$ are presented as a function of the non-dimensional stagger x/L_1 in Figure 7.

Our results are compared with the near-field slender body theory results by Wang [3]. One can see on the figure that the near-field theory gives higher prediction of force and moments than the present 3D method, for the B – type slender body. This is consistent with figure 6. It also shows that the near-field results demonstrate obvious 2D effect because the interaction forces and moments exist only in the range where two bodies have sections in interaction transversely. This is a shortage compared with 3D method and the far-field slender body theory.

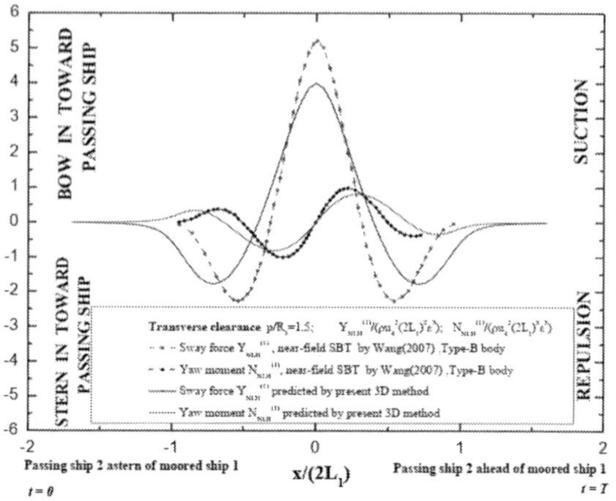

Figure 7: Calm-water results for the interaction sway force and yaw moment on the moored slender body 1 due to passage of an identical slender body 2. Calculated values by present 3D method, the near-field results by Wang [3].

4. NUMERICAL STUDY

The case study performed in the present work concerns underway lightering operation in calm and deep water. The 'Aframax' and 'KVLCC2' (Moeri tanker) [16] tanker type of ships with the main particulars according to Figure 8 are used.

	'KVLCC2'	'Aframax'
Hull particulars		
Length between perpendiculars, L_{PP}	320.0 m	230.0 m
Beam, B	58.0 m	42.0 m
Draft (even keel), T	20.8 m (full load)	7.5 m (ballast)
Block coefficient, C_B	0.8097	0.78
Displacement	312571.2 t	53251.1 t
Centre of gravity, COG (x_g, y_g, z_g)	(0.0, 0.0, -8.06) m	(0.0, 0.0, -2.88) m
Centre of buoyancy, COB (x_b, y_b, z_b)	(0.0, 0.0, -10.87) m	(0.0, 0.0, -3.6) m
Metacentric height (transverse), G_{MT}	24.312 m	17.59 m
Metacentric height (longitudinal), G_{ML}	387.402 m	442.49 m
Radius of gyration, roll k_{xx}	17.66 m	12.6 m
Radius of gyration, pitch/yaw k_{yy}/k_{zz}	83.38 m	56.35 m
Nominal (design) ship speed, U	16 knots	16 knots

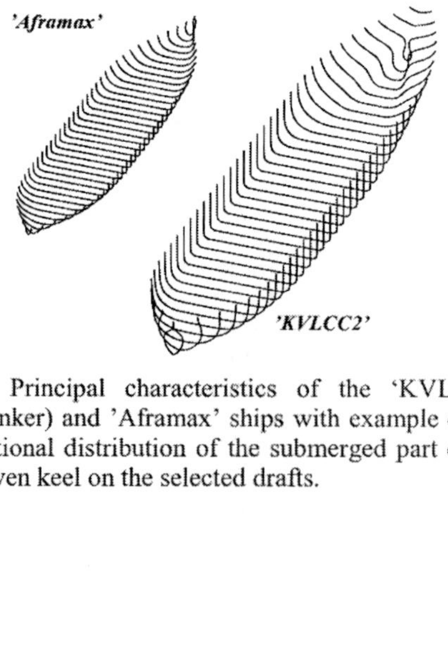

Figure 8: Principal characteristics of the 'KVLCC2' (Moeri tanker) and 'Aframax' ships with example of the cross sectional distribution of the submerged part of the hulls at even keel on the selected drafts.

Figure 9 shows ships trajectories and main maneuvering parameters on above mentioned ships involved in theoretically investigated underway lightering maneuver. The maneuver is performed on a part of circular preselected path with origin at point A (1100.0 m, 10400 m) characterized by the radius r = 10000.0 m and desired transient port headings y_d. The starting position of the lightering ship A is at (0.0 m, – 200.0 m) while, for the ship to be lightered B is (800.0 m, 400.0 m) in the Earth-fixed coordinate system $OX_EY_EZ_E$ (see Figure 9). The motion control system ('autopilots', see subsection 2.1)

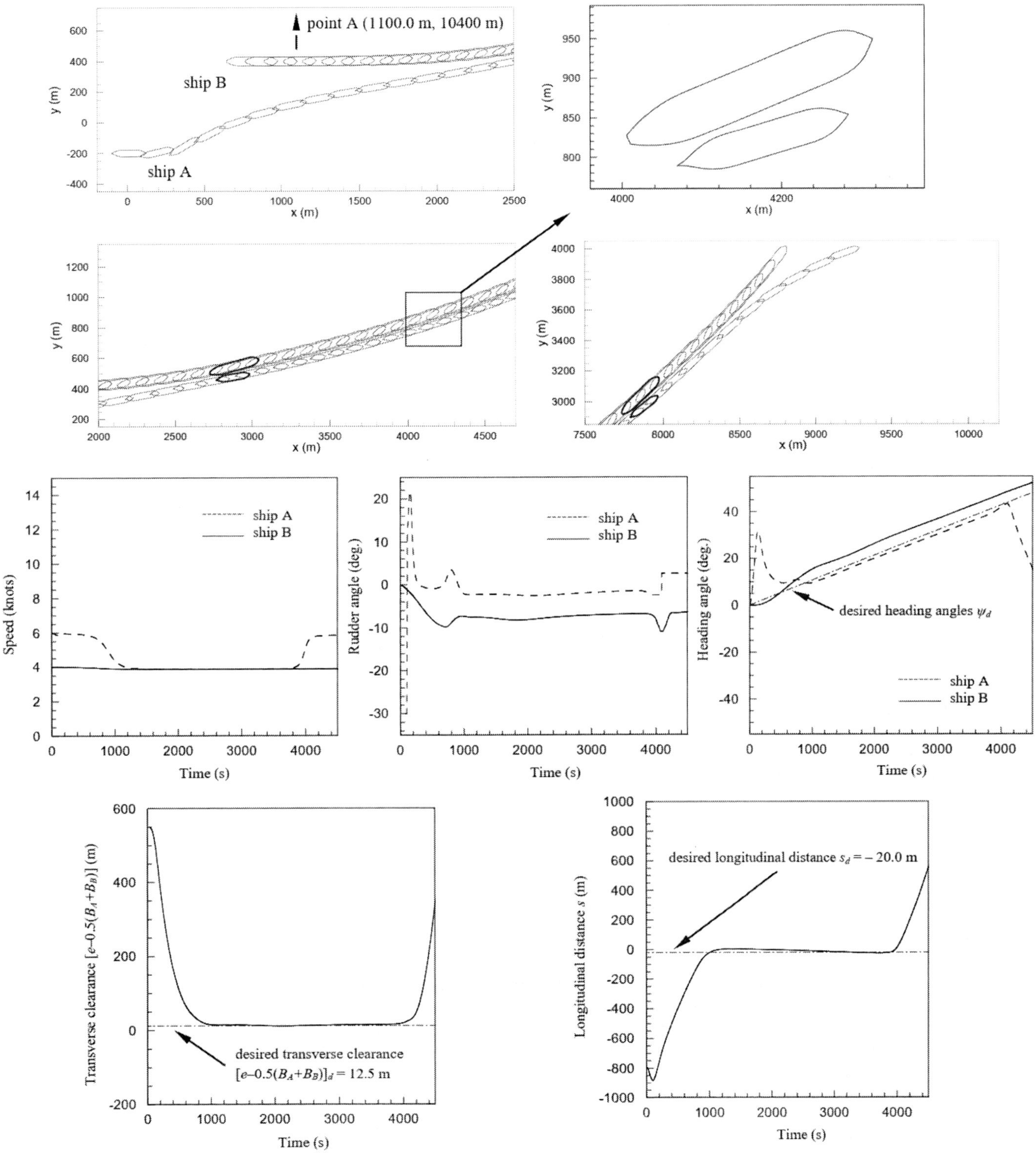

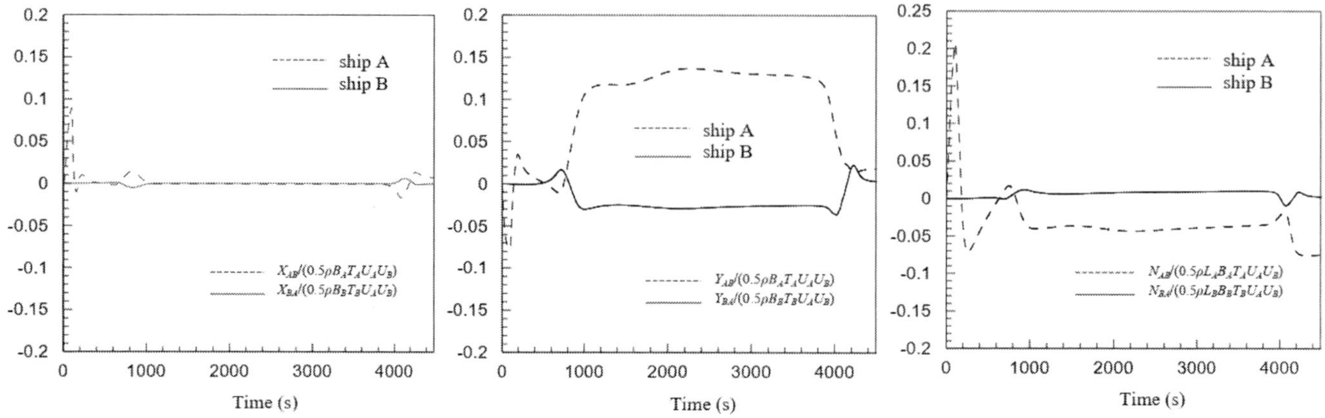

Figure 9: The underway lightering maneuver between the 'Aframax' and 'KVLCC2' (Moeri tanker) ships advancing in calm water on a part of desired circular path (origin at point A (1100.0 m, 10400.0 m) not shown) of radius r = 10000.0 m. The starting positions: ship A (0.0 m, – 200.0 m), ship B (800.0 m, 400.0 m). Note: Part of the figure where 4700.0 m < x < 7500 m and 1100.0 m < y < 2850.0 m is not shown since, both ships are in abeam configuration and on a desired course. B_A – beam of the 'Aframax' hull on main frame, B_B – beam of the 'KVLCC2' (Moeri tanker) hull on main frame, 1 knot = 0.5144444 m/s.

is deployed on both ships from the beginning of the simulation. The lightering ship, i.e. ship A is assumed to advance with the approach speed of 6 knots and to adhere on desired transient course angles ψ_d with ship to be lightered, i.e., ship B during the underway lightering operation. The approach speed of the ship A is set to 6 knots. The engagement of motion control after certain time interval ensures the abeam configuration of involved ships by matching their forward speeds and, as well as, the desired longitudinal distance s_d = – 20.0 m and transverse clearance of $[e - 0.5(B_A + B_B)]_d$ = 12.5 m defined between the facing sides of both hulls. Here, B_A is the beam of the lightering ship A, while B_B is the beam of the ship B (ship which needs to be lightered).

The abeam configuration is achieved by following the zero collision risk tolerance. This means that the motion controllers are now the replacement for the experienced helmsman and deck crew on both ships which are responsible for the execution of lightering operation. As can be seen from the Figure 9 the lightering ship A catches up with the ship B at x position of approximately 4100.0 m given in respect to the global Earth-fixed coordinate system $OX_EY_EZ_E$. At the same time the relative longitudinal position s between both ships decreased to the desired value of – 20.0 m, i.e., both ships are abeam and zero relative speed, i.e., $(u_A - u_B) = 0$ m/s together with the desired transversal clearance $[e - 0.5(B_A + B_B)]_d$ of 12.5 meters is achieved. Both ships now start to experience significant hydrodynamic interaction loads shown on the lower part of Figure 9.

As the same figure shows, the initial development of the hydrodynamic interaction loads (surge and sway forces and, yaw moments) is characterized by the certain peak values on the lightering ship A (time t <≈ 400.0 s). These, in fact do not present the time development of the 'real' hydrodynamic interaction effects because the two ships are still quite away from each other as can be seen from their maneuvering parameters and trajectories on the figure. Therefore, the reason of the peaks is due to nonlinear added mass terms of the ship A associated with the surge, sway and yaw velocities of the same ship as the lightering ship A executes the maneuver in order to align with the lightering ship on the same heading (see section 3). After this initial transient maneuvering phase of the lightering ship A the distances and speeds on involved ships are closing up and the hydrodynamic interaction loads start to show their real behavior in time. First, both ships experience an initial repulsion phase which is shortly after followed by the attraction phase characterized with the peak values of interaction surge and sway forces and the yaw moments tend to swing the bow of the each ship from each other (see lower part of Figure 9).

In order to reduce the simulation time, the authors assumed that the duration time of the underway lightering maneuver is 75 minutes like it can be seen from Figure 9. During the phase where the underway lightering is in progress, the hydrodynamic interaction loads shows constant behaviour which may be expected due to calm water operational conditions.

In the last (separation) stage of the underway lightering maneuver, corresponding to x position of about 7900.0 m, lightering ship A slowly starts to increase her speed in respect to the ship B, which remains at steady speed and target heading. The possibility of collision increases since both ships are now experiencing the repulsive interaction surge and sway forces and the interaction yaw moments which are pointing in the same direction. In practice, this situation is recognised as a critical one due to increased possibility of steel-to-steel contact between the stern part of the hull of a lightering ship A and facing side of the hull of ship B [10]. However, the mentioned collision scenario is completely avoided here due to engagement of experienced personnel (the motion control module – 'autopilots').

When the relative speed ($u_A - u_B$) between the lightering ship A and ship B reaches approximately 1 knot, a rudder on lightering ship A is deflected to starboard so that she actually disengages herself from the transient heading configuration with the ship B.

The underway lightering manoeuvre in calm and deep water presented above showed that the calm water hydrodynamic interaction loads between two ships estimated with Xiang and Faltinsen 3D theory [1] can not be neglected when the two ships are involved in the close proximity manoeuvre(s). Furthermore, with application of the motion control module we exemplify the requirement for 'experienced' helmsmen and deck crews which need to be fulfilled in order to accomplish the underway lightering operation with zero collision risk tolerance.

5. CONCLUSION

Theoretical investigation of underway lightering manoeuvre in calm and deep water has been carried out by using a unified seakeeping and manoeuvring model of two interacting ships by Skejic and Faltinsen [4, 5] and Skejic [6].

The unified model is developed on the modular concept. This means that the propulsion and resistance, rudder loads and nonlinear viscous loads are accounted for through modules commonly used within the manoeuvring research field. In addition to the mentioned modules, the hydrodynamic interaction effects between two ships advancing with moderate low speeds on general desired course in calm and deep water are estimated with a full 3D method based on Xiang and Faltinsen theory [1]. The applicability of the theory is verified on the example of submerged ellipsoid advancing in close proximity of the wall ([2], [17]), and the case scenario of moored slender body passed by another slender body [3].

The underway lightering manoeuvre is simulated, as closely as possible, on a realistic way by following [10]. It has been demonstrated that the importance of hydrodynamic interaction effects predicted by the above theory can not be neglected when two bodies are operating in vicinity of each other. In this context, the critical stages of the lightering operation are exemplified and discussed from the navigational safety point of view.

6. ACKNOWLEDGEMENTS

The work was carried out jointly at CeSOS and MARINTEK. The authors express gratitude to the financial supports from Research Council of Norway through CeSOS and the KMB project 'Investigating Hydrodynamic Aspects and Control Strategies for Ship-to-Ship Operations' at MARINTEK.

7. REFERENCES

1. Xiang, X. and Faltinsen, O.M., 'Maneuvering of two Interacting Ships in Calm Water' In Proceedings of PRADS2010, the 11th International Symposium on Practical Design of Ships and other Floating Structures, Rio de Janeiro, Brazil, 2010.

2. Newman, J.N., 'The force and moment on a slender body of revolution moving near a wall', David Taylor Model Basin, Rep. 2127, Hydromechanics Laboratory, Washington D.C., USA, 1965.

3. Wang, Q.X., 'An analytical solution for two slender bodies of revolution translating in very close proximity', Journal of Fluid Mechanics, 582, pp. 223-251, 2007.

4. Skejic, R. and Faltinsen, O.M., 'A unified seakeeping and maneuvering analysis of two interacting ships' In Proceedings of the 2nd International Conference on Marine Research and Transportation, Ischia, Italy, pp. 209–218, 2007. http://www.icmrt07.unina.it/Proceedings/Papers/A/68.pdf.

5. Skejic, R. and Faltinsen, O.M., 'A unified seakeeping and maneuvering analysis of ships in regular waves' Journal of Marine Science and Technology 13:371–394, 2008. http://www.springerlink.com/content/88373xm3g12h4067/fulltext.pdf.

6. Skejic, R. 'Maneuvering and Seakeeping of a Single Ship and of Two Ships in Interaction' Doctoral thesis, Department of Marine Technology, Faculty of Engineering Science and Technology, Norwegian University of Science and Technology, Trondheim, Norway, 2008.

7. Skejic, R. and Berg, T.E., 'Hydrodynamic Interaction Effects during Lightering Operation in Calm Water – Theoretical Aspects' In Proceedings of the International Conference on Marine Simulations and Ship Maneuverability - MARSIM 2009, Panama City, Panama, 2009.

8. Skejic, R. and Berg, T.E., 2010, 'Combined Seakeeping and Maneuvering analysis of a Two Ship Lightering Operation' In Proceedings of the 29th International Conference on Ocean, Offshore and Arctic Engineering – OMAE 2010, Shanghai, China, 2010.

9. Tuck, E.O. and Newman, J.N., 'Hydrodynamic interaction between ships' In Proceedings of the 10th Symposium on Naval Hydrodynamics, Cambridge, MA, USA, pp. 35–70, 1974.

10. 'Ship-to-Ship (STS) Transfer Guide (Petroleum)', 4th edition, OCIMF, 2005.

11. Skejic, R., Breivik, M., Fossen, T.I., Faltinsen, O.M., 'Modeling and Control of Underway Replenishment

Operations in Calm Water' 8th IFAC International Conference on Manoeuvring and Control of Marine Craft, Guarujá (SP), Brazil, 2009.

12. ITTC 'The Specialist Committee on Esso Osaka' Final Report and Recommendations to the 23rd ITTC. In Proceedings of the 23rd ITTC, Vol. 2, Italian Ship Model Basin (INSEAN), Venice, pp 581–617,2002. http://ittc.sname.org/proc23/Esso%20Osaka.pdf.

13. Imlay, F.H. 'The Complete Expressions for Added Mass of a Rigid Body Moving in an Ideal Fluid' DTMB Report 1528, Washington D.C., USA, 1961.

14. Salvesen, N., Tuck, E.O. and Faltinsen, O.M., 'Ship motions and sea loads' Trans SNAME 78:250–287, 1970.

15. Faltinsen, O.M., *Sea Loads on Ships and Offshore Structures*, Cambridge, England: Cambridge University Press, 1990.

16. SIMMAN 2008 'Workshop on Verification and Validation of Ship Manoeuvring Simulation Methods' Preprints of Workshop Proceedings, Vol. 1: Summary, Test Cases, Methods and Papers, Editors: Stern, F. and Agdrup, K., Copenhagen, Denmark, April 14th -16th 2008. http://www.simman2008.dk/KVLCC/KVLCC2/kvlcc2_geometry.html.

17. Wang, Q. X. , 'Analysis of a slender body moving near a curved-ground', Phys. Fluids 17 (9): Art. No. 097102, 2005.

8. AUTHORS BIOGRAPHY

Xu Xiang is currently PhD candidate in marine hydrodynamics at CeSOS, NTNU. His PhD project is maneuvering of two interacting ships in waves. He was employed as a research assistant and later research engineer at the seakeeping and maneuverability basin of China Ship Scientific Research Center (CSSRC) between 2003 and 2007.

Renato Skejic obtained his PhD degree in marine hydrodynamics at CeSOS, NTNU. He is currently employed as a research scientist at the Norwegian Marine Technology Research Institute – MARINTEK. His current research interests include ship hydrodynamics, especially considering combined seakeeping and maneuvering of single and multiple bodies operating in various environmental conditions including ice infested waters.

Odd M. Faltinsen is Professor in marine hydrodynamics at CeSOS and Department of Marine Technology, NTNU. He was employed by DNV from 1968 to 1974 before becoming a professor at NTNU. Faltinsen has been member of 5 ITTC committees including 2 as chairman, and 3 ISSC Committees. He is a foreign member of the National Academy of Engineering, USA and the Chinese Academy of Engineering and has educated 50 PhD students.

Tor E. Berg obtained his Dr.ing. degree in marine hydrodynamics at the Department of Naval Architecture at the Norwegian Institute of Technology (now NTNU). He is currently employed as a principal research engineer at MARINTEK and Professor II at NTNU. His current research and project activities include topics related to ship maneuvering, ship-to-ship interaction, marine emergency operations and safety at sea in Arctic waters.

STUDY ON THE EFFECTS OF UNSTEADY SHIP TO SHIP INTERACTION BY CFD METHOD

H Yang, China Ship Scientific Research Center, Wuxi, China
B S Wu, China Ship Scientific Research Center, Wuxi, China
QM Miao, China Ship Scientific Research Center, Wuxi, China
X Xiang, Centre for Ships and Ocean Structures (CeSOS), Norwegian University of Science and Technology, Norway
T E Berg, MARINTEK, Norway
XF Kuang, China Ship Scientific Research Center, Wuxi, China

SUMMARY

The present paper presents the study of hydrodynamic interaction loads between two ships by CFD method. A series of different lateral distances between two ships sailing in the deep and shallow water are considered to study the hydrodynamic interaction.
One ship passing a moored ship is simulated using two identical Wigley hulls. The results give close results with the BEM method. Then, the benchmark model KVLCC2 was used to study the hydrodynamic interaction. The hydrodynamic interaction loads were predicted to investigate the effects of relative positions and water depth. Finally, the future requirements and recommendations concerning the accurate prediction of the hydrodynamic interaction loads are discussed from the actual sea-operation's point of view.

NOMENCLATURE

The nomenclatures used in this paper were shown below.

ρ	Density of water (kg m^{-3})
L	Ship length (m)
B	Ship breadth (m)
T	Ship draft (m)
h	Water depth (m)
V	Forward speed (m s^{-1})
X	Surge force (N)
Y	Sway force (N)
N	Yaw moment (N m)
C_x	Surge force coefficient
Cy	Sway force coefficient
C_n	Yaw moment coefficient
y^+	Yplus

1. INTRODUCTION

The lightering operation between two ships typically takes place in open sea under low speed and small lateral distance. The application of underway lightering of crude oil is increasing. The complex flow between two ships will increase the possibility of the collision accident when the lateral distance is sufficiently small. The main objective of the research project "Investigating Hydrodynamic Aspects and Control Strategies for Ship-to-Ship Operations" is to increase knowledge and understanding of the complex flow between two ships operating in close proximity which is organized by Norwegian Research Council through MARINTEK. A reliable method is needed to predict the hydrodynamic performance of two ships sailing closely. Newton [1] experimentally investigated interaction forces and moments between two ships at different longitudinal and transversal clearances in deep water. Tuck and Newman [2] calculated the interaction loads between two slender bodies of revolution by a theoretical way. Brix [3] estimated the forces and moments acting on a ship during an overtaking maneuver based on a semi-empirical approach. With the rapid development of the numerical techniques and computer hardware, the CFD (Computation Fluid Dynamics) method becomes a feasible way to study the complex flow problem between two ships. Hamn-Ching Chen [4] study Ship-Ship Interactions in Shallow Water and Restricted Waterway by Chimera RANS (Reynolds Averaged Navier-Stokes) Method.
The unsteady ship to ship interaction surge and sway force and yaw moment were predicted by CFD method through the commercial software FLUENT for the present paper. The CFD results were compared with the results by a 3-D potential flow method (Xiang and Faltinsen [5]) for one ship passing a moored ship with good agreement for the surge, sway forces and yaw moment on the moored ship. Then, the hydrodynamic forces and moment on two ships at different relative lateral distances in deep and shallow waters were computed by CFD method to study the effects of the relative lateral distance and water depth.

2. NUMERICAL METHOD DESCRIPTIONS

2.1 COORDINATE SYSTEM AND THE EXPRESSION OF NUMERICAL RESULTS

Two ships are named Ship I ($I=1$, 2). Two sets of coordinate systems are used. The earth-fixed coordinate system $OXYZ$ and two moving coordinate systems fixed to the center of gravity (CoG) of each ship $o_i x_i y_i z_i$. The moving systems are always parallel to the earth fixed system. The signs of convention for the hydrodynamic forces and positive directions are referred to Figure 1.

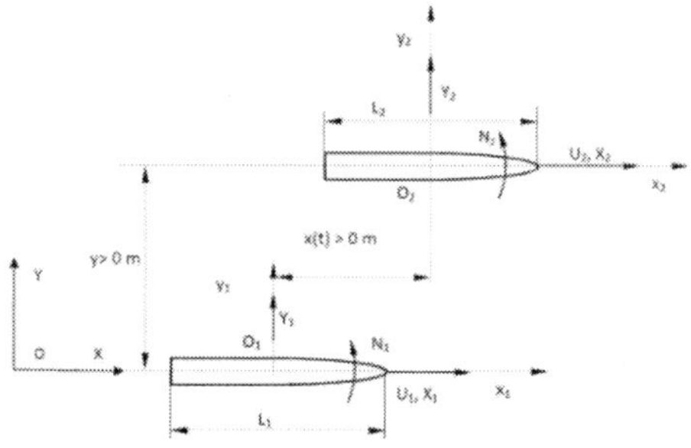

Figure 1: signs conventions: definition of longitudinal and lateral distance x, y; positive direction of interaction forces X_1, Y_1, X_2, Y_2 and interaction moments N_1, N_2, t is time

The surge and sway forces and yaw moment are also named as X, Y, N for each ship, respectively. They are non-dimensionized as follows:

$$C_x = \frac{X}{0.5\rho V^2 L^2} \quad C_y = \frac{Y}{0.5\rho V^2 L^2} \quad C_n = \frac{N}{0.5\rho V^2 L^3}$$

2.2 GOVERNING EQUATIONS

The governing equations for incompressible, turbulent flow are the continuity equation for mass conservation and RANS equations for momentum transport.

1) Continuity equation

$$\frac{\partial \overline{u_i}}{\partial x_i} = 0 \qquad (1)$$

2) RANS equation

$$\rho\frac{\partial \overline{u_i}}{\partial t} + \rho\overline{u_j}\frac{\partial \overline{u_i}}{\partial x_j} = \rho\overline{F_i} - \frac{\partial \overline{p}}{\partial x_i} + \frac{\partial}{\partial x_j}\left(\mu\frac{\partial \overline{u_i}}{\partial x_j} - \rho\,\overline{u_i' u_j'}\right) \qquad (2)$$

3) Turbulence model

The RNG k−ε turbulence model and wall function are suitable for prediction of the maneuvering hydrodynamics of full-ship model (Yang [6]). Turbulent kinetic energy (k) transport equation is

$$\frac{\partial}{\partial t}(\rho k) + \frac{\partial}{\partial x_i}(\rho k u_i) = \frac{\partial}{\partial x_j}\left[\left(\mu + \frac{\mu_t}{\sigma_k}\right)\frac{\partial k}{\partial x_j}\right] + G_k - \rho\varepsilon \qquad (3)$$

And the equation of dissipation of turbulent energy (ε) is

$$\frac{\partial}{\partial t}(\rho\varepsilon) + \frac{\partial}{\partial x_i}(\rho\varepsilon u_i) = \frac{\partial}{\partial x_j}\left[\left(\mu + \frac{\mu_t}{\sigma_\varepsilon}\right)\frac{\partial \varepsilon}{\partial x_j}\right] + C_{1\varepsilon}\frac{\varepsilon}{\kappa}G_k - C_{2\varepsilon}\rho\frac{\varepsilon^2}{\kappa} - R_\varepsilon \qquad (4)$$

Where, $\mu_t = \rho C_\mu \frac{k^2}{\varepsilon}$, $G_k = -\rho\overline{u_i' u_j'}\frac{\partial u_j}{\partial x_i}$,

$R_\varepsilon = \frac{C_\mu \rho \eta^3 (1-\eta/\eta_0)}{1+\beta\eta^3}\frac{\varepsilon^2}{k}$, $\eta = Sk/\varepsilon$;

And $\eta_0, \beta, C_{1\varepsilon}, C_{2\varepsilon}, C_\mu, \sigma_k, \sigma_\varepsilon$ are the empirical constants.

2.3 BOUNDARY CONDITION

2.3(a) Inlet boundary is two hull lengths ahead of bow, where the velocity is prescribed for the steady motion case.
2.3(b) Outlet boundary is three hull lengths after stern, where pressure outlet is set for the simulation.
2.3(c) Outer boundary is two hull lengths away from hull, where the velocity inlet boundary is applied.
2.3(d) Hull surface: non-slip condition is applied.
2.3(e) Free surface: symmetry boundary is applied to ignore the effect of the free surface.

3. COMPARATIVE STUDY OF CFD AND POTENTIAL FLOW PREDICTIONS

A comparative study was carried out for two identical Wigley hulls via CFD method and a 3-D potential flow method, Xiang and Faltinsen[5]. The domain and grids set is referred to Figure 2.

In our case ship model length L=3m, the breadth B=0.3m. We let ship B be stationary (moored) and ship A advances with constant speed V=0.43m/s, corresponding to F_n=0.079 and R_e=1.29×10^6. The non-dimensional lateral distance between the two ships is y/B=1.2. The surge and sway force coefficients and yaw moment coefficient on the moored ship B are presented as a function of the non-dimensional stagger x/L in Figure 3 to Figure 5.

Good agreements are obtained. When x/L = -0.3. As the moving ship approaches, the moored ship experiences a negative surge force, a positive sway force, and a negative yaw moment. When the two ships are abreast, surge forces and yaw moments are near zero, while sway force is up to positive peak value, i.e. the moored ship is pulled in toward the moving ship. Further on, as the ships depart, for example when x/L=0.6, surge force turns to positive, sway forces turns to negative, and yaw moment turns to positive.

The agreement is good for both forces and the yaw moment coefficients on the moored Wigley ship. This means that in such case the cross-flow is not significant for the moored ship, i.e., there is not much contribution from the viscous force, thus the BEM method gives close results with the CFD method.

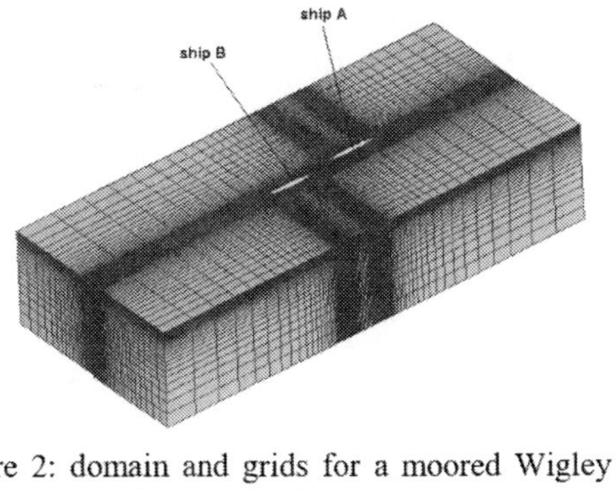

Figure 2: domain and grids for a moored Wigley ship passing by another identical Wigley ship

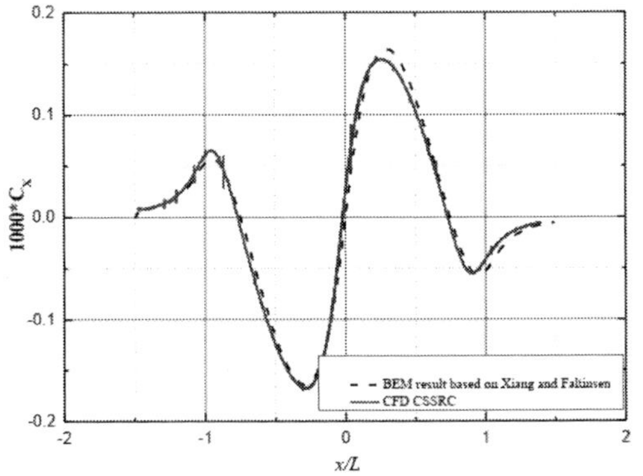

Figure 3: Surge force coefficient C_x on one moored Wigley hull passed by an identical one, clearance between centerlines Y=y/B=1.2: BEM results compared with CSSRC CFD results

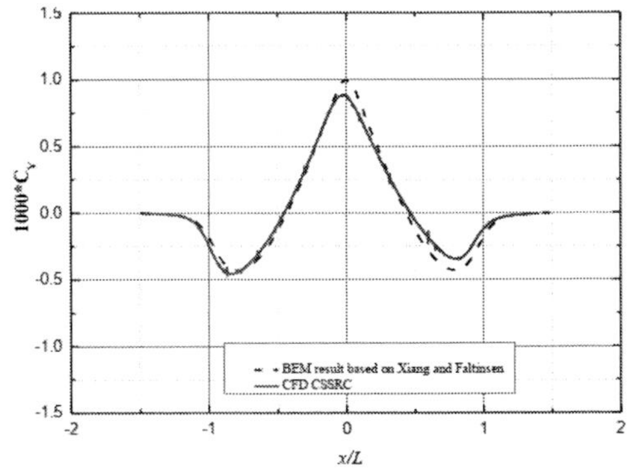

Figure 4: Sway force coefficient C_y on one moored Wigley hull passed by an identical one, clearance between centerlines Y=y/B=1.2: BEM results compared with CSSRC CFD results

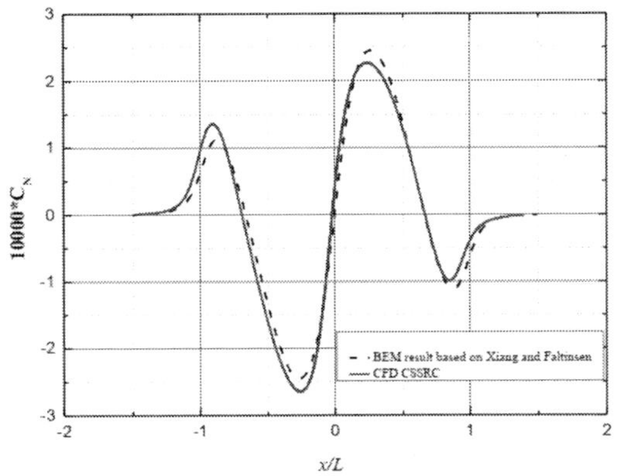

Figure 5: Yaw moment coefficient C_n on one moored Wigley hull passed by an identical one, clearance between centerlines Y=y/B=1.2: BEM results compared with CSSRC CFD results

4. EFFECTS OF LATERAL DISTANCE BETWEEN TWO SHIPS ON THE HYDRODYNAMIC LOADS

Two identical benchmark models KVLCC2 were chose to study the effects of the lateral distance when two ships advance parallel with the same speed. A series of different lateral distances(full scale: y=62m, 68m, 83m, 108m and 158m) were studied to evaluate the hydrodynamic interaction between two ships. The main dimensions of KVLCC2 were shown in Table1.

Table1 principal dimensions of KVLCC2 hull

	L(m)	B(m)	T(m)	C_b
Full scale	320.0	58.0	20.8	0.8098

4.1 GRIDS OF COMPUTATION

Multi-block structural grid is adopted in this paper, and the grid number is about 1.3millions.The first element thickness equates to a y^+~50. The topology is body-fitted, and the grids near the hull are refined. Because of the symmetry of flow, only half of the computation domain is considered. The domain and grids are shown in Figure 6.

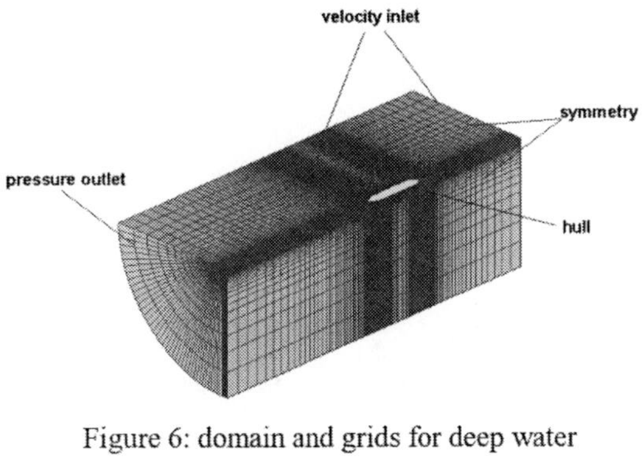

Figure 6: domain and grids for deep water

4.2 COMPUTATION RESULTS AND ANALYSIS

The interaction sway force and yaw moment at different lateral distances when two ships advance parallel with the same speed were shown in Table 2 and Figure 7 and Figure 8.

It is obvious that the sway force becomes remarkably larger while the lateral distance decreases, i.e., a suction force is developed between two ships. It is interesting to find that the yaw moment tends to turn two ships bow in when the lateral distance less than 1.17*B*, while stern in at other situation.

The speed has almost no effects on the sway force coefficient, and little on the yaw moment coefficient.

Table 2 the sway force and yaw moment coefficients between two ships under two speeds.

Lateral distance y(m)	U=4kn		U=6kn	
	Sway force, C_y	Yaw moment, C_n	Sway force, C_y	Yaw moment, C_n
62	-3.81×10^{-3}	-1.02×10^{-4}	-3.88×10^{-3}	-1.40×10^{-4}
68	-3.06×10^{-3}	-0.07×10^{-4}	-3.08×10^{-3}	-0.20×10^{-4}
83	-1.85×10^{-3}	0.57×10^{-4}	-1.85×10^{-3}	0.48×10^{-4}
108	-1.03×10^{-3}	0.73×10^{-4}	-1.02×10^{-3}	0.64×10^{-4}
158	-4.79×10^{-4}	0.70×10^{-4}	-4.69×10^{-4}	0.63×10^{-4}

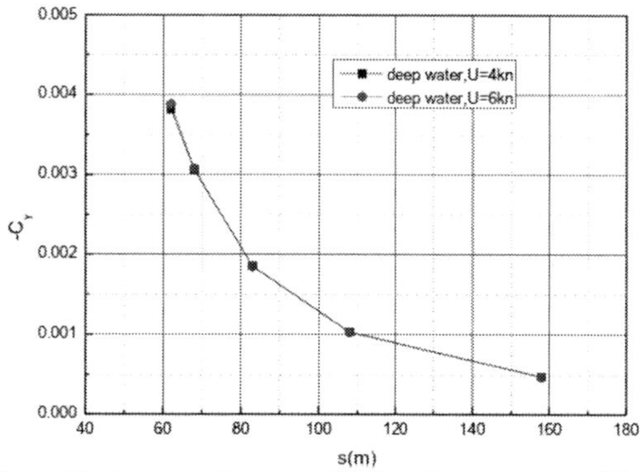

Figure 7: the sway force coefficient between two ships at different lateral distances in deep water

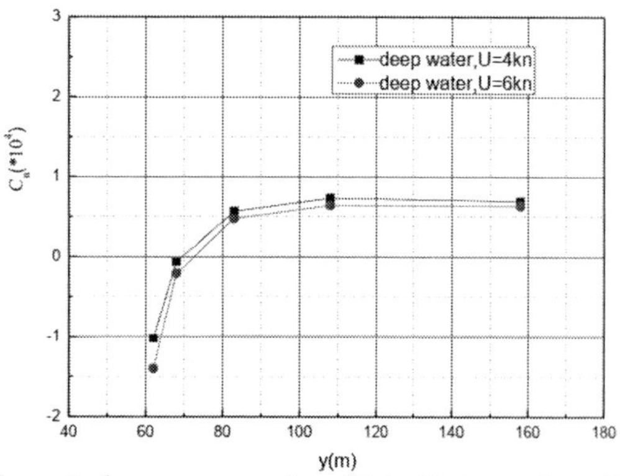

Figure 8: the yaw moment coefficient between two ships at different lateral distances in deep water

5. EFFECTS OF SHIP'S DRAFT AND UNDER KEEL CLEARANCE

The ship's draft and under keel clearance was referred to the definition by PIANC [6]:

Deep water $h/T > 3.0$

Medium water $1.5 < h/T < 3.0$

Shallow water $1.2 < h/T < 1.5$

Here, h is the water depth, T the draft。

The depth to draft ratio of 1.2(h/T=1.2) was chosen to study the hydrodynamic performance of two ships in shallow water. The computation grid was shown in 9.

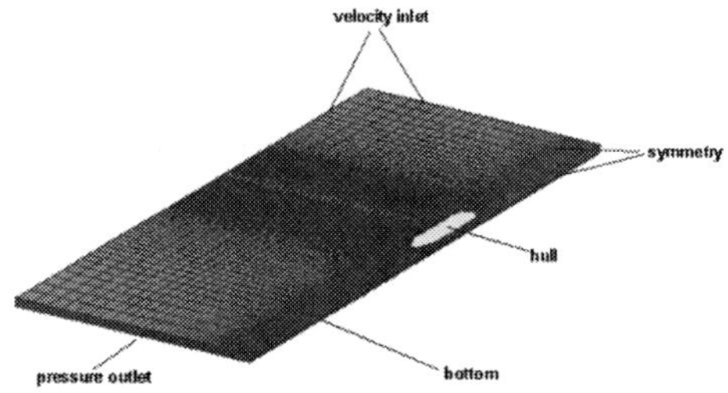

Figure 9: domain and grids for shallow water

The interaction sway force and yaw moment coefficients at different lateral distances when two ships advance parallel with the same speed were shown in Table3 and Figure 10 and Figure 11.

As expected, the depth to ship's draft ratio appears to be a very important parameter, which can be concluded from Figure 10 and Figure 11: if the ships' draft kept constant, interaction forces increase significantly as water depth decreases.

It is also found that the effect of advance speed is more remarkable than that in the deep water case.

Table 3 the sway force and yaw moment coefficients on one of the two ships at different relative lateral positions in shallow water (h/T=1.2)

Lateral distance y(m)	U=4kn		U=6kn	
	Sway force, C_y	Yaw moment, C_n	Sway force, C_y	Yaw moment, C_n
62	-3.97×10^{-3}	8.40×10^{-4}	-4.38×10^{-3}	7.12×10^{-4}
68	-3.42×10^{-3}	9.35×10^{-4}	-3.70×10^{-3}	8.17×10^{-4}
83	-2.16×10^{-3}	8.61×10^{-4}	-2.41×10^{-3}	7.69×10^{-4}
108	-1.29×10^{-3}	6.31×10^{-4}	-1.51×10^{-3}	5.62×10^{-4}
158	-7.79×10^{-4}	3.28×10^{-4}	-9.77×10^{-4}	2.87×10^{-6}

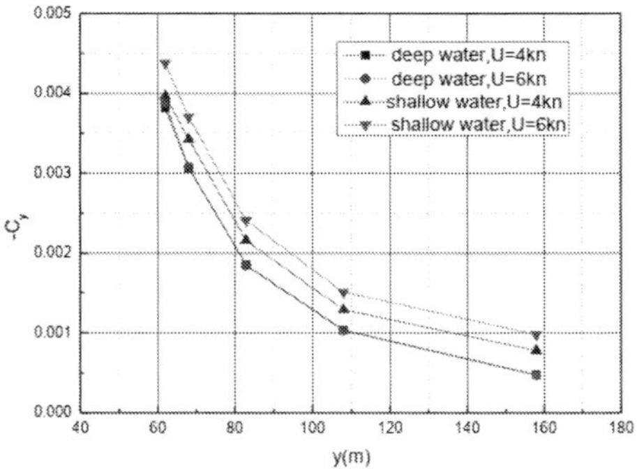

Figure 10: the sway force coefficient on one of the two ships advancing parallel at different lateral distance in shallow and deep water

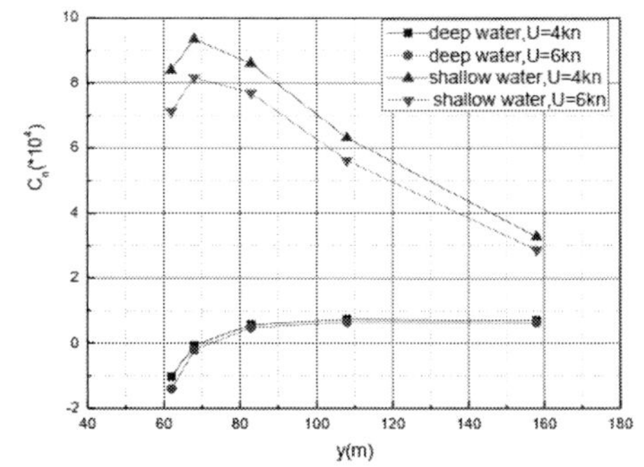

Figure 11: the yaw moment coefficient on one of the two ships advancing parallel at different lateral distance in shallow and deep water

6. CONCLUSION

Numerical investigation of hydrodynamic interaction forces and moment between two ships in deep and shallow water has been carried out by CFD method based on RANS. It has been demonstrated that the importance of hydrodynamic interaction effects cannot be neglected when two ships operating in vicinity of each other. The relative position and water depth effects on the interaction hydrodynamic forces and moment are shown to be important for two ships advancing in close proximity. Although a great deal of further study is required, this paper clearly demonstrates the potential of using the RANS method for ship-ship interaction problems.

7. REFERENCES

1. Newton, R.N. Some notes on interaction effects between ships close aboard in deep water. First Symposium on Naval Maneuverability, Washington DC, pp. 1-24, 1960.

2. Tuck, E.O. and Newman, J.N. Hydrodynamic interaction between ships. In Proceedings of the 10th Symposium on Naval Hydrodynamics, Cambridge, MA, USA, pp. 35–70, 1974.

3. Brix, J. (editor). Manoeuvring Technical Manual. Seehafen Verlag GmbH, Hamburg, 1993.

4. Hamn-Ching Chen, Woei-Min Lin and Wei-Yuan Hwang.Validation and Application of Chimera RANS Method for Ship-Ship Interactions in Shallow Water and Restricted Waterway. 24th Symposium on Naval Hydrodynamics, Fukuoka, JAPAN, 2002.

5. Xiang, X. and Faltinsen, O.M. 'Maneuvering of two Interacting Ships in Calm Water' In Proceedings of PRADS2010, the 11th International Symposium on Practical Design of Ships and other Floating Structures, Rio de Janeiro, Brazil, 2010.

6. Pianc.Capability of ship manoeuvring simulation models for approach channels and fairways in harbours. Report of working group no.20 of Permanent Technical Committee II, 1992.

7. SIMMAN 2008 'Workshop on Verification and Validation of Ship Manoeuvring Simulation Methods' Preprints of Workshop Proceedings, Vol. 1: Summary, Test Cases, Methods and Papers, Editors: Stern, F. and Agdrup, K., Copenhagen, Denmark, April 14th -16th 2008. http://www.simman2008.dk/KVLCC/KVLCC2/kvlcc2_geometry.html.

8. ACKNOWLEDGEMENTS

The authors express gratitude to the financial supports from CSSRC and the KMB project 'Investigating Hydrodynamic Aspects and Control Strategies for Ship-to-Ship Operations' at MARINTEK.

9. AUTHORS BIOGRAPHY

Han Yang is currently research engineer at marine hydrodynamics department of CSSRC. His main research interest includes CFD applications in ship maneuverability prediction, ship-ship interaction and so on.

Bao-shan Wu is currently research professor and vice director of technical of CSSRC. His main research interests include ship hydrodynamics and Uncertainty Analysis in experimental and computational fluid dynamics.

Quan-Ming Miao obtained his PhD degree in marine hydrodynamics in University of Hongkong. He is currently research professor and vice director of chief-engineertechnical of CSSRC. His main research interests include ship and offshore hydrodynamics, ocean engineering technology and offshore wind turbines.

Xu Xiang is currently PhD candidate in marine hydrodynamics at CeSOS, NTNU. His PhD project is maneuvering of two interacting ships in waves. He was employed as a research assistant and later research engineer at the seakeeping and maneuverability basin of China Ship Scientific Research Center (CSSRC) between 2003 and 2007.

Tor E. Berg obtained his Dr.ing. degree in marine hydrodynamics at the Department of Naval Architecture at the Norwegian Institute of Technology (now NTNU). He is currently employed as a principal research engineer at MARINTEK and Professor II at NTNU. His current research and project activities include topics related to ship maneuvering, ship-to-ship interaction, marine emergency operations and safety at sea in Arctic waters.

Xiao-feng Kuang is currently senior research engineer at marine hydrodynamics department of CSSRC. His main research interests include ship and offshore hydrodynamics, ocean engineering technology.

HYDRODYNAMIC INTERACTION OF TWO THIN SHIPS WITH RUDDER IN CLOSE PROXIMITY

H. Yasukawa and S. Yoshida, Hiroshima University, Japan

SUMMARY

The hydrodynamic force characteristics of two Wigley hulls moving side by side in close proximity with steering are investigated experimentally and theoretically. Captive model tests are carried out to measure the hydrodynamic forces acting on two hulls and two rudders with changing lateral clearance between ships, hull drift angle and rudder angle. The experimental results are compared with numerical analysis using the nonlinear lifting surface theory. We found that when a ship moves obliquely to be close to other ship, suction force and bow-in moment act on both ships and become larger with decrease of the lateral clearance between ships. The lifting-surface based calculation method can capture the hydrodynamic force characteristics of two thin ship models with rudder.

NOMENCLATURE

B	Ship breadth
C_b	Block coefficient
D	Ship depth
d	Ship draft
F'_N	Rudder normal force coefficient
F_n	Froude number based on ship length
L	Ship length
N'	Yaw moment coefficient around midship
U	Ship speed
V	Displacement volume
X'	Surge force coefficient acting on ship
Y'	Lateral force coefficient acting on ship
y_0	Lateral clearance between two ships
β	Hull drift angle
δ	Rudder angle
ρ	Water density
suffix 1	left positioning ship
suffix 2	right positioning ship

1. INTRODUCTION

Application of ship-to-ship transfer operation is an alternative way to transfer crude oil, liquefied oil, liquefied gas etc. Alongside position of ship to each other closely, it becomes possible for ships to load and unload the oils either in stationary or underway conditions. This offers advantages especially saving cost and time landed in the harbor since the ships do no need to do so.

To prevent this accident and ensure a safe navigation during the transfer operation, it is required to grasp characteristics of hydrodynamic force acting on the ships that moving side by side. Many researchers have studied both using analytical and experimental model of the hydrodynamic force characteristics on two ships sailing side by side. The results obtained that suction force and bow-out yaw moment act on two ships moving in parallel with same speed. However, this analysis is still insufficient since effect of controlling devices for example rudder was not included in the previous studies.

In this paper, hydrodynamic interaction effect of rudder with two ships sailing side by side in close proximity is dealt with. Since the hydrodynamic interaction of actual ships is complex problems, we employ several simplifications as follows:

- Wigley parabolic hull is selected as the hull form with non-propellered ship and a rectangular shaped rudder is used.
- Two ships model with same dimensions and shapes.
- There is no stagger between two ships in longitudinal direction.

We conducted tank test and numerical analysis to capture the hydrodynamic interactions between two Wigley hulls and two rudders. In this experiment, the hydrodynamic forces acting on two hull models and two rudder models in close proximity were measured with changing the parameters such as the lateral clearance between ships, hull drift angle and rudder angle. In the same way, the hydrodynamic forces acting on two ship hulls and two rudders were calculated using a method based on the nonlinear lifting surface theory. Through the comparison between experiment and calculation, the characteristics of the hydrodynamic interaction are investigated.

2. OUTLINE OF CAPTIVE MODEL TEST

2.1 SHIP MODEL

Tables 1 and 2 show principal dimensions of the main hull and the rudder used in the model test. Wigley parabolic hull was employed as the hull form with non-propellered ship. Also a rectangular typed rudder model with NACA0020 wing section was used. Figure 1 shows arrangement of Wigley hull model with rudder.

Table 1: Principal dimensions of ship model

Length L	1200 (mm)
Depth D	200 (mm)
Breadth B	120 (mm)
draft d	75 (mm)
Volume V	0.0048 (m^3)
block coefficient C_b	0.444

Table 2: Principal dimensions of rudder model

Chord length	50 (mm)
Span	150 (mm)
Aspect ratio (below the waterline)	1.5
Gap between rudder and hull	5 (mm)
Wing section	NACA0020

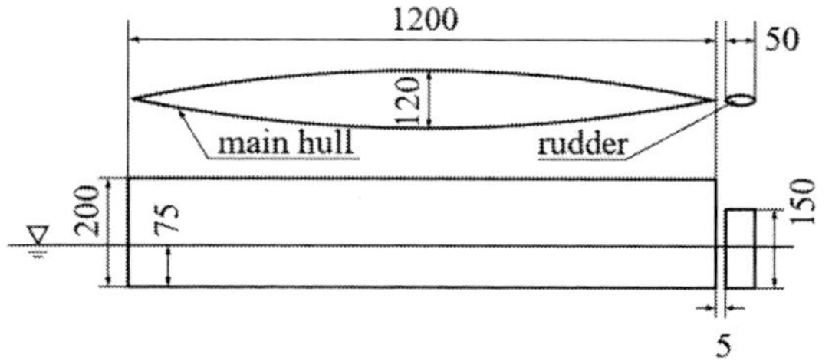

Figure 1: Arrangement of Wigley hull model with rudder (unit:mm)

2.2 MEASUREMENT

Captive model tests were conducted in towing tank of Hiroshima University, Japan. Two ship models are arranged side by side as shown in Figure 2, and forces and moment acting on the ship hulls and rudders were measured. In this paper, a ship of the left hand side in Figure 2 is called Ship1 and the right hand side is Ship2. A 3 components dynamo-meter installed on the midship of the hull model was used to measures the surge force X, lateral force Y and yaw moment around the midship of ship N, and a component dynamo-meter equipped on aft part of the ship model was used to measure the rudder normal force F_N. Totally, four dynamometers were used to measure the hydrodynamic forces acting on two hulls and two rudders.

Definition of the positive direction of forces and moment acting on a ship model, and the hull drift angle β and rudder angle δ of each ship model is shown in Figure 3. Measured values of surge force X, lateral force Y, yaw moment N and rudder normal force F_N are non-dimensionalized as follows:

$$X', Y', F'_N = \frac{X, Y, F_N}{(1/2)\rho L d U^2}, \quad N' = \frac{N}{(1/2)\rho L^2 d U^2}$$

The prime means the non-dimensionalized value.

To distinguish two ships, suffix 1 and 2 is used for the notation. For example, rudder angle and drift angle of Ship1 are shown as δ_1 and β_1 respectively, and rudder angle and drift angle of Ship2 are shown as δ_2 and β_2 respectively.

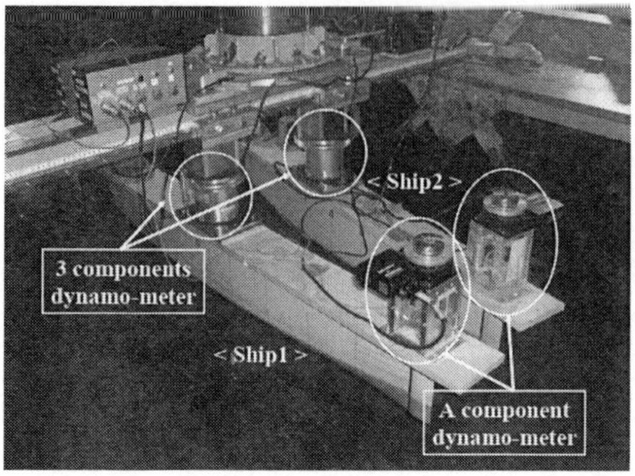

Figure 2: A scenery of captive model test

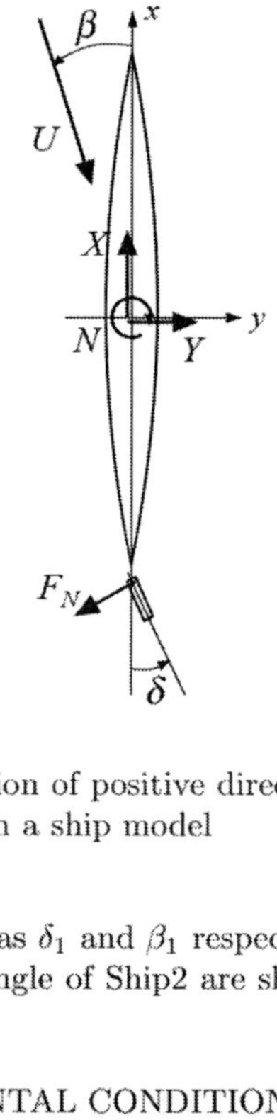

Figure 3: Definition of positive direction of forces and moment acting on a ship model

2.3 EXPERIMENTAL CONDITIONS

All the tests were carried out under the condition of the speed $U = 0.565$m/s, which is corresponding to Froude number $F_n = 0.165$.

2.3(a) Rudder force test

In the test, under the condition of $\beta_1 = \beta_2 = 0$ and $\delta_1 = 0$, we measure the hydrodynamic forces with only changing δ_2 in the range of $-15 \sim 15$deg by 5deg, as shown in Figure 4 (left). Then, the lateral clearance between two ships y_0 changed in three patterns, y_0=60mm, 140mm and 230mm ($y_0/B = 0.5$, 1.17 and 1.92).

Also, the hydrodynamic forces were measured while

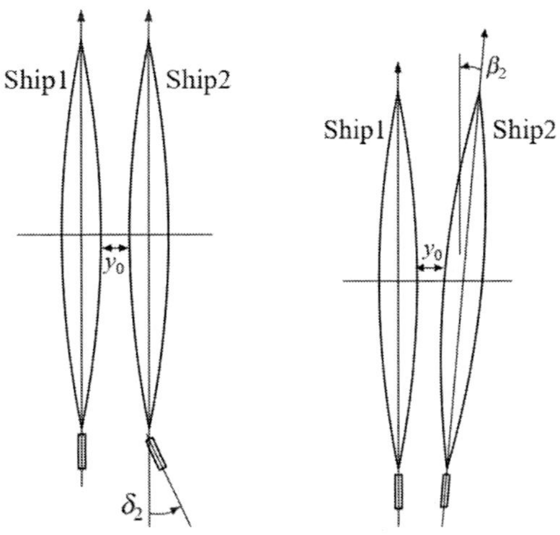

Figure 4: Kind of captive model tests (left: rudder force test, right: oblique towing test)

keeping $\beta_1 = \beta_2 = 0$ and $\delta_2 = 0$ with only changing δ_1 in the range of $-15 \sim 15$deg by 5deg. y_0 changed in two patterns, y_0=60mm and 140mm ($y_0/B = 0.5$ and 1.17). The experimental conditions in $y_0/B = 0.5$ and 1.17 are essentially the same as the conditions mentioned in the previous paragraph because Ship1/2 was only replaced with Ship2/1. Therefore, the absolute values of the measured hydrodynamic forces should be same for Ship1 and Ship2. We can check the accuracy through comparison of two measurements.

2.3(b) Oblique towing test

In the test, the hydrodynamic forces are measured when $\delta_1 = \delta_2 = 0$ and $\beta_1 = 0$ with only changing β_2 in the range of $-10 \sim 10$deg by 5deg, as shown in Figure 4 (right). Then, y_0 changed in two patterns, $y_0 = 140$mm and 230mm ($y_0/B = 1.17$ and 1.92).

3. OUTLINE OF HYDRODYNAMIC FORCE CALCULATIONS

In the calculations, we employ the assumption that thickness of ship and rudder is negligible. The reason for using the thin ship/rudder assumption is due to the easiness for understanding the physical phenomenon. Then, the ship hull and rudder are regarded as two flat plate wings with different aspect ratios and a method based on the nonlinear lifting surface theory is applicable for calculating the hydrodynamic forces acting on the flat wings (ships and rudders). In the past, some papers where a ship with rudder was dealt with as two flat plate wings were published by applying the lifting surface theory, for example, Hess[1], Fujino et al.[2] and Yasukawa[3]. In this study, we expand those methods to treat the problem of hydrodynamic forces acting on four flat wings (two hulls and two rudders).

For the numerical treatments of the lifting surface theory, QCM (Quasi-Continuous Method) presented by Lan[4] was applied. The outline of the calculation method is as follows:

1. Distribute horse shoe vortex with unit strength on the center planes of the flat wings.
2. Calculate the velocity components of the collocation points on the center planes of the flat wings induced by the distributed horse shoe vortices using Biot-Savart law.
3. Determine the strength of the horse shoe vortices from the boundary conditions of the wings.
4. Calculate the hydrodynamic forces acting on the wings using Kutta-Joukowski's theorem.

Trailing vortices on the hull (small aspect wing) are assumed to release from the hull center plane with half angle of the hull drift angle according to Bollay's wing theory[5]. For the rudder, it is assumed to flow out from the trailing edge to the direction of uniform flow. Figure 5 shows the vortex model for hull and rudder of a thin ship. In addition, the free-surface is treated as a rigid wall. Note that the present method can not deal with the suction force between ships moving in parallel since the thickness of the hull and rudder is neglected.

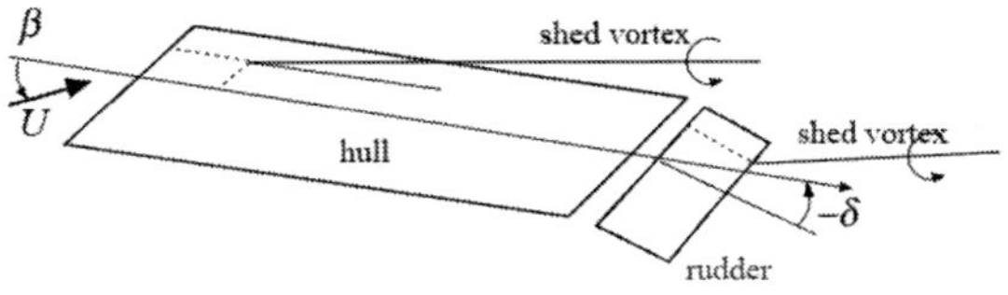

Figure 5: Vortex model for hull and rudder of a thin ship

4. HULL AND RUDDER INTERACTION OF A SHIP

Before dealing with the hydrodynamic interaction problem between two ships moving side by side, the hull-rudder interaction problem of a ship was experimentally and theoretically investigated. In 1979, Fujino et al.[2] carried out similar study on the hydrodynamic interaction between the two flat plates which were regarded as ship hull and rudder. In this study, a Wigley hull model and a rudder model with NACA0020 wing section were used in the experiments. The hydrodynamic forces acting on the hull and rudder were measured when rudder angle δ changed in the range of $-15 \sim 15$deg by 5deg and when hull drift angle β also changed in the range of $-10 \sim 10$deg by 5deg. This test was conducted by using two models of Ship1 and Ship2.

Figure 6 shows comparison of lateral force, yaw moment and rudder normal force of a ship versus rudder angle δ. Calculation results of the hydrodynamic forces acting on the hull and rudder were plotted in

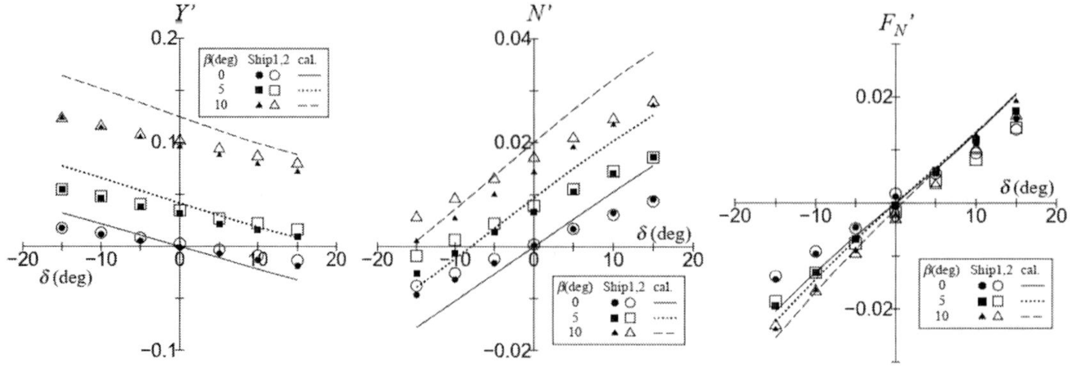

Figure 6: Comparison of lateral force, yaw moment and rudder normal force of a ship versus rudder angle

the figure together with experimental results in non-dimensionalized form. Experimental values should be same for Ship1 and Ship2. However, some difference in yaw moment N' is observed when δ is negative. Except that, the experimental results are reasonable. Calculation results are qualitatively agreed with the experiments, although some discrepancy is observed: inclinations of the calculated Y' and N' versus rudder angle δ are larger than the experiments.

Figure 7 shows comparison of lateral force, yaw moment and rudder normal force of a ship versus hull drift angle β. Note that the rudder normal force F_N' is ten times magnified in the figure. Experimental values of lateral force Y' are almost the same for Ship1 and Ship2. The calculated results of Y' and N' are qualitatively agreed well with experiments. Quantitatively, the results are close to the experiments. Discrepancy is observed in the experiments of rudder normal force F_N' between Ship1 and Ship2. The reason is due to the fact that measured values are small. The calculated F_N' is roughly agreed with the experiment.

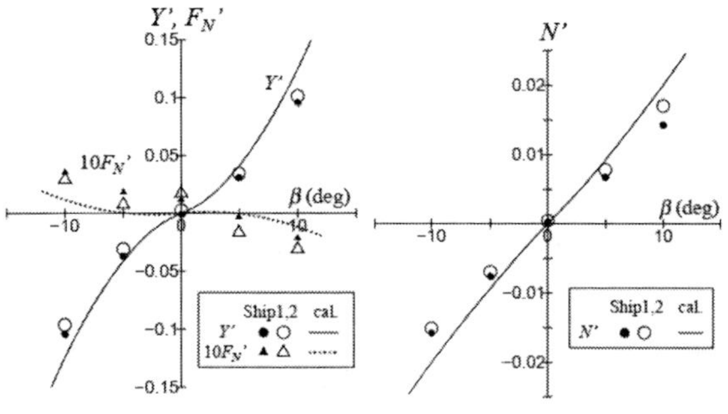

Figure 7: Comparison of lateral force, yaw moment and rudder normal force of a ship versus hull drift angle

5. HYDRODYNAMIC INTERACTION OF TWO SHIPS IN CLOSE PROXIMITY

5.1 SUCTION FORCES

First, we consider the lateral force and yaw moment acting on ships when two ships are moving in parallel with same speed. Rudder angle and hull drift angle are set to be zero. Then, magnitudes of the lateral force and yaw moment should be same for two ships and the direction should be opposite because the ships are same dimensions and same shapes.

Figure 8 shows the experimental results of suction force (lateral force) and bow-out moment (yaw moment) between two ships. Actually, some fluctuation in the measured values was observed since the order of magnitude

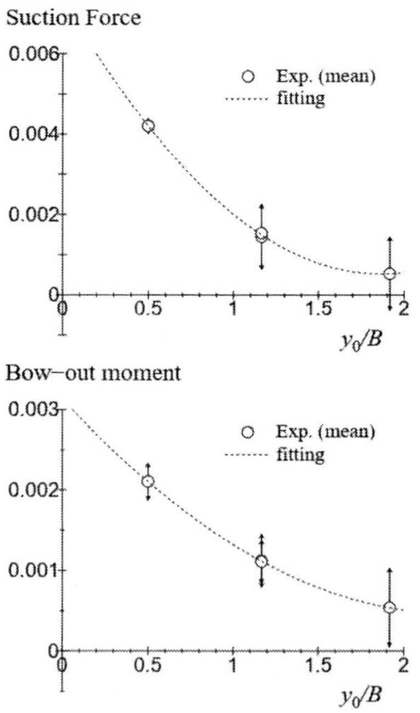

Figure 8: Suction force and bow-out moment between ships

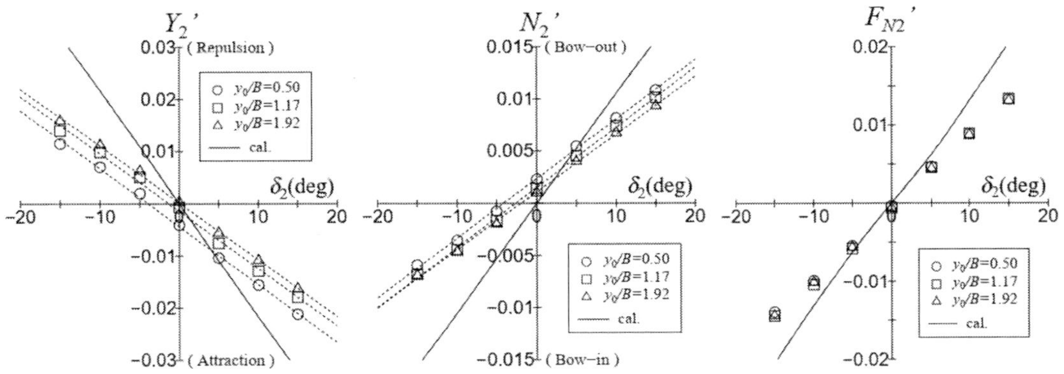

Figure 9: Lateral force, yaw moment and rudder normal force acting on Ship2 versus rudder angle of Ship2

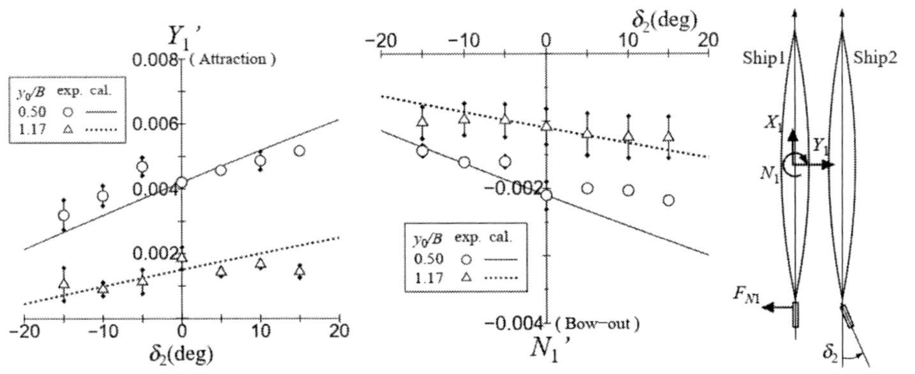

Figure 10: Lateral force and yaw moment acting on Ship1 versus rudder angle of Ship2

of the measured values is small. In the figure, the range of the fluctuation and the averaged value are shown. As the horizontal axis in the figure, the non-dimensional lateral clearance between ships y_0/B is taken, and the zero means touch of the ships. We see that the suction force and bow-out moment act on two ships and become larger with decrease of y_0/B. On the other hand, they tend toward zero with increase of y_0/B. This tendency is agreed with the tendency of the hydrodynamic forces acting on a ship moving straight in the proximity of vertical bank.

5.2 EFFECT OF RUDDER STEERING ON HYDRODYNAMIC FORCES

Figure 9 shows the results of lateral force (Y_2'), yaw moment (N_2') and rudder normal force (F_{N2}') acting on Ship2 versus the rudder angle of Ship2 (δ_2). The measured lateral force (Y_2') shifts downwards with decrease of lateral clearance between ships (y_0/B), and we see the suction force of a certain value is added to Ship2. The inclination of Y_2' to δ_2 is almost the same for any y_0/Bs. The yaw moment (N_2') has the same tendency as Y_2'. N_2' shifts upwards with decrease of y_0/B and the bow-out moment of a certain value is added to Ship2. Also, the inclination of N_2' to δ_2 is almost the same for any y_0/Bs. Namely, the hydrodynamic force characteristics of Ship2 do not change so much by steering of Ship2 and only some shift occurs by adjacency of Ship1. The calculations support the tendency: there is no difference in the result of inclinations of Y_2' and N_2' to δ_2 with change of y_0/B. The experimental values of rudder normal force (F_N') are almost the same for any y_0/Bs and the tendency is the same in the calculations.

Next, Figure 10 shows lateral force (Y_1') and yaw moment (N_1') acting on Ship1 when the rudder angle of Ship2 (δ_2) changes. Calculated results are also shown in Figure 10. The calculation method can not deal with the steady suction force and bow-out moment acting on two ships as mentioned in chapter 3, so that the hydrodynamic forces become zero at $\delta_2 = 0$. To correct the defect of the calculation results, we added the experimental values of steady suction force and bow-out moment shown in Figure 8 to the results. The corrected calculations are shown in Figure 10. Y_1' increases and N_1' decreases with increase of δ_2. This means that the attraction force and bow-out moment act on Ship1 when Ship2 takes the positive rudder angle to leave from Ship1. This tendency becomes more remarkable when the lateral clearance between ships is small. The calculation method can capture the tendency of the hydrodynamic force changes. Here, there is no figure of the rudder normal force of Ship1 (F_{N1}) versus δ_2 since

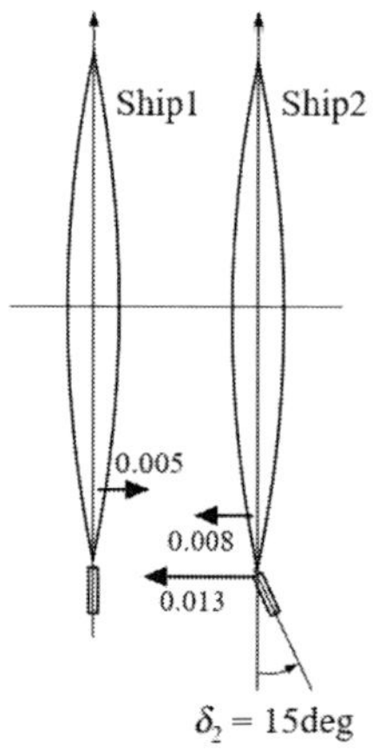

Figure 11: A schematic diagram showing lateral force coefficient components acting on Ship1 and Ship2 ($\delta_2 =$ 15deg, $y_0/B = 0.5$)

the order of magnitude is significantly small.

Figure 11 shows a schematic diagram showing lateral force coefficient components acting on two ships in the condition of $y_0/B = 0.5$ and $\delta_2 =$ 15deg. This figure was made based on experimental result. Due to steering of Ship2, lateral force is induced to the rudder of Ship2 with magnitude of 0.013. Then, flow at back side of the rudder is accelerated. Due to the influence of the flow acceleration, the flow at aft part of port side of Ship2 is also accelerated and an additional lateral force is induced to the hull of Ship2 with magnitude of 0.008. Since the magnitude of the steady suction force component between ships is 0.004, we see that the magnitude of the additional force component caused by the rudder steering is 0.004 (=0.008-0.004), and the order of magnitude is about 30%(=0.004/0.0013*100) of the rudder force. Due to the influence of the flow acceleration at aft part of port side of Ship2, flow at aft part of starboard side of Ship1 is also accelerated and an additional lateral force is induced to Ship1's hull with magnitude of 0.005. Then, the magnitude of the additional force component caused by the rudder steering becomes 0.001(=0.005-0.004) by reducing the steady suction force component, and the order of magnitude is about 8%(=0.001/0.0013*100) of the rudder force of Ship2. Thus, we found that additional lateral force component was induced to the aft part of Ship1's hull due to steering of Ship2. The direction of the additional force component is the suction side. However, the situation is not so serious because acting point of the additional force component is located at aft part of the hull and it means the bow-out yaw moment.

5.3 EFFECT OF HULL DRIFT ON HYDRODYNAMIC FORCES

Figure 12 shows lateral force (Y_2'), yaw moment (N_2') and rudder normal force (F_{N2}') acting on Ship2 when the drift angle of Ship2 (β_2) changes. There is no large difference of Y_2', N_2' and F_{N2}' with change of y_0/B in the experiments. The experimental results of Y_2' and N_2' include the steady suction force and bow-out moment between ships. However, the scale of the figure is too large to see the difference clearly. In the calculations, also, we can not notice the difference of the hydrodynamic forces by change of y_0/B in the figure. The hydrodynamic force characteristics of Ship2 versus hull drift angle of Ship2 are not influenced so much by the lateral clearance between two ships.

Figure 13 shows lateral force (Y_1') and yaw moment (N_1') acting on Ship1 when the hull drift angle of Ship2 (β_2) changes. In the region of negative β_2, Y_1' and N_1' increase with decrease of β_2. Namely, lateral force of suction side and yaw moment of bow-in side act on Ship1 when Ship2 moves obliquely with negative drift angle (the direction where Ship2's bow is close to Ship1). This tendency is more remarkable when y_0/B is small. The situation where the region of negative β_2 may be serious because the suction force and bow-in moment act on Ship2, too. The calculation method can capture the tendency of such the hydrodynamic force characteristics. Here, there is no figure of the rudder normal force of Ship1 (F_{N1}) versus β_2 since the order of magnitude is significantly small.

Figure 14 shows a schematic diagram showing lateral force coefficient components acting on two ships in the conditions of $\beta_2 =$ 0deg and -10deg of $y_0/B = 1.17$. This figure was made based on experimental result. In case of $\beta_2 =$ 0deg, lateral force acts on the stern part of the ship with the magnitude of 0.002. Namely, we see that suction force and bow-out moment act on both ships. In case of $\beta_2 = -10$deg, large lateral force acts on the fore part of Ship2's hull with the magnitude of 0.102 due to the influence of hull drift. This is operated as attraction force and bow-in moment to Ship1. Also, rudder force acts on Ship2 with magnitude of 0.003. When the bow part of Ship2 is close to Ship1, flow near the starboard side at fore part of Ship1 is accelerated and an additional lateral force is induced to Ship1's hull with the magnitude of 0.005. The acting point of the additional force is located at fore part of the hull, so that the bow-in moment acts on Ship1. Thus, the condition such as $\beta_2 = -10$deg is in danger of ship collision because the suction force and bow-in moment act on both ships at the same time.

6. CONCLUDING REMARKS

The hydrodynamic force characteristics of two Wigley hulls moving side by side in close proximity with steering were investigated experimentally and theoretically. Captive model tests were carried out to measure the hydrodynamic forces acting on two hulls and two rudders with changing lateral clearance between two ships, hull drift angle and rudder angle. Then, this experimental result was compared with numerical analysis using the non-linear lifting surface theory. The present study is concluded as follows:

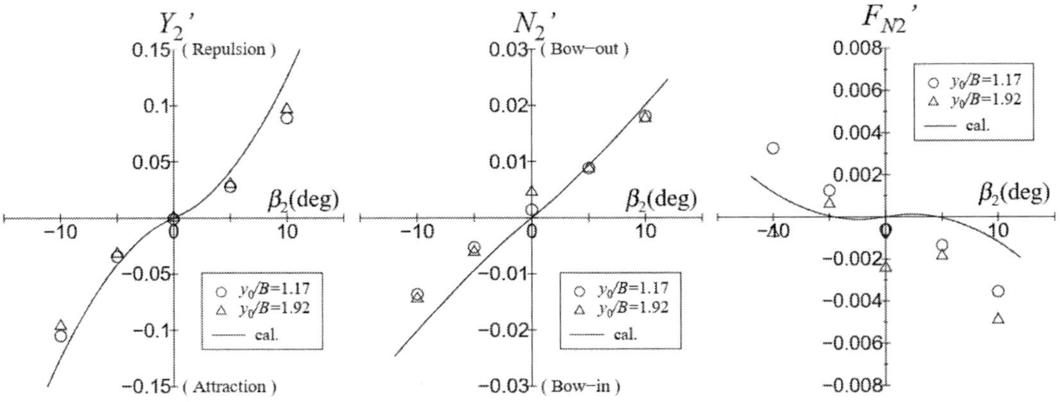

Figure 12: Lateral force, yaw moment and rudder normal force acting on Ship2 versus drift angle of Ship2

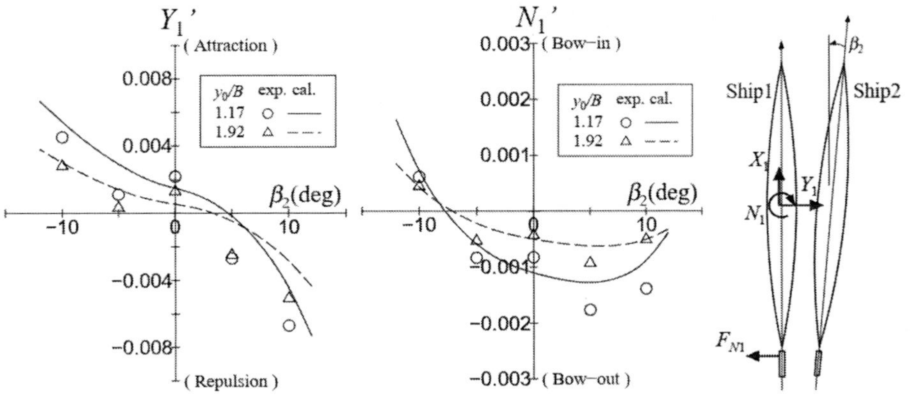

Figure 13: Lateral force and yaw moment acting on Ship1 versus drift angle of Ship2

1. Steady suction force and bow-out moment act on ships moving in parallel and become larger with decrease of lateral clearance between ships. The hydrodynamic force changes on Ship2 versus steering of Ship2 are not influenced by the lateral clearance between ships. When Ship2 steers so as to leave from Ship1, suction force and bow-out moment act on Ship1 and become larger with decrease of the lateral clearance.

2. The hydrodynamic force changes on Ship2 versus hull drift angle of Ship2 are not influenced by the lateral clearance between ships. When Ship2 moves obliquely with negative drift angle so as to be close to Ship1, suction force and bow-in moment act on Ship1 and become larger with decrease of the lateral clearance. This situation may be serious because the suction force and bow-in moment act on Ship2, too.

3. The lifting-surface based calculation method can capture the hydrodynamic force characteristics of two thin ship models with rudder.

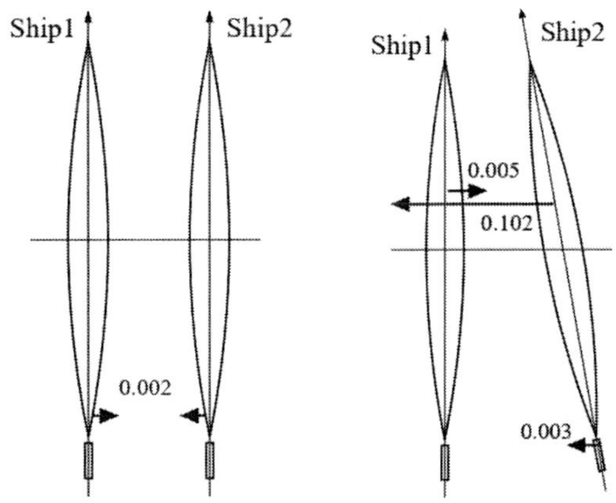

$\beta_2 = 0$deg $\beta_2 = -10$deg

Figure 14: A schematic diagram showing lateral force coefficient components acting on Ship1 and Ship2 ($y_0/B = 1.17$)

REFERENCES

[1] Hess, F.: *Lifting-Surface Theory Applied to Ship-Rudder Systems*, International Shipbuilding

Progress, Vol.25, No.292 (1978), pp.299-305.

[2] Fujino, M., Numata, T. and Motora, S.: *A Fundamental Study on Hull to Rudder Interaction*, J. Society of Naval Architects of Japan, Vol.146 (1979), pp.213-221 (in Japanese).

[3] Yasukawa, H.: *Hydrodynamic Interactions among Hull, Rudder and Propeller of a Turning Thin Ship*, Trans. West-Japan Soc. Naval Arch., No.84 (1992), pp.59-83.

[4] Lan, C. E.: *A Quasi-Vortex-Lattice Method in Thin Wing Theory*, J. Aircraft, Vol.11, No.9 (1974), pp.518-527.

[5] Bollay, W.: *A Non-Linear Wing Theory and Its Application to Rectangular Wings of Small Aspect Ratio*, Z.A.M.M., Bd.19, Nr.1 (1939), pp.21-35.

Authors Biography

Hironori YASUKAWA holds the current position of professor in department of vehicle and environmental systems, Hiroshima University. He is responsible for research on the ship dynamics and hydrodynamics. Now, he is maneuvering committee member of ITTC.

Seiko YOSHIDA is master course student in Hiroshima University. She is responsible for a development of calculation method for hydrodynamic force predictions in restricted shallow water.

NUMERICAL INVESTIGATION ON SHIP-SHIP HYDRODYNAMIC INTERACTION IN RESTRICTED WATERS

ZHANG Chen-xi and **ZOU Zao-jian**, School of Naval Architecture, Ocean and Civil Engineering, Shanghai Jiao Tong University, China

SUMMARY

Adjacent ships moving relatively such as meeting and overtaking, and operating in side-by-side or in tandem array will induce hydrodynamic interaction which has significant influence on ship maneuverability. In this paper, a numerical study on the hydrodynamic interaction between ships undergoing meeting and overtaking in restricted waters is carried out. The unsteady RANS equation in conjunction with a RNG k-ε turbulence model is solved. The relative motion of ships is dealt with by using the dynamic mesh method and sliding interface technique. Numerical simulation of the three-dimensional unsteady viscous flow around the ships is conducted, and the hydrodynamic interaction is calculated. The results at different water depth and distance to bank are compared to investigate the influence of bottom clearance and bank on the ship-to-ship interaction.

1. INTRODUCTION

A ship sailing close to other ships or fixed obstacles will experience a particular force and moment caused by the hydrodynamic interaction between these objects. This phenomenon has received attention for a long time, but some new circumstances make it more frequent and complicated. The increasing number of ships leads to a high density of ships in waterways, so the interaction become unavoidable. These interaction scenarios include meeting and passing of ships, moored ships affected by passing ships etc. On the other hand, the great progress in navy and ocean resource development brings about some other complicated interaction cases such as tandem lightering, interaction with marine structure etc.

The hydrodynamic interaction has significant influence on ship maneuverability which is closely related to safe operation and effective control of ships. As the ship dimensions become larger, the effect of the bank and water bottom on the hydrodynamic interaction between ships sailing in restricted water such as a harbor or a channel has been amplified. During the last decade, people paid more and more attention to the maritime safety and ocean environment protection. As a result, prediction of ship maneuverability and related investigation of hydrodynamic interaction between ships in restricted water have become hot topic and research focus.

Some experiments were conducted to study the hydrodynamic interaction between ships. Newton [1] investigated two ship models overtaking in deep water. Müller [2] presented both meeting and overtaking results of two ships in a narrow channel. Dand [3] carried out test of two ships on parallel courses. Vantorre et al. [4] reported a comprehensive result of two ships' interaction with different models and conditions.

This work is financially supported by the National Natural Science Foundation of China (Grant No: 50779033) and the National High Technology Research and Development Program (863 Program) of China (Grant No: 2007AA11Z250).

Other researchers also developed many numerical methods using different theories. Tuck & Newman [5] calculated hydrodynamic forces between two ships in deep water using the method of matched asymptotic expansions based on slender-body theory. Yeung [6] applied a similar approach to solve the problem of two ships interaction in shallow water. Kijima & Yasukawa [7] considered the hydrodynamic behavior of ships during meeting and passing in narrow water channel. Kaplan & Sankaranarayanan [8] analyzed the interaction of two ships passing and overtaking in a shallow asymmetric channel. Varyani et al. [9][10] obtained interactive forces between two or more ships in restricted waters. Chen et al. [11] used Chimera RANS method to compute multi-ship interaction in confined water. They also analyzed the results to determine the effect of ship speed, ship type, bottom clearance, etc. Lee & Lee [12] predicted the hydrodynamic forces of two ships sailing in confined water under wind-effect.

In this article, the commercial software FLUENT is used to simulate the viscous flow around two ships moving on parallel courses in restricted waterway. The three-dimensional unsteady viscous flow and the hydrodynamic interaction are calculated. Numerical results including both meeting and overtaking are compared to investigate the bottom clearance effect and bank effect of interactive force and moment.

2. PROBLEM FORMULATION

As shown in Figure 1, three right-handed coordinate systems are used to describe the problem. Coordinate system o-xyz fixed in space, o_1-$x_1y_1z_1$ and o_2-$x_2y_2z_2$ fixed on Ship1 and Ship2, with x-axis pointing to the moving direction of ship. The z-axis of all three systems is pointing upward. The constant speeds of two ships are U_1 and U_2 ($U_2>U_1$ when overtaking), and their directions are both parallel the x-axis of the space fixed system. Sp and ST are transverse and longitudinal distance between two ships, h is the water depth. Sp_i (i=1, 2) is the distance between ship i and bank. $\mathrm{W=Sp_1+Sp+Sp_2}$ is the width of the channel.

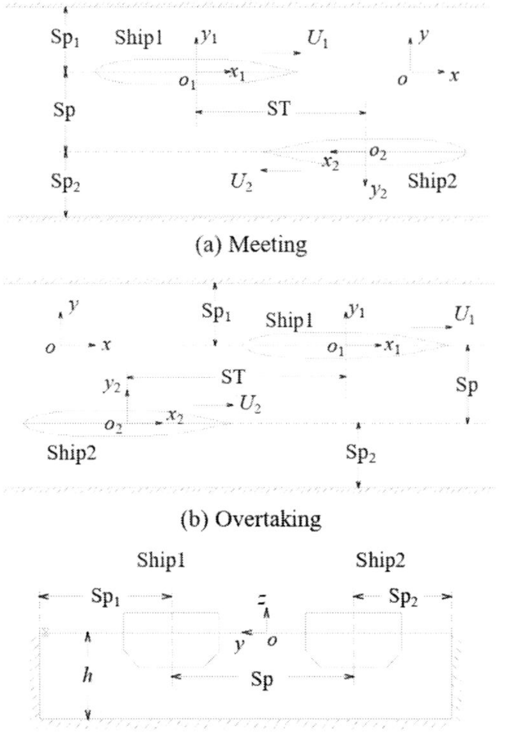

(a) Meeting

(b) Overtaking

Figure 1: Coordinate system

The elevation of free surface is assumed to be negligible for the small Froude number. Fluid around ships is single phase, 3-D, viscous and incompressible. Using Reynolds-Averaged method, the governing equations for this flow problem are:

$$\frac{\partial u_i}{\partial x_i} = 0 \tag{1}$$

$$\rho\frac{\partial u_i}{\partial t} + \rho u_j \frac{\partial u_i}{\partial x_j} = -\frac{\partial p}{\partial x_i} + \frac{\partial}{\partial x_j}\left(\mu\frac{\partial u_i}{\partial x_j} - \rho\overline{u'_i u'_j}\right) \tag{2}$$

where x_i (i= 1, 2, 3) is the ith component of the fixed coordinate system, ρ is the density of fluid, u_i is mean velocity component, p is the mean pressure, μ is the viscosity and $-\rho\overline{u'_i u'_j}$ is the Reynolds stress.

The RNG k-ε model is adopted to solve the Reynolds stress. The transportation equations of turbulence kinetic energy k and its rate of dissipation ε are:

$$\frac{\partial(\rho k)}{\partial t} + \frac{\partial(\rho k u_i)}{\partial x_i} = \frac{\partial}{\partial x_j}\left[\alpha_k \mu_{eff}\frac{\partial k}{\partial x_j}\right] + G_k - \rho\varepsilon \tag{3}$$

$$\frac{\partial(\rho \varepsilon)}{\partial t} + \frac{\partial(\rho \varepsilon u_i)}{\partial x_i} = \frac{\partial}{\partial x_j}\left[\alpha_\varepsilon \mu_{eff}\frac{\partial \varepsilon}{\partial x_j}\right] + \frac{C^*_{1\varepsilon}\varepsilon}{k}G_k - C_{2\varepsilon}\rho\frac{\varepsilon^2}{k} \tag{4}$$

where G_k represents the generation of turbulence kinetic energy due to the mean velocity gradients, μ_t is the turbulence viscosity, they are expressed as:

$$G_k = \mu_t\left(\frac{\partial u_i}{\partial x_j} + \frac{\partial u_j}{\partial x_i}\right)\frac{\partial u_i}{\partial x_j}, \quad \mu_t = \rho C_\mu \frac{k^2}{\varepsilon} \tag{5}$$

The other terms have the following value and expression:

$$\mu_{eff} = \mu + \mu_t, \quad C_\mu = 0.0845, \quad \alpha_k = \alpha_\varepsilon = 1.39,$$

$$C^*_{1\varepsilon} = C_{1\varepsilon} - \frac{\eta(1-\eta/\eta_0)}{1+\beta\eta^3}, \quad C_{1\varepsilon} = 1.42, \quad C_{2\varepsilon} = 1.68,$$

$$\eta_0 = 4.377, \beta = 0.012, \quad \eta = \left(2E_{ij}\cdot E_{ij}\right)^{1/2}\frac{k}{\varepsilon},$$

$$E_{ij} = \frac{1}{2}\left(\frac{\partial u_i}{\partial x_j} + \frac{\partial u_j}{\partial x_i}\right)$$

3. NUMERICAL METHOD

3.1 BOUNDARY CONDITIONS

The domain is chosen as a cube, and the length should be long enough to make sure that the outflow boundary is far away from the ship. Here the length is 10 times ship length. The boundary conditions of ship meeting are shown in Figure 2. Considering that the free surface is assumed to be rigid in the analysis, the top surface of the domain is set as symmetry.

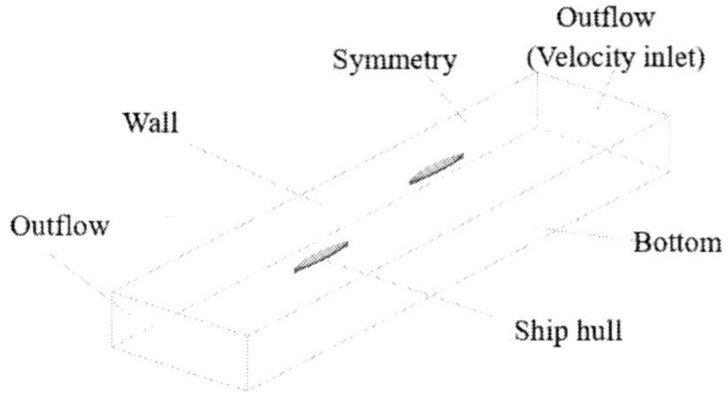

Figure 2: Boundary conditions

Different from meeting condition, to simulate overtaking condition in real case need more computational time and larger domain. For this reason, in the overtaking condition, the slower ship (Ship1) is fixed, while the space-fixed coordinate system is moving. An inflow speed which has the same magnitude but in opposite direction as Ship1's speed is set to simulate the ships' motion as shown in Figure 2 in bracket. The overtaking ship will move with the relative speed of two ships. Also the water bottom and the bank move with the same speed and direction as inflow, so they are stationary in space-fixed coordinate system.

3.2 DYNAMIC MESH METHOD

In this paper, dynamic mesh method was used to simulate the relative motion between two ships. There are three kinds of methods in FLUENT: smoothing, layering and remeshing. The smoothing method requires that the motion is small in magnitude. Remeshing method need a very small time step to avoid a negative cell volume, and it is hard to control the quality of regenerated mesh during the procedures. In this paper, the layering method is chosen to update the mesh.

The basic principle of layering is to split or collapse cells near the moving boundary with defined movement. Figure 3 shows the sketch of layering, where H is the height of jth layer, α_s and α_c are the split factor and collapse factor, h_0 is the ideal height defined by user. If $H > (1+\alpha_s)\, h_0$, Layer j will split. If $H < \alpha_c h_0$, Layer j will collapse.

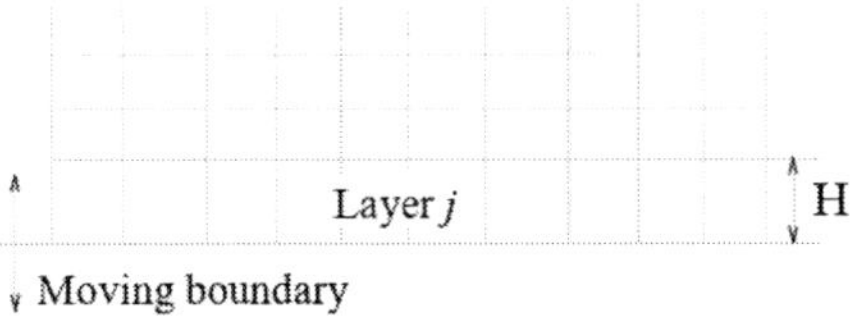

Figure 3: Sketch of layering

When using this method, boundaries between two moving zones or between moving zone and stationary zone are set as interface. Interface is usually used when calculating non-conformal grid of two zones. It uses interpolate method to calculate the flux cross two faces. The zones in Figure 4 which contain the ships are set as moving zone moving at the speed of ship, and the dashed line indicates the interface. Boundaries in solid ellipse and dashed ellipse are places where grids split and collapse.

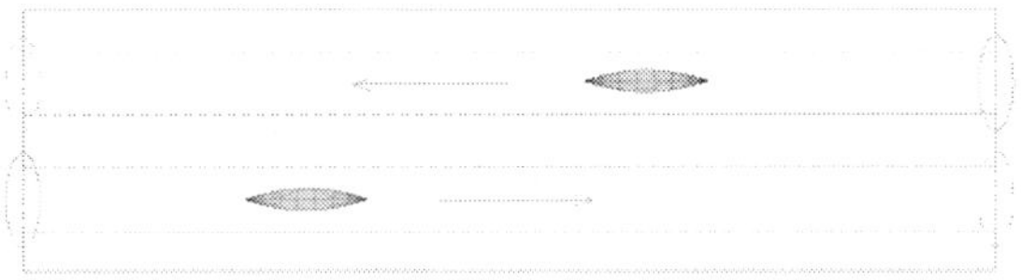

Figure 4: Definition of moving zone in meeting condition

Figure 5 shows the moving zone of overtaking condition. Here the boundary of dynamic mesh zone is set apart from the outflow boundary to avoid reverse flow. As meeting condition, boundaries in solid ellipse and dashed ellipse are places where grids split and collapse.

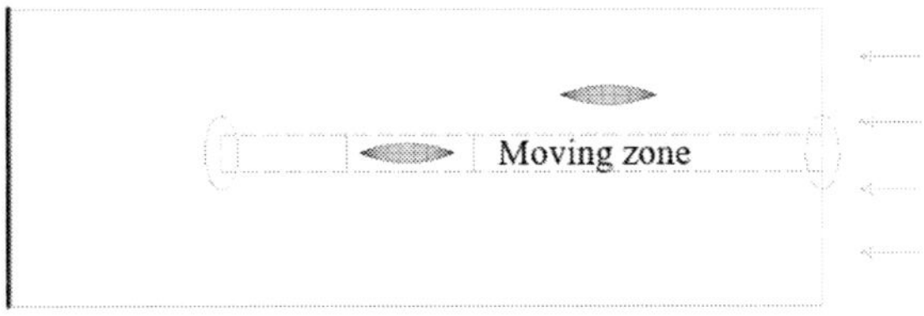

Figure 5: Definition of moving zone in overtaking condition

3.3 USING OF UDF

User-defined function, or UDF, is a function programmed by a user that can be dynamically linked with the FLUENT solver to enhance the standard features of the code. In this paper, UDF is used to set the speed of ship, to define the gravity center of each ship and to calculate the force and moment on each ship. UDF is also used to output the results as text file to edit.

4. NUMERICAL RESULTS AND DISCUSSION

4.1 SHIP TYPE AND GRID GENERATION

Two ships in meeting and overtaking condition at different water depth and distance to the bank are calculated. The parameters of ships are shown in Table 1.

Table 1: main parameters of ships

Ship Type	KCS	OSAKA
L (m)	230.000	237.000
B (m)	32.200	30.400
D (m)	10.800	10.800
C_B	0.65	0.831
λ	64	64

where L is the ship length, B is the ship breadth, D is the ship draught, C_B is the block coefficient, λ is the scaling factor. Two KCS ships are used to calculate in meeting condition, and different ships are used in overtaking condition. The speed of both ships in meeting condition is 8 knot, while in overtaking condition the speed is 4 knot (KCS) and 8 knot. The lateral force, yaw moment and longitudinal distance are expressed in the non-dimensional forms as:

$$C_{Fi} = \frac{F_i}{\frac{1}{2}\rho U_1 U_2 B_i D_i} \qquad C_{Mi} = \frac{M_i}{\frac{1}{2}\rho U_1 U_2 B_i D_i L_i}$$

$$ST' = \frac{ST}{\frac{1}{2}(L_1 + L_2)}$$

where i=1, 2 represent Ship1 and Ship2, and the initial ST' is defined negative. Here the mixed mesh types are used to generate the grid. Tetrahedral mesh scheme is used around the stern where the geometry is complicated. Hex and Cooper mesh scheme are used in other zones. Mesh on KCS ship hull is shown in Figure 6.

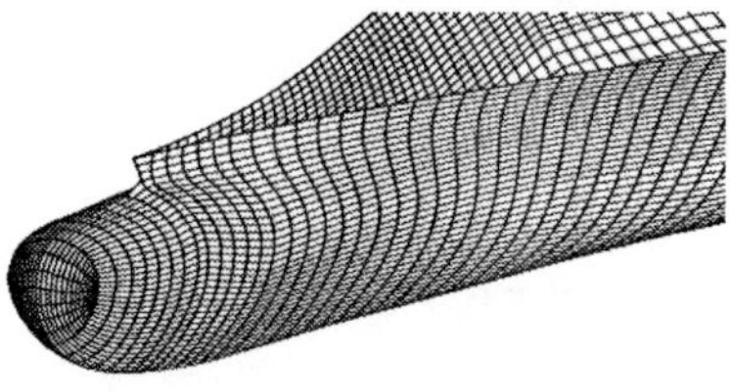

Figure 6a: Grids on bow

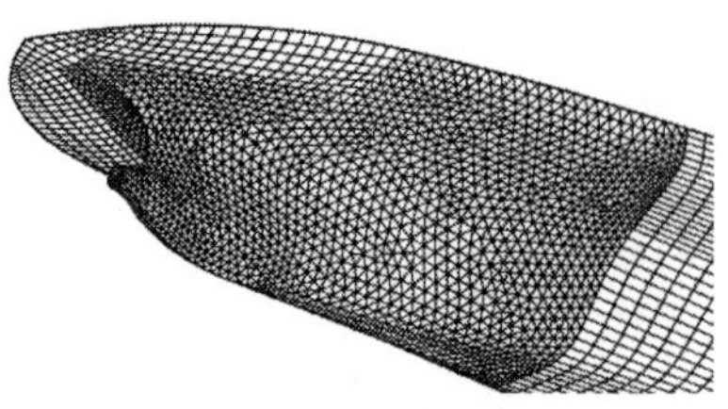

Figure 6b: Grids on stern

4.2 SOLUTION SETTINGS

The unsteady part in governing equations is discretized by the first-order implicit scheme. The algorithm SIMPLE is chosen to solve the velocity-pressure coupling problem. Momentum, turbulent kinetic energy, turbulent dissipation rate are discretized by second-order upwind scheme. Time step is set as 0.1s to make sure that the product of the time step times velocity is smaller than the mesh size in moving zone.

4.3 EFFECT OF BOTTOM CLEARANCE

The water depth selected for comparison is shown in Table 2. Here the bank effect is not taken into account.

Table 2: water depth variation

	Sp	*h/D*
Meeting	0.4*L*	1.2,1.3,1.5,1.8,2.0
Overtaking	0.4*L*	1.3,1.4,1.5,1.7,2.0

Figure 7 and Figure 8 show the lateral force and yaw moment in meeting and overtaking conditions. Here only the results of overtaken ship in overtaking conditions are given, because the force and moment act on that ship is larger than the other one. It is seen that as the *h/D* decreases the hydrodynamic force and moment increase. In meeting conditions, when *h/D* is larger than 1.8, the effect of bottom clearance becomes small.

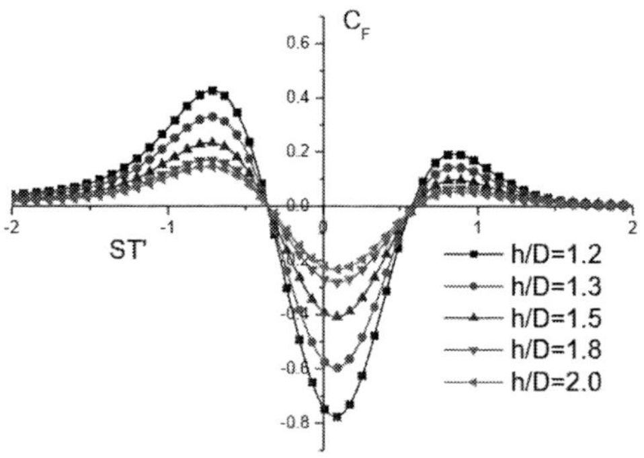

Figure 7a: Sway coefficients changing with water depth in meeting condition

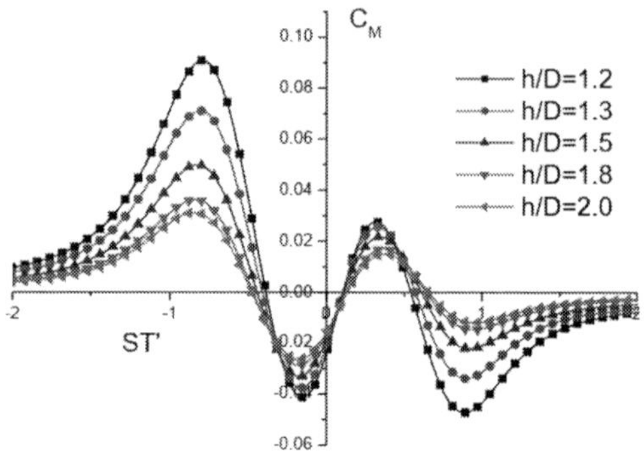

Figure 7b: Yaw coefficients changing with water depth in meeting condition

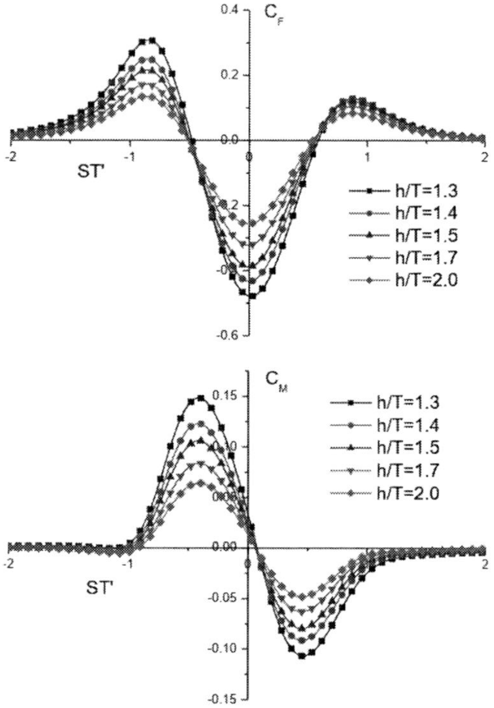

Figure 8: Sway and yaw coefficients changing with water depth in overtaking condition

4.4 EFFECT OF THE BANK

As it is well known, a lateral force and a yaw moment will act on the ship when it sails near the bank. The direction of the lateral force has been found to be towards the near bank when *h/D* is greater than 1.2. For *h/D* is 1.1 and lower the lateral force may be away from the bank. Whatever the *h/D* is, the yaw moment will turning the bow away from the bank.

In this section the bank effect to the hydrodynamic interaction between two ships is investigated. Table 3 and Table 4 show the cases of bank influence in meeting and overtaking conditions. The *h/D* is 1.267 in all cases, and ∞ means no bank exists.

Table 3: parametric variation of bank effects in meeting condition

Sp	Sp_1	Sp_2	W
0.5L	0.25L,0.35L,0.5L 0.65L,0.75L,1.0L	Sp_1	1.0L,1.2L,1.5L, 1.8L,2.0L,2.5L

Table 4: parametric variation of bank effects in overtaking condition

Case	Sp	Sp_1	Sp_2	W
1	0.4L	∞	0.3L,0.4L,0.5L 0.6L,0.8L,1.0L	∞
2	0.4L	0.3L,0.4L,0.5L 0.6L,0.8L,1.0L	∞	∞

Figure 9 shows the effect of channel width on lateral force and yaw moment acting on ship. It is seen that when the width is larger than 1.5L the bank effects on the hydrodynamic force is small. The effect is more obvious in yaw moment. With the decrease of W, the yaw moment decreases, but the reduction keeps approximately constant. In these cases, the bank effects can be neglected when the width is larger than 1.8L.

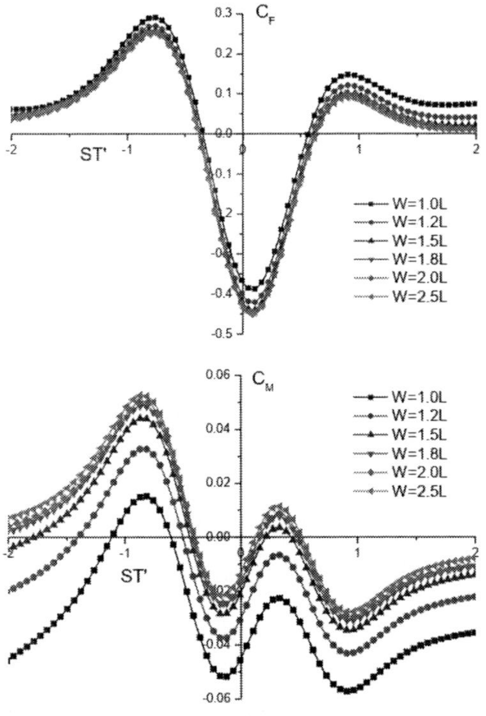

Figure 9: Sway and yaw coefficients changing with channel width in meeting condition

Figure 10 shows the effect of bank at side of Ship1 on the hydrodynamic forces during overtaking. The second peaks of lateral force and yaw moment decrease as Sp1 decreases. It is seen that when Sp1 is larger than 0.6L, its effect is very small.

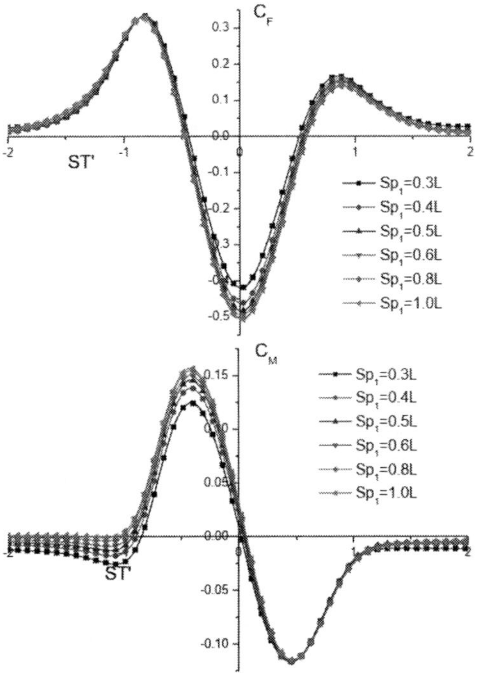

Figure 10: Sway and yaw coefficients changing with Sp1 in overtaking condition

Figure 11 shows the effect of bank at side of Ship2 on the hydrodynamic forces during overtaking. The first two peaks of lateral force and yaw moment increase as Sp2 decreases. It is seen that when Sp2 is larger than 0.8L, its effect is very small.

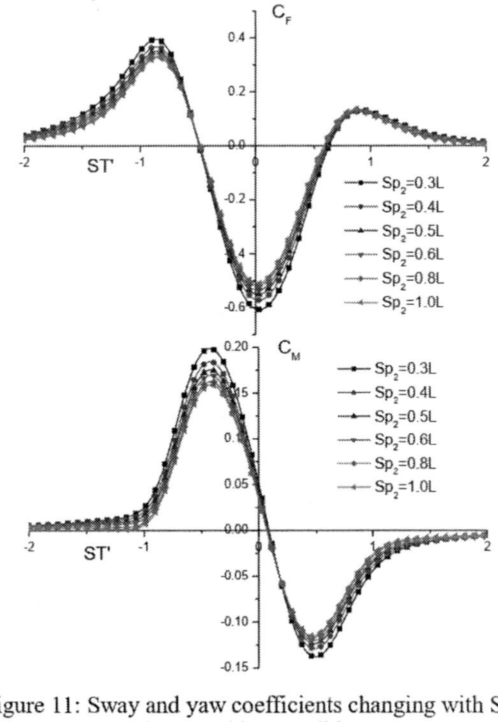

Figure 11: Sway and yaw coefficients changing with Sp2 in overtaking condition

5. CONCLUSION

A numerical study on the unsteady hydrodynamic interaction between two ships meeting and overtaking in a restricted waterway is carried out by using FLUENT. UDF is used to define the movement of ships, and the dynamic mesh method is adopted to update the mesh. Lateral force and yaw moment act on ships during these processes are calculated. Both the effect of bottom clearance and bank on the interactive force and moment are investigated. The main conclusions are as follows:

- The bottom clearance does have a major influence on lateral force and yaw moment in both meeting and overtaking conditions. The magnitude of hydrodynamic force and moment increases as the clearance decreases.
- It is more dangerous to overtake a ship when there is a bank on the side of overtaking ship than on the side of overtaken ship. No matter under what kind of bank condition, the bank effects on the ship-ship interaction can be neglected if the distance between ship and bank is larger than 0.8L.
- The present study can qualitatively estimate the ship-to-ship interaction in restricted water. It can provide certain guidance on safe maneuvering and control of ships navigating in restricted water.

6. REFERENCE

1. NEWTON, R.N., 'Some notes on interaction effects between ships close aboard in deep water', First Symposium on Naval Manoeuvrability, 1960.

2. MÜLLER, E., 'Untersuchungen über die gegenseitige Kursbeeinflussung von Schiffen auf Binnenwasserstrassen', Schiff und Hafen, 1967.

3. DAND, I.W., 'Some measurements of interaction between ship models passing on parallel courses', National Maritime Institute, 1981.

4. VANTORRE, M., LAFORCE, E., VERZHBITSKAYA E., 'Model test based formulations of ship-ship interaction forces for simulation purposes' IMSF-28th Annual General Meeting, 2001.

5. TUCK, E.O., NEWMAN, J.N, 'Hydrodynamic interactions between ships', 10th ONR Symposium on Naval Hydrodynamics, 1974.

6. YEUNG, R.W., 'On the interactions of slender ships in shallow water', Journal of Fluid Mechanics, 1978.

7. KIJIMA, K., YASUKAWA, H., 'Manoeuvrability of ships in Narrow Waterway', J.S.N.A., 1984.

8. KAPLAN, P., SANKARANARAYANAN, K., 'Theoretical analysis of generalized hydrodynamic interaction forces on ships in shallow channels', SNAME Trans, 1991.

9. VARYANI, K.S., MCGREGOR, R.C., WOLD, P., 'Interactive forces and moments between several ships meeting in confined waters', Control Eng., 1998.

10. VARYANI, K.S., MCGREGOR, R.C., WOLD, P., 'Identification of trends in extremes of sway-yaw interference for several ships meeting in restricted waters', Ship Technology Research, 2002.

11. CHEN, H.C., LIU, T., HUANG, E.T., 'Application of Chimera RANS Method for Multiple-Ship Interactions in a Navigation Channel', Proceedings of the Twelfth International Offshore and Polar Engineering Conference, 2002.

12. LEE C.K., LEE S.G., 'Hydrodynamic forces between vessels and safe maneuvering under wind-effect in confined waters', Journal of Mechanical Science and Technology, 2007.

7. AUTHORS BIOGRAPHY

Chen-xi Zhang is a master student at the School of Naval Architecture, Ocean and Civil Engineering, Shanghai Jiao Tong University. His current research interest is numerical study on ship hydrodynamics in restricted water.

Zao-jian Zou is a Professor at the School of Naval Architecture, Ocean and Civil Engineering, Shanghai Jiao Tong University. His research interest covers ship manoeuvrability and control, numerical towing tank and application of artificial intelligence in ship hydrodynamics.

CFD PREDICTION AND VALIDATION OF SHIP-BANK INTERACTION IN A CANAL

L Zou, Chalmers University of Technology, Sweden
L Larsson, Chalmers University of Technology, Sweden
G Delefortrie, Flanders Hydraulics Research, Belgium
E Lataire, Maritime Technology Division Ghent University, Belgium

SUMMARY

This paper utilizes CFD (Computational Fluid Dynamics) methods to investigate the bank effects on a tanker moving straight ahead at low speed in a canal characterized by surface piercing banks. For varying water depths and ship-to-bank distances, the sinkage and trim as well as the viscous hydrodynamic forces on the hull are predicted mainly by a steady state RANS (Reynolds Averaged Navier-Stokes) solver, in which the double model approximation is adopted to simulate the flat free surface. A potential flow method is also applied to evaluate the effect of the free surface and viscosity on the solutions. In addition, focus is placed on V&V (Verification and Validation) based on a grid convergence study and comparison with EFD (Experimental Fluid Dynamics) data, as well as the exploration of the modelling error in RANS computations to enable more accurate and reliable predictions of the bank effects.

NOMENCLATURE

α	Constant (-)
δ_{RE}	Discretization error (-)
ν	Kinematic viscosity (N·s·m^{-2})
ρ	Density of water (kg·m^{-3})
σ	Sinkage (m)
$\bar{\sigma}_{ij}$	Average stress (N·m^{-2})
τ	Trim (°)
ϕ_0	Extrapolated solution at zero grid size (-)
ϕ_i	i th grid solution (-)
ω	Specific dissipation rate (s^{-1})
A_w	Water plane area (m^2)
B	Beam (m)
C_F	Frictional resistance coefficient (-)
C_{PV}	Viscous pressure resistance coefficient (-)
D	Experimental data (-)
E	Comparison error (-)
F_i	Body force (N·kg^{-1})
Fr	Froude number, $Fr=U/\sqrt{gL_{PP}}$ (-)
g	Acceleration of gravity (m·s^{-2})
h	Water depth (m)
h_1	Finest grid step (-)
h_i	i th grid step (-)
I_w	Longitudinal moment of inertia of the water plane area (m^4)
k	Turbulent kinetic energy, $k=\overline{u_i' u_i'}/2$ (m^2·s^{-2})
K	Roll moment (N·m)
K'	Non-dimensional roll moment (-)
L_{PP}	Ship length between perpendiculars (m)
M	Trim moment (N·m)
N	Yaw moment (N·m)
N'	Non-dimensional yaw moment (-)
n_g	Available grids number (-)
$\bar{p}$	Average pressure (N·m^{-2})
p	Order of accuracy (-)
p_{th}	Theoretical order of accuracy (-)
r	Grid refinement ratio (-)
Re	Reynolds number, $Re=U{\cdot}L_{PP}/\nu$ (-)
R_F	Frictional resistance (N)
$R_{ij}=R_{ji}$	Reynolds stresses (N·m^{-2})
R_{PV}	Viscous pressure resistance (N)
S	Numerical solution (-)
S_W	Wetted hull surface (m²)
T	Draft (m)
U	Ship speed (m·s^{-1})
U_D	Data uncertainty (-)
$\bar{u}_{i(j)}$	Average velocity (m·s^{-1})
u_i'	Fluctuating velocity (m·s^{-1})
UKC	Under keel clearances (m)
U_{num}	Numerical uncertainty (-)
U_S	Discretization uncertainty (-)
U_{Val}	Validation uncertainty (-)
X	Longitudinal force (N)
X'	Non-dimensional longitudinal force (-)
Y	Sway force (N)
Y'	Non-dimensional sway force (-)
y_B	Ship-to-bank distance (m)
y^+	Non-dimensional wall distance (-)
Z	Sinking force (N)

1. INTRODUCTION

It is well known that when a ship is passing through restricted waterways, such as canals, narrow channels, interaction occurs between the ship and the banks. Additional hydrodynamic forces and moments generated by the vicinity of the bank act on the hull and influence the ship motion. The explanation of these bank effects is that when the distance between the hull and the waterway boundary is narrowed, the flow is accelerated and the pressure accordingly decreased, which induces the variation in the hydrodynamic characteristics. The produced hydrodynamic forces, especially in shallow water, may considerably affect the manoeuvring performance of the ship, making it difficult to steer. The ship may collide with the side wall or run aground due to the so-called 'squat' phenomenon. From this point of view, bank effects are extremely important for ship navigation. In the past decades, many investigations on bank effects have been carried out, both experimentally and numerically. A notable event was, the International

Conference on Ship Manoeuvring in Shallow and Confined Water: Bank Effects [1], in which a broad concern about this problem was expressed and many interesting papers were presented.

However, historical investigations have mostly relied on some kind of experimental tools, such as model tests and empirical or semi-empirical formulae, which normally treat the bank effect as a function involving hull-bank distance, water depth, ship speed, hull form, bank geometry, propeller performance, etc. During the 1970s, Norrbin at SSPA Sweden carried out experimental research and then, based on the experiments, proposed empirical formulae to estimate the hydrodynamic forces for flooded [2], vertical [3] and sloping [3] banks. Li et al. [4] continued Norrbin's investigations and tested the bank effect in extreme conditions for three different hull forms (tanker, ferry and catamaran). The influence of ship speed, propeller loading and bank inclination was evaluated. Ch'ng et al. [5] conducted a series of model tests and developed an empirical formula to estimate the bank-induced sway force and yaw moment for a ship handling simulator. In recent years, comprehensive model tests in a towing tank have been carried out at Flanders Hydraulics Research (FHR), Belgium, to build up mathematical models for bank-effect investigations and to provide data for computation validation. Vantorre et al. [6] discussed the influence of water depth, lateral distance, forward speed and propulsion on the hydrodynamic forces and moments based on a systematic captive model test program for three ship models moving along a vertical surface-piercing bank. Empirical formulae for the prediction of ship-bank interaction forces were proposed. From extensive tests, Lataire et al. [7] developed a mathematical model for the estimation of the hydrodynamic forces, moments and motions taking the ship speed, propulsion and ship/bank geometry into consideration. In addition, two parameters defining the distance to a bank of irregular geometry and the equivalent blockage, indicating the influence of water particles on the bank effect, were established.

Although experimental tools and empirical formulae are widely used for bank-effect prediction, they have their shortcomings because of the approximation. For example, empirical formulae are only suitable for cases with similar hull form and conditions. Otherwise, the prediction is barely reliable. To establish a mathematical model a significant amount of systematic and expensive model tests are always required. However, the most important weakness of these tools is their inability to provide detailed information of the flow field, which can explain the flow mechanism behind the bank effect. In view of this, people resort to using numerical methods to deal with the phenomena of bank effects. Among existing numerical methods, the potential flow method (based on the inviscid flow assumption) is the most common one. Newman [8] applied the Green function to predict the interaction force between a ship and an adjacent rectangular canal wall. Miao et al. [9] studied the case of a ship travelling in a rectangular channel and the lateral force, yaw moment and wave pattern were estimated based on Dawson's method. It was shown that the applied potential flow method was able to predict reasonable hydrodynamic forces for a water depth to draught ratio larger than 1.5, but it failed for the ratio smaller than 1.5. Lee et al. [10] applied the potential flow method to estimate the hydrodynamic forces as a function of the water depth and the spacing between the ship and a wedge-shaped bank. With respect to the application of viscous flow method, Lo et al. [11] studied the bank effect on the KRISO (Korea Research Institute of Ships & Ocean Engineering) 3600 TEU Container Ship (KCS) model using CFD software based on Navier-Stokes equations. The effect of vessel speed and distance to bank on the magnitude and temporal variation of the yaw angle and sway force were reported. Some details of the predicted flow field are available in this work. Wang et al. [12] recently studied the bank effects using a RANS method to predict viscous hydrodynamic forces on a Series 60 hull at varying water depth ratio (1.5, 3.0 and 10.0) and ship-bank distance.

From available documentation, it is seen that investigations on bank effects by viscous methods are very rare; only a few reports referring to this type of computation have been presented. In the framework of an ongoing project aiming at extensive CFD investigation of the hydrodynamic forces on the hull in restricted waterways, the present work intends to study the bank effects in the case when a ship is close to the seabed and/or a canal wall. By means of numerical methods, quantitative predictions of the most interesting hydrodynamic quantities, such as the sinkage and trim and the hydrodynamic forces on the hull are obtained. Since this is a validation study emphasis is placed on formal Verification and Validation (V&V) and an investigation of modelling errors.

2. GEOMETRY, TEST CONDITIONS AND DATA

2.1 THE HULL

A model scale tanker hull, known as the KVLCC2 (2nd version of the KRISO Very Large Crude-oil Carrier) was used in the investigation. This hull form is extensively used in the CFD ship hydrodynamics community as it is one of the benchmark models adopted by the ITTC (International Towing Tank Conference), the series of CFD workshops on Ship Hydrodynamics (2000, 2005, 2010) and the Workshop on Verification and Validation of Ship Manoeuvring Simulation Methods (SIMMAN 2008 and 2012). The geometry is available and obtained from [13]. In the present work a 1/75 scale model was used, with principal dimensions: hull length L_{PP}=4.267m, beam B=0.773m, draft T=0.277m. It should be mentioned that there was a horn-type rudder with a NACA0018 wing section attached to the hull.

2.2 TEST DESCRIPTION AND CONDITIONS

To validate the computations, the computational conditions were carefully selected according to the bank-effect measurements performed at Flanders Hydraulics Research (FHR), Belgium.

During the test series in the shallow water towing tank at FHR (co-operation with the Maritime Technology Division of Ghent University) the following bank configuration was built in, see Figure 1:

- A vertical wall followed by a bank with slope 1:1 on one side of the tank;
- A bank with slope 1:4 followed by a bank with slope 1:3 on the other side of the tank.
- Each bank was 30 m long.

Figure 1: Built in configuration during captive manoeuvring tests at FHR (The arrow indicates the direction of motion and the circled bank is the vertical one in Figure 2 below)

The KVLCC2 was tested at three different under keel clearances (*UKC*), namely 50%, 35% and 10% of draft. Straight line tests were conducted at different longitudinal velocities (6~12 knots full scale) and different lateral positions. The lateral positions were chosen according to the water depth. Whenever possible the middle of the ship sailed above the toe of the banks and the lateral distance was varied in steps of 0.5*B*. In any case the minimal distance either between ship and bank or between ship and bottom was 20mm. Tests were conducted both at self propulsion and at zero propeller rate, but always with fitted propeller.

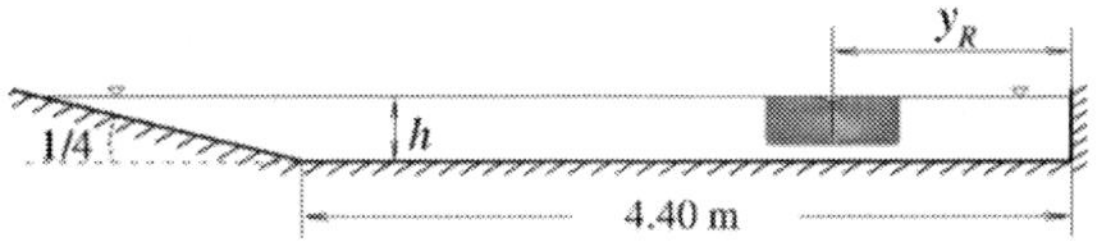

Figure 2: Cross-section of the canal, seen in the direction of motion

In the computations presented here only a subset of all test conditions was considered. The canal configuration is seen in Figure 2, where the view is in the opposite direction as compared to Figure 1. Thus, the hull is moving close to the vertical bank, and the considered non-dimensional bank distances, y_R/B, and water depth ratios, h/T, are presented in Table 1. As can be seen, there are six combined conditions, some of which rather extreme, which have put the computational tools to a severe test. No waves were considered since the tanker moved at a low speed U=0.356m/s (6 knots full scale) (Froude number Fr=0.055; Reynolds number Re=1.513× 10^6), thus the double model approximation was adopted and no sinkage and trim was allowed. In addition, propeller performance was not included in the computations, but the drag of the fixed propeller has been deducted from the measured resistance in the results below.

Table 1: Matrix of test conditions

h/T \ y_R/B	0.669	0.758	1.258	1.758
1.5 (*UKC*=50%*T*)		o		
1.35 (*UKC*=35%*T*)	o	o	o	o
1.1 (*UKC*=10%*T*)		o		

3. NUMERICAL METHOD DESCRIPTION

3.1 GOVERNING EQUATIONS

In this paper, the viscous flow around the KVLCC2 tanker is assumed incompressible and the numerical problem is described by the following steady RANS equations coupled with the time-averaged continuity equation:

$$\frac{\partial}{\partial x_j}(\bar{u}_i\bar{u}_j) = -\frac{1}{\rho}\frac{\partial \bar{p}}{\partial x_i} + F_i + \frac{1}{\rho}\frac{\partial}{\partial x_j}\left(\bar{\sigma}_{ji} + R_{ji}\right) \qquad (1)$$

$$\frac{\partial \bar{u}_i}{\partial x_i} = 0 \qquad (2)$$

here $\bar{u}_{i(j)}, \bar{p}$ and $\bar{\sigma}_{ij}$ denote the average velocity, pressure and stress; ρ is the water density; F_i represents the body force, which is regarded as a constant term; and $R_{ji} = R_{ij} = -\rho\overline{u_i'u_j'}$ denotes the Reynolds stresses. To compute these Reynolds stresses, the turbulence model EASM (Explicit Algebraic Stress Model) [14], originally developed by Gatski and Speziale [15] is applied.

A finite volume solver, XCHAP, in the SHIPFLOW4.4 software [16] is applied to solve the steady RANS equations. In the solver, the discretization of convective terms is implemented by a Roe scheme and for the diffusive fluxes central differences are applied. To approach second order accuracy, a flux correction is adopted. An ADI (Alternating Direction Implicit scheme) is utilized to solve the equations.

3.2 COMPUTATIONAL SETUP AND BOUNDARY CONDITIONS

Due to the asymmetry of the bank geometry and so the flow field, the computational domain has to cover the flow field around the whole hull in the canal. A schematic diagram indicating the hull/rudder geometry, the coordinate system and the computational domain is given in Figure 3. As presented in the figure, the computational domain is made up by seven boundaries: hull surface, inflow plane, outflow plane, flat free surface, seabed boundary, as well as two side banks. The inflow plane is located at one hull length in front of the hull and the outflow plane is at one and a half hull lengths behind the hull. The flat free surface is considered at z=0, and the seabed and the two side banks are placed at specific locations according to the test conditions in Table 1.

As to the adopted boundary conditions in the computations, a fixed velocity, k, ω and zero pressure gradient are set at the inflow boundary; zero gradient of velocity, k, ω and fixed pressure are set at the outflow boundary; a no-slip condition (velocity components, k and pressure gradient are zero) is satisfied on the hull (no wall function is introduced and thus y^+<1 is employed instead); a slip condition (normal velocity component, normal gradient of all other flow quantities are zero) is used at the flat free surface (z=0), the seabed and the side banks.

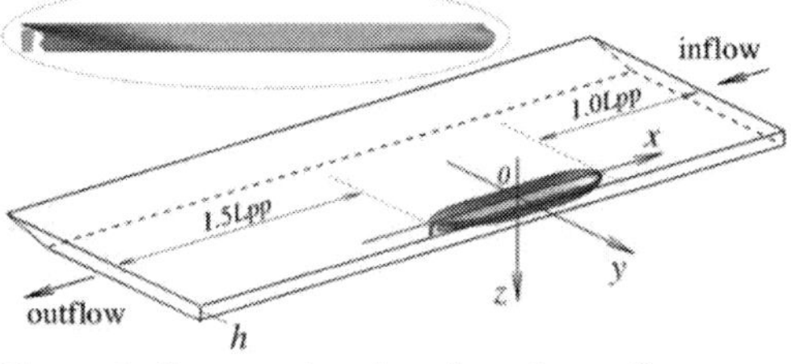

Figure 3: Computation domain and coordinate system

The coordinate system is defined as a body-fixed and right-handed Cartesian system. Its origin is at the intersection of the ship centreline and the mid-ship section. The axes x, y, z are directed towards the bow, to starboard and downwards, respectively.

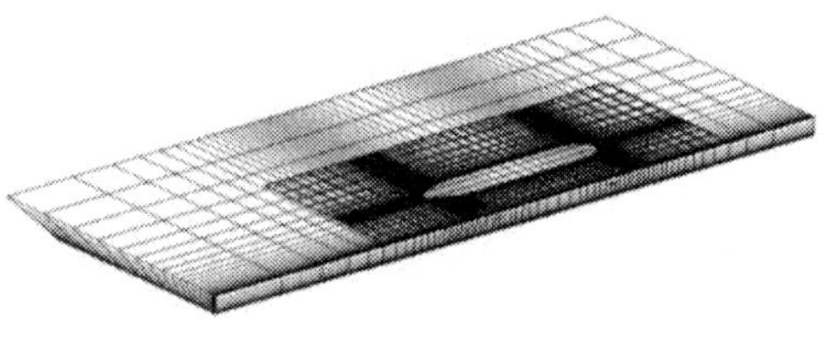

Figure 4: Sketch of grid distribution
(Coarse grids, for clearer illustration)

4. GRID GENERATION

An overlapping grid is used for the computations. With the capability of handling overlapping grids in SHIPFLOW, the computation with more complicated geometries (such as sloping banks, appendages, etc.) is possible and more flexible. As illustrated by Figure 4, the overlapping grid in the present work is mainly built up by two segments: a cylindrical H-O grid (for hull geometry) and a rectilinear H-H grid (for canal geometry). The former grid is immersed in the latter. The body-fitted H-O grid produced by the module XGRID in SHIPFLOW covers the main flow field around the hull, in which two clusters of grid points are concentrated around the bow and stern regions so as to resolve the flow field more precisely. A small radius ($0.5L_{PP}$) of the cylindrical grid is used to save grid points. On the other hand, a 'box' of the rectilinear grid is employed to take care of the remaining part of the domain and to define the grids over the boundary of flat inflow plane, outflow plane, seabed and side banks. The rectilinear grid is created in the mesh generator ANSYS ICEMCFD 12.1 and imported into SHIPFLOW. Moreover, an internally generated grid in SHIPFLOW is used to describe the rudder behind the hull.

5. GRID CONVERGENCE STUDY (ACCURACY VERSUS COMPUTING EXPENSE)

When it comes to numerical simulations, there is always a concern about the accuracy, or the balance between the computing expense and accuracy. Formal Verification and Validation (V&V) methods have been accepted gradually as a useful tool to quantify the numerical and modelling errors in CFD method and its solution. In the numerical method the hydrodynamic problem is formulated by mathematical equations which generally have to be discretized in space and time. From a theoretical point of view, the discretization error should approach zero when the number of grid points tend to infinity, however since the number of grid points is limited there is always a discretization error. To estimate this error and to obtain more reliable results from a numerical point of view, verification of the computation is essential, and a so-called convergence study is always required. In steady flow simulation, only a grid convergence study is necessary.

The present work relies on a preliminary grid convergence study (Verification) carried out for a similar computation, where the KVLCC2 tanker moves straight-ahead in a canal with surface piercing and vertical side walls under a very extreme condition: water depth ratio h/T=1.12 and ship-bank distance $0.6B$. The method applied to estimate the numerical error or uncertainty follows the proposal by Eça and Hoekstra on the assumption that the theoretical order of accuracy of the method is 2 (p_{th}=2) and the iterative and round-off errors are negligible. The detailed procedure is based on [17, 18, 19] and personal communication. It should be noted that this method is characterized by using a Least Squares Root approach (curve fit) to take the scatter in the numerical solutions into account and applying a safety factor for the numerical uncertainty estimation. The discretization error estimator on the basis of Richardson Extrapolation is given as:

$$\delta_{RE} = \phi_i - \phi_0 = \alpha h_i^p \qquad (3)$$

where ϕ_i is the solution of a quantity on the i th grid (i=1, 2...n_g, n_g: available number of grids), ϕ_0 is the extrapolated solution to zero step size, α is a constant, h_i represents the step size (grid spacing) of the i th grid and p is the observed order of accuracy.

In the grid convergence study, a uniform grid refinement ratio $r=h_{i+1}/h_i=\sqrt[4]{2}$ was utilized to create the systematically similar grids, where h_{i+1} and h_i are the grid spacing of two successively refined grids. Like in previous work [20], scatter in numerical solutions existed. Therefore six systematically refined grids were introduced in the investigation so as to enable a curve fit by the Least Squares Root method. From the finest to the coarsest density, the grid point numbers are given in Table 2, where the grid refinement ratio is denoted by h_i/h_1 and h_1 corresponds to the finest grid. C_F, C_{PV} and X', Y', K', N' were selected for the error and uncertainty estimation. C_F, C_{PV} represent the frictional resistance coefficient and viscous pressure resistance coefficient; X', Y', K', N' stand for the non-dimensional longitudinal and sway force and the roll and yaw moment on the hull respectively. The non-dimensional quantities are defined as follows:

$$C_F=\frac{R_F}{0.5\rho U^2 S_W},\ C_{PV}=\frac{R_{PV}}{0.5\rho U^2 S_W},$$
$$X'=\frac{X}{0.5\rho U^2 L_{PP}T},\ Y'=\frac{Y}{0.5\rho U^2 L_{PP}T}, \quad (4)$$
$$K'=\frac{K}{0.5\rho U^2 L_{PP}T^2},\ N'=\frac{N}{0.5\rho U^2 L_{PP}^2 T}$$

where S_W is the wetted hull surface.

Table 2: Grid sizes in grid convergence study

No.	Grid Points (Million)	h_i/h_1 (i=1, 2, ..., 6)
grid1	~7.94	1.0
grid2	~4.59	1.189
grid3	~2.82	1.414
grid4	~1.71	1.682
grid5	~1.00	2.0
grid6	~0.61	2.378

In the study, the Least Squares Root approach was first used to get the observed order of accuracy p in the solutions, and then the errors and uncertainties were estimated based on the comparison between the observed order of accuracy p and the theoretical one p_{th}. The estimated discretization uncertainties U_S of C_F, C_{PV} and X', Y', K', N' are listed in Table 3, where the uncertainties at three grid densities are given to indicate the grid dependent behaviour of these quantities. '1' denotes the finest grid1 and '2', '3' are the subsequent grid2 and grid3. S represents the simulated result. It should be noticed that the coarsest grid6 was dropped from the curve fit as this grid is too coarse to give reasonable results.

Table 3: Discretization uncertainties of C_F, C_{PV} and X', Y', K', N'

	C_F	C_{PV}	X'	Y'	K'	N'
$\lvert U_S\%S\rvert_1$	7.69	14.53	3.32	24.45	4.95	4.54
$\lvert U_S\%S\rvert_2$	9.29	14.59	3.34	26.21	4.91	6.67
$\lvert U_S\%S\rvert_3$	12.17	14.45	3.33	29.91	4.90	8.54

It is seen in Table 3 that the resistance component C_F converges for an increasing number of grid points as the uncertainties become smaller and smaller; while the resistance component C_{PV}, the longitudinal force X' and the roll moment K' are independent of the grid density, as their uncertainties are almost constant. The sway force Y' and yaw moment N' indicate converged tendencies as well, but Y' converges at a slow rate and its uncertainty is still large even with finer grids. After considering the computing expense and the accuracy in solutions, similar density of grid3 is selected to perform the further computations in this paper.

6. RESULTS

This section presents the results of the computations for varying water depth and ship-to-bank distance. In addition to the hydrodynamic forces X', Y' and moments K', N' defined above, the mean sinkage and trim on the KVLCC2 tanker hull are presented. The mean sinkage σ and trim τ are obtained from the formulae (positive sinkage downwards and positive trim bow-up):

$$\sigma = Z/\rho g A_w \quad (5)$$
$$\tau = M/\rho g I_w \quad (6)$$

Here ρ is the water density; Z is the sinking force; M is the trim moment; A_w represents the water plane area and I_w denotes the longitudinal moment of inertia of the water plane area about the centre of floatation.

In this work, a steady state potential flow panel method XPAN in SHIPFLOW was used as well; for a method description reference is made to [16]. The motivation was mainly to investigate the influence of neglecting the free surface in the RANS method. In XPAN a non-linear free surface boundary condition is satisfied. It is thus possible to investigate the effect of waves, especially in extreme shallow water condition and in the vicinity of a vertical bank. It is also possible to compare the accuracy of the bank effect simulation by two different numerical methods, the potential flow and the RANS method. Since the former neglects the water viscosity, it is interesting to see how this simplification affects the accuracy.

All the predicted hydrodynamic quantities were compared with the test data from FHR. It should be noticed that the measured data was obtained from a confidential project, so that the absolute values of all the data and results are hidden in the following figures. Only a zero value is given for reference. Besides, the measured negative thrust from the non-rotating propeller is subtracted from the total longitudinal force to enable a direct comparison between computations and measurements.

6.1 SINKAGE AND TRIM

The computed sinkage and trim results from both the potential flow method (free sinkage and trim, without rudder) and the RANS method are plotted in the same figures, coupled with the experimental data. The dashed lines indicate the zero magnitude. Figure 5 shows the sinkage and trim results versus the water depth ratio h/T at the same small ship-to-bank distance y_R/B=0.758. In general, the sinkage and trim rise with the decrease of water depth, and especially at a water depth less than h/T=1.35, the sinkage and trim change more sharply revealing a significant shallow water effect. The results versus the ship-bank distance at the same water depth h/T=1.35 are presented in Figure 6. Similar tendencies are displayed: when the hull moves closer to the bank, the sinkage and trim increase.

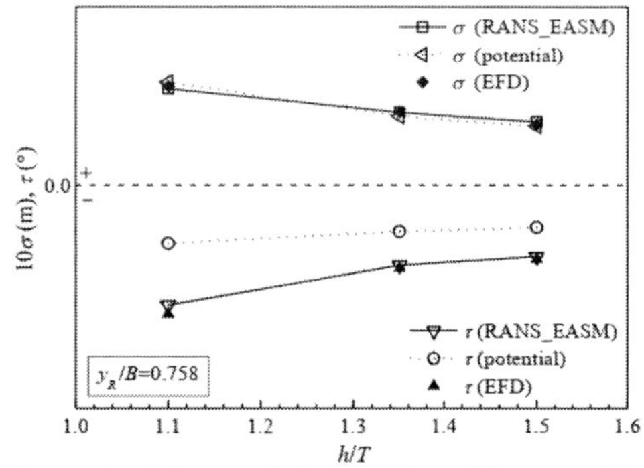

Figure 5: Sinkage σ (m) and Trim τ (°) versus h/T

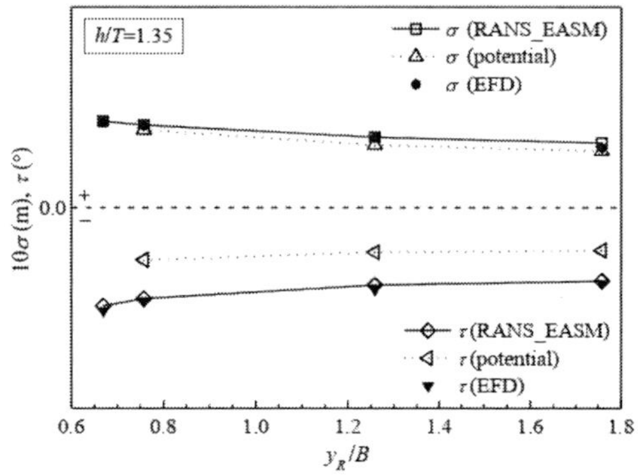

Figure 6: Sinkage σ (m) and Trim τ (°) versus y_R/B

Comparing the results from the potential flow and RANS methods, it is shown in Figure 5 and 6 that the RANS method worked very well in predicting the sinkage and trim, while the potential flow method managed to capture the primary tendency. The sinkage results agree well with the RANS simulations and the measured data. But the code failed to compute the sinkage and trim at the closest ship-bank distance and the quantitative estimation of the trim is not satisfactory. The predicted trim is only half of the measured value. This may be mainly due to the fact that there is significant flow separation along the hull which influences the pressure at the stern. Thus a higher pressure difference between the bow and stern is produced. The pronounced flow separation around the stern at h/T=1.1 and y_R/B=0.758 is clearly predicted by RANS method, as indicated from the non-dimensional axial velocity u_X/U contours in Figure 7. However the potential flow method is unable to simulate these important characteristic bank effects. This can be further verified by the predicted pressure distribution on the hull surface. Normalizing the pressure by $0.5\rho U^2$, the pressure coefficient distributions from potential flow and RANS methods are displayed in Figure 8, where the pressure difference between the bow and stern is almost invisible in the result of potential flow method.

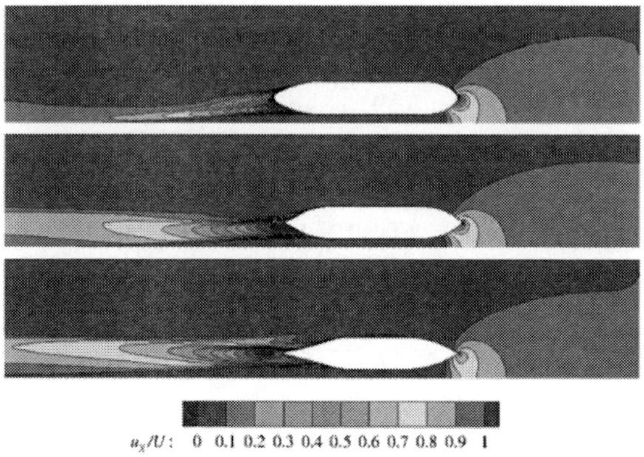

Figure 7: RANS predicted axial velocity u_X/U contours around the tanker in the horizontal plane
(h/T=1.1, y_R/B=0.758)
[From top to bottom: z/L_{PP}=0, -0.032, -0.06]

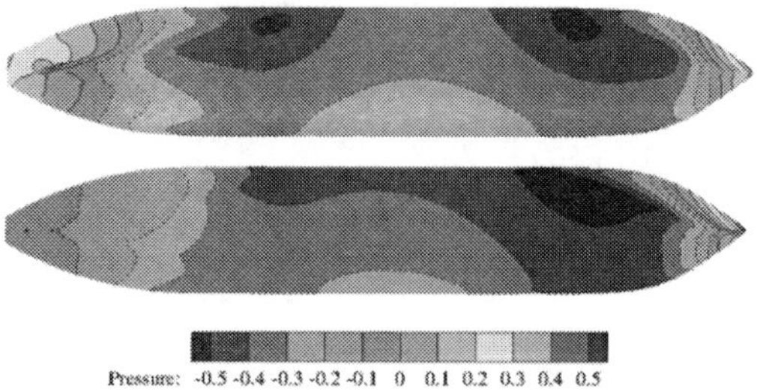

Figure 8: Pressure distribution on the hull surface
(h/T=1.1, y_R/B=0.758)
(Bottom view) upper: potential flow; lower: RANS

6.2 VALIDATION: HYDRODYNAMIC FORCES AND MOMENTS

The predicted forces and moments by the RANS method are shown in the following figures, where only the zero level is given, as mentioned above. However no results from the potential flow method are presented here as it is indicated from the predicted sinkage and trim and the pressure distribution that viscous effects cannot be neglected. The results for the X', Y' forces and the K', N' moments at a specific ship-bank distance, y_R/B=0.758, versus water depth ratio are shown in Figure 9a and 9b. Results at a specific water depth ratio h/T =1.35 versus ship-bank distance are shown in Figures 10a and 10b.

Comparing with the measurements, the tendencies of the forces and moments are captured well. As seen in Figure 9a and 9b, when the hull bottom approaches the seabed, the X' force and K' moment become larger, while the Y' force (a suction force towards the bank) behaves in a different way. It increases between h/T=1.5 and 1.35, but drops rapidly between h/T=1.35 and 1.1. The N' moment shows a monotonic increase for diminishing water depth as well, but in contrast with the measured data the predicted yaw moment changes from a bow in to a bow out moment between h/T =1.5 and 1.35. However, the magnitudes of the yaw moment are very small. In Figure 10a and 10b, with varying ship-to-bank distance at h/T =1.35, the hull is attracted to the bank while the bow is pushed away.

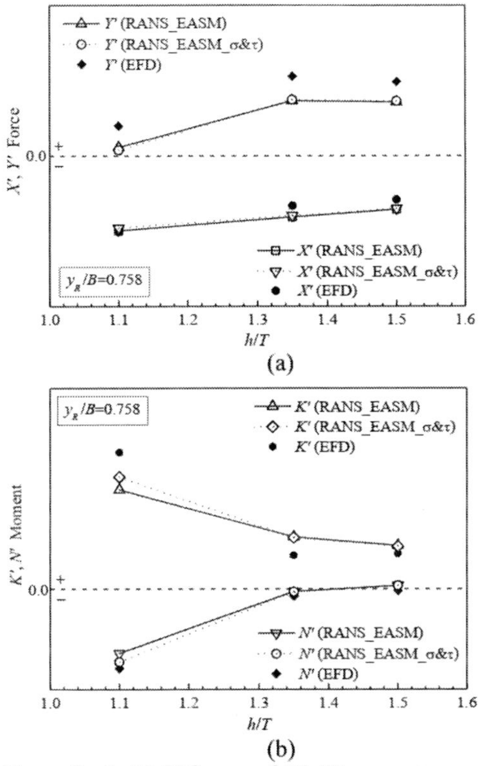

Figure 9a, b: X', Y' force and K', N' moment versus h/T

While the tendencies of the predicted forces and moments correspond well with those of the measured data the absolute level is not well predicted in some cases. The most obvious difference between computations and measurements is seen in the Y' force, for which there is a more or less constant shift to lower predicted values. This tendency is seen clearly in Figure 9a and 10a. To more quantitatively investigate the absolute accuracy a formal validation study was made. The validation procedure was based on the approach adopted in the 3rd Workshop on CFD Uncertainty Analysis [21]. Two parameters were introduced in this procedure, one is the validation comparison error denoted as $E = S\text{-}D$ and the other is the validation uncertainty defined as $U_{Val}^2 = U_{num}^2 + U_D^2$. Here S and D represent the simulated solution and experimental data respectively. U_{num} is the numerical uncertainty, which is approximated as the discretization uncertainty U_S in the grid convergence study (since the iterative uncertainty is neglected), and U_D is the data uncertainty in the measurements. Here it is estimated as the standard deviation of measured variables in repeated captive shallow water model tests at FHR for the SIMMAN 2012 Workshop [22]. In this way, the precision error is accounted for, but not the bias error.

If $|E| \leq U_{Val}$, the error is within the "noise level" and not much can be said about the modelling error; but if $|E| >> U_{Val}$, the sign and magnitude of E could be used as to improve the modelling. Considering the available discretization and data uncertainties, the condition h/T =1.1 at y_R/B=0.758 was selected for the validation of the X', Y', K', N' prediction. The measured data (D) was used to normalize the estimated uncertainties and errors and the results are presented in Table 4.

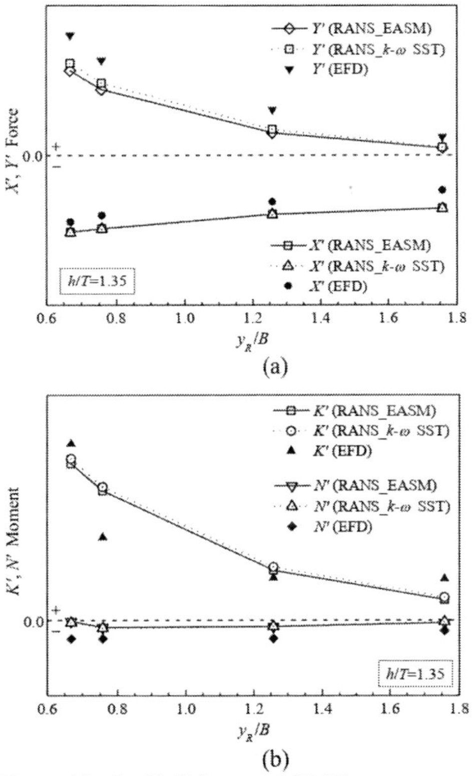

Figure 10a, b: X', Y' force and K', N' moment versus y_R/B

Table 4: Validation results of X', Y', K', N'

	X'	Y'	K'	N'
$\lvert U_D\% D\rvert$	2.20	9.40	3.55	1.39
$\lvert U_S\% D\rvert$	3.27	8.38	3.57	6.92
$\lvert U_{Val}\% D\rvert$	3.94	12.59	5.03	7.06
$\lvert E\% D\rvert$	1.85	71.99	27.19	18.94

It is seen that three quantities Y', K' and N', exhibit a larger comparison error than the validation uncertainty, which implies that there are significant modelling errors, in the computations and/or the measurements. The neglected bias error in the measured uncertainty corresponds to a modelling error and has not been investigated in the present work. However systematic investigations of the modelling in the computations have been carried out and will be reported below.

6.3 DISCUSSION: THE MODELLING ERROR

There are several potential sources of modelling errors in the computations. Examples are: neglect of free surface, non-free sinkage and trim, turbulence modelling, boundary condition on the solid wall, and absence of propeller.

As mentioned before, the free surface effect was neglected in RANS computations. Its influence on hydrodynamic quantities is a concern. As shown in Figure 5 and 6, the computation of sinkage and trim indicates that without considering the free surface in RANS computations, the predicted sinkage and trim correspond very well with the measurements and the results of potential flow computations which take the free surface and free sinkage and trim into account. This shows that it is acceptable to neglect the free surface effect when predicting the sinkage and trim at the present low speed. However, the effect of the sinkage and trim on the other forces and moments, especially at the small water depths, is needed to be investigated. Thus, further computations with a given sinkage and trim from prior computations were carried on. The dotted lines in Figure 9a and 9b indicate the results with the initial sinkage and trim (σ&τ). From the comparisons, it is interesting to see that the influence of sinkage and trim is generally very small. However, it cannot be neglected for the K' and N' moments at the very shallow water depth (h/T=1.1), where the correction from sinkage and trim increases the $|U_{Val}\%D|$ in N' moment to 7.99% and reduces the $|E\%D|$ to 7.92%, while for other quantities the improvement is not large enough to reduce E to the U_{Val} level, so there must be other significant modelling issues.

As for the turbulence modelling, the EASM model was used in the systematic computations. This turbulence model is capable of predicting more accurate wake profiles behind the hull in the deep water condition [23], but in the present case the hydrodynamic quantities were under-predicted with this model. Therefore another turbulence model, the Menter k-ω SST model [24] was applied to evaluate the influence of the turbulence modelling. The focus was placed on the computations with the variation of the bank distance at the same water depth h/T=1.35. Results are presented by dotted lines in Figure 10a and 10b for better comparison with the results of EASM model. As exhibited by the figures, there is a slight improvement in the prediction of sway force Y' (the maximum decrease in $|E\%D|$ is around 8%), but the improvement is very small compared to the difference between the computed and measured results. It should be pointed out that the massive separation mentioned above near the stern on the bank side of the hull may be influenced by the turbulence modelling. The prediction of this separation is crucial for the prediction of the forces.

Moreover, a slip condition was satisfied at the walls (seabed, side bank) in the computations, but when the hull moves over/along the seabed/side bank, a boundary layer is developed on it due to the disturbance velocities from the hull. Therefore, the slip condition might be inadequate and a moving no-slip condition could be more suitable. Preliminary computations have indicated some effect, but this will be further investigated.

The last modelling error stated here is the absence of the non-rotating propeller. The evaluation of its influence on the flow field at the stern, and accordingly its contribution to hydrodynamic forces, will also be investigated in future work.

7. CONCLUSIONS

The investigation of bank effects by CFD methods is a challenging work due to the importance and complexity of the hydrodynamic problem itself and the difficulty in modelling the physical problem by numerical tools. The present work attempts to compute the bank effects by applying a viscous flow method to predict the sinkage and trim and the hydrodynamic forces for varying water depths and ship-to-bank distances. The results show that it is possible to predict sinkage and trim very accurately in bank-effect investigations. Tendencies of the viscous hydrodynamic forces and moments are also well predicted, but the absolute level of some quantities, notably the sway force, is less accurate. A formal validation analysis showed that there are modelling errors in the computations and/or the measured data. Due to time restrictions the latter could not be investigated, but three types of computational modelling errors were discussed, and shown to have some effect on the results. However, the improvement could not fully explain the differences between the computed and measured results. Further work to investigate the differences is underway.

8. ACKNOWLEDGEMENTS

The computing resources were provided by C3SE, Chalmers Centre for Computational Science and Engineering. All model tests were carried out in the Towing Tank for Manoeuvres in Shallow Water (co-operation Flanders Hydraulic Research - Ghent University). The authors thank Professor Marc Vantorre for his review of this article.

9. REFERENCES

1. http://www.bankeffects.ugent.be/index.html.

2. Norrbin, N., 'Bank Effects on a Ship Moving Through a Short Dredged Channel', *10th ONR Symposium on Naval Hydrodynamics*, Cambridge, MA, USA, 1974.

3. Norrbin, N., 'Bank Clearance and Optimal Section Shape for Ship Canals', *26th PIANC International Navigation Congress*, Brussels, Belgium, Section 1, Subject 1, pp. 167-178, 1985.

4. Li, D. Q., Leer-Andersen, M., Ottosson, P., Trägårdh, P., 'Experimental Investigation of Bank Effects under Extreme Conditions', *Practical design of ships and other floating structures*, Oxford, Elsevier Science Ltd: 541-546, 2001.

5. Ch'ng, P. W., Doctors, L. J., and Renilson, M. R., 'A Method of Calculating the Ship-Bank Interaction Forces and Moments in Restricted Water', *International Shipbuilding Progress*, Vol. 40, No.421, pp 7-23, 1993.

6. Vantorre, M., and Delefortrie, G., et al., 'Experimental Investigation of Ship-Bank Interaction Forces', *Proc. of International Conference on Marine Simulation and Ship Manoeuvrability, MARSIM'03*, Kanazawa, Japan, 2003.

7. Lataire, E., Vantorre, M., Eloot, K., 'Systematic model tests on ship-bank interaction effects', *International conference on Ship manoeuvring in shallow and confined water: bank effects(2009)*, Antwerp, Belgium, May 2009.

8. Newman, J. N., 'The Green Function for Potential Flow in a Rectangular Channel', *Journal of Engineering Mathematics, Anniversary Issue*, pp. 51–59, 1992.

9. Miao, Q. M., Xia, J. Z., and Chwang, A. T., et al., 'Numerical Study of Bank Effects on a Ship Travelling in a Channel', *Proc. of 8th International Conference on Numerical Ship Hydrodynamics*, Busan, Korea, 2003.

10. Lee, C. K., and Lee, S. G., 'Investigation of Ship Manoeuvring with Hydrodynamic Effects Between Ship and Bank', *Journal of Mechanical Science and Technology*, 22(6): 1230-1236, 2008.

11. Lo, D. C., Su, Dong-Taur and Chen, Jan-Ming, 'Application of Computational Fluid Dynamics Simulations to the Analysis of Bank Effects in Restricted Waters', *Journal of Navigation*, 62, pp 477-491, 2009.

12. Wang, H. M., Zou, Z. J., Xie, Y. H., Kong, W. L., 'Numerical Study of Viscous Hydrodynamic Forces on a Ship Navigating near Bank in Shallow Water', *Proceedings of the Twentieth (2010) International Offshore and Polar Engineering Conference*, Beijing, China, June 20-25, 2010.

13. http://www.gothenburg2010.org/kvlcc2_gc.html.

14. Deng, G.B., Queutey, P., and Visonneau, M., 'Three-Dimensional Flow Computation with Reynolds Stress and Algebraic Stress Models', *Engineering Turbulence Modelling and Experiments* 6, Rodi, W., and Mulas, M., eds. ELSEVIER, 389-398, 2005.

15. Gatski, T. B., and Speziale, C. G., 'On Explicit Algebraic Stress Models for Complex Turbulent Flows', *Journal of Fluid Mech.*, 254, 59-78, 1993.

16. Broberg, L., Regnström, B., Östberg, M., 'SHIPFLOW Theoretical Manual', *FLOWTECH International AB*, Gothenburg, Sweden, 2007.

17. Eça, L., and Hoekstra, M., 'An Evaluation of Verification Procedures for CFD Applications', *24th Symposium on Naval Hydrodynamics*, Fukuoka, Japan, 2002.

18. Eça, L., Vaz, G., Hoekstra, M., 'Code Verification, Solution Verification and Validation in RANS Solvers', *Proceedings of ASME 29th International Conference OMAE2010*, Shanghai, China, June 6-11, 2010.

19. Eça, L., Vaz, G., Hoekstra, M., 'A Verification and Validation Exercise for the Flow Over a Backward Facing Step', *V European Conference on Computational Fluid Dynamics, ECCOMAS CFD 2010*, Eds. Pereira J.C.F., Sequeira A., Lisbon, June 2010.

20. Zou, L., Larsson, L., Orych, M., 'Verification and Validation of CFD Predictions for a Manoeuvring Tanker', *Journal of Hydrodynamics*, Ser. B 22 (5, Supplement 1): 438-445, 2010.

21.http://maretec.ist.utl.pt/html_files/CFD_workshops/Workshop_2008.htm.

22. Delefortrie, G., Eloot, K., Mostaert, F., 'SIMMAN 2012: Execution of model tests with KCS and KVLCC2. Version 2_0, WL Rapporten, 846_01', *Flanders Hydraulics Research*, Antwerp, Belgium, 2011.

23. Zou, L., Larsson, L., 'CFD Prediction of the Local Flow around the KVLCC2 Tanker in Fixed Condition', *Proceedings of A Workshop on Numerical Ship Hydrodynamics*, Gothenburg, Sweden, 2010.

24. Menter, F.R., 'Zonal Two Equation *k-ω* Turbulence Models for Aerodynamic Flows', *24th Fluid Dynamics Conference*, Orlando, AIAA paper 93-2906, 1993.

10. AUTHORS BIOGRAPHY

Lu Zou is a PhD student at the Department of Shipping and Marine Technology, Chalmers University of Technology, Sweden. Her PhD project is mainly focused on the CFD prediction of the hydrodynamic forces and moments on a manoeuvring ship in restricted water. Some of her research interest is in the V&V for CFD simulations.

Lars Larsson is a Professor of Hydrodynamics at Chalmers University of Technology. Formerly head of the Department of Naval Architecture and Director of the Rolls-Royce University Technology Centre for

Computational Hydrodynamics. Chairman and former Managing Director of the CFD software company FLOWTECH International AB. Author of more than 120 papers on CFD in hydrodynamics, and the author of two books.

Guillaume Delefortrie, PhD, is junior expert nautical research at Flanders Hydraulics Research, Belgium. His MSc project dealt with the modelling of bank effects. He is presently in charge for the experimental research in ship hydrodynamics at Flanders Hydraulics Research and the development of mathematical manoeuvring models in shallow and confined waters.

Evert Lataire, MSc, is assistant at the Maritime Technology Division of Ghent University, Belgium. He performs PhD research on bank effects and co-organised the 1st International Conference on Manoeuvring in Shallow and Confined Waters: Bank Effects in 2009.

flanders
HYDRAULICS RESEARCH
KNOWLEDGE CENTRE MANOEUVRING
IN SHALLOW AND CONFINED WATER

MARITIME TECHNOLOGY
GHENT UNIVERSITY

The Knowledge Centre Manoeuvring in Shallow and Confined Water aims to consolidate, extend and disseminate knowledge on the behaviour of ships in shallow and confined water. The dissemination activities involve the participation in and organization of conferences and workshops, the hosting of the website www.shallowwater.be and releasing a newsletter four times a year.

The Knowledge Centre is founded on a collaboration between Flanders Hydraulics Research with its unique experimental facilities and the Maritime Technology Division at Ghent University, which provides research assistance. The experimental facilities comprise a towing tank for manoeuvres in shallow water, a lock model of the Panama Canal and three manoeuvring simulators of which one simulator dedicated to inland shipping was inaugurated in December 2010.

Research activities have recently involved the study of bank effects, shallow water, ship hydrodynamics in a lock and the concept of a nautical bottom; the execution of self-propulsion tests for the SIMMAN Workshop; simulation studies in the framework of admittance policy and onboard ship motion measurement using very accurate RTK-GPS based equipment.

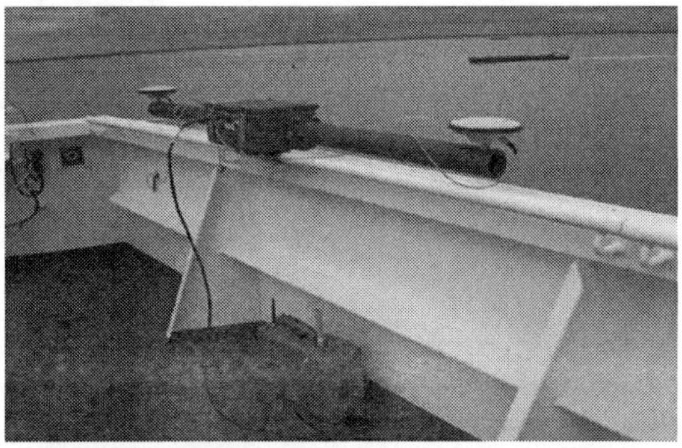

Extensive model testing has been carried out in the towing tank to study ship-ship interaction and ship-tugboat interaction. The analysis of the model tests, which were carried under the research project entitled 'Investigating hydrodynamic aspects and control strategies for ship-to-ship operations' co-ordinated by MARINTEK and financially supported by the Research Council of Norway, has significantly enhanced knowledge of the complex flows arising between vessels that sail closely to each other.

In collaboration with other partners, the Knowledge Centre organizes a biannual international conference on manoeuvring in shallow and confined water. The proceedings of the first conference, which pays special attention to bank effects, can be ordered from the R.I.N.A. publications website.

Department of Marine Technology

The Department of Marine Technology is a part of the Faculty of Engineering Science and Technology at NTNU and graduates 90-100 masters of technology students every year. The department collaborates closely with MARINTEK at the Marine Technology Centre at Tyholt. Currently around 100 PhD-students are hosted by the department.

The departments education and research activities are closely linked to the challenges and opportunities created by the sea. With less than a thousandth of the world's inhabitants, Norway is world class in terms of marine science, marine technology and maritime industrial enterprise. Marine technology encompasses activities tied to our largest exporting industries:

- Oil and gas extraction
- Fisheries technology and aquaculture
- Marine engineering and the associated industries.

The Centre for Ships and Ocean Structures (CeSOS) is hosted by the department, with approx. 40 PhD-students and researchers from a number of countries.

The department are organized in two groups;
The Marine Constructions Group which educates and conducts research in the fields of hydrodynamics, load effect analyses and control systems related to all types of marine structures; including ships, oil platforms, and pipelines. The research group works closely with the department's other research groups, MARINTEK and with SINTEF Building and Infrastructure.

The Marine Systems Group as teach and do research on all major aspects of marine systems, such as machinery, maritime transport systems, offshore oil and gas installations, fisheries and aquaculture facilities and offshore energy production systems. The group focuses on system design and operation, as well as interaction and adaptation to the surroundings in a total life cycle perspective.

The hydrodynamic laboratories at the Marine Technology Centre are jointly used by the department, CeSOS, and MARINTEK. NTNU uses the laboratories for research and teaching, while MARINTEK uses the laboratories for contract research.

Students have access to the laboratories for demonstrations and exercises in topics related to hydrodynamics, and use the laboratories for projects and masters theses. In particular, both the small towing tank and the student workshop are available for students outside of organized teaching activities.

Research activities for advanced ship and offshore marine operations

MARINTEK is among the world's largest independent research institutes in the maritime and offshore sectors. More than 200 researchers work in extensive laboratory facilities covering hydrodynamics, strength, engine technology and operations. MARINTEK is certified to ISO 9001 and will in 2011 be an accredited organisation for measuring emissions to air from engines and turbines in laboratory and field. Operational safety and efficiency are two basic questions when planning and performing complex operations at sea. Here are some examples of our achievements and activities.

Hull and propulsive performance

Since the opening of the towing tank in 1939, MARINTEK has played an important role in reducing ship resistance and improving propulsive efficiency. Today, both state-of-the-art laboratories and analytical tools are applied to all aspects of our research. Hull optimization research programs have reduced fuel conspumption and emissions significantly. In addition to evaluation and modification of hull designs, MARINTEK can be a key partner in the testing and development of novel fuel saving devices related to ship propulsion. Focus must always be on testing the ship in realistic conditions close to the actual operating environment.

VeSim/ShipX

MARINTEK develops Software tools for evaluation of ship designs such as the ShipX program suite including tools for performance prediction (speed and powering in calm water and in a seaway), evaluation of seakeeping performance and manoeuvring and station-keeping. The VeSim ship simulator can be applied for "real-time" or faster simulations of ship motions and manoeuvring in a seaway including the effects from wind, waves and current. Propeller and engine dynamics can be modeled and sensor signals can be provided so that real control systems can be tested.

Offloading simulation

Various configurations of transfer of oil or gas has been studied by model tests in the Ocean Laboratory and by computer simulation, using the software SIMO and Riflex. DP positioning and dynamic forces in the connection lines have been addresssed.

CFD

Computational Fluid Dynamics (CFD) is used to study the performance of different design alternatives prior to model test. CFD is also used for investigating local flow phenomena, such as flow separation, wake field, generation of vorticities, thruster/hull interaction, propeller performance, in order to gain a better insight into the physics involved.

Regulations and standards

As part of our research activities we are active in legislative and standards work, in IMO as well as in international standard organisations.

SPT - MARINE TRANSFER SOLUTIONS
Best Companies to Work for in Texas 2010
SPT CONQUEROR
SPTMTS.COM